NEURAL CREST
CELLS

NEURAL CREST CELLS

CELLS

Evolution, Development and Disease

Edited by

PAUL A. TRAINOR

AMSTERDAM • BOSTON • HEIDELBERG • LONDON
NEW YORK • OXFORD • PARIS • SAN DIEGO
SAN FRANCISCO • SINGAPORE • SYDNEY • TOKYO

Academic Press is an imprint of Elsevier

Academic Press is an imprint of Elsevier
32 Jamestown Road, London NW1 7BY, UK
225 Wyman Street, Waltham, MA 02451, USA
525 B Street, Suite 1800, San Diego, CA 92101-4495, USA

Notice

No responsibility is assumed by the publisher for any injury and/or damage to persons
or property as a matter of products liability, negligence or otherwise, or from any use or
operation of any methods, products, instructions or ideas contained in the material herein.
Because of rapid advances in the medical sciences, in particular, independent verification
of diagnoses and drug dosages should be made.

British Library Cataloguing-in-Publication Data
A catalogue record for this book is available from the British Library

Library of Congress Cataloging-in-Publication Data
A catalog record for this book is available from the Library of Congress

ISBN: 978-0-12-401730-6

For information on all Academic Press publications visit
our website at www.store.elsevier.com

Cover Figure
The cover figure was kindly provided by Amanda Barlow and is of a late E9.5 DAPI nuclear stained
(blue) mouse embryo illustrating the distribution of neural crest cells (green) and the commencement
of their differentiation into neurons (red).

Typeset by MPS Limited, Chennai, India
www.adi-mps.com

Working together
to grow libraries in
developing countries

www.elsevier.com • www.bookaid.org

Contents

I

NEURAL CREST CELL EVOLUTION AND DEVELOPMENT

14. Neural Crest Cells and Pigmentation

ALBERTO LAPEDRIZA, KLEIO PETRATOU AND
ROBERT N. KELSH

15. Neural Crest Cells in Vascular Development

SOPHIE E. WISZNIAK AND QUENTEN P. SCHWARZ

16. Neural Crest Cells and Cancer: Insights into Tumor Progression

DAVALYN R. POWELL, JENEAN H. O'BRIEN,
HEIDE L. FORD AND KRISTIN BRUK ARTINGER

III

TISSUE ENGINEERING AND REPAIR

21. Using Induced Pluripotent Stem Cells as a Tool to Understand Neurocristopathies

JOHN AVERY, LAURA MENENDEZ, MICHAEL L.
CUNNINGHAM, HAROLD N. LOVVORN III AND STEPHEN
DALTON

Preface

Neural crest cells comprise a migratory, stem and progenitor cell population and are synonymous with vertebrate evolution and development. Although thought to have been first identified by William His in 1868, the term neural crest is attributed to Arthur Milnes Marshall in recognition of the cells anatomical origins. Over the past 145 years, neural crest cells have held the fascination of developmental and evolutionary biologists alike by providing a unique paradigm with which to study various developmental processes such as morphogenetic induction, migratory behaviors, and fate determination.

During the first half of the twentieth century, the majority of neural crest cell research was undertaken in amphibian embryos as reviewed in Horstadius' well-known 1950 monograph. The 1960s saw the introduction of tritiated thymidine cell labeling techniques employed by Weston and Chibon to visualize the migration of neural crest cells throughout the developing amphibian and chick embryos. This was followed shortly thereafter in 1969 through Nicole Le Douarin's seminal introduction of the quail-chick marking system. While similar in principle to the earlier generation of amphibian chimeras by German embryologists such as Andres and Wagner, this now enabled embryologists to distinguish neural crest cells of one species from the surrounding tissue of another species. With this technique, generations of scientists reliably marked and studied the properties of neural crest cells, the principles of which were detailed in Nicole Le Douarin's

1982 book "The Neural Crest." These chimera approaches remain a fundamental staple of neural crest cell research today.

The year 1983 marked another seminal year in the neural crest cell field. Drew Noden illustrated the remarkable properties of cranial neural crest cells and Carl Gans and Glen Northcutt published their "New Head" hypothesis which posited the central importance of neural crest cells in the endless variation of vertebrate craniofacial features throughout evolution. In the late 1980s, new vital dyes cell labeling techniques pioneered by Marianne Bronner and Scott Fraser opened the door to visualizing neural crest cells in any species, but particularly fish and mice. This afforded the opportunity in combination with advances in imaging to follow the dynamics of neural crest cells in real time. Together with advances in molecular techniques, all of these techniques collectively facilitated the study of comparative neural crest cell development, fate, and evolution, highlighting the interplay between patterning and plasticity and spurring the quest to identify the evolutionary origins of neural crest cells. These principles and developments were captured in Brian Hall's 1999 "The Neural Crest in Evolution and Development" and Jean-Pierre Saint Jeannet's 2006 "Neural Crest Induction and Differentiation" books.

In the twenty-first century, much of our focus now revolves around the contributions of neural crest cells to congenital disorders and diseases which are collectively termed neurocristopathies. Understanding the true genetic and cellular etiology and pathogenesis of individual neurocristopathies offers the potential for developing therapeutic avenues for their clinical prevention. Furthermore, continuing advances have uncovered and characterized the pluripotent stem cell-like characteristics of neural crest cells during embryogenesis and their persistence into adulthood. Thus, there is tremendous excitement in the potential for neural crest cells to be used in tissue engineering and regenerative medicine. This book "Neural Crest Cells: Evolution, Development and Disease" ambitiously tries to capture the classic principles and recent advances in our comprehension of the roles of neural crest cells in vertebrate evolution and development. Much of this has come from the application of new genetic, imaging and systems biology approaches exploring the gene regulatory control of neural crest cell formation, migration, and differentiation. This book also illustrates how this foundational knowledge influences our understanding of the central roles neural crest cells play in the pathogenesis of congenital disorders and diseases and the potential for neural crest cells to be used in tissue engineering and regenerative medicine to treat disorders and disease.

I am extremely grateful to all of the authors for sharing their time, knowledge, and expertise in contributing chapters to this book. I also want to thank all the past and present members of my laboratory as well as colleagues and friends in the neural crest cell field for continually making science and life stimulating and fun.

Acknowledgment

I am indebted to Patrick Tam and Robb Krumlauf for introducing me to neural crest cells and for their scientific training and continual support, mentorship, and friendship.

"Learn from yesterday, live for today, hope for tomorrow. The important thing is to not stop questioning."
Albert Einstein

Contributors

Sinu Jasrapuria-Agrawal Department of Biochemistry and Cell Biology, Rice University, Houston, TX, USA

Kristin Bruk Artinger Department of Craniofacial Biology, School of Dental Medicine, University of Colorado Anschutz Medical Campus, Aurora, CO, USA

John Avery Department of Biochemistry and Molecular Biology, University of Georgia, USA

Chang-Joon Bae Department of Basic Science and Craniofacial Biology, College of Dentistry, New York University, New York, USA

Amanda J. Barlow Department of Surgery, University of Wisconsin, USA

Jo Begbie Department of Physiology, Anatomy and Genetics, University of Oxford, Oxford, UK

Erin Betters Yale University, New Haven, CT, USA

Marianne E. Bronner Division of Biology, California Institute of Technology, Pasadena, CA, USA

Yang Chai Center for Craniofacial Molecular Biology, Herman Ostrow School of Dentistry, University of Southern California, Los Angeles, CA, USA

Michael L. Cunningham Department of Pediatrics, University of Washington, Seattle, USA; Seattle Children's Research Institute, Seattle, USA

Stephen Dalton Department of Biochemistry and Molecular Biology, University of Georgia, USA

Elisabeth Dupin Institut de la Vision Research Center, Department of Developmental Biology, Paris, France

Anthony B. Firulli Riley Heart Research Center, Herman B Wells Center for Pediatric Research Division of Pediatrics Cardiology, Departments of Anatomy, Biochemistry, and Medical and Molecular Genetics, Indiana University Medical School, IN, USA

Jennifer L. Fish Department of Orthopaedic Surgery, University of California at San Francisco, San Francisco, CA, USA

Stephen J. Fleenor Department of Physiology, Anatomy and Genetics, University of Oxford, Oxford, UK

Heide L. Ford Department of Pharmacology, School of Medicine, University of Colorado Anschutz Medical Campus, Aurora, CO, USA

Martín I. Garcia-Castro Department of Molecular Cell & Developmental Biology, Yale University, New Haven, CT, USA

Robert N. Kelsh Department of Biology and Biochemistry and Centre for Regenerative Medicine, University of Bath, Bath, UK

Alberto Lapedriza Department of Biology and Biochemistry and Centre for Regenerative Medicine, University of Bath, Bath, UK

Nicole M. Le Douarin Institut de la Vision Research Center, Department of Developmental Biology, Paris, France

Pierre Le Pabic Department of Developmental and Cell Biology, University of California, Irvine, Irvine, CA, USA

Alan W. Leung Yale University, New Haven, CT, USA

Harold N. Lovvorn III Department of Pediatric Surgery, Vanderbilt University Medical Center, USA

Peter Y. Lwigale Department of Biochemistry and Cell Biology, Rice University, Houston, TX, USA

Roberto Mayor Department of Cell and Developmental Biology, University College London, UK

Laura Menendez Department of Biochemistry and Molecular Biology, University of Georgia, USA

Barbara Murdoch Department of Biology, Eastern Connecticut State University, CT, USA

Jenean H. O'Brien Department of Pharmacology, School of Medicine, University of Colorado Anschutz Medical Campus, Aurora, CO, USA

Rangarajan Padmanabhan Department of Animal and Avian Sciences, University of Maryland, College Park, MD, USA

Carolina Parada Center for Craniofacial Molecular Biology, Herman Ostrow School of Dentistry, University of Southern California, Los Angeles, CA, USA

Kleio Petratou Department of Biology and Biochemistry and Centre for Regenerative Medicine, University of Bath, Bath, UK

Davalyn R. Powell Department of Craniofacial Biology, School of Dental Medicine, University of Colorado Anschutz Medical Campus, Aurora, CO, USA

Andrew Prendergast Department of Biological Structure, University of Washington, USA

David W. Raible Department of Biological Structure, University of Washington, USA

Jean-Pierre Saint-Jeannet Department of Basic Science and Craniofacial Biology, College of Dentistry, New York University, New York, USA

Pedro A. Sanchez-Lara Children's Hospital Los Angeles, Department of Pediatrics & Pathology and Laboratory Medicine, Keck School of Medicine, University of Southern California, Los Angeles, CA, USA; Center for Craniofacial Molecular Biology, Ostrow School of Dentistry, University of Southern California, Los Angeles, CA, USA

Lisa Sandell Birth Defects Center—MCCB, School of Dentistry, University of Louisville, Louisville KY, USA

Thomas F. Schilling Department of Developmental and Cell Biology, University of California, Irvine, CA, USA

Richard A. Schneider Department of Orthopaedic Surgery, University of California at San Francisco, San Francisco, CA, USA

Quenten P. Schwarz Centre for Cancer Biology, SA Pathology, Frome Road, Adelaide, Australia

Paul Sharpe Department of Craniofacial Development and Stem Cell Biology, Dental Institute, Kings College London, Guy's Hospital, London, UK

Pablo H. Strobl-Mazzulla Laboratory of Developmental Biology, Instituto de Investigaciones Biotecnológicas-Instituto Tecnológico de Chascomús (CONICET-UNSAM), Chascomús, Argentina

Lisa A. Taneyhill Department of Animal and Avian Sciences, University of Maryland, College Park, MD, USA

Eric Theveneau Department of Cell and Developmental Biology, University College London, UK

Paul A. Trainor Stowers Institute for Medical Research, Kansas City, MO, USA; Department of Anatomy and Cell Biology, University of Kansas School of Medicine, Kansas City, KS, USA

Joshua W. Vincentz Riley Heart Research Center, Herman B Wells Center for Pediatric Research Division of Pediatrics Cardiology, Departments of Anatomy, Biochemistry, and Medical and Molecular Genetics, Indiana University Medical School, Indianapolis, IN, USA

Kristin E. Noack Watt Stowers Institute for Medical Research, Kansas City, MO; Department of Anatomy and Cell Biology, University of Kansas School of Medicine, Kansas City, KS, USA

Sophie E. Wiszniak Centre for Cancer Biology, SA Pathology, Frome Road, Adelaide, Australia

Hu Zhao Center for Craniofacial Molecular Biology, Ostrow School of Dentistry, University of Southern California, Los Angeles, CA, USA

NEURAL CREST CELL EVOLUTION AND DEVELOPMENT

The Neural Crest, a Fourth Germ Layer of the Vertebrate Embryo: Significance in Chordate Evolution

Nicole M. Le Douarin and Elisabeth Dupin

Institut de la Vision Research Center, Department of Developmental Biology, UMR INSERM S968/ CNRS 7210/ UPMC, 17 rue Moreau 75012 Paris, France

OUTLINE

Neural Crest Cells.
DOI: http://dx.doi.org/10.1016/B978-0-12-401730-6.00001-6

GENERAL CHARACTERISTICS OF THE NEURAL CREST

The neural crest (NC) is capable of giving rise to nearly all cell types that are produced by ectoderm and mesoderm (neurons and glia, muscles, bone, cartilage, connective tissues, endocrine cells, etc.). A few exceptions exist, however, such as the hemangioblasts, precursors of both the blood and vascular endothelial cells, or the excretory system, which are exclusive products of the mesoderm. This is why, after Karl Von Baer formulated the germ layer theory at the dawn of the nineteenth century, the notion of a fourth germ layer was put forward to designate this particular structure of the vertebrate embryo discovered by Wilhelm His in 1868 [1] (see also [2,3]). This "germ layer," whose appearance is postponed with respect to that of the three others that are laid during gastrulation, has another character: its component cells issued from the ectoderm are deployed within the developing embryo through interstitial migration.

The canonical migratory character of the NC cells (NCC) is at the origin of their presence in virtually all tissues of the adult body. Their initial migration that gives rise to the peripheral and enteric nervous systems (PNS and ENS) is extended by the colonization of the body by Schwann cells lining the nerves that travel within the tissues together with the blood vessels, thus making the NCC nearly ubiquitous. NCC migration involves at its

origin a process widespread in metazoans designated "epithelial-to-mesenchymal transition" (EMT), defined as "the events that convert epithelial cells into individual migratory cells that can invade the extracellular matrix" [4]. Such a mechanism is a general feature of embryogenesis and takes place in several organogenetic and morphogenetic processes. Moreover, EMT is crucial for cancer progression and metastasis [5,6]. Dispersion of the NCC has been one of the models in which the genetic control of this important problem of cell biology has been investigated.

The NC is a vertebrate innovation, and it is interesting to look for the existence of possible precursors of this structure in the other members of chordates and to consider the role that it may have played in the evolution of this phylum. Among the benefits that were conferred to vertebrates by the NC, one can cite the coordination of their physiological functions via the PNS and a better adaptation capacity to the variations of their environment. For example, the vasomotoricity of the peripheral blood vessels, which plays a major role in thermal regulation, is controlled by the PNS, and the protective screen against the deleterious effects of UV radiations is provided by NC-derived pigment cells that colonize the skin. NC derivatives forming the PNS control various other physiological regulations. They participate in the construction of the cardiovascular system [7–10], and, as recently shown and further developed in this chapter, the cephalic NC plays a major role in the development of the most recent structures of the brain—namely, the forebrain and midbrain [11].

Among the remarkable characters of vertebrates are their extraordinary radiations in both aquatic and terrestrial environments since they have appeared some 550 Million years (Myr) ago and the developmental complexity of their brain that culminate in human primates. Brain expansion was accompanied by the appearance of sense organs that developed either from the neural plate or from ectodermal placodes, another vertebrate innovation to which is added the first osseous tissues, which were NC derived. These spectacular changes led vertebrates to switch from filter feeding to a predatory behavior and greatly contributed to their extraordinary evolutionary success.

In this chapter we will discuss the views that emerged during the last decades from a variety of approaches going from embryological, to genetic, to paleontological studies on the role that the NC played in the emergence of vertebrate traits within the group of chordates. We will also consider when and how the still undetermined ancestors of vertebrates did acquire this structure to which invasiveness and pluripotency have conferred a role, much more significant than previously believed, in the development and evolution of this phylum.

TRANSITION FROM INVERTEBRATES CHORDATES TO VERTEBRATES

In the late nineteenth and twentieth centuries the origin of vertebrates was a major research theme. Some authors thought that their ancestors were segmented protostomes, like annelids [12–14], nemertines [15,16], or arthropods [17–19]. One of the arguments in

favor of this view was that the body plan of these ancestors with their ventral nervous system was highly comparable to a vertebrate embryo that would have been remodeled after rolling onto its back. Modern molecular developmental biology has confirmed this dorsal–ventral inversion at the genetic level [20]; however, the hypothesis of a deuterostomian origin of chordates and vertebrates is the one that is now unanimously favored. For various anatomical and physiological reasons the hemichordate enteropneusts [21] and the protochordates, among which essentially the cephalochordates [22–24], are considered to be the most plausible ancestors of the evolutionary successful phylum of vertebrates.

A large amount of work has been devoted in the last decades to the hypothesis that evolution of vertebrates has proceeded from a putative protochordate ancestor similar in many respects to the extant *Amphioxus* [25]. However, recently, whole genome phylogenetic analyses placed the tunicates urochordates rather than cephalochordates as the sister clade of vertebrates [26,27] (Figure 1.1). Nevertheless, most of the interest devoted to the transition from invertebrate chordates (and essentially a putative *Amphioxus-like* ancestor) to vertebrates was prompted by the seminal articles written in 1983 by Carl Gans and Glenn Northcutt [28,29]. These authors provided a reinterpretation of vertebrate origins in the light of various considerations including novel embryological evidence about the contribution of the NC to vertebrate development. At that time, this subject was being unraveled owing to the development of appropriate cell-marking techniques: the use of tritiated thymidine [30–32] first, and later the construction of heterospecific chimeras between two species of birds, the chick and the Japanese quail [33–36], which enabled the tracing of the migratory NCC throughout development up to the adult stage.

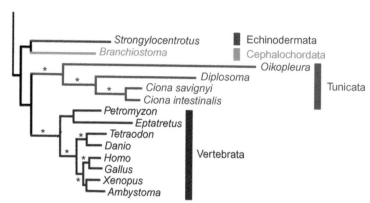

FIGURE 1.1 Phylogenetic relationship between chordates. Phylogenetic analyses of genomic data from 14 different chordates [26] strongly support the sister-group relationship between vertebrates and tunicates, arguing against the previous view that cephalochordates (*Amphioxus*) would be the closest living relatives of vertebrates. Cephalochordates surprisingly grouped with echinoderms, a hypothesis that needs to be tested with additional data. Source: *Adapted from [26], with permission.*

The data provided by these experimental approaches in amniote vertebrates (avians) completed the studies that had been carried out in fish and amphibians during the first half of the twentieth century and were reviewed in the well-acknowledged monograph written in 1950 by Sven Hörstadius [2]. The advantage of the quail-chick marker technique over those used in the previous investigations was its stability. It allowed for observation of the contribution of the NC to mature tissues and to show that it was much more considerable than previously estimated ([35,36] for references).

THE MOLECULAR CONTROL OF EMT STUDIED IN NCC DEVELOPMENT

Among EMT models, the best known at the genetic level are the formation of the ventral furrow in *Drosophila* [37] and the ingression of primary mesenchymal cells in the sea urchin [38]. In vertebrate embryos, primary EMT takes place during gastrulation and in the NC. The primary transcriptional regulators of EMT are Snail zinc-finger proteins and Twist bHLH proteins, which are essential for mesoderm formation in most species [4,39]. *Snail* and *Twist* genes are phylogenetically

conserved, however their interactions and temporal sequence of expression may differ between species. Twist acts as an activator and is a determinant of EMT mostly in invertebrates, while, in amniotes Snail proteins play a major role in the gastrulation process. Snail1 in mouse (and Snail2 in chick) regulates cell polarity, delamination and motility by acting as a repressor of E-cadherin, adherens junction genes, and basal polarity genes (e.g., *occludin* and *claudin*).

In the NC, regulation of the EMT also depends on Snail and Twist proteins, but in a more flexible manner compared to gastrulation. While defects in individual genes lead to very strong EMT phenotypes at gastrulation, a high degree of cooperation and plasticity is in play during NC development [4,39]. Moreover, the transcriptional regulators of EMT and those that control NC specification largely overlap (e.g., *Snail*, *SoxE*, *FoxD3* genes). Because the induction, survival, maintenance of multipotency and emigration of early NCC are intimately correlated processes, most of the genes involved in the control of the EMT regulate some other aspects of NC development [40]. In the chick, NC delamination requires *Snail2* [41]; Snail2, in cooperation with phosphorylated Sox9 induces full EMT in the neural epithelium [42–44]. In *Xenopus*, both Snail1 and Snail2 control the NC induction and

EMT [45]. In mouse, the role played by Twist and Snail proteins in the NC EMT is less clear. Null mutations of *Twist1* caused severe cranial defects without the abolition of NCC migration [46,47]. In fact, *Twist1* is expressed in the mesoderm and in the cephalic NC only after the onset of NCC migration, which is consistent with the defects in cranial NC skeletal derivatives observed after inactivation of *Twist1* specifically in the NC [48]. Surprisingly, the inactivation of *Snail1* and *Snail2*, alone or in combination in murine NC, did not impair NC formation and delamination [49]. Another repressor of E-cadherin, Smad-interacting protein-1 (Sip1/Zeb2/Zfhx1b), appears to be crucial for mouse NC EMT [50]. *Zfhx1b* is expressed in premigratory and migratory mouse NCC and its conditional mutation in the NC altered the development of melanocytes, cranial skeleton, sympathetic ganglia, adrenal gland, and ENS [51].

In jawless vertebrates (hagfish and lamprey), the expression of *Twist* and *Snail* genes homologs markedly differ from that of gnathostomes. *Twist* expression in the lamprey is initiated only in the migratory NCC, not at the premigratory stage [51]. The lamprey and hagfish *Snail* homologs are present in the neural plate at early developmental stages, but later, they are absent in the premigratory NC and become ubiquitously expressed in the mesoderm and the ectoderm [51,52].

In the cephalochordates (*Amphioxus* and ascidians) lacking a NC, no EMT takes place in the neural ectoderm. *Snail* expression is restricted to the mesoderm, and later to the ectoderm bordering the neural plate. The neural plate similarly does not express *Twist*, which is present only in the mesoderm [53]. Altogether, these findings suggest that EMT key regulators were first expressed in the mesoderm and have been recruited by the neuroepithelium/neural plate border cells facilitating delamination and migration of NCC in vertebrates.

ROLE OF THE NC IN THE DEVELOPMENT OF THE VERTEBRATE HEAD

One of the major results brought about by the experiments aimed at deciphering the contribution of the NC to the vertebrate body was that the vertebrate head, which is largely a chordate innovation, is mainly derived from ectoderm, with a major participation of the NC. In some species, the whole head skeleton (except for the occipital region and part of the otic capsule), the connective and adipose tissues (including the connective tissue associated with the muscle fibers together with the tendons), the dermis of the face and ventral part of the neck, and various structures associated with sense organs are NC derived [7,35,54,55] (Table 1.1).

This and various other considerations led Gans and Northcutt [28] to put forward the "new head" concept, according to which the acquisition of a *neural crest* and of the ectodermal *placodes* allowed the development of a *head* endowed with sense organs that did not exist in protochordates. The brain, sense organs, and the facial skeleton with the jaw as organ of predation provided this novel group of animals with the capacity to adopt a life style different from that of their filter-feeder ancestors, the *Amphioxus*-like animals: vertebrates became predators.

This profound change in lifestyle required other major evolutions in anatomical and physiological systems that were pointed out by Northcutt and Gans [29]. Compared to their predecessors, these authors are notable for considering the evolution of the various organ systems from protochordates to vertebrates, not only individually, but also by examining the correlations that exist between them.

The filter-feeder protochordates possess a poorly developed cephalic region: consisting of merely rudimentary sense organs and no

TABLE 1.1 Summary of the Differentiated Cell Types Derived from the Cephalic and Trunk NC in Amniote Vertebrates

		NC-Derived Cell Types		
	Neural Cell Types	Pigment Cells	Endocrine Cells	Mesenchymal Cell Types
Cephalic NC	**Neurons:** PNS cranial sensory ganglia and parasympathetic (ciliary) ganglia ENS ganglia **Glial cells:** Satellite cells of PNS ganglia ENS glial cells Schwann cells of PNS nerves Ensheathing olfactory cells	Melanocytes (skin and inner ear)	Carotid body cells C cells (thyroid gland/ ultimobranchial body)	*In head and neck:* Osteocytes and chondrocytes (Craniofacial skeleton) Smooth muscle cells (vascular wall) Heart conotruncus Pericytes (brain vessels) Meninges (forebrain) Connective cells in muscles and glands Dermal cells Tendons Adipocytes Keratocytes (corneal endothelium and stroma) Odontoblasts (*Mammals*)
Trunk NC	**Neurons:** PNS sensory DRG, sympathetic and parasympathetic ganglia **Glial cells:** Satellite cells of PNS ganglia Schwann cells of PNS nerves	Melanocytes (skin)	Adrenal-medullary cells	

skeletal tissues. Therefore several parallel changes occurred in the protochordate-to-vertebrate transition. First, metabolic, circulatory, and respiratory modifications had to occur to move from the filter-feeding life style to the more active behavioral requirements of predators. Sensory systems had to be acquired in association with increased size and complexity of the brain for effective predation. Circulatory and respiratory changes also had to take place, particularly at the level of the pharynx, to improve oxygenation as required by more extensive muscle activity.

In the ammocoetes (the larval form of the lamprey), representative of primitive forms of extant vertebrates, a hypomeric striated musculature is present that increases the flux of water in the pharynx, a feature absent in *Amphioxus*. This musculature is provided with an innervation (i.e., cranial nerves V, VII, IX,

X) of NC origin (see [28,35] for references). Moreover, the branchial apparatus is supported by cartilaginous bars of NC origin [56–58]. In enteropneust hemichordates and protochordates, water goes through the pharynx due to the beat of cilia that line the pharyngeal endoderm, whereas gas exchanges in the branchial apparatus are much more efficient in extant anamniote fish owing to specializations of the epithelium forming the gills that increase the surface for these exchanges.

It is noteworthy that a recent article from the research group at the University of Tübingen led by Nüsslein-Volhard [59] has shown that, within the gills of the extant teleosts (zebrafish), pillar cells, which mechanically support the filaments of the gill apparatus and form gill-specific capillaries, are NC derived. These cells play a critical role in the efficiency of blood oxygenation by increasing the respiratory surface area. Pillar cells are present in lower vertebrates: hagfish, chondrichthyans, and osteognathostomes but not in cephalochordates (see [59] and references therein) and are therefore a vertebrate innovation. The ontogeny of these pillar cells is not well documented. However, the fact that they express smooth muscle myosin [60] suggests that they may be a particular type of differentiation of the smooth muscle cells derived from the NC in amniote embryos, which participate in the blood vessel walls [7,8,61].

Blood circulation is driven in vertebrates by a muscular pump (the heart) also derived from the mesodermal hypomere with the participation of the NC (the so-called cardiac NC) in septation. Moreover, the large arteries exiting from the heart have a musculo-connective wall surrounding the vascular endothelium, which is NC derived [7−10].

The pressure and gas levels of the blood are regulated in vertebrates by vasoreceptors (type 1 and type 2 cells of the carotid body) derived from the NC [35,62].

THE EVOLUTIONARY ORIGIN OF THE VERTEBRATE NERVOUS SYSTEM

The most spectacular changes that distinguish vertebrates from other chordates took place in the nervous system. The primitive condition of the nervous system in metazoans is that of dispersed nerve cells within the epithelium lining the external surface of the body and in the interstitial space between ectoderm and endoderm as present in *Hydra*. During evolution, the use of muscles made nerve cells necessary for locomotion; hence the switch from widespread neural plexuses to more concentrated and organized neural centers. Such a process accounts for the formation of the dorsal nerve cord seen in prochordates and vertebrates. It is noticeable that, in the latter, neural

plexuses have been conserved in the gut wall where they are so abundant that they are sometimes qualified as a "second brain." This designation is also based upon the fact that, in contrast to the PNS, the ENS harbors complete reflex systems, including sensory and motor neurons connected by interneurons as in the CNS. In the vertebrate gut, the ENS forms two distinct plexuses, and the dispersion of their precursor cells takes place from the NC, in which they are concentrated before invading the developing gut [63].

In some species of enteropneust hemichordates, the nervous system consists of a short hollow neural tube that exists only in the collar section and opens caudally and rostrally through neuropores. The inner layer of this tubular structure is ciliated and the neurons, mostly of the motor type with also some interneurons [64], are only present within its ventral and lateral sides. In addition, sensory and some additional motor neurons are distributed outside the neural tube within epidermal plexuses that are particularly developed in relationship with the oral system of food acquisition.

The nervous system of extant protochordates involves, as in vertebrates, the formation of a dorsal hollow nerve cord including motoneurons and interneurons. In cephalochordates (*Amphioxus*), the brain is merely a simple vesicle divided into rostral and caudal halves by an "infundibular" recess. The rostral part is formed by a simple layer of ependymal cells and does not possess neurons. The caudal part contains neurons ventrally and laterally, whereas the dorsal roof exists as a thin membrane devoid of neurons. The motor neurons are located ventrally, while a dorsolateral zone is sensory. Plexuses are present in the atrium and pharyngeal region, the alimentary tract, and beneath the somatic musculature of the trunk ([28] and references therein).

It appears, therefore, that the evolution from protochordates to vertebrates was characterized by a hypercentralization of the nervous system and the complex development of the brain in relation with the appearance of paired sense organs that developed either from the neural plate itself (optic vesicles) or from the cephalic placodes. Moreover, the sense organs of protochordates and enteropneusts are either absent or poorly developed. According to Lacalli [65], in *Amphioxus* the discrete sensory modules reside within the CNS and include several distinct structures: a frontal ventral—medial "eye" built of ciliary receptors, a lamellar body located dorsally in the anterior part of the neural tube, rhabdomeric photoreceptors called *Joseph cells* forming a dorsal column, and numerous rhabdomeric dorsal pigmented ocelli (designated as *Hesse organs*) positioned in the ventral neural tube along the whole body.

In urochordates, unpaired sense organs are dispersed in the body [64]. The development and structure of the CNS in the tadpole larva of the ascidian *Ciona intestinalis* has been the subject of detailed and careful investigations and provided significant results: with its 330—350 cells, the CNS of *Ciona* offers a chordate nervous system in miniature. Like in vertebrates, it develops from a neural plate whose anterior-most part does not roll up but contributes to the dorsal-anterior epidermis with adhesive organs, head sensory neurons and "pharynx" [66—68]. The posterior part of the CNS is divided into a rostral "sensory vesicle" comprising 215 cells with the sensory receptor systems, including the pigmented ocellus [68], and anterior to it, a pigmented otolith [69]. The caudal part of *Ciona* CNS is a nerve cord (65 cells, mostly ependymal). Between anterior and posterior CNS lies a visceral ganglion made up of about 45 cells. A slender neck region connects the sensory vesicle to the visceral ganglion (Figure 1.2).

Ascidian orthologs of vertebrate developmental genes are expressed in the CNS in an anterior—posterior pattern that closely parallels the one that regionalizes the vertebrate brain: the tripartite organization of the neurons corresponding to the three regions of the vertebrate fore-, mid-, and hindbrain is found in both *Ciona* and *Halocynthia* [70,71] (Figure 1.2). It is characterized by the regionalized expression, from rostral to caudal, of *Otx*, *Pax-2/5/8*, and *Hox* genes. Homology with the vertebrate anterior spinal cord can also be recognized: the anterior region of the ascidian caudal nerve corresponds to the anterior vertebrate spinal cord on the basis of *Hox5* expression; the posterior visceral ganglion corresponds to rhombomeres 5—8 (r5—r8) on the basis of *Hox3* expression; and *Hox1* labels both the anterior visceral ganglion in ascidians and r4 in vertebrates. From gene expression patterns, one can deduce that the visceral ganglion of urochordates is similar to the descending brain stem reticular-spinal neurons of vertebrates.

The neck region of the ascidian larva located between the *Otx* territory of the sensory vesicle and the *Hox3*-expressing region of the anterior visceral ganglion expresses *Pax2/5/8* and *Fgf5/17/18* [72,73], making it resemble the vertebrate *Isthmus*, although in a different rostro—caudal sequence [74]. The posterior vesicle is similar to the metencephalon, both expressing *Fgf9/16/20* and *engrailed*. The only marker of telencephalon that has been identified in ascidians is *emx* and it is expressed by the anterior epidermal territory [70] (Figure 1.2).

These findings led to the conclusion that urochordates, like cephalochordates [25], have no structure equivalent to a telencephalon. The gene expression profiles observed in these primitive animals thus point to the possibility that CNS emergence proceeded in several steps, by the transformation of an ancestral diffuse network into a centralized neuronal assembly. It seems therefore that the enormous developmental complexity of the brain and the

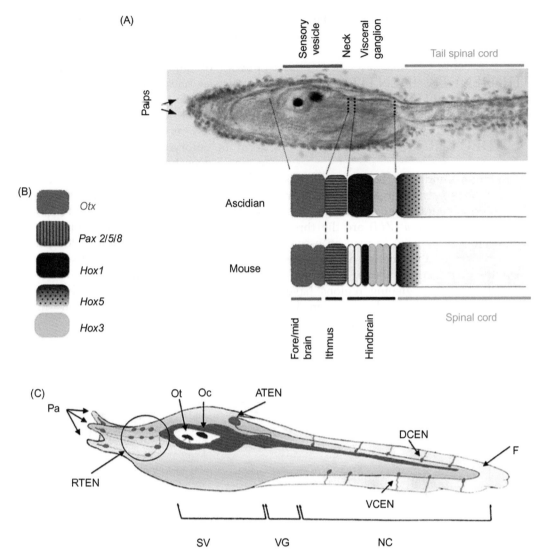

FIGURE 1.2 **Organization of the ascidian nervous system.** (A) Lateral view of the anterior section of a *Ciona intestinalis* larva. (B) Schematic representation of the expression pattern of conserved genes between ascidians and vertebrates. The name of the vertebrate territories thought to be homologous to the sensory vesicle (SV), neck, visceral ganglion (VG), and tail nerve cord (NC) is indicated. (C) Positions in the larva of the CNS (blue), epidermal sensory neurons of the PNS (red), and pigmented ocellus (Oc) and otolith (Ot). Pa, palps; RTEN, rostral trunk epidermal neurons; ATEN, apical trunk epidermal neurons; VCEN, ventral caudal epidermal neurons; DCEN, dorsal caudal epidermal neurons; F, Fin. *Source:* *Reproduced from [70], with permission.*

appearance of the telencephalon are exclusive vertebrate characters. However, the overall plan of the brain was already present in their urochordate ancestors and even in more primitive animals such as the hemichordate *Saccoglossus kowalevskii*. In this animal, a genetic scaffold is already present in the rostral part of the ectoderm in which several landmarks of the vertebrate brain can be recognized [75], as described below.

THE GENETIC IDENTITY OF VERTEBRATE SECONDARY BRAIN ORGANIZERS AS LANDMARKS FOR CNS EVOLUTION IN DEUTEROSTOMES

The midbrain–hindbrain boundary (MHB also designated as *Isthmus* organizer) together with the anterior neural ridge (ANR) and the *zona limitans intrathalamica* (ZLI) are the three signaling centers recognized to play an essential role in patterning the anterior neural plate in vertebrates. As described above, only unobtrusive traces of such signaling regions have been detected in extant protochordates on the basis of their gene expression profile [67,76–79].

A recent study of the early stages of development of an enteropneust hemichordate, *Saccoglossus kowalevskii* [75], has shown that ancestors of the three signaling centers of vertebrates, identified through the genes they express, are already present in the rostral ectoderm in embryos of this species. *Fgf8/17/18* (a single gene homologous to vertebrate *Fgf8, Fgf17, Fgf18*), *sfrp1/5, hh*, and *wnt1* are expressed in a vertebrate-like arrangement in hemichordate ectoderm [75] (Figure 1.3).

These results thus tend to support the view that these genetic programs belonged to an ancestral regulatory scaffold used for body patterning in deuterostomes. This genetic information was retained in the vertebrate phylum where it was applied in brain

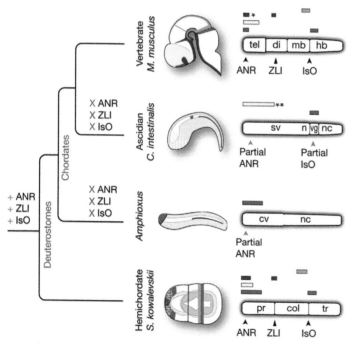

FIGURE 1.3 **Evolutionary gain and loss of ANR, ZLI, and *Isthmus* organizer (IsO)-like genetic programs.** Schematic diagrams depicting the expression of *Fgf8, Sfrp1, Shh*, and *Wnt1* homologs in the mouse brain and the ectoderm of *Ciona intestinalis, Amphioxus*, and the hemichordate, *S. kowalevskii*. Embryos and CNS regions are oriented with the anterior side to the left. Diagrams depict only expression domains that are related to signaling components of vertebrate CNS signaling centers. (cv, cerebral vesicle; n, neck; nc, nerve cord; sv, sensory vesicle; vg, visceral ganglion). Single asterisk indicates that *Shh* is expressed in the medial ganglionic eminence, near the ANR. Double asterisk indicates that *sfrp1/5* is expressed in the *Ciona intestinalis* anterior ectoderm from the 64-cell stage up to neurulation but is then downregulated in the anterior ectoderm and CNS (yellow stripes). Source: *Reproduced from [75], with permission.*

patterning. Cephalochordates, at least as far as their brain is concerned, are derived chordates in which brain development involved the loss of characters that were parts of an ancient genetic regulatory scaffold, preceding the morphological innovation of vertebrates (Figure 1.3).

It is striking that the presence of these signaling centers in hemichordates was not exploited for the construction of complex morphological structures in their neural anlagen. It is only during the early and successive steps of vertebrate evolution that these ancestral landmarks of the rostral ectoderm were called upon to pattern novel neural structures in a conserved regulatory framework that, in hemichordates, seems to have been merely devolved to ectodermal patterning.

When this preestablished regulatory network was first involved in regulating the rostral region of the neural plate and whether it was associated with the onset of vertebrate novelties remain unresolved questions. In the extant invertebrate chordates, these signaling center components were nearly completely lost, and, in the absence of knowledge about the brain structure of the early vertebrate ancestors, it is difficult to assess when this preestablished regulatory network was first applied to brain patterning.

Thus, basal chordates have not retained all ancestral characters and have undergone substantial independent evolutionary changes. This has long been accepted for urochordates, but cephalochordates were considered as the most informative extant group for reconstructing ancestral chordate characters. This is likely to be true for trunk structures, but clearly not for head and brain.

Whether the ANR, *ZLI*, and *Isthmus* genetic programs are unique deuterostome features or have even deeper bilaterian origins has not yet been addressed. Broad phylogenetic sampling including morphologically divergent out-groups will be instrumental to identify gene regulatory innovations responsible for evolutionary changes in body plans.

ARE PRECURSORS OF THE NC PRESENT IN NON-VERTEBRATE CHORDATES?

As stressed by Gans and Northcutt in their seminal 1983 article [28], embryological, paleontological, and physiological considerations all point to the decisive role of the NC in the evolution of vertebrates. Hence, there is considerable interest in identifying forerunners of this structure during development and in the adults of species considered as the closest relatives of vertebrates: the invertebrate chordates [26] (see Figure 1.1). A migratory cell population, originating from the vicinity of the neural plate, was identified in the mangrove tunicate and proposed to be a rudimentary precursor of the NC [80]. However, other studies on several ascidians provided the evidence that these cells were of mesodermal origin [81].

Several genes expressed in the NC lineage of vertebrates were found to be activated in *Ciona intestinalis* in cells belonging to a cephalic melanocytic lineage derived from blastomere *a9.49* and arising from the neural plate border. These cells expressed not only neural plate border genes but also a number of NC specification genes such as *Id*, *Snail*, *Ets*, and *FoxD* [82–87]. Moreover, the posterior cells of this lineage form the gravity-sensing otolith and the melanocytes of the light-sensitive ocellus [88]. Abitua et al. [89] have recently shown that these two cell types are derived from a single progenitor. Their differentiation depends upon *Wnt* signaling according to a pathway observed in vertebrates where Mitf (microphthalmia-associated transcription factor) directly activates target genes involved in melanogenesis of NC-derived melanocytes. Both precursor cells express *Mitf* before neurulation. Subsequently, the posterior cell receives a localized Wnt signal and activates *FoxD* (the ascidian homolog of *FoxD3* of vertebrates), which attenuates *Mitf*, leading to a faint pigmentation in the ocellus. In the anterior cell, *Mitf* expression is sustained and leads to the densely pigmented otolith [89].

Interestingly, cells of the *a9.49* lineage can be reprogrammed into *ectomesenchyme* by the targeted misexpression of *Twist* [89], a regulatory gene involved in EMT and formation of mesenchymal cells by the vertebrate cephalic NC. This suggests that the NC regulatory network predated the divergence of tunicates and vertebrates. The authors propose that the co-option of a mesenchymal determinant such as *Twist* into the neural ectoderm was critical in the acquisition of the migratory properties of the border cells of the neural plate and the formation of the vertebrate *new head*. These observations suggest that the mesenchymal properties of the NCC were the last to appear during vertebrate evolution, as previously proposed by Gans and Northcutt [28] and Shimeld and Holland [90].

THE MULTIPLE ROLES OF THE NC IN THE CONSTRUCTION OF THE VERTEBRATE HEAD AND BRAIN

As mentioned previously, the participation of the NC to the head skeleton had been demonstrated by the pioneers, whose experimental animals were essentially Amphibians and whose work extended over the first half of the twentieth century (see [2] for references). At that time, however, neither the extent of the participation of the NC to the cephalic skeleton, nor its contribution to soft tissues of the head had been established (see [35] for a review). In addition, apart from some extirpation experiments performed in chick embryos ([91–93]), no investigations on NCC fates concerned amniote embryos before the late 1960s (see [35,36] for references). An analysis based on the use of the quail-chick marker system disclosed the respective contribution of the cephalic NC, the cephalic paraxial mesoderm, and the five first somites to the head skeleton. It turned out that in birds, except for the occipital and the otic (partly) domains, the head skeleton is nearly entirely NC derived [54,94] (Figure 1.4).

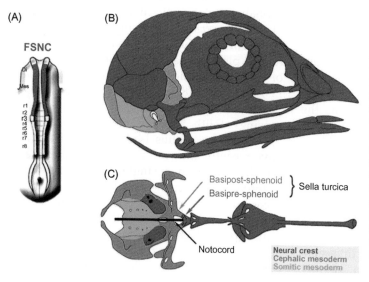

FIGURE 1.4 **The major part of the vertebrate skull originates from the NC.** (A) Schematic drawing of the avian embryo at 5 somite stage showing the NF, including the FSNC (extending from the diencephalon to r2; in red), which is at the origin of most of the head skeleton. The NC from r3–r8 (yellow) contributes to the hyoid cartilage. The posterior limit of the NC capable of differentiating into mesenchymal derivatives corresponds to the level of somite 4 included. The trunk NC begins caudal to somite 4. Lateral (B) and basal (C) views of the chondrocranium in E10-avian embryo, showing the triple origin of the head skeleton [54], from the NC (red), the cephalic mesoderm (blue), and the anterior somitic mesoderm (green). Source: *Reproduced from [95], with permission.*

This picture does not fully apply to all vertebrate species studied. In mammals, the parietal bone was found to be mesodermally derived [96,97]. In the more primitive forms of vertebrates in which these types of cell lineage studies can be performed, some variations in the origin of cranial bones have been found (see [98] for an overview). It is noteworthy that the homologies between bones in different vertebrate groups is in some cases imprecise. The NC origin of the facial skeleton and of most of the cranium, however, appears as a rule.

In addition to these skeletal tissues, the cephalic NC (extending from the mid-diencephalon to the level of somite 4) yields other types of mesenchymal derivatives such as connective tissue, tendons, adipose, and smooth muscle cells (Table 1.1). The mesoderm-derived striated myofibrils of the masticatory muscles are associated with connective tissue of NC origin and the muscles are attached to the bones by NC-derived tendons [7,99,100].

Moreover, cephalic NCC give rise to the forebrain meninges, while these structures covering the remainder of the CNS are derived from the mesoderm. The dermis of the face and ventral part of the neck contains connective and adipose tissues that arise from the NC. The wall of the cephalic blood vessels is NC in origin, but their endothelium is derived from the mesoderm [8]. The contribution of NC to the conotruncus of the heart has also been fully demonstrated [7–10] (Table 1.1). Of note is the fact that the capacity of the NC to yield mesenchymal derivatives is restricted to the cephalic NC in amniote vertebrates (see [7,54] and references therein). An exception to this rule however concerns the turtle exoskeleton whose ventral part (plastron) is claimed by Gilbert and colleagues [101–103] to be derived from the trunk NC. A particular behavior of a late-emigrating population of the trunk NC, which remains for a while in contact to the dorsal neural tube in a sort of "migrating staging area" before reaching the ventral side of the embryo, is considered to be responsible for the development of the plastron bones [103]. This is a remarkable exception that has no equivalent in any other vertebrate group. Although none of the molecular markers (HNK1, p75, FoxD3, and PDGFRα) used to follow the dorsal ventral migration of the cells that form these bones is strictly NCC specific, the fact that they are expressed together is a strong argument in favor of the NC origin of turtle plastron.

HOX GENE EXPRESSION IN THE CEPHALIC NC

The area of the neural fold (NF)/premigratory NC, endowed with the capacity to yield skeletal tissues in amniote vertebrates, is divided into two domains differing by their expression of genes of the *Hox* clusters: a rostral domain (from the mid-diencephalon to r3 included), in which *Hox* genes are not expressed, and a caudal part (including r3, partly to r7 included) reaching the level of somite 4 in the avian embryo, in which *Hox* genes situated at the 3′ end of the *Hox* clusters (*Hoxa2*, *Hoxa3*, *Hoxb4*) are expressed. Expression of *Hox* genes is shared by the NCC of the entire trunk region [104–106].

A particular area of the anterior NC domain was shown to produce the cells responsible for the construction of the entire facial skeleton. It was, for this reason, designated "facial skeletogenic neural crest" (FSNC). The problem was then raised as to whether the fact that *Hox* genes are not expressed in the rostral domain of the NC is, or not, critical for its role in the construction of the vertebrate head.

This question could be addressed through two experimental designs. (1) Removal of the FSNC (*Hox*-negative domain of the cephalic NC): if the rostral *Hox*-negative and caudal *Hox*-positive NCC are both equally capable to yield the facial skeleton and other derivatives in the head, the regeneration capacities of the precocious embryonic tissues should provide the necessary cells and no or only mild malformations should be observed. (2) The second experimental design consisted in triggering the forced expression of *Hox* genes (*Hoxa2*, *Hoxa3*, *Hoxb4*) in the rostral domain of the NC designated FSNC.

Experiment 1 showed that no regeneration of the excised *Hox*-negative domain of the NC from the more caudal *Hox*-positive area took place. This resulted in the absence of head

skeleton and (unexpectedly) in severe malformations of the pre-otic brain resulting in *exencephaly*. In contrast, transplantation of part (one third of the entire length) of the FSNC from quail to chick following ablation of the FSNC in the recipient was sufficient to rescue the phenotype, while grafting of NF from the *Hox*-positive region of the head was insufficient (Figure 1.5). This demonstrated that the regeneration capacity of the FSNC at this stage is restricted to the *Hox*-negative domain of the head and absent in the *Hox*-positive domain that gives rise to the hyoid bone and cartilage.

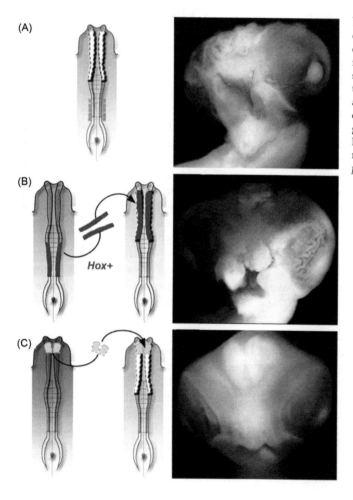

(A)

(B)

Hox+

(C)

FIGURE 1.5 **Detrimental effect of *Hox* gene expression on head development.** (A) Bilateral extirpation of the FSNC leads to the absence of facial structures and anencephaly. (B) Bilateral substitution of quail r4–r8 *Hox*-positive NCC to the FSNC results in anencephaly and complete absence of upper face and lower jaw. (C) In embryos after excision of the FSNC, a bilateral graft of the posterior diencephalic *Hox*-negative NC is sufficient to restore normal head development. Source: *Reproduced from [11], with permission.*

It is striking to see that the regeneration took place inside the *Hox*-negative domain, which cannot be rescued by the caudal part of the NC in which *Hox* genes are expressed [107]. This result shows that head and trunk are, from an early embryonic stage, distinct entities whose development obeys different patterning rules. This supports the view put forward by Gans and Northcutt according to which the vertebrate head would be an *addition* to an initial chordate body plan rather that the transformation of its most rostral end [28,29].

MALFORMATIONS OF THE BRAIN INDUCED BY FSNC REMOVAL

A few hours after the FSNC surgical ablation, the dorsal closure of the neural tube took place at the level where the NF had been removed (i.e., from the mid-diencephalon to r3 excluded), as it does in intact embryos. However, this closure was only transitory. When observed at 3.5 days, the neural tube was not closed dorsally any more and the telencephalon did not develop normally (Figure 1.6). Moreover, neither the thalamus nor the optic tectum was present in the operated embryos. Interestingly, expression of *Wnt1* and *Wnt8b* in dorsal brain vesicles was lost and the expression domain of *Shh*, which is medioventrally restricted in the brain of control embryos, extended over the whole neural epithelium after FSNC ablation [107]. Therefore, removal of the *Hox*-negative domain of the NC prevents the development of the alar plates of the neural anlagen, causing a "ventralization" of the anterior neural structures arising from the basal plates.

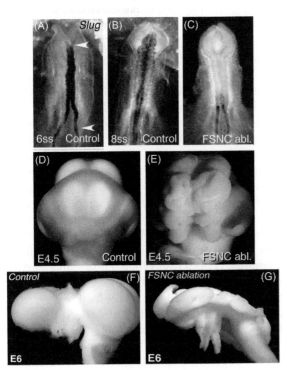

FIGURE 1.6 **Ablation of FSNC causes defects of the craniofacial mesenchyme and the anterior brain.** In control chick embryos, expression of *Slug/Snail2* delineates the NC at premigratory stage (A, 6-somite stage) and early migratory stage (B, 8-somite stage). No expression of *Slug* remains at 8-somite stage after surgical removal of FSNC; note that the neural tube is closed at the level of the excision (C). (D and E) Ventral views of E4.5 chick embryos show defects of the craniofacial region after ablation of FSNC (E), as compared to the control (D). (F, G; whole-mount preparations of the brain) At E6, ablation of the FSNC results in reduction and absence of closure of the anterior brain (G) as compared to the control (F). Source: *Adapted from [107].*

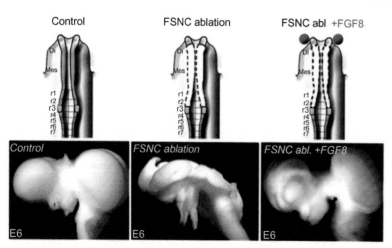

FIGURE 1.7 Exogenous FGF8 restores brain development after ablation of the FSNC. Whole-mount brain preparations dissected out from control, FSNC-deprived, and FGF8-supplemented embryos at E6. After ablation of FSNC, brain is exencephalic and partitions into telencephalon, diencephalon, and mesencephalon are no longer recognizable. When stimulated with exogenous FGF8 in ANR or BA1 ectoderm, neural tube closure occurs and brain regionalization is restored. Source: *Reproduced from [11], with permission.*

In addition, the development of the so-called secondary brain organizers (i.e., the ANR and the *Isthmus*) was severely affected by removal of the FSNC. These structures have been shown to regulate the early patterning of brain vesicles through the signaling molecule Fgf8 [108–111]. In operated embryos, the production of Fgf8 was severely reduced. Addition of Fgf8-soaked beads rescued the phenotype resulting from FSNC ablation. Not only did skeletal facial structures develop, but the encephalic vesicles also showed nearly normal morphology and gene expression [112] (Figure 1.7). The origin of the NCC responsible for rescuing the craniofacial skeleton and connective tissues in the operated embryos was found to be from r3 whose cells invading the *Hox*-free environment of the anterior head lost their initial *Hoxa2* expression. As shown in quail-chick chimeras, in the presence of exogenous Fgf8, r3-NCC in FSNC-ablated host embryos could give rise to all the anterior head structures normally generated by the FSNC [112]. In contrast, cells from r4 did not regenerate a NCC contingent able to loose their expression of *Hoxa2* and to colonize the rostral NC domain following its ablation.

EFFECT OF FORCED EXPRESSION OF *HOX* GENES IN THE FSNC ON FACE AND BRAIN DEVELOPMENT

The second set of experiments that consisted in forcing the expression of *Hoxa2*, *Hoxa3*, or *Hoxb4* genes in the cells of the FSNC domain of the cephalic NF led to the same conclusion: that the head structures could not develop from NCC expressing genes of the *Hox* clusters [113]. The phenotypes of the embryos so-treated were similar to those resulting from the surgical ablation of the FSNC. The forced expression of *Hoxa2* in FSNC thus resulted in complete loss of the facial skeleton together with severe defects in the brain [113]. *Hoxa2* is known to act as a selector gene for second arch identity, and its ectopic expression in branchial arch 1(BA1) NC led to transformation of the upper jaw into hyoid-like skeletal elements [114–116]. When *Hoxa3* and *Hoxb4* were overexpressed individually in the FSNC, the effects were milder. In contrast, expression of both together perfectly reproduced the malformations resulting from *Hoxa2* alone [113].

It thus appears from these experiments that (1) the expression of *Hox* genes by NCC in the most rostral part of the head is not compatible with NCC differentiation into skeletal tissues (either membrane bones or cartilages), (2) the cephalic NC has a more important function in building the vertebrate head than by just providing the mesenchymal cells at the origin of cephalic skeletal and connective tissues (it exerts a strong regulatory role in brain development), and (3) expression of *Hox* genes in the cells of the FSNC prevents their development into skeletal tissues and induces severe defects in the pre-otic brain. These effects are similar to those produced by the surgical removal of the cephalic NC.

The next question concerned the mechanisms through which the cephalic NC regulates the level of Fgf8 in the secondary brain organizers that are known to be critical for brain development at this stage.

ROLE OF THE CEPHALIC NC IN THE REGULATION OF Fgf8 PRODUCTION BY THE ANR AND ISTHMUS

In addition to Fgf8 expression in the secondary brain organizers, other secreted factors such as Bmps have been observed [111,117]. It has been shown that Bmp4 downregulates *Fgf8* expression in the developing brain, including the ANR of chick and mouse [118]. This indicates that forebrain morphogenesis is the result of a cross-regulation between Bmps, Fgfs, and Shh signaling centers. During the early stages of neurulation, cephalic NCC express genes encoding the Bmp-antagonists Gremlin and Noggin [119]. The decrease in Fgf8 production by secondary brain organizers, observed after FSNC ablation, thus results from the increase of Bmps due to the absence of the Bmp-antagonists normally produced by the NCC [120]. Similarly, downregulation of Fgf8 in the ANR and *Isthmus* by RNA interference or by increasing the amount of

Bmp4 in the forebrain leads to a strong reduction in forebrain size accompanied by a disorganization of the inter-hemispheric structures, absence of olfactory bulbs and nerves, and atrophy of choroid plexus. In contrast, after downregulation of *Gremlin* and *Noggin* in the FSNC by RNA interference, the level of *Fgf8* transcripts is severely decreased in the ANR and *Isthmus*, resulting in significant atrophy of the telencephalon, thalamus, and optic tectum [121]. Gain of function of *Gremlin* and *Noggin*, in contrast, leads to hypertrophy of the pre-otic brain [121].

It is therefore very striking to see that the cephalic NC not only provides the head with most of its skeleton, but also that it plays a critical role in the development of the neural primordium, which is at the origin of the telencephalon and of the thalamus and optic tectum [11,121].

THE EVOLUTION OF SKELETAL STRUCTURES IN THE VERTEBRATE PHYLUM

It has been amply documented that the first mineralized skeletal tissues that appeared in the vertebrate *phylum* were of NC origin. The pteraspidomorphs agnathans, which appeared in the early Ordovician (-470 Myr), possessed an ossified integument armor bearing tubercles made up of *dentine*. Dentine is considered as a marker of NC origin when it is detected in primitive forms of vertebrate fossils [122–124]. Although possessing a probably cartilaginous endoskeleton likely derived from the somites, they seemed to have been devoid of well-characterized vertebrae [125,126].

Later on (after -430 Myr) in osteostracans, the NC-derived exoskeleton was present like in gnathostomes, together with an endoskeleton made up of true cellular bone of somitic origin. In extant teleost fish (and probably also jawed fish possessing scales with a dentinous layer), the NC-derived skeleton is reduced to the skull and facial bones and cartilages, while the endoskeleton is fully developed [127]. The question as to whether the osseous rays of their caudal and dorsal fins are of NC or mesoderm origin is debated. In zebrafish, Smith and colleagues [124] found that a migration of trunk NCC is taking place to the caudal fin, and they proposed that these cells contribute to the lepidotrichia. However, this assertion did not rest upon lineage tracing. Recent genetic labeling of NCC in zebrafish yields results that are still controversial: for Lee et al. [128,129] the bony rays of the fins are mesodermally derived, a view shared by Shimada et al. [130] in medaka; in contrast, Kague et al. [98] provide convincing evidence of a contribution of labeled NCC to the lepidotrichia of the caudal fin, when observed in the fully differentiated state.

Vertebrate skeletal evolution is thus characterized by the increasing development of mesodermal bone and cartilage with the regression of the primitive exoskeleton of NC origin. The latter was essentially preserved in the head and its extension closely accompanied that of the brain. Concomitantly to the substitution of the trunk exoskeleton by the vertebral column derived from the sclerotomal part of the somites, the trunk NC has lost its participation to the truncal mesenchymal cell types in amniote embryos. It was clearly documented by the labeling studies carried out in birds that, *in vivo*, the trunk NC does not yield mesenchymal derivatives in normal development (see [35,36] for references). Similar conclusions were reached in mouse [97,131]. However, as mentioned above, in turtles recent studies have demonstrated the contribution of the trunk NC to the shell [101–103]. Mesenchymal potentialities are not totally absent in the TNC, as shown by heterotopic NC transplantation from trunk to head in quail-chick chimeras [132]. In such circumstances, a few mesenchymal cells (e.g., cells

lining the blood vessels but no skeletal elements) were seen to arise from the posterior NC levels. This was in certain conditions only, when the trunk NCC was mixed with cephalic NCC [132]. This suggested that the environment provided by the cephalic NC created a permissive milieu for the expression of latent mesenchymal potentialities still present in the trunk NCC.

Later, the demonstration of a remnant of ectomesenchymal potentialities in trunk NC was clearly provided through *in vitro* cultures where trunk NCC exhibited the capacity to yield mesenchymal derivatives, including skeletal tissues, in birds and mammals [133–138]. This shows that an "evolution memory" still exists in the truncal part of the NC, which makes these cells able to yield mesenchymal derivatives when challenged with appropriate culture conditions.

CONCLUSIONS AND PERSPECTIVES

This chapter, which briefly reviewed our knowledge about the characteristics of the NC, a specific structure of the vertebrate embryo, has also highlighted some new developments of the recent research concerning the role it may have played in vertebrate evolution. Vertebrates are characterized by their remarkable radiation since their appearance 570 Myr ago, their capacity to occupy most of the available environments of the planet, and their impressive diversification. Most of these capabilities are related to the development of their brain, which reached its pinnacle state in human primates.

Analyses of the migration and fate of the NCC in the avian embryo has provided evidence that the vertebrate head is nearly entirely made up of cells of the ectoderm, among which the NC is the main provider. This led to the notion that the appearance of this structure in a chordate ancestor has facilitated the formation of a "new head" [28] in which the brain and sense organs have developed and profoundly influenced the behavior of these animals.

The critical event in the evolution of the vertebrate *phylum* was therefore thought to be the appearance of the NC in a chordate ancestor. This prompted a large amount of research to test this hypothesis, in particular the search for the forerunner of a NC in more closely related extant protochordates. In this chapter, some of the recent results that have enlightened the origin of the NC and its role in the development of the vertebrates—namely, of their head—have been reported and discussed. One of the highlights of the abundant literature on the evolution of the NC in recent times was the observation of cells in *Ciona intestinalis*, which, by their migratory behavior and gene expression pattern, prefigure the vertebrate NC [89].

Another interesting observation mentioned in this review is the existence in primitive organisms like *Saccoglossus kowalevskii*, a hemichordate, of a genetic program belonging to an ancestral regulatory network devoted to the patterning of the rostral part of the body, which is used in vertebrates to construct the brain [75]. The fact that it is present in hemichordates and, with some loss and variations, in protochordates indicates once more the pertinence of the remark by François Jacob who evoked the "tinkering" of evolution, which uses "the Old" to make "the New."

Another point, which opens new research avenues is the role played by the cephalic NC in brain development. In addition to providing the forebrain with its meninges and the entire brain with most of its skeletal case, the NC was shown to regulate the amount of Fgf8 at a stage of development when the brain imperatively needs this growth factor to develop. The cephalic NC can therefore be considered as a third "secondary brain organizer" in addition

to the ANR and *Isthmus*. It controls these two structures by strictly regulating the amount of Fgf8 they produce. This therefore reinforces the view that the NC has been an important asset in vertebrate evolution. It is particularly striking to see that the selective removal of the rostral NF, which is at the origin of the skull and facial skeleton, negatively affects the development of a large part of the brain—namely, the telencephalon, the site of the cognitive abilities.

A problem, which unfolds in the field of NC research, is the distinction between its cephalic and truncal domains. From the studies carried out *in vivo* in amniote embryos, the notion was solidly established that the trunk NC is devoid of the capacity to yield ectomesenchymal derivatives. The finding produced by Scott Gilbert's laboratory about the NC origin of the plastron of turtle is of great interest. The production of osseous tissues (lepidotrichia) by trunk NC in the extant teleost fins is far from reaching unanimity among researchers interested in this problem. It has been recently shown in our laboratory [138] that a large proportion of trunk NCC of the avian embryo, provided with appropriate culture conditions, are able to yield a large variety of fully differentiated ectomesenchyme-derived cell types (bone, cartilage, smooth muscle, and adipose cells) together with neural (neurons, glia) and pigment cells. One can therefore suspect that these potentialities are inhibited in some way during embryonic development. The identification of the molecular cues able to play this role would be a great advance in the field of NC evolutionary biology. There is a consensus that the first skeletal tissues that have appeared in chordates are NC in origin. These NC bony derivatives have strongly regressed in the trunk while the endoskeleton developed. Is there a genetic relationship between the mechanisms responsible for the disappearance of the NC-derived exoskeleton of primitive vertebrates and the simultaneous development of the mesodermal endoskeleton of more evolved forms of vertebrates? Answering these questions would be an important step in our knowledge of vertebrate evolution.

Abbreviations

ANR anterior neural ridge
BA branchial arch
EMT epithelial-to-mesenchymal transition
ENS enteric nervous system
FSNC facial skeletogenic neural crest
MHB midbrain—hindbrain boundary
Million years (Myr)
NC neural crest
NCC neural crest cell
NF neural fold
PNS peripheral nervous system
R rhombomere
Shh Sonic Hedgehog
ZLI zona limitans intrathalamica

Acknowledgments

Work in the authors' laboratory is supported by the Centre National de la Recherche Scientifique (CNRS) and the Institut National du Cancer (INCa). The authors thank Michèle Scaglia for help in preparing the manuscript.

References

[1] His W. Untersuchung über die ersteanlage des Wirbeltierleibes. Die erste entwickling des Hühnchens. Leipzig: Vogel; 1868.

[2] Hörstadius S. The neural crest: its properties and derivatives in the light of experimental research. London: Oxford University Press; 1950.

[3] Hall BK. The neural crest in development and evolution. New York, NY: Springer-Verlag; 1999.

[4] Acloque H, Adams MS, Fishwick K, Bronner-Fraser M, Nieto MA. Epithelial—mesenchymal transitions: the importance of changing cell state in development and disease. J Clin Invest 2009;119:1438—49.

[5] Savagner P. The epithelial—mesenchymal transition (EMT) phenomenon. Ann Oncol Engl 2010;21:vii89—92.

[6] Nieto MA. The ins and outs of the epithelial to mesenchymal transition in health and disease. Annu Rev Cell Dev Biol 2011;27:347—76.

[7] Le Lièvre C, Le Douarin N. Mesenchymal derivatives of the neural crest: analysis of chimeric quail and chick embryos. J Embryol Exp Morphol 1975;34:125—54.

[8] Etchevers HC, Vincent C, Le Douarin NM, Couly GF. The cephalic neural crest provides pericytes and smooth muscle cells to all blood vessels of the face and forebrain. Development 2001;128:1059—68.

[9] Kirby ML, Gale TF, Stewart DE. Neural crest cells contribute to normal aorticopulmonary septation. Science 1983;220:1059—61.

[10] Kirby ML, Waldo KL. Neural crest and cardiovascular patterning. Circ Res 1995;77:211—5.

[11] Le Douarin NM, Couly G, Creuzet SE. The neural crest is a powerful regulator of pre-otic brain development. Dev Biol 2012;366:74—82.

[12] Dohrn FA. Der Ursprung der Wirbelthiere und das Princip des Functionswechsels, genealogische Skizzen. Leipzig: Wilhelm Engelmann; 1875.

[13] Semper C. *Arbeiten aus dem Zoologisch-Zootomischen Institut in Würzburg* (Volume Bd.2). Würzburg: Stahel'schen Buch & Kunsthandlung; 1875.

[14] Delsman HC. The ancestry of vertebrates as a means of understanding the principal features of their structure and development. Weltevreden, Java Visser & Company; 1922.

[15] A.A.W. Hubrecht. The relation of the nemertea to the vertebrata. Q J Microsc Sci 1887;27:605—44.

[16] Willmer P. Invertebrate relationships; patterns in animal evolution. Cambridge University Press, Cambridge; 1990.

[17] Perrier E. La Philosophie zoologique avant darwin. Paris: Felix Alcan; 1884.

[18] Gaskell WH. The origin of vertebrates. London: Longmans, Green, and Co; 1908.

[19] Patten W. The evolution of the vertebrates and their kin. Philadelphia, PA: P. Blakiston's Son & Co; 1912.

[20] De Robertis EM. Evolutionary biology commentary: the molecular ancestry of segmentation mechanisms. Proc Natl Acad Sci USA 2008;105:16411—2.

[21] Bateson W. The ancestry of the Chordata. Q J Microsc Sci 1886;26:535—71.

[22] Darwin C. The descent of man, and selection in relation to sex. 1st ed. London: John Murray; 1871.

[23] Willey A. Amphioxus and the ancestry of the vertebrates. New York; London: Macmillan and Co; 1894.

[24] Sewertzoff AN. Morphologische gesetzmäßigkeiten der Evolution. Jena: Gustav Fischer; 1931.

[25] Holland LZ, Holland ND. Evolution of neural crest and placodes: amphioxus as a model for the ancestral vertebrate? J Anat 2001;199:85—98.

[26] Delsuc F, Brinkmann H, Chourrout D, Delsuc HP. Tunicates and not cephalochordates are the closest living relatives of vertebrates. Nature 2006;439:965—8.

[27] Shimeld SM, Donoghue PC. Evolutionary crossroads in developmental biology: cyclostomes (lamprey and hagfish). Development 2012;139:2091−9.

[28] Gans C, Northcutt G. Neural crest and the origin of vertebrates: a new head. Science 1983;220:268−73.

[29] Northcutt G, Gans C. The genesis of neural crest and epidermal placodes: a reinterpretation of vertebrate origins. Q Rev Biol 1983;58:1−28.

[30] Weston JA. A radiographic analysis of the migration and localization of trunk neural crest cells in the chick. Dev Biol 1963;6:279−310.

[31] Chibon P. Analyse par la méthode de marquage nucléaire à la thymidine tritiée des derivés de la crête neurale céphalique chez l'Urodèle Pleurodeles waltlii Michah. CR Acad Sci Paris 1964;259:3624−7.

[32] Chibon P. Marquage nucléaire par la thymidine tritiée des dérivés de la crête neurale chez l'Amphibien Urodèle Pleurodeles waltilii Michah. J Embryol Exp Morphol 1967;18:343−58.

[33] Le Douarin N. Particularités du noyau interphasique chez la Caille japonaise (Coturnix coturnix japonica). Utilisation de ces particularités comme "marquage biologique" dans des recherches sur les interactions tissulaires et les migrations cellulaires au cours de l'ontogenèse. Bull Biol Fr Belg 1969;103:435−52.

[34] Le Douarin N. Comparative ultrastructural study of the interphasic nucleus in the quail (Coturnix coturnix japonica) and the chicken (Gallus gallus) by the regressive EDTA staining method. CR Acad Sci 1971;272:2334−7.

[35] Le Douarin N. The neural crest. Cambridge: Cambridge University Press; 1982.

[36] Le Douarin NM, Kalcheim C. The neural crest. 2nd ed. Cambridge: Cambridge University Press; 1999.

[37] Baum B, Settleman J, Quinlan MP. Transitions between epithelial and mesenchymal states in development and disease. Semin Cell Dev Biol 2008;19: 294−308.

[38] Wu SY, Ferkowicz M, McClay DR. Ingression of primary mesenchyme cells of the sea urchin embryo: a precisely timed epithelial mesenchymal transition. Birth Defects Res C Embryo Today 2007;81:241−52.

[39] Lim J, Thiery JP. Epithelial−mesenchymal transitions: insights from development. Development 2012;139: 3471−86.

[40] Kerosuo L, Bronner-Fraser M. What is bad in cancer is good in the embryo: importance of EMT in neural crest development. Semin Cell Dev Biol 2012;23:320−32.

[41] Nieto MA, Sargent MG, Wilkinson DG, Cooke J. Control of cell behavior during vertebrate development by Slug, a zinc finger gene. Science 1994;264: 835−9.

[42] Cheung M, Chaboissier MC, Mynett A, Hirst E, Schedl A, Briscoe J. The transcriptional control of trunk neural crest induction, survival, and delamination. Dev Cell 2005;8:179−92.

[43] Sakai D, Suzuki T, Osumi N, Wakamatsu Y. Cooperative action of Sox9, Snail2 and PKA signaling in early neural crest development. Development 2006;133:1323−33.

[44] Liu JA, Wu MH, Yan CH, Chau BK, So H, Ng A, et al. Phosphorylation of Sox9 is required for neural crest delamination and is regulated downstream of BMP and canonical Wnt signaling. Proc Natl Acad Sci USA 2013;110:2882−7.

[45] Shi J, Severson C, Yang J, Wedlich D, Klymkowsky MW. Snail2 controls mesodermal BMP/Wnt induction of neural crest. Development 2011;138:3135−45.

[46] Chen ZF, Behringer RR. twist is required in head mesenchyme for cranial neural tube morphogenesis. Genes Dev 1995;9:686−99.

[47] Soo K, O'Rourke MP, Khoo PL, Steiner KA, Wong N, Behringer RR, et al. Twist function is required for the morphogenesis of the cephalic neural tube and the differentiation of the cranial neural crest cells in the mouse embryo. Dev Biol 2002;247:251−70.

[48] Bildsoe H, Loebel DA, Jones VJ, Chen YT, Behringer RR, Tam PP. Requirement for Twist1 in frontonasal and skull vault development in the mouse embryo. Dev Biol 2009;331:176−88.

[49] Murray SA, Gridley T. Snail1 gene function during early embryo patterning in mice. Cell Cycle 2006;5:2566−70.

[50] Van de Putte T, Maruhashi M, Francis A, Nelles L, Kondo H, Huylebroeck D, et al. Mice lacking ZFHX1B, the gene that codes for Smad-interacting protein-1, reveal a role for multiple neural crest cell defects in the etiology of Hirschsprung disease-mental retardation syndrome. Am J Hum Genet 2003;72:465−70.

[51] Van de Putte T, Francis A, Nelles L, van Grunsven LA, Huylebroeck D. Neural crest-specific removal of Zfhx1b in mouse leads to a wide range of neurocristopathies reminiscent of Mowat-Wilson syndrome. Hum Mol Genet 2007;16:1423−36.

[52] Nikitina NV, Bronner-Fraser M. Gene regulatory networks that control the specification of neural-crest cells in the lamprey. Biochim Biophys Acta 2009;1789:274−8.

[53] Yu JK. The evolutionary origin of the vertebrate neural crest and its developmental gene regulatory network—insights from amphioxus. Zoology (Jena) 2010;113:1−9.

[54] Couly G, Coltey P, Le Douarin NM. The triple origin of skull in higher vertebrates: a study in quail-chick chimeras. Development 1993;117:409−29.

[55] Creuzet S, Vincent C, Couly G. Neural crest contribution to eye development, periocular structures and eyelids. Int J Dev Biol 2005;49:161–71.

[56] Landacre FL. The fate of the neural crest in the head of the Urodeles. J Comp Neurol 1921;20:309–411.

[57] Newth DR. On the neural crest of the lamprey embryo. J Embryol Exp Morphol 1956;4:358–75.

[58] Le Lièvre C. Rôle des cellules mésectodermiques issues des crêtes neurales céphaliques dans la formation des arcs branchiaux et du squelette viscéral. J Embryol Exp Morphol 1974;31:453–77.

[59] Mongera A, Singh AP, Levesque MP, Chen YY, Konstantinidis P, Nüsslein-Volhard C. Genetic lineage labeling in zebrafish uncovers novel neural crest contributions to the head, including gill pillar cells. Development 2013;140:916–25.

[60] Smith DG, Chamley-Campbell J. Localization of smooth-muscle myosin in branchial pillar cells of snapper (Chrysophys auratus) by immunofluorescence histochemistry. J Exp Zool 1981;215:121–4.

[61] Yoshida T, Vivatbutsiri P, Morriss-Kay G, Saga Y, Iseki S. Cell lineage in mammalian craniofacial mesenchyme. Mech Dev 2008;125:797–808.

[62] Le Douarin N, Le Lièvre C, Fontaine J. Recherches expérimentales sur l'origine embryologique du corps carotidien chez les Oiseaux. CR Acad Sci Paris 1972;275:583–6.

[63] Le Douarin N, Teillet MA. The migration of neural crest cells to the wall of the digestive tract in avian embryo. J Embryol Exp Morphol 1973;30:31–48 (Citation classics).

[64] Bullock TH, Horridge GA. Structure and function in the nervous system of vertebrates. W.H. Freeman and Co; San Francisco; 1965.

[65] Lacalli TC. Sensory systems in amphioxus: a window on the ancestral chordate condition. Brain Behav Evol 2004;64:148–62.

[66] Nishida H. Cell lineage analysis in ascidian embryos by intracellular injection of a tracer enzyme. III. Up to the tissue restricted stage. Dev Biol 1987;121:526–54.

[67] Meinertzhagen IA, Lemaire P, Okamura Y. The neurobiology of the ascidian tadpole larva: recent developments in an ancient chordate. Annu Rev Neurosci 2004;27:453–85.

[68] Dilly PN. Studies on the receptors in the cerebral vesicle of the ascidian tadpole, 2. The ocellus. Q J Microsc Sci 1964;105:13–20.

[69] Dilly PN. Studies on the receptors in the cerebral vesicle of the ascidian tadpole, 1. The otolith. Q J Microsc Sci 1962;103:393–8.

[70] Wada MR, Ohtani Y, Shibata Y, Tanaka KJ, Tanimoto N, Nishikata T. An alternatively spliced gene encoding a Y-box protein showing maternal expression and tissue-specific zygotic expression in the ascidian embryo. Dev Growth Differ 1998;40:631–40.

[71] Lemaire P, Bertrand V, Hudson C. Early steps in the formation of neural tissue in ascidian embryos. Dev Biol 2002;252:151–69.

[72] Imai KS, Satoh N, Satou YA. Twist-like bHLH gene is a downstream factor of an endogenous FGF and determines mesenchymal fate in the ascidian embryos. Development 2003;130:4461–72.

[73] Jiang D, Smith WC. An ascidian engrailed gene. Dev Genes Evol 2002;212:399–402.

[74] Wurst W, Bally-Cuif L. Neural plate patterning: upstream and downstream of the isthmic organizer. Nat Rev Neurosci 2001;2:91–9.

[75] Pani AM, Mullarkey E, Aronowicz J, Assimacopoulos S, Grove EA, Lowe CJ. Ancient deuterostome origins of vertebrate brain signalling centres. Nature 2012;483:289–94.

[76] Wicht H, Lacalli TC. The nervous system of amphioxus: structure, development, and evolutionary significance. Can J Zool 2005;83:122–50.

[77] Lacalli T. Prospective protochordate homologs of vertebrate midbrain and MHB, with some thoughts on MHB origins. Int J Biol Sci 2006;2:104–9.

[78] Holland LZ, Albalat R, Azumi K, Benito-Gutiérrez E, Blow MJ, Bronner-Fraser M, et al. The amphioxus genome illuminates vertebrate origins and cephalochordate biology. Genome Res 2008;18:1100–11.

[79] Holland LZ. Chordate roots of the vertebrate nervous system: expanding the molecular toolkit. Nat Rev Neurosci 2009;10:736–46.

[80] Jeffery WR, Strickler AG, Yamamoto Y. Migratory neural crest-like cells form body pigmentation in a urochordate embryo. Nature 2004;43:696–9.

[81] Jeffery WR. Ascidian neural crest-like cells: phylogenetic distribution, relationship to larval complexity, and pigment cell fate. J Exp Zool B 2006;306B:470–80.

[82] Jeffery WR, Chiba T, Krajka FR, Deyts C, Satoh N, Joly JS. Trunk lateral cells are neural crest-like cells in the ascidian Ciona intestinalis: insights into the ancestry and evolution of the neural crest. Dev Biol 2008;324:152–60.

[83] Tassy O, Dauga D, Daian F, Sobral D, Robin F, Khoueiry P, et al. The ANISEED database: digital representation, formalization, and elucidation of a chordate developmental program. Genome Res 2010;20:1459–68.

[84] Russo MT, Donizetti A, Locascio A, D'Aniello S, Amoroso A, Aniello F, et al. Regulatory elements controlling Ci-msxb tissue-specific expression during Ciona intestinalis embryonic development. Dev Biol 2004;267:517–28.

[85] Imai KS, Levine M, Satoh N, Satou Y. Regulatory blueprint for a chordate embryo. Science 2006;312: 1183–7.

[86] Wada H, Makabe K. Genome duplications of early vertebrates as a possible chronicle of the evolutionary history of the neural crest. Int J Biol Sci 2006;2:133–41.

[87] Squarzoni P, Parveen F, Zanetti L, Ristoratore F, Spagnuolo A. FGF/MAPK/Ets signaling renders pigment cell precursors competent to respond to Wnt signal by directly controlling Ci-Tcf transcription. Development 2011;138:1421–32.

[88] Nishida H, Satoh N. Determination and regulation in the pigment cell lineage of the ascidian embryo. Dev Biol 1989;132:355–67.

[89] Abitua PB, Wagner E, Navarrete IA, Levine M. Identification of a rudimentary neural crest in a nonvertebrate chordate. Nature 2012;492:104–7.

[90] Shimeld SM. Holland PWH. Vertebrate innovations. Proc Natl Acad Sci USA 2000;97:4449–52.

[91] Van Campenhout E. Le développement du système nerveux sympathique chez le Poulet. Arch Biol 1931;42:479–507.

[92] Van Campenhout E. Further experiments on the origin of the enteric nervous system in the chick. Physiol Zool 1932;5:333–53.

[93] Van Campenhout E. Le développement du système nerveux crânien chez le poulet. Arch Biol 1937;48:611–6.

[94] Le Lièvre C. Participation of neural crest-derived cells in the genesis of the skull in birds. J Embryol Exp Morphol 1978;47:7–17.

[95] Dupin E, Calloni GW, Le Douarin NM. The cephalic neural crest of amniote vertebrates is composed of a large majority of precursors endowed with neural, melanocytic, chondrogenic and osteogenic potentialities. Cell Cycle 2010;9:238–49.

[96] Le Douarin NM, Dupin E. The neural crest in vertebrate evolution. Curr Opin Genet Dev 2012;22:381–9.

[97] Morriss-Kay GM. Derivation of the mammalian skull vault. J Anat 2001;199:143–51.

[98] Jiang X, Iseki S, Maxson RE, Sucov HM, Morriss-Kay GM. Tissue origins and interactions in the mammalian skull vault. Dev Biol 2002;241:106–16.

[99] Kague E, Gallagher M, Burke S, Parsons M, Franz-Odendaal T, Fisher S. Skeletogenic fate of zebrafish cranial and trunk neural crest. PLoS One 2012;7:e47394.

[100] Matsuoka T, Ahlberg PE, Kessaris N, Iannarelli P, Dennehy U, Richardson WD, et al. Neural crest origins of the neck and shoulder. Nature 2005;436:347–55.

[101] Grenier J, Teillet MA, Grifone R, Kelly RG, Duprez D. Relationship between neural crest cells and cranial mesoderm during head muscle development. PLoS One 2009;4:e4381.

[102] Clark K, Bender G, Murray BP, Panfilio K, Cook S, Davis R, et al. Evidence for the neural crest origin of turtle plastron bones. Genesis 2001;31:111–7.

[103] Cebra-Thomas JA, Betters E, Yin M, Plafkin C, McDow K, Gilbert SF. Evidence that a late-emerging population of trunk neural crest cells forms the plastron bones in the turtle Trachemys scripta. Evol Dev 2007;9:267–77.

[104] Cebra-Thomas JA, Terrell A, Branyan K, Shah S, Rice R, Gyi L, et al. Late-emigrating trunk neural crest cells in turtle embryos generate an osteogenic ectomesenchyme in the plastron. Dev Dyn 2013;31.

[105] Prince V, Lumsden A. Hoxa-2 expression in normal and transposed rhombomeres: independent regulation in the neural tube and neural crest. Development 1994;120:911–23.

[106] Couly G, Grapin-Botton A, Coltey P, Le Douarin NM. The regeneration of the cephalic neural crest, a problem revisited: the regenerating cells originate from the contralateral or from the anterior and posterior neural fold. Development 1996;122:3393–407.

[107] Kontges G, Lumsden A. Rhombencephalic neural crest segmentation is preserved throughout craniofacial ontogeny. Development 1996;122:3229–42.

[108] Creuzet SE, Martinez S, Le Douarin NM. The cephalic neural crest exerts a critical effect on forebrain and midbrain development. Proc Natl Acad Sci USA 2006;103:14033–8.

[109] Shimamura K, Rubenstein JL. Inductive interactions direct early regionalization of the mouse forebrain. Development 1997;124:2709–18.

[110] Houart C, Westerfield M, Wilson SW. A small population of anterior cells patterns the forebrain during zebrafish gastrulation. Nature 1998;391: 788–92.

[111] Martinez S, Crossley PH, Cobos I, Rubenstein JL, Martin GR. FGF8 induces formation of an ectopic isthmic organizer and isthmocerebellar development via a repressive effect on Otx2 expression. Development 1999;126:1189–200.

[112] Crossley PH, Martinez S, Martin GR. Midbrain development induced by FGF8 in the chick embryo. Nature 1996;380:66–8.

[113] Creuzet S, Schuler B, Couly G, Le Douarin NM. Reciprocal relationships between Fgf8 and neural crest cells in facial and forebrain development. Proc Natl Acad Sci USA 2004;101:4843–7.

[114] Creuzet S, Couly G, Vincent C, Le Douarin NM. Negative effect of Hox gene expression on the development of the neural crest-derived facial skeleton. Development 2002;129:4301–13.

[115] Grammatopoulos GA, Bell E, Toole L, Lumsden A, Tucker AS. Homeotic transformation of branchial

arch identity after Hoxa2 overexpression. Development 2000;127:5355–65.

[116] Pasqualetti M, Ori M, Nardi I, Rijli FM. Ectopic Hoxa2 induction after neural crest migration results in homeosis of jaw elements in *Xenopus*. Development 2000;127:5367–78.

[117] Hunter MP, Prince VE. Zebrafish hox paralogue group 2 genes function redundantly as selector genes to pattern the second pharyngeal arch. Dev Biol 2002;247:367–89.

[118] Crossley PH, Martinez S, Ohkubo Y, Rubenstein JL. Coordinate expression of Fgf8, Otx2, Bmp4, and Shh in the rostral prosencephalon during development of the telencephalic and optic vesicles. Neuroscience 2001;108:183–206.

[119] Ohkubo Y, Chiang C, Rubenstein JL. Coordinate regulation and synergistic actions of BMP4, SHH and FGF8 in the rostral prosencephalon regulate morphogenesis of the telencephalic and optic vesicles. Neuroscience 2002;111:1–17.

[120] Bardot B, Lecoin L, Huillard E, Calothy G, Marx M. Expression pattern of the drm/gremlin gene during chicken embryo development. Mech Dev 2001;101: 263–5.

[121] Tzahor E, Kempf H, Mootoosamy RC, Poon AC, Abzhanov A, Tabin CJ, et al. Antagonists of Wnt and BMP signaling promote the formation of vertebrate head muscle. Genes Dev 2003;17:3087–99.

[122] Creuzet SE. Regulation of pre-otic brain development by the cephalic neural crest. Proc Natl Acad Sci USA 2009;106:15774–9.

[123] Smith MM. Putative skeletal neural crest cells in early late ordovician vertebrates from Colorado. Science 1991;251:301–3.

[124] Janvier P. Early vertebrates. Oxford: Clarendon Press; 1996.

[125] Smith M, Hickman A, Amanze D, Lumsden A, Thorogood P. Trunk neural crest origin of caudal fin mesenchyme in the zebrafish *Brachydanio rerio*. Proc R Soc Biol Lond B 1994;256:137–45.

[126] Ota KG, Fujimoto S, Oisi Y, Kuratani S. Identification of vertebra-like elements and their possible differentiation from sclerotomes in the hagfish. Nat Commun 2011;2:373.

[127] Janvier P. Comparative anatomy: all vertebrates do have vertebrae. Curr Biol 2011;21:R661–3.

[128] Kimmel CB, Miller CT, Keynes RJ. Neural crest patterning and the evolution of the jaw. J Anat 2001;199:105–20.

[129] Lee RT, Thiery JP, Carney TJ. Dermal fin rays and scales derive from mesoderm, not neural crest. Curr Biol 2013;23:R336–7.

[130] Lee RT, Knapik EW, Thiery JP, Carney TJ. An exclusively mesodermal origin of fin mesenchyme demonstrates that zebrafish trunk neural crest does not generate ectomesenchyme. Development 2013;140: 2923–32.

[131] Shimada A, Kawanishi T, Kaneko T, Yoshihara H, Yano T, Inohaya K, et al. Trunk exoskeleton in teleosts is mesodermal in origin. Nat Commun 2013;4:1639.

[132] Chai Y, Jiang X, Ito Y, Bringas Jr. P, Han J, Rowitch DH, et al. Fate of the mammalian cranial neural crest during tooth and mandibular morphogenesis. Development 2000;127:1671–9.

[133] Nakamura H, Ayer-le Lievre CS. Mesectodermal capabilities of the trunk neural crest of birds. J Embryol Exp Morphol 1982;70:1–18.

[134] McGonnell IM, Graham A. Trunk neural crest has skeletogenic potential. Curr Biol 2002;12:767–71.

[135] Ido A, Ito K. Expression of chondrogenic potential of mouse trunk neural crest cells by FGF2 treatment. Dev Dyn 2006;235:361–7.

[136] John N, Cinelli P, Wegner M, Sommer L. Transforming growth factor beta-mediated Sox10 suppression controls mesenchymal progenitor generation in neural crest stem cells. Stem Cells 2011;29:689–99.

[137] Calloni GW, Glavieux-Pardanaud C, Le Douarin NM, Dupin E. Sonic Hedgehog promotes the development of multipotent neural crest progenitors endowed with both mesenchymal and neural potentials. Proc Natl Acad Sci USA 2007;104:19879–84.

[138] Coelho-Aguiar JM, Le Douarin NM, Dupin E. Environmental factors unveil dormant developmental capacities in multipotent progenitors of the trunk neural crest. Dev Biol 2013 (http://dx.doi.org/10.1016/j.ydbio.2013.09.030).

Induction and Specification of Neural Crest Cells: Extracellular Signals and Transcriptional Switches

Chang-Joon Bae and Jean-Pierre Saint-Jeannet

Department of Basic Science and Craniofacial Biology, College of Dentistry, New York University, New York, USA

INTRODUCTION

The neural crest (NC) is a transient embryonic cell population originating from the ectoderm, in a region lateral to the prospective neural plate. NC cells (NCCs) carry the remarkable ability to migrate in the embryo and differentiate into a vast number of derivatives. Evolutionarily, the NC played an essential role in the emergence of two of the defining features of the vertebrates: an advanced craniofacial skeleton and specialized paired sensory organs [1]. NCC are multipotent with the ability to give rise to cell types as diverse as neurons, glia, smooth muscle cells, melanocytes, chondrocytes, and odontoblasts. Because of these unique properties, the NC has

Neural Crest Cells.
DOI: http://dx.doi.org/10.1016/B978-0-12-401730-6.00002-8

attracted the attention of developmental biologists for over a century. It is an excellent model for exploring fundamental mechanisms underlying embryonic induction, cell migration, cell fate determination, and differentiation. More recent work has also focused on the stem cell-like properties of the NC and its potential use in regenerative medicine (reviewed in [2,3]).

Several human diseases have been associated with abnormal development of the NC and its derivatives. They are collectively known as neurocristopathies. These diseases can be the result of defects in NC specification, migration, proliferation, survival, or differentiation. Because of its contribution to multiple lineages, abnormal development of the NC often results in a wide array of clinical manifestations, affecting multiple organ systems. They include conditions such as Waardenburg-Shah syndrome (aganglionic megacolon, hypopigmentation, and deafness), DiGeorge syndrome (thymic hypoplasia, craniofacial and heart defects) or Treacher-Collins syndrome (cleft palate, micrognathia, and deafness). The characterization of the regulatory inputs controlling NC formation is therefore critical to understand how these processes may be altered in neurocristopathies.

The formation of the NC is a multistep process regulated by a complex set of signaling events. The process is initiated at gastrulation by the induction of NC progenitors at the lateral edges of the neural plate, a region where neural and non-neural ectoderm meets. As the neural plate closes into a tube, NC progenitors occupy the most dorsal aspect of the neural tube. NCC eventually delaminate from the neuroepithelium and migrate along specific routes throughout the embryo (Figure 2.1). Upon reaching their final destination, NCC differentiate into specific cell types depending on their origin, history, and final position in the embryo. Here, we review the molecular players directing NC induction and specification, comparing information from four model organisms: mouse, chick, *Xenopus*, and zebrafish.

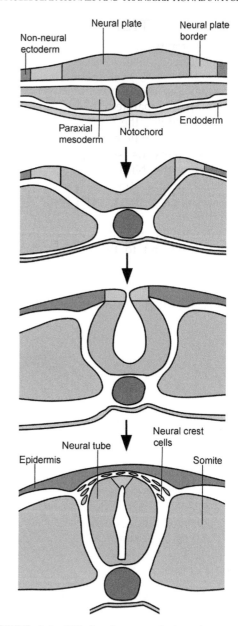

FIGURE 2.1 **NC development during the morphogenetic movements of neurulation.** At the beginning of neurulation, the NC is located at the boundary between the neural plate and the non-neural ectoderm, in a region known as the neural plate border (NPB). As the neural plate folds into a tube, the NC-forming region ends up at the dorsal aspect of the neural tube. In most vertebrates, neural crest delamination and migration starts upon neural tube closure.

NC INDUCTION AT THE NPB

At the end of gastrulation, the ectoderm of the vertebrate embryo can be divided into three major domains: the non-neural ectoderm and the neural plate separated by a region known as the NPB. While the non-neural ectoderm and neural plate will develop into epidermis and central nervous system, respectively, in most species the NPB gives rise to the NC but also to another transitory cell population, the pre-placodal ectoderm (PE). The PE eventually segregates into cranial placodes that contribute to the paired sense organs (nose, ear, and lens) and to the cephalic peripheral nervous system (cranial ganglia). In anamniotes, the NPB also gives rise to primary sensory neurons (Rohon–Beard neurons) in the trunk region, and to hatching gland cells anteriorly (Figure 2.2). Topographically, the PE is positioned lateral to the NC within the head, except for the most anterior region, where the PE directly abuts the neural plate.

Besides their common origin at the NPB, NC and PE share a number of important features. Some of the same signals involved in the generation of the NC have also been implicated in the induction of the PE [4–7]. Moreover, NC and PE have a similar ability to delaminate from the epithelial structure

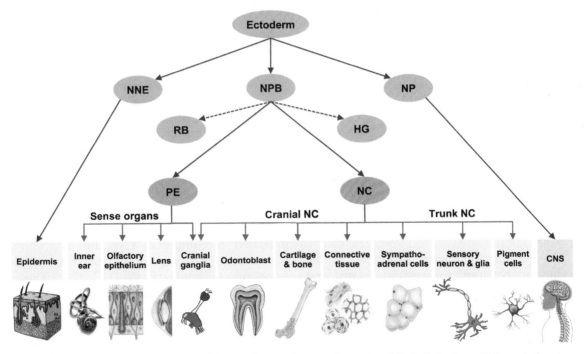

FIGURE 2.2 **The three subdivisions of the ectoderm at the neurula stage and their derivatives.** At the end of gastrulation, the ectoderm is subdivided into three discrete domains: the neuronal plate (NP), non-neuronal ectoderm (NNE), and NPB. NP and NNE give rise to the central nervous system (CNS) and epidermis, respectively, while the NPB give rise to neural crest (NC) and PE. Cranial NC differentiates into odontoblasts, cartilage, bone, and connective tissues and contributes to cranial ganglia, while trunk NC differentiates symphatho-adrenal cells, sensory neuron and glia, and pigment cells. PE ultimately segregates into individual cranial placodes to form the paired sense organs (olfactory epithelium,

from which they originate, and differentiate into a large array of cell types, including sensory neurons, glia, and supporting cells [8]. However, the NC also has a set of unique characteristics: it develops a much broader repertoire of cell types as compared to the PE, including pigment cells, cartilage, and smooth muscle cells (Figure 2.2), and it will also migrate over greater distances. In addition, unlike cranial placodes, NCC are not exclusively restricted to the head region; they arise from the entire length of the neural tube, starting from the mesencephalon.

The induction of the NC is a classical example of embryonic induction. A group of cells located at the NPB segregate from neighboring cells in response to inductive signals produced by surrounding tissues. As a consequence of these inductive interactions, a first set of transcription factors is activated that define the NPB (NPB specifier genes). In turn, these factors activate another set of genes more restricted to the NC territory, known as NC specifiers. The NC specifiers are thought to regulate the expression of downstream NC effector genes implicated in the control of NCC migration and differentiation (reviewed in [9,10]).

Timing and Tissues

The NC-forming region is located at the boundary between the neural plate and the non-neural ectoderm, and sitting atop the mesoderm. Because of their position relative to the NC, each one of these tissues has been proposed as a source of NC-inducing signal. There are some differences in the relative contribution of these tissues to NC induction among vertebrates. This is presumably related to differences in the timing of NC induction, as well as differences in the precise mechanisms of neurulation employed by each one of these organisms (Figure 2.3). For example, the

sequential activation of NPB and NC specifier genes occurs in a matter of hours in *Xenopus* and zebrafish [13,21,22], while the same process takes much longer in birds [23]. The activation of NPB and NC specifier genes occurs at gastrulation in fish and frogs, while it is initiated at the onset of neurulation in chick and mouse embryos [11−20]. Another difference is in the timing of NCC migration. For example, unlike other species, NCC migration in the mouse is initiated prior to neural tube closure, and shortly after NC progenitors have acquired their identity [20,24].

By analyzing the expression of NC specifier genes such as *Snail2*, *Foxd3*, or *Sox8/9/10* (reviewed in [25,26]), it has been possible to monitor NC induction. It is now well accepted that NC induction is a multistep process initiated during gastrulation and persisting at least until neural tube closure. Two steps can be distinguished: first, the induction of NC progenitors at the gastrula stage, followed by the maintenance of these progenitors at the neurula stage. This was demonstrated experimentally in both *Xenopus* and chick embryos. For example, NPB explants downregulate NC-specific genes, which indicates that further signaling events are required to elicit NC fate [27,28].

Transplantation experiments performed in the early 1990s using *Axolotl* embryos demonstrated the importance of the interaction between the neural plate and the non-neural ectoderm to generate the NC [29]. These findings were subsequently confirmed and expanded in *Xenopus* and chick embryos [30,31]. For example, explants of neural plate do not generate NCC when cultured *in vitro*; however, grafts of neural plate explants into the adjacent non-neural ectoderm induces expression of NC genes at their boundary. Furthermore, lineage tracing studies have demonstrated that under these experimental conditions NCC are derived from both tissues, the neural and non-neural ectoderm [30,31].

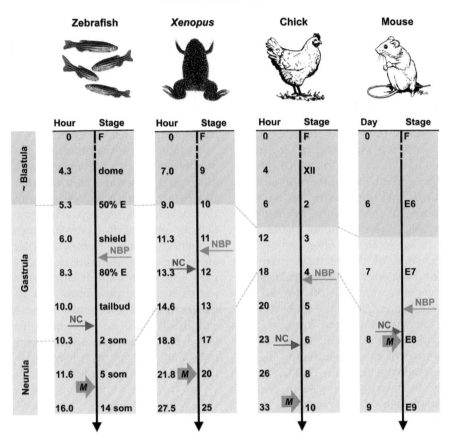

FIGURE 2.3 **Timing of NC development in zebrafish, _Xenopus_, chick, and mouse embryos.** The initial expression of NPB and NC specifier genes differs across species. Note that the expression of NPB and NC specifier genes is initiated at the end of gastrulation in anamniotes (zebrafish and _Xenopus_) and at the beginning of neurulation in amniotes (chick and mouse), respectively. In zebrafish, _Foxd3_ and _Ap2a_ are first expressed at early gastrula stage [11], while _Snail2_, _Sox9/10_, and other NC specifiers appear in NC toward the end of the gastrulation [12]. In _Xenopus_, NPB specifiers such as _Pax3_, _Zic1_, and _Msx1_ are first expressed at stage 11.5 [13,14], followed immediately by the expression of NC specifiers (_Snail2_, _Foxd3_, and _Sox8/9_) at stage 12, even before gastrulation is completed [15]. In chick, _Pax7_ is expressed in the prospective NPB region at an early stage (stage 4), together with other NPB specifier genes (_Msx1_, _Pax3_, and _Dlx5_), which exhibit broader expression domains [16]. The first NC specifier (_Foxd3_) expression takes place after stage 6 [17]. NC migration begins between stage 9 and stage 10. In mouse, most NPB specifiers begin expression at approximately E7.5. NC migration begins at E8. NC specifiers (_Sox9/10_ and _Foxd3_) are first detected at E7.75, immediately before NC migration is initiated [18–20]. F, fertilization; M, initiation of NC migration.

However, recent work in _Xenopus_ has challenged this view. By grafting neural plate explants into ventral ectoderm, Schlosser and colleagues [32] have shown that NC-specific genes were exclusively restricted to the graft (neural plate), while PE-specific genes were induced on the non-neural side. Therefore, it has been proposed that competence to form NC is confined to the neural plate, while the non-neural ectoderm is only competent to generate PE [32], suggesting that NC induction is directly dependent on the induction of neural plate tissue. This work also implies that NC is induced in response to non-neural-derived signals, whereas the PE would require neural plate-derived signals. This possibility is consistent with transplantation experiments in chicken embryos showing that epidermal/neural plate interactions are sufficient to generate neural crest derivatives [30].

Early work in Uroledes demonstrated that lateral plate and paraxial mesoderm explants transplanted into the blastocoel of gastrula-stage embryos had the ability to induce ectopic NCC [33]. Since then, a number of studies have confirmed the importance of paraxial mesoderm for NC induction in *Xenopus* [34–36]. The NC specifier gene *Snail2* was activated in naïve ectoderm after recombination with explants of dorsolateral marginal zone (DLMZ) isolated at the gastrula stage [34,36]. Conversely, removal of the DLMZ inhibited *Snail2* expression in the embryo [35]. More recent fate map studies in *Xenopus* have attempted to identify the specific region of the mesoderm source of the NC-inducing signal. These studies indicate that the DLMZ at the gastrula stage eventually give rise to part of the intermediate mesoderm, which is positioned directly underneath the NC-forming region at the neurula stage [28]. They further provided evidence that signals from the DLMZ participate in NC induction during gastrulation, and that signals derived from the intermediate mesoderm were involved in the maintenance of the NC identity at the neurula stage [28].

In chick, recombination experiments between somitic mesoderm and neural plate explants can elicit formation of NCC [30]. However, there is also evidence that NC induction can occur in epiblast explants isolated at the gastrula stage, independent of the formation of mesoderm [16]. Zebrafish embryos, in which mesoderm formation and involution have been blocked by disruption of Nodal signaling, can still form NC progenitors, suggesting that the mesoderm and its derived signals are dispensable for NC formation in fish [37]. However, in these embryos

the DLMZ adjacent to the presumptive NC territory is normally specified [38,39] and expresses *Wnt8*, a known NC-inducing signal [40]. Therefore, a role for mesoderm in NC induction cannot completely be ruled out based on these studies. Rather, the work suggests that the involuting mesoderm may not be required for the maintenance of NC progenitors.

In mouse, the contribution of the ectoderm and mesoderm to NC induction has not been addressed experimentally due to the difficulties in accessing the corresponding tissues in the developing embryo. In mouse mutants lacking part of the paraxial mesoderm, NC induction appears largely unaffected [41,42]. However, these mutants do not show a complete loss of mesoderm, and therefore we cannot exclude the possibility that the remaining tissues may have retained some NC-inducing activity. There is no information on the importance of the interaction between the neural plate and non-neural ectoderm in the generation of the NC in mammals.

Signaling Factors

At least three distinct classes of signaling molecules have been implicated in NC induction: the bone morphogenetic proteins (Bmp), the Wnt family of glycoproteins, and fibroblast growth factors (Fgf). While they all appear to have some role in NC formation, their relative importance in this process varies among species. Here, we summarize the past decade of work analyzing the involvement of these signaling pathways in NC induction.

Bone Morphogenetic Protein Signaling

According to the current model of NC induction, the diffusion of Bmp antagonists produced by the axial mesoderm, including chordin, noggin, and follistatin, creates a gradient of Bmp activity in the overlying ectoderm, such that NC forms at levels of Bmp signaling intermediate to those required for formation of the neural plate (low Bmp signaling) and non-neural ectoderm (high Bmp signaling). This model is primarily based on work performed in frog and fish. In *Xenopus*, modulation of Bmp signaling in ectoderm explants by graded expression of noggin or expression of a dominant-negative Bmp receptor can induce epidermal, NC, and neural plate fates in a concentration-dependent manner [35,43,44]. A similar gradient model was proposed based upon genetic analysis of Bmp signaling pathway mutants in zebrafish [45,46]. More recent work suggested that this gradient could be interpreted at the level of the downstream effectors of the Bmp pathway, Smad1/5. In this study, attenuation of Smad5 expression by injection of increasing amounts of Smad5 antisense oligonucleotides resulted first in an increase and then a loss of NC progenitors [47].

Earlier work in chick suggested that Bmps expressed in the neural folds and adjacent ectoderm might function as NC inducers. Conditioned medium from Bmp4- or Bmp7-transfected cells induced NC markers in neural plate explants, mimicking the NC-inducing activity of the non-neural ectoderm [48,49]. These observations were, however, challenged by a subsequent study [50]. While recombinant Bmp4 was a potent NC inducer in a culture medium containing additives (F12-N2), Bmp4 NC-inducing activity was completely lost when using a chemically defined medium (DMEM) [50]. These observations suggest that other signals synergize with Bmp4 to induce NC in chick neural plate explants. In the chick, the timing and pattern of expression of noggin, chordin, and follistatin are not completely consistent with a role in NC induction [51−53]. However, there is evidence that Bmp attenuation in the chick epiblast might be mediated through Fgf signaling rather than Bmp antagonists [54−56]. In *Xenopus*, active Fgf signaling is also required for neutralization of the ectoderm and NC induction by Bmp antagonists [57−60], although this activity has not been directly linked to a change in Bmp signaling levels.

A requirement for Bmp antagonists in the generation of NC is not as clearly established in mouse embryos. For example, chordin/noggin double mutant mouse embryos form an excess of NC progenitors [61,62], which is opposite to what the gradient model predicts. In *Bmp2* null mutant mouse embryos, NCC are induced and express several early NC markers, however these cells failed to migrate, suggesting that Bmp2 is required for the production of migratory NCC [63,64]. Mouse mutant embryos for *Bmp4* or for other components of the Bmp signaling pathway die early, before NC specification [65−71]. Conditional knockouts, using the Wnt1-Cre line, to specifically delete gene function in the NC lineage have primarily demonstrated a requirement for Bmp signaling later in NC development [72−75]. However, it is important to point out that *Wnt1* is first expressed in the dorsal neural tube at a time point slightly downstream of these early inductive events, and therefore the importance of Bmp signaling in NC induction in mouse embryos remains an open question.

A couple of recent reports indicate that the sole inhibition of Bmp signaling cannot account for all aspects of NC induction in *Xenopus* and chick embryos [15,76]. Both studies demonstrated that Bmp signaling may have a distinct temporal requirement during NC induction (Figure 2.4): first, inhibition of Bmp signaling (in combination with a canonical Wnt; discussed below) at the gastrula stage, followed by a period of active Bmp and Wnt signaling to maintain the NC population at the neurula stage [15,76]. This two-step model, as well as the gradient model, strongly argues for an important role of Bmp signaling in the early steps of NC induction. However, there is also strong evidence that Bmp signaling alone is not sufficient to activate expression of NC

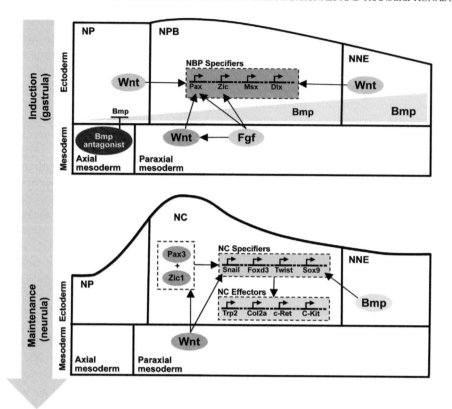

FIGURE 2.4 **Spatiotemporal regulation of NPB and NC specifier genes by signaling molecules.** NC induction begins at the NPB, which is mediated by several inducing signals such as Fgf and canonical Wnt from paraxial mesoderm and non-canonical Wnt from the adjacent NNE and NP, and intermediate level of Bmp activity in NPB. These signals redundantly stimulate the expression of individual NPB specifiers, including genes of the *Pax*, *Zic*, *Msx*, and *Dlx* families. Among these genes *Pax3* and *Zic1* in turn activate the expression of NC specifiers such as *Snail1/2*, *Sox8/9/10*, *Foxd3*, *Twist*, *Myc*, *Id3*, and *Tfap2a* in presumptive NC. These NC specifier genes activate NC effectors (*Trp2*, *Col2a*, *c-Ret*, and *c-Kit*), which regulate the migration and differentiation of NCC along distinct lineages. This diagram is primarily based on work performed in *Xenopus* and may not apply to all organisms. For clarity, only a subset of NPB/NC specifiers and NC effectors are shown.

markers [50,77–79], and therefore it must act in concert with other signals to induce the NC.

Wnt Signaling

A large body of evidence strongly indicates that activation of canonical Wnt signaling is critical to specify the NC in fish, frog, and chick. The canonical Wnt pathway involves binding of the extracellular Wnt ligands to frizzled proteins (receptor) and LRP/arrow (co-receptor), which in turn signals through the cytosolic

adaptor protein Disheveled (Dsh), leading to inhibition of GSK-3β and subsequent stabilization of β-catenin. Wnt-dependent gene expression is activated by β-catenin together with Tcf/Lef DNA binding factors (reviewed in [80]). Interfering with any components of the canonical Wnt pathway in the ectoderm, either using dominant negative forms of Frizzled 3 (Fz3), Frizzled 7 (Fz7), and their co-receptor Lrp6, or morpholino-mediated knockdown of Fz3, Fz7, Lrp6, Kermen, Dsh and beta-catenin are all sufficient to block NC formation in the whole embryo [81–86]. Active canonical Wnt signaling is necessary for NC induction, but it is not sufficient to induce NC in naïve ectoderm, as it also requires attenuation of Bmp signaling.

In frog and fish, the source of the Wnt signal has been proposed to reside primarily in the paraxial mesoderm underlying the NC-forming region, while in chick, it is believed to originate from the non-neural ectoderm [40,50,87]. Wnt8 has been proposed as the endogenous NC-inducing signal in zebrafish, where it is expressed in a region immediately adjacent to the NC-forming region [40,88], and Wnt8 morpholino-mediated knockdown prevented expression of early NC markers [50]. In frogs, Wnt ligands have been proposed to reside in the non-neural ectoderm (Wnt7b) [89] and the paraxial mesoderm (Wnt8 [87] and Wnt3a [90]), as the spatiotemporal expression of these ligands appears to be compatible with NC induction. In *Xenopus*, knockdown of Wnt3a or Wnt8 prevented the expression of NPB and NC specifier genes [14,60,91]. In chick, the non-neural ectoderm expresses Wnt6 and has been proposed as the Wnt ligand involved in NC induction [50]. Consistent with this, Wnt6 overexpression enhanced NC markers expression, and Wnt6 siRNA-mediated knockdown reduced NC markers expression [92].

Mouse embryos with targeted inactivation of the β-catenin gene in the dorsal neural tube (Wnt1-Cre) have severe defects in several craniofacial skeletal elements of NC origin [93]. In these embryos, it is believed that signaling through β-catenin is required for survival and/or differentiation of cranial NCC [93]. Loss of Wnt1 and Wnt3a, both expressed in the dorsal neural tube, resulted in defects in a broad range of NC derivatives, and it has been proposed that Wnt signaling in the dorsal neural tube is primarily required for the expansion of NC progenitors rather than for

NC induction [94,95]. More recently, it has been shown that sustained Wnt activity in mouse NC progenitors had little effect on the population size and instead regulated fate decisions [96]. All gene perturbation experiments in the mouse point to a role for canonical Wnt signaling in lineage specification and differentiation rather than induction (reviewed in [97]). However, an earlier role for Wnt signaling in NC induction cannot be completely ruled out, since Wnt ligands can act redundantly. Moreover, most studies thus far have targeted the Wnt signaling pathway in the dorsal neural tube using the Wnt1-Cre line, which may not interfere with early NC induction events.

A recent study has also implicated non-canonical Wnt signaling in NC induction [98]. Non-canonical Wnt ligands do not stabilize β-catenin or activate Tcf-dependent transcription; rather, they regulate changes in cell shape and motility through the activation of small Rho GTPases and Rho-associated kinases [99]. This branch of the Wnt signaling pathway has been more typically implicated in the regulation of NCC migration [100–102]. The non-canonical Wnt ligand, Wnt11R, is expressed in the neural plate immediately adjacent to the NC-forming region [103]. Wnt11R loss of function resulted in a marked reduction of *Foxd3* and *Sox8* expression and a partial loss of NPB genes such as *Pax3* and *Tfap2a* [98]. The authors propose that non-canonical Wnts are acting by changing the localization and activity of the polarity kinase Par-1. Indeed, Par-1 is required for NC specification and can rescue NC-specific gene expression in embryos depleted of non-canonical Wnt ligands [98]. These observations are linking changes in cell polarity to cell fate specification, thereby raising the intriguing possibility that the cell shape changes occurring at the NPB during neural plate elevation and closure may be an integral part of the process of NC induction. This is not the first study proposing a role for non-canonical Wnt signaling in NC induction. In chick, Wnt6 expressed in the non-neural ectoderm has been shown to mediate its NC-inducing activity independent of β-catenin [92].

Fibroblast Growth Factor Signaling

Historically, the first evidence for a role of Fgf signaling in NC induction came from work in *Xenopus*. In two experimental systems, dissociated cells [104] and ectoderm explants [43], basic Fgf (bFgf) in combination with attenuation of Bmp signaling was shown to induce pigment cells and *Snail2* expression, respectively. More definitive evidence came from the overexpression of a dominant-negative Fgf receptor, which blocked expression of *Snail2* in intact *Xenopus* embryos without affecting surrounding tissues [105]. Moreover, based on recombination experiments, it has been proposed that the non-neural ectoderm was the source of the Fgf signal [105].

More recent work has implicated Fgf8 as an inducer of NPB cells. Interference with Fgf8 signaling in the embryo prevented the expression of NPB genes [21,60]. Fgf8 is also capable of transiently inducing NC genes in an isolated explant assay without the supplement of Bmp antagonist; however, only a limited repertoire of genes were activated under these conditions [36]. Fgf8 is expressed in the mesoderm at gastrulation and has been proposed as the paraxial mesoderm-derived signal responsible for NC induction in *Xenopus* [36]. This proposal is based on the observation that explants of DLMZ, which normally induce NC markers in the ectoderm [34], were unable to induce NC when recombined with ectoderm explants made refractory to Fgf signaling by expression with a dominant-negative Fgf receptor (XFD). Because intact Fgf signaling is required for neutralization of the ectoderm by Bmp antagonists [57–59], an alternative interpretation would be that NC induction was blocked, not as a result of the inability of a DLMZ-derived Fgf ligand to signal in the ectoderm, but rather because

the neutralization of these explants was impaired by expression of XFD [57]. Consistent with this possibility, the MAPK inhibitor, U0126, blocks neutralization by Chordin [59,60]. Consequently, it has been proposed that Fgf8 NC-inducing activity was likely to be indirect through the activation of Wnt8 in the paraxial mesoderm. While Wnt8 can rescue the expression of NC genes in Fgf8 morphants, Fgf8 is unable to rescue NC progenitors in Wnt8 or β-catenin-depleted embryos. Moreover, Fgf8 upregulates Wnt8 expression in the embryos, and Fgf8 morphants lack Wnt8 expression in the mesoderm [60].

In chicken, recent work indicates that Fgf activity is required in the ectoderm at gastrulation for NC specification. Expression of dominant-negative FGF receptor 1 at the gastrula stage prevented *Snail2* and *Pax7* expression in NC progenitors [56]. Interestingly, subsequent work indicated that activation of Fgf signaling in the epiblast upregulates both Bmp and Wnt signaling [106]. There is no direct evidence of Fgf signaling involvement in mouse NC induction.

A recent study in zebrafish has provided important information on the mechanism of integration of these signaling pathways in the generation of the NC [107]. In this work, the authors showed that the NPB genes *Zic3* and *Pax3a* possess two enhancers differently regulated by Bmp, Wnt, and Fgf signaling. *Zic3* and *Pax3a* each have an enhancer regulated by both Wnt and Fgf, as well as an enhancer regulated either by Wnt (*Zic3*) only or by Fgf (*Pax3a*) only. Changes in the relative influence/dominance of these enhancers may account for some of the differences in the importance of Wnt and Fgf signaling in NC induction in various species [107].

Other Signaling Pathways

Signaling through the membrane-bound protein Delta and its receptor Notch has also been implicated in NC formation in several species. In zebrafish, Delta has been reported to be required for trunk NC formation, as Delta-deficient embryos form supernumerary Rohon—Beard sensory neurons in the trunk at the expense of NCC [108]. Interestingly, cranial NCC were unaffected in these embryos, suggesting that trunk and cranial NC are regulated by different mechanisms [108]. In *Xenopus*, constitutive activation of Notch signaling by expression of a Notch construct lacking the ligand-binding domain resulted in a dramatic expansion of neural tissues and prevented expression of epidermal and NC markers [109]. Using hormone inducible dominant-negative constructs, another study demonstrated that the timing of Notch activation was critical for NC formation. Activation of Notch at gastrulation resulted in a dramatic expansion of the NC territory without affecting surrounding tissues [110]. Notch-mediated expansion of NC progenitors is thought to occur through the activation of the transcription factor Hairy2, which in turn suppresses *Bmp4* expression in the ectoderm [110]. Similarly in birds, Delta is involved in regulating *Bmp4* expression levels in the ectoderm and thus is believed to be indirectly required for NC induction [110,111]. In mouse, Notch signaling has been primarily associated with NC migration, proliferation, and differentiation [112—114].

Retinoic acid (RA) signaling has been implicated in NC development as well. In *Xenopus*, RA can induce *Snail2* expression in explants of anterior neural plate [115]. However, because RA functions as a posteriorizing signal in the neural tube, it was thought that *Snail2* induction in these explants was secondary to the posteriorizing activity of RA [116]. Another study using *Xenopus* animal caps recombined with chick mesoderm or NP explants demonstrated that induction of Pax3 occurred independently of RA signaling, suggesting that this pathway is not directly implicated in NC induction [117]. In vitamin A-deficient quail embryos, cranial NC progenitors form properly, yet shortly after migration is initiated these cells undergo massive apoptosis [118]. Double knockout mice for the RA-degrading enzymes Cyp26a1 and Cyp26c1, which have excess RA signaling, showed normal expression of NC markers, although these embryos were affected by abnormal NC migration [119]. In the triple knockout mice for Cyp26a1/Cyp26c1, and the RA-synthesizing enzyme (retinaldehyde dehydrogenase 2; Raldh2), which presumably completely lack RA signaling, NCC were properly specified, and the NC migration defect was largely rescued in these embryos [119]. These observations suggest that endogenous levels of RA signaling in the mouse may not be essential for NC induction or migration.

Endothelin-1/Endothelin-A receptor signaling can regulate NC formation in *Xenopus*, and this pathway has been more specifically implicated in the induction and maintenance of NC progenitors [120]. The receptor (*EdnrA*) is detected at the NPB, and the ligand, Endothelin-1 (ET-1), is produced by the mesoderm underlying the NC territory. Interference with ET-1 signaling causes a loss of NC-specific genes and induces apoptosis indicative of a role in the maintenance of the NC progenitor pool at the neurula stage. Epistatic analysis indicated that this signaling pathway was acting downstream of the NPB gene *Msx1* and upstream of the NC specifiers, *Sox9* and *Sox10* [120]. In mouse, ET-1 signaling is primarily required for the later phases of NC development. ET-1 is expressed in the epithelial layer of the branchial arches as well as in the mesodermal core, while the receptor is expressed in the arches mesenchyme [121,122]. Disruption of the pathway resulted in hypoplastic branchial arches due to apoptosis of the NC-derived arches mesenchyme [123].

Finally, Indian hedgehog (Ihh) signaling has been proposed to regulate multiple aspects of NC development in *Xenopus* [124]. Interference with Ihh signaling using morpholino anti-sense oligonucleotides, dominant-negative constructs, and chemical inhibitors causes a loss of both NPB and NC genes, suggesting an early role in NC specification [124]. In addition, Ihh has a later function in promoting NCC migration through both autocrine or paracrine mechanisms [124]. Another member of the hedgehog family, Sonic hedgehog (Shh), has also been shown to regulate NC migration in chick [125]. In mouse, targeted deletion of the Shh receptor, Smo, in the neural crest lineage showed an important role for this signaling pathway in the patterning and growth of NC-derived facial skeletal elements [126]. There is no evidence that hedgehog signaling is involved in NC specification in amniotes.

GENE REGULATORY NETWORK INVOLVED IN NC SPECIFICATION

In response to signaling events mediated by Bmp, Wnt, and Fgf, distinct sets of transcriptions factors are sequentially activated at the lateral edge of the neural plate. A first set of genes, known as NPB specifiers, is initially broadly activated at the NPB, and their expression domain typically comprises the prospective NC tissue as well as other subdomains of the NPB. These transcription factors include several homeobox-containing proteins, Pax3/7, Mxs1/2, Dlx5, Tfap2a, and Gbx2, as well as zinc finger-containing factors of the Zic family. In turn, these factors activate a second set of genes more restricted to the NC territory, known as NC specifiers, which include, among others, genes of the Snail, Sox, and Fox family of transcription factors. These NC specifiers are thought to regulate the expression of downstream NC effector genes implicated in the control of NCC migration and differentiation. The proper expression of these three sets of genes in time and space is central to the

specification of NC progenitors (Figure 2.4; reviewed in [9,127,128]). Here, we describe a subset of NPB and NC specifiers that have been identified in various species, and how they currently fit in the gene regulatory network underlying NC formation.

NPB Specifier Genes

In *Xenopus*, NPB specifiers include *Msx1*, *Pax3*, *Zic1*, *Hairy2*, *Gbx2*, *Tfap2a*, and *Meis3* (Figure 2.4; for a comprehensive list of NPB specifiers, see [128–130]). They are all expressed in a region of the ectoderm that includes the prospective NC, and loss-of-function studies have demonstrated that they are required for NC formation in the embryo [13,14,21,131–135].

Among these factors, Tfap2a has been proposed as a "master regulator" of the NC regulatory cascade, as it is required for the expression of other NPB specifiers such as *Msx1*, *Pax3*, *Zic1*, and *Hairy2* [134]. Consistent with this view, genome-wide analyses of chromatin marking patterns and transcription factors occupancy have shown that human NC enhancers are primarily occupied by TFAP2A [136], confirming that this transcription factor is a key regulator of NC fate. It is also important to mention that this regulation by Tfap2a is not unidirectional, since most NPB genes are also involved in maintaining each other's expression or may act at multiple steps during NC development [134]. In zebrafish, Tfap2a function has been primarily associated with NC diversification [22], while the requirement of Tfap2a at the chick NPB formation has not been evaluated [23]. In mouse, *Tfap2a* expression begins at E7 (Figure 2.3), ahead of the activation of NC specifier genes [137], as predicted for a NPB specifier gene. Targeted deletion of *Tfap2a* resulted in perinatal lethality. Mutant mice showed severe craniofacial defects and failure of cranial neural tube closure, associated with increased apoptosis in the midbrain/hindbrain region [138]. Interestingly, the expression of *Pax3* was largely unaffected in these mutants [139], suggesting that *Pax3* is acting upstream of Tfap2a, or that both factors are functioning in parallel pathways.

The transcription factors Pax3 and Zic1 are especially relevant to NPB formation since they are not only necessary but also sufficient to promote the formation of multiple NPB cell types. Gain-of-function experiments in the embryos have shown that Pax3 and Zic1 can promote hatching gland and PE fates [21], respectively, while their combined activity is essential to specify the NC [107]. Moreover, by manipulating the expression levels of Pax3 and Zic1 in naïve ectoderm explants, it is possible to generate a pure population of NC progenitors, in the absence of other NPB cell types [21], further demonstrating the importance of the cooperation between Pax3 and Zic1 in promoting NC fate [21]. Interestingly, cells derived from ectoderm explants expressing Pax3 and Zic1 have the ability to migrate and produce a full repertoire of NC derivatives when transplanted into the embryo [140]. These observations suggest that a limited number of factors are sufficient to initiate the NC developmental program.

While Pax3 is essential for NPB formation in frogs, its paralog, Pax7, plays a critical role in NC formation in chick. Pax7 is the only factor identified so far with broad expression in the NPB at early stages that has been directly involved in NC specification in chick [16,141]. In mouse, Pax3 and Zic genes expression are initiated around E7.5 (Figure 2.3), a timing of expression that is consistent with a potential role as NPB specifiers [142–144]. In the *Pax3* mouse mutant *Splotch*, the pool of NC progenitors fails to expand and to complete its migration, resulting in scarce cardiac NC derivatives [144,145]. Homozygous *Splotch* mice die *in utero* with persistent truncus arteriosus and pharyngeal arch patterning defects [144,146].

Because NPB specifiers are the first group of genes activated in response to extracellular signals, it is expected that their cis-regulatory sequences contain response elements for the signaling pathways implicated in NC induction (Figure 2.4). It is only recently that these potential direct interactions have been analyzed at the molecular level. In zebrafish, the two NPB genes *Zic3* and *Pax3a* each possess two intronic enhancers, which are differently regulated by Bmp, Wnt, and Fgf signaling [107]. This study provides the first evidence that *Pax* and *Zic* gene families are directly activated by NC-inducing signals. Another NPB specifier, *Gbx2*, which is essential for the positioning of the neural folds and required for the expression of NC specifiers, is an immediate direct target of canonical Wnt signaling in *Xenopus* [132]. Moreover, rescue experiments have shown that Gbx2 act upstream of *Pax3* and *Msx1* [132], thereby positioning Gbx2 as one of the earliest factors mediating the Wnt NC-inductive signal. In *Xenopus* and zebrafish, Msx1 expression at the NPB is activated by attenuation of Bmp signaling [147]. Consistent with these observations, in mouse a Bmp-responsive enhancer has been identified upstream of the *Msx2* gene, which appears to respond to intermediate levels of Bmp signal [148]. *Msx1/2* double knockout mice have delayed NCC migration in the pre- and post-otic regions and elevated number of apoptotic cells in the NC populations contributing to the cranial ganglia and the first pharyngeal arch [149].

NC Specifier Genes

NPB specifier genes, which are broadly expressed at the NPB, are thought to activate a subset of genes with a more restricted expression in the NC lineage, known as NC specifier genes. Several families of transcription factors have been identified as NC specifiers, including c-Myc, Id3, Snail1/2, Foxd3, Sox8/9/10, and Twist (Figure 2.4; for a comprehensive list of NC specifiers, see [128−130]). Typically, these genes are activated in NC progenitors before they initiate their migration, and their expression is either maintained or downregulated as NCC migrate in the periphery and differentiate along specific lineages. Their expression pattern in NC progenitors is well conserved across species, but there are also some variations in the precise timing of their expression.

For example, the SoxE family of transcription factors, which include Sox8, Sox9, and Sox10, has well-established roles in NC development (reviewed in [150−153]). While they all show expression in NC progenitors, they also show differences in their onset and sequence of expression in various organisms. In Xenopus, Sox8 is the first SoxE family member detected in the prospective NC at the late gastrula stage [154], immediately followed by Sox9 [155], while Sox10 is activated slightly later at the early neurula stage [156,157]. In zebrafish, Sox8 is not expressed in the NC; it is Sox9b that is first detected in NC progenitors [158], with a later onset of expression for Sox10 [159]. In contrast to other species, Sox10 is not maintained in migrating NCC in zebrafish. In chick and mouse embryos, the expression of Sox9 and Sox10 is initiated before Sox8 expression [160,161], and, at least in chick, Sox9 is the first SoxE gene expressed in NC progenitors [162,163].

Morpholino-mediated knockdown of Sox8, Sox9, or Sox10 in Xenopus results in the loss of expression of several NC specifiers, such as Snail2, Foxd3, and Twist [154,155,157]. However, because Snail2, Sox9, Foxd3, and Twist precede Sox10 expression in NC progenitors, Sox10 is more likely to be involved in maintaining the expression of these genes [156]. In zebrafish, Sox10 mutations do not prevent NC specification; instead, NCC fail to migrate and undergo apoptosis [159]. Similarly, Sox10 knockout mice exhibited massive NCC death in the trunk prior

to or shortly after delamination [164]. In chick and mouse, it has also been shown that Sox9 functions in NC progenitors formation, delamination, and in the development of specific NC derivatives [163,165].

Regarding the upstream regulators of SoxE genes, very little information is currently available. In Xenopus, Sox9 expression is activated by Tfap2a [166], and in mouse, putative Tfap2a binding motifs have been identified within the Sox9 cis-regulatory region [167], suggesting that Tfap2a may directly regulate Sox9 expression. There is also evidence that Sox8 and Sox9 may directly regulate Sox10 expression in Xenopus [154]. A more recent study, using a combination of chromatin Immunoprecipitation (ChIP) and reporter assay, has demonstrated that Sox9, Ets1, and cMyb directly bind and activate a Sox10 enhancer in the chicken cranial NC [168].

The winged-helix transcription factor Foxd3 is expressed in the premigratory NC and is an important regulator of NC development in all vertebrates. In mouse embryo, Foxd3 is required for maintenance of multipotent NC progenitors. A NC-specific deletion of a floxed allele of Foxd3 resulted in a broad loss or severe reduction of multiple NC derivatives [24]. Foxd3 is also likely to control NC survival because these mutants have an increase in NC apoptosis. In zebrafish, knockdown of Foxd3 indicated that it is required for the differentiation of a subset of NC lineages but did not appear to be involved in NC induction or migration [169]. Zebrafish Foxd3sym1 homozygous mutants start with normal numbers of premigratory NCC [170]. Foxd3sym1 mutants have increased NCC apoptosis in a region of the hindbrain corresponding to third NC stream, indicating a role for Foxd3 in the survival of at least a subpopulation of NCC [170]. In chicken embryos, misexpression of Foxd3 within the dorsal neural tube causes an expansion of the NC domain [17,171]. Similarly, in Xenopus Foxd3 overexpression in the embryo or in explants induced a broad array of NC-specific genes [172]. Expression of a dominant-negative Foxd3 construct inhibited NC differentiation in the embryo, a phenotype that can be rescued by *Snail2* expression [172].

The cooperative activity of Zic1 and Pax3, in combination with inputs from Wnt signaling, has been proposed to regulate the NC expression of Foxd3 in *Xenopus* [13]. A more recent study has identified two Foxd3 enhancers (NC1 and NC2) that drive differential expression in the chick cranial (NC1) and trunk (NC2) NC. Mutational analysis, *in vivo* ChIP, and morpholino knockdowns demonstrated that the transcription factors Pax7 and Msx1/2 cooperate with the NC specifier gene Ets1 to bind to the cranial NC1 regulatory element. At the trunk level, Pax7 and Msx1/2 function together with the NPB specifier Zic1, which directly binds to the NC2 enhancer [161]. This work provides strong evidence directly linking *Foxd3* expression in the NC territory to the activation of the NPB specifiers *Pax3/7*, *Zic1*, and *Msx1/2*.

The two members of the Snail family of zinc finger transcription factors, Snail1 and Snail2, have a critical role in NC development. Functional studies in chick and *Xenopus* demonstrate that overexpression of these genes resulted in an expansion of the NC territory, while their inhibition prevented NC formation and migration [173–175]. In *Xenopus*, *Snail1* is expressed in the prospective NC slightly earlier than *Snail2*, and epistatic analysis indicated that *Snail1* was an upstream regulator of *Snail2* during NC development [175,176]. Indirect evidence suggests that *Snail2* expression might be regulated by *Zic1* and *Pax3* in *Xenopus*, while *Msx1* may carry that function in zebrafish [13,147]. *Snail2* null mouse are viable and have normal NC formation, migration, and differentiation [177]. Homozygous null mutation in *Snail1* is embryonic lethal because of its early role in gastrulation [178]. An NC-specific deletion of the *Snail1* gene demonstrated that

Snail1 was not essential for NC formation and delamination [179]. The double knockout of both *Snail1* and *Snail2* demonstrated that this gene family is not required for the generation, delamination, or migration of the NCC in mouse embryos [179], indicating that the Snail gene family does not have an evolutionary conserved function during NC formation.

The protooncogene *c-Myc* and its downstream effector *Id3* have a fairly broad expression domain at the NPB in *Xenopus*. It includes both the prospective NC and PE [180–182]. Because c-Myc and Id3 expression is initiated prior to most NC specifiers, such as *Snail2* and *Sox9*, it has been proposed that *c-Myc* and *Id3* may function as a bridge between NPB and NC specifier genes. Consistent with this possibility, knockdown of Id3 results in the downregulation of early NC specifier genes *Snail2*, *Sox10*, *Foxd3*, and *Twist* [181,182]. Forced expression of Id3 in *Xenopus* NC prevented the development of most NC derivatives, presumably by maintaining NCC in a progenitor state beyond their normal course and thereby preventing their timely differentiation [181].

CONCLUSIONS AND PERSPECTIVES

In this chapter, we have summarized the major signaling pathways involved in NC induction and have presented a sample of the molecular players—NPB and NC specifiers—directing NC specification, comparing information from different model organisms.

Studies in fish, frog, and chicken point to some differences in the source and nature of the signaling molecules involved in the induction of NC progenitors in vertebrates. This presumably reflects differences in the timing of neural and NC induction, as well as differences in the precise mechanisms of neurulation employed by each one of these organisms.

NC induction in mammals is not as well understood as in other species, due to the difficulties of manipulating these processes early in the embryo. The development of new tools to specifically remove gene function in the presumptive NC territory, at an earlier stage than what can be achieved using the Wnt1-Cre line, will be critical to evaluate the level conservation of these mechanisms and pathways in the induction of mammalian NCC. In that respect, Pax3-Cre-mediated deletions have not yet provided information different from what has been learned using Wnt1-Cre.

NC specification is guided by the careful orchestration of regulatory circuits that can be assembled in a gene regulatory network. This regulatory cascade is temporally and spatially regulated by factors (NPB and NC specifiers), often acting in an iterative manner, which adds to the complexity of the network. While a great deal has been learned in the last few years on the expression and the role of these factors individually, the challenge is now to better understand how they interact with one another. Perturbation experiments in several model organisms have been extremely useful to delineate the interactions between the components of this network. However, so far only a small number of these interactions have been validated to discriminate direct from indirect regulations (reviewed in [128,183]). This is an essential step in establishing the functional linkage between these factors and defining their position within the NC gene regulatory network. The validation of these interactions through the identification of cis-regulatory sequences using ChIP and ChIP-Sequence is the focus of extensive work in many laboratories around the world. This effort will provide invaluable information to assemble this gene regulatory network. The comprehensive characterization of the regulatory inputs controlling NC formation is essential to understand how these processes may be altered in pathological situations.

Acknowledgments

We thank Dr. Jane McCutcheon for comments on the manuscript. We would like to apologize to colleagues whose work is not cited here, due to space limitations. Work in J-P S-J's lab is supported by a grant from the National Institutes of Health (RO1-DE014212).

References

[1] Northcutt RG, Gans C. The genesis of neural crest and epidermal placodes: a reinterpretation of vertebrate origins. Q Rev Biol 1983;58(1):1–28.

[2] Crane JF, Trainor PA. Neural crest stem and progenitor cells. Annu Rev Cell Dev Biol 2006;22:267–86.

[3] Achilleos A, Trainor PA. Neural crest stem cells: discovery, properties and potential for therapy. Cell Res 2012;22(2):288–304.

[4] Brugmann SA, Pandur PD, Kenyon KL, Pignoni F, Moody SA. Six1 promotes a placodal fate within the lateral neurogenic ectoderm by functioning as both a transcriptional activator and repressor. Development 2004;131(23):5871–81.

[5] Glavic A, Maris Honore S, Gloria Feijoo C, Bastidas F, Allende ML, Mayor R. Role of BMP signaling and the homeoprotein Iroquois in the specification of the cranial placodal field. Dev Biol 2004;272(1):89–103.

[6] Ahrens K, Schlosser G. Tissues and signals involved in the induction of placodal Six1 expression in *Xenopus laevis*. Dev Biol 2005;288(1):40–59.

[7] Litsiou A, Hanson S, Streit A. A balance of FGF, BMP and WNT signalling positions the future placode territory in the head. Development 2005;132(18):4051–62.

[8] Baker CV, Bronner-Fraser M. Vertebrate cranial placodes I. Embryonic induction. Dev Biol 2001;232(1):1–61.

[9] Meulemans D, Bronner-Fraser M. Gene-regulatory interactions in neural crest evolution and development. Dev Cell 2004;7(3):291–9.

[10] Sauka-Spengler T, Bronner-Fraser M. Insights from a sea lamprey into the evolution of neural crest gene regulatory network. Biol Bull 2008;214(3):303–14.

[11] Wang WD, Melville DB, Montero-Balaguer M, Hatzopoulos AK, Knapik EW. Tfap2a and Foxd3 regulate early steps in the development of the neural crest progenitor population. Dev Biol 2011;360(1):173–85.

[12] Li W, Cornell RA. Redundant activities of Tfap2a and Tfap2c are required for neural crest induction and development of other non-neural ectoderm derivatives in zebrafish embryos. Dev Biol 2007;304(1):338–54.

[13] Sato T, Sasai N, Sasai Y. Neural crest determination by co-activation of Pax3 and Zic1 genes in *Xenopus* ectoderm. Development 2005;132(10):2355–63.

[14] Monsoro-Burq AH, Wang E, Harland R. Msx1 and Pax3 cooperate to mediate FGF8 and WNT signals during *Xenopus* neural crest induction. Dev Cell 2005;8(2):167–78.

[15] Steventon B, Carmona-Fontaine C, Mayor R. Genetic network during neural crest induction: from cell specification to cell survival. Semin Cell Dev Biol 2005;16(6):647–54.

[16] Basch ML, Bronner-Fraser M, Garcia-Castro MI. Specification of the neural crest occurs during gastrulation and requires Pax7. Nature 2006;441(7090):218–22.

[17] Kos R, Reedy MV, Johnson RL, Erickson CA. The winged-helix transcription factor FoxD3 is important for establishing the neural crest lineage and repressing melanogenesis in avian embryos. Development 2001;128(8):1467–79.

[18] Zhao Q, Eberspaecher H, Lefebvre V, De Crombrugghe B. Parallel expression of Sox9 and Col2a1 in cells undergoing chondrogenesis. Dev Dyn 1997;209(4):377–86.

[19] Kuhlbrodt K, Herbarth B, Sock E, Hermans-Borgmeyer I, Wegner M. Sox10, a novel transcriptional modulator in glial cells. J Neurosci 1998;18(1):237–50.

[20] Labosky PA, Kaestner KH. The winged helix transcription factor Hfh2 is expressed in neural crest and spinal cord during mouse development. Mech Dev 1998;76(1-2):185–90.

[21] Hong CS, Saint-Jeannet JP. The activity of Pax3 and Zic1 regulates three distinct cell fates at the neural plate border. Mol Biol Cell 2007;18(6):2192–202.

[22] Arduini BL, Bosse KM, Henion PD. Genetic ablation of neural crest cell diversification. Development 2009;136(12):1987–94.

[23] Khudyakov J, Bronner-Fraser M. Comprehensive spatiotemporal analysis of early chick neural crest network genes. Dev Dyn 2009;238(3):716–23.

[24] Teng L, Mundell NA, Frist AY, Wang Q, Labosky PA. Requirement for Foxd3 in the maintenance of neural crest progenitors. Development 2008;135(9):1615–24.

[25] Huang X, Saint-Jeannet JP. Induction of the neural crest and the opportunities of life on the edge. Dev Biol 2004;275(1):1–11.

[26] Heeg-Truesdell E, LaBonne C. A slug, a fox, a pair of sox: transcriptional responses to neural crest inducing signals. Birth Defects Res C Embryo Today 2004;72(2):124–39.

[27] Basch ML, Selleck MA, Bronner-Fraser M. Timing and competence of neural crest formation. Dev Neurosci 2000;22(3):217–27.

[28] Steventon B, Araya C, Linker C, Kuriyama S, Mayor R. Differential requirements of BMP and Wnt signalling during gastrulation and neurulation define two steps in neural crest induction. Development 2009;136(5):771–9.

[29] Moury JD, Jacobson AG. The origins of neural crest cells in the axolotl. Dev Biol 1990;141(2):243–53.

[30] Selleck MA, Bronner-Fraser M. Origins of the avian neural crest: the role of neural plate–epidermal interactions. Development 1995;121(2):525–38.

[31] Mancilla A, Mayor R. Neural crest formation in *Xenopus laevis*: mechanisms of Xslug induction. Dev Biol 1996;177(2):580–9.

[32] Pieper M, Ahrens K, Rink E, Peter A, Schlosser G. Differential distribution of competence for panplacodal and neural crest induction to non-neural and neural ectoderm. Development 2012;139(6):1175–87.

[33] Raven K. Induction by medial and lateral pieces of archenteron roof, with special reference to the determination of the neural crest. Acta Neerl Norm Pathol 1945;55:348–62.

[34] Bonstein L, Elias S, Frank D. Paraxial-fated mesoderm is required for neural crest induction in *Xenopus* embryos. Dev Biol 1998;193(2):156–68.

[35] Marchant L, Linker C, Ruiz P, Guerrero N, Mayor R. The inductive properties of mesoderm suggest that the neural crest cells are specified by a BMP gradient. Dev Biol 1998;198(2):319–29.

[36] Monsoro-Burq AH, Fletcher RB, Harland RM. Neural crest induction by paraxial mesoderm in *Xenopus* embryos requires FGF signals. Development 2003;130(14):3111–24.

[37] Ragland JW, Raible DW. Signals derived from the underlying mesoderm are dispensable for zebrafish neural crest induction. Dev Biol 2004;276(1):16–30.

[38] Gritsman K, Zhang J, Cheng S, Heckscher E, Talbot WS, Schier AF. The EGF-CFC protein one-eyed pinhead is essential for nodal signaling. Cell 1999;97(1):121–32.

[39] Thisse C, Thisse B. Antivin, a novel and divergent member of the TGFbeta superfamily, negatively regulates mesoderm induction. Development 1999;126(2):229–40.

[40] Lewis JL, Bonner J, Modrell M, Ragland JW, Moon RT, Dorsky RI, et al. Reiterated Wnt signaling during zebrafish neural crest development. Development 2004;131(6):1299–308.

[41] Yoshikawa Y, Fujimori T, McMahon AP, Takada S. Evidence that absence of Wnt-3a signaling promotes neuralization instead of paraxial mesoderm development in the mouse. Dev Biol 1997;183(2):234–42.

[42] Chapman DL, Papaioannou VE. Three neural tubes in mouse embryos with mutations in the T-box gene Tbx6. Nature 1998;391(6668):695–7.

[43] Mayor R, Morgan R, Sargent MG. Induction of the prospective neural crest of *Xenopus*. Development 1995;121(3):767–77.

[44] Morgan R, Sargent MG. The role in neural patterning of translation initiation factor eIF4AII; induction of neural fold genes. Development 1997;124(14):2751–60.

[45] Nguyen M, Park S, Marques G, Arora K. Interpretation of a BMP activity gradient in *Drosophila* embryos depends on synergistic signaling by two type I receptors, SAX and TKV. Cell 1998;95 (4):495–506.

[46] Nguyen VH, Trout J, Connors SA, Andermann P, Weinberg E, Mullins MC. Dorsal and intermediate neuronal cell types of the spinal cord are established by a BMP signaling pathway. Development 2000;127 (6):1209–20.

[47] Schumacher JA, Hashiguchi M, Nguyen VH, Mullins MC. An intermediate level of BMP signaling directly specifies cranial neural crest progenitor cells in zebrafish. PLoS One 2011;6(11):e27403.

[48] Liem Jr. KF, Tremml G, Roelink H, Jessell TM. Dorsal differentiation of neural plate cells induced by BMP-mediated signals from epidermal ectoderm. Cell 1995;82(6):969–79.

[49] Liem Jr. KF, Tremml G, Jessell TM. A role for the roof plate and its resident TGFbeta-related proteins in neuronal patterning in the dorsal spinal cord. Cell 1997;91 (1):127–38.

[50] Garcia-Castro MI, Marcelle C, Bronner-Fraser M. Ectodermal Wnt function as a neural crest inducer. Science 2002;297(5582):848–51.

[51] Streit A, Lee KJ, Woo I, Roberts C, Jessell TM, Stern CD. Chordin regulates primitive streak development and the stability of induced neural cells, but is not sufficient for neural induction in the chick embryo. Development 1998;125(3):507–19.

[52] Storey KG, Crossley JM, De Robertis EM, Norris WE, Stern CD. Neural induction and regionalisation in the chick embryo. Development 1992;114(3):729–41.

[53] Levin M, Pagan S, Roberts DJ, Cooke J, Kuehn MR, Tabin CJ. Left/right patterning signals and the independent regulation of different aspects of situs in the chick embryo. Dev Biol 1997;189(1):57–67.

[54] Wilson SI, Graziano E, Harland R, Jessell TM, Edlund T. An early requirement for FGF signalling in the acquisition of neural cell fate in the chick embryo. Curr Biol 2000;10(8):421–9.

[55] Streit A, Berliner AJ, Papanayotou C, Sirulnik A, Stern CD. Initiation of neural induction by FGF signalling before gastrulation. Nature 2000;406(6791):74–8.

[56] Stuhlmiller TJ, Garcia-Castro MI. FGF/MAPK signaling is required in the gastrula epiblast for avian neural crest induction. Development 2012;139(2):289–300.

[57] Launay C, Fromentoux V, Shi DL, Boucaut JC. A truncated FGF receptor blocks neural induction by endogenous *Xenopus* inducers. Development 1996;122 (3):869–80.

[58] Delaune E, Lemaire P, Kodjabachian L. Neural induction in *Xenopus* requires early FGF signalling in addition to BMP inhibition. Development 2005;132 (2):299–310.

[59] Kuroda H, Fuentealba L, Ikeda A, Reversade B, De Robertis EM. Default neural induction: neuralization of dissociated *Xenopus* cells is mediated by Ras/MAPK activation. Genes Dev 2005;19(9):1022–7.

[60] Hong CS, Park BY, Saint-Jeannet JP. Fgf8a induces neural crest indirectly through the activation of Wnt8 in the paraxial mesoderm. Development 2008;135 (23):3903–10.

[61] Bachiller D, Klingensmith J, Kemp C, Belo JA, Anderson RM, May SR, et al. The organizer factors Chordin and Noggin are required for mouse forebrain development. Nature 2000;403(6770):658–61.

[62] Anderson RM, Stottmann RW, Choi M, Klingensmith J. Endogenous bone morphogenetic protein antagonists regulate mammalian neural crest generation and survival. Dev Dyn 2006;235(9):2507–20.

[63] Kanzler B, Foreman RK, Labosky PA, Mallo M. BMP signaling is essential for development of skeletogenic and neurogenic cranial neural crest. Development 2000;127(5):1095–104.

[64] Correia AC, Costa M, Moraes F, Bom J, Novoa A, Mallo M. Bmp2 is required for migration but not for induction of neural crest cells in the mouse. Dev Dyn 2007;236(9):2493–501.

[65] Beppu H, Kawabata M, Hamamoto T, Chytil A, Minowa O, Noda T, et al. BMP type II receptor is required for gastrulation and early development of mouse embryos. Dev Biol 2000;221(1):249–58.

[66] Fujiwara T, Dehart DB, Sulik KK, Hogan BL. Distinct requirements for extra-embryonic and embryonic bone morphogenetic protein 4 in the formation of the node and primitive streak and coordination of left-right asymmetry in the mouse. Development 2002;129(20):4685–96.

[67] Gu Z, Reynolds EM, Song J, Lei H, Feijen A, Yu L, et al. The type I serine/threonine kinase receptor ActRIA (ALK2) is required for gastrulation of the mouse embryo. Development 1999;126(11):2551–61.

[68] Mishina Y, Crombie R, Bradley A, Behringer RR. Multiple roles for activin-like kinase-2 signaling during mouse embryogenesis. Dev Biol 1999;213 (2):314–26.

[69] Mishina Y, Suzuki A, Ueno N, Behringer RR. Bmpr encodes a type I bone morphogenetic protein receptor that is essential for gastrulation during mouse embryogenesis. Genes Dev 1995;9(24):3027–37.

[70] Dunn NR, Winnier GE, Hargett LK, Schrick JJ, Fogo AB, Hogan BL. Haploinsufficient phenotypes in Bmp4 heterozygous null mice and modification by mutations in Gli3 and Alx4. Dev Biol 1997;188(2):235−47.

[71] Stottmann RW, Klingensmith J. Bone morphogenetic protein signaling is required in the dorsal neural folds before neurulation for the induction of spinal neural crest cells and dorsal neurons. Dev Dyn 2011;240 (4):755−65.

[72] Dudas M, Sridurongrit S, Nagy A, Okazaki K, Kaartinen V. Craniofacial defects in mice lacking BMP type I receptor Alk2 in neural crest cells. Mech Dev 2004;121(2):173−82.

[73] Kaartinen V, Dudas M, Nagy A, Sridurongrit S, Lu MM, Epstein JA. Cardiac outflow tract defects in mice lacking ALK2 in neural crest cells. Development 2004;131(14): 3481−90.

[74] Stottmann RW, Choi M, Mishina Y, Meyers EN, Klingensmith J. BMP receptor IA is required in mammalian neural crest cells for development of the cardiac outflow tract and ventricular myocardium. Development 2004;131(9):2205−18.

[75] Wang J, Nagy A, Larsson J, Dudas M, Sucov HM, Kaartinen V. Defective ALK5 signaling in the neural crest leads to increased postmigratory neural crest cell apoptosis and severe outflow tract defects. BMC Dev Biol 2006;6:51.

[76] Patthey C, Edlund T, Gunhaga L. Wnt-regulated temporal control of BMP exposure directs the choice between neural plate border and epidermal fate. Development 2009;136(1):73−83.

[77] Wilson PA, Lagna G, Suzuki A, Hemmati-Brivanlou A. Concentration-dependent patterning of the *Xenopus* ectoderm by BMP4 and its signal transducer Smad1. Development 1997;124(16):3177−84.

[78] Saint-Jeannet JP, He X, Varmus HE, Dawid IB. Regulation of dorsal fate in the neuraxis by Wnt-1 and Wnt-3a. Proc Natl Acad Sci USA 1997;94(25):13713−8.

[79] LaBonne C, Bronner-Fraser M. Neural crest induction in *Xenopus*: evidence for a two-signal model. Development 1998;125(13):2403−14.

[80] MacDonald BT, Tamai K, He X. Wnt/beta-catenin signaling: components, mechanisms, and diseases. Dev Cell 2009;17(1):9−26.

[81] Tamai K, Semenov M, Kato Y, Spokony R, Liu C, Katsuyama Y, et al. LDL-receptor-related proteins in Wnt signal transduction. Nature 2000;407(6803):530−5.

[82] Deardorff MA, Tan C, Saint-Jeannet JP, Klein PS. A role for frizzled 3 in neural crest development. Development 2001;128(19):3655−63.

[83] Abu-Elmagd M, Garcia-Morales C, Wheeler GN. Frizzled7 mediates canonical Wnt signaling in neural crest induction. Dev Biol 2006;298(1):285−98.

[84] Hassler C, Cruciat CM, Huang YL, Kuriyama S, Mayor R, Niehrs C. Kremen is required for neural crest induction in *Xenopus* and promotes LRP6-mediated Wnt signaling. Development 2007;134(23):4255−63.

[85] Gray RS, Bayly RD, Green SA, Agarwala S, Lowe CJ, Wallingford JB. Diversification of the expression patterns and developmental functions of the dishevelled gene family during chordate evolution. Dev Dyn 2009;238(8):2044−57.

[86] Wu J, Yang J, Klein PS. Neural crest induction by the canonical Wnt pathway can be dissociated from anterior-posterior neural patterning in *Xenopus*. Dev Biol 2005;279(1):220−32.

[87] Bang AG, Papalopulu N, Goulding MD, Kintner C. Expression of Pax-3 in the lateral neural plate is dependent on a Wnt-mediated signal from posterior nonaxial mesoderm. Dev Biol 1999;212(2):366−80.

[88] Lekven AC, Thorpe CJ, Waxman JS, Moon RT. Zebrafish wnt8 encodes two wnt8 proteins on a bicistronic transcript and is required for mesoderm and neurectoderm patterning. Dev Cell 2001;1(1):103−14.

[89] Chang C, Hemmati-Brivanlou A. Cell fate determination in embryonic ectoderm. J Neurobiol 1998;36 (2):128−51.

[90] McGrew LL, Lai CJ, Moon RT. Specification of the anteroposterior neural axis through synergistic interaction of the Wnt signaling cascade with noggin and follistatin. Dev Biol 1995;172(1):337−42.

[91] Elkouby YM, Elias S, Casey ES, Blythe SA, Tsabar N, Klein PS, et al. Mesodermal Wnt signaling organizes the neural plate via Meis3. Development 2010;137 (9):1531−41.

[92] Schmidt C, McGonnell IM, Allen S, Otto A, Patel K. Wnt6 controls amniote neural crest induction through the non-canonical signaling pathway. Dev Dyn 2007;236(9):2502−11.

[93] Brault V, Moore R, Kutsch S, Ishibashi M, Rowitch DH, McMahon AP, et al. Inactivation of the beta-catenin gene by Wnt1-Cre-mediated deletion results in dramatic brain malformation and failure of craniofacial development. Development 2001;128(8):1253−64.

[94] Ikeya M, Lee SM, Johnson JE, McMahon AP, Takada S. Wnt signalling required for expansion of neural crest and CNS progenitors. Nature 1997;389(6654): 966−70.

[95] Hari L, Brault V, Kleber M, Lee HY, Ille F, Leimeroth R, et al. Lineage-specific requirements of beta-catenin in neural crest development. J Cell Biol 2002;159 (5):867−80.

[96] Lee HY, Kleber M, Hari L, Brault V, Suter U, Taketo MM, et al. Instructive role of Wnt/beta-catenin in sensory fate specification in neural crest stem cells. Science 2004;303(5660):1020−3.

[97] Jones NC, Trainor PA. Role of morphogens in neural crest cell determination. J Neurobiol 2005;64(4):388–404.

[98] Ossipova O, Sokol SY. Neural crest specification by noncanonical Wnt signaling and PAR-1. Development 2011;138(24):5441–50.

[99] Komiya Y, Habas R. Wnt signal transduction pathways. Organogenesis 2008;4(2):68–75.

[100] Carmona-Fontaine C, Matthews H, Mayor R. Directional cell migration in vivo: Wnt at the crest. Cell Adh Migr 2008;2(4):240–2.

[101] De Calisto J, Araya C, Marchant L, Riaz CF, Mayor R. Essential role of non-canonical Wnt signalling in neural crest migration. Development 2005;132(11):2587–97.

[102] Matthews HK, Marchant L, Carmona-Fontaine C, Kuriyama S, Larrain J, Holt MR, et al. Directional migration of neural crest cells in vivo is regulated by Syndecan-4/Rac1 and non-canonical Wnt signaling/RhoA. Development 2008;135(10):1771–80.

[103] Matthews HK, Broders-Bondon F, Thiery JP, Mayor R. Wnt11r is required for cranial neural crest migration. Dev Dyn 2008;237(11):3404–9.

[104] Kengaku M, Okamoto H. Basic fibroblast growth factor induces differentiation of neural tube and neural crest lineages of cultured ectoderm cells from *Xenopus* gastrula. Development 1993;119(4):1067–78.

[105] Mayor R, Guerrero N, Martinez C. Role of FGF and noggin in neural crest induction. Dev Biol 1997;189(1):1–12.

[106] Yardley N, Garcia-Castro MI. FGF signaling transforms non-neural ectoderm into neural crest. Dev Biol 2012;372(2):166–77.

[107] Garnett AT, Square TA, Medeiros DM. BMP, Wnt and FGF signals are integrated through evolutionarily conserved enhancers to achieve robust expression of Pax3 and Zic genes at the zebrafish neural plate border. Development 2012;139(22):4220–31.

[108] Cornell RA, Eisen JS. Delta signaling mediates segregation of neural crest and spinal sensory neurons from zebrafish lateral neural plate. Development 2000;127(13):2873–82.

[109] Coffman CR, Skoglund P, Harris WA, Kintner CR. Expression of an extracellular deletion of Xotch diverts cell fate in *Xenopus* embryos. Cell 1993;73(4):659–71.

[110] Glavic A, Silva F, Aybar MJ, Bastidas F, Mayor R. Interplay between Notch signaling and the homeoprotein Xiro1 is required for neural crest induction in *Xenopus* embryos. Development 2004;131(2):347–59.

[111] Endo Y, Osumi N, Wakamatsu Y. Bimodal functions of Notch-mediated signaling are involved in neural crest formation during avian ectoderm development. Development 2002;129(4):863–73.

[112] Humphreys R, Zheng W, Prince LS, Qu X, Brown C, Loomes K, et al. Cranial neural crest ablation of Jagged1 recapitulates the craniofacial phenotype of Alagille syndrome patients. Hum Mol Genet 2012;21(6):1374–83.

[113] Mead TJ, Yutzey KE. Notch pathway regulation of neural crest cell development in vivo. Dev Dyn 2012;241(2):376–89.

[114] De Bellard ME, Ching W, Gossler A, Bronner-Fraser M. Disruption of segmental neural crest migration and ephrin expression in delta-1 null mice. Dev Biol 2002;249(1):121–30.

[115] Villanueva S, Glavic A, Ruiz P, Mayor R. Posteriorization by FGF, Wnt, and retinoic acid is required for neural crest induction. Dev Biol 2002;241(2):289–301.

[116] Papalopulu N, Kintner C. A posteriorising factor, retinoic acid, reveals that anteroposterior patterning controls the timing of neuronal differentiation in *Xenopus* neuroectoderm. Development 1996;122(11):3409–18.

[117] Bang AG, Papalopulu N, Kintner C, Goulding MD. Expression of Pax-3 is initiated in the early neural plate by posteriorizing signals produced by the organizer and by posterior non-axial mesoderm. Development 1997;124(10):2075–85.

[118] Halilagic A, Zile MH, Studer M. A novel role for retinoids in patterning the avian forebrain during presomite stages. Development 2003;130(10):2039–50.

[119] Uehara M, Yashiro K, Mamiya S, Nishino J, Chambon P, Dolle P, et al. CYP26A1 and CYP26C1 cooperatively regulate anterior-posterior patterning of the developing brain and the production of migratory cranial neural crest cells in the mouse. Dev Biol 2007;302(2):399–411.

[120] Bonano M, Tribulo C, De Calisto J, Marchant L, Sanchez SS, Mayor R, et al. A new role for the Endothelin-1/Endothelin-A receptor signaling during early neural crest specification. Dev Biol 2008;323(1):114–29.

[121] Clouthier DE, Hosoda K, Richardson JA, Williams SC, Yanagisawa H, Kuwaki T, et al. Cranial and cardiac neural crest defects in endothelin-A receptor-deficient mice. Development 1998;125(5):813–24.

[122] Yanagisawa H, Yanagisawa M, Kapur RP, Richardson JA, Williams SC, Clouthier DE, et al. Dual genetic pathways of endothelin-mediated intercellular signaling revealed by targeted disruption of endothelin converting enzyme-1 gene. Development 1998;125(5):825–36.

[123] Thomas T, Kurihara H, Yamagishi H, Kurihara Y, Yazaki Y, Olson EN, et al. A signaling cascade involving endothelin-1, dHAND and msx1 regulates development of neural-crest-derived branchial arch mesenchyme. Development 1998;125(16):3005–14.

[124] Aguero TH, Fernandez JP, Lopez GA, Tribulo C, Aybar MJ. Indian hedgehog signaling is required for proper formation, maintenance and migration of *Xenopus* neural crest. Dev Biol 2012;364(2):99–113.

[125] Testaz S, Jarov A, Williams KP, Ling LE, Koteliansky VE, Fournier-Thibault C, et al. Sonic hedgehog restricts adhesion and migration of neural crest cells independently of the Patched- Smoothened-Gli signaling pathway. Proc Natl Acad Sci USA 2001;98(22):12521–6.

[126] Jeong J, Mao J, Tenzen T, Kottmann AH, McMahon AP. Hedgehog signaling in the neural crest cells regulates the patterning and growth of facial primordia. Genes Dev 2004;18(8):937–51.

[127] Sauka-Spengler T, Bronner-Fraser M. A gene regulatory network orchestrates neural crest formation. Nat Rev Mol Cell Biol 2008;9(7):557–68.

[128] Betancur P, Bronner-Fraser M, Sauka-Spengler T. Assembling neural crest regulatory circuits into a gene regulatory network. Annu Rev Cell Dev Biol 2010;26:581–603.

[129] Nelms BL, Pfaltzgraff ER, Labosky PA. Functional interaction between Foxd3 and Pax3 in cardiac neural crest development. Genesis 2011;49(1):10–23.

[130] Milet C, Monsoro-Burq AH. Neural crest induction at the neural plate border in vertebrates. Dev Biol 2012;366(1):22–33.

[131] Nichane M, Ren X, Souopgui J, Bellefroid EJ. Hairy2 functions through both DNA-binding and non DNA-binding mechanisms at the neural plate border in *Xenopus*. Dev Biol 2008;322(2):368–80.

[132] Li B, Kuriyama S, Moreno M, Mayor R. The posteriorizing gene Gbx2 is a direct target of Wnt signalling and the earliest factor in neural crest induction. Development 2009;136(19):3267–78.

[133] Collins CA, Gnocchi VF, White RB, Boldrin L, Perez-Ruiz A, Relaix F, et al. Integrated functions of Pax3 and Pax7 in the regulation of proliferation, cell size and myogenic differentiation. PLoS One 2009;4(2):e4475.

[134] de Croze N, Maczkowiak F, Monsoro-Burq AH. Reiterative AP2a activity controls sequential steps in the neural crest gene regulatory network. Proc Natl Acad Sci USA 2011;108(1):155–60.

[135] Gutkovich YE, Ofir R, Elkouby YM, Dibner C, Gefen A, Elias S, et al. *Xenopus* Meis3 protein lies at a nexus downstream to Zic1 and Pax3 proteins, regulating multiple cell-fates during early nervous system development. Dev Biol 2010;338(1):50–62.

[136] Rada-Iglesias A, Bajpai R, Prescott S, Brugmann SA, Swigut T, Wysocka J. Epigenomic annotation of enhancers predicts transcriptional regulators of human neural crest. Cell Stem Cell 2012;11(5):633–48.

[137] Mitchell PJ, Timmons PM, Hebert JM, Rigby PW, Tjian R. Transcription factor AP-2 is expressed in neural crest cell lineages during mouse embryogenesis. Genes Dev 1991;5(1):105–19.

[138] Zhang J, Hagopian-Donaldson S, Serbedzija G, Elsemore J, Plehn-Dujowich D, McMahon AP, et al. Neural tube, skeletal and body wall defects in mice lacking transcription factor AP-2. Nature 1996;381 (6579):238–41.

[139] Schorle H, Meier P, Buchert M, Jaenisch R, Mitchell PJ. Transcription factor AP-2 essential for cranial closure and craniofacial development. Nature 1996;381 (6579):235–8.

[140] Milet C, Maczkowiak F, Roche DD, Monsoro-Burq AH. Pax3 and Zic1 drive induction and differentiation of multipotent, migratory, and functional neural crest in *Xenopus* embryos. Proc Natl Acad Sci USA 2013;110(14):5528–33.

[141] Otto A, Schmidt C, Patel K. Pax3 and Pax7 expression and regulation in the avian embryo. Anat Embryol (Berl) 2006;211(4):293–310.

[142] Inoue T, Hatayama M, Tohmonda T, Itohara S, Aruga J, Mikoshiba K. Mouse Zic5 deficiency results in neural tube defects and hypoplasia of cephalic neural crest derivatives. Dev Biol 2004;270(1):146–62.

[143] Inoue T, Ota M, Mikoshiba K, Aruga J. Zic2 and Zic3 synergistically control neurulation and segmentation of paraxial mesoderm in mouse embryo. Dev Biol 2007;306(2):669–84.

[144] Conway SJ, Henderson DJ, Copp AJ. Pax3 is required for cardiac neural crest migration in the mouse: evidence from the splotch (Sp2H) mutant. Development 1997;124(2):505–14.

[145] Conway SJ, Bundy J, Chen J, Dickman E, Rogers R, Will BM. Decreased neural crest stem cell expansion is responsible for the conotruncal heart defects within the splotch (Sp(2H))/Pax3 mouse mutant. Cardiovasc Res 2000;47(2):314–28.

[146] Epstein JA, Li J, Lang D, Chen F, Brown CB, Jin F, et al. Migration of cardiac neural crest cells in Splotch embryos. Development 2000;127(9):1869–78.

[147] Tribulo C, Aybar MJ, Nguyen VH, Mullins MC, Mayor R. Regulation of Msx genes by a Bmp gradient is essential for neural crest specification. Development 2003;130(26):6441–52.

[148] Brugger SM, Merrill AE, Torres-Vazquez J, Wu N, Ting MC, Cho JY, et al. A phylogenetically conserved cis-regulatory module in the Msx2 promoter is sufficient for BMP-dependent transcription in murine and *Drosophila* embryos. Development 2004;131(20): 5153–65.

[149] Ishii M, Han J, Yen HY, Sucov HM, Chai Y, Maxson Jr. RE. Combined deficiencies of Msx1 and Msx2 cause impaired patterning and survival of the cranial neural crest. Development 2005;132(22):4937–50.

[150] Hong CS, Saint-Jeannet JP. Sox proteins and neural crest development. Semin Cell Dev Biol 2005;16 (6):694–703.

[151] Haldin CE, LaBonne C. SoxE factors as multifunctional neural crest regulatory factors. Int J Biochem Cell Biol 2010;42(3):441–4.

[152] Lee YH, Saint-Jeannet JP. Sox9 function in craniofacial development and disease. Genesis 2011;49 (4):200–8.

[153] Kelsh RN. Sorting out Sox10 functions in neural crest development. Bioessays 2006;28(8):788–98.

[154] O'Donnell M, Hong CS, Huang X, Delnicki RJ, Saint-Jeannet JP. Functional analysis of Sox8 during neural crest development in Xenopus. Development 2006;133 (19):3817–26.

[155] Spokony RF, Aoki Y, Saint-Germain N, Magner-Fink E, Saint-Jeannet JP. The transcription factor Sox9 is required for cranial neural crest development in Xenopus. Development 2002;129(2):421–32.

[156] Aoki Y, Saint-Germain N, Gyda M, Magner-Fink E, Lee YH, Credidio C, et al. Sox10 regulates the development of neural crest-derived melanocytes in Xenopus. Dev Biol 2003;259(1):19–33.

[157] Honore SM, Aybar MJ, Mayor R. Sox10 is required for the early development of the prospective neural crest in Xenopus embryos. Dev Biol 2003;260 (1):79–96.

[158] Li M, Zhao C, Wang Y, Zhao Z, Meng A. Zebrafish sox9b is an early neural crest marker. Dev Genes Evol 2002;212(4):203–6.

[159] Dutton KA, Pauliny A, Lopes SS, Elworthy S, Carney TJ, Rauch J, et al. Zebrafish colourless encodes sox10 and specifies non-ectomesenchymal neural crest fates. Development 2001;128(21):4113–25.

[160] Southard-Smith EM, Kos L, Pavan WJ. Sox10 mutation disrupts neural crest development in Dom Hirschsprung mouse model. Nat Genet 1998;18 (1):60–4.

[161] Simoes-Costa MS, McKeown SJ, Tan-Cabugao J, Sauka-Spengler T, Bronner ME. Dynamic and differential regulation of stem cell factor FoxD3 in the neural crest is Encrypted in the genome. PLoS Genet 2012;8(12):e1003142.

[162] Cheng Y, Cheung M, Abu-Elmagd MM, Orme A, Scotting PJ. Chick sox10, a transcription factor expressed in both early neural crest cells and central nervous system. Brain Res Dev Brain Res 2000;121 (2):233–41.

[163] Cheung M, Briscoe J. Neural crest development is regulated by the transcription factor Sox9. Development 2003;130(23):5681–93.

[164] Mollaaghababa R, Pavan WJ. The importance of having your SOX on: role of SOX10 in the development of neural crest-derived melanocytes and glia. Oncogene 2003;22(20):3024–34.

[165] Cheung M, Chaboissier MC, Mynett A, Hirst E, Schedl A, Briscoe J. The transcriptional control of trunk neural crest induction, survival, and delamination. Dev Cell 2005;8(2):179–92.

[166] Luo T, Lee YH, Saint-Jeannet JP, Sargent TD. Induction of neural crest in Xenopus by transcription factor AP2alpha. Proc Natl Acad Sci USA 2003;100 (2):532–7.

[167] Bagheri-Fam S, Barrionuevo F, Dohrmann U, Gunther T, Schule R, Kemler R, et al. Long-range upstream and downstream enhancers control distinct subsets of the complex spatiotemporal Sox9 expression pattern. Dev Biol 2006;291(2):382–97.

[168] Betancur P, Sauka-Spengler T, Bronner MA. Sox10 enhancer element common to the otic placode and neural crest is activated by tissue-specific paralogs. Development 2011;138(17):3689–98.

[169] Lister JA, Cooper C, Nguyen K, Modrell M, Grant K, Raible DW. Zebrafish Foxd3 is required for development of a subset of neural crest derivatives. Dev Biol 2006;290(1):92–104.

[170] Stewart RA, Arduini BL, Berghmans S, George RE, Kanki JP, Henion PD, et al. Zebrafish foxd3 is selectively required for neural crest specification, migration and survival. Dev Biol 2006;292(1):174–88.

[171] Dottori M, Gross MK, Labosky P, Goulding M. The winged-helix transcription factor Foxd3 suppresses interneuron differentiation and promotes neural crest cell fate. Development 2001;128(21):4127–38.

[172] Sasai N, Mizuseki K, Sasai Y. Requirement of FoxD3-class signaling for neural crest determination in Xenopus. Development 2001;128(13):2525–36.

[173] LaBonne C, Bronner-Fraser M. Induction and patterning of the neural crest, a stem cell-like precursor population. J Neurobiol 1998;36(2):175–89.

[174] del Barrio MG, Nieto MA. Overexpression of Snail family members highlights their ability to promote chick neural crest formation. Development 2002;129 (7):1583–93.

[175] Aybar MJ, Nieto MA, Mayor R. Snail precedes slug in the genetic cascade required for the specification and migration of the Xenopus neural crest. Development 2003;130(3):483–94.

[176] Linker C, Bronner-Fraser M, Mayor R. Relationship between gene expression domains of Xsnail, Xslug, and Xtwist and cell movement in the prospective neural crest of Xenopus. Dev Biol 2000;224(2):215–25.

[177] Jiang R, Lan Y, Norton CR, Sundberg JP, Gridley T. The Slug gene is not essential for mesoderm or neural crest development in mice. Dev Biol 1998;198 (2):277–85.

[178] Murray SA, Gridley T. Snail1 gene function during early embryo patterning in mice. Cell Cycle 2006; 5(22):2566–70.

[179] Murray SA, Gridley T. Snail family genes are required for left-right asymmetry determination, but not neural crest formation, in mice. Proc Natl Acad Sci USA 2006;103(27):10300–4.

[180] Bellmeyer A, Krase J, Lindgren J, LaBonne C. The protooncogene c-myc is an essential regulator of neural crest formation in *Xenopus*. Dev Cell 2003;4 (6):827–39.

[181] Light W, Vernon AE, Lasorella A, Iavarone A, LaBonne C. *Xenopus* Id3 is required downstream of Myc for the formation of multipotent neural crest progenitor cells. Development 2005;132(8):1831–41.

[182] Kee Y, Bronner-Fraser M. To proliferate or to die: role of Id3 in cell cycle progression and survival of neural crest progenitors. Genes Dev 2005;19(6):744–55.

[183] Prasad MS, Sauka-Spengler T, LaBonne C. Induction of the neural crest state: control of stem cell attributes by gene regulatory, post-transcriptional and epigenetic interactions. Dev Biol 2012;366(1):10–21.

The Cell Biology of Neural Crest Cell Delamination and EMT

Lisa A. Taneyhill and Rangarajan Padmanabhan

Department of Animal and Avian Sciences, University of Maryland, College Park, MD 20742, USA

INTRODUCTION

In 1982, Newgreen and Gibbins suggested that four conditions must be fulfilled for neural crest cells (NCCs) to emigrate from the neural tube: (i) absence of physical barriers to emigration, such as the basal lamina, (ii) an appropriate substratum for migration, (iii) loss of intercellular adhesion, and (iv) acquisition of migratory competence [1] (Figure 3.1). Although the first condition with respect to

the basal lamina as a physical barrier was later proved incorrect, the remaining conditions together describe the epithelial-to-mesenchymal transition (EMT), in which a stationary cell undergoes a series of events to become migratory (reviewed in [2]). NCC have traditionally been considered one of the best models in which to study this process, being used from the earliest days of EMT research [3]. Although all NCC undergo EMT to become migratory, axial level differences in

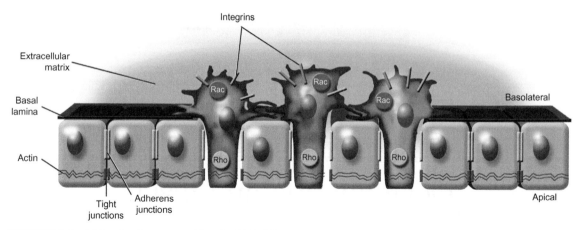

FIGURE 3.1 Molecular changes associated with NCC EMT. Premigratory NCC undergo several changes during EMT to allow them to emigrate from the neural tube. Apically localized tight junctions and adherens junctions are dismantled, followed by the rearrangement of cytoskeletal factors to coordinate subsequent motility. Emigrating NCC exit from the dorsal region of the neural tube through a region devoid of basal lamina. Membrane proteins such as integrins facilitate migration into and through the extracellular matrix.

delamination and EMT exist. For example, cranial NCC delaminate and undergo an *en masse* EMT all at once, migrating in regions devoid of somitic mesoderm (except in the case of post-otic NCC). Trunk NCC, however, undergo EMT individually, delaminating from the dorsal neural tube in a progressive, drip-like fashion and then migrating through the somites [4–11]. In mouse and *Xenopus*, cranial NCC delamination and EMT occur before neural tube closure [6,12], while in birds, delamination and EMT happen concurrently with neural fold fusion [13,14]. These distinct delamination and EMT mechanisms can be ascribed to specific molecular pathways operating at these different axial levels, as observed in the chick. For instance, trunk NCC delamination and EMT in the chick are closely coupled to the cell cycle [13, 15], with NCC delaminating in S phase, although this entry is not sufficient to promote delamination. Furthermore, recent work using photoconvertible fluorescent proteins to label small populations of trunk NCC and track their exit from the neural tube reveals that about half of the premigratory NCC population does not undergo cell division prior to

exit [16]. In chick midbrain NCC, however, Ets-1 serves to disconnect the cell cycle and delamination, thus enabling simultaneous NCC delamination and EMT, irrespective of cell cycle position. Moreover, somite dissociation controls chick rostral trunk NCC delamination and EMT by relieving noggin-mediated BMP4 inhibition [17]. Remarkably, despite these apparent axial level differences, the molecular changes that accompany EMT, including the dismantling of cellular junctions and cytoskeletal rearrangements, are still well conserved. This chapter will specifically discuss the cell biological aspects of NCC delamination and EMT, highlighting findings at multiple axial levels (Figures 3.2 and 3.3). The reader is advised to look elsewhere for information pertaining to gene expression changes that encompass NCC EMT [18].

LOSS OF INTERCELLULAR ADHESION

Adherens Junctions

Adherens junctions (AJs) are broadly defined as cell–cell junctions that function to

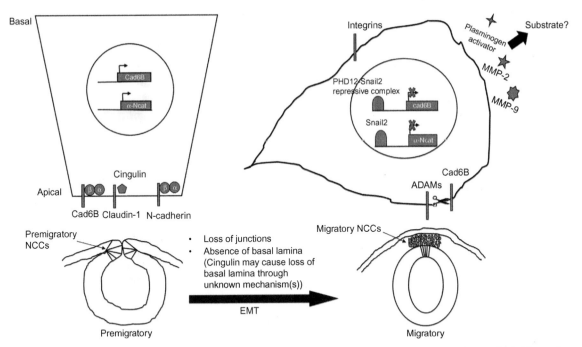

FIGURE 3.2 Midbrain NCC EMT. Premigratory NCC are located in the dorsal neural tube. These cells, held together in part by Cad6B [27, 29–31], express claudin-1 [74] and N-cadherin [26, our unpublished data], to which cingulin [73] and αN-catenin ([57] αN-cat; through β-catenin) could also interact. The basal lamina surrounds the neural tube on its basolateral surface (not shown), but it is not present in the dorsal region of the neural tube prior to NCC emigration [73,74], possibly due to secreted proteases that facilitate its degradation. As NCC transition to a migratory state, they lose intercellular AJs and TJs (through transcriptional and post-translational mechanisms) [28, 30, 31, 41, 73, 74 our unpublished data]. Migratory NCC express mesenchymal proteins such as integrins [158–166] and secrete proteases [122–124, 126–130] that help NCC migrate through the complex extracellular matrix.

hold adjacent cells together. Ultrastructurally, they are characterized by dense plaques at regions of cell–cell contact, principally comprised of transmembrane molecules and actin filaments. Termed zonula adherens in polarized epithelial cells, AJs form the "adhesion belt" by interacting with circumferential F-actin, thereby linking cells into a continuous sheet and separating the apical and basolateral membranes [19,20]. AJs consist of transmembrane cadherin and nectin/afadin molecules [19,21], and here we focus on cadherins, a large family of calcium-dependent transmembrane proteins that form primarily homophilic interactions with cadherins of adjacent cells. Cadherins are divided into the classical type I

cadherins (e.g., E-cadherin) and type II cadherins (e.g., Cadherin-5). The cadherin cytoplasmic domain harbors motifs to permit binding of two catenin proteins, β-catenin and p120-catenin [22]. β-catenin in turn interacts with α-catenin, which can interact directly with and stabilize the actin cytoskeleton through the formation of α-catenin homodimers [22].

Cadherins

CHICK

Cadherin expression begins early in chick embryos with E-cadherin in the ectoderm [23]. During neurulation, E-cadherin expression is

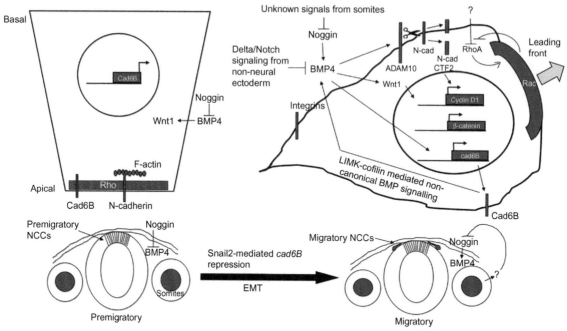

FIGURE 3.3   Trunk NCC EMT. Premigratory NCC are maintained in epithelial state by N-cadherin (N-cad) [28], Cad6B [27, 29], and Rho [97, 107], ensuring a stable actin cytoskeleton. During EMT, unknown signals from somites inhibit Noggin activity [17], thereby triggering BMP4-mediated Wnt1 signaling and activation of *cyclin D1* and *β-catenin* transcription [13, 15]. BMP4 also stimulates ADAM10 and γ-secretase mediated N-cadherin proteolysis, releasing an N-cadherin cytoplasmic fragment (CTF2) that further augments *cyclin D1* transcription [28]. Unknown mechanisms operate to remove Cad6B from later migratory NCC. A positive feedback loop exists between BMP4 and Cad6B [34, 43, 44], which is possibly broken by Delta/Notch signaling from the non-neural ectoderm [45]. Unknown signals inhibit Rho and stimulate Rac activity at the leading front [107], which further inhibits Rho and ensures directed migration of NCC.

lost in the newly formed neural tube and "switched" to N-cadherin [24]. Cadherin switching (in which one cadherin replaces the expression of another) is a phenomenon that occurs during processes underlying normal embryonic development, including cell segregation and, in many cases, EMT. Stationary epithelial tumor cells also deploy this process to become metastatic (for review, see [25]). This switch is critical with respect to NCC because the levels of N-cadherin play an important role in regulating NCC emigration. Microinjection of an N-cadherin function-blocking antibody disrupts neural tube architecture and leads to ectopic NCC emigration [26], while N-cadherin overexpression impedes NCC emigration

[27,28]. Prospective NCC located in the dorsal neural tube are postulated to segregate from other cells within the newly formed neural tube through an N-cadherin to Cadherin-6B (Cad6B) switch [27,29]. It is thought that this differential expression prevents mixing of dorsal NCC and other neuroepithelial cells [29]. Cad6B is considered to be a definitive marker of premigratory NCC [27,29–31]. Although Cad6B is expressed in all prospective NCC, the levels of expression (mRNA and protein) are qualitatively much higher in the trunk compared to the midbrain (our unpublished data).

Upon neural tube closure, chick midbrain premigratory NCC down-regulate N-cadherin protein (and transcripts to some extent, our unpublished data), while Cad6B (transcripts and protein) is up-regulated. During chick trunk NCC EMT, however, *N-cadherin* transcripts remain high while protein levels decline, suggesting that N-cadherin protein down-regulation is likely happening post-transcriptionally [28]. Indeed, Shoval et al. [28] showed that BMP-mediated signaling activates the A Disintegrin and Metalloproteinase (ADAM) family member ADAM10, which proteolytically cleaves the N-cadherin extracellular domain. The remaining truncated N-cadherin protein is subsequently processed within its cytoplasmic domain by an unknown protease (most likely γ-secretase; see [32,33]), releasing a C terminal N-cadherin fragment (CTF2). CTF2 translocates to the nucleus and directly or indirectly stimulates transcription of genes important for modulating NCC EMT, such as the G1/S phase-specific cyclin, *cyclin D1* [28], which correlates with the occurrence of trunk NCC EMT during the G1/S phase of the cell cycle [13, 15]. N-cadherin processing is crucial because maintenance of N-cadherin, as observed in overexpression experiments, inhibits NCC delamination, migration and signaling [27,28,34]. As such, N-cadherin, and its proteolytic products, is postulated to play both adhesive and signaling roles in NCC migration [28,34–39]. Although this mechanism of N-cadherin regulation cannot necessarily be extrapolated to NCC residing in other axial regions of the chick (primarily due to differences in NCC delamination and EMT), recent work suggests that N-cadherin processing also occurs in the head (our unpublished data).

NCC delamination from the chick midbrain, but not the hindbrain and trunk, is associated with Cad6B down-regulation from premigratory NCC. Cad6B protein is not lost at these other axial levels until NCC have completely segregated from the neural tube, once again highlighting the differences in the molecular mechanisms of emigration among axial levels. Cad6B down-regulation happens at the transcriptional and post-translational levels. Fairchild and Gammill have shown that Tetraspanin18 functions to protect cranial Cad6B protein from down-regulation [40], and Cad6B is processed by ADAM10 and ADAM19 at this axial level (our unpublished data). *Cad6B* repression in the midbrain is mediated by Snail2, a transcription factor expressed by premigratory and migratory cranial NCC that binds to E-boxes (Snail2 binding sites) within the *Cad6B* promoter [31]. Morpholino (MO)-mediated depletion of Snail2 from premigratory NCC reduces the number of migratory NCC, and this phenotype can be partially rescued by depletion of both Snail2 and Cad6B [31]. The exact molecular mechanism of this repression by Snail2 was only recently discovered [30]. Epigenetic modification of the *Cad6B* promoter occurs through the activity of a protein complex consisting of PHD12, Sin3A/HDAC, and Snail2. Together, this complex deacetylates lysines on histone H3 within the *Cad6B* promoter during NCC EMT, thereby repressing *Cad6B* transcription [41]. Although Snail2 associates with E-boxes within the *cad6B* promoter in trunk premigratory NCC, it is not clear whether Snail2 is sufficient to repress trunk *cad6B*, as *Cad6B* transcripts possess a similar spatiotemporal expression pattern to Snail2 protein in this region [31]. Regardless, down-regulation of chick midbrain Cad6B transcripts and protein is crucial because Cad6B overexpression or knockdown inhibits or augments NCC emigration and migration, respectively [30]. Interestingly, a recent study suggests the opposite function for Cad6B in the chick trunk. Park et al. [34] showed that Cad6B triggers de-epithelialization, with Cad6B knockdown decreasing the number of migratory NCC. These conflicting phenotypes can be explained, in part, by (1) axial differences in NCC EMT and (2) different experimental methods and metrics for data analysis.

It is hypothesized that post-translational mechanisms most likely exist to down-regulate

Cad6B in midbrain premigratory NCC because of the relatively short time period between loss of *Cad6B* transcripts and Cad6B protein [31]. Two potential means by which to clear Cad6B protein from premigratory NCC are proteolysis, as observed for Cad6B, N-cadherin and Cadherin-11 (below), and/or endocytosis, as noted with several classical cadherins [42]. Recent live-imaging studies of chick trunk NCC delamination in slice culture have given important insights to cadherin down-regulation. Ahlstrom et al. showed that although most ($>90\%$) of the premigratory NCC initiated migration after "normal" down-regulation of cell–cell junctions, a small percentage of cells broke away and left behind parts of their adhesive complexes within the neural tube. This suggests that the down-regulation of adhesive complexes may not be necessary prior to NCC EMT, although in both instances AJs were technically "lost" from emigrating cells (either through molecular down-regulation or mechanical separation of the apical portion of the cell possessing the AJs). It is not known if a similar phenotype is observed during chick cranial NCC EMT.

Similar to N-cadherin, Cad6B is likely to have adhesive and signaling roles during NCC migration. Recent work by Park et al. [34,43] has shed light on the functional role of Cad6B in the chick trunk. These studies demonstrate that Cad6B augments BMP4 signaling through a LIMK/cofilin-mediated, non-canonical BMP receptor pathway, thereby directly affecting NCC delamination. Interestingly, a prior report showed that BMP4 increases *Cad6B* transcripts in the trunk [44]. A model for chick trunk NCC delamination has now emerged in which BMP4 controls ADAM10-mediated N-cadherin cleavage and stimulates *cad6B* transcription. Cad6B then increases BMP4 signaling to ensure sufficient processing of N-cadherin in order to initiate delamination. It remains to be determined what breaks this cycle, however, as only newly emigrating, but not later migratory NCC that have completely separated from the neural tube, are Cad6B-positive [27,29,34]. This may require the inhibition of Delta/Notch signaling, which has been shown to activate *Bmp4* in the avian epidermis [45].

XENOPUS

Xenopus NCC induction is similar to chick in that an E- to N-cadherin switch occurs in the neural plate [7]. Unlike chick, though, *Xenopus* embryos do not express Cad6B. Instead, they express Cadherin-11, whose expression pattern seems to be the sum total of chick Cad6B (premigratory) and chick Cadherin-7 (migratory) expression. Cadherin-11 is expressed in premigratory, cephalic and certain trunk migratory NCC populations [46]. Like chick Cad6B, Cadherin-11 modulation is crucial for *Xenopus* cranial NCC migration, as Cadherin-11 knockdown and overexpression augments or prevents NCC migration, respectively [47]. This latter effect can be rescued by overexpressing ADAM13 [48], which separates the Cadherin-11 extracellular and intracellular domains by proteolysis. ADAM13 knockdown inhibits NCC migration and increases Cadherin-11 protein levels. This NCC phenotype can be rescued by re-introduction of ADAM13 or by injecting an mRNA that encodes the N-terminal portion of Cadherin-11 containing the first three cadherin repeats. These results indicate that a balance of full-length and cleaved Cadherin-11 protein is required for proper cranial NCC migration in *Xenopus*. Similar to N-cadherin, Cadherin-11 has dual adhesive and signaling roles [49].

ZEBRAFISH AND MOUSE

Parachute, the zebrafish homolog of N-cadherin, affects neural tube morphogenesis, but its role in NCC formation and migration has not been characterized [50]. N-cadherin trafficking, however, is important for normal NCC migration in zebrafish [38]. The Wnt target gene ovo1 controls plasma membrane N-cadherin levels in pigment precursor NCC. Ovo1 knockdown increases N-cadherin endocytosis due to de-repression of *rab GTPases* involved in N-cadherin trafficking, resulting in the depletion of membrane-bound N-cadherin, accumulation of N-cadherin within intracellular puncta, and aberrant NCC migration [38]. These results further underscore that a fine balance must be maintained between intracellular and membrane-bound N-cadherin.

Mouse NCC express Cadherin-6, the mouse ortholog of chick Cad6B. The expression pattern of Cadherin-6, like Cadherin-11 in *Xenopus*, appears to be the sum total of chick Cad6B and Cadherin-7 expression [51]. Mouse mesodermal cells and migratory NCC also express Cadherin-11 [52–54]. The role of Cadherin-6 or Cadherin-11 in mouse NCC EMT and migration has yet to be explored. In the mouse, Cadherin-6 plays a role in the mesenchymal-to-epithelial transition that occurs during the formation of the renal vesicle, leading to the loss of nephrons upon its deletion [55], while Cadherin-11 appears to function together with N-cadherin to regulate epithelial structure within the somites [56].

Catenins

The function of catenins in NCC development is largely unknown *in vivo* except for recent work by Jhingory et al. [57]. Here, the authors characterized the function of αN-catenin, the neural isoform of α-catenin, during chick midbrain NCC emigration and migration. The αN-catenin protein is found throughout the apical region of the neural tube (including premigratory NCC), a pattern reminiscent of proteins associated with AJs, but is severely diminished dorsally within the neural tube as NCC delaminate and undergo EMT [57]. Newly emigrating and early migratory NCC are devoid of αN-catenin transcripts and protein. αN-catenin influences NCC emigration by controlling levels of the AJ proteins

N-cadherin and Cad6B. This regulation results in a statistically significant expansion or reduction in the migratory NCC population upon αN-catenin depletion or overexpression, respectively, with concomitant precocious loss or maintenance of N-cadherin and Cad6B. Importantly, *αN-catenin* expression within premigratory and migratory NCC is controlled, in part, by Snail2-mediated transcriptional repression [17]. In keeping with this observation, MO-mediated depletion of both Snail2 and αN-catenin reestablishes normal NCC emigration/migration [57]. Therefore, Snail2 plays a critical role in chick cranial premigratory NCC by directly repressing *cad6B* and *αN-catenin* to initiate EMT.

The role of β-catenin in NCC delamination and EMT has been studied *in vitro* in chick trunk NCC explants [58]. While β-catenin is localized to the nucleus and cytoplasm in NCC close to the explant, it (along with N-cadherin) is found only in transient cell contacts made by NCC that have migrated farther away from the explant. This distribution is not influenced by changes in cell–cell adhesion. Ectopically activating β-catenin in premigratory and migratory NCC through addition of lithium chloride or Wnt abrogates NCC migration, decreases cell proliferation, and reduces the number of cells expressing Notch and Delta. Thus, transient activation of β-catenin may affect trunk NCC delamination *in vitro* through multiple mechanisms [58]. Since β-catenin is similarly observed in the nuclei of chick trunk neural folds [59], it is likely to possess important signaling functions in the formation and delamination of NCC *in vivo*.

Although the preceding studies have shed light on the importance of AJ proteins during NCC EMT, many questions still remain. What upstream pathways control cadherin and catenin transcriptional and post-translational regulation? Which catenins interact with premigratory NCC cadherins? Exploring these questions will provide additional insight into the function of AJ proteins during NCC EMT.

Tight Junctions

Tight junctions (TJs) are present at the most luminal side of the lateral plasma membrane [20]. Also called zonula occludens, the major function of TJs is to regulate the diffusion of solutes through the plasma membranes of apposing cells with size and charge selectivity [60]. Molecularly, TJs are similar to AJs in that they are comprised of transmembrane and cytoplasmic proteins. Claudins constitute the key transmembrane protein for TJ formation and function, forming homophilic interactions [60]. Occludin and tricellulin are other transmembrane proteins that are incorporated into, or closely localized to, claudin-based TJs, and function to modulate TJ formation [61,62]. ZO-1, ZO-2, and ZO-3 are the major cytoplasmic proteins that interact with the cytoplasmic domain of claudins, occludin, tricellulin, and JAM-A [63−65]. Cingulin is another cytoplasmic protein associated with TJs [66] through its interaction with ZO proteins [67,68]. ZO and cingulin proteins also associate with the actin cytoskeleton and thus contribute to TJ structural integrity [69−71].

Although the expression pattern of occludins and ZO-1 during neural tube formation has been characterized [72], any role in NCC formation and delamination remains undocumented. Occludin and ZO-1 localize to the apical region of the neural tube, including dorsally where premigratory NCC reside. As neurulation ceases and NCC migration initiates in chick, occludin is lost in the neural tube and is absent in migratory NCC. Expression of ZO-1, however, increases from the neural plate to the closed neural tube stage. Two recent studies, however, now reveal the roles of the TJ components claudin-1 and cingulin in chick midbrain NCC delamination and are discussed below [73,74].

Claudin-1

Claudin-1 protein is present in the apical neural tube of Hamburger−Hamilton (HH) stage 8 (HH8) through HH10 embryos, representing time points at which NCC are transitioning from a premigratory to migratory state. Levels of claudin-1 protein decrease in the dorsal neural folds well prior to NCC delamination. Migratory NCC also lack claudin-1 protein [74]. Claudin-1 protein influences NCC emigration, as MO-mediated knockdown of claudin-1 results in a statistically significant increase in the number of migratory NCC. Moreover, claudin-1 knockdown causes premature emigration of Sox10-positive NCC in embryos as early as HH8. Conversely, claudin-1 overexpression results in a statistically significant reduction in the number of migratory NCC cells. Interestingly, neither knockdown nor overexpression of claudin-1 causes any changes in the distribution of important NCC EMT markers such as Cad6B and laminin, suggesting that claudin-1 does not control EMT by modulating components of AJs and the basal lamina [74]. It is possible that claudin-1 exerts its effects differently, such as at the level of NCC induction.

Cingulin

Similar to claudin-1, cingulin is localized to the apical neural tube of HH8 embryos. One stage later, cingulin protein becomes diffusely expressed and diminished apically within the dorsal neural tube, even after NCC have delaminated and initiated migration [73]. Interestingly, delaminating and migratory NCC do not express cingulin, in keeping with the lack of TJs found in mesenchymal cell types [75]. Proper cingulin expression is critical for appropriate NCC delamination. MO-mediated depletion of cingulin significantly increases the migratory NCC population, as assessed by several molecular markers, at both early and late time points after MO introduction.

This increase is not due to changes in the premigratory NCC population. Cingulin depletion also causes a loss of dorsolateral basal lamina at stages not only prior to NCC delamination but later after NCC migration has ceased. Consistent with this observation, NCC continue to delaminate from these dorsolateral regions with compromised basal lamina [73].

Surprisingly, cingulin overexpression also expands the migratory NCC population. Unlike cingulin knockdown, however, this phenotype can be attributed to an increase in the number of premigratory NCC. This overexpressed cingulin protein is not only present at the apical side of the neural tube (where it normally localizes) but also in the nucleus and along the lateral edges of neuroepithelial cells. The nuclear localization of a TJ component has been observed previously for both cingulin [76,77] and symplekin [78], with the latter functioning to polyadenylate RNAs and regulate transcription [79–81]. As such, the possibility remains that forced expression of cingulin leads to the formation of unique and novel transcription factor complexes that in turn regulate the expression of premigratory NCC genes. Overexpression of cingulin also causes ectopic delamination of electroporated ventrolateral neuroepithelial cells, correlated with a loss of laminin from these delamination sites. Similar to cingulin knockdown embryos, loss of laminin persists at much later stages after cessation of midbrain NCC migration. The comparable phenotype upon cingulin knockdown and overexpression likely reflects (1) the requirement for cingulin-containing complexes to form with the correct stoichiometry and/or (2) indirect effects due to aberrant localization of cingulin within premigratory NCC, away from its normal site of activity.

The ectopic delamination of cells overexpressing cingulin from ventrolateral regions of the neural tube suggests a role of cytoskeleton-modulating elements in this process. Indeed, cingulin is found to regulate the levels of RhoA in the ventrolateral neural tube, with cingulin knockdown or overexpression increasing or decreasing RhoA levels, respectively, as seen *in vitro* [82,83]. RhoA downregulation is consistent with the observed loss of laminin and ectopic delamination of cells from the ventrolateral regions of the neural tube, similar to what is observed during gastrulation [84]. These results could be attributed to the presence of the correct RhoA GEFs to mediate loss of RhoA activity exclusively in this region. Cingulin, however, has no effect on RhoA levels in the dorsal neural tube, suggesting that the enhanced NCC migration phenotype observed upon cingulin overexpression and knockdown is probably independent of cingulin-mediated RhoA modulation, perhaps due to the absence of appropriate RhoA GEFs in the dorsal neural tube. These data are also in keeping with the normal down-regulation of RhoA that occurs in the neural folds after neural tube closure [85]. Neither cingulin knockdown nor overexpression causes any changes in the expression of cell junction markers whose down-regulation is associated with EMT, such as Cad6B, claudin-1, and ZO-1, suggesting that cingulin control of NCC emigration is independent of its function in TJs or AJs. These results collectively support a model in which cingulin could function to prevent premature breakdown of the basal lamina [86]. Cingulin may accomplish this through (1) inhibiting protease(s) that act to degrade the basal lamina (see later section), (2) modulating cytoskeletal elements [67,71], and/or (3) associating with other proteins that interact with components of the basal lamina.

Despite these recent discoveries, a dearth of information still exists with respect to TJ proteins and their importance during NCC delamination and EMT. Of particular note will be whether cross-talk exists between components of AJs and TJs, and the temporal order in which these junctions are dismantled during

NCC EMT. Moreover, it will be of great interest to uncover whether orthologs of these chick TJ proteins play similar roles during NCC EMT in other species. The answers to these questions will further enhance our knowledge of how premigratory NCC become primed for emigration out of the dorsal neural tube.

ACQUISITION OF MIGRATORY CAPACITY

Rho GTPases and the Cytoskeleton

Initiation of cell migration involves complex changes in cellular architecture, including the extension of a leading edge protrusion or lamellipodium, the establishment of new adhesion sites at the front, cell body contraction, and detachment of adhesions at the cell rear [87]. Although a variety of molecules have been implicated in cell migration, the Rho family of GTPases play a crucial role in regulating the various pathways controlling cell migration [88]. Rho GTPases are small G-protein signaling molecules that constitute a subfamily of the Ras superfamily. Twenty-two mammalian genes encode 3 Rho isoforms, 3 Rac isoforms, Cdc42, and several other proteins belonging to this family [89]. With respect to cell migration, the consensus is that Cdc42 regulates the direction of migration, Rac induces polarized membrane protrusions at the leading edge of the cell through stimulation of actin polymerization and focal adhesion complexes, and Rho promotes actin-myosin contraction in the cell body and at the rear (reviewed in [89]). In general, spatiotemporal expression of Rho and Rac is mutually exclusive in the migrating cell. Rho is always expressed at the rear of the migrating cell (inhibiting Rac) and Rac is expressed at the leading edge (inhibiting Rho), ensuring that protrusions are always at the front end [89].

Upon BMP induction in the chick trunk, *RhoB* expression is observed in premigratory NCC and persists in early, but not later, migrating NCC located close to the neural tube [44]. In contrast, *RhoA* is detected throughout the neural tube and surrounding mesoderm. Upon neural tube closure, *RhoA* is concentrated laterally and ventrally in the neural tube and is absent in the dorsal neural folds; *RhoC* is expressed robustly in the notochord and at very low levels in the neural tube. These results suggest that only RhoB appears to be involved in chick trunk NCC EMT [44]. Blocking Rho activation using the *Clostridium botulinum* C3 exotoxin inhibits NCC delamination [90]. Although RhoB possesses the proper spatiotemporal pattern to function during NCC EMT, it is important to note that C3 exotoxin is a general inhibitor of Rho activity, and, as such, this phenotype cannot be ascribed specifically to RhoB [44].

Studies from time-lapse imaging of zebrafish hindbrain NCC EMT demonstrate that inhibiting Rho-kinase (ROCK), a downstream effector of Rho GTPases, through ROCKi (3-[4−pyridyl] indole) leads to a reduction in cell blebbing activity and, subsequently, the number of NCC undergoing EMT [91]. In addition, overexpression of a wild-type or dominant active *RhoB* in the trunk of HH10 to HH12 chick embryos causes a severe distortion of the neural tube due to massive cell death. When *RhoB* is overexpressed along with *sox9* in the neural tube, however, cells acquire a mesenchymal phenotype and undergo EMT, which is correlated with a reduction in neural tube laminin expression [92]. RhoB and Sox9 thus appear to cooperate and play complementary roles: RhoB triggers EMT, conferring a mesenchymal phenotype, and Sox9 protects migrating cells from apoptosis and renders them with NCC features [92]. Given these data, it is clear that Rho GTPases play a positive regulatory role in triggering hindbrain and trunk NCC EMT in zebrafish and chick, respectively.

In contrast, other experiments performed in chick trunk NCC, both *in vitro* and *in vivo*, reveal that Rho GTPase inhibition (C3 exotoxin [90], ROCK inhibitor Y27632 [93]) enhances NCC emigration, while Rho GTPase activation (lysophosphatidic acid (LPA) treatment [94–96]) inhibits NCC emigration [97]. Active RhoA and RhoB proteins localize to premigratory NCC plasma membranes and are down-regulated upon delamination. Interestingly, inhibition of ADAM10-mediated N-cadherin proteolytic cleavage (by GI254023X [28]) also impedes NCC migration [28], and these effects are reversed upon Y27632 treatment, suggesting that down-regulation of RhoA and N-cadherin is tightly coupled. This is consistent with the observation that treatment with Y27632, C3, or the use of dominant negative RhoB causes premature loss of N-cadherin from premigratory trunk NCC, both *in vitro* and *in vivo*. Conversely, *in vivo* treatment with LPA maintains N-cadherin in the dorsal neural tube at axial levels where N-cadherin is normally absent. These contradictory results may be attributed to experiments using relatively high levels of C3 exotoxin [44], which could have had extraneous influences on unrelated molecular pathways. With respect to studies performed in zebrafish hindbrain using ROCKi [91], the conflicting results could once again be explained by the ROCK inhibitors used (Y27632 is a more potent ROCK inhibitor than ROCKi [98]), as well as differences in species, axial levels, methods, and the stage of NCC migration examined. Rupp and Kulesa have also noted no function for RhoA in the delamination of post-otic NCC, again highlighting that genetic, chemical and axial level differences can potentially affect resulting phenotypes [99].

Studies on *Xenopus* Rho GTPases have provided additional insight into the roles of Rho GTPases in initiating NCC EMT. As is the case with mouse embryos [100], *Xenopus* express *rhoB* in migratory, but not premigratory, NCC [101], along with *rhoV* in premigratory and *rhoU* in migratory NCC (discussed later) [102,103]. *RhoV* is induced in the prospective NCC territory as a canonical Wnt response gene but is absent in migrating NCC [103]. Knockdown or overexpression of rhoV impairs or augments expression of *snai2*, *twist*, and *sox9*, resulting in a dramatic loss of NCC-derived cranial structures or an expansion of the NCC territory, respectively [103]. Interestingly, RhoV is required to maintain E-cadherin at AJs during zebrafish epiboly, with loss of RhoV causing E-cadherin to shift to intracellular vesicles [104]. This mirrors the prior findings by Groysman et al. where RhoA and RhoB are required for proper maintenance of N-cadherin in chick trunk NCC [97]. In addition, *Xenopus* express *rnd1*, another member of the Rho GTPase family, in cranial NCC [105], but its role in NCC EMT has yet to be explored. *RhoU* and *rhoV* have also been observed in chick premigratory NCC, and similar to *Xenopus*, *rhoU* has been noted on migrating chick cranial NCC [106]. Not surprisingly, knockdown of RhoU reduces chick cranial NCC migration [102], suggesting that similar mechanisms may exist in these two species.

Studies by Shoval and Kalcheim [107] have elucidated the function of Rac GTPases and their interaction with Rho GTPases in chick trunk NCC EMT. Rac1 is expressed only in migrating trunk NCC lamellopodia but not in the epithelial NCC progenitors, suggesting a specific role in NCC migration. Expression of dominant negative Rac1 in the chick trunk leads to normal NCC specification and delamination but not migration. Instead, these NCC cluster adjacent to the dorsal neural tube and express HNK-1 [107]. Similarly, expression of constitutively active Rac1 prevents normal NCC migration, but this is due to the failure of these NCC to down-regulate N-cadherin and undergo a complete EMT. Thus, Rac1 is required for normal trunk NCC EMT, along with N-cadherin (and RhoA) down-regulation

[107]. These results have been corroborated by recent studies on the collective migration of *Xenopus* cranial NCC, where RhoA and Rac1 act antagonistically, generating directed protrusions in migrating NCC [36,108].

Myosins also play a role during NCC migration. The unconventional myosin Myosin-X (Myo10) is expressed both in *Xenopus* premigratory and migratory cranial NCC [109,110]. MO-mediated knockdown of myo10 impairs NCC migration, leads to head cartilage hypoplasia [109,110], and delays induction of NCC marker genes [110]. In the zebrafish hindbrain, Berndt et al. [91] showed that myosin II positively regulates NCC blebbing and EMT through the use of the myosin II inhibitor blebbistatin. In contrast to Myo10, however, blebbistatin-mediated inhibition of myosin II does not significantly reduce lamellopodia and filopodia (only blebbing behavior) [91]. This difference in inhibition could be attributed to the species, the properties of the myosin analyzed, axial differences with respect to EMT, the stage of migrating NCC examined, and the approaches employed to inhibit myosin.

Although the functions of Rho GTPases have been well characterized, several important questions remain unanswered, and the future is very exciting. What are the molecular players involved in the down-regulation of Rho and activation of Rac in delaminating NCC? Moreover, considering that chick midbrain NCC delaminate *en masse*, how do Rho GTPases function at this axial level? The answers to these questions will shed further light on the role of cytoskeletal molecules during NCC EMT.

APPROPRIATE SUBSTRATE FOR MIGRATION

Breakdown of Basal Lamina

The basal lamina is a fibrous protein matrix that surrounds the neural tube and is also localized to the basal surface of the ectoderm. The major constituent of basal lamina at cranial and trunk axial levels is laminin, in addition to fibronectin, collagen I, and collagen IV [111−113]. Initially, the basal lamina is continuous with the ectoderm. During neurulation, the neural plate invaginates and neural folds elevate to form NCC at the hinge region between the neural and non-neural ectoderm, thus disrupting the continuous basal lamina. Upon formation of the neural tube, the basal lamina re-seals around the neural tube and the ectoderm, except in the dorsal neural tube where premigratory NCC reside. Importantly, the basal lamina is not completely reconstituted in the head until all NCC have emigrated [9,73]. It was suggested initially that the basal lamina could serve as a physical barrier for migration, with its breakdown required for emigration [114]. This was substantiated by the observation that the basal lamina is impenetrable to NCC emigration [115]. Martins-Green and Erickson [116], however, later showed that the absence of basal lamina is a necessity but the basal lamina itself does not constitute a physical barrier for NCC emigration. Electron microscopic examination of mouse, chick, salamander, and zebrafish embryos has shown that the basal lamina is not present in the region that overlies NCC, at least 10 hours before EMT in the trunk and just before EMT in the cranial region [5,117−121]. Nevertheless, in the head, NCC emigration and loss of basal lamina seem to be directly correlated, although detailed studies are required to delineate the exact relationship between these two processes. Interestingly, NCC emigration and basal lamina degradation are independent, as expression of phosphorylatable Ets-1 stimulates NCC delamination and emigration and is associated with basal lamina degradation, while expression of a non-phosphorylatable Ets-1 only results in NCC emigration [13]. Basal lamina degradation in the trunk, however, is not correlated with NCC

EMT [116], possibly because sustained NCC delamination occurs at this axial level. Given that *Ets-1* is not detected in the trunk, it is tempting to speculate that Ets-1 function could explain the difference in the role of the basal lamina in NCC EMT in the head and trunk.

NCC can secrete a variety of proteases to potentially degrade the basal lamina. Chick cranial and trunk NCC synthesize plasminogen activator during migration *in vitro* [122–124], which may in turn degrade several cell surface proteins and extracellular matrix (ECM) components encountered during migration (reviewed in [125]). Recently, a number of matrix metalloproteases (MMPs) and ADAMs have been discovered to play important roles in NCC migration. MMP-2 and its natural inhibitor, TIMP-2, are expressed during avian cardiac NCC migration [126–128]. Recent experiments reveal that leader migrating cranial NCC express *Mmp-2* while trailing cells do not [129]. MMP-2 is crucial for normal NCC EMT, as MMP-2 inhibitors (BB-94 [128], KB8301 [127]) cause aberrant migration of cardiac and trunk NCC, respectively. Similarly, MMP-2 knockdown perturbs chick NCC EMT but has no effects on migration of NCC that have already separated from the neural tube [128]. MMP-9 is another protease expressed in delaminating and migrating NCC at both the cranial and trunk levels [130]. Blocking MMP-9 chemically dramatically reduces NCC delamination and migration without perturbing specification or survival, suggesting a role for MMP-9 in NCC detachment from the neural tube as well as in their migration. Conversely, MMP-9 overexpression or addition of MMP-9-conditioned media enhances NCC migration and stimulates precocious emigration *in vitro* and *in vivo*. Substrates for MMP-9 include laminin and to a lesser extent N-cadherin [130]. Considering that Ets-1 has been shown to up-regulate several MMPs *in vitro* and *in vivo* (reviewed in [131]), and that it is expressed in the chick head [13], it is possible that Ets-1

could stimulate chick cranial NCC delamination by up-regulating expression of MMPs.

ADAM13 is one of the best-characterized ADAMs in the context of NCC EMT, and it performs multiple functions aside from Cadherin-11 cleavage, as depleting cadherin-11 is not sufficient to rescue the ADAM13 knockdown [132]. ADAM13 is processed by γ-secretase to release a cytoplasmic fragment that translocates to the nucleus [133], and this fragment regulates the expression of multiple genes in cranial NCC [133]. Consistent with this, expression of a mutant ADAM13 lacking its cytoplasmic domain cannot rescue the ADAM13 knockdown phenotype (inhibition of cranial NCC migration), suggesting that the nuclear translocation of the cytoplasmic domain is crucial for promoting cranial NCC EMT *in vivo*. Moreover, ADAM13 can bind, cleave, and remodel a fibronectin substrate *in vivo*, thereby promoting NCC migration [134], and is required cell autonomously [132]. Interestingly, ADAM13 may execute its functions through cooperation with ADAM19, as the ADAM13 and ADAM19 double knockdown more strongly inhibits NCC migration than the single knockdowns, and the ADAM19 cytoplasmic domain rescues the phenotype of ADAM13 knockdown embryos [133]. ADAM19 was also identified to play a role in inducing *Xenopus* cranial NCC [135], but its role in NCC EMT is not clear. In the chick head, *ADAM19* and γ-secretase are expressed in premigratory and migratory NCC, and ADAM10 is expressed by only premigratory NCC. Both ADAMs function to process Cad6B (our unpublished data).

ECM

Experiments in axolotl embryos revealed that grafting ECM-coated filters underneath the ectoderm in contact with premigratory NCC leads to premature NCC emigration, prompting speculation that the ECM could serve as one of the signals stimulating NCC EMT [120]. This property is not conserved, however, as avian trunk NCC explants cultured on ECM substrates fail to precociously emigrate [1]. Many of the ECM components required for NCC migration are already present in the correct spatiotemporal pattern prior to NCC delamination such that extensive changes in ECM composition are unnecessary [4,112]. As such, the direct extracellular environment of premigratory NCC is permissive for emigration long before emigration occurs. In the chick, this environment consists of proteins such as fibronectin [136], laminin [137], tenascin/cytotactin [138], collagens [111], proteoglycans [139,140], vitronectin [141], and hyaluronan [142–144]. Initial work on hyaluronan in axolotl revealed that although NCC can preferentially migrate on a hyaluronan matrix (compared to fibronectin) *in vitro*, this positive effect is not observed *in vivo* and could be mitigated in different embryo backgrounds [145]. Comparable *in vitro* studies in chick demonstrated that the hyaluronan binding proteins aggrecan and versican do not support NCC attachment and migration, with aggrecan serving to strongly inhibit NCC motility on permissive matrix molecules [144]. Two recent reports, however, have shed new light on the role of hyaluronan in supporting NCC migration *in vivo*. In *Xenopus*, the hyaluronan synthase Xhas2 influences trunk NCC migration, with MO-mediated knockdown resulting in the random movement of NCC and impaired migration [146]. Additional work in *Xenopus* at the cranial axial level later revealed that hyaluronan and its cell surface receptor, CD44, regulate *Xenopus* NCC migration [142]. *Xhas1* and *CD44* are expressed in migrating cranial NCC, while *Xhas2* is observed in the pharyngeal endoderm into which these NCC migrate. Single MO-mediated depletion of Xhas1, Xhas2, or CD44 results in delayed NCC migration manifesting as craniofacial defects,

with simultaneous knockdown of Xhas1 and Xhas2 exacerbating these phenotypes. The importance of the ECM to migration has also been demonstrated through the use of function-blocking antibodies against ECM components, which severely perturb NCC emigration/migration [147–149].

These different ECM components raise the question whether NCC show any change in preference of ECM upon acquisition of the mesenchymal state. Considering the delamination phase alone, initial experiments showed that chick trunk NCC explants attach readily to fibronectin over a broad concentration range. Preference to attach to laminin, however, is initially reduced but increases over time and concentration [150–152]. Furthermore, initial emigrating cranial NCC (but not the ones following them nor trunk NCC) have the ability to synthesize and construct fibronectin matrices [136]. Similar to chick trunk NCC, *Xenopus* cranial NCC show a greater preference for fibronectin over laminin during migration, although both molecules can support efficient adhesion [153].

Integrins

Although a variety of integrins are expressed by delaminating/early migratory chick trunk NCC—$\alpha v\beta 1$, $\alpha v\beta 3$, $\alpha v\beta 5$ (vitronectin, [141]); $\alpha 1\beta 1$, $\alpha 3\beta 1$, $\alpha 6\beta 1$, $\alpha 7\beta 1$, $\alpha v\beta 3$ (laminin-1), [151,154,155]; $\alpha 3\beta 1$, $\alpha 4\beta 1$, $\alpha 5\beta 1$, $\alpha 8\beta 1$, $\alpha v\beta 1$, $\alpha v\beta 3$, $\beta 8$ integrin (fibronectin), [156]; and $\alpha 1\beta 1$ (collagen I), [157]—not all are involved in NCC migration. It was noted that early migratory, but not premigratory, chick trunk NCC express $\alpha 4\beta 1$ *in vitro* [156], and that $\alpha 4\beta 1$ function is dispensable for initial NCC delamination but not NCC migration and survival [156]. Explants of chick trunk NCC demonstrate that migratory NCC also express αv, a receptor for ECM components such as fibronectin and vitronectin [141].

Although there is abundant data on the expression profiles of integrins *in vitro*, very little is known about their expression *in vivo*. One of the most prominent integrins expressed at various axial levels of chick, mouse, and *Xenopus* are the β1-integrins, which bind fibronectin, collagens, and laminin. These are crucial for cranial NCC delamination, as they are expressed in premigratory and migratory cranial and trunk NCC, and a function-blocking antibody severely perturbs cranial NCC migration *in vitro* and *in vivo* [158–162]. Premigratory and migratory chick cranial and trunk NCC also express the fibronectin receptor α4 [163,164]. Competitive inhibition of α4 function by CS-1 peptide results in abnormal NCC migration, suggesting that α4 is important for normal delamination of at least chick cranial NCC *in vivo* [163,165]. α6, a laminin receptor, is expressed in the neural tube and in cranial premigratory NCC but is downregulated during NCC migration [166], although migrating cranial NCC do express α6 *in vitro* [167]. α1, a receptor for laminin and collagen, is also weakly expressed by migrating trunk NCC, and its levels do not change significantly during EMT [168]. Integrin α7, another laminin receptor, is only observed on a subpopulation of migrating trunk NCC [155]. Apart from these four α subunits, chick cranial premigratory and migratory NCC also express β3 and α4β3, which can bind to fibronectin and vitronectin [141,169]. *Xenopus* cranial premigratory and migratory NCC express α5β1 [153], a fibrinogen receptor, but not α6β1, a laminin receptor [170]. A function-blocking antibody against α5β1 prevents migration of cranial NCC explants grown on fibronectin [153]. Consistent with this observation, transcripts encoding the α5 subunit are enriched in the premigratory and migratory cranial NCC populations [171]. Not surprisingly, there are some species differences between chick and *Xenopus* integrin expression profiles (reviewed in [172]). In the mouse, cranial migratory NCC express α4 [173] and possibly α5 [174], but it is not clear whether they are also expressed by premigratory NCC. Although not observed in delaminating NCC, β1-integrins are expressed in mouse migrating enteric NCC and their role in their migration has been well characterized [175–178]. Conditional knockout of *β1-integrins* results in a Hirschsprung-like phenotype, accompanied by abnormal ganglia organization due to impaired NCC motility on the extracellular matrix and increased cell aggregation, likely mediated by cadherins [178]. Intriguingly, conditional knockout of *N-cadherin* in migrating enteric NCC delays NCC colonization of the gut (although to a lesser extent than that observed with the *β1-integrin* knock-out), but this delay is reversed later in development [176]. The *N-cadherin/β1-integrin* double conditional knockout mouse then revealed additional defects in the colonization of the distal gut due to impaired NCC locomotory behavior and directionality, suggesting a cooperative effect between N-cadherin and β1-integrins [176]. Recent studies in mice also demonstrate that β1-integrin in enteric NCC interacts with Phactr4, a novel regulator of enteric NCC migration. Phactr4 decreases Rho/ROCK-mediated integrin signaling and regulates cytoskeletal dynamics through cofilin phosphorylation [175].

Based upon our current knowledge of the ECM through which chick NCC migrate and their integrin profiles, it appears that delaminating NCC express a set of integrins that binds to fibronectin to facilitate migration. Unfortunately, our knowledge of *in vivo* integrin expression in the chick, as well as in other species, is woefully inadequate. Similar to the cadherin switch, is there an integrin switch during NCC delamination? What other integrins are expressed by delaminating NCC in *Xenopus*, mouse, and fish? The answers to these questions will surely help pave the way for additional functional studies on integrins across species.

CONCLUSIONS AND PERSPECTIVES

In summary, premigratory NCC must undergo multiple changes to delaminate from the neuroepithelium and undergo EMT. Important questions still linger, however, with respect to the molecules that orchestrate cranial vs. trunk NCC EMT, including how junction components are temporally down-regulated, and the role of Rho GTPases, myosins, and integrins during NCC EMT and migration *in vivo*. With this information, a molecular blueprint for EMT will emerge that is likely to be highly conserved with other normal developmental and aberrant EMTs, shedding light on these important processes.

References

[1] Newgreen D, Gibbins I. Factors controlling the time of onset of the migration of neural crest cells in the fowl embryo. Cell Tissue Res 1982;224(1):145–60.

[2] Nieto MA. The ins and outs of the epithelial to mesenchymal transition in health and disease. Annu Rev Cell Dev Biol 2011;27:347–76.

[3] Hay ED. An overview of epithelio-mesenchymal transformation. Acta Anat 1995;154(1):8–20.

[4] Thiery JP, Duband JL, Delouvee A. Pathways and mechanisms of avian trunk neural crest cell migration and localization. Dev Biol 1982;93(2):324–43.

[5] Erickson CA, Weston JA. An SEM analysis of neural crest migration in the mouse. J Embryol Exp Morphol 1983;74:97–118.

[6] Sadaghiani B, Thiebaud CH. Neural crest development in the *Xenopus laevis* embryo, studied by interspecific transplantation and scanning electron microscopy. Dev Biol 1987;124(1):91–110.

[7] Davidson LA, Keller RE. Neural tube closure in *Xenopus laevis* involves medial migration, directed protrusive activity, cell intercalation and convergent extension. Development 1999;126(20):4547–56.

[8] Krispin S, Nitzan E, Kassem Y, Kalcheim C. Evidence for a dynamic spatiotemporal fate map and early fate restrictions of premigratory avian neural crest. Development 2010;137(4):585–95.

[9] Duband JL. Diversity in the molecular and cellular strategies of epithelium-to-mesenchyme transitions:

[10] Jesuthasan S. Contact inhibition/collapse and pathfinding of neural crest cells in the zebrafish trunk. Development 1996;122(1):381–9.

[11] Theveneau E, Mayor R. Neural crest delamination and migration: from epithelium-to-mesenchyme transition to collective cell migration. Dev Biol 2012;366 (1):34–54.

[12] Nichols DH. Neural crest formation in the head of the mouse embryo as observed using a new histological technique. J Embryol Exp Morphol 1981;64:105–20.

[13] Theveneau E, Duband JL, Altabef M. Ets-1 confers cranial features on neural crest delamination. PloS One 2007;2(11):e1142.

[14] Duband JL, Thiery JP. Distribution of fibronectin in the early phase of avian cephalic neural crest cell migration. Dev Biol 1982;93(2):308–23.

[15] Burstyn-Cohen T, Kalcheim C. Association between the cell cycle and neural crest delamination through specific regulation of G1/S transition. Dev Cell 2002;3 (3):383–95.

[16] McKinney MC, Fukatsu K, Morrison J, McLennan R, Bronner ME, Kulesa PM. Evidence for dynamic rearrangements but lack of fate or position restrictions in premigratory avian trunk neural crest. Development 2013;140(4):820–30.

[17] Sela-Donenfeld D, Kalcheim C. Inhibition of noggin expression in the dorsal neural tube by somitogenesis: a mechanism for coordinating the timing of neural crest emigration. Development 2000;127(22): 4845–54.

[18] Sauka-Spengler T, Bronner-Fraser M. A gene regulatory network orchestrates neural crest formation. Nat Rev Mol Cell Biol 2008;9(7):557–68.

[19] Meng W, Takeichi M. Adherens junction: molecular architecture and regulation. Cold Spring Harb Perspect Biol 2009;1(6):a002899.

[20] Farquhar MG, Palade GE. Junctional complexes in various epithelia. J Cell Biol 1963;17:375–412.

[21] Takai Y, Ikeda W, Ogita H, Rikitake Y. The immunoglobulin-like cell adhesion molecule nectin and its associated protein afadin. Ann Rev Cell Dev Biol 2008;24:309–42.

[22] Pokutta S, Weis WI. Structure and mechanism of cadherins and catenins in cell–cell contacts. Ann Rev Cell Dev Biol 2007;23:237–61.

[23] Edelman GM, Gallin WJ, Delouvee A, Cunningham BA, Thiery JP. Early epochal maps of two different cell adhesion molecules. Proc Natl Acad Sci USA 1983; 80(14):4384–8.

[24] Hatta K, Takeichi M. Expression of N-cadherin adhesion molecules associated with early morphogenetic

events in chick development. Nature 1986;320(6061): 447–9.

[25] Wheelock MJ, Shintani Y, Maeda M, Fukumoto Y, Johnson KR. Cadherin switching. J Cell Sci 2008;121 (Pt 6):727–35.

[26] Bronner-Fraser M, Wolf JJ, Murray BA. Effects of antibodies against N-cadherin and N-CAM on the cranial neural crest and neural tube. Dev Biol 1992;153 (2):291–301.

[27] Nakagawa S, Takeichi M. Neural crest emigration from the neural tube depends on regulated cadherin expression. Development 1998;125(15):2963–71.

[28] Shoval I, Ludwig A, Kalcheim C. Antagonistic roles of full-length N-cadherin and its soluble BMP cleavage product in neural crest delamination. Development 2007;134(3):491–501.

[29] Nakagawa S, Takeichi M. Neural crest cell-cell adhesion controlled by sequential and subpopulation-specific expression of novel cadherins. Development 1995;121(5):1321–32.

[30] Coles EG, Taneyhill LA, Bronner-Fraser M. A critical role for Cadherin6B in regulating avian neural crest emigration. Dev Biol 2007;312(2):533–44.

[31] Taneyhill LA, Coles EG, Bronner-Fraser M. Snail2 directly represses cadherin6B during epithelial-to-mesenchymal transitions of the neural crest. Development 2007;134(8):1481–90.

[32] Marambaud P, Shioi J, Serban G, Georgakopoulos A, Sarner S, Nagy V, et al. A presenilin-1/gamma-secretase cleavage releases the E-cadherin intracellular domain and regulates disassembly of adherens junctions. EMBO J 2002;21(8):1948–56.

[33] Marambaud P, Wen PH, Dutt A, Shioi J, Takashima A, Siman R, et al. A CBP binding transcriptional repressor produced by the PS1/epsilon-cleavage of N-cadherin is inhibited by PS1 FAD mutations. Cell 2003;114(5):635–45.

[34] Park KS, Gumbiner BM. Cadherin 6B induces BMP signaling and de-epithelialization during the epithelial mesenchymal transition of the neural crest. Development 2010;137(16):2691–701.

[35] Xu X, Li WE, Huang GY, Meyer R, Chen T, Luo Y, et al. Modulation of mouse neural crest cell motility by N-cadherin and connexin 43 gap junctions. J Cell Biol 2001;154(1):217–30.

[36] Theveneau E, Marchant L, Kuriyama S, Gull M, Moepps B, Parsons M, et al. Collective chemotaxis requires contact-dependent cell polarity. Dev Cell 2010;19(1):39–53.

[37] Monier-Gavelle F, Duband JL. Control of N-cadherin-mediated intercellular adhesion in migrating neural crest cells in vitro. J Cell Sci 1995;108(Pt 12):3839–53.

[38] Piloto S, Schilling TF. Ovo1 links Wnt signaling with N-cadherin localization during neural crest migration. Development 2010;137(12):1981–90.

[39] Carmona-Fontaine C, Matthews HK, Kuriyama S, Moreno M, Dunn GA, Parsons M, et al. Contact inhibition of locomotion in vivo controls neural crest directional migration. Nature 2008;456(7224):957–61.

[40] Fairchild CL, Gammill LS. Tetraspanin18 is a FoxD3-responsive antagonist of cranial neural crest epithelial-to-mesenchymal transition that maintains cadherin-6B protein. J Cell Sci 2013;126(Pt 6):1464–76.

[41] Strobl-Mazzulla PH, Bronner ME. A PHD12-Snail2 repressive complex epigenetically mediates neural crest epithelial-to-mesenchymal transition. J Cell Biol 2012;198(6):999–1010.

[42] Kowalczyk AP, Nanes BA. Adherens junction turn-over: regulating adhesion through cadherin endocytosis, degradation, and recycling. Subcell Biochem 2012;60:197–222.

[43] Park KS, Gumbiner BM. Cadherin-6B stimulates an epithelial mesenchymal transition and the delamination of cells from the neural ectoderm via LIMK/cofilin mediated non-canonical BMP receptor signaling. Dev Biol 2012;366(2):232–43.

[44] Liu JP, Jessell TM. A role for rhoB in the delamination of neural crest cells from the dorsal neural tube. Development 1998;125(24):5055–67.

[45] Endo Y, Osumi N, Wakamatsu Y. Bimodal functions of Notch-mediated signaling are involved in neural crest formation during avian ectoderm development. Development 2002;129(4):863–73.

[46] Vallin J, Girault JM, Thiery JP, Broders F. Xenopus cadherin-11 is expressed in different populations of migrating neural crest cells. Mech Dev 1998; (1–2):171–4.

[47] Borchers A, David R, Wedlich D. Xenopus cadherin-11 restrains cranial neural crest migration and influences neural crest specification. Development 2001;128 (16):3049–60.

[48] McCusker C, Cousin H, Neuner R, Alfandari D. Extracellular cleavage of cadherin-11 by ADAM metalloproteases is essential for Xenopus cranial neural crest cell migration. Mol Biol Cell 2009;20(1):78–89.

[49] Kashef J, Kohler A, Kuriyama S, Alfandari D, Mayor R, Wedlich D. Cadherin-11 regulates protrusive activity in Xenopus cranial neural crest cells upstream of Trio and the small GTPases. Genes Dev 2009;23(12):1393–8.

[50] Lele Z, Folchert A, Concha M, Rauch GJ, Geisler R, Rosa F, et al. Parachute/n-cadherin is required for morphogenesis and maintained integrity of the zebrafish neural tube. Development 2002;129(14):3281–94.

[51] Inoue T, Chisaka O, Matsunami H, Takeichi M. Cadherin-6 expression transiently delineates specific

rhombomeres, other neural tube subdivisions, and neural crest subpopulations in mouse embryos. Dev Biol 1997;183(2):183−94.

[52] Kimura Y, Matsunami H, Inoue T, Shimamura K, Uchida N, Ueno T, et al. Cadherin-11 expressed in association with mesenchymal morphogenesis in the head, somite, and limb bud of early mouse embryos. Dev Biol 1995;169(1):347−58.

[53] Hoffmann I, Balling R. Cloning and expression analysis of a novel mesodermally expressed cadherin. Dev Biol 1995;169(1):337−46.

[54] Simonneau L, Kitagawa M, Suzuki S, Thiery JP. Cadherin 11 expression marks the mesenchymal phenotype: towards new functions for cadherins? Cell Adh Commun 1995;3(2):115−30.

[55] Mah SP, Saueressig H, Goulding M, Kintner C, Dressler GR. Kidney development in cadherin-6 mutants: delayed mesenchyme-to-epithelial conversion and loss of nephrons. Dev Biol 2000;223 (1):38−53.

[56] Horikawa K, Radice G, Takeichi M, Chisaka O. Adhesive subdivisions intrinsic to the epithelial somites. Dev Biol 1999;215(2):182−9.

[57] Jhingory S, Wu CY, Taneyhill LA. Novel insight into the function and regulation of αN-catenin by Snail2 during chick neural crest cell migration. Dev Biol 2010;344(2):896−910.

[58] de Melker AA, Desban N, Duband JL. Cellular localization and signaling activity of β-catenin in migrating neural crest cells. Dev Dyn 2004;230(4):708−26.

[59] Garcia-Castro MI, Marcelle C, Bronner-Fraser M. Ectodermal Wnt function as a neural crest inducer. Science 2002;297(5582):848−51.

[60] Furuse M. Molecular basis of the core structure of tight junctions. Cold Spring Harb Perspect Biol 2010;2 (1):a002907.

[61] Furuse M, Hirase T, Itoh M, Nagafuchi A, Yonemura S, Tsukita S, et al. Occludin: a novel integral membrane protein localizing at tight junctions. J Cell Biol 1993;123(6 Pt 2):1777−88.

[62] Ikenouchi J, Furuse M, Furuse K, Sasaki H, Tsukita S, Tsukita S. Tricellulin constitutes a novel barrier at tricellular contacts of epithelial cells. J Cell Biol 2005;171 (6):939−45.

[63] Stevenson BR, Siliciano JD, Mooseker MS, Goodenough DA. Identification of ZO-1: a high molecular weight polypeptide associated with the tight junction (zonula occludens) in a variety of epithelia. J Cell Biol 1986;103(3):755−66.

[64] Gumbiner B, Lowenkopf T, Apatira D. Identification of a 160-kDa polypeptide that binds to the tight junction protein ZO-1. Proc Natl Acad Sci USA 1991;88 (8):3460−4.

[65] Haskins J, Gu L, Wittchen ES, Hibbard J, Stevenson BR. ZO-3, a novel member of the MAGUK protein family found at the tight junction, interacts with ZO-1 and occludin. J Cell Biol 1998;141(1):199−208.

[66] Citi S, Sabanay H, Jakes R, Geiger B, Kendrick-Jones J. Cingulin, a new peripheral component of tight junctions. Nature 1988;333(6170):272−6.

[67] Cordenonsi M, D'Atri F, Hammar E, Parry DA, Kendrick-Jones J, Shore D, et al. Cingulin contains globular and coiled-coil domains and interacts with ZO-1, ZO-2, ZO-3, and myosin. J Cell Biol 1999;147 (7):1569−82.

[68] D'Atri F, Nadalutti F, Citi S. Evidence for a functional interaction between cingulin and ZO-1 in cultured cells. J Biol Chem 2002;277(31):27757−64.

[69] Fanning AS, Jameson BJ, Jesaitis LA, Anderson JM. The tight junction protein ZO-1 establishes a link between the transmembrane protein occludin and the actin cytoskeleton. J Biol Chem 1998;273(45):29745−53.

[70] Wittchen ES, Haskins J, Stevenson BR. Protein interactions at the tight junction. Actin has multiple binding partners, and ZO-1 forms independent complexes with ZO-2 and ZO-3. J Biol Chem 1999;274(49):35179−85.

[71] D'Atri F, Citi S. Cingulin interacts with F-actin in vitro. FEBS Lett 2001;507(1):21−4.

[72] Aaku-Saraste E, Hellwig A, Huttner WB. Loss of occludin and functional tight junctions, but not ZO-1, during neural tube closure—remodeling of the neuroepithelium prior to neurogenesis. Dev Biol 1996;180 (2):664−79.

[73] Wu CY, Jhingory S, Taneyhill LA. The tight junction scaffolding protein cingulin regulates neural crest cell migration. Dev Dyn 2011;240(10):2309−23.

[74] Fishwick KJ, Neiderer TE, Jhingory S, Bronner ME, Taneyhill LA. The tight junction protein claudin-1 influences cranial neural crest cell emigration. Mech Dev 2012;3.

[75] Ikenouchi J, Matsuda M, Furuse M, Tsukita S. Regulation of tight junctions during the epithelium−mesenchyme transition: direct repression of the gene expression of claudins/occludin by Snail. J Cell Sci 2003;116(Pt 10):1959−67.

[76] Citi S, Cordenonsi M. Tight junction proteins. Acta Biochim Biophys 1998;1448(1):1−11.

[77] Nakamura T, Blechman J, Tada S, Rozovskaia T, Itoyama T, Bullrich F, et al. huASH1 protein, a putative transcription factor encoded by a human homologue of the Drosophila ash1 gene, localizes to both nuclei and cell−cell tight junctions. Proc Natl Acad Sci USA 2000;97(13):7284−9.

[78] Keon BH, Schafer S, Kuhn C, Grund C, Franke WW. Symplekin, a novel type of tight junction plaque protein. J Cell Biol 1996;134(4):1003−18.

[79] Takagaki Y, Manley JL. Complex protein interactions within the human polyadenylation machinery identify a novel component. Mol Cell Biol 2000;20(5):1515–25.

[80] Barnard DC, Ryan K, Manley JL, Richter JD. Symplekin and xGLD-2 are required for CPEB-mediated cytoplasmic polyadenylation. Cell 2004;119(5):641–51.

[81] Kavanagh E, Buchert M, Tsapara A, Choquet A, Balda MS, Hollande F, et al. Functional interaction between the ZO-1-interacting transcription factor ZONAB/DbpA and the RNA processing factor symplekin. J Cell Sci 2006;119(Pt 24):5098–105.

[82] Aijaz S, D'Atri F, Citi S, Balda MS, Matter K. Binding of GEF-H1 to the tight junction-associated adaptor cingulin results in inhibition of Rho signaling and G1/S phase transition. Dev Cell 2005;8(5):777–86.

[83] Citi S, Paschoud S, Pulimeno P, Timolati F, De Robertis F, Jond L, et al. The tight junction protein cingulin regulates gene expression and RhoA signaling. Ann N Y Acad Sci 2009;1165:88–98.

[84] Nakaya Y, Sukowati EW, Wu Y, Sheng G. RhoA and microtubule dynamics control cell-basement membrane interaction in EMT during gastrulation. Nat Cell Biol 2008;10(7):765–75.

[85] Kinoshita N, Sasai N, Misaki K, Yonemura S. Apical accumulation of Rho in the neural plate is important for neural plate cell shape change and neural tube formation. Mol Biol Cell 2008;19(5):2289–99.

[86] Wu CY, Taneyhill LA. Annexin a6 modulates chick cranial neural crest cell emigration. PloS One 2012;7(9):e44903.

[87] Ridley AJ, Schwartz MA, Burridge K, Firtel RA, Ginsberg MH, Borisy G, et al. Cell migration: integrating signals from front to back. Science 2003;302(5651):1704–9.

[88] Raftopoulou M, Hall A. Cell migration: Rho GTPases lead the way. Dev Biol 2004;265(1):23–32.

[89] Jaffe AB, Hall A. Rho GTPases: biochemistry and biology. Ann Rev Cell Dev Biol 2005;21:247–69.

[90] Aktories K, Hall A. Botulinum ADP-ribosyltransferase C3: a new tool to study low molecular weight GTP-binding proteins. Trends Pharmacol Sci 1989;10(10):415–8.

[91] Berndt JD, Clay MR, Langenberg T, Halloran MC. Rho-kinase and myosin II affect dynamic neural crest cell behaviors during epithelial to mesenchymal transition in vivo. Dev Biol 2008;324(2):236–44.

[92] Cheung M, Chaboissier MC, Mynett A, Hirst E, Schedl A, Briscoe J. The transcriptional control of trunk neural crest induction, survival, and delamination. Dev Cell 2005;8(2):179–92.

[93] Ishizaki T, Uehata M, Tamechika I, Keel J, Nonomura K, Maekawa M, et al. Pharmacological properties of Y-27632, a specific inhibitor of rho-associated kinases. Mol Pharmacol 2000;57(5):976–83.

[94] Ridley AJ, Hall A. Distinct patterns of actin organization regulated by the small GTP-binding proteins Rac and Rho. Cold Spring Harb Symp Quant Biol 1992;57:661–71.

[95] Ren XD, Kiosses WB, Schwartz MA. Regulation of the small GTP-binding protein Rho by cell adhesion and the cytoskeleton. EMBO J 1999;18(3):578–85.

[96] Weiner JA, Fukushima N, Contos JJ, Scherer SS, Chun J. Regulation of Schwann cell morphology and adhesion by receptor-mediated lysophosphatidic acid signaling. J Neurosci 2001;21(18):7069–78.

[97] Groysman M, Shoval I, Kalcheim C. A negative modulatory role for rho and rho-associated kinase signaling in delamination of neural crest cells. Neural Dev 2008;3:27.

[98] Yarrow JC, Totsukawa G, Charras GT, Mitchison TJ. Screening for cell migration inhibitors via automated microscopy reveals a Rho-kinase inhibitor. Chem Biol 2005;12(3):385–95.

[99] Rupp PA, Kulesa PM. A role for RhoA in the two-phase migratory pattern of post-otic neural crest cells. Dev Biol 2007;311(1):159–71.

[100] Henderson DJ, Ybot-Gonzalez P, Copp AJ. RhoB is expressed in migrating neural crest and endocardial cushions of the developing mouse embryo. Mech Dev 2000;95(1–2):211–4.

[101] Vignal E, de Santa Barbara P, Guemar L, Donnay JM, Fort P, Faure S. Expression of RhoB in the developing Xenopus laevis embryo. Gene Expr Patterns 2007;7(3):282–8.

[102] Fort P, Guemar L, Vignal E, Morin N, Notarnicola C, de Santa Barbara P, et al. Activity of the RhoU/Wrch1 GTPase is critical for cranial neural crest cell migration. Dev Biol 2011;350(2):451–63.

[103] Guemar L, de Santa Barbara P, Vignal E, Maurel B, Fort P, Faure S. The small GTPase RhoV is an essential regulator of neural crest induction in Xenopus. Dev Biol 2007;310(1):113–28.

[104] Tay HG, Ng YW, Manser E. A vertebrate-specific Chp-PAK-PIX pathway maintains E-cadherin at adherens junctions during zebrafish epiboly. PloS One 2010;5(4):e10125.

[105] Wunnenberg-Stapleton K, Blitz IL, Hashimoto C, Cho KW. Involvement of the small GTPases XRhoA and XRnd1 in cell adhesion and head formation in early Xenopus development. Development 1999;126(23):5339–51.

[106] Notarnicola C, Le Guen L, Fort P, Faure S, de Santa Barbara P. Dynamic expression patterns of RhoV/Chp and RhoU/Wrch during chicken embryonic development. Dev Dyn 2008;237(4):1165−71.

[107] Shoval I, Kalcheim C. Antagonistic activities of Rho and Rac GTPases underlie the transition from neural crest delamination to migration. Dev Dyn 2012;241 (7):1155−68.

[108] Matthews HK, Marchant L, Carmona-Fontaine C, Kuriyama S, Larrain J, Holt MR, et al. Directional migration of neural crest cells in vivo is regulated by Syndecan-4/Rac1 and non-canonical Wnt signaling/RhoA. Development 2008;135(10):1771−80.

[109] Hwang YS, Luo T, Xu Y, Sargent TD. Myosin-X is required for cranial neural crest cell migration in Xenopus laevis. Dev Dyn 2009;238(10):2522−9.

[110] Nie S, Kee Y, Bronner-Fraser M. Myosin-X is critical for migratory ability of Xenopus cranial neural crest cells. Dev Biol 2009;335(1):132−42.

[111] Duband JL, Thiery JP. Distribution of laminin and collagens during avian neural crest development. Development 1987;101(3):461−78.

[112] Sternberg J, Kimber SJ. Distribution of fibronectin, laminin and entactin in the environment of migrating neural crest cells in early mouse embryos. J Embryol Exp Morphol 1986;91:267−82.

[113] Tuckett F, Morriss-Kay GM. The distribution of fibronectin, laminin and entactin in the neurulating rat embryo studied by indirect immunofluorescence. J Embryol Exp Morphol 1986;94:95−112.

[114] Newgreen DF, Gibbins IL, Sauter J, Wallenfels B, Wutz R. Ultrastructural and tissue-culture studies on the role of fibronectin, collagen and glycosaminoglycans in the migration of neural crest cells in the fowl embryo. Cell Tissue Res 1982;221(3):521−49.

[115] Erickson CA. Behavior of neural crest cells on embryonic basal laminae. Dev Biol 1987;120(1):38−49.

[116] Martins-Green M, Erickson CA. Basal lamina is not a barrier to neural crest cell emigration: documentation by TEM and by immunofluorescent and immunogold labelling. Development 1987;101(3):517−33.

[117] Raible DW, Wood A, Hodsdon W, Henion PD, Weston JA, Eisen JS. Segregation and early dispersal of neural crest cells in the embryonic zebrafish. Dev Dyn 1992;195(1):29−42.

[118] Martins-Green M. Origin of the dorsal surface of the neural tube by progressive delamination of epidermal ectoderm and neuroepithelium: implications for neurulation and neural tube defects. Development 1988;103(4):687−706.

[119] Tosney KW. The early migration of neural crest cells in the trunk region of the avian embryo: an electron microscopic study. Dev Biol 1978;62(2):317−33.

[120] Lofberg J, Nynas-McCoy A, Olsson C, Jonsson L, Perris R. Stimulation of initial neural crest cell migration in the axolotl embryo by tissue grafts and extracellular matrix transplanted on microcarriers. Dev Biol 1985;107(2):442−59.

[121] Newgreen DF, Erickson CA. The migration of neural crest cells. Int Rev Cytol 1986;103:89−145.

[122] Valinsky JE, Le Douarin NM. Production of plasminogen activator by migrating cephalic neural crest cells. EMBO J 1985;4(6):1403−6.

[123] Erickson CA, Isseroff RR. Plasminogen activator activity is associated with neural crest cell motility in tissue culture. J Exp Zool 1989;251(2):123−33.

[124] Agrawal M, Brauer PR. Urokinase-type plasminogen activator regulates cranial neural crest cell migration in vitro. Dev Dyn 1996;207(3):281−90.

[125] Andreasen PA, Egelund R, Petersen HH. The plasminogen activation system in tumor growth, invasion, and metastasis. Cell Mol Life Sci 2000;57 (1):25−40.

[126] Cai DH, Brauer PR. Synthetic matrix metalloproteinase inhibitor decreases early cardiac neural crest migration in chicken embryos. Dev Dyn 2002;224 (4):441−9.

[127] Cai DH, Vollberg Sr. TM, Hahn-Dantona E, Quigley JP, Brauer PR. MMP-2 expression during early avian cardiac and neural crest morphogenesis. Anat Rec 2000;259(2):168−79.

[128] Duong TD, Erickson CA. MMP-2 plays an essential role in producing epithelial-mesenchymal transformations in the avian embryo. Dev Dyn 2004;229 (1):42−53.

[129] McLennan R, Dyson L, Prather KW, Morrison JA, Baker RE, Maini PK, et al. Multiscale mechanisms of cell migration during development: theory and experiment. Development 2012;139(16):2935−44.

[130] Monsonego-Ornan E, Kosonovsky J, Bar A, Roth L, Fraggi-Rankis V, Simsa S, et al. Matrix metalloproteinase 9/gelatinase B is required for neural crest cell migration. Dev Biol 2012;364(2):162−77.

[131] Hsu T, Trojanowska M, Watson DK. Ets proteins in biological control and cancer. J Cell Biochem 2004;91 (5):896−903.

[132] Cousin H, Abbruzzese G, McCusker C, Alfandari D. ADAM13 function is required in the 3 dimensional context of the embryo during cranial neural crest cell migration in Xenopus laevis. Dev Biol 2012;368 (2):335−44.

[133] Cousin H, Abbruzzese G, Kerdavid E, Gaultier A, Alfandari D. Translocation of the cytoplasmic domain of ADAM13 to the nucleus is essential for Calpain8-a expression and cranial neural crest cell migration. Dev Cell 2011;20(2):256−63.

[134] Alfandari D, Cousin H, Gaultier A, Smith K, White JM, Darribere T, et al. Xenopus ADAM 13 is a metalloprotease required for cranial neural crest-cell migration. Curr Biol 2001;11(12):918−30.

[135] Neuner R, Cousin H, McCusker C, Coyne M, Alfandari D. *Xenopus* ADAM19 is involved in neural, neural crest and muscle development. Mech Dev 2009;126(3-4):240−55.

[136] Newgreen D, Thiery JP. Fibronectin in early avian embryos: synthesis and distribution along the migration pathways of neural crest cells. Cell Tissue Res 1980;211(2):269−91.

[137] Krotoski DM, Domingo C, Bronner-Fraser M. Distribution of a putative cell surface receptor for fibronectin and laminin in the avian embryo. J Cell Biol 1986;103(3):1061−71.

[138] Tan SS, Crossin KL, Hoffman S, Edelman GM. Asymmetric expression in somites of cytotactin and its proteoglycan ligand is correlated with neural crest cell distribution. Proc Natl Acad Sci USA 1987;84 (22):7977−81.

[139] Perris R, Krotoski D, Lallier T, Domingo C, Sorrell JM, Bronner-Fraser M. Spatial and temporal changes in the distribution of proteoglycans during avian neural crest development. Development 1991;111 (2):583−99.

[140] Erickson CA. Control of pathfinding by the avian trunk neural crest. Development 1988;103 (Suppl):63−80.

[141] Delannet M, Martin F, Bossy B, Cheresh DA, Reichardt LF, Duband JL. Specific roles of the α V β 1, α V β 3 and α V β 5 integrins in avian neural crest cell adhesion and migration on vitronectin. Development 1994;120(9):2687−702.

[142] Casini P, Nardi I, Ori M. Hyaluronan is required for cranial neural crest cells migration and craniofacial development. Dev Dyn 2012;241(2):294−302.

[143] Perissinotto D, Iacopetti P, Bellina I, Doliana R, Colombatti A, Pettway Z, et al. Avian neural crest cell migration is diversely regulated by the two major hyaluronan-binding proteoglycans PG-M/versican and aggrecan. Development 2000;127(13):2823−42.

[144] Perris R, Perissinotto D, Pettway Z, Bronner-Fraser M, Morgelin M, Kimata K. Inhibitory effects of PG-H/aggrecan and PG-M/versican on avian neural crest cell migration. FASEB J 1996; 10(2):293−301.

[145] Epperlein HH, Radomski N, Wonka F, Walther P, Wilsch M, Muller M, et al. Immunohistochemical demonstration of hyaluronan and its possible involvement in axolotl neural crest cell migration. J Struct Biol 2000;132(1):19−32.

[146] Ori M, Nardini M, Casini P, Perris R, Nardi I. XHas2 activity is required during somitogenesis and precursor cell migration in *Xenopus* development. Development 2006;133(4):631−40.

[147] Poole TJ, Thiery JP. Antibodies and a synthetic peptide that block cell-fibronectin adhesion arrest neural crest cell migration in vivo. Prog Clin Biol Res 1986;217B:235−8.

[148] Bronner-Fraser M. Distribution and function of tenascin during cranial neural crest development in the chick. J Neurosci Res 1988;21(2−4):135−47.

[149] Bronner-Fraser M, Lallier T. A monoclonal antibody against a laminin-heparan sulfate proteoglycan complex perturbs cranial neural crest migration in vivo. J Cell Biol 1988;106(4):1321−9.

[150] Delannet M, Duband JL. Transforming growth factor-β control of cell-substratum adhesion during avian neural crest cell emigration in vitro. Development 1992;116(1):275−87.

[151] Desban N, Duband JL. Avian neural crest cell migration on laminin: interaction of the α1β1 integrin with distinct laminin-1 domains mediates different adhesive responses. J Cell Sci 1997;110(Pt 21): 2729−44.

[152] Rovasio RA, Delouvee A, Yamada KM, Timpl R, Thiery JP. Neural crest cell migration: requirements for exogenous fibronectin and high cell density. J Cell Biol 1983;96(2):462−73.

[153] Alfandari D, Cousin H, Gaultier A, Hoffstrom BG, DeSimone DW. Integrin α5β1 supports the migration of *Xenopus* cranial neural crest on fibronectin. Dev Biol 2003;260(2):449−64.

[154] Lallier T, Deutzmann R, Perris R, Bronner-Fraser M. Neural crest cell interactions with laminin: structural requirements and localization of the binding site for α1 β1 integrin. Dev Biol 1994;162(2):451−64.

[155] Kil SH, Bronner-Fraser M. Expression of the avian α 7-integrin in developing nervous system and myotome. Int J Dev Neurosci 1996;14(3):181−90.

[156] Testaz S, Duband JL. Central role of the α4β1 integrin in the coordination of avian truncal neural crest cell adhesion, migration, and survival. Dev Dyn 2001;222(2):127−40.

[157] Perris R, Syfrig J, Paulsson M, Bronner-Fraser M. Molecular mechanisms of neural crest cell attachment and migration on types I and IV collagen. J Cell Sci 1993;106(Pt 4):1357−68.

[158] Bronner-Fraser M. Alterations in neural crest migration by a monoclonal antibody that affects cell adhesion. J Cell Biol 1985;101(2):610−7.

[159] Duband JL, Rocher S, Chen WT, Yamada KM, Thiery JP. Cell adhesion and migration in the early vertebrate embryo: location and possible role of the

putative fibronectin receptor complex. J Cell Biol 1986;102(1):160–78.

[160] Kil SH, Lallier T, Bronner-Fraser M. Inhibition of cranial neural crest adhesion in vitro and migration in vivo using integrin antisense oligonucleotides. Dev Biol 1996;179(1):91–101.

[161] Testaz S, Delannet M, Duband J. Adhesion and migration of avian neural crest cells on fibronectin require the cooperating activities of multiple integrins of the (β)1 and (β)3 families. J Cell Sci 1999;112 (Pt 24):4715–28.

[162] Gawantka V, Ellinger-Ziegelbauer H, Hausen P. β1-integrin is a maternal protein that is inserted into all newly formed plasma membranes during early *Xenopus* embryogenesis. Development 1992;115(2):595–605.

[163] Kil SH, Krull CE, Cann G, Clegg D, Bronner-Fraser M. The α4 subunit of integrin is important for neural crest cell migration. Dev Biol 1998;202(1):29–42.

[164] Stepp MA, Urry LA, Hynes RO. Expression of α4 integrin mRNA and protein and fibronectin in the early chicken embryo. Cell Adh Commun 1994;2(4):359–75.

[165] Dufour S, Duband JL, Humphries MJ, Obara M, Yamada KM, Thiery JP. Attachment, spreading and locomotion of avian neural crest cells are mediated by multiple adhesion sites on fibronectin molecules. EMBO J 1988;7(9):2661–71.

[166] Bronner-Fraser M, Artinger M, Muschler J, Horwitz AF. Developmentally regulated expression of α6 integrin in avian embryos. Development 1992;115 (1):197–211.

[167] Strachan LR, Condic ML. Neural crest motility and integrin regulation are distinct in cranial and trunk populations. Dev Biol 2003;259(2):288–302.

[168] Duband JL, Belkin AM, Syfrig J, Thiery JP, Koteliansky VE. Expression of α1 integrin, a laminin-collagen receptor, during myogenesis and neurogenesis in the avian embryo. Development 1992;116(3):585–600.

[169] Pietri T, Thiery JP, Dufour S. Differential expression of β3 integrin gene in chick and mouse cranial neural crest cells. Dev Dyn 2003;227(2):309–13.

[170] Lallier TE, Whittaker CA, DeSimone DW. Integrin α6 expression is required for early nervous system development in *Xenopus laevis*. Development 1996;122(8):2539–54.

[171] Joos TO, Whittaker CA, Meng F, DeSimone DW, Gnau V, Hausen P. Integrin α5 during early development of *Xenopus laevis*. Mech Dev 1995;50(2–3):187–99.

[172] McKeown SJ, Wallace AS, Anderson RB. Expression and function of cell adhesion molecules during neural crest migration. Dev Biol 2012;373(2):244–57.

[173] Pinco KA, Liu S, Yang JT. α4 integrin is expressed in a subset of cranial neural crest cells and in epicardial progenitor cells during early mouse development. Mech Dev 2001;100(1):99–103.

[174] Goh KL, Yang JT, Hynes RO. Mesodermal defects and cranial neural crest apoptosis in α5 integrin-null embryos. Development 1997;124(21):4309–19.

[175] Zhang Y, Kim TH, Niswander L. Phactr4 regulates directional migration of enteric neural crest through PP1, integrin signaling, and cofilin activity. Genes Dev 2012;26(1):69–81.

[176] Broders-Bondon F, Paul-Gilloteaux P, Carlier C, Radice GL, Dufour S. N-cadherin and β1-integrins cooperate during the development of the enteric nervous system. Dev Biol 2012;364(2):178–91.

[177] Breau MA, Dahmani A, Broders-Bondon F, Thiery JP, Dufour S. β1 integrins are required for the invasion of the caecum and proximal hindgut by enteric neural crest cells. Development 2009;136 (16):2791–801.

[178] Breau MA, Pietri T, Eder O, Blanche M, Brakebusch C, Fassler R, et al. Lack of β1 integrins in enteric neural crest cells leads to a Hirschsprung-like phenotype. Development 2006;133(9):1725–34.

Neural Crest Cell Migration: Guidance, Pathways, and Cell–Cell Interactions

Eric Theveneau and Roberto Mayor

Department of Cell and Developmental Biology, University College London, UK

OUTLINE

PATHWAYS OF NEURAL CREST CELL MIGRATION: AN OVERVIEW

The neural crest cells (NCC) are induced at the interface between the neuroepithelium and the prospective epidermis [1]. Shortly after induction, NCC delaminate from their surrounding tissue by activating an epithelium-to-mesenchyme transition (EMT) program and migrate extensively throughout the embryo [2]. They depart as a continuous cell population, progressively dispersing away from the neuroepithelium, and organize into discrete streams. Along the way toward their target

Neural Crest Cells.
DOI: http://dx.doi.org/10.1016/B978-0-12-401730-6.00004-1

region, NCC are influenced by multiple positive and negative signals present in their local environment. They also respond to direct physical interactions with one another and various other tissues along the way. Therefore, the overall directionality of NC cell migration results from a complex integration of cooperative behaviors due to cell–cell interactions, chemotactic signals, and distribution of extracellular matrix.

NCC migration follows stereotypical routes. In the head (Figure 4.1A, B), early migratory

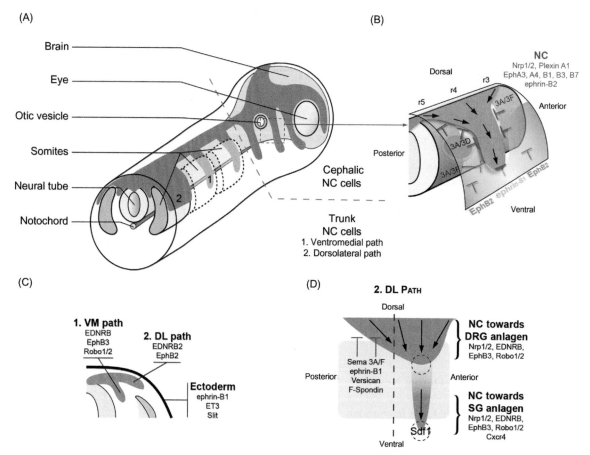

FIGURE 4.1 **Molecular control of neural crest cell migration.** (A) Neural crest cells (NCC) come out of the dorsal region of the neural tube and quickly split into discrete streams. In the head, three main subpopulations are generated. The anterior-most cephalic NCC surround the eye, while another two streams are generated around the otic vesicle. In the trunk, NCC are seen migrating as a continuous wave under the skin (dorsolateral path) and in a segmented manner throughout toward the ventral region of the embryo (mediolateral path). (B–D) Examples based on data obtained on chick NCC of how guidance is locally controlled by a given set of signaling molecules. (B) Formation of the second stream of cephalic NCC by cells originating from the rhombomeres 3 to 5 (r3–r5). (C) Eph/ephrin, endothelin, and Slit/Robo signaling determine which trunk NCC will enter the dorsolateral path. (D) NC migration is restricted to the anterior sclerotome and targeted to the peripheral ganglia anlagen (DRG, SG) by a combination of semaphorins, ephrins, Sdf1, and extracellular matrix signaling. DRG, dorsal root ganglion; SG, sympathetic ganglion; r, rhombomere.

NCC rapidly organize in three main subpopulations that will further split into additional subgroups as migration proceeds. The cephalic NCC giving rise to neurons and glial cells of the cranial ganglia stop dorsally, while NCC forming bones and cartilages continue further ventrally to invade the branchial arches [3,4]. Some NCC delaminating from the caudal rhombencephalon migrate even further.

A subpopulation of the NCC arising from rhombomeres 4 to 7 migrates to the heart and is known as the cardiac NC [5]. The NCC that colonize the gut are called enteric NCC. Two subpopulations are involved: the vagal NCC that depart from the neural tube adjacent to somites 1−7, and the sacral NCC that delaminate from the caudal trunk, after somite 28 in chick. The vagal NCC colonize the entire gut, whereas the sacral NC only contribute to the most posterior third of the gut innervation [6]. The overall pattern of cephalic NCC is very well conserved across species. In the trunk, variations in timing and trajectories are observed, but two main pathways can be defined: a dorsolateral pathway mostly used by pigment cell precursors and a ventromedial route primarily colonized by glial and neuronal precursors (Figure 4.1C). In chick and mouse embryos, NC migration first starts in between the somites along the intersomitic vessels, and then NCC pass through the anterior half of the sclerotome and along the basement membrane of the dermomyotome (Figure 4.1D). These NCC will form the sympathetic and dorsal root ganglia, glial cells along the roots of the spinal cord, and the boundary cap cells [4,7]. In mouse, both dorsolateral and ventromedial routes are invaded simultaneously, while in chick the dorsolateral path is invaded 24 h after the onset of NC cell migration. In zebrafish, trunk NCC first migrate along the middle part of the somites and the neural tube and aligned with slow muscle cells [8,9]. After a short delay, NCC migration is observed along the dorsolateral

pathway between the epidermis and the somite [8]. *Xenopus* trunk NCC pass in between the caudal part of the somite and the neural tube [10]. Few *Xenopus* NCC use the lateral pathway. Interestingly, in chick and mouse the dorsolateral path is used solely by NCC committed to the melanoblastic lineage, but fish and frog pigment cells use both routes [10,11]. Most of the differences observed across species are related to differences in somite organization and differentiation. In fish and frogs, the relative size of the sclerotome is reduced compared to amniotes and plays no major role in restricting NC cell migration.

CEPHALIC NEURAL CREST CELL MIGRATION

Inhibitors

The cephalic NC population separates into different groups due to Eph/ephrins and Semaphorin/Neuropilin/Plexin signaling pathways. The distribution of ephrins and Eph receptors within NCC and their surrounding tissues varies across species [12−17]. In *Xenopus*, NC and the adjacent mesoderm have a similar Eph/ephrin signature [15], but in chick the NC, the ectoderm, and the neural tube show complementary patterns [14,17]. Inhibition of ephrin/Eph signaling reported in the literature has two main consequences. It leads to NCC being targeted to the wrong migratory stream, or it promotes ectopic migration into otherwise NC-free regions. Therefore, Eph/ephrin signaling is important to generate NC-free regions and to target individual subpopulations of NCC to a given migratory route [12−17].

In addition, cephalic NCC express neuropilin 1 and 2 (Nrp1/2) [18−23]. Nrp1/2 associate with plexinA receptors and bind secreted class 3 semaphorins [24−26]. Inhibition of semaphorin signaling leads to ectopic NC cell migration, indicating that they also contribute to the formation of NC-free regions [19–23]. For instance, knockout mice lacking Nrp1 or Nrp2 have NCC migrating into the mesenchyme adjacent to odd-numbered rhombomeres. Similar effects have been obtained in zebrafish using antisense Morpholinos and in chicken by overexpressing soluble version of Nrp [20–23]. The role of Sema/Plexin/Nrp signaling at later stages of cephalic NC migration is more ambiguous. NCC invade the branchial arches, where several semaphorins are expressed [22]. NC may either lose the ability to sense semaphorins or may then use semaphorins as positive guidance cues as seen in axonal guidance [27,28]. Along these lines, evidence of positive guidance of post-otic NCC by Semaphorin-3C have been reported [29] and thus support a dual role for semaphorin signaling in patterning NC cell migration. Semaphorins are important to partition NC population into subgroups and may then attract some specific NCC to particular locations.

Some tissues/organs acting as a physical barrier also help shape the NC streams. For instance, the otic placode is adjacent to the rhombencephalon in amniotes and forces chick and mouse NCC to migrate anteriorly and posteriorly after delamination to join cells coming out of nearby rhombomeres. In fish and frogs, however, the otic placode is at a distance, and NCC can start migrating as a continuous sheet and separate as streams later on. The otic region expresses semaphorins [19,30], but evidence for an inhibitory effect is lacking. Indeed, heterotopically grafted otic vesicles attract NCC [31]. Similarly, mice mutants for kreisler lack rhombomere 5 and have a misplaced otic placode that leads a rerouting of NC migration [32]. Thus, the physical obstacle created by the otic vesicle is the most likely explanation for the change of direction of NCC in its vicinity. An example of how multiple signals cooperate to restrict cephalic NC cell migration is shown in Figure 4.1B.

Positive Regulators of Cephalic Neural Crest Cell Migration

Expression of numerous growth factors including VEGF, FGFs, and PDGFs have been reported in the face, neck, and branchial arches [4,33–44]. FGF and VEGF signaling are important for the homing of NCC in the second branchial arch (BA2) [42,43,45]. FGF2 and FGF8 have been proposed as chemotactic factors for mesencephalic and cardiac NCC, respectively [41,44]. Similarly, FGF8 might act as a chemoattractant driving NC cell migration toward the frontonasal process [46] since ectopic sources of FGF8 can drive NC migration toward different targets. However, in most of these cases formal demonstration of FGF signaling as a chemotactic cue is still awaited. Roles for PDGF signaling in NC development have been extensively documented [34]. Typical PDGF-related loss-of-function phenotypes include cleft palate, cranial bones malformations, and cardiac defects [33,34,39,47–51]. However, we know little about how PDGF signaling influences NC development, as some of these phenotypes could be equally due to chemotaxis, chemokinesis, survival, or differentiation defects.

The chemokine Stromal cell-derived factor 1 (Sdf1 or CXCL12) is expressed in the cephalic region at the time of NC migration, and its main receptor, Cxcr4, is expressed by the migratory NCC [52–55]. Sdf1 is a well-characterized attractant involved in gastrulation in Xenopus [56], germ cells, and posterior lateral line migration in zebrafish [57–61], as well as lymphocytes, stem cells, and cancer cells [62–68]. In Xenopus, Sdf1 is required for cephalic NC migration in vivo [55]. Cxcr4 signaling acts by activation of the small GTPase Rac1, which in turn stabilizes cell protrusions. Sdf1 is expressed all along the migratory routes, and its inhibition leads to an arrest of NC migration and not to cell dispersion [54,55], as it would be expected for a chemoattractant. Similar observations have been made for other positive regulators of NC migration, and altogether current data supporting clear chemotaxis *in vivo* remain scarce.

One striking example is that of the enteric NC migration along the digestive track [69]. The concept of attractant organized as a shallow gradient over long distances would theoretically apply to a situation like that of the enteric NCC that have to colonize a structure of immense size, the gut. However, vagal and sacral NCC migration along the gut in opposite directions seems to rely on the same signals, making it unlikely that any gradient driving long-distance migration exists along the gut. Furthermore, as enteric NC migrate equally well rostrally as they do caudally along explants of embryonic midgut, there is currently no functional evidence for gradients along the gut that promote the caudally directed migration of vagal NC [70]. A simpler explanation would be that the positive regulators of NC migration (Sdf1/VEGFs/FGFs/PDGFs/GDNF) act by promoting chemokinesis (random motility) instead of attraction. This random migration of individual cells could lead to directional migration of the whole population, as proposed for gastrulating lateral mesoderm in chick [71].

TRUNK NEURAL CREST CELL MIGRATION

The Ventromedial Pathway

In amniotes trunk NCC migrate through the anterior part of each somite and express EphA/B receptors whereas the posterior somite expresses ephrin-B ligands [17,72—75]. Blocking Eph/ephrin signaling is sufficient to promote migration through the posterior sclerotome (Figure 4.1D). Semaphorin 3A and 3F are also expressed in the posterior half of the somites (Figure 4.1D) and inhibiting Sema3F/Nrp2 signaling leads to unsegmented migration

of trunk NCC. However, this is not sufficient to affect the final patterning of the dorsal ganglia [76]. It does lead to ectopic NCC in the intersomitic space and disturbs the distribution of sympathetic ganglia [77]. Double inhibition of Sema3A and 3F induces a complete loss of NCC segmentation and a fusion of peripheral ganglia [77]. The role of Eph/ephrin and semaphorin pathways in trunk NC cell migration in *Xenopus* and zebrafish embryos has not been assessed.

Routes of NC migration are lined with permissive extracellular matrix (ECM) such as fibronectin and laminins [78—82]. Of particular interest for trunk NC segmentation are F-spondin [83] and versicans [84—86], which are present in the caudal sclerotome and around the notochord (Figure 4.1D). Inhibition of F-spondin with blocking antibodies promotes ectopic NC cell migration through the caudal half of the somite *in vivo* [83], which suggests that F-spondin has primarily an inhibitory role. Data on versican's function are less clear. Versican is mainly distributed around NC streams and is required for neural crest migration *in vivo* but acts as a repellent in *in vitro* migration assays [84,86].

Studies on Sdf1 in trunk NC migration did not show any major role but rather revealed its function into guiding specific subpopulations (Figure 4.1D). For instance, in mouse Cxcr4 is present in NCC that form the dorsal root ganglia [87], whereas in chick it is found in precursors of the sympathetic ganglia [88,89]. In these two species, affecting Cxcr4 expression and/or Sdf1 distribution perturbs the formation of the dorsal root and sympathetic ganglia, respectively. Other pathways also influence ganglia formation. In zebrafish, for instance, neuregulin/ErbB signaling pathway is required for proper migration of trunk NCC toward the anlagen of the DRG [90,91], whereas in chick, N-cadherin and ephrin-B1 are important for the final aggregation of sympathetic precursors attracted by Sdf1 [88,89,92].

Finally, Robo receptors are expressed by trunk NCC using the ventromedial path whereas Slit ligands are found around the gut [93,94]. Slit/Robo signaling prevents trunk NCC from colonizing the gut but allows NCC from the vagal region (caudal hindbrain) to do so, thus acting as checkpoint to let enteric NCC access the gut.

The Dorsolateral Pathway

Migration along the dorsolateral path is regulated by Eph/ephrin, Slit/Robo, and endothelin signaling (Figure 4.1C). In chick, the first wave of trunk NCC express EphB3 and are repelled by ephrin-B1 present in the dorsolateral path, which forces these cells to take the ventromedial route. However, pigment cells precursors express EphB2 and thus respond positively to ephrin-B1, allowing them to migrate under the ectoderm [74]. Similarly, neuronal and glial precursors are prevented to enter the dorsal migration pathway because they express endothelin receptor B (EDNRB), whereas melanoblasts expressing endothelin receptor B2 (EDNRB2) can do so [95–98]. Finally, trunk NCC expressing Robo1 and 2 also avoid the dorsolateral path, and interfering with Robo function is sufficient to induce premature migration under the skin [94].

In amniotes, prospective pigment cells can enter the dorsolateral path [11,99]. However, in *Xenopus*, only few melanocytes use that route. Most pigment cells use the ventromedial pathway together with glial and neuronal precursors and move laterally going around the somites toward the epidermis [10]. In zebrafish, some pigment cells show specific preferences for one path or another, but melanocytes use both ventral and dorsal paths equally [11].

Finally, in mouse, melanocytes are guided from the skin to the hair follicle [11] in an Sdf1-dependent manner [100]. In the chick, where melanocytes colonize both feathers without completely abandoning the skin, additional mechanisms may be required to maintain a pool of pigment cells in the epidermis, or only a fraction of the melanocytes may be expressing Cxcr4.

CELL–CELL INTERACTIONS DURING NEURAL CREST CELL MIGRATION

The directionality of NCC cannot emerge only from a balance of positive and negative cues promoting invasion of specific areas of the embryo. Inhibitors cannot give directionality per se. The putative NC attractants identified so far do not have very accurate expression patterns. Therefore, their distribution alone does not account for the overall directional migration of the NCC. The migratory routes are long, and a chemoattractant expressed in a target tissue would only work if stable long-distance gradients were established along the paths to guide the cells. Evidence for such *in vivo* gradients is lacking. Thus, it is more likely that directionality is controlled on a local scale and that guidance cues act at short range to modulate it.

Neural Crest–Neural Crest Interactions and Cell Polarity

Long-lasting cell–cell contacts and transient cell collisions mediating Contact inhibition of locomotion (CIL) have a direct effect on cell polarity. CIL is the process by which a cell stops migrating upon collision with another cell [101,102]. The physical contact between two neighboring cells inhibits cell protrusions. When cells are at low density, this local inhibition of cell protrusions upon contact promotes a change of polarity and directionality. At high cell density, only cells exposed to a free space (i.e., an area devoid of other cells) can establish and stabilize cell protrusions. Cells located at the border of the population have a clear front—back polarity where the back identity is imposed by the contact with other cells. Therefore, cells at the outskirt tend to migrate away from the cluster. This explains the overall radial dispersion observed when NCC are cultured on 2D permissive matrices (Figure 4.2A). However, *in vivo*, where not all

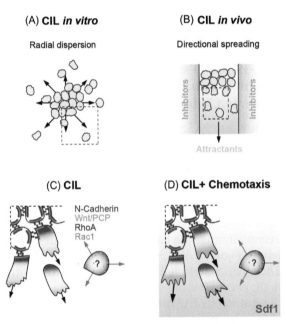

(A) CIL *in vitro*

Radial dispersion

(B) CIL *in vivo*

Directional spreading

(C) CIL

N-Cadherin
Wnt/PCP
RhoA
Rac1

(D) CIL+ Chemotaxis

Sdf1

FIGURE 4.2 **Contact inhibition of locomotion (CIL) promotes cell dispersion and cooperates with chemotaxis to control directional cell migration.** (A) *In vitro.* Cells are polarized by CIL with other cells leading to radial dispersion. (B) *In vivo.* NCC are polarized via CIL, but, in addition, they are exposed to negative signals restricting their migration (inhibitors, shades of orange) and positive signals (attractant, shades of khaki). Altogether, these signals promote directional migration. (C, D) Molecular pathways involved in *Xenopus* NCC. (C) CIL acts via N-cadherin and the noncanonical Wnt/PCP pathway that activates RhoA. Cell protrusions are mostly oriented outward (Rac1). (D) Attractants such as Sdf1 are able to reinforce the CIL-induced cell polarity by increasing Rac1 activity and further stabilizing cell protrusions. Note that in both cases, CIL alone and CIL + chemotaxis, cells that lose contact with other cells fail to polarize properly. Such isolated cells undergo random migration even when exposed to a chemotactic gradient.

areas are equally available for migration, this behavior can lead to the directional migration of the whole group [103] (Figure 4.2B). CIL had been proposed as a driving force of NC migration in the mid-1980s [104], but the actual demonstration of its importance for *in vivo* migration of zebrafish and *Xenopus* NC took 33 years [105].

In the chick embryo, migratory cephalic NCC make short- and long-range cell—cell interactions [106—109] that directly influence cell trajectories, confirming previous *in vitro* observations [104]. These NCC form chains similarly to heart fibroblast cultured *in vitro*, which exhibit CIL [110]. After contact, chick NCC either move away from each other or resume migration toward the leader cell after a short pause. The contact is systematically followed by a retraction of the cell protrusions, which is consistent with CIL. Cells migrating in chains have a higher directionality than cells wandering as single cells, suggesting that cell—cell contacts promote the directional migration of the NC population. The molecules responsible for the CIL-like behavior observed in chick cephalic NCC remain unknown.

The role of CIL in NC cell migration *in vivo* was deciphered using *Xenopus* and zebrafish cephalic NCC [105]. CIL has two main roles. Upon contact, NCC collapse their cell protrusions and repolarize to move away from each other. At high cell density, CIL prevents cell overlapping by inhibiting the cell protrusions at the region of cell—cell contact, thus restricting the protrusions to the free edge (Figure 4.2A). In *Xenopus*, CIL is mediated by N-cadherin and the noncanonical Wnt/planar cell polarity pathway (PCP) (Figure 4.2C [55,105,111—113]). N-cadherin reduces Rac1 activity, whereas Wnt/PCP induces that of RhoA at the cell—cell contacts [55,105,114]. RhoA is involved in cell contractility, formation of stress fibers, and at high level defines the back identity of migratory cells, while Rac1 is involved in membrane ruffling and cytoskeleton polymerization, and is required to form and stabilize cell protrusions [115–117]. Inhibiting N-cadherin and/or Wnt/PCP impairs CIL in NCC. Consequently, NCC produce protrusions on top of each other and fail to repolarize upon collision [55,105]. Due to its effect on cell protrusion, CIL promotes an overall dispersion of the cells.

Neural Crest–Neural Crest Interactions Promote Collective Guidance

By controlling cell polarity, CIL has a direct influence on the chemotactic abilities of the cells [55]. Groups of *Xenopus* NCC are well guided by gradients of Sdf1. Cells polarize according to their cell–cell contacts and CIL-dependent local inhibition of cell protrusions. Cxcr4 signaling promotes Rac1 activity, and thus protrusions facing high concentration of Sdf1 are stabilized (Figure 4.2D). This favors migration toward the source of Sdf1. By contrast, single NCC poorly chemotax [55]. If single cells are cultured at high cell density, allowing frequent collisions, individual NCC respond efficiently to chemotactic signals. Blocking CIL by affecting N-cadherin inhibits this collective chemotaxis [55]. Therefore, the interplay between CIL and chemotaxis can explain the directional migration of NCC toward region of Sdf1 expression. N-cadherin and Sdf1 signaling are both required for NC cell migration in fish, but the possible connection between the two has not been assessed in this species [54,118].

Migratory cephalic NCC have gap junctions [119–124] that play a role in polarity, survival, and guidance. For instance, gap junctions are required for NCC to respond to semaphorins [119]. Results on N-cadherin/Sdf1 and gap junctions/semaphorin suggest that the competence to respond to external cues is contact dependent. Thus, groups of cells would have an advantage over single cells, and that would favor collective against solitary cell migration

(for discussion, see [125]). Although NCC benefit from cell–cell interactions, they undergo an epithelial–mesenchymal transition at the onset of migration and are thus unable to maintain stable cell–cell junctions. In spite of this, NCC remains in close proximity throughout migration, and most subpopulations have a relatively steady spatial organization over time. These observations suggest that additional mechanisms are involved in maintaining the cells together.

CIL-Based Dispersion Is Counterbalanced by Mutual Attraction to Promote Collective Cell Migration

How are NC maintained together in spite of cell dispersion promoted by CIL? *Xenopus* NCC attract each other from a distance by autocrine/paracrine chemotaxis mediated by complement factor C3a and its receptor C3aR, which are co-expressed in NCC [126]. NC explants attract each other, whereas single cells leaving a group eventually migrate back toward the main population (Figure 4.3). C3a is secreted by each NCC. Thus, regions of high cell density are areas with high concentration of C3a. Rac1 is activated downstream of C3aR, thus polarizing the cells when the pathway is triggered. The fact that NCC attract one another prevents cell dispersion by favoring inward migration with respect to the NC population, thus counterbalancing CIL, which promotes outward migration. In addition, since cells are attracted to each other, the occurrence of collisions is increased, each event repolarizing the cells. In a system where CIL dominates, rapid dispersion is the main outcome. When cell–cell interactions are lost, CIL no longer affects cell polarity. Therefore, by preventing massive dispersion, co-attraction maintains cell density and positively feedbacks into CIL by increasing the probability of cell collisions. Despite the fact that co-attraction has not been formally tested in other animal models,

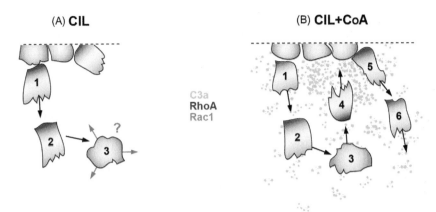

FIGURE 4.3   Co-attraction (CoA) counterbalances CIL and maintains cell density in migratory NCC. (A) CIL alone leads to cell dispersion. Cells are polarized based on the region of cell–cell contact (1), migrate away from the group (2), and subsequently lose the contact-dependent polarity (3), making them likely to continue moving in any direction (gray arrows, question mark). (B) Each NC cell expresses the complement factor C3a (which is secreted, represented as green dots) and its cognate receptor C3aR (a G protein-coupled receptor). Regions of higher concentration of NCC contain higher concentration of C3a, which acts as a local attractant. Cells are polarized owing to their cell–cell interactions (1) and thus move away from each other (2). The contact-based polarity is lost (3). Signaling via C3a/C3aR activates Rac1 and is sufficient to repolarize cells that have left the group (4) such that they migrate back toward regions of high cell density. Upon collision with another cell, each NC cell is repolarized owing to CIL (5) and thus moves away (6). This balancing act (CIL versus CoA) allows a mesenchymal population like the NC to maintain a high cell density over time.

time-lapse movies in zebrafish and chick embryos have revealed that single NCC exiting a migratory stream rejoin the other cells soon after [106,109,127]. This observation, together with the fact C3 is expressed in cephalic crest in chick (Bronner and Mayor unpublished) and mouse (Lambris and Mayor unpublished) embryos, opens the possibility that co-attraction might be a common feature of NC cell migration.

Finally, inhibitors surrounding each NC stream and the inability of NCC from different subpopulations to mix also prevent excessive dispersion and together with cell–cell adhesion and CIL ensure a high level of cell–cell interactions (for discussion, see [128]).

Dialogue Between Neural Crest Cells and Their Local Environment

One important aspect about the neural crest is that it influences the development of the tissues it interacts with during its migration. Not only do NCC receive signals from other cells in order to reach specific targets, they also have a direct effect on some of these tissues. Several examples have been described. The most anterior part of the forebrain is, as mentioned above, a source of FGF8 that is essential for NC migration and survival. It turns out that FGF8 production is NC-dependent. In absence of NCC, FGF8 expression from the anterior neural ridge is lost while Sonic Hedgehog expression expands. These local changes of gene expression have severe consequences on the overall patterning of the forebrain and midbrain that result in exencephaly [129]. These results indicate that NCC have a direct impact on the expression of one of their positive migratory cues and that they directly influence the development of the brain around which they provide protection by later differentiating as cartilage and bone.

Another example involving the cephalic NCC concerns the interaction between NC and placodal cells. NC and placodes are induced adjacent to one another within the neural plate border region and conjointly form the ganglia of the cephalic peripheral nervous system [128,130,131]. Placode-derived neurons are known to use the migratory NCC as a guide to reach the anlagen of the cephalic ganglia, but the early phases of NC and placode development were thought to happen mostly independently. Time-lapse movies in zebrafish revealed that the first wave of cephalic NC that contributes to the cartilage and bones of the face directly influence the position of the placodal domains within the ectoderm [132].

Further studies performed in *Xenopus and zebrafish* have shown that the pre-placodal domain is a source of Sdf1, which, as described above, controls cephalic NC migration [133]. Intriguingly, placode cells attract NCC from a distance by secreting Sdf1, whereas NC and placodes repel each other when in direct physical contact. This repulsive contact between NC and placodes is mediated by contact inhibition of locomotion in a N-cadherin/Wnt-PCP–dependent manner [133]. When NCC invade the region where Sdf1-positive placodes are located, placodal cells are repelled, thus shifting the source of Sdf1 further ventrally. Importantly, impairing chemotaxis or CIL in this system had consequences on both NC and placodes. More precisely, blocking placodes development was sufficient to stop NC migration, whereas inhibiting NC migration prevented the formation of discrete placodes from the original pre-placodal domain [133].

Interestingly, PDGF-A expression within the cephalic ectoderm and GDNF expression in the gut are located at the front of NC migration and progressively shift ventrally and posteriorly when NC migration proceeds [51,134]. These observations suggest that interactions between migratory NCC and the tissues producing the positive cues might be a general mechanism driving directional movement.

Other interesting examples of NC-environment interactions have been described. For instance, trunk NCC that migrate underneath the dermomyotome trigger myogenesis by transiently activating Notch signaling in muscle precursors [135]. Thus, NC migration is patterned by the antero-posterior polarity of the somites, while migratory NCC control the onset of myogenesis in the dorsal part of the somites. Another case is that of the putative interaction between trunk NCC and the neuromasts deposited by the lateral line. Neuromasts are mechanosensors located along the body of aquatic vertebrates involved spatial awareness. NCC provide glia to these structures. Extirpation of the trunk NCC have shown that in absence of trunk NCC the lateral line still migrates properly from anterior to posterior but that it fails to deposit neuromasts [136], suggesting another kind of NC–placodes interaction required for the development of the peripheral nervous system.

CONCLUSION AND PERSPECTIVES

The NC population is a heterogeneous group of multipotent progenitors that migrate throughout the embryo following precise routes. The separation of the NC population into subgroups and the subsequent guidance of these individual groups require an intricate array of signaling molecules, including local inhibitors, attractants, and cell adhesion molecules. During migration, each NC cell has to integrate a vast amount of information at any given time in order to proceed successfully. Some of the works discussed above have hinted at the interplay between different types of signaling. For instance, cell–cell adhesion influences how NCC interpret chemotactic cues and repellents suggesting that all these different signaling pathways may share common downstream effectors that they regulate in opposite manner. Much work remains to be done on this aspect to fully understand how cell polarity and directionality are regulated over time during NCC migration. Importantly, research on NC migration has highlighted the fact that mesenchymal cells can cooperate and are capable of undergoing collective cell migration—a mode of migration previously thought to require stable cell–cell junction between migratory cells and thus implicitly restricted to epithelial cell types. Data on other mesenchymal tissues such as the gastrulating mesoderm [137–139] support the notion that collective cell migration is indeed a common feature in mesenchymal tissues. This realization modifies the common view about cell populations undergoing EMT. It strongly suggests that metastatic cancer cells that have downregulated their cell–cell adhesions may still be able to cooperate in order to migrate collectively. Tumors are now considered as organs recapitulating developmental processes in a noncontrolled manner [140–142]. Strikingly, the various steps of tumor progression (cells separating from the original tissue, EMT, cell migration, reversion to an epithelial phenotype) are very similar to NC development. This means that NCC are one of the best models to study the mechanisms controlling cell guidance over time and how they may lead to tumor formation and dispersion when they are triggered inappropriately.

Acknowledgments

This investigation was supported by grants from MRC, BBSRC, and Wellcome Trust to RM and the Wellcome Trust Value in People award to ET.

References

[1] Milet C, Monsoro-Burq AH. Neural crest induction at the neural plate border in vertebrates. Dev Biol 2012; 366(1):22–33.

[2] Theveneau E, Mayor R. Neural crest delamination and migration: from epithelium-to-mesenchyme transition to collective cell migration. Dev Biol 2012.

[3] Hall B. The neural crest and neural crest cells in vertebrate development and evolution. 2nd ed. New York, NY: Springer; 2008.

[4] Le Douarin N, Kalcheim C. The neural crest. 2nd ed. Cambridge; New York, NY: Cambridge University Press; 1999.

[5] Kirby ML, Hutson MR. Factors controlling cardiac neural crest cell migration. Cell Adh Migr 2010;4(4).

[6] Le Douarin NM, Teillet MA. The migration of neural crest cells to the wall of the digestive tract in avian embryo. J Embryol Exp Morphol 1973;30(1):31–48.

[7] Vermeren M, Maro GS, Bron R, McGonnell IM, Charnay P, Topilko P, et al. Integrity of developing spinal motor columns is regulated by neural crest derivatives at motor exit points. Neuron 2003;37(3): 403–15.

[8] Raible DW, Wood A, Hodsdon W, Henion PD, Weston JA, Eisen JS. Segregation and early dispersal of neural crest cells in the embryonic zebrafish. Dev Dyn 1992;195(1):29–42.

[9] Honjo Y, Eisen JS. Slow muscle regulates the pattern of trunk neural crest migration in zebrafish. Development 2005;132(20):4461–70.

[10] Collazo A, Bronner-Fraser M, Fraser SE. Vital dye labelling of *Xenopus laevis* trunk neural crest reveals multipotency and novel pathways of migration. Development 1993;118(2):363–76.

[11] Kelsh RN, Harris ML, Colanesi S, Erickson CA. Stripes and belly-spots—a review of pigment cell morphogenesis in vertebrates. Semin Cell Dev Biol 2009;20(1):90–104.

[12] Adams RH, Diella F, Hennig S, Helmbacher F, Deutsch U, Klein R. The cytoplasmic domain of the ligand ephrinB2 is required for vascular morphogenesis but not cranial neural crest migration. Cell 2001; 104(1):57–69.

[13] Davy A, Aubin J, Soriano P. Ephrin-B1 forward and reverse signaling are required during mouse development. Genes Dev 2004;18(5):572–83.

[14] Mellott DO, Burke RD. Divergent roles for Eph and ephrin in avian cranial neural crest. BMC Dev Biol 2008;8:56.

[15] Smith A, Robinson V, Patel K, Wilkinson DG. The EphA4 and EphB1 receptor tyrosine kinases and ephrin-B2 ligand regulate targeted migration of branchial neural crest cells. Curr Biol 1997;7(8):561–70.

[16] Helbling PM, Tran CT, Brandli AW. Requirement for EphA receptor signaling in the segregation of *Xenopus* third and fourth arch neural crest cells. Mech Dev 1998;78(1–2):63–79.

[17] Baker RK, Antin PB. Ephs and ephrins during early stages of chick embryogenesis. Dev Dyn 2003;228 (1):128–42.

[18] Koestner U, Shnitsar I, Linnemannstons K, Hufton AL, Borchers A. Semaphorin and neuropilin expression during early morphogenesis of Xenopus laevis. Dev Dyn 2008;237(12):3853–63.

[19] Eickholt BJ, Mackenzie SL, Graham A, Walsh FS, Doherty P. Evidence for collapsin-1 functioning in the control of neural crest migration in both trunk and hindbrain regions. Development 1999;126(10):2181–9.

[20] Osborne NJ, Begbie J, Chilton JK, Schmidt H, Eickholt BJ. Semaphorin/neuropilin signaling influences the positioning of migratory neural crest cells within the hindbrain region of the chick. Dev Dyn 2005;232 (4):939–49.

[21] Yu HH, Moens CB. Semaphorin signaling guides cranial neural crest cell migration in zebrafish. Dev Biol 2005;280(2):373–85.

[22] Gammill LS, Gonzalez C, Bronner-Fraser M. Neuropilin 2/semaphorin 3F signaling is essential for cranial neural crest migration and trigeminal ganglion condensation. Dev Neurobiol 2007;67(1):47–56.

[23] Schwarz Q, Vieira JM, Howard B, Eickholt BJ, Ruhrberg C. Neuropilin 1 and 2 control cranial ganglion genesis and axon guidance through neural crest cells. Development 2008;135(9):1605–13.

[24] Kruger RP, Aurandt J, Guan KL. Semaphorins command cells to move. Nat Rev Mol Cell Biol 2005;6 (10):789–800.

[25] Jackson RE, Eickholt BJ. Semaphorin signalling. Curr Biol 2009;19(13):R504–7.

[26] Eickholt BJ. Functional diversity and mechanisms of action of the semaphorins. Development 2008;135 (16):2689–94.

[27] Castellani V, Chedotal A, Schachner M, Faivre-Sarrailh C, Rougon G. Analysis of the L1-deficient mouse phenotype reveals cross-talk between Sema3A and L1 signaling pathways in axonal guidance. Neuron 2000;27(2):237–49.

[28] Falk J, Bechara A, Fiore R, Nawabi H, Zhou H, Hoyo-Becerra C, et al. Dual functional activity of semaphorin 3B is required for positioning the anterior commissure. Neuron 2005;48(1):63–75.

[29] Toyofuku T, Yoshida J, Sugimoto T, Yamamoto M, Makino N, Takamatsu H, et al. Repulsive and attractive semaphorins cooperate to direct the navigation of cardiac neural crest cells. Dev Biol 2008;321(1):251–62.

[30] Bao ZZ, Jin Z. Sema3D and Sema7A have distinct expression patterns in chick embryonic development. Dev Dyn 2006;235(8):2282–9.

[31] Sechrist J, Scherson T, Bronner-Fraser M. Rhombomere rotation reveals that multiple mechanisms contribute to the segmental pattern of hindbrain neural crest migration. Development 1994;120(7):1777–90.

[32] Manzanares M, Trainor PA, Nonchev S, Ariza-McNaughton L, Brodie J, Gould A, et al. The role of kreisler in segmentation during hindbrain development. Dev Biol 1999;211(2):220–37.

[33] Tallquist MD, Soriano P. Cell autonomous requirement for PDGFRalpha in populations of cranial and cardiac neural crest cells. Development 2003;130(3):507–18.

[34] Smith CL, Tallquist MD. PDGF function in diverse neural crest cell populations. Cell Adh Migr 2010;4 (4):561–6.

[35] Richarte AM, Mead HB, Tallquist MD. Cooperation between the PDGF receptors in cardiac neural crest cell migration. Dev Biol 2007;306(2):785–96.

[36] Ho L, Symes K, Yordan C, Gudas LJ, Mercola M. Localization of PDGF A and PDGFR alpha mRNA in Xenopus embryos suggests signalling from neural ectoderm and pharyngeal endoderm to neural crest cells. Mech Dev 1994;48(3):165–74.

[37] Orr-Urtreger A, Bedford MT, Do MS, Eisenbach L, Lonai P. Developmental expression of the alpha receptor for platelet-derived growth factor, which is deleted in the embryonic lethal Patch mutation. Development 1992;115(1):289–303.

[38] Orr-Urtreger A, Lonai P. Platelet-derived growth factor-A and its receptor are expressed in separate, but adjacent cell layers of the mouse embryo. Development 1992;115(4):1045–58.

[39] Schatteman GC, Morrison-Graham K, van Koppen A, Weston JA, Bowen-Pope DF. Regulation and role of PDGF receptor alpha-subunit expression during embryogenesis. Development 1992;115(1):123–31.

[40] Takakura N, Yoshida H, Ogura Y, Kataoka H, Nishikawa S. PDGFR alpha expression during mouse embryogenesis: immunolocalization analyzed by whole-mount immunohistostaining using the monoclonal anti-mouse PDGFR alpha antibody APA5. J Histochem Cytochem 1997;45(6):883–93.

[41] Sato A, Scholl AM, Kuhn EB, Stadt HA, Decker JR, Pegram K, et al. FGF8 signaling is chemotactic for cardiac neural crest cells. Dev Biol 2011;17.

[42] Trokovic N, Trokovic R, Partanen J. Fibroblast growth factor signalling and regional specification of the pharyngeal ectoderm. Int J Dev Biol 2005;49 (7):797–805.

[43] McLennan R, Teddy JM, Kasemeier-Kulesa JC, Romine MH, Kulesa PM. Vascular endothelial growth factor (VEGF) regulates cranial neural crest migration in vivo. Dev Biol 2010;339(1):114–25.

[44] Kubota Y, Ito K. Chemotactic migration of mesencephalic neural crest cells in the mouse. Dev Dyn 2000; 217(2):170–9.

[45] McLennan R, Kulesa PM. Neuropilin-1 interacts with the second branchial arch microenvironment to mediate chick neural crest cell dynamics. Dev Dyn 2010;239(6):1664–73.

[46] Creuzet SE, Martinez S, Le Douarin NM. The cephalic neural crest exerts a critical effect on forebrain and midbrain development. Proc Natl Acad Sci USA 2006;103(38):14033–8.

[47] Soriano P. The PDGF alpha receptor is required for neural crest cell development and for normal patterning of the somites. Development 1997;124(14):2691–700.

[48] Morrison-Graham K, Schatteman GC, Bork T, Bowen-Pope DF, Weston JAA. PDGF receptor mutation in the mouse (Patch) perturbs the development of a non-neuronal subset of neural crest-derived cells. Development 1992;115(1):133–42.

[49] Robbins JR, McGuire PG, Wehrle-Haller B, Rogers SL. Diminished matrix metalloproteinase 2 (MMP-2) in ectomesenchyme-derived tissues of the Patch mutant mouse: regulation of MMP-2 by PDGF and effects on mesenchymal cell migration. Dev Biol 1999;212(2): 255–63.

[50] Stoller JZ, Epstein JA. Cardiac neural crest. Semin Cell Dev Biol 2005;16(6):704–15.

[51] Eberhart JK, He X, Swartz ME, Yan YL, Song H, Boling TC, et al. MicroRNA Mirn140 modulates Pdgf signaling during palatogenesis. Nat Genet 2008;40 (3):290–8.

[52] Yusuf F, Rehimi R, Dai F, Brand-Saberi B. Expression of chemokine receptor CXCR4 during chick embryo development. Anat Embryol (Berl) 2005;210(1):35–41.

[53] Rehimi R, Khalida N, Yusuf F, Dai F, Morosan-Puopolo G, Brand-Saberi B. Stromal-derived factor-1 (SDF-1) expression during early chick development. Int J Dev Biol 2008;52(1):87–92.

[54] Olesnicky Killian EC, Birkholz DA, Artinger KB. A role for chemokine signaling in neural crest cell migration and craniofacial development. Dev Biol 2009;333(1):161–72.

[55] Theveneau E, Marchant L, Kuriyama S, Gull M, Moepps B, Parsons M, et al. Collective chemotaxis requires contact-dependent cell polarity. Dev Cell 2010;19(1):39–53.

[56] Fukui A, Goto T, Kitamoto J, Homma M, Asashima M. SDF-1 alpha regulates mesendodermal cell migration during frog gastrulation. Biochem Biophys Res Commun 2007;354(2):472–7.

[57] Boldajipour B, Mahabaleshwar H, Kardash E, Reichman-Fried M, Blaser H, Minina S, et al. Control of chemokine-guided cell migration by ligand sequestration. Cell 2008;132(3):463–73.

[58] Blaser H, Eisenbeiss S, Neumann M, Reichman-Fried M, Thisse B, Thisse C, et al. Transition from non-motile behaviour to directed migration during early PGC development in zebrafish. J Cell Sci 2005;118(Pt 17): 4027–38.

[59] Doitsidou M, Reichman-Fried M, Stebler J, Koprunner M, Dorries J, Meyer D, et al. Guidance of primordial germ cell migration by the chemokine SDF-1. Cell 2002;111(5):647–59.

[60] David NB, Sapede D, Saint-Etienne L, Thisse C, Thisse B, Dambly-Chaudiere C, et al. Molecular basis of cell migration in the fish lateral line: role of the chemokine receptor CXCR4 and of its ligand, SDF1. Proc Natl Acad Sci USA 2002;99(25):16297–302.

[61] Haas P, Gilmour D. Chemokine signaling mediates self-organizing tissue migration in the zebrafish lateral line. Dev Cell 2006;10(5):673–80.

[62] Kucia M, Reca R, Miekus K, Wanzeck J, Wojakowski W, Janowska-Wieczorek A, et al. Trafficking of normal stem cells and metastasis of cancer stem cells involve similar mechanisms: pivotal role of the SDF-1-CXCR4 axis. Stem Cells 2005;23(7):879–94.

[63] Kucia M, Ratajczak J, Ratajczak MZ. Bone marrow as a source of circulating CXCR4 + tissue-committed stem cells. Biol Cell 2005;97(2):133–46.

[64] Kucia M, Jankowski K, Reca R, Wysoczynski M, Bandura L, Allendorf DJ, et al. CXCR4-SDF-1 signalling, locomotion, chemotaxis and adhesion. J Mol Histol 2004;35(3):233–45.

[65] Koizumi K, Hojo S, Akashi T, Yasumoto K, Saiki I. Chemokine receptors in cancer metastasis and cancer cell-derived chemokines in host immune response. Cancer Sci 2007;98(11):1652–8.

[66] Dewan MZ, Ahmed S, Iwasaki Y, Ohba K, Toi M, Yamamoto N. Stromal cell-derived factor-1 and CXCR4 receptor interaction in tumor growth and metastasis of breast cancer. Biomed Pharmacother 2006;60(6):273–6.

[67] Aiuti A, Webb IJ, Bleul C, Springer T, Gutierrez-Ramos JC. The chemokine SDF-1 is a chemoattractant for human CD34 + hematopoietic progenitor cells and provides a new mechanism to explain the mobilization of CD34 + progenitors to peripheral blood. J Exp Med 1997;185(1):111–20.

[68] Bleul CC, Fuhlbrigge RC, Casasnovas JM, Aiuti A, Springer TA. A highly efficacious lymphocyte chemoattractant, stromal cell-derived factor 1 (SDF-1). J Exp Med 1996;184(3):1101–9.

[69] Sasselli V, Pachnis V, Burns AJ. The enteric nervous system. Dev Biol 2012;366(1):64–73.

[70] Anderson RB, Bergner AJ, Taniguchi M, Fujisawa H, Forrai A, Robb L, et al. Effects of different regions of the developing gut on the migration of enteric neural crest-derived cells: a role for Sema3A, but not Sema3F. Dev Biol 2007;305(1):287–99.

[71] Benazeraf B, Francois P, Baker RE, Denans N, Little CD, Pourquie O. A random cell motility gradient downstream of FGF controls elongation of an amniote embryo. Nature 2010;466(7303):248–52.

[72] Krull CE, Lansford R, Gale NW, Collazo A, Marcelle C, Yancopoulos GD, et al. Interactions of Eph-related receptors and ligands confer rostrocaudal pattern to trunk neural crest migration. Curr Biol 1997;7(8):571–80.

[73] De Bellard ME, Ching W, Gossler A, Bronner-Fraser M. Disruption of segmental neural crest migration and ephrin expression in delta-1 null mice. Dev Biol 2002; 249(1):121–30.

[74] Santiago A, Erickson CA. Ephrin-B ligands play a dual role in the control of neural crest cell migration. Development 2002;129(15):3621–32.

[75] Wang HU, Anderson DJ. Eph family transmembrane ligands can mediate repulsive guidance of trunk neural crest migration and motor axon outgrowth. Neuron 1997;18(3):383–96.

[76] Gammill LS, Gonzalez C, Gu C, Bronner-Fraser M. Guidance of trunk neural crest migration requires neuropilin 2/semaphorin 3F signaling. Development 2006;133(1):99–106.

[77] Schwarz Q, Maden CH, Davidson K, Ruhrberg C. Neuropilin-mediated neural crest cell guidance is essential to organise sensory neurons into segmented dorsal root ganglia. Development 2009;136 (11):1785–9.

[78] Coles EG, Gammill LS, Miner JH, Bronner-Fraser M. Abnormalities in neural crest cell migration in laminin alpha5 mutant mice. Dev Biol 2006;289(1):218–28.

[79] Bronner-Fraser M. Distribution and function of tenascin during cranial neural crest development in the chick. J Neurosci Res 1988;21(2–4):135–47.

[80] Duband JL, Thiery JP. Distribution of laminin and collagens during avian neural crest development. Development 1987;101(3):461–78.

[81] Brauer PR, Bolender DL, Markwald RR. The distribution and spatial organization of the extracellular matrix encountered by mesencephalic neural crest cells. Anat Rec 1985;211(1):57–68.

[82] Perris R, Perissinotto D. Role of the extracellular matrix during neural crest cell migration. Mech Dev 2000;95(1–2):3–21.

[83] Debby-Brafman A, Burstyn-Cohen T, Klar A, Kalcheim C. F-Spondin, expressed in somite regions avoided by neural crest cells, mediates inhibition of distinct somite domains to neural crest migration. Neuron 1999;22(3):475–88.

[84] Dutt S, Kleber M, Matasci M, Sommer L, Zimmermann DR. Versican V0 and V1 guide migratory neural crest cells. J Biol Chem 2006;281(17):12123–31.

[85] Perissinotto D, Iacopetti P, Bellina I, Doliana R, Colombatti A, Pettway Z, et al. Avian neural crest cell migration is diversely regulated by the two major hyaluronan-binding proteoglycans PG-M/versican and aggrecan. Development 2000;127(13):2823–42.

[86] Perris R, Perissinotto D, Pettway Z, Bronner-Fraser M, Morgelin M, Kimata K. Inhibitory effects of PG-H/aggrecan and PG-M/versican on avian neural crest cell migration. FASEB J 1996;10(2):293–301.

[87] Belmadani A, Tran PB, Ren D, Assimacopoulos S, Grove EA, Miller RJ. The chemokine stromal cell-derived factor-1 regulates the migration of sensory neuron progenitors. J Neurosci 2005;25(16):3995–4003.

[88] Kasemeier-Kulesa JC, McLennan R, Romine MH, Kulesa PM, Lefcort F. CXCR4 controls ventral migration of sympathetic precursor cells. J Neurosci 2010;30 (39):13078–88.

[89] Saito D, Takase Y, Murai H, Takahashi Y. The dorsal aorta initiates a molecular cascade that instructs sympatho-adrenal specification. Science 2012;336(6088): 1578–81.

[90] Heermann S, Schwab MH. Molecular control of Schwann cell migration along peripheral axons: keep moving! Cell Adh Migr 2012;7(1):18–22.

[91] Honjo Y, Kniss J, Eisen JS. Neuregulin-mediated ErbB3 signaling is required for formation of zebrafish dorsal root ganglion neurons. Development 2008;135 (15):2615–25.

[92] Kasemeier-Kulesa JC, Bradley R, Pasquale EB, Lefcort F, Kulesa PM. Eph/ephrins and N-cadherin coordinate to control the pattern of sympathetic ganglia. Development 2006;133(24):4839–47.

[93] De Bellard ME, Rao Y, Bronner-Fraser M. Dual function of Slit2 in repulsion and enhanced migration of trunk, but not vagal, neural crest cells. J Cell Biol 2003;162(2):269–79.

[94] Jia L, Cheng L, Raper J. Slit/Robo signaling is necessary to confine early neural crest cells to the ventral migratory pathway in the trunk. Dev Biol 2005;282 (2):411–21.

[95] Harris ML, Hall R, Erickson CA. Directing pathfinding along the dorsolateral path - the role of EDNRB2 and EphB2 in overcoming inhibition. Development 2008;135(24):4113–22.

[96] Pla P, Alberti C, Solov'eva O, Pasdar M, Kunisada T, Larue L. Ednrb2 orients cell migration towards the dorsolateral neural crest pathway and promotes melanocyte differentiation. Pigment Cell Res 2005;18 (3):181–7.

[97] Lee HO, Levorse JM, Shin MK. The endothelin receptor-B is required for the migration of neural crest-derived melanocyte and enteric neuron precursors. Dev Biol 2003;259(1):162–75.

[98] Shin MK, Levorse JM, Ingram RS, Tilghman SM. The temporal requirement for endothelin receptor-B signalling during neural crest development. Nature 1999;402(6761):496–501.

[99] Harris ML, Erickson CA. Lineage specification in neural crest cell pathfinding. Dev Dyn 2007;236(1):1–19.

[100] Belmadani A, Jung H, Ren D, Miller RJ. The chemokine SDF-1/CXCL12 regulates the migration of melanocyte progenitors in mouse hair follicles. Differentiation 2009;77(4):395–411.

[101] Abercrombie M, Dunn GA. Adhesions of fibroblasts to substratum during contact inhibition observed by interference reflection microscopy. Exp Cell Res 1975;92(1):57–62.

[102] Abercrombie M, Heaysman JE. Observations on the social behaviour of cells in tissue culture. I. Speed of movement of chick heart fibroblasts in relation to their mutual contacts. Exp Cell Res 1953;5(1):111–31.

[103] Mayor R, Carmona-Fontaine C. Keeping in touch with contact inhibition of locomotion. Trends Cell Biol 2010;20(6):319–28.

[104] Erickson CA. Control of neural crest cell dispersion in the trunk of the avian embryo. Dev Biol 1985;111(1):138–57.

[105] Carmona-Fontaine C, Matthews HK, Kuriyama S, Moreno M, Dunn GA, Parsons M, et al. Contact inhibition of locomotion in vivo controls neural crest directional migration. Nature 2008;456(7224):957–61.

[106] Teddy JM, Kulesa PM. In vivo evidence for short- and long-range cell communication in cranial neural crest cells. Development 2004;131(24):6141–51.

[107] Kulesa PM, Fraser SE. Neural crest cell dynamics revealed by time-lapse video microscopy of whole embryo chick explant cultures. Dev Biol 1998;204(2):327–44.

[108] Kulesa PM, Fraser SE. In ovo time-lapse analysis of chick hindbrain neural crest cell migration shows cell interactions during migration to the branchial arches. Development 2000;127(6):1161–72.

[109] Kulesa P, Bronner-Fraser M, Fraser S. In ovo time-lapse analysis after dorsal neural tube ablation shows rerouting of chick hindbrain neural crest. Development 2000;127(13):2843–52.

[110] Ambrose EJ. The movements of fibrocytes. Exp Cell Res 1961;(Suppl. 8):54–73.

[111] Carmona-Fontaine C, Matthews H, Mayor R. Directional cell migration in vivo: Wnt at the crest. Cell Adh Migr 2008;2(4):240–2.

[112] De Calisto J, Araya C, Marchant L, Riaz CF, Mayor R. Essential role of non-canonical Wnt signalling in neural crest migration. Development 2005;132(11):2587–97.

[113] Matthews HK, Broders-Bondon F, Thiery JP, Mayor R. Wnt11r is required for cranial neural crest migration. Dev Dyn 2008;237(11):3404–9.

[114] Theveneau E, Mayor R. Integrating chemotaxis and contact-inhibition during collective cell migration: small GTPases at work. Small GTPases 2010;1(2):1–5.

[115] Ridley AJ, Schwartz MA, Burridge K, Firtel RA, Ginsberg MH, Borisy G, et al. Cell migration: integrating signals from front to back. Science 2003;302(5651):1704–9.

[116] Ridley AJ, Paterson HF, Johnston CL, Diekmann D, Hall A. The small GTP-binding protein rac regulates growth factor-induced membrane ruffling. Cell 1992;70(3):401–10.

[117] Ridley AJ, Hall A. Distinct patterns of actin organization regulated by the small GTP-binding proteins Rac and Rho. Cold Spring Harb Symp Quant Biol 1992;57:661–71.

[118] Piloto S, Schilling TF. Ovo1 links Wnt signaling with N-cadherin localization during neural crest migration. Development 2010;137(12):1981–90.

[119] Xu X, Francis R, Wei CJ, Linask KL, Lo CW. Connexin 43-mediated modulation of polarized cell movement and the directional migration of cardiac neural crest cells. Development 2006;133(18):3629–39.

[120] Bannerman P, Nichols W, Puhalla S, Oliver T, Berman M, Pleasure D. Early migratory rat neural crest cells express functional gap junctions: evidence that neural crest cell survival requires gap junction function. J Neurosci Res 2000;61(6):605–15.

[121] Waldo KL, Lo CW, Kirby ML. Connexin 43 expression reflects neural crest patterns during cardiovascular development. Dev Biol 1999;208(2):307–23.

[122] Huang GY, Cooper ES, Waldo K, Kirby ML, Gilula NB, Lo CW. Gap junction-mediated cell-cell communication modulates mouse neural crest migration. J Cell Biol 1998;143(6):1725–34.

[123] Lo CW, Cohen MF, Huang GY, Lazatin BO, Patel N, Sullivan R, et al. Cx43 gap junction gene expression and gap junctional communication in mouse neural crest cells. Dev Genet 1997;20(2):119–32.

[124] Wang X, Veruki ML, Bukoreshtliev NV, Hartveit E, Gerdes HH. Animal cells connected by nanotubes can be electrically coupled through interposed gap-junction channels. Proc Natl Acad Sci USA 2010;107(40):17194–9.

[125] Theveneau E, Mayor R. Can mesenchymal cells undergo collective cell migration? The case of the neural crest. Cell Adh Migr 2011;5:6.

[126] Carmona-Fontaine C, Theveneau E, Tzekou A, Woods M, Page K, Tada M, et al. Complement fragment C3a controls mutual cell attraction during collective cell migration. Dev Cell 2011;21:6.

[127] Kulesa PM, Lu CC, Fraser SE. Time-lapse analysis reveals a series of events by which cranial neural crest cells reroute around physical barriers. Brain Behav Evol 2005;66(4):255–65.

[128] Theveneau E, Mayor R. Collective cell migration of the cephalic neural crest: the art of integrating information. Genesis 2011;49(4):164–76.

[129] Creuzet SE. Regulation of pre-otic brain development by the cephalic neural crest. Proc Natl Acad Sci USA 2009;106(37):15774–9.

[130] Pieper M, Eagleson GW, Wosniok W, Schlosser G. Origin and segregation of cranial placodes in *Xenopus laevis*. Dev Biol 2011;360(2):257–75.

[131] Streit A. The cranial sensory nervous system: specification of sensory progenitors and placodes. Harvard Stem Cell Institute 2008; StemBook [Internet].

[132] Culbertson MD, Lewis ZR, Nechiporuk AV. Chondrogenic and gliogenic subpopulations of neural crest play distinct roles during the assembly of epibranchial ganglia. PLoS One 2011;6(9):e24443.

[133] Theveneau E, Steventon B, Scarpa E, Garcia S, Trepat X, Streit A, et al. Chase-and-run between adjacent cell populations promotes directional collective migration. Nat Cell Biol 2013;15:7.

[134] Natarajan D, Marcos-Gutierrez C, Pachnis V, de Graaff E. Requirement of signalling by receptor tyrosine kinase RET for the directed migration of enteric nervous system progenitor cells during mammalian embryogenesis. Development 2002;129 (22):5151–60.

[135] Rios AC, Serralbo O, Salgado D, Marcelle C. Neural crest regulates myogenesis through the transient activation of NOTCH. Nature 2011.

[136] Hörstadius SO. The neural crest; its properties and derivatives in the light of experimental research. London, New York: Oxford University Press; 1950.

[137] Winklbauer R, Selchow A, Nagel M, Angres B. Cell interaction and its role in mesoderm cell migration during *Xenopus* gastrulation. Dev Dyn 1992;195(4):290–302.

[138] Weber GF, Bjerke MA, Desimone DW. A mechanoresponsive cadherin-keratin complex directs polarized protrusive behavior and collective cell migration. Dev Cell 2012;22(1):104–15.

[139] Ulrich F, Krieg M, Schotz EM, Link V, Castanon I, Schnabel V, et al. Wnt11 functions in gastrulation by controlling cell cohesion through Rab5c and E-cadherin. Dev Cell 2005;9(4):555–64.

[140] Egeblad M, Nakasone ES, Werb Z. Tumors as organs: complex tissues that interface with the entire organism. Dev Cell 2010;18(6):884–901.

[141] Hanahan D, Weinberg RA. Hallmarks of cancer: the next generation. Cell 2011;144(5):646–74.

[142] Thiery JP, Acloque H, Huang RY, Nieto MA. Epithelial–mesenchymal transitions in development and disease. Cell 2009;139(5):871–90.

Epigenetic Regulation of Neural Crest Cells

Pablo H. Strobl-Mazzulla[a] and Marianne E. Bronner[b]

[a]Laboratory of Developmental Biology, Instituto de Investigaciones Biotecnológicas-Instituto Tecnológico de Chascomús (CONICET-UNSAM), Chascomús, Argentina [b]Division of Biology 139−74, California Institute of Technology, Pasadena, CA, USA

INTRODUCTION

The neural crest (NC), often referred to as the fourth germ layer, is a transient embryonic cell population that has broad differentiation potential. It is well established that a highly orchestrated gene regulatory network plays an important role in formation of the NC, as reviewed in [1−6]. However, an emerging body of evidence suggests that post-transcriptional and epigenetic contributions also play central roles in NC development and that aberrant epigenetic modifications may contribute to diseases such as cancer. Consequently, understanding the normal epigenetic mechanisms involved in NC cell (NCC) development will provide important clues regarding the mistakes that may lead to abnormal development or loss of the differentiated state.

The term epigenetic was initially coined by Waddington [7] as "the branch of biology which studies the causal interactions between

Neural Crest Cells.
DOI: http://dx.doi.org/10.1016/B978-0-12-401730-6.00005-3

89

genes and their products, which bring the phenotype into being." He described cellular differentiation as a process largely governed by changes in the "epigenetic landscape" rather than alterations in genetic inheritance. In this context, epigenetics is actually defined as "the study of any potentially stable and, ideally, heritable change in gene expression or cellular phenotype that occurs without changes in Watson-Crick base-pairing of DNA" [8].

Epigenetic phenomena are linked by the fact that DNA is not "naked" in eukaryotes, but exists as an intimate complex with specialized proteins called histones, which together with DNA comprise chromatin. Initially, chromatin was considered to be a passive packaging molecule. However, now it is clear that there exists a plethora of covalent and noncovalent modifications. These include, but are not limited to, post-translational histone modifications, chromatin-remodeling events that mobilize or alter nucleosome structure, dynamic shuffling of new histones or variants in and out of nucleosomes, and the targeting role of small noncoding RNAs. The DNA itself is also susceptible to being modified covalently by methylation, mostly occurring at cytosine nucleotides. Many, but not all, of these modifications and chromatin changes are reversible and depend upon intrinsic and external stimuli [9], which ultimately regulate the accessibility and/or procession of the transcriptional machinery [10]. Although, most of these epigenetic modifications are unlikely to be propagated through the germ line, they can remain stable through several cell divisions, thus reflecting a type of cellular memory.

Recently, there have been numerous publications dedicated to epigenetic mechanisms, which are gaining recognition as a key factor in the fine-tuning regulation of gene expression and cell differentiation, particularly in cancer and stem cells. In the neural crest as well, there is emerging data in this promising field describing the "epigenetic landscape" involved in NC development (Figure 5.1). In this review,

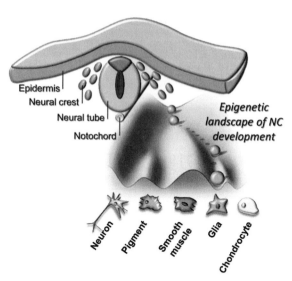

FIGURE 5.1   Schematic diagram representing the "epigenetic landscape" of NC development.

we will describe the current state of knowledge regarding the actors involved in the "writing," "erasing," and/or "reading" of the "epigenetic code" necessary to achieve the cellular memory that dictates gene expression programs in NC development (Figure 5.2).

HISTONE MODIFICATIONS

Gene expression occurs in a chromatin context. Therefore, wrapping the genome is known to be a critical feature of regulating gene expression. The nucleosome is the main unit of chromatin, comprised of a histone octamer that contains two molecules each of the four core histones (H2A, H2B, H3, and H4), around which is wrapped 147 bp of genomic DNA. Adjacent nucleosomes are connected by histone H1−bound linker DNA. In addition to these canonical histone proteins, many variant forms of histones exist in different organisms. These are able to regulate chromatin structure and have been associated with different transcriptional states [12].

	Modification	Gene	Target genes	Role on NC	Reference
Chromatin remodeling complexes		Brg1 (SWI/SNF)		Induction and differentiation	(Erogly et al., 2006)
		CHD7-SWI/SNF	Sox9, Twist, Slug	Specification	(Bajpai et al., 2010)
		WSTF		Migration and/or maintenance	(Yoshimura et al., 2009; Barnett et al., 2012)
Histone methylation	H4K20me1 H3K9me1/2	PHF8	MSX1/MSXB	Cranial development	(Phillips et al., 2006)
	H3K9me3	JmjD2A	Sox10, Snail2	Specification	(Strobl-Mazzulla et al., 2010)
	H3K27me3	Aebp2-PRC2	Ret, Gdnf, Ednrb, Sox10, Bmp4, Pax3, Snail2, Phox2b, Zfhx1b	Specification and migration	(Kim et al., 2011)
Histone variants	H3.3			Ectomesenchyme potential to cranial NC cells	(Cox et al., 2012)
Histone acetylation		HDAC1	FoxD3	Melanophore development	(Ignatius et al., 2008)
		HDAC8	Otx2, Lhx1	Cranial development	(Haberland et al., 2009)
		HDAC3		Cardiac differentiation	(Singh et al., 2011)
		HDAC4		Craniofacial development	(Delaurier et al., 2012)
	H3KAc	PHD12-Sin3A/HDAC	Cad6B	Migration	(Strobl-Mazzulla and Bronner, 2012)
DNA methylation		DNMT3B DNMT3A	Sox2, Sox3	Cranial development Repress neural fate	(Jin et al., 2008) (Hu et al., 2012)

FIGURE 5.2 **Summary of the current state of knowledge of epigenetic contributions to NC development.** *Source: Adapted from Ref. [11].*

The core histones are highly conserved basic proteins with globular domains and flexible N-terminal "tails" that protrude from the nucleosome and are susceptible to a variety of post-translational modifications [13–15]. Acetylation and methylation of core histones, notably H3 and H4, were among the first covalent modifications to be described, and they have been proposed to correlate, respectively, with positive and negative changes in transcriptional activity. Since the pioneering studies of Allfrey and coworkers [16], 130 sites for post-transcriptional histone modifications have been identified and characterized; these include histone propionylation, butyrylation, formylation, phosphorylation, ubiquitination, sumoylation, deimination, citrullination, proline isomerization, ADP ribosylation, tyrosine hydroxylation, and lysine crotonylation [17]. Most of these modifications act as landmarks to recruit additional factors, including chromatin-remodeling factors and the transcriptional machinery necessary to throw the switch on the "on/off" state of gene transcription.

Even though all of these histone modifications have potential implications in transcriptional regulation, only methylation and acetylation have been studied during NC development to date. It should be noted that sumoylation and other modifications also can affect transcriptional regulation in NCC by directly modifying transcription factors and their interacting partners [18,19].

Methylation

Histone methylation is one of the best-studied histone modifications and is associated with both transcriptional activation and repression. Histone methylation primarily occurs in histone tails, mostly on lysine residues. Although it would be tempting to propose that methylation might act to directly regulate chromatin structure, there is no evidence for lysine methylation directly affecting chromatin dynamics. Rather, methylation has been shown to be involved in the recruitment of various modifiers of chromatin and transcriptional activators or repressors [15]. Until now, at least three protein motifs—the chromodomain, the tudor domain, and the WD40-repeat domain—have been identified as capable of specifically interacting with methylated lysine residues (for review see [20]). This specific recruitment step appears to play an important role in gene regulation and the concomitant biological outcomes that are associated with different methylation events.

Moreover, there are additional levels of complexity, since some residues susceptible to methylation can adopt one of three different methylation states. Also, the binding affinity of the recruited proteins is affected by their particular methylation states and these ultimately mediate unique cellular responses [21–23].

Histone methylation and demethylation are catalyzed by specific histone methyltransferase and demethylases, respectively. As a general rule, trimethylation of H3K4 (H3K4me3) is catalyzed by histone methyltranferases of the Trithorax-group (TrxG) proteins and associated with active transcription [24–26]. In contrast, trimethylation of H3K27 (H3K27me3) is catalyzed by Polycomb-group (PcG) proteins and associated with transcriptional repression [27–29]. Similar to H3K4me3, trimethylated H3 lysine 36 (H3K36me3) is frequently located in transcriptionally active euchromatic regions, with the former predominantly found in the promoter and the second in the bodies of genes. On the other hand,

trimethylated H3 lysines 9 and 27 (H3K9me3 and H3K27me3) are repressive marks associated with heterochromatin or in the proximity of the transcription start site (TSS) of euchromatically repressed genes [30]. Most modifications are distributed in distinct patterns within the upstream region, the core promoter close to the TSS, the 5′ end of the open reading frame (ORF) and the 3′ end of the ORF. Indeed, the location of a modification is tightly regulated and is crucial for its effect on transcription (Figure 5.3).

In the developing embryo and embryonic stem cells, early progenitor cells have a broad developmental potential. This correlates with cis-regulatory elements of developmental genes existing in a "poised" state characterized by bivalent histone methylation states [31,32]. A recent study has used ChIP-seq analysis to characterize the genome-wide chromatin patterns associated with the regulatory regions of human embryonic stem cells differentiated into neural crest cells [33]. This study identified over 4300 genomic elements marked by an active enhancer signature defined by the occupancy of p300, with simultaneous enrichment of H3K27ac and H3K4me1 at flanking regions and the absence of H3K4me3. Moreover, 79% of the elements were exclusively identified in human cells differentiated into NCC, but not in human embryonic cells or human neurectodermal cells. This type of approach helps to efficiently identify all functional enhancer regions for genes potentially important in all aspects of NC development. In contrast to enhancers, considerably less cell type-specific variation was observed in chromatin marking patterns at proximal promoter regions [33]. Thus, the establishment of the correct "epigenetic code," removing or adding repressive or activating marks at different genomic positions, plays a key role during development. Failure in either step can cause dysregulation of gene expression, which has been strongly linked to a variety of diseases [34–38].

Changes in histone methylation clearly have a critical effect on the timing of NC development.

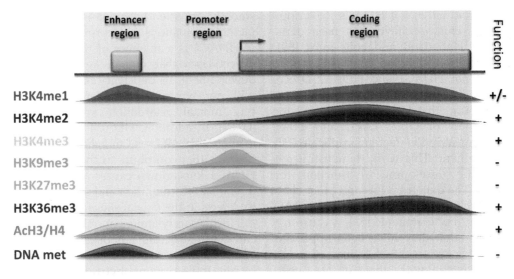

FIGURE 5.3 **Correlation between epigenetic modifications and gene transcription.** The distribution of epigenetic marks is mapped onto an arbitrary gene, based on genome-wide data. ± indicates cases where the association with active transcription is ambiguous.

For example, the histone demethylase, KDM4A (also known as JmjD2A), which is responsible for demethylation of both H3K9me3 and H3K36me3, is required for proper spatiotemporal activation of several key NC specifier genes, such us *Sox9*, *Sox10*, *FoxD3*, and *Snail2* [39]. In avian embryos, KDM4A is initially found throughout the neural plate, but as neurulation proceeds, its expression becomes restricted to the dorsal aspect of the neural tube at the time of NC specification, disappearing later from migrating NCC. By demethylating H3K9me3, KDM4A removes a repressive mark from the transcriptional start site of both Sox10 and Snail2, thus allowing activation of their expression in premigratory NCC [39]. These findings highlight the importance of epigenetics regulation setting up the developmental program for NCC specification.

Acetylation

Histone acetylases HATs [40]; acetylate-specific lysine residues on histones, whereas histone deacetylases HDACs [41]; do the opposite.

Acetylation of lysine residues on histones located on promoter regions (Figure 5.3) allows opening of the chromatin structure, thus facilitating access of the transcriptional machinery to the target gene [42,43] as well as the creation of binding sites for bromodomain proteins, which function as transcriptional activators [44]. Specifically, the acetylation of lysines neutralizes their positive charge, reducing the strength of the binding between the histone and the negatively charged DNA.

There are 11 HDAC enzymes encoded in the mammalian genome, which based on similarity and function can be grouped into four different classes: class I (HDAC1, 2, 3, and 8), IIa (HDAC4, 5, 7, and 9), IIb (HDAC6 and 10), and IV (Hdac11) [45]. The class I HDACs are widely expressed during embryogenesis and organogenesis [46–51] and can regulate several genes involved in different developmental pathways. On this basis, one might assume that they play global and redundant roles as modulators of gene expression. Surprisingly, however, knockout studies reveal specific and nonredundant roles for individual class I HDACs in various cell lineages. This specificity has been demonstrated by Haberland and colleagues [50], who showed that deletion by Wnt1-cre of HDAC8 but not HDAC1 and HDAC2 affects cranial NC cell differentiation into facial skeleton. They found that HDAC8 is required for the suppression of Otx2, Lhx1, and other homeobox transcription factors implicated in skull development. This work demonstrates a highly specific developmental function for HDAC8 [50]. The data from this and other class I HDACs knockout studies indicate that HDAC inhibitors may be teratogenic [52], a cause for concern given that they have been approved recently for the treatment of human diseases [53,54]. Consistent with this, it has been reported that the fetuses of mothers who take the HDAC inhibitor valproic acid, an anticonvulsant and mood-stabilizing drug, during the first trimester of pregnancy have a significantly increased risk of developing craniofacial abnormalities [55,56]. On the other hand, valproic acid has been shown to have anticancer properties for the NC-derived human tumor cell lines G361 melanoma, U87MG glioblastoma, and SKNMC Askin tumor [57].

Although HDAC1 is not essential for cranial NC differentiation, studies in zebrafish demonstrate a requirement for this enzyme for proper differentiation and migration of neural crest-derived melanophores and their precursors. Specifically, HDAC1 is required to repress FoxD3 expression to permit the induction of MITFA resulting in melanogenesis by a subset of neural crest-derived cells [58]. Moreover, a recent study demonstrates that HDAC3 plays a critical and specific regulatory role in the neural crest-derived smooth muscle lineage and in the formation of the cardiac outflow tract [59]. The authors suggest that HDAC3 may be required for Notch-mediated upregulation of Jagged1 and subsequent smooth muscle differentiation in the aortic arch arteries. Thus, HDAC inhibitors may be useful for modulation of vascular

smooth muscle behavior—for example, in patient populations with coronary disease.

HDAC4 is a member of the class II histone deacetylases that play a key role in human development. Its deficiency has been associated with non-syndromic oral clefts and brachydactyly mental retardation syndrome (BDMR) [60,61]. In addition, HDAC4 is widely present in migratory NC cells in zebrafish, particularly in those that contribute to the palatal skeleton. HDAC4 knockdown causes a severe reduction in cranial NC migration and shortened face due to defects in the palatal skeleton [62].

Histone acetylation not only acts directly on chromatin structure but also facilitates binding of bromodomain proteins, which function to recruit transcriptional activators or repressors [44]. Along these lines, our recent work showed that a plant-homeodomain 12 (PHD12) protein, containing two PHD domains and a bromodomain, is expressed by premigratory NC cells [63]. Interestingly, PHD12 interacts with Sin3A, which in turn interacts with Snail2, recruiting the complex to the Cad6b regulatory region. This results in a dramatic histone H3 lysine deacetylation on the Cad6b transcription start site necessary for its repression and for the subsequent NC epithelial-to-mesenchymal transition. Thus, PHD12 reads the epigenetic code (acetylated H3) and Snail2 the DNA marks (E-boxes) to recruit a repressive complex. This dual specifictiy confers an additional fine level of regulation during NC development [63].

Histone Variants

In addition to the canonical histone proteins, many variant histone forms exist in different organisms [12]. These are usually present as single-copy genes and are expressed throughout the cell cycle. Unlike the major histone subtypes, the variant genes contain introns and their transcripts are often polyadenylated, giving them the capacity to be post-transcriptionally regulated.

Some variants exchange with preexisting core histones during development and differentiation, and are therefore referred to as replacement histones. This replacement is an easy way for cells to "erase" certain epigenetic marks and other chromatin-associated proteins. However, little is known about the transcriptional activity present at certain loci where such histone variants are localized during development. Interestingly, recent studies in zebrafish have demonstrated an essential role for the histone variant H3.3 to confer ectomesenchyme potential to cranial NC cells [64]. Specifically, a zebrafish dominant mutation in h3f3a, one of five variant histone H3.3 genes, eliminates head skeleton and a subset of pigment cells while other cranial and trunk NC derivatives remain intact. Futures studies are likely to reveal the target genes of this or other histone replacements and how these might regulate the transcriptional profile of different NC progenitor populations.

CHROMATIN MODIFIERS

During NC development, cells gradually restrict their differentiation potential to produce specialized cell types, tissues, and organs. Recent studies have demonstrated that unique chromatin states are associated with retained or restricted differentiation potential [65]. It seems likely that a similar situation occurs during NC cell development.

The chromatin state is generally regulated by chromatin-remodeling complexes involving enzymes that transiently disrupt the association between DNA and histones in an ATP-dependent manner. This, in turn, may induce conformational changes in nucleosomes and control different degrees of condensation to the chromatin. Even though little is known about mechanisms of chromatin regulation during NC formation, a recent study has shown that CHD7 (chromodomain helicase DNA-binding domain), which catalyzes nucleosome sliding on DNA,

is essential for activation of core components of the NC transcriptional circuitry, including Sox9, Twist, and Snail2 [66]. Specifically, CHD7 is able to associate with PBAF a SWI/SNF family chromatin-remodeling complex, and both co-occupy a distal enhancer marked by H3K4me1 orchestrating NC gene expression programs. Importantly, mutations of the CHD7 gene were identified as the major cause of CHARGE syndrome [67], an autosomal-dominant malformation syndrome in which several organs and systems are variably involved. The clinical characteristics include coloboma of the eye, heart malformations, atresia of choanae, retardation of growth and development, genital hypoplasia, and ear anomalies. This syndrome affects about 1 child of 10,000 born and leads to death before the age of 5 in approximately 30% of the cases.

Williams syndrome transcription factor (WSTF) is one of approximately 25 contiguous genes, presented on chromosome 7, that are haplodeficient in individuals with the complex developmental disorder Williams syndrome (WS) [68]. These patients exhibit developmental delays, infantile hypercalcemia, mental retardation, and characteristic malformations of craniofacial, heart, and neural structures. WSTF is a core component of three functionally distinct chromatin-remodeling complexes, including ISWI, which has been implicated in DNA repair, replication, transcriptional activation, and repression, and also appears to have histone H2A kinase activity [69]. WSTF is expressed in a wide variety of tissues [68], including NC cells [70], during development. Consistent with this, WSTF-null mice exhibit heart and craniofacial defects [71]. Moreover, WSTF-depleted *Xenopus* embryos have a severe defect in NC migration and/or maintenance [72], though induction and specification appear normal. Taken together, these results support the idea that NC defects, resulting from WSTF haploinsufficiency, may be one of the major contributors to symptoms of individuals with WS.

DNA METHYLATION

DNA methylation is one of the most important epigenetic regulatory mechanisms during mammalian development, including events such as genomic imprinting, X-chromosome inactivation, genomic stability, and regulation of gene expression [73,74]. After fertilization, repressive parental DNA methylation is erased or gradually lost, reprogramming the cells of the early embryo to enable reacquisition of pluripotency [75]. As development proceeds, new DNA methylation marks are established to gradually restrict the cell's potential [75–77]. Generally, most mammalian genomes are highly methylated except at active or "poised" promoters, enhancers, and CpG islands, where DNA methylation has a repressive effect (Figure 5.3). DNA methyltransferases (DNMTs) are the enzymes in charge of DNA methylation, catalyzing transfer of a methyl group to the cytosine residues on DNA [78]. There are three DNMT families in vertebrates: DNMT1, DNMT2, and DNMT3 [79]. DNMT1 is implicated in maintaining existing methylation patterns as well as in the regulation of histone methylation [80]. The functional importance of DNMT2 in vertebrates remains largely unclear. However, a recent study in zebrafish has shown its involvement in cytoplasmic RNA methylation [80]. On the other hand, both DNMT3A and 3B have been shown to be essential for *de novo* methylation throughout embryonic cell differentiation [76,79,81]. Although DNMT3A and DNMT3B have overlapping functions, they seem to play distinctive roles during development. Interestingly, the knockout of both genes in mES cells reveals common as well as distinct DNA targets [81,82]. Whereas DNMT3A-null mice die several weeks after birth, DNMT3B-null embryos have rostral neural tube defects and growth impairment [82] but are viable. Recent work in avian embryos reveals that DNMT3A acts as a molecular switch mediating the neural tube-to-neural crest fate transition [83]. DNMT3A promotes neural crest specification by repressing, via promoter DNA methylation, neural genes like Sox2 and Sox3 in the neural folds. Moreover, DNMT3A knockdown causes ectopic Sox2 and Sox3 expression in the NC territory, which in turn represses neural crest genes.

On the other hand, DNMT3B also seems to be important for normal NC development. High levels of DNMT3B are observed in neuroepithelial cells of the early neural plate and dorsal spinal cord [84–86]. Moreover, mutations in human DNMT3B result in ICF (immunodeficiency, centromeric instability, and facial anomalies) syndrome, in which patients exhibit facial abnormalities [87], suggesting a defect related to abnormal NC development. Likewise, mice carrying DNMT3B mutations, similar to ICF syndrome, survive to term and exhibit craniofacial anomalies that resemble human patients [88]. Furthermore, depletion of DNMT3B in hES cells results in a hypomethylation of pericentromeric regions and enhanced expression of regional specifier genes that are important for neural identity and neural crest lineages [89]. However, a recent report in mice using NC-specific conditional deletion of DNMT3B found normal neural crest migration and differentiation into craniofacial and cardiac structures [90]. One possibility is that the NC defect obtained in other DNMT3B mutant embryos [82,88] may be due to a requirement for this protein in neighboring cell types.

CONCLUSIONS AND PERSPECTIVES

Abnormal NC development results in a myriad of birth defects and diseases termed neurocristopathies. Some neurocristopathies occur as tumors, and others as malformations or lesions. Tumors derived from NC cells include melanoma, neuroblastoma, neurofibroma, medullary thyroid cancer, and pheochromocytoma. Neural crest malformations include albinism, cleft lip and/or palate, aganglionic megacolon (Hirschsprung's disease), conotruncal heart malformations, congenital nevi, and craniofacial abnormalities associated with fetal alcohol syndrome. Although some of these have been well-studied clinically, the complex molecular mechanisms associated with these conditions remain largely unknown.

NCC display enormous plasticity, and it is becoming evident that epigenetic mechanisms play a major role in conferring this plasticity during development. Presently, we are uncovering the "tip of the iceberg" in terms of understanding the contributions of chromatin regulatory mechanisms in NC development. This promises to be an important and fertile area of future investigation. Moreover, knowledge regarding epigenetic contributions toward particular cell fate choices during NC differentiation may have far-reaching clinical implications for treating different neurocristopathies.

Acknowledgments

Parts of the work described here were funded by DE16459 and HD037105 grants to MEB and a Fogarty grant (R03 DE022521) to MEB and PS-M.

References

[1] Aybar MJ, Mayor R. Early induction of neural crest cells: lessons learned from frog, fish and chick. Curr Opin Genet Dev 2002;12(4):452−8.

[2] Betancur P, Bronner-Fraser M, Sauka-Spengler T. Assembling neural crest regulatory circuits into a gene regulatory network. Annu Rev Cell Dev Biol 2010;26:581−603.

[3] Mayor R, Young R, Vargas A. Development of neural crest in Xenopus. Curr Top Dev Biol 1999;43:85−113.

[4] Meulemans D, Bronner-Fraser M. Gene−regulatory interactions in neural crest evolution and development. Dev Cell 2004;7(3):291−9.

[5] Sauka-Spengler T, Bronner-Fraser M. A gene regulatory network orchestrates neural crest formation. Nat Rev Mol Cell Biol 2008;9(7):557−68.

[6] Steventon B, Carmona-Fontaine C, Mayor R. Genetic network during neural crest induction: from cell specification to cell survival. Semin Cell Dev Biol 2005;16 (6):647−54.

[7] Waddington CH. The strategy of the genes; a discussion of some aspects of theoretical biology. London; 1957.

[8] Goldberg AD, Allis CD, Bernstein E. Epigenetics: a landscape takes shape. Cell 2007;128(4):635−8.

[9] Jaenisch R, Bird A. Epigenetic regulation of gene expression: how the genome integrates intrinsic and environmental signals. Nat Genet 2003;33(Suppl): 245−54.

[10] Sims 3rd RJ, Mandal SS, Reinberg D. Recent highlights of RNA-polymerase-II-mediated transcription. Curr Opin Cell Biol 2004;16(3):263−71.

[11] Strobl-Mazzulla PH, Marini M, Buzzi A. Epigenetic landscape and miRNA involvement during neural crest development. Dev Dyn 2012;241(12):1849−56.

[12] Kamakaka RT, Biggins S. Histone variants: deviants? Genes Dev 2005;19(3):295−310.

[13] Berger SL. The complex language of chromatin regulation during transcription. Nature 2007;447(7143): 407−12.

[14] Gibney ER, Nolan CM. Epigenetics and gene expression. Heredity (Edinb) 2010;105(1):4−13.

[15] Kouzarides T. Chromatin modifications and their function. Cell 2007;128(4):693−705.

[16] Allfrey VG, Faulkner R, Mirsky AE. Acetylation and methylation of histones and their possible role in the regulation of RNA synthesis. Proc Natl Acad Sci USA 1964;51:786−94.

[17] Tan M, Luo H, Lee S, Jin F, Yang JS, Montellier E, et al. Identification of 67 histone marks and histone lysine crotonylation as a new type of histone modification. Cell 2011;146(6): 1016−28.

[18] Luan Z, Liu Y, Stuhlmiller TJ, Marquez J, Garcia-Castro MI. SUMOylation of Pax7 is essential for neural crest and muscle development. Cell Mol Life Sci CMLS 2013;70(10):1793−806.

[19] Taylor KM, Labonne C. SoxE factors function equivalently during neural crest and inner ear development and their activity is regulated by SUMOylation. Dev Cell 2005;9(5):593−603.

[20] Martin C, Zhang Y. The diverse functions of histone lysine methylation. Nat Rev Mol Cell Biol 2005;6 (11):838−49.

[21] Jenuwein T, Allis CD. Translating the histone code. Science 2001;293(5532):1074−80.

[22] Strahl BD, Allis CD. The language of covalent histone modifications. Nature 2000;403(6765):41−5.

[23] Turner BM. Histone acetylation and an epigenetic code. Bioessays 2000;22(9):836−45.

[24] Barski A, Cuddapah S, Cui K, Roh TY, Schones DE, Wang Z, et al. High-resolution profiling of histone methylations in the human genome. Cell 2007;129 (4):823−37.

[25] Cheung I, Shulha HP, Jiang Y, Matevossian A, Wang J, Weng Z, et al. Developmental regulation and

individual differences of neuronal H3K4me3 epigenomes in the prefrontal cortex. Proc Natl Acad Sci USA 2010;107(19):8824−9.

[26] Pan G, Tian S, Nie J, Yang C, Ruotti V, Wei H, et al. Whole-genome analysis of histone H3 lysine 4 and lysine 27 methylation in human embryonic stem cells. Cell Stem Cell 2007;1(3):299−312.

[27] Liu T, Rechtsteiner A, Egelhofer TA, Vielle A, Latorre I, Cheung MS, et al. Broad chromosomal domains of histone modification patterns in C. elegans. Genome Res 2011;21(2):227−36.

[28] Schwartz YB, Kahn TG, Nix DA, Li XY, Bourgon R, Biggin M, et al. Genome-wide analysis of Polycomb targets in Drosophila melanogaster. Nat Genet 2006;38 (6):700−5.

[29] Tolhuis B, de Wit E, Muijrers I, Teunissen H, Talhout W, van Steensel B, et al. Genome-wide profiling of PRC1 and PRC2 Polycomb chromatin binding in Drosophila melanogaster. Nat Genet 2006;38(6):694−9.

[30] Simon JA, Kingston RE. Mechanisms of polycomb gene silencing: knowns and unknowns. Nat Rev Mol Cell Biol 2009;10(10):697−708.

[31] Bernstein BE, Mikkelsen TS, Xie X, Kamal M, Huebert DJ, Cuff J, et al. A bivalent chromatin structure marks key developmental genes in embryonic stem cells. Cell 2006;125(2):315−26.

[32] Rada-Iglesias A, Bajpai R, Swigut T, Brugmann SA, Flynn RA, Wysocka J. A unique chromatin signature uncovers early developmental enhancers in humans. Nature 2011;470(7333):279−83.

[33] Rada-Iglesias A, Bajpai R, Prescott S, Brugmann SA, Swigut T, Wysocka J. Epigenomic annotation of enhancers predicts transcriptional regulators of human neural crest. Cell Stem Cell 2012;11(5):633−48.

[34] Herz HM, Shilatifard A. The JARID2-PRC2 duality. Genes Dev 2010;24(9):857−61.

[35] Swigut T, Wysocka J. H3K27 demethylases, at long last. Cell 2007;131(1):29−32.

[36] Nottke A, Colaiacovo MP, Shi Y. Developmental roles of the histone lysine demethylases. Development 2009; 136(6):879−89.

[37] Lindeman LC, Andersen IS, Reiner AH, Li N, Aanes H, Ostrup O, et al. Prepatterning of developmental gene expression by modified histones before zygotic genome activation. Dev Cell 2011;21(6):993−1004.

[38] Ho L, Crabtree GR. Chromatin remodelling during development. Nature 2010;463(7280):474−84.

[39] Strobl-Mazzulla PH, Sauka-Spengler T, Bronner-Fraser M. Histone demethylase JmjD2A regulates neural crest specification. Dev Cell 2010;19(3):460−8.

[40] Carrozza MJ, Utley RT, Workman JL, Cote J. The diverse functions of histone acetyltransferase complexes. Trends Genet TIG 2003;19(6):321−9.

[41] Hsieh J, Nakashima K, Kuwabara T, Mejia E, Gage FH. Histone deacetylase inhibition-mediated neuronal differentiation of multipotent adult neural progenitor cells. Proc Natl Acad Sci USA 2004;101(47):16659−64.

[42] Ekwall K. Genome-wide analysis of HDAC function. Trends Genet TIG 2005;21(11):608−15.

[43] Wang Z, Zang C, Cui K, Schones DE, Barski A, Peng W, et al. Genome-wide mapping of HATs and HDACs reveals distinct functions in active and inactive genes. Cell 2009;138(5):1019−31.

[44] Grunstein M. Histone acetylation in chromatin structure and transcription. Nature 1997;389(6649):349−52.

[45] Yang XJ, Seto E. Lysine acetylation: codified crosstalk with other posttranslational modifications. Mol Cell 2008;31(4):449−61.

[46] Knutson SK, Chyla BJ, Amann JM, Bhaskara S, Huppert SS, Hiebert SW. Liver-specific deletion of histone deacetylase 3 disrupts metabolic transcriptional networks. EMBO J 2008;27(7):1017−28.

[47] Dovey OM, Foster CT, Cowley SM. Histone deacetylase 1 (HDAC1), but not HDAC2, controls embryonic stem cell differentiation. Proc Natl Acad Sci USA 2010;107(18):8242−7.

[48] Montgomery RL, Davis CA, Potthoff MJ, Haberland M, Fielitz J, Qi X, et al. Histone deacetylases 1 and 2 redundantly regulate cardiac morphogenesis, growth, and contractility. Genes Dev 2007;21(14):1790−802.

[49] Ye F, Chen Y, Hoang T, Montgomery RL, Zhao XH, Bu H, et al. HDAC1 and HDAC2 regulate oligodendrocyte differentiation by disrupting the beta-catenin-TCF interaction. Nat Neurosci 2009; 12(7):829−38.

[50] Haberland M, Mokalled MH, Montgomery RL, Olson EN. Epigenetic control of skull morphogenesis by histone deacetylase 8. Genes Dev 2009;23(14):1625−30.

[51] Lagger G, O'Carroll D, Rembold M, Khier H, Tischler J, Weitzer G, et al. Essential function of histone deacetylase 1 in proliferation control and CDK inhibitor repression. EMBO J 2002;21(11):2672−81.

[52] Menegola E, Di Renzo F, Broccia ML, Giavini E. Inhibition of histone deacetylase as a new mechanism of teratogenesis. Birth Defects Res Part C Embryo Today Rev 2006;78(4):345−53.

[53] Duvic M, Vu J. Vorinostat: a new oral histone deacetylase inhibitor approved for cutaneous T-cell lymphoma. Expert Opin Invest Drugs 2007;16(7):1111−20.

[54] Kavanaugh SM, White LA, Kolesar JM. Vorinostat: a novel therapy for the treatment of cutaneous T-cell lymphoma. Am J Health-Syst Pharm AJHP: Off J Am Soc Health-Syst Pharm 2010;67(10):793−7.

[55] Alsdorf R, Wyszynski DF. Teratogenicity of sodium valproate. Expert Opin Drug Safety 2005;4(2):345−53.

[56] Wyszynski DF, Nambisan M, Surve T, Alsdorf RM, Smith CR, Holmes LB, et al. Increased rate of major

malformations in offspring exposed to valproate during pregnancy. Neurology 2005;64(6): 961–5.

[57] Papi A, Ferreri AM, Rocchi P, Guerra F, Orlandi M. Epigenetic modifiers as anticancer drugs: effectiveness of valproic acid in neural crest-derived tumor cells. Anticancer Res 2010;30(2):535–40.

[58] Ignatius MS, Moose HE, El-Hodiri HM, Henion PD. colgate/hdac1 repression of foxd3 expression is required to permit mitfa-dependent melanogenesis. Dev Biol 2008;313(2):568–83.

[59] Singh N, Trivedi CM, Lu M, Mullican SE, Lazar MA, Epstein JA. Histone deacetylase 3 regulates smooth muscle differentiation in neural crest cells and development of the cardiac outflow tract. Circ Res 2011;109 (11):1240–9.

[60] Park JW, Cai J, McIntosh I, Jabs EW, Fallin MD, Ingersoll R, et al. High throughput SNP and expression analyses of candidate genes for non-syndromic oral clefts. J Med Genet 2006;43(7): 598–608.

[61] Williams SR, Aldred MA, Der Kaloustian VM, Halal F, Gowans G, McLeod DR, et al. Haploinsufficiency of HDAC4 causes brachydactyly mental retardation syndrome, with brachydactyly type E, developmental delays, and behavioral problems. Am J Hum Genet 2010;87(2):219–28.

[62] Delaurier A, Nakamura Y, Braasch I, Khanna V, Kato H, Wakitani S, et al. Histone deacetylase-4 is required during early cranial neural crest development for generation of the zebrafish palatal skeleton. BMC Dev Biol 2012;12(1):16.

[63] Strobl-Mazzulla PH, Bronner ME. A PHD12-Snail2 repressive complex epigenetically mediates neural crest epithelial-to-mesenchymal transition. J Cell Biol 2012;198(6):999–1010.

[64] Cox SG, Kim H, Garnett AT, Medeiros DM, An W, Crump JG. An essential role of variant histone H3.3 for ectomesenchyme potential of the cranial neural crest. PLoS Genet 2012;8(9):e1002938.

[65] Mohn F, Schubeler D. Genetics and epigenetics: stability and plasticity during cellular differentiation. Trends Genet TIG 2009;25(3):129–36.

[66] Bajpai R, Chen DA, Rada-Iglesias A, Zhang J, Xiong Y, Helms J, et al. CHD7 cooperates with PBAF to control multipotent neural crest formation. Nature 2010;463(7283):958–62.

[67] Vissers LE, van Ravenswaaij CM, Admiraal R, Hurst JA, de Vries BB, Janssen IM, et al. Mutations in a new member of the chromodomain gene family cause CHARGE syndrome. Nat Genet 2004;36 (9):955–7.

[68] Lu X, Meng X, Morris CA, Keating MT. A novel human gene, WSTF, is deleted in Williams syndrome. Genomics 1998;54(2):241–9.

[69] Barnett C, Krebs JE. WSTF does it all: a multifunctional protein in transcription, repair, and replication. Biochem Cell Biol 2011;89(1):12–23.

[70] Cus R, Maurus D, Kuhl M. Cloning and developmental expression of WSTF during Xenopus laevis embryogenesis. Gene Expr Patterns 2006;6(4):340–6.

[71] Yoshimura K, Kitagawa H, Fujiki R, Tanabe M, Takezawa S, Takada I, et al. Distinct function of 2 chromatin remodeling complexes that share a common subunit, Williams syndrome transcription factor (WSTF). Proc Natl Acad Sci USA 2009;106(23): 9280–5.

[72] Barnett C, Yazgan O, Kuo HC, Malakar S, Thomas T, Fitzgerald A, et al. Williams syndrome transcription factor is critical for neural crest cell function in Xenopus laevis. Mech Dev 2012;129(9-12):324–38.

[73] Bird A. DNA methylation patterns and epigenetic memory. Genes Dev 2002;16(1):6–21.

[74] Ooi SK, O'Donnell AH, Bestor TH. Mammalian cytosine methylation at a glance. J Cell Sci 2009;122(Pt 16):2787–91.

[75] Mayer W, Niveleau A, Walter J, Fundele R, Haaf T. Demethylation of the zygotic paternal genome. Nature 2000;403(6769):501–2.

[76] Reik W. Stability and flexibility of epigenetic gene regulation in mammalian development. Nature 2007;447 (7143):425–32.

[77] Borgel J, Guibert S, Li Y, Chiba H, Schubeler D, Sasaki H, et al. Targets and dynamics of promoter DNA methylation during early mouse development. Nat Genet 2010;42(12):1093–100.

[78] Cheng X, Blumenthal RM. Mammalian DNA methyltransferases: a structural perspective. Structure 2008;16(3):341–50.

[79] Goll MG, Bestor TH. Eukaryotic cytosine methyltransferases. Ann Rev Biochem 2005;74:481–514.

[80] Rai K, Nadauld LD, Chidester S, Manos EJ, James SR, Karpf AR, et al. Zebra fish Dnmt1 and Suv39h1 regulate organ-specific terminal differentiation during development. Mol Cell Biol 2006;26 (19):7077–85.

[81] Chen T, Ueda Y, Dodge JE, Wang Z, Li E. Establishment and maintenance of genomic methylation patterns in mouse embryonic stem cells by Dnmt3a and Dnmt3b. Mol Cell Biol 2003;23(16):5594–605.

[82] Okano M, Bell DW, Haber DA, Li E. DNA methyltransferases Dnmt3a and Dnmt3b are essential for de novo methylation and mammalian development. Cell 1999;99(3):247–57.

[83] Hu N, Strobl-Mazzulla P, Sauka-Spengler T, Bronner ME. DNA methyltransferase3A as a molecular switch mediating the neural tube-to-neural crest fate transition. Genes Dev 2012;26(21):2380–5.

[84] Hirasawa R, Sasaki H. Dynamic transition of Dnmt3b expression in mouse pre- and early post-implantation embryos. Gene Expr Patterns 2009;9(1):27−30.

[85] Watanabe D, Suetake I, Tada T, Tajima S. Stage- and cell-specific expression of Dnmt3a and Dnmt3b during embryogenesis. Mech Dev 2002;118(1-2):187−90.

[86] Watanabe D, Uchiyama K, Hanaoka K. Transition of mouse de novo methyltransferases expression from Dnmt3b to Dnmt3a during neural progenitor cell development. Neuroscience 2006;142(3):727−37.

[87] Jin B, Tao Q, Peng J, Soo HM, Wu W, Ying J, et al. DNA methyltransferase 3B (DNMT3B) mutations in ICF syndrome lead to altered epigenetic modifications and aberrant expression of genes regulating development, neurogenesis and immune function. Hum Mol Genet 2008;17(5):690−709.

[88] Ueda Y, Okano M, Williams C, Chen T, Georgopoulos K, Li E. Roles for Dnmt3b in mammalian development: a mouse model for the ICF syndrome. Development 2006;133(6):1183−92.

[89] Martins-Taylor K, Schroeder DI, Lasalle JM, Lalande M, Xu RH. Role of DNMT3B in the regulation of early neural and neural crest specifiers. Epigenetics 2012;7(1):71−82.

[90] Jacques-Fricke BT, Roffers-Agarwal J, Gammill LS. DNA methyltransferase 3b is dispensable for mouse neural crest development. PLoS One 2012;7(10): e47794.

Neural Crest-Mediated Tissue Interactions During Craniofacial Development: The Origins of Species-Specific Pattern

Jennifer L. Fish and Richard A. Schneider

Neural Crest Cells.
DOI: http://dx.doi.org/10.1016/B978-0-12-401730-6.00007-7

INTRODUCTION

From Where Does Craniofacial Pattern Come?

The craniofacial complex represents one of the most highly diversified and evolutionarily adapted anatomical aspects of vertebrates. Individual components within the craniofacial complex can change rapidly over time and frequently are finely tuned to meet ecological and functional demands with phenomenal precision. Many studies on the origins of species-specific pattern in the craniofacial complex have focused on the role of external factors such as the natural environment or feeding behavior in directing the course of evolution. In contrast, less is known about molecular and cellular mechanisms that generate species-specific differences. However, recent work in this area is beginning to offer a clearer picture of how the craniofacial complex acquires species-specific pattern and, importantly, such work offers fundamental insights on how developmental programs become modified internally to provide the variation necessary for evolution.

Much has been postulated about the source of pattern during craniofacial development. In a way, asking the question, "From where does pattern come?" deserves a comedic answer like, "Who's on first, what's on second, and I don't know is on third." This is because the ability to name and define the source of craniofacial pattern has long been confounded by numerous and often unidentified players that switch positions and roles throughout embryonic time and space. On the one hand, pattern is an emergent property of dynamic and complex developmental systems, which by their very nature are constantly changing, progressing, interacting, and morphing. Thus, we should not expect to find the source of pattern in any single player, place, or event. But, on the other hand, pattern is something that repeats itself with a high degree of fidelity, can be inherited, and evolves. Thus,

there should be a way to identify the major determinants of pattern and see where, when, and how they become implemented and/or altered over time. Confusion also comes from the fact that pattern is a relative term, and there are two main types that need to be considered in the context of craniofacial development. First, there is general anatomical pattern derived from mechanisms that provide axial orientation (e.g., dorsal—ventral, rostral—caudal, proximal—distal, medial—lateral), histological nature (e.g., bone, cartilage, tendon, muscle), and structural identity (e.g., upper jaw versus lower jaw, eye versus ear, feather versus scale). Second, there is species-specific pattern derived from mechanisms that primarily superimpose parameters for size and shape onto general anatomical pattern. This chapter focuses mainly on the basis for species-specific pattern, but there is also some attention given to how such mechanisms differ from, relate to, or become integrated with processes that underlie general anatomical pattern.

Origins and Functions of Craniofacial Tissues

Like other parts of the vertebrate body plan, the craniofacial complex is composed of epithelial and mesenchymal tissues derived from three germ layers (i.e., ectoderm, mesoderm, and endoderm). Reciprocal and unidirectional interactions among these tissues allow the craniofacial complex to achieve structural and functional integration [1] (Figure 6.1). Epithelia and mesenchyme play distinct patterning and structural roles during craniofacial morphogenesis. Epithelia are arranged as polarized sheets of tightly connected cells and include the endoderm-derived lining of the pharynx and ectoderm-derived neural tube, which are both important signaling centers and sources of craniofacial pattern [3–20]. Epithelial tissues derived from surface (i.e., non-neural) ectoderm include placodes that produce tissues like olfactory and inner ear epithelium as well as cranial nerve ganglia [21–27]. Ectoderm-derived epithelia also include epidermis, which becomes stratified into multiple layers [28–30]. For example, the upper layer of mature epidermis is composed of the nonliving stratum corneum, which typically produces keratinized portions of structures such as hair, horns, feathers, scales, beaks, and egg teeth [6,22,31–34]. A thin basement membrane separates epidermis from underlying dermis, which is derived from mesenchyme.

Mesenchyme is defined as loosely associated stellate-shaped cells, which in the trunk and caudal regions of the head arise from mesoderm and in the face and portions of the neck mainly come from cranial neural crest [35–41]. Neural crest-derived mesenchyme originates within the dorsal margins of the neural tube during neurulation and migrates extensively through out the craniofacial complex [42,43]. Neural crest-derived

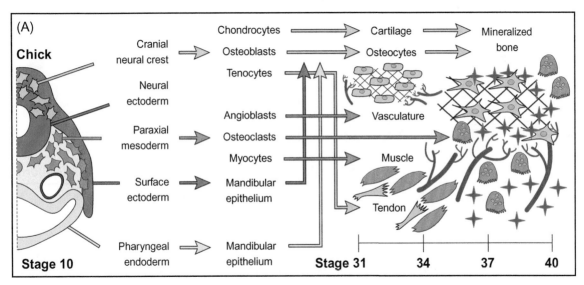

FIGURE 6.1 **Lineages, cell types, and tissue interaction in the craniofacial complex.** (A) Schematic hemi-transverse section through the pharyngeal region of a stage 10 chick showing germ layers, major cell lineages, and tissue interactions during mandible development. NCC (light blue) give rise to chondrocytes that make cartilage, osteocytes that make bone, and tenocytes that make tendon. Paraxial mesoderm (orange) gives rise to angioblasts that make blood vessels, osteoclasts that resorb bone, and myocytes that make muscle. Surface ectoderm (brown) and pharyngeal endoderm (yellow) generate mandibular epithelia that interact (arrows) with neural crest derivatives and provide general anatomical pattern. Overt differentiation of craniofacial tissues occurs by stages 31–40. (B) At stage 9.5 (dorsal view) NCC (light blue) migrate from the forebrain (fb), midbrain (mb), and hindbrain rhombomeres (r). NCC migrate alongside paraxial mesoderm (m; orange). (C) By HH25, the frontonasal (fn), maxillary (mx), mandibular (ma), and hyoid (hy) primordia (sagittal view) are surrounded by surface ectoderm (se), pharyngeal endoderm (pe), and forebrain neuroepithelium (fb) and contain contributions from neural crest, nasal placode (np), and cranial ganglia (V, VII, IX). Mesoderm (m) that produces skeletal tissues is distributed caudally. (D) The facial primordia give rise to the upper and lower portions of the beak as shown in a quail embryo. (E) By HH40 in chick, NCC produce the facial and jaw skeletons (light blue) whereas mesoderm forms the caudal cranial vault and skull base (orange). (F) The boundary in the skull between bones and cartilages derived from neural crest (light blue) versus mesoderm (orange) is highly conserved as shown in the mouse. *Source: Panels B–E modified from Ref. [2]; F modified from Ref. [1].*

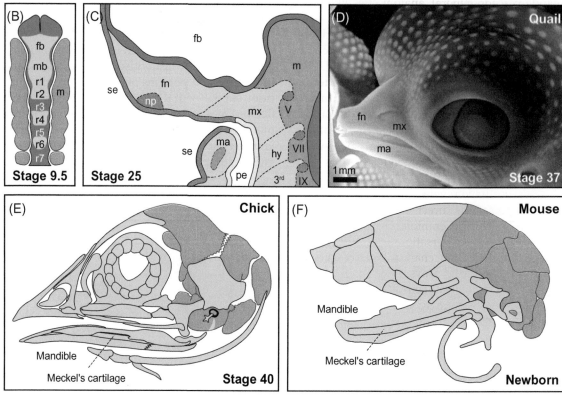

FIGURE 6.1   (Continued)

mesenchyme gives rise to a variety of tissues, including bone and cartilage of the skull, and muscle connective tissues such as tendon [35,44–47]. Neural crest cells (NCC) also make sensory neurons and glia, Schwann cells, pericytes, and pigment-producing melanocytes that infiltrate epidermis and become the source of color in hair, skin, beaks, and feathers [48–53]. Like neural crest-derived mesenchyme, mesodermal mesenchyme gives rise to bone and cartilage, but only in caudal portions of the head [54–57]. Yet mesoderm also uniquely generates components such as vascular endothelium, osteoclasts, and voluntary muscles [38,45,58,59].

For the most part, epithelia that surround the developing jaws and face (e.g., frontonasal,

maxillary, mandibular primordia) appear to supply positional cues and maintenance factors required for patterned outgrowth of individual components. For example, signaling by ectodermal epithelium around the frontonasal primordium (i.e., mid-face and upper face) is required for proper expansion and orientation of skeletal elements along the proximodistal, mediolateral, and dorsoventral axes [60–62]. Likewise, endodermal epithelium that lines the pharynx is necessary for growth and orientation of bone and cartilage in the jaw skeleton [13,63]. When either of these epithelia is surgically rotated, the underlying neural crest-derived skeleton follows suit. Thus, epithelia of both ectodermal and endodermal origin function as local sources of signals for general anatomical pattern that elicit and/or maintain programmatic responses from underlying neural crest-derived mesenchyme [64–71].

Tissue Interactions and the Evolution of Species-Specific Pattern

A broad range of experiments designed to study and identify critical signaling interactions among cranial NCC and neighboring embryonic tissues have reveled that active and persistent conversations direct most, if not all, of the key events during craniofacial morphogenesis. Altering these conversations on the molecular or cellular level can change three-dimensional pattern in profound ways, which helps explain why the craniofacial complex is both highly evolvable and extremely susceptible to developmental defects [72]. While epithelia and mesenchyme typically work together to produce composite structures, mesenchyme, and, in particular, cranial neural crest-derived mesenchyme seems to act as the dominant source of species-specific patterning information [1,73–81]. In this capacity, cranial neural crest has likely functioned as a responsive target of natural selection and a driving force for generating species-specific variation [82]. Moreover, in order for species to evolve, their underlying developmental programs must display plasticity [83–85], and this appears to be a defining characteristic of programs that utilize mesenchyme, especially cranial neural crest-derived mesenchyme [70,77,82,86]. Such plasticity has clearly enhanced the evolutionary potential (i.e., adaptability) of the pharyngeal and rostral portions of the vertebrate head [87–93], as well as the lateral wall of the skull [94,95].

Although there is substantial conservation in the general anatomical organization of neural crest derivatives, such as bone and cartilage, these structures also exhibit great diversity in species-specific size and shape. In fact, some of the most dramatic and innovative evolutionary changes to the craniofacial skeleton have occurred in those rostral regions derived primarily from neural crest mesenchyme. As noted above, cells that form the back and base of the skull come from mesoderm, and thus there is a boundary that separates the portion of the skull derived from neural crest and that from mesoderm (Figure 6.1). This boundary may represent a fundamental evolutionary landmark that distinguishes the rostral "new" head and caudal "old" head of vertebrates [46,87,96]. While the interface between neural crest and mesodermal mesenchyme is identifiable only when using lineage-specific markers such as vital dyes, radioactive and fluorescent labels, interspecific transplantations, or reporter gene constructs [37,43,54–56,97–103], the way these dual mesenchymal lineages become distributed in the craniofacial complex remains remarkably regular and consistent [1,104,105]. Interestingly, disruption of the mesoderm/neural crest boundary at the coronal suture causes craniosynostosis [106]. Ventrally, this boundary runs through the hypophyseal fenestra on either side of the pituitary gland [1,46] and delineates one of the most highly conserved regions of the skull across vertebrates [95]. Thomas Huxley concluded in his famous lecture "On the Theory of the Vertebrate Skull" that the "pituitary body may be regarded as marking the organic centre, as it were, of the skull—its relations to the axial cranial bones being the same, as far as I am aware, in all *Vertebrata*" [107]. Likewise, Goodrich [108] argued that the hypophyseal fenestra marks "a point of comparison in the head of all Craniates", a claim substantiated by other notable anatomists [109,110]. In stark contrast, the greatest size and shape changes that have occurred during vertebrate evolution appear to reside rostral to the hypophyseal fenestra in the neural crest-derived region of the jaws and skull, as most clearly exemplified by the varied faces of dogs and beaks of birds [82,111]. As described below, a common denominator that has enabled vertebrates to achieve astounding diversity in the size, shape, arrangement, and color of structures in their craniofacial complex is undeniably the cranial neural crest. Whether the topic is the jaw skeleton and its associated musculature, or integument and its various appendages, neural crest plays a pivotal role in providing species-specific patterning information primarily by mediating interactions with neighboring and participating tissues.

ORIGINS OF SPECIES-SPECIFIC PATTERN

Comparative Early Development of Neural Crest

When, where, and how during development do species-specific differences arise? For derivatives of neural crest, mechanisms that contribute to species-specific pattern—in particular, size and shape—could first come in the form of raw cellular materials as building blocks, and this could be deeply influenced by evolutionary changes to initial population size, timing and routes of migration, spatial allocation, and rates of proliferation. Such early cell biological determinants could ultimately bias, direct, or constrain the range of later morphological outcomes when NCC begin to assemble and grow at their final sites of differentiation. Species-specific pattern could also arise from external signals that differentially regulate neural crest along the way, as well as through regulatory mechanisms intrinsic to NCC themselves that define and control morphological outcomes by dominating interactions with adjacent tissues.

Cranial neural crest migration has been extensively described using multiple methods in placental mammals [112—120], marsupials [121], avians [42,52,122—132], amphibians [100,133—140], and bony fish [141—144]. These analyses indicate that cranial neural crest migration patterns are highly conserved in vertebrates [145,146]. Cranial neural crest migrates into the pharyngeal arches as three major streams (mandibular, hyoid, and branchial), and they also extend more rostrally into the frontonasal prominence. Conservation of cranial neural crest migration extends to the lamprey, an agnathan vertebrate, which shares cranial neural crest migration and Hox gene expression patterns with jawed vertebrates [147—149]. These data suggest that major migration patterns of cranial neural crest are highly conserved and may be constrained by developmental programs associated with coordination of muscle, nerve, and skeletal elements related to the evolution of the vertebrate head [87,91,130].

While patterns of cranial neural crest migration from the neural tube appear to be highly conserved across vertebrates, there is significant species-specific variation in the timing of the onset of migration [146,150]. In mouse and rat, cranial NCC start to migrate at headfold stages long before completion of neural tube closure. This is in contrast to chick, zebrafish, *Xenopus*, and axolotl, where the neural folds close prior to neural crest emigration [138,146,150—152]. Cranial neural crest migration is even earlier in marsupials, where the onset begins before any somites appear or regional differentiation exists in the neural tube [121,153]. Evolutionary alterations in the timing of cranial neural crest migration may be related to species precocity at birth and ultimately affect differential growth and evolutionary integration of neural crest derivatives with surrounding structures [154—158].

Cranial NCC are distinct from trunk NCC in that they normally generate bone and cartilage (i.e., ectomesenchymal derivatives), in addition to non-ectomesenchyme derivatives, such as neurons, glia, and pigment cells [43,51,70,159—164]. How the ectomesenchymal lineage becomes specified and differentiated from non-ectomesenchymal lineages is still greatly debated. A likely scenario involves species-specific differences in the deployment of an ectomesenchymal-specific gene regulatory network (GRN). Comparative developmental and gene expression analyses reveal that early chordates possessed migratory neural tube cells with the ability to generate non-ectomesenchymal derivatives (e.g., neurons and glia). More recent comparative data from chordates and jawed vertebrates indicate that cranial neural crest evolved an ectomesenchymal lineage by repeated co-option of genes expressed in other cell types, particularly mesoderm [165–170]. The gradual co-option of genes, especially transcription factors involved in proto-cartilage differentiation, may have been assembled into a cartilage GRN that is operated by cranial neural crest [168]. This suggests that the ectomesenchymal versus non-ectomesenchmyal lineage "switch" involves activation of a GRN, and that the timing of its onset and relative distribution of its activation within the total cranial neural crest population, may be subject to evolutionary modification within jawed vertebrates.

Cultured avian cranial NCC have been shown to generate both ectomesenchymal and non-ectomesenchymal derivatives clonally, pointing to a role for neural crest as a stem cell-like progenitor [171]. The fact that cranial neural crest can do so in culture supports the hypothesis that they are initially multipotent and that subsequent lineage restriction is mediated by signals that remain close to the neural tube (and thus would be absent from culture conditions). However, lineage-tracing experiments in zebrafish embryos have failed to identify a multipotent progenitor, instead signifying that cranial neural crest progenitors are fate-restricted prior to their migration from the neural tube [142]. Other experiments in zebrafish embryos indicate that specification of the ectomesenchymal lineage initiates at the onset of cranial neural crest delamination and is related to variance in activation of bone morphogenetic protein (BMP) signaling among cranial neural crest populations. In particular, the laterally positioned, earliest migrating

cranial neural crest appear to be specified as ectomesenchyme as they migrate away from BMP sources in the neural tube, whereas later-migrating cranial neural crest are specified as non-ectomesenchyme through a BMP-specific pathway [172]. Similarly, data from mouse embryos suggest that cranial neural crest ectomesenchyme derives from a subpopulation of early-migrating cranial neural crest, which are also laterally positioned in the dorsal neural fold and are the first to delaminate and migrate away from the neural tube [173]. In contrast, heterotopic transplants in avians indicate that late-migrating neural crest have equivalent potential to early-migrating neural crest [161]. Trunk NCC also have the potential to generate bone and cartilage when provided with the proper environmental signals [162]. These data suggest that neural crest maintain their mulitpotent potential through migratory stages, and that cell fate is ultimately determined by exposure to the right signal, rather than a lack of competence to respond.

Tissue Interactions that Program the Neural Crest

As NCC migrate to their ultimate destinations, they become more restricted and acquire certain patterning biases through a variety of sequential interactions with surrounding tissues. For example, physical contact with neuroepithelium at the time of migration and early exposure to molecular signals that regionalize the brain, such as fibroblast growth factor 8 (FGF8) at the midbrain/hindbrain boundary, provide NCC with positional information necessary for subsequent morphogenesis in the pharyngeal arches [17,174–176]. Similarly, interactions with paraxial mesoderm at the level of the neural tube [177,178], surface ectoderm and endoderm in the pharyngeal arches [13,70,104,139,179,180], and forebrain and facial ectoderm along the frontonasal process

[11,19,20,60–62,181–184] all affect the fate of NCC and are necessary to maintain proper histogenesis and morphogenesis of their derivatives. This is especially true for skeletogenesis, where endoderm and surface ectoderm provide vital signals, especially FGFs, BMPs, and sonic hedgehog (SHH) that positively regulate bone and cartilage [16,98,185–189].

Other patterning determinants appear to emerge from inside neural crest themselves and function through mechanisms that are spatially distinct. In caudal regions of the neural tube, NCC arise from different hindbrain rhombomeres where combinations of *Hox* genes and other transcription factors pattern the hyoid and subsequent arches [177,190]. In contrast, rostral NCC that populate the frontonasal, maxillary, and mandibular primordia arise from a *Hox*-free zone [13,174,191,192]. When these rostral populations of NCC are rotated 180° to transpose frontonasal and mandibular precursors, they can still produce normal facial and jaw skeletons, demonstrating that at the earliest stages of development cranial NCC have the capacity to generate almost any other structure within their range of derivatives [193]. If the *Hox* code is partially eliminated in populations of neural crest destined to form the hyoid arch either by surgically replacing hyoid-bound NCC with non-Hox-expressing mandibular or frontonasal neural crest, or by targeted knockout (i.e., *Hoxa2* mutants), then mandibular skeletal structures form in place of hyoid arch elements [193–195]. Conversely, forced expression of *Hoxa2* in mandibular arch neural crest precursors produces hyoid rather than mandibular skeletal structures [196,197]. Further demonstrating the importance of neural crest and neuroepithelial tissue interactions, *Hoxa2* is downregulated by FGF8, and when *Fgf8* is expressed ectopically in the hindbrain, hyoid arch structures are disrupted [198].

That NCC rely upon intrinsic transcription factor modules for establishing pattern is evidenced by experiments manipulating the *Dlx* and *endothelin* pathways, which transform maxillary versus mandibular identity [199–202]. Similarly, the maxillary primordia can be turned into a frontonasal process by exposing NCC to retinoic acid and the BMP antagonist Noggin [181], presumably by altering expression of key transcription factors within NCC themselves. Such data reveal that neural crest–mediated pattern is achieved in a site-specific manner because NCC establish the identity of the local environment by activating intrinsic patterns of gene expression and by regulating domains of secreted molecules within adjacent tissues. The net result is that cranial NCC at different axial levels exhibit differential degrees of plasticity and responsiveness during their generation and migration.

Neural Crest and the Origins of Species-Specific Pattern

As described above, ongoing interactions between NCC and surrounding tissues underlie proper histogenesis and morphogenesis of the craniofacial complex. These interactions are hierarchical, involve progenitor populations acting in both autonomous and non-autonomous manners, and have outcomes that are driven by the unique abilities of NCC, especially in regard to propagating species-specific pattern. In fact, the degree to which NCC convey intrinsic patterning information has been made most evident by comparing and manipulating domains of gene expression and by using surgically created chimeric embryos that exploit differences between species. For example, differential domains of gene expression in neural crest correlate with species-specific variations in size and shape of beaks among various bird embryos, including Darwin's finches, ducks, and cockatiels, and mimicking these species-specific expression patterns in chicken embryos affects cell proliferation and skeletal outgrowth, as well as producing variations in beak size and shape [203–205]. For example, augmenting spatiotemporal domains of *Bmp4* in mesenchyme of chick can expand the depth and width of the beak like that observed in duck embryos [204,206]. Similarly, overexpression of *Bmp4* in mice leads to expansion of nasal cartilage and loss, or reduction, in bone [207]. In cichlids that normally develop elongated jaws, *Bmp4* overexpression results in a shorter, deeper mandible than normal, phenocopying morphological differences associated with distinct feeding strategies [208]. These data highlight both similarity in gene function across taxa (broadening of the upper jaw) and differences in gene function in time and space (*Bmp4* affect on cartilage versus bone).

Besides candidate gene approaches, there have also been attempts to screen large numbers of genes simultaneously using high throughput genome-based strategies [2]. In an effort to identify novel genes that account for species-specific differences in beak morphology among Darwin's finches, a DNA microarray strategy was employed with tissues harvested from the Galápagos Islands [209]. Results show that a protein called calmodulin, which is a mediator of cellular calcium signaling, is expressed at higher levels in long and pointed finch species. When the calmodulin-dependent pathway is experimentally upregulated in frontonasal mesenchyme of chicken embryos, the upper beak becomes elongated. Thus, neural crest-derived mesenchyme mediates species-specific pattern by establishing species-specific spatial domains and levels of gene expression.

Schotte and Spemann first uncovered the origins of species-specific pattern in the 1930s by transplanting mouth-forming tissues between frogs and newts [210–212]. These classic experiments showed that general anatomical aspects of the mouth were directed by local signals, but that species-specific pattern was imposed by intrinsic information within donor cells. Apparently, Spemann explained his discovery like this, "The ectoderm says to the inducer, 'you tell me to make a mouth; all right, I'll do so, but I can't make your kind of mouth; I can make my own and I'll do that'" [213]. Subsequent transplant experiments between salamanders and frogs, and between mice and chicks, have also supported the notion that species-specific pattern (in this case for jaws and teeth) is largely driven by neural crest [68,73,74,214,215]. Equivalent conclusions have been made using divergent species of birds [75,76,79,81]. In particular, the quail-duck chimeric transplant system has offered insights into the molecular and cellular basis for species-specific pattern and has helped define mechanistic contributions of NCC to tissue interactions that are necessary for the structural and functional integration of the craniofacial complex (Figure 6.2). Making chimeras between quail and duck embryos capitalizes on three features that set these species of birds apart [215]. First, quail and duck embryos are distinct in terms of their overall size and shape, which offers a direct means to assess if morphological outcomes are mediated by donor- or host-derived tissues. Second, quail and duck embryos have highly divergent maturation rates (17 days versus 28 days), which provides a way to evaluate effects of donor cells on the host by looking for species-specific changes to the timing of gene expression, tissue interactions, histogenesis, and other events during development. Third, as with the quail-chick chimeric system [217], there is an anti-quail antibody (Q¢PN) that does not recognize duck cells but which allows donor and host contributions to be distinguished from one another permanently (i.e., Q¢PN-positive versus Q¢PN-negative). Therefore, the quail-duck chimeric system by design enables the role of various cell populations to be characterized during development and provides a potent tool for identifying tissue interactions that generate species-specific pattern in the craniofacial complex.

Transplanting neural crest-derived mesenchyme destined to form the beak between quail and duck embryos sheds light on how neural crest serves as a source of species-specific pattern [79]. In these experiments, quail NCC give rise to short, blunt quail-like beaks on duck hosts ("quck"), whereas duck NCC produce long, broad duck-like bills on quail hosts ("duail"). Additional analyses established that donor mesenchyme executes autonomous molecular programs (i.e., transcription factors) and regulates gene expression (i.e., secreted molecules) in host epithelium, a finding that was further substantiated through ensuing

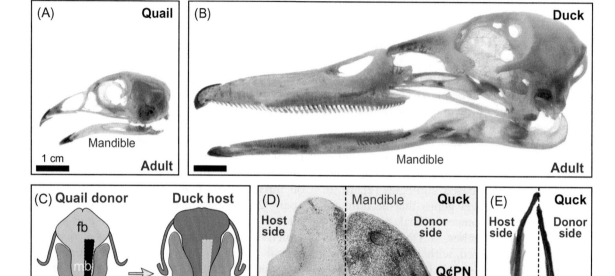

FIGURE 6.2 **The quail-duck chimeric system and the origins of species-specific pattern.** (A, B) Japanese quail and white Pekin duck show species-specific differences in the size and shape of the jaw and skull. (C) To make chimeras, neural crest is grafted from the hindbrain (hb), midbrain (mb), and forebrain (fb) of quail to duck. (D) Coronal section through the mandible of a chimeric quck (rostral at top). Quail donor cells can be detected with the Q¢PN antibody (black). (E) In quck mandibles, the quail donor-derived Meckel's cartilage is shorter and straighter than that observed for the contralateral duck host-derived Meckel's cartilage, which is larger and curved. Secondary cartilage (asterisk) is found along the jaw on the duck side but not quail. (F) In quck, jaw muscles come from the host whereas skeletal and connective tissues come from donor neural crest. Jaw anatomy on the host side is like that found in duck. The mandibular adductor muscle, which closes the jaw, is broader and inserts laterally on the surangular bone. Secondary cartilage forms within the muscle insertion (asterisk). (G) On the donor side, the mandibular adductor is narrower, inserts dorsally along the surangular, and, as in quail, does not contain secondary cartilage. (H) Cranial feather buds arise via interactions between dermis and epidermis. At stage 33, there is little histological evidence for feathers, but by stage 36, feather buds contain a placode and a dermal condensation as they begin to rise above the integument. (I, J) Compared to those of quail, duck feather buds are smaller and spaced closer together. (K) Faster-developing, quail-like feathers form on the donor side of quck. *Source: Panels A, B modified from Ref. [80]; C, D, F, G modified from Ref. [216]; H—K modified from Ref. [77].*

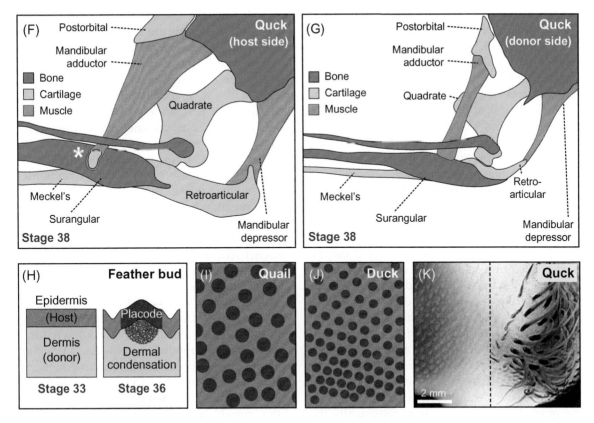

FIGURE 6.2   (Continued)

investigations of species-specific feather formation [77]. Equivalent anatomical modifications
were obtained in a separate series of experiments where NCC that generate the proximal
jaw skeleton were transplanted between quail
and duck embryos [81]. Just like that observed
for beak pattern, the size and shape of elements
around the jaw joint, which differ considerably
between these species due to their specialized
feeding behaviors, was like that of the quail
donor. Thus, quail-duck transplant experiments
demonstrate that neural crest-derived mesenchyme is a fundamental source for species-
species pattern that controls size and shape of
the avian beak as well as other elements in the
craniofacial skeleton. But what remains to be

identified are precise mechanisms through
which NCC accomplish such an undoubtedly
complicated task. Research efforts to accomplish this goal using the quail-duck chimeric
system are described below.

Mechanisms of Species-Specific Pattern in the Neural Crest-Derived Skeleton

Cartilage, like any other patterned tissue,
forms through the interconnected processes of
histogenesis and morphogenesis. Cartilage histogenesis comprises the differentiation of mesenchyme in two stages. First, pre-chondrogenic
cells distinguish themselves by undergoing

condensation, and second, they begin overt chondrification, whereby they secrete extracellular matrix [218,219]. Cartilage morphogenesis involves the establishment of general anatomical pattern (e.g., relative position and orientation) as well as species-specific pattern (e.g., size and shape). While the process of cartilage histogenesis has remained highly conserved across vertebrates [220], changes to the cartilage morphogenetic program have enabled remarkable evolutionary diversification, especially in the jaw skeleton. In this regard, Meckel's cartilage, which encompasses much of the lower jaw, is an excellent example for studying acquisition of size and shape.

When making quck chimeras, NCC can be transplanted unilaterally from quail donors into stage-matched duck hosts, and donor cells will ultimately fill one side of the duck host mandible. This type of experimental approach maintains the nonsurgical side as an internal control [77,81] and facilitates a direct comparison of quail donor-derived tissue and duck host-–derived tissue in the same chimeric facial skeleton. In quck embryos, Meckel's cartilage on the quail donor-derived side is consistently the size and shape that is appropriate for a quail, whereas the duck host-derived Meckel's cartilage is the same size and shape as in duck controls. Also, quail donor mesenchyme maintains its faster rate of growth within the relatively slower duck host environment, and Meckel's cartilage on the quail donor side becomes more advanced in terms of stage-specific size and shape than that found on the relatively younger contralateral duck host side. Thus, implementation of species-specific size and shape in the jaw is due to quail donor-directed programs for cartilage morphogenesis.

Further analyses show that on the histological level, neural crest mesenchyme establishes cartilage size and shape by regulating the program for cartilage differentiation. Chondrocytes on the quail donor side of quck mandibles synthesize cartilage matrix on the time frame of quail donor cells, and these donor-dependent shifts in histogenesis are evident from the very beginning of mesenchymal condensation. Both *sox9*, which is the earliest known marker of chondrogenic condensations [218,221–223], and *col2a1*, which is directly regulated by *sox9* [224], are expressed prematurely by quail donor cells relative to duck host mesenchyme of the contralateral side. Moreover, FGF signaling, which functions upstream of *sox9* and chondrogenesis [223,225–230] is also controlled by mandibular mesenchyme, which upregulates the receptor *Fgfr2* prematurely in quail donor cells. Thus, by establishing the timing of FGF signaling as well as expression of *sox9* and *col2a1*, neural crest mesenchyme transmits information for stage-specific and species-specific size and shape to Meckel's cartilage.

The discovery that neural crest mesenchyme drives significant aspects of cranial chondrogenesis can be well integrated with known functions of surrounding epithelia. Whereas epithelia derived from ectoderm and endoderm may specify sites of chondrogenic condensations [104,231–233], which do not appear to differ grossly between quail and duck, neural crest-derived mesenchyme consequently responds by implementing intrinsic stage-specific and species-specific programs of histogenesis and morphogenesis that govern cartilage size and shape presumably through differential expression of target genes. These results are in line with the permissive role found for epithelia during mandibular osteogenesis, where neural crest mesenchyme controls the timing of critical signaling interactions required for intramembranous ossification [187], and the observation that many chondrogenic signals, including FGFs, are continuously expressed by epithelia prior to and during the arrival of neural crest-derived mesenchyme in the mandible [180,234–238]. Thus, initiation and implementation of skeletogenic programs may be controlled by precisely when and where neural crest mesenchyme activates expression of pathway-specific receptors.

Mechanisms of Species-Specific Pattern in the Craniofacial Musculature

In addition to bone and cartilage, cranial neural crest mesenchyme also generates muscle connective tissues, including ligaments, tendons, fascia, and epi- and endomysia [45]. In contrast, craniofacial musculature is derived from mesodermal mesenchyme, which flanks the neural tube [38,56,193,239]. Therefore, the quail-duck chimeric system offers a way to test the extent to which donor neural crest mesenchyme provides species-specific pattern to host muscle [80]. A variety of experimental strategies—including mutant screens in zebrafish [240], extirpations in amphibians [241,242], analyses of quail-chick chimeras [193,243], and gene misexpression experiments in chick [196] and *Xenopus* [197]—have revealed that cranial neural crest mesenchyme is essential for muscle development [1,57,99,130,244,245]. Specifically, the migration, differentiation, and patterning of myogenic mesenchyme relies on interactions with surrounding tissues, especially those derived from cranial neural crest [37,47,57,177, 193,240–243,246–250].

In terms of species-specific muscle pattern, quail and duck have unique jaw morphologies in conjunction with their species-specific modes of feeding (Figure 6.2). Quail peck at seed on the ground whereas duck use leverage to strain sediment, and these behavioral differences are reflected in the size, shape, and attachment sites of their muscles [251,252]. Quail-duck chimeras demonstrate that neural crest mesenchyme provides species-specific pattern to duck host jaw muscles [80]. In chimeric quck embryos, quail donor cells generate skeletal and muscular connective tissues, and as a result duck host jaw muscles become more elongated rostrally and attached dorsally like those of quail. Significantly, in these experiments donor neural crest mesenchyme does not appear to alter early programs for host myogenic specification or differentiation, which are initiated autonomously by the host. What donor neural crest mesenchyme does do is alter spatial and temporal patterns of expression of connective tissue markers like *Tcf4*, which is a transcription factor that operates downstream of the Wnt/β-catenin signaling pathway and is critical for skeletal muscle development [253–256]. Such neural crest—mediated changes ultimately affect muscle shape and attachment sites, which in turn can provide mechanical forces necessary for species-specific differences in functional morphology such as those associated with formation of secondary cartilage in the jaw skeleton of duck but not quail [216]. Thus, by executing autonomous molecular programs and by interacting with their neighbors, neural crest-derived connective tissues convey species-specific patterning information to muscles in the craniofacial complex. Such observations illuminate the mechanistic basis for how elements in the jaw complex have so intimately co-evolved and are oftentimes mutually compromised in cases of congenital defects.

Mechanisms of Species-Specific Pattern in the Craniofacial Integument

The integument offers a classic example of the importance of tissue interactions during vertebrate development [32,257–261]. In the craniofacial complex, the integument consists of epidermis derived from ectodermal epithelium and dermis that forms from either mesoderm-derived or neural crest-derived mesenchyme. Integumentary structures include hair, feathers, scales, horns, and nails, and the most critical developmental process underlying their morphogenesis is a series of reciprocal signaling interactions between dermis and epidermis. In birds, feather formation proceeds much like that observed for other epidermal appendages where the first morphological evidence of initiation is the aggregation of mesenchyme into a layer of dense dermis beneath the epithelium [262–264]. Adjacent epithelium then thickens into an epidermal placode, mesenchyme forms a dermal condensation, the placode and mesenchyme rise above the integumentary surface, and both tissues undergo proliferation, cell movements, and differentiation [211,261]. Presumably, dermis releases a primary signal, which instructs the epithelium to begin making a placode, and while the identity of the first dermal signal is not known, likely candidates include molecules in the BMP and FGF families [261,265–267].

Tissue recombination experiments in birds have demonstrated that the time of appearance, location, size, number, and morphological identity of feathers is determined by dermis [257,262,268–276]. For example, recombining differently staged dermis and epidermis from wild-type and featherless chicken mutants demonstrates that dermis is endowed with an ability to induce epidermal placodes, but this capacity is quickly lost without proper epidermal interactions [277]. In other situations, however, epithelium can function as a source of pattern, dictating whether epidermis generates scales or feathers [276,278] or determining late-stage branching patterns [279,280].

Japanese quail have cranial feathers that are relatively large, widely spaced, and pigmented, whereas those of white Pekin duck are smaller, closely arranged, and unpigmented [32]. Exchanging NCC between quail and duck embryos challenges host epidermis to respond to species-specific differences in signals that are propagated by donor neural crest-derived dermis [77]. In resulting chimeras, donor neural crest alters the spatial pattern and changes the time at which host cranial feathers form. To accomplish this task, donor neural crest elaborates a set of molecular instructions intrinsic to its own genome and induces a donor-specific pattern of gene expression and histogenesis, which overrides epidermal programs in the host. In particular, donor neural crest alters

expression of members and targets of the BMP, SHH, and Delta/Notch pathways. Such results demonstrate the essential role played by neural crest-derived dermis and the plasticity found within overlying epidermis. In so doing, this research provides insight into developmental mechanisms that can account for the remarkable evolutionary diversity observed in feathers and other derivatives of the integument [82].

CONCLUSIONS AND PERSPECTIVES

The craniofacial complex exhibits both significant conservation in general anatomy and great diversity in species-specific size and shape. This duality reflects the ability of the craniofacial complex to meet fundamental requirements for survival (e.g., feeding, breathing, sensing) and at the same time generate sufficient morphological variation for adaptive evolution. Notably, most of the diversity in the craniofacial complex occurs in those structures that arise from the cranial neural crest. Here, we have detailed molecular and cellular programs driving the development of neural crest-derived structures, as well as the requisite interactions with neighboring tissues (e.g., epithelia and mesodermal mesenchyme). Data from multiple experimental systems performed in taxa representing significant vertebrate diversity show that neural crest mesenchyme controls species-specific pattern. In terms of the mechanistic origins of species-specific pattern, the experiments presented here paint a picture of permissive epithelia providing general anatomic pattern (e.g., make an upper jaw), whereas neural crest mesenchyme provides instructive information for size and shape (e.g., make a quail-like upper jaw versus a duck-like one). Thus, the initiation and implementation of the histogenic and morphogenetic programs that build the craniofacial complex are dependent upon precisely when, where, and for how long neural crest mesenchyme activates intrinsic molecular programs such as the expression of pathway-specific receptors as well as a variety of transcription factors. Quite importantly, the ability of neural crest mesenchyme to keep track simultaneously of stage-specific and species-specific pattern offers a potent mechanism that links development and evolution in the musculoskeletal and integumentary systems, reveals how development can play a generative role in the evolution of species-specific pattern [281], and lends support to theories on heterochrony, which postulate that species-specific transformations in size and shape can arise through changes in the timing of development events [282–285]. During the evolution of the "new head" and its wide-ranging elaboration, epithelial instructions for general anatomical pattern have remained highly conserved, reflecting strong selection for preserving functional abilities (e.g., making an upper and lower jaw that can articulate). In contrast, molecular and cellular programs within the neural crest foster greater plasticity, which allows for the generation of variation and, eventually, species-specific adaptations.

In this context, one major open question for future research is how exactly variation in molecular and cellular programs employed by the neural crest relates to morphological variation. Answering this question will have important implications for understanding evolution as well as the etiologies of craniofacial defects. Although much work has pointed to a role for heterochrony in generating morphological variation, what remains unclear is how to incorporate the multidimensional effects of complex changes in the spatial distribution and levels of gene expression [286]. Further strategies, particularly those pinpointing systems-level differences in gene regulation within neural crest of divergent taxa, have the potential to clarify these issues. Once such differences are elucidated, manipulation via genetic engineering or tissue recombination will allow for key mechanisms in

neural crest molecular and cellular programs to be identified.

References

[1] Noden DM, Schneider RA. Neural crest cells and the community of plan for craniofacial development: historical debates and current perspectives. Adv Exp Med Biol 2006;589:1–23.

[2] Schneider RA. How to tweak a beak: molecular techniques for studying the evolution of size and shape in Darwin's finches and other birds. Bioessays 2007;29(1):1–6.

[3] Shimamura K, Hartigan DJ, Martinez S, Puelles L, Rubenstein JL. Longitudinal organization of the anterior neural plate and neural tube. Development 1995;121(12):3923–33.

[4] Ekker SC, Ungar AR, Greenstein P, von Kessler DP, Porter JA, Moon RT, et al. Patterning activities of vertebrate hedgehog proteins in the developing eye and brain. Curr Biol 1995;5(8):944–55.

[5] Sasai Y, De Robertis EM. Ectodermal patterning in vertebrate embryos. Dev Biol 1997;182(1):5–20.

[6] Pera E, Stein S, Kessel M. Ectodermal patterning in the avian embryo: epidermis versus neural plate. Development 1999;126(1):63–73.

[7] Veitch E, Begbie J, Schilling TF, Smith MM, Graham A. Pharyngeal arch patterning in the absence of neural crest. Curr Biol 1999;9(24):1481–4.

[8] Piotrowski T, Nusslein-Volhard C. The endoderm plays an important role in patterning the segmented pharyngeal region in zebrafish (Danio rerio). Dev Biol 2000;225(2):339–56.

[9] Miller CT, Schilling TF, Lee K, Parker J, Kimmel CB. Sucker encodes a zebrafish Endothelin-1 required for ventral pharyngeal arch development. Development 2000;127(17):3815–28.

[10] Withington S, Beddington R, Cooke J. Foregut endoderm is required at head process stages for anteriormost neural patterning in chick. Development 2001;128(3):309–20.

[11] Schneider RA, Hu D, Rubenstein JL, Maden M, Helms JA. Local retinoid signaling coordinates forebrain and facial morphogenesis by maintaining FGF8 and SHH. Development 2001;128(14):2755–67.

[12] Olivera-Martinez I, Thelu J, Teillet MA, Dhouailly D. Dorsal dermis development depends on a signal from the dorsal neural tube, which can be substituted by Wnt-1. Mech Dev 2001;100(2):233–44.

[13] Couly G, Creuzet S, Bennaceur S, Vincent C, Le Douarin NM. Interactions between Hox-negative cephalic neural crest cells and the foregut endoderm

in patterning the facial skeleton in the vertebrate head. Development 2002;129(4):1061–73.

[14] Anderson RM, Lawrence AR, Stottmann RW, Bachiller D, Klingensmith J. Chordin and noggin promote organizing centers of forebrain development in the mouse. Development 2002;129(21):4975–87.

[15] Graham A. Development of the pharyngeal arches. Am J Med Genet A 2003;119(3):251–6.

[16] Crump JG, Maves L, Lawson ND, Weinstein BM, Kimmel CB. An essential role for Fgfs in endodermal pouch formation influences later craniofacial skeletal patterning. Development 2004;131(22):5703–16.

[17] Alvarado-Mallart RM. The chick/quail transplantation model: discovery of the isthmic organizer center. Brain Res Brain Res Rev 2005;49(2):109–13.

[18] Wada N, Javidan Y, Nelson S, Carney TJ, Kelsh RN, Schilling TF. Hedgehog signaling is required for cranial neural crest morphogenesis and chondrogenesis at the midline in the zebrafish skull. Development 2005;132(17):3977–88.

[19] Marcucio RS, Young NM, Hu D, Hallgrimsson B. Mechanisms that underlie co-variation of the brain and face. Genesis 2011;49(4):177–89.

[20] Chong HJ, Young NM, Hu D, Jeong J, McMahon AP, Hallgrimsson B, et al. Signaling by SHH rescues facial defects following blockade in the brain. Dev Dyn 2012;241(2):247–56.

[21] Bancroft M, Bellairs R. Placodes of the chick embryo studied by SEM. Anat Embryol 1977;151(1):97–108.

[22] Couly G, Le Douarin NM. The fate map of the cephalic neural primordium at the presomitic to the 3-somite stage in the avian embryo. Development 1988;103(Suppl):101–13.

[23] Webb JF, Noden DM. Ectodermal placodes: contributions to the development of the vertebrate head. Am Zool 1993;33(4):434–47.

[24] Ladher RK, O'Neill P, Begbie J. From shared lineage to distinct functions: the development of the inner ear and epibranchial placodes. Development 2010;137 (11):1777–85.

[25] Francis-West PH, Ladher RK, Schoenwolf GC. Development of the sensory organs. Sci Prog 2002;85 (Pt 2):151–73.

[26] Baker CV, Bronner-Fraser M. Vertebrate cranial placodes I. Embryonic induction. Dev Biol 2001;232 (1):1–61.

[27] Baker CV, Stark MR, Marcelle C, Bronner-Fraser M. Competence, specification and induction of Pax-3 in the trigeminal placode. Development 1999;126 (1):147–56.

[28] Romanoff AL. Avian embryo. New York, NY: The Macmillan Company; 1960.

[29] Hamilton HL. Lillie's development of the chick: an introduction to embryology. 3rd ed. New York, NY: Holt, Rinehart and Winston; 1965.

[30] Yasui K, Hayashi Y. Morphogenesis of the beak of the chick embryo: histological, histochemical and auto-radiographic studies. Embryologia (Nagoya) 1967;10 (1):42–74.

[31] Kingsbury JW, Allen VG, Rotheram BA. The histological structure of the beak in the chick. Anat Rec 1953;116(1):95–115.

[32] Lucas AM, Stettenheim PR. Avian anatomy: integument. Washington, DC: United States Department of Agriculture; 1972.

[33] Sawyer RH, O'Guin WM, Knapp LW. Avian scale development. X. Dermal induction of tissue-specific keratins in extraembryonic ectoderm. Dev Biol 1984;101(1):8–18.

[34] Yu M, Yue Z, Wu P, Wu DY, Mayer JA, Medina M, et al. The developmental biology of feather follicles. Int J Dev Biol 2004;48(2-3):181–91.

[35] Noden DM. The control of avian cephalic neural crest cytodifferentiation. I. Skeletal and connective tissues. Dev Biol 1978;67(2):296–312.

[36] Noden DM. Origins and patterning of craniofacial mesenchymal tissues. J Craniofac Genet Dev Biol 1986;2:15–31.

[37] Noden DM. Interactions and fates of avian craniofacial mesenchyme. Development 1988;103:121–40.

[38] Couly GF, Coltey PM, Le Douarin NM. The developmental fate of the cephalic mesoderm in quail-chick chimeras. Development 1992;114(1):1–15.

[39] Olivera-Martinez I, Coltey M, Dhouailly D, Pourquie O. Mediolateral somitic origin of ribs and dermis determined by quail-chick chimeras. Development 2000;127(21):4611–7.

[40] Olivera-Martinez I, Thelu J, Dhouailly D. Molecular mechanisms controlling dorsal dermis generation from the somitic dermomyotome. Int J Dev Biol 2004;48(2-3):93–101.

[41] Matsuoka T, Ahlberg PE, Kessaris N, Iannarelli P, Dennehy U, Richardson WD, et al. Neural crest origins of the neck and shoulder. Nature 2005;436 (7049):347–55.

[42] Tosney KW. The segregation and early migration of cranial neural crest cells in the avian embryo. Dev Biol 1982;89(1):13–24.

[43] Hall BK, Hörstadius S. The neural crest. London: Oxford University Press; 1988.

[44] Le Lièvre CS, Le Douarin NM. Mesenchymal derivatives of the neural crest: analysis of chimaeric quail and chick embryos. J Embryol Exp Morphol 1975;34 (1):125–54.

[45] Noden DM. The embryonic origins of avian cephalic and cervical muscles and associated connective tissues. Am J Anat 1983;168:257–76.

[46] Couly GF, Coltey PM, Le Douarin NM. The triple origin of skull in higher vertebrates: a study in quail-chick chimeras. Development 1993;117:409–29.

[47] Grenier J, Teillet MA, Grifone R, Kelly RG, Duprez D. Relationship between neural crest cells and cranial mesoderm during head muscle development. PLoS One 2009;4(2):e4381.

[48] Noden DM. The control of avian cephalic neural crest cytodifferentiation: II. Neural Tissues Dev Biol 1978;67:313–29.

[49] Rawles ME. Origin of melanophores and their role in the development of color pattern in vertebrates. Physiol Rev 1948;28(4):383–408.

[50] Cramer SF. The origin of epidermal melanocytes. Implications for the histogenesis of nevi and melanomas. Arch Pathol Lab Med 1991;115(2):115–9.

[51] Le Douarin NM, Dupin E. Cell lineage analysis in neural crest ontogeny. J Neurobiol 1993;24(2):146–61.

[52] Bronner-Fraser M. Neural crest cell formation and migration in the developing embryo. Faseb J 1994;8 (10):699–706.

[53] Hirobe T. Structure and function of melanocytes: microscopic morphology and cell biology of mouse melanocytes in the epidermis and hair follicle. Histol Histopathol 1995;10(1):223–37.

[54] Jiang X, Iseki S, Maxson RE, Sucov HM, Morriss-Kay GM. Tissue origins and interactions in the mammalian skull vault. Dev Biol 2002;241(1):106–16.

[55] McBratney-Owen B, Iseki S, Bamforth SD, Olsen BR, Morriss-Kay GM. Development and tissue origins of the mammalian cranial base. Dev Biol 2008;322(1):121–32.

[56] Evans DJ, Noden DM. Spatial relations between avian craniofacial neural crest and paraxial mesoderm cells. Dev Dyn 2006;235(5):1310–25.

[57] Noden DM, Trainor PA. Relations and interactions between cranial mesoderm and neural crest populations. J Anat 2005;207(5):575–601.

[58] Noden DM. Embryonic origins and assembly of blood vessels. Am Rev Respir Dis 1989;140 (4):1097–103.

[59] Couly G, Coltey P, Eichmann A, Le Douarin NM. The angiogenic potentials of the cephalic mesoderm and the origin of brain and head blood vessels. Mech Dev 1995;53(1):97–112.

[60] Hu D, Marcucio RS, Helms JA. A zone of frontonasal ectoderm regulates patterning and growth in the face. Development 2003;130(9):1749–58.

[61] Foppiano S, Hu D, Marcucio RS. Signaling by bone morphogenetic proteins directs formation of an ectodermal signaling center that regulates craniofacial development. Dev Biol 2007;312(1):103–14.

[62] Hu D, Marcucio RS. Unique organization of the frontonasal ectodermal zone in birds and mammals. Dev Biol 2009;325(1):200–10.

[63] Haworth KE, Wilson JM, Grevellec A, Cobourne MT, Healy C, Helms JA, et al. Sonic hedgehog in the pharyngeal endoderm controls arch pattern via regulation of Fgf8 in head ectoderm. Dev Biol 2007;303 (1):244–58.

[64] Richman JM, Tickle C. Epithelia are interchangeable between facial primordia of chick embryos and morphogenesis is controlled by the mesenchyme. Dev Biol 1989;136(1):201–10.

[65] Langille RM, Hall BK. Pattern formation and the neural crest. In: Hanken J, Hall BK, editors. The skull. Chicago, IL: University of Chicago Press; 1993. p. 77–111.

[66] Tucker AS, Yamada G, Grigoriou M, Pachnis V, Sharpe PT. Fgf-8 determines rostral-caudal polarity in the first branchial arch. Development 1999;126(1):51–61.

[67] Ferguson CA, Tucker AS, Sharpe PT. Temporospatial cell interactions regulating mandibular and maxillary arch patterning. Development 2000;127(2):403–12.

[68] Mitsiadis TA, Cheraud Y, Sharpe P, Fontaine-Perus J. Development of teeth in chick embryos after mouse neural crest transplantations. Proc Natl Acad Sci USA 2003;100(11):6541–5.

[69] Santagati F, Rijli FM. Cranial neural crest and the building of the vertebrate head. Nat Rev Neurosci 2003;4(10):806–18.

[70] Le Douarin NM, Creuzet S, Couly G, Dupin E. Neural crest cell plasticity and its limits. Development 2004;131(19):4637–50.

[71] Wilson J, Tucker AS. Fgf and Bmp signals repress the expression of Bapx1 in the mandibular mesenchyme and control the position of the developing jaw joint. Dev Biol 2004;266(1):138–50.

[72] Jheon AH, Schneider RA. The cells that fill the bill: neural crest and the evolution of craniofacial development. J Dent Res 2009;88(1):12–21.

[73] Andres G. Untersuchungen an Chimären von Triton und Bombinator. Genetica 1949;24:387–534.

[74] Wagner G. Untersuchungen an Bombinator-Triton-Chimaeren. Roux' Archiv für Entwicklungsmechanik der Organismen 1959;151:136–58.

[75] Sohal GS. Effects of reciprocal forebrain transplantation on motility and hatching in chick and duck embryos. Brain Res 1976;113(1):35–43.

[76] Yamashita T, Sohal GS. Development of smooth and skeletal muscle cells in the iris of the domestic duck, chick and quail. Cell Tissue Res 1986;244(1):121–31.

[77] Eames BF, Schneider RA. Quail-duck chimeras reveal spatiotemporal plasticity in molecular and histogenic programs of cranial feather development. Development 2005;132(7):1499–509.

[78] Eames BF, Schneider RA. The genesis of cartilage size and shape during development and evolution. Development 2008;135(23):3947–58.

[79] Schneider RA, Helms JA. The cellular and molecular origins of beak morphology. Science 2003;299 (5606):565–8.

[80] Tokita M, Schneider RA. Developmental origins of species-specific muscle pattern. Dev Biol 2009;331 (2):311–25.

[81] Tucker AS, Lumsden A. Neural crest cells provide species-specific patterning information in the developing branchial skeleton. Evol Dev 2004;6 (1):32–40.

[82] Schneider RA. Developmental mechanisms facilitating the evolution of bills and quills. J Anat 2005;207 (5):563–73.

[83] West-Eberhard MJ. Phenotypic plasticity and the origins of diversity. Ann Rev Ecol Syst 1989;20:249–78.

[84] West-Eberhard MJ. Developmental plasticity and evolution. Oxford; New York: Oxford University Press; 2003.

[85] Schlosser G, Wagner GP. Modularity in development and evolution. Chicago, IL: University of Chicago Press; 2004.

[86] Trainor PA, Melton KR, Manzanares M. Origins and plasticity of neural crest cells and their roles in jaw and craniofacial evolution. Int J Dev Biol 2003;47(7-8):541–53.

[87] Gans C, Northcutt RG. Neural crest and the origin of vertebrates: a new head. Science 1983;220:268–74.

[88] Northcutt RG, Gans C. The genesis of neural crest and epidermal placodes. Quart Rev Biol 1983;58 (1):1–27.

[89] Gans C. Craniofacial growth, evolutionary questions. Development 1988;103(Suppl):3–15.

[90] Noden DM. Craniofacial development: new views on old problems. Anat Rec 1984;208:1–13.

[91] Northcutt RG. The new head hypothesis revisited. J Exp Zool B Mol Dev Evol 2005;304B(4):274–97.

[92] Smith KK. The evoultion of the mammalian pharynx. Zool J Linn Soc 1992;104:313–49.

[93] Smith KK. The form of the feeding apparatus in terrestrial vertebrates: studies of adaptation and constraint. In: Hanken J, Hall BK, editors. The skull. Chicago, IL: University of Chicago Press; 1993. p. 150–96.

[94] Smith KK, Schneider RA. Have gene knockouts caused evolutionary reversals in the mammalian first arch? BioEssays 1998;20(3):245–55.

[95] Schneider RA. Neural crest can form cartilages normally derived from mesoderm during development of the avian head skeleton. Dev Biol 1999;208 (2):441–55.

[96] Hunt P, Whiting J, Nonchev S, Sham MH, Marshall H, Graham A, et al. The branchial Hox code and its implications for gene regulation, patterning of the nervous system and head evolution. Development 1991;2(Suppl):63–77.

[97] Le Douarin N, Kalcheim C. The neural crest. 2nd ed. Cambridge, UK; New York, NY: Cambridge University Press; 1999.

[98] Hall BK. The neural crest in development and evolution. New York, NY; Berlin: Springer; 1999.

[99] Francis-West PH, Robson L, Evans DJ. Craniofacial development: the tissue and molecular interactions that control development of the head. Adv Anat Embryol Cell Biol 2003;169(III-VI):1–138.

[100] Gross JB, Hanken J. Cranial neural crest contributes to the bony skull vault in adult Xenopus laevis: insights from cell labeling studies. J Exp Zool B Mol Dev Evol 2005;304(2):169–76.

[101] Morriss-Kay GM. Derivation of the mammalian skull vault. J Anat 2001;199(Pt 1-2):143–51.

[102] Gross JB, Hanken J. Segmentation of the vertebrate skull: neural-crest derivation of adult cartilages in the clawed frog, Xenopus laevis. Integr Comp Biol 2008;48(5):681–96.

[103] Gross JB, Hanken J, Oglesby E, Marsh-Armstrong N. Use of a ROSA26:GFP transgenic line for long-term Xenopus fate-mapping studies. J Anat 2006;209 (3):401–13.

[104] Thorogood P. Differentiation and morphogenesis of cranial skeletal tissues. In: Hanken J, Hall BK, editors. The skull. Chicago, IL: University of Chicago Press; 1993. p. 112–52.

[105] Gross JB, Hanken J. Review of fate-mapping studies of osteogenic cranial neural crest in vertebrates. Dev Biol 2008;317(2):389–400.

[106] Merrill AE, Bochukova EG, Brugger SM, Ishii M, Pilz DT, Wall SA, et al. Cell mixing at a neural crest-mesoderm boundary and deficient ephrin-Eph signaling in the pathogenesis of craniosynostosis. Hum Mol Genet 2006;15(8):1319–28.

[107] Huxley TH. On the theory of the vertebrate skull. Proc R Soc London 1858;9:381–457.

[108] Goodrich ES. Studies on the structure and development of vertebrates. Chicago, IL: University of Chicago Press; 1930.

[109] de Beer GR. The development of the vertebrate skull. Chicago, IL: University of Chicago Press; 1937.

[110] Romer AS. Osteology of the reptiles. Chicago, IL: University of Chicago Press; 1956.

[111] Coppinger R, Schneider R. Evolution of working dogs. In: Serpell J, editor. The domestic dog. Cambridge: Cambridge University Press; 1995. p. 21–47.

[112] Nichols DH. Neural crest formation in the head of the mouse embryo as observed using a new histological technique. J Embryol Exp Morphol 1981;64:105–20.

[113] Nichols DH. Formation and distribution of neural crest mesenchyme to the first pharyngeal arch region of the mouse embryo. Am J Anat 1986;176(2):221–31.

[114] Erickson CA, Loring JF, Lester SM. Migratory pathways of HNK-1-immunoreactive neural crest cells in the rat embryo. Dev Biol 1989;134(1):112–8.

[115] Erickson CA, Reedy MV. Neural crest development: the interplay between morphogenesis and cell differentiation. Curr Top Dev Biol 1998;40:177–209.

[116] Osumi-Yamashita N, Ninomiya Y, Doi H, Eto K. The contribution of both forebrain and midbrain crest cells to the mesenchyme in the frontonasal mass of mouse embryos. Dev Biol 1994;164(2):409–19.

[117] Serbedzija GN, Bronner-Fraser M, Fraser SE. Vital dye analysis of cranial neural crest cell migration in the mouse embryo. Development 1992;116:297–307.

[118] Serbedzija GN, McMahon AP. Analysis of neural crest cell migration in Splotch mice using a neural crest-specific LacZ reporter. Dev Biol 1997;185(2):139–47.

[119] Trainor PA, Tam PP. Cranial paraxial mesoderm and neural crest cells of the mouse embryo: co-distribution in the craniofacial mesenchyme but distinct segregation in branchial arches. Development 1995;121(8):2569–82.

[120] Yoshida T, Vivatbutsiri P, Morriss-Kay G, Saga Y, Iseki S. Cell lineage in mammalian craniofacial mesenchyme. Mech Dev 2008;125(9-10):797–808.

[121] Vaglia JL, Smith KK. Early differentiation and migration of cranial neural crest in the opossum, monodelphis domestica. Evol Dev 2003;5(2):121–35.

[122] Johnston MC. A radioautographic study of the migration and fate of cranial neural crest cells in the chick embryo. Anat Rec 1966;156(2):143–55.

[123] Le Douarin NM, Teillet MM. Experimental analysis of the migration and differentiation of neuroblasts of the autonomic nervous system and of neurectodermal mesenchymal derivatives, using a biological cell marking technique. Dev Biol 1974;41:162–84.

[124] Noden DM. An analysis of the migratory behavior of avian cephalic neural crest cells. Dev Biol 1975;42:106–30.

[125] Duband JL, Thiery JP. Distribution of fibronectin in the early phase of avian cephalic neural crest migration. Dev Biol 1982;93:308–23.

[126] Bronner-Fraser M, Stern C. Effects of mesodermal tissues on avian neural crest cell migration. Dev Biol 1991;143(2):213–7.

[127] Lumsden A, Sprawson N, Graham A. Segmental origin and migration of neural crest cells in the hindbrain region of the chick embryo. Development 1991;113:1281–91.

[128] Sechrist J, Serbedzija GN, Scherson T, Fraser SE, Bronner-Fraser M. Segmental migration of the hindbrain neural crest does not arise from its segmental generation. Development 1993;118:691–703.

[129] Birgbauer E, Sechrist J, Bronner-Fraser M, Fraser S. Rhombomeric origin and rostrocaudal reassortment of neural crest cells revealed by intravital microscopy. Development 1995;121(4):935–45.

[130] Köntges G, Lumsden A. Rhombencephalic neural crest segmentation is preserved throughout craniofacial ontogeny. Development 1996;122(10):3229–42.

[131] Kulesa PM, Fraser SE. In ovo time-lapse analysis of chick hindbrain neural crest cell migration shows cell interactions during migration to the branchial arches. Development 2000;127(6):1161–72.

[132] Wada N, Nohno T, Kuratani S. Dual origins of the prechordal cranium in the chicken embryo. Dev Biol 2011;356(2):529–40.

[133] Platt J. The development of the cartilaginous skull and of the branchial and hypoglossal musculature in *Necturus*. Morphol Jahrb 1898;25:377–467.

[134] Stone LS. Further experiments on the extirpation and transplantation of mesectoderm in *Amblystoma punctatum*. J Exp Zool 1926;44:95–131.

[135] de Beer GR. The differentiation of neural crest cells into visceral cartilages and odontoblasts in amblystoma, and a re-examination of the germ-layer theory. Proc R Soc Ser B Biol Sci 1947;134(876):377–98.

[136] Sadaghiani B, Thiebaud CH. Neural crest development in the *Xenopus laevis* embryo, studied by interspecific transplantation and scanning electron microscopy. Dev Biol 1987;124(1):91–110.

[137] Mitgutsch C, Haas A, Olsson L. Cranial neural crest migration in discoglossid frogs. J Morphol 2001;248(3):262.

[138] Falck P, Hanken J, Olsson L. Cranial neural crest emergence and migration in the Mexican axolotl (*Ambystoma mexicanum*). Zool (Jena) 2002;105(3):195–202.

[139] Cerny R, Meulemans D, Berger J, Wilsch-Brauninger M, Kurth T, Bronner-Fraser M, et al. Combined intrinsic and extrinsic influences pattern cranial neural crest migration and pharyngeal arch morphogenesis in axolotl. Dev Biol 2004;266(2):252–69.

[140] Epperlein H, Meulemans D, Bronner-Fraser M, Steinbeisser H, Selleck MA. Analysis of cranial neural crest migratory pathways in axolotl using cell markers and transplantation. Development 2000;127(12):2751–61.

[141] Raible DW, Wood A, Hodsdon W, Henion PD, Weston JA, Eisen JS. Segregation and early dispersal of neural crest cells in the embryonic zebrafish. Dev Dyn 1992;195(1):29–42.

[142] Schilling TF, Kimmel CB. Segment and cell type lineage restrictions during pharyngeal arch development in the zebrafish embryo. Development 1994;120:483–94.

[143] Halloran MC, Berndt JD. Current progress in neural crest cell motility and migration and future prospects for the zebrafish model system. Dev Dyn 2003;228 (3):497–513.

[144] Ericsson R, Joss J, Olsson L. The fate of cranial neural crest cells in the Australian lungfish, Neoceratodus forsteri. J Exp Zool B Mol Dev Evol 2008;310(4):345–54.

[145] Kulesa P, Ellies DL, Trainor PA. Comparative analysis of neural crest cell death, migration, and function during vertebrate embryogenesis. Dev Dyn 2004;229 (1):14–29.

[146] Theveneau E, Mayor R. Neural crest delamination and migration: from epithelium-to-mesenchyme transition to collective cell migration. Dev Biol 2012;366(1):34–54.

[147] Horigome N, Myojin M, Ueki T, Hirano S, Aizawa S, Kuratani S. Development of cephalic neural crest cells in embryos of Lampetra japonica, with special reference to the evolution of the jaw. Dev Biol 1999;207(2):287–308.

[148] McCauley DW, Bronner-Fraser M. Neural crest contributions to the lamprey head. Development 2003;130(11):2317–27.

[149] Kuratani S, Adachi N, Wada N, Oisi Y, Sugahara F. Developmental and evolutionary significance of the mandibular arch and prechordal/premandibular cranium in vertebrates: revising the heterotopy scenario of gnathostome jaw evolution. J Anat 2013;222(1):41–55.

[150] Aybar MJ, Mayor R. Early induction of neural crest cells: lessons learned from frog, fish and chick. Curr Opin Genet Dev 2002;12(4):452–8.

[151] Alfandari D, Cousin H, Marsden M. Mechanism of Xenopus cranial neural crest cell migration. Cell Adh Migr 2010;4(4):553–60.

[152] Hanken J, Gross JB. Evolution of cranial development and the role of neural crest: insights from amphibians. J Anat 2005;207(5):437–46.

[153] Smith KK. Early development of the neural plate, neural crest and facial region of marsupials. J Anat 2001;199(Pt 1–2):121–31.

[154] Smith KK. Comparative patterns of craniofacial development in eutherian and metatherian mammals. Evolution 1997;51(5):1663–78.

[155] Smith KK. Heterochrony revisited: the evolution of developmental sequences. Biol J Linn Soc 2001;73 (2):169–86.

[156] Smith KK. The evolution of mammalian development. Bull Museum Comp Zool 2001;156(1):119–35.

[157] Smith KK. Sequence heterochrony and the evolution of development. J Morphol 2002;252(1):82–97.

[158] Smith KK. Craniofacial development in marsupial mammals: developmental origins of evolutionary change. Dev Dyn 2006;235(5):1181–93.

[159] Hörstadius S. The neural crest; its properties and derivatives in the light of experimental research. London; New York, NY: Oxford University Press; 1950.

[160] Bronner-Fraser M. Origins and developmental potential of the neural crest. Exp Cell Res 1995;218 (2):405–17.

[161] Baker CV, Bronner-Fraser M, Le Douarin NM, Teillet MA. Early- and late-migrating cranial neural crest cell populations have equivalent developmental potential in vivo. Development 1997;124 (16):3077–87.

[162] McGonnell IM, Graham A. Trunk neural crest has skeletogenic potential. Curr Biol 2002;12(9):767–71.

[163] Abzhanov A, Tzahor E, Lassar AB, Tabin CJ. Dissimilar regulation of cell differentiation in mesencephalic (cranial) and sacral (trunk) neural crest cells in vitro. Development 2003;130(19):4567–79.

[164] Lwigale PY, Conrad GW, Bronner-Fraser M. Graded potential of neural crest to form cornea, sensory neurons and cartilage along the rostrocaudal axis. Development 2004;131(9):1979–91.

[165] Yu JK, Holland ND, Holland LZ. An amphioxus winged helix/forkhead gene, AmphiFoxD: insights into vertebrate neural crest evolution. Dev Dyn 2002;225(3):289–97.

[166] Sauka-Spengler T, Meulemans D, Jones M, Bronner-Fraser M. Ancient evolutionary origin of the neural crest gene regulatory network. Dev Cell 2007;13 (3):405–20.

[167] Meulemans D, Bronner-Fraser M. Insights from amphioxus into the evolution of vertebrate cartilage. PLoS One 2007;2(8):e787.

[168] Meulemans D, Bronner-Fraser M. Central role of gene cooption in neural crest evolution. J Exp Zool B Mol Dev Evol 2005;304(4):298–303.

[169] Meulemans D, Bronner-Fraser M. Gene-regulatory interactions in neural crest evolution and development. Dev Cell 2004;7(3):291–9.

[170] Cattell M, Lai S, Cerny R, Medeiros DM. A new mechanistic scenario for the origin and evolution of vertebrate cartilage. PLoS One 2011;6(7):e22474.

[171] Baroffio A, Dupin E, Le Douarin NM. Clone-forming ability and differentiation potential of migratory neural crest cells. Proc Natl Acad Sci USA 1988;85 (14):5325–9.

[172] Das A, Crump JG. Bmps and id2a act upstream of Twist1 to restrict ectomesenchyme potential of the cranial neural crest. PLoS Genet 2012;8(5):e1002710.

[173] Breau MA, Pietri T, Stemmler MP, Thiery JP, Weston JA. A nonneural epithelial domain of embryonic cranial neural folds gives rise to ectomesenchyme. Proc Natl Acad Sci USA 2008;105(22):7750−5.

[174] Couly G, Grapin-Botton A, Coltey P, Ruhin B, Le Douarin NM. Determination of the identity of the derivatives of the cephalic neural crest: incompatibility between Hox gene expression and lower jaw development. Development 1998;125(17):3445−59.

[175] Nakamura H, Katahira T, Matsunaga E, Sato T. Isthmus organizer for midbrain and hindbrain development. Brain Res Brain Res Rev 2005;49(2):120−6.

[176] Trainor PA, Ariza-McNaughton L, Krumlauf R. Role of the isthmus and FGFs in resolving the paradox of neural crest plasticity and prepatterning. Science 2002;295(5558):1288−91.

[177] Trainor P, Krumlauf R. Plasticity in mouse neural crest cells reveals a new patterning role for cranial mesoderm. Nat Cell Biol 2000;2(2):96−102.

[178] Ladher RK, Wright TJ, Moon AM, Mansour SL, Schoenwolf GC. FGF8 initiates inner ear induction in chick and mouse. Genes Dev 2005;19(5):603−13.

[179] Graham A, Begbie J, McGonnell I. Significance of the cranial neural crest. Dev Dyn 2004;229(1):5−13.

[180] Shigetani Y, Nobusada Y, Kuratani S. Ectodermally derived FGF8 defines the maxillomandibular region in the early chick embryo: epithelial−mesenchymal interactions in the specification of the craniofacial ectomesenchyme. Dev Biol 2000;228(1):73−85.

[181] Lee SH, Fu KK, Hui JN, Richman JM. Noggin and retinoic acid transform the identity of avian facial prominences. Nature 2001;414(6866):909−12.

[182] Song Y, Hui JN, Fu KK, Richman JM. Control of retinoic acid synthesis and FGF expression in the nasal pit is required to pattern the craniofacial skeleton. Dev Biol 2004;276(2):313−29.

[183] Marcucio RS, Cordero DR, Hu D, Helms JA. Molecular interactions coordinating the development of the forebrain and face. Dev Biol 2005;.

[184] Young NM, Chong HJ, Hu D, Hallgrimsson B, Marcucio RS. Quantitative analyses link modulation of sonic hedgehog signaling to continuous variation in facial growth and shape. Development 2010;137 (20):3405−9.

[185] Hall BK. Tissue interactions and the initiation of osteogenesis and chondrogenesis in the neural crest-derived mandibular skeleton of the embryonic mouse as seen in isolated murine tissues and in recombinations of murine and avian tissues. J Embryol Exp Morphol 1980;58(6):251−64.

[186] Helms JA, Schneider RA. Cranial skeletal biology. Nature 2003;423(6937):326−31.

[187] Merrill AE, Eames BF, Weston SJ, Heath T, Schneider RA. Mesenchyme-dependent BMP signaling directs the timing of mandibular osteogenesis. Development 2008;135(7):1223−34.

[188] Walshe J, Mason I. Fgf signalling is required for formation of cartilage in the head. Dev Biol 2003;264 (2):522−36.

[189] David NB, Saint-Etienne L, Tsang M, Schilling TF, Rosa FM. Requirement for endoderm and FGF3 in ventral head skeleton formation. Development 2002;129(19):4457−68.

[190] Couly G, Le Douarin NM. Head morphogenesis in embryonic avian chimeras: evidence for a segmental pattern in the ectoderm corresponding to the neuromeres. Development 1990;108(4):543−58.

[191] Hunt P, Krumlauf R. Deciphering the hox code: clues to patterning branchial regions of the head. Cell 1991;66:1075−8.

[192] Hunt P, Wilkinson D, Krumlauf R. Patterning the vertebrate head: murine Hox 2 genes mark distinct subpopulations of premigratory and migrating cranial neural crest. Development 1991;112(1):43−50.

[193] Noden DM. The role of the neural crest in patterning of avian cranial skeletal, connective, and muscle tissues. Dev Biol 1983;96:144−65.

[194] Gendron-Maguire M, Mallo M, Zhang M, Gridley T. Hoxa-2 mutant mice exhibit homeotic transformation of skeletal elements derived from cranial neural crest. Cell 1993;75(7):1317−31.

[195] Rijli FM, Mark M, Lakkaraju S, Dierich A, Dolle P, Chambon P. A homeotic transformation is generated in the rostral branchial region of the head by disruption of *Hoxa-2*, which acts as a selector gene. Cell 1993;75(7):1333−49.

[196] Grammatopoulos GA, Bell E, Toole L, Lumsden A, Tucker AS. Homeotic transformation of branchial arch identity after Hoxa2 overexpression. Development 2000;127(24):5355−65.

[197] Pasqualetti M, Ori M, Nardi I, Rijli FM. Ectopic Hoxa2 induction after neural crest migration results in homeosis of jaw elements in *Xenopus*. Development 2000;127(24):5367−78.

[198] Creuzet S, Couly G, Vincent C, Le Douarin NM. Negative effect of Hox gene expression on the development of the neural crest-derived facial skeleton. Development 2002;129(18):4301−13.

[199] Depew MJ, Lufkin T, Rubenstein JL. Specification of jaw subdivisions by Dlx genes. Science 2002;298 (5592):381−5.

[200] Kuraku S, Takio Y, Sugahara F, Takechi M, Kuratani S. Evolution of oropharyngeal patterning mechanisms

involving Dlx and endothelins in vertebrates. Dev Biol 2010;341(1):315–23.

[201] Miller CT, Yelon D, Stainier DY, Kimmel CB. Two endothelin 1 effectors, hand2 and bapx1, pattern ventral pharyngeal cartilage and the jaw joint. Development 2003;130(7):1353–65.

[202] Sato T, Kurihara Y, Asai R, Kawamura Y, Tonami K, Uchijima Y, et al. An endothelin-1 switch specifies maxillomandibular identity. Proc Natl Acad Sci USA 2008;105(48):18806–11.

[203] Abzhanov A, Protas M, Grant BR, Grant PR, Tabin CJ. Bmp4 and morphological variation of beaks in Darwin's finches. Science 2004;305(5689):1462–5.

[204] Wu P, Jiang TX, Suksaweang S, Widelitz RB, Chuong CM. Molecular shaping of the beak. Science 2004;305 (5689):1465–6.

[205] Wu P, Jiang TX, Shen JY, Widelitz RB, Chuong CM. Morphoregulation of avian beaks: comparative mapping of growth zone activities and morphological evolution. Dev Dyn 2006;235(5):1400–12.

[206] Grant PR, Grant BR, Abzhanov A. A developing paradigm for the development of bird beaks. Biol J Linn Soc 2006;88(1):17–22.

[207] Bonilla-Claudio M, Wang J, Bai Y, Klysik E, Selever J, Martin JF. Bmp signaling regulates a dose-dependent transcriptional program to control facial skeletal development. Development 2012;139(4):709–19.

[208] Albertson RC, Streelman JT, Kocher TD, Yelick PC. Integration and evolution of the cichlid mandible: the molecular basis of alternate feeding strategies. Proc Natl Acad Sci USA 2005;102(45):16287–92.

[209] Abzhanov A, Kuo WP, Hartmann C, Grant BR, Grant PR, Tabin CJ. The calmodulin pathway and evolution of elongated beak morphology in Darwin's finches. Nature 2006;442(7102):563–7.

[210] Spemann H. Embryonic development and induction. New Haven, CT: Yale University Press; 1938.

[211] Olivera-Martinez I, Viallet JP, Michon F, Pearton DJ, Dhouailly D. The different steps of skin formation in vertebrates. Int J Dev Biol 2004;48(2-3):107–15.

[212] Kuroda H, Wessely O, De Robertis EM. Neural induction in Xenopus: requirement for ectodermal and endomesodermal signals via chordin, noggin, beta-catenin, and cerberus. PLoS Biol 2004;2(5):E92.

[213] Harrison RG. Some difficulties of the determination problem. Am Nat 1933;67(711):306–21.

[214] Lumsden AG. Spatial organization of the epithelium and the role of neural crest cells in the initiation of the mammalian tooth germ. Development 1988;103 (Suppl 15):155–69.

[215] Lwigale PY, Schneider RA. Other chimeras: quail-duck and mouse-chick. Methods Cell Biol 2008;87:59–74.

[216] Solem RC, Eames BF, Tokita M, Schneider RA. Mesenchymal and mechanical mechanisms of secondary cartilage induction. Dev Biol 2011;356(1):28–39.

[217] Le Douarin NM, Dieterlen-Lievre F, Teillet M. Quail-chick transplantations. In: Bronner-Fraser M, editor. Methods in avian embryology. San Diego, CA: Academic Press; 1996. p. 23–59.

[218] Eames BF, de la Fuente L, Helms JA. Molecular ontogeny of the skeleton. Birth Defects Res C Embryo Today 2003;69(2):93–101.

[219] Hall BK. Bones and cartilage: developmental and evolutionary skeletal biology. San Diego, CA: Elsevier Academic Press; 2005.

[220] Eames BF, Allen N, Young J, Kaplan A, Helms JA, Schneider RA. Skeletogenesis in the swell shark Cephaloscyllium ventriosum. J Anat 2007;210(5):542–54.

[221] Healy C, Uwanogho D, Sharpe PT. Expression of the chicken Sox9 gene marks the onset of cartilage differentiation. Ann N Y Acad Sci 1996;785:261–2.

[222] Zhao Q, Eberspaecher H, Lefebvre V, De Crombrugghe B. Parallel expression of Sox9 and Col2a1 in cells undergoing chondrogenesis. Dev Dyn 1997;209(4):377–86.

[223] Eames BF, Sharpe PT, Helms JA. Hierarchy revealed in the specification of three skeletal fates by Sox9 and Runx2. Dev Biol 2004;274(1):188–200.

[224] Bell DM, Leung KK, Wheatley SC, Ng LJ, Zhou S, Ling KW, et al. SOX9 directly regulates the type-II collagen gene. Nat Genet 1997;16(2):174–8.

[225] Bobick BE, Thornhill TM, Kulyk WM. Fibroblast growth factors 2, 4, and 8 exert both negative and positive effects on limb, frontonasal, and mandibular chondrogenesis via MEK-ERK activation. J Cell Physiol 2007;211(1):233–43.

[226] Govindarajan V, Overbeek PA. FGF9 can induce endochondral ossification in cranial mesenchyme. BMC Dev Biol 2006;6:7.

[227] Murakami S, Kan M, McKeehan WL, de Crombrugghe B. Up-regulation of the chondrogenic Sox9 gene by fibroblast growth factors is mediated by the mitogen-activated protein kinase pathway. Proc Natl Acad Sci USA 2000;97(3):1113–8.

[228] Petiot A, Ferretti P, Copp AJ, Chan CT. Induction of chondrogenesis in neural crest cells by mutant fibroblast growth factor receptors. Dev Dyn 2002;224 (2):210–21.

[229] Healy C, Uwanogho D, Sharpe PT. Regulation and role of Sox9 in cartilage formation. Dev Dyn 1999;215 (1):69–78.

[230] de Crombrugghe B, Lefebvre V, Behringer RR, Bi W, Murakami S, Huang W. Transcriptional mechanisms of chondrocyte differentiation. Matrix Biol 2000;19 (5):389–94.

[231] Bee J, Thorogood P. The role of tissue interactions in the skeletogenic differentiation of avian neural crest cells. Dev Biol 1980;78:47−66.

[232] Thorogood P. Morphogenesis of cartilage. In: Hall BK, editor. Cartilage New York: Academic Press, Inc; 1983. p. 223−53.

[233] Thorogood P. Mechanisms of morphogenetic specification in skull development. In: Wolff JR, Severs J, Berry M, editors. Mesenchymal−epithelial interactions in neural development. Berlin: Springer-Verlag; 1987. p. 141−52.

[234] Francis-West PH, Tatla T, Brickell PM. Expression patterns of the bone morphogenetic protein genes Bmp-4 and Bmp-2 in the developing chick face suggest a role in outgrowth of the primordia. Dev Dyn 1994;201(2):168−78.

[235] Wall NA, Hogan BL. Expression of bone morphogenetic protein-4 (BMP-4), bone morphogenetic protein-7 (BMP-7), fibroblast growth factor-8 (FGF-8) and sonic hedgehog (SHH) during branchial arch development in the chick. Mech Dev 1995;53(3):383−92.

[236] Mina M, Wang YH, Ivanisevic AM, Upholt WB, Rodgers B. Region- and stage-specific effects of FGFs and BMPs in chick mandibular morphogenesis. Dev Dyn 2002;223(3):333−52.

[237] Ashique AM, Fu K, Richman JM. Signalling via type IA and type IB bone morphogenetic protein receptors (BMPR) regulates intramembranous bone formation, chondrogenesis and feather formation in the chicken embryo. Int J Dev Biol 2002;46(2):243−53.

[238] Havens BA, Rodgers B, Mina M. Tissue-specific expression of Fgfr2b and Fgfr2c isoforms, Fgf10 and Fgf9 in the developing chick mandible. Arch Oral Biol 2006;51(2):134−45.

[239] Wachtler F, Jacob M. Origin and development of the cranial skeletal muscles. Bibl Anat 1986;29:24−46.

[240] Schilling TF, Piotrowski T, Grandel H, Brand M, Heisenberg CP, Jiang YJ, et al. Jaw and branchial arch mutants in zebrafish I: branchial arches. Development 1996;123(4):329−44.

[241] Olsson L, Falck P, Lopez K, Cobb J, Hanken J. Cranial neural crest cells contribute to connective tissue in cranial muscles in the anuran amphibian, Bombina orientalis. Dev Biol 2001;237(2):354−67.

[242] Ericsson R, Cerny R, Falck P, Olsson L. Role of cranial neural crest cells in visceral arch muscle positioning and morphogenesis in the Mexican axolotl, Ambystoma mexicanum. Dev Dyn 2004;231(2):237−47.

[243] Noden DM. Patterning of avian craniofacial muscles. Dev Biol 1986;116:347−56.

[244] Schnorrer F, Dickson BJ. Muscle building; mechanisms of myotube guidance and attachment site selection. Dev Cell 2004;7(1):9−20.

[245] Noden DM, Francis-West P. The differentiation and morphogenesis of craniofacial muscles. Dev Dyn 2006;235(5):1194−218.

[246] Hall EK. Experimental modifications of muscle development in Amblystoma puncatum. J Exp Zool 1950;113:355−77.

[247] Trainor PA, Sobieszczuk D, Wilkinson D, Krumlauf R. Signalling between the hindbrain and paraxial tissues dictates neural crest migration pathways. Development 2002;129(2):433−42.

[248] Tzahor E, Kempf H, Mootoosamy RC, Poon AC, Abzhanov A, Tabin CJ, et al. Antagonists of Wnt and BMP signaling promote the formation of vertebrate head muscle. Genes Dev 2003;17(24):3087−99.

[249] Borue X, Noden DM. Normal and aberrant craniofacial myogenesis by grafted trunk somitic and segmental plate mesoderm. Development 2004;131(16):3967−80.

[250] Rinon A, Lazar S, Marshall H, Buchmann-Moller S, Neufeld A, Elhanany-Tamir H, et al. Cranial neural crest cells regulate head muscle patterning and differentiation during vertebrate embryogenesis. Development 2007;134(17):3065−75.

[251] Zweers G. Structure, movement, and myography of the feeding apparatus of the mallard (Anas platyrhynchos L.). A study in functional anatomy. Netherlands J Zool 1974;24(4):323−467.

[252] Soni VC. The role of kinesis and mechanical advantage in the feeding apparatus of some partridges and quails. Annals Zool 1979;15:103−10.

[253] Anakwe K, Robson L, Hadley J, Buxton P, Church V, Allen S, et al. Wnt signalling regulates myogenic differentiation in the developing avian wing. Development 2003;130(15):3503−14.

[254] Bonafede A, Kohler T, Rodriguez-Niedenfuhr M, Brand-Saberi B. BMPs restrict the position of pre-muscle masses in the limb buds by influencing Tcf4 expression. Dev Biol 2006;299(2):330−44.

[255] Miller KA, Barrow J, Collinson JM, Davidson S, Lear M, Hill RE, et al. A highly conserved Wnt-dependent TCF4 binding site within the proximal enhancer of the anti-myogenic Msx1 gene supports expression within Pax3-expressing limb bud muscle precursor cells. Dev Biol 2007;311(2):665−78.

[256] Kardon G, Harfe BD, Tabin CJA. Tcf4-positive mesodermal population provides a prepattern for vertebrate limb muscle patterning. Dev Cell 2003;5 (6):937−44.

[257] Dhouailly D. Dermo-epidermal interactions between birds and mammals: differentiation of cutaneous appendages. J Embryol Exp Morphol 1973;30 (3):587−603.

[258] Dhouailly D. Formation of cutaneous appendages in dermo-epidermal recombinations between reptiles,

birds and mammals. Roux Arch Dev Biol 1975;177 (4):323—40.

[259] Widelitz RB, Jiang TX, Noveen A, Ting-Berreth SA, Yin E, Jung HS, et al. Molecular histology in skin appendage morphogenesis. Microsc Res Tech 1997;38(4):452—65.

[260] Widelitz RB, Chuong CM. Early events in skin appendage formation: induction of epithelial placodes and condensation of dermal mesenchyme. J Invest Dermatol Symp Proc 1999;4(3):302—6.

[261] Pispa J, Thesleff I. Mechanisms of ectodermal organogenesis. Dev Biol 2003;262(2):195—205.

[262] Wessells NK. Morphology and proliferation during early feather development. Dev Biol 1965;12 (1):131—53.

[263] Brotman HF. Abnormal morphogenesis of feather structures and pattern in the chick embryo integument. II. Histological description. J Exp Zool 1977;200(1):107—24.

[264] Mayerson PL, Fallon JF. The spatial pattern and temporal sequence in which feather germs arise in the white Leghorn chick embryo. Dev Biol 1985;109 (2):259—67.

[265] Tao H, Yoshimoto Y, Yoshioka H, Nohno T, Noji S, Ohuchi H. FGF10 is a mesenchymally derived stimulator for epidermal development in the chick embryonic skin. Mech Dev 2002;116(1-2):39—49.

[266] Mandler M, Neubuser A. FGF signaling is required for initiation of feather placode development. Development 2004;131(14):3333—43.

[267] Song HK, Lee SH, Goetinck PF. FGF-2 signaling is sufficient to induce dermal condensations during feather development. Dev Dyn 2004;231(4):741—9.

[268] Cairns JM, Saunders JW. The influence of embryonic mesoderm on the regional specification of epidermal derivatives in the chick. J Exp Zool 1954;127:221—48.

[269] Saunders JW, Gasseling MT. The origin of pattern and feather germ tract specificity. J Exp Zool 1957;135:503—28.

[270] Rawles ME. Tissue interactions in scale and feather development as studied in dermal-epidermal recombinations. J Embryol Exp Morphol 1963;11:765—89.

[271] Dhouailly D. Analysis of the factors in the specific differentiation of the neoptile feathers in the duck and chicken. J Embryol Exp Morphol 1967;18 (3):389—400.

[272] Dhouailly D. The determination of specific differentiation of neoptile and teleoptile feathers in the chick

and the duck. J Embryol Exp Morphol 1970;24 (1):73—94.

[273] Linsenmayer TF. Control of integumentary patterns in the chick. Dev Biol 1972;27(2):244—71.

[274] Dhouailly D, Sawyer RH. Avian scale development. XI. Initial appearance of the dermal defect in scaleless skin. Dev Biol 1984;105(2):343—50.

[275] Song HK, Sawyer RH. Dorsal dermis of the scaleless (sc/sc) embryo directs normal feather pattern formation until day 8 of development. Dev Dyn 1996;205 (1):82—91.

[276] Prin F, Dhouailly D. How and when the regional competence of chick epidermis is established: feathers vs. scutate and reticulate scales, a problem en route to a solution. Int J Dev Biol 2004;48 (2—3):137—48.

[277] Viallet JP, Prin F, Olivera-Martinez I, Hirsinger E, Pourquié O, Dhouailly D. Chick Delta-1 gene expression and the formation of the feather primordia. Mech Dev 1998;72(1—2):159—68.

[278] Widelitz RB, Jiang TX, Lu J, Chuong CM. beta-catenin in epithelial morphogenesis: conversion of part of avian foot scales into feather buds with a mutated beta-catenin. Dev Biol 2000;219(1):98—114.

[279] Harris MP, Fallon JF, Prum RO. Shh-Bmp2 signaling module and the evolutionary origin and diversification of feathers. J Exp Zool 2002;294(2):160—76.

[280] Yu M, Wu P, Widelitz RB, Chuong CM. The morphogenesis of feathers. Nature 2002;420(6913):308—12.

[281] Alberch P. The generative and regulatory roles of development in evolution. In: Mossakowski D, Roth G, editors. Environmental adaptation and evolution: a theoretical and empirical approach. Stuttgart: G. Fischer-Verlag; 1982. p. 19—36.

[282] de Beer GR. Embryology and evolution. Oxford: Clarendon Press; 1930.

[283] Smith KK. Time's arrow: heterochrony and the evolution of development. Int J Dev Biol 2003;47(7-8):613—21.

[284] Alberch P, Gould SJ, Oster GF, Wake DB. Size and shape in ontogeny and phylogeny. Paleobiology 1979;5(3):296—317.

[285] Hall BK. Developmental processes underlying heterochrony as an evolutionary mechanism. Canadian J Zool 1984;62:1—7.

[286] Depew MJ, Simpson CA. 21st century neontology and the comparative development of the vertebrate skull. Dev Dyn 2006;235(5):1256—91.

NEURAL CREST CELL DIFFERENTIATION AND DISEASE

Neural Crest Cells in Craniofacial Skeletal Development

Thomas F. Schilling and Pierre Le Pabic

Department of Developmental and Cell Biology, University of California,
Irvine, 4109 Natural Sciences II, Irvine, CA

INTRODUCTION

In addition to many other derivatives, neural crest cells (NCC) in the cranial region form most of the craniofacial skeleton. Vertebrates have evolved a remarkable variety of craniofacial shapes and functions, from filter feeding sieves to powerful, toothy jaws. Defects in the craniofacial apparatus underlie some of the most common human birth defects, such as cleft palate and craniosynostosis. Therefore, it is crucial to understand the

Neural Crest Cells.
DOI: http://dx.doi.org/10.1016/B978-0-12-401730-6.00008-9

factors that specify skeletogenic NCC, guide them to their correct destinations in the head, and induce them to differentiate into bones of the right size and shape.

Such factors are best understood if the skeletal elements are considered as mosaic assemblies of modules with intrinsic properties, such as (1) NC versus mesodermally derived elements, (2) segments along the anterior−posterior (A−P) axis in the pharyngeal region, (3) dorsal−ventral (D−V) domains within each pharyngeal segment, and (4) oral−aboral subdivisions within the developing palate [1]. Gene functions implicated in skeletogenic NC point to several extremely important signaling pathways in cranial NC development, and here we try and consider each of these within a broader context of craniofacial gene regulatory networks (GRNs).

SKELETOGENIC NC SPECIFICATION AND ANTERIOR−POSTERIOR PATTERNING

The Ectomesenchymal Lineage and its NC Origins

Skeletogenic NCC are often called ectomesenchyme (EM), referring to their ectodermal origin. Over a century of embryological studies either ablating or labeling premigratory NCC have demonstrated a NC origin for this unusual skeletogenic population [2−4], but how closely related is it to other NC lineages (e.g., neural, glial, or pigment cells)? In both avian and zebrafish embryos, a subpopulation of migrating cells downregulates early NC markers (e.g., *Sox10*, *FoxD3*) and upregulates *Dlx2* when entering the pharyngeal arches, and fails to do so if deprived of Fgf signaling, suggesting that EM and non-EM populations share early NC-specific gene expression programs that only later diverge after migration

(Figure 7.1A) [5]. A growing list of mutants disrupt both EM and non-EM lineages [6−8]. Cranial NCC maintained *in vitro* form chondrocytes and osteocytes in addition to non-EM derivatives [9,10]. Long-term NC labeling *in vivo* also marks both EM and non-EM populations [11]. Finally, under some conditions EM cells can be driven to adopt non-EM fates, and vice versa; loss of Twist1 in zebrafish disrupts skeletogenesis and leads to an excess of non-EM fates [12], while conversely loss of FoxD3 disrupts pigment and glial cells but leads to excess EM [13]. Thus, although some EM cells in mice arise from a more lateral domain of the non-neural ectoderm than the bulk of the NC [14], these results indicate a close lineal relationship between EM and the rest of the NC.

Does EM exist outside of the cranial region? Many extinct vertebrate species (e.g., placoderms, ostracoderms) had dermal bones covering their entire bodies, but such bones are restricted to the skull vault in living species [15]. This raises an interesting evolutionary question: Did these bones form from NC, similar to their cranial counterparts? It also raises the developmental question: Does NC in the trunk in living species retain any skeletogenic potential? A NC contribution to the ventral part of the turtle shell—the plastron—has been suggested using molecular markers and DiI labeling, lending support to this theory [16]. Micromass cultures of avian trunk NCC can form cartilage, demonstrating that they have such potential *in vitro* [17]. Previous, DiI cell labeling studies in zebrafish have suggested that NCC give rise to mesenchyme of the larval median fins, which form bony rays [18], but have not ruled out a possible mesodermal origin. Recent long-term fate-mapping studies in zebrafish using *sox10:CRE* have also suggested such contributions of NC to median fins [19]. However, other studies tracking NC- and mesoderm-specific transgenes suggest that

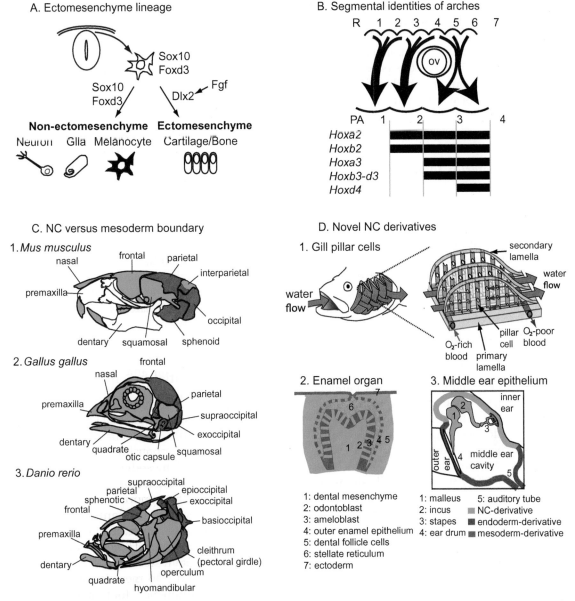

FIGURE 7.1 **Cranial neural crest specification and migration.** (A) Cranial NC migration and some of the molecules that specify distinct lineages, including the EM. An NCC leaving the dorsal neural tube, shown in transverse section, can form but is not limited to sensory neurons or glial cells, pigment cells, or skeletal derivatives. (B) Three streams of cranial NC arising from particular rhombomeres (R) of the hindbrain are shown migrating into the pharyngeal arches (PA) in lateral view. Expression domains of Hox genes that specify segmental identities of NCC in the arches are represented by horizontal black bars. (C) NC (blue) versus mesodermal (red) contributions to the skull in mouse, chick, and fish (redrawn after Noden [20]; Gross and Hanken [21]; Cubbage and Mabee [22]). (D) Newly discovered cranial NC derivatives: (1) gill pillar cells, (2) enamel organ, and (3) middle ear epithelium.

median fin mesenchyme and fin rays (as well as scales) are all mesoderm-derived [23]. Median fin mesenchyme has also been shown to arise from somites in amphibians [24]. Thus, while they may still have the potential to do so, the jury is still out on whether or not trunk NCC normally form any part of the trunk skeleton.

Anterior—Posterior Identities of Cranial NCC

Cranial NC migrates in three major streams into the pharyngeal region to give rise to variable numbers of arches depending on the species (e.g., four in mammals, seven in zebrafish) (Figure 7.1B). How are segmental differences between arches conferred within these streams? One long-standing idea is that Hox genes give positional identity to NCC based on their segmental origins from specific rhombomeres in the hindbrain. Arch 2 (hyoid) NC arises from rhombomeres 3, 4, and 5 (R3-5) and expresses Hox paralogue group 2 genes, while more posterior (branchial) arches arise from R5-7 and express Hox 3 and 4 paralogues (Figure 7.1B). In contrast, NCC in arch 1 (mandibular) do not express Hox genes despite originating from a Hox-positive region (R2/3) of the hindbrain as well as the Hox-negative midbrain [25,26]. Mutant mice lacking *Hoxa2* display homeotic transformations of arch 2 to a more mandibular identity [27—30], while conversely *Hoxa2* overexpression in Xenopus or zebrafish embryos causes arch 1 to acquire arch 2 characteristics [29,31,32]. This role as a selector gene of segmental identity appears to be NC autonomous, suggesting a role for Hox genes in modulating the response of each arch to local epithelial signals [33,34]. Epigenetic regulation of Hoxa2 expression has been shown to play a crucial role in arch 2, since in mutants in zebrafish *moz*

(Myst3, a histone acetyltransferase) or *brf1* (a Trithorax group member), arch 2 similarly acquires arch 1 identity [35,36]. Likewise, mouse mutants in *histone deacetylase 8* (*Hdac8*) upregulate *Otx2* and *Lhx1* and display specific skeletal defects in the rostral, Hox-negative domain [37]. Hox-gene functions appear to be more redundant in the posterior arches, but simultaneous inactivation of all HoxA cluster genes in NCC in mice leads to transformations of arches 2—4 to an arch 1 identity, providing some of the first genetic evidence for segmental homology throughout the pharyngeal series [38]. Taken together, these results demonstrate the modular organization of the pharyngeal region during the critical period of Hox-dependent patterning: each pharyngeal arch has a mandibular ground pattern that may be modified somewhat independently from other arches through modifications in its combinatorial expression of Hox genes.

Defining the Spatial Extent of the EM and its Derivatives

Most of the skeleton (axial, appendicular, posterior cranial) arises from embryonic mesoderm. Where precisely is the boundary between NC- and mesoderm-derived bones? This may have important implications for understanding the regional specificity of skeletal changes both in human malformations and during craniofacial evolution. Classic fate-mapping studies using chick-quail chimeras have shown that ventrally the entire pharyngeal skeleton is NC derived, except for the columella (dual NC/mesoderm origin). Dorsally, the anterior avian skull is NC derived, while the posterior skull is mesodermal, with a boundary within the frontal bone (Figure 7.1C) [39,40]. These results have been largely confirmed in mice either using a *Wnt1* promoter driving Cre recombinase (*Wnt1:CRE*) crossed to a Rosa26 reporter to mark NCC genetically [41,42] or a mesoderm-specific *Mesp1-CRE* line driving Rosa26 reporter expression [43]. However, in contrast to the chick, the boundary between NC and mesoderm in the mouse corresponds to the coronal suture between the frontal and parietal bones. These studies raise questions as to the homology of bones with lineal differences (i.e., the murine frontal is completely NC while the avian frontal bone has a dual NC/mesoderm origin; the murine interparietal bone also has a dual origin) and further support the argument for homology for others (i.e., the avian columella and murine stapes both have a dual NC/mesodermal origin) [44,45].

Such long-term fate-mapping studies have not only delineated the boundaries of cranial NC derivatives but have also revealed new derivatives. For example, cranial NCC in zebrafish labeled with *sox10:ER^{T2}-CRE; ß-actin: switch*, form the gill pillar cells that mechanically support gill filaments (Figure 7.1D) [46]. A similar *Sox10-iCreER^{T2}; R26R* system in mice has revealed NC contributions to brain pericytes [47] and to the otic vesicle, the dual origins of which have been confirmed using *Wnt1:CRE; R26R, Pax3:CRE; R26R,* and *Hoxb1: CRE; R26R* combinations [48]. Similar studies using *P0-CRE; R26R* have provided evidence that, in addition to dentin-secreting odontoblasts, cranial NCC also contribute to the enamel forming cells of teeth (Figure 7.1D) [49]. Surprisingly, NCC also contribute to the epithelial lining of the auditory tube of the mammalian middle ear that surrounds the middle ear bones (Figure 7.1D) and defects in this population may cause conductive hearing loss in humans [50,51]. Taken together, these studies suggest that the full spectrum of potential cell types and tissues to which NC can contribute is more extensive than previously appreciated.

PATTERNING WITHIN CRANIOFACIAL SKELETAL PRIMORDIA

After cranial NCC migrate they become surrounded by epithelia, and numerous studies have demonstrated the critical role played by local epithelial–mesenchymal interactions in craniofacial patterning. For example, mandibular endoderm rotations in quail-chick chimeras result in matching rotations of the corresponding skeleton [52], while loss of endoderm in mutant zebrafish results in complete loss of the pharyngeal skeleton [53]. In addition, quail-duck chimeras have demonstrated that the EM possesses the ability to translate epithelial signals into species-specific shape and size [54]. Consequently, understanding how pattern arises in the EM boils down largely to unraveling these epithelial–mesenchymal interactions. Similar to the pharyngeal region, the modular nature of the face is patent during embryonic development as demonstrated by the physically distinct facial prominences (Figure 7.2A), each of which is a semi-autonomous unit that eventually becomes integrated into the craniofacial complex. With the new techniques available for genetic labeling of cells and conditional perturbations of gene functions (particularly *Wnt1:CRE* to disrupt gene functions in NCC in mice), huge strides have been made in this area in the last few years. Here, we discuss some of the major signaling pathways and their influences on three subdivisions of the EM: (1) the pharyngeal arches, (2) the frontonasal process and palate, and (3) the dermal bones of the skull vault.

A GRN of D–V Pharyngeal Patterning

As discussed above, experimental manipulation of Hox-gene expression has demonstrated the modularity of the pharyngeal arch series

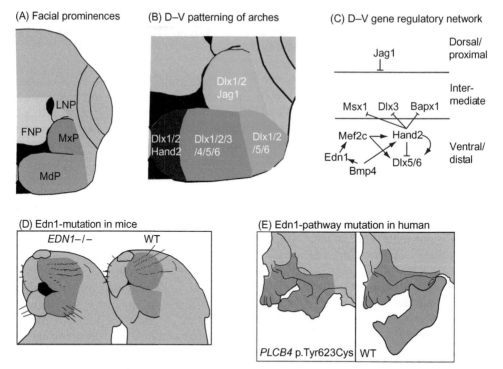

FIGURE 7.2   **Patterning skeletogenic neural crest in the developing face.** (A) Facial prominences, including upper and lower jaws. Anterior view of the left side of an embryonic mouse. FNP: Frontonasal Prominence, LNP: Lateral Nasal Process, MdP: Mandibular Prominence, MxP: Maxillary Prominence. (B) Close up of MdP and MxP showing expression domains of dorsal−ventral (D−V) patterning genes in the NC of the upper and lower jaws. (C) GRN controlling D−V pharyngeal arch patterning. (D) Lower jaw to upper jaw homeotic transformation in Edn1 mutant mice (redrawn after Ozeki et al. [55]). (E) Lower to upper jaw homeotic transformation in human patient with auriculocondylar syndrome caused by a PLCB4 mutation (redrawn after Rieder et al. [56]).

along the A−P axis. Here, we discuss the modular organization of pharyngeal arches along their D−V axis. Each pharyngeal segment consists of ectodermal and endodermal epithelia ensheathing a cylindrical wall of NC mesenchyme itself surrounding a mesodermal core [57]. In 1998 the Yanigasawa group showed requirements for the type-A Endothelin receptor (EdnrA) in formation of the lower jaw (the ventral module of the mandibular arch) in mice [58]. Since then it has become clear that the ligand, Endothelin-1 (Edn1), is secreted by ventral pharyngeal epithelia and promotes ventral cell fates in the arches (distal in mice). It does so, in part, by inducing a nested pattern of key transcription factors, including Hand2 and Dlx3/5/6, which specify distinct D−V domains (Figure 7.2B and C) [59,60]. Mutant screens in zebrafish and mice have identified several essential components of this Edn1 signaling system, such as Galpha (q)/Galpha(11), phospholipase-b3 (Plcb3), and Mef2c [61,62]. Loss-of-function mutations in these genes typically lead to loss of much of the lower jaw and in more severe cases show homeotic transformations of ventral to more dorsal mandibular fates [63]. Edn1 loss of function in mice causes similar ventral-to-dorsal transformations (Figure 7.2D) [55]. Recent studies of auriculocondylar syndrome have revealed the first evidence of human homeotic transformations along the D−V axis of the mandibular arch, due to mutations in *PLCB4* and *GNAI3*, both components of Edn1 signaling (Figure 7.2E) [56].

How does Edn1 signaling impart ventral identities to mandibular cells? In part, it does so by localized production of the ligand, Edn1, in a subset of pharyngeal cell types (ectoderm, endoderm, mesoderm) and of these only ectoderm-derived Edn1 appears to be essential [64,65]. Based on the homeotic transformations discussed above, Edn1 might act as a diffusible morphogen, promoting ventral versus dorsal NCC fates in a concentration-dependent manner [63]. In support of this model, a knock-in of *Edn1* into the *EdnrA* locus, which results in constitutive *Ednra* activation throughout the craniofacial mesenchyme, induces the transformation of dorsal, maxillary into ventral, mandibular structures [66]. Thus, at least at some stages, dorsal NCC within the arch are competent to respond to Edn1, and Edn1 alone is sufficient to drive them to a ventral identity. In mice, this critical period appears to be restricted to a period of about 2 days (E8.0–E10.0) when cranial NCC are migrating into and proliferating within the arches [67].

In fact, Edn1 initiates a dynamic set of gene regulatory interactions within an arch that further subdivides the NC along the D−V axis (Figure 7.2B and C) [59]. The transcription factors Hand2, Mef2c, Dlx5, and Dlx6 all define the ventral arch NC domain and are required for its specification [61,68,69]. Dlx5/6 initially induce Hand2, through a branchial arch-specific enhancer, while later Hand2 feeds back to suppress Dlx5/6. In mice, this down-regulation is required to repress Runx2 and osteogenesis (Hand2 interacts physically with Runx2 to suppress DNA binding) and allows a subset of ventral NCC to give rise to connective tissue of the tongue [70,71]. Hand2 function is partially redundant with Hand1 in this regard [72]. Mef2c induces transcription of all three of the other genes and binds an arch-specific enhancer in the Dlx5/6 locus [69]. Bapx1, Dlx3, and Msx1 (Msxe in zebrafish) define a more dorsal domain that corresponds, at least in part, to the region that forms the jaw joint [73–75]. Disruption of Bapx1 in zebrafish eliminates the jaw joint, and zebrafish mutants in Barx1 lead to defects in mandibular development, including an ectopic, more ventrally located joint within Meckels cartilage [76,77]. Thus, spatial and temporal dynamics in the D−V GRN lead to additional "intermediate" domains (modules) along the D−V axis [74].

With this D−V modularity in mind, we can begin to understand the influences of other signals on the pharyngeal skeleton. Bmp4 has a similar ventralizing activity to Edn1, as was revealed recently using conditional gain- and loss-of-function approaches in zebrafish to bypass earlier well-known roles in NC specification, proliferation, survival, and differentiation (Figure 7.2C) [75,78]. Both Bmp and Edn1 are required for the proper ventral and intermediate domains of gene expression along the D−V axis, and the Bmp inhibitor, Grem1, counteracts this ventralizing activity of Bmp in the dorsal domain. Initially both signals can also induce these genes and rescue the loss-of-function phenotypes, but at later stages Bmp strongly induces *hand2* and *dlx*, while Edn1 cannot rescue *hand2* expression in the absence of Bmp signaling. Shh is also required for pharyngeal D−V patterning as loss of the essential signal transducer for Shh Smoothened (Smo) in zebrafish disrupts *dlx5/6* and other D−V patterning genes in the arches. Restoring Smo in endoderm rescues this expression, suggesting that in this case endodermal cells, rather than NC, respond to Shh [79]. Wnt signaling is also restricted to ventral arch epithelia and NC [80]. These results suggest that while there are multiple ventralizing signals in the arches, each signal plays a distinct role in the induction of a subset of genes along the D−V axis.

Bmp4 overexpression by 36 hpf in zebrafish or E10.5 in mice no longer influences D−V patterning but induces genes involved in osteogenesis and mesenchymal stem cell self-renewal [81].

While Edn1 and Bmp4/7 are busy promoting ventral arch fates, Notch signaling promotes dorsal identities [82]. The Notch ligand, Jagged 1 (Jag1), is expressed in dorsal arch NCC in both zebrafish and mice during D−V patterning, and loss-of-function mutations in fish lead to loss of dorsal domain and a dorsal expansion of genes expressed in both ventral and intermediate domains [82]. Deleting *Jag1* from cranial NC in mice also leads to midfacial hypoplasia [83]. Like Bmps, Jagged/Notch signaling also plays later roles in chondrogenesis and osteogenesis [84].

An unexpected new player involved in the D−V pharyngeal patterning network is a receptor for serotonin 2B (5-hydroxytryptamine 2B; 5-HTR2B). 5-HT is best known as a monoamine neurotransmitter in the nervous system, but it also plays roles as a growth and differentiation factor. 5-HT as well as dopamine can stimulate NCC migration [85]. *5-Htr2b* overexpression in *Xenopus* embryos causes ectopic *Hand2* and *Bapx1* expression, while loss of function eliminates Bapx1 and jaw joint formation, indicating effects on the intermediate domain. In addition 5-Htr2b depletion interacts genetically with Plcb3, suggesting interactions with Edn1 signaling [86].

Midfacial Integration

Under normal conditions, craniofacial development results in a functionally and morphologically integrated structure such as the human face. To do so, facial prominences fuse at the midline following the period of semi-autonomous development described above. Midfacial clefting such as cleft palate is a common consequence of a failure in this integration process and often results from reduced Shh signaling, such as in holoprosencephaly (HPE) [87]. Conversely, excessive Shh signaling through mutations in GLI3 or PTC causes midline expansion such as in Greig cephalopolysyndactily and Gorlin syndrome [88]. Hh proteins must be properly processed in secreting cells, and recent studies in mice have shown that mutations in a Hh acyltransferase (Hhat) cause HPE and agnathia [89]. Distinct requirements for Hh signaling in the arches, versus the palate, have also been revealed by depletion of a zebrafish inositol phosphate kinase (Ip6k2) that appears critical for signaling downstream of Smo but upstream of *gli* transcription [90].

The broad effect of Shh on midfacial integration seems to result from the concerted action of several signaling centers, one of which is the frontonasal ectoderm zone (FEZ). The FEZ model proposes that a specialized midfacial ectoderm coordinates frontonasal and palate development (Figure 7.3A). The FEZ forms at the boundary between *Fgf8* and *Shh* expression and transplantation to ectopic locations in avian embryos reorganizes NCC leading to beak duplications [93]. Shh signaling from the ventral forebrain as well as Bmp2/4 in the frontonasal process itself are critical for inducing the FEZ [94−96].

In both zebrafish and chick embryos, two distinct mediolateral populations of NCC give rise to the palate and both require Shh (Figure 7.3B) [91,97−99]. Trabecular and ethmoid cartilages in zebrafish are homologous to the mammalian nasal capsule and primary palate, though fish lack a secondary, soft palate [79,100]. Similar to studies of the FEZ in chick, bidirectional Shh signaling between neural and oral epithelium and NC-derived palate progenitors are critical for primary palate formation in zebrafish [97,98]. A *glid:mCherry* reporter for Hh signaling is first detected in this region around 48 h postfertilization, after

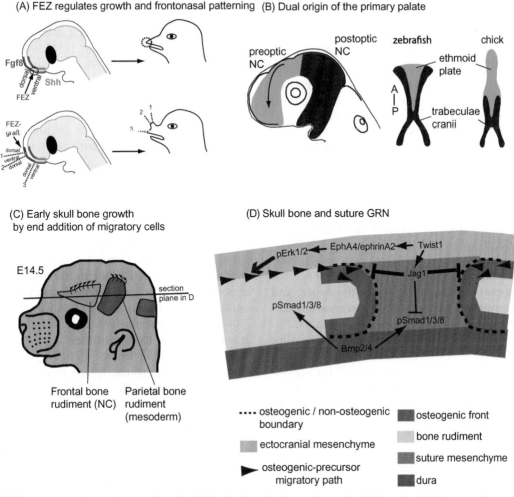

FIGURE 7.3 **Growth and patterning of NC in the developing palate and skull vault.** (A) Frontonasal ectodermal zone (FEZ) model for D–V patterning of the frontonasal process by Fgf8 and Shh in chick and beak duplication resulting from a FEZ grafted into a more dorsal location. Lateral views of the heads of chick embryos at approximately stage 25 and resulting chicks after hatching. (B) Dual origin for the primary palate based on lineage tracing in both zebrafish and chick (redrawn after Wada et al. [91]). Left panel shows a lateral view, anterior to the left, of the head of a stage 13 chick embryo. Right panels show dorsal views, anterior to the top, of the early ethmoid and trabecular cartilages of zebrafish and chick, derived either from preoptic (light blue) or postoptic (dark blue) NC. (C) End addition of cells to growing membrane bones of the skull vault. Early condensations of progenitors are shown in lateral view in mouse at E14.5, including the NC-derived frontal bone (blue) and the mesoderm-derived parietal and occipital bones (red). Arrows indicate end addition of osteoblasts at the growing mesenchymal front (redrawn after Yen et al. [92]). (D) GRN controlling suture development. Section through frontal and parietal bones illustrated in (C).

cranial NCC have completed migration and just prior to the onset of cartilage differentiation [101]. *miR-140* overexpression or pharmacological inhibition of Shh signaling results in

clefting and a failure of NCC to form the ethmoid [99]. The secondary palate in mammals arises by outgrowth of paired maxillary processes, which also relies on Shh—loss of Smo in palatal NCC disrupts skeletal mesenchyme proliferation, which in turn disrupts patterning in the adjacent epithelia [102].

Dlx5/6 are not only required for lower jaw patterning, but also for the dorsal nasal capsule [103]. Loss of Dlx5 leads to downregulation of *Fgf7* and expansion of *Shh* toward nasal, partially rescuing cleft palate in *Msx1* mutants [104]. Similarly, Hand2 is not only required for lower jaw development but also specifies the upper jaw and palate, where it is induced by Bmp signaling. However, here Hand2 appears to be required in the palatal epithelium to maintain NC proliferation and palate extension [105]. In the developing palate, Bmpr1a is required to maintain *Msx1*, *Fgf10*, and *Shh* to promote palate growth [106]. Thus, many of the same developmental regulatory genes are required in both upper and lower jaws but appear to have somewhat different modes of action in the two domains.

Primary cilia are critical organelles required for transducing the Hh signal, which also play roles in Fgf, Wnt, and Pdgf signaling. Over the past several years a growing body of evidence points to a new category of craniofacial malformation in humans caused by cilia defects, so-called ciliopathies. These include oro-facial-digital syndrome, Joubert syndrome, Bardet−Biedel syndrome (BBS), Meckel−Gruber syndrome, Ellis−van Creveld syndrome, and Sensenbrenner syndrome [107,108]. Shh-responding cells concentrate Patched (Ptc) receptors and other signal transduction machinery in cilia [109]. Both Shh and cilia are required for closure of the buccohypophyseal canal, pituitary formation, and morphogenesis at the midline of the cranial base [110]. BBS, for example, is due to defective intraflagellar transport within cilia, and one of the hallmarks of these patients is midfacial defects [111]. Other cilia proteins with functions in chondrogenesis and osteogenesis include Pkd1 and Pkd2, which interestingly are activated in response to mechanical stress [110,112].

Modular Patterning of Tongue and Teeth

Both Shh and Wnt signaling are also essential for tongue development [113]. While not commonly appreciated, NCC of the ventral mandibular arch form the connective tissues of the tongue, defects in which cause secondary palate defects, which are by far the most common form of cleft palate in humans. As mentioned above, Hand2 suppression of osteogenesis in ventral NCC allows formation of tongue connective tissue [70,71]. Mice mutant for the *TGFbeta activating kinase 1* (*Tak1*), which regulates Smad-independent signaling within the NC, overexpress *Fgf10* in NCC of the tongue and later display cleft palate and micrognathia [114]. Thus, development of the tongue and its NC contributions also has important implications for understanding the causes of craniofacial skeletal malformations.

A similar modularity is observed in the development of teeth, another important EM derivative [115]. Bmp4 signaling restricts expression of the transcription factor Barx1 to dorsal mesenchyme that gives rise to molars, which expands into the more ventral incisor domain and causes incisor-to-molar transformations when Bmp signaling is suppressed [116]. Inactivation of Bmp4 in NC arrests mandibular but not maxillary tooth formation, presumably because it normally suppresses tooth inhibitors in the lower jaw, such as the Wnt inhibitor Dkk2, and induces Msx1 to give tooth potential [117]. Likewise, loss of Bmpr1a causes mandibular hypoplasia and loss of lower incisors, as well as cleft palate, [106,118]. TGFbeta and Wnt signaling interact genetically to promote NC-derived odontoblast differentiation and dentin formation [118]. Dlx3 promotes ossification of mandibular and frontal bones and directly regulates dentin production in NC-derived odontoblasts [119,120]. Disrupted patterning along the D−V (buccallingual) axis of jaws can lead to supernumerary teeth, and suppression of Shh and Wnt signaling disrupts both tooth patterning and tooth number.

Gene Regulatory Interactions in the Developing Skull Vault

New insights into growth and patterning of the dermal bones of the skull vault in mice, and presumably in all mammals, has come recently from lineage tracing of frontal and parietal bone rudiments. Prior to the radial pattern of mineralization observed at the final positions of these bones above the brain, osteo-progenitor cells migrate dorsally from a more ventrolateral area where they already have formed frontal and parietal bone rudiments (Figure 7.3C) [43,121]. Migrating precursor cells in the supraorbital ridge continually add to the leading edges of the developing bone. As embryonic bones become mineralized miniatures of their adult forms, continued growth then involves a signaling system leading to osteogenesis at the sutural margins, where long-term skull growth mainly occurs. Because of the prevalence of premature suture closure—craniosynostosis in humans—great efforts have been made to understand suture development. Of particular interest for human health, the sagittal and coronal sutures make the greatest contributions to skull growth and their premature closure dramatically impairs brain growth. Developmentally, craniosynostosis involves failure of signals that drive growth and differentiation at the sutural margins, and the finding that certain sutures (e.g., coronal in mice) form as juxtapositions between NC and mesoderm has provided an additional level of complexity to this system [41].

How are such boundaries specified? Experimental studies in animals and genetic mapping of craniosynostosis in humans have identified key roles for both diffusible signals (Fgfs, Wnts, Bmps, retinoic acid [RA]), as well as contact-dependent signaling (ephrin/Eph,

Jagged/Notch). Developmental studies in mice have begun to reveal functional links between some of these factors (Figure 7.3D). For example, Msx2 functions downstream of, and in opposition to, the Wnt-target Twist1 in regulating the Eph/ephrin-dependent integrity of the osteogenic/non-osteogenic boundary at the coronal suture [122]. Twist1 regulates several ephrins (*Efna2/4*) and *Eph receptor A4* (*EphA4*) in a strip of cells termed "ectocranial" mesenchyme overlying frontal and parietal bone rudiments [121]. Guided migration of osteogenic cells along the ectocranial mesenchyme toward the osteogenic front and their exclusion from non-osteogenic domains requires EfnaA2/EphA4 signaling (Figure 7.3D). Twist1 also regulates Jag1-Notch signaling and *Jag1* mutations, which cause craniosynostosis in human Alagille syndrome, leading to fusions of the coronal suture in mice [92]. Jag1 is required (1) to repress the osteogenic fate in sutural cells in part by repressing Bmp-responses and (2) to maintain the boundary between osteogenic and non-osteogenic compartments (Figure 7.3D). Importantly, in addition to the coronal suture, synostosis also occurs in the squamosal, lambdoid, occipito-interparietal, and interfrontal sutures of Twist1-Jag1 mutants, implying that similar mechanisms may operate in most sutures [92].

As implied above, Bmp signaling also contributes to calvarial development and its activity is regulated in a time-dependent manner by Msx1/2. First, Msx1/2 prevents the Bmp-dependent differentiation of the early migrating mesenchyme associated with the frontal bone rudiment [123]. Second, Msx1/2 promotes proliferation, differentiation and survival of the osteogenic lineage at later stages [124,125]. A requirement for tightly regulated Bmpr1a within the NC-derived frontal bone rudiment has been confirmed based on the premature fusion of the metopic suture (separating left and right frontals) and the reduced width of the frontals in mice expressing a constitutively active *Bmpr1a* (*CA-Bmpr1a*) transgene in the NC-derived skull [126]. These studies also report elevated Fgf signaling in CA-Bmpr1a + frontals, thus linking Bmp- to Fgf signaling—FGF receptor mutations have been reported in at least six distinct syndromes whose defining feature is craniosynostosis [127].

Translating Patterning into Growth and Survival

As the proper patterning of individual craniofacial modules results in skeletal elements with specific morphology, patterning is conceptually bound to result in tightly controlled spatial and temporal patterns of NCC proliferation, survival, migration or differentiation. However, such links remain mostly unknown because many of the same signals involved in patterning often influence several of these processes at once, and unraveling these interconnected functions is one of the major challenges of craniofacial skeletal biology.

Much of what we know about the GRNs that regulate EM growth and survival come from studies of the human 22q11.2 deletion—or DiGeorge syndrome (DGS)—which comprises multiple organ defects, including craniofacial anomalies affecting the entire series of pharyngeal arches. Craniofacial defects variably include cleft palate, micrognathia, and/or middle ear defects. This syndrome most commonly results from the hemizygous loss of a 3 Mb region of human chromosome 22 containing 35 genes. *TBX1* has been a primary candidate for the cause of DGS, as it is included in the 22q11 interval and heterozygous loss of *Tbx1* in mice phenocopies the craniofacial phenotype of DGS [128]. *Tbx1* is expressed in pharyngeal endoderm and regulates the endodermal expression of signals required for EM survival and patterning in zebrafish [129]. In fact, Tbx1

is required for expression of many of the critical epithelial signals implicated in arch development, including Fgf8, Bmp4, Shh, and RA (Figure 7.4A) [130–132]. In addition, *Tbx1* expression is induced by many of these same signals [133–135]. Shh produced by the endoderm appears to kick-start lower-jaw development by inducing *Fgf8* and *Bmp4* expression [136]—Fgf8 induction occurs through sphingosine-1-phosphate signaling [137] and is required for *Tbx1* expression and NC survival [134,138–143]. Tbx1, in turn, promotes *Shh* expression in the ventral endoderm through *Tbx2/3* [144]. Notably, *Tbx2/3* and *Msx2/3* interact with *Hdac3* to repress NC proliferation [145]. Several microRNAs are also required to regulate Shh signaling and NC survival [146–148].

Further scrutiny of the human 22q11 interval has led to a more detailed knowledge of the GRN involved. As some DGS individuals have intact *TBX1* alleles, other genes within the 22q11 interval have been hypothesized to contribute to DGS etiology. These include the adaptor protein CRKL and ERK2 (Figure 7.4B) [131,149]. Fgf8 induces tyrosine phosphorylation of FGF receptors 1 and 2 (Fgfr1/2) and their binding to CrkL, which in turn is required for normal cellular responses to Fgf8, including survival and migration [131], as a result of active ERK1/2 pathway signaling. Notably, *ERK2* microdeletions have been detected in a minority of DGS patients, and the NC-targeted loss of either ERK1/2, their upstream regulators MEK1/2 or B-Raf/C-Raf, or their downstream transcription factor SRF, in mice causes cardiac and craniofacial abnormalities similar to DGS (Figure 7.4B) [149]. In addition, NC-specific inactivation of another pathway intermediate downstream of CRKL, the protein phosphatase SHP2, leads to decreased phosphorylated ERK (pERK) activity and a DGS-like phenotype [150]. Fgf signaling also induces a *Vestigial-like* gene within the pharyngeal endoderm, which is required for NC survival [151].

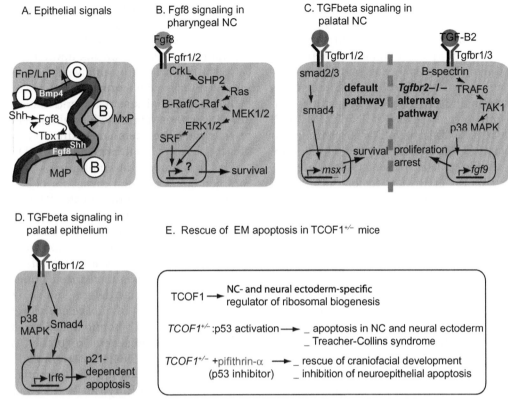

FIGURE 7.4 **Genetic pathways controlling ectomesenchymal growth and survival.** (A) Epithelial signals supporting Tbx1 expression and growth of the facial prominences. Anterior view of the left side of the face. Shh in pharyngeal endoderm (purple), Fgf8 in ectoderm (green), and Bmp4 (red). Circled letters indicate signaling pathways illustrated in (B–D). (B) Fgf8 signaling promotes EM cell survival. Orange square represents a pharyngeal NCC. (C) Canonical Tgfbeta signaling is a default pathway promoting NCC survival in the palate (left side), while an alternate pathway is activated through TAK1 in Tgfbr2-deficient EM (right side). (D) Canonical (Smad4) and non-canonical (p38) Tgfbr1/2-dependent palatal epithelium apoptosis. (E) Regulation of cell survival by TCOF and rescue of apoptosis by p53 inhibition.

Another body of research that has begun to uncover the molecular mechanisms of EM proliferation and survival has focused on the etiology of cleft palate, a birth defect in many cases caused by NC insufficiency. Here also, TGFbeta studies highlight the new insights coming from tissue-specific gene inactivation (Figure 7.4C). In the canonical Tgfbeta/BMP-signaling pathway, ligand binding to a heteromeric complex of type I and type II receptors (Tgfbr1/2) activates Tgfbr1, which in turn regulates a transcriptional response in part through Smad4. However, while NC-specific ablation of either Tgfbr1/2 or Smad4 results in cleft palate, different cellular mechanisms are regulated in these mutants as Tgfbr1 and Smad4 are required for NC survival, while Tgfbr2 promotes proliferation (Figure 7.4C) [152–155]. This apparent discrepancy in phenotypes was recently shown to result from the NC-activation of an alternate, Smad-independent, Tgfbr1/3 pathway in the absence of Tgfbr2, where signal transduction operates through Traf6/Tak1/p38 Mapk [156]. This non-canonical TGFb-mediated p38 Mapk signaling represses Fgf9/Pitx2-dependent expression of *Cyclin D1/D3*, resulting in decreased proliferation of palatal mesenchyme [157]. Interestingly, conditional expression of the Fgf-signaling modulator *Spry1* in NC decreases proliferation and increases apoptosis in association with lowered levels of Tfap2a, Msx1/2, and Dlx5/6, leading to midfacial clefting and cleft palate [158]. In contrast to their requirements in the palatal mesenchyme, Tgfbr1/2 both promote epithelial fusion of palatal shelves through apoptosis [153–159], and they do so by activating *Irf6*-dependent expression of *p21* by redundant transduction through both Smad4 and p38 Mapk pathways (Figure 7.4D) [157,159].

How do Shh and Fgf signaling ultimately regulate the EM cell cycle? In contrast to TGFbr1/2, few studies have examined these questions. Small GTP-ases Rac1 and Cdc42 are required cell-autonomously in postmigratory NC to promote proliferation [160]. This study also suggests that while the NCC cycle is Fgf responsive during migration, it later becomes more responsive to Egf signaling at postmigratory stages. Consequently, distinct factors regulate NCC cycle at different stages.

ENVIRONMENTAL FACTORS IN CRANIOFACIAL DEVELOPMENT

So far we have emphasized genetic regulation of the EM, but a growing body of evidence suggests that these cells are also highly sensitive to environmental factors and stress. These appear to feed in to p53-dependent cell survival pathways, including those described above, with very distinct craniofacial defects. This raises the exciting prospect that some craniofacial disorders, even congenital syndromes, may be ameliorated if such stresses can be minimized.

Important insights have come from the studies of mouse models for Treacher Collins syndrome, which is caused by mutations in the human nucleolar phosphoprotein Treacle (*TCOF1*) [161]. Haploinsufficiency for *Tcof1* in mice stabilizes p53 leading to cell cycle arrest in neuroepithelial and NCC and their subsequent apoptosis (Figure 7.4E). Inhibition of p53 rescues this cell death and the craniofacial defects in $Tcof1^{+/-}$ mutants (Figure 7.4D) [162].

This has important implications for other common environmental causes for NC defects, such as prenatal hypoxia, folate defiency, and fetal alcohol syndrome. Incubation of avian embryos in various hypoxic conditions induces apoptosis in NCC [163]. Depletion of *reduced folate carrier* (*Rfc*) in *Xenopus* or loss-of-function mutants in mice leads to reduced numbers of NCC and craniofacial defects, while overexpression of *Rfc* or of *5-methyltetrahydrofolate* (*Mthf*) expands NC populations, suggesting this depends on epigenetic regulation of histone methylation [164]. Interestingly, gain-of-function mutations in the essential Wnt receptor cofactor, LRP6, can be rescued with dietary folate supplements, which attenuates elevated Wnt signaling and restores proliferation in NC progenitors [165].

It is well known that cranial NCC are also extremely sensitive to alcohol and defects in EM are hallmarks of fetal alcohol spectrum disorders. Ethanol induces apoptosis in NCC *in vitro* and *in vivo* reduces Tgfbeta1 expression and disrupts formation of parietal bones [166]. Induction of endogenous antioxidants through activation of nuclear factor-erythroid 2-related factor 2 (Nrf2) prevents oxidative stress and apoptosis in ethanol-exposed mice, revealing interactions between EtOH-induced cell death and hypoxia [167]. Other causes of early EM death may feed into these stress pathways. For example, conditional loss of Bmpr1a in NC-derived cells leads to p53 activation, defects in nasal and frontal bone development, and ultimately palatal clefting [168]. Overexpression of

Smad7, an inhibitory Smad common to multiple TGFbeta pathways, induces NCC death and craniofacial skeletal defects [169]. Finally, loss-of-function mutations in *Tfap2a* also elevate apoptosis in zebrafish and mouse models [6], and mutations in *TFAP2A* cause branchio-oculo-facial syndrome in humans [170].

EM cells also respond to mechanical stresses. Classical experiments have shown defects in jaw development following paralysis of jaw muscles [171]. Consistent with these results, paralyzed zebrafish mutants lacking acetylcholine receptors or mice lacking differentiated skeletal muscle, show a failure in mediolateral intercalation of chondrocytes and stacking [172]. Responses to mechanical stress in chondrogenesis and osteogenesis appear to require the functions of primary cilia, since mutations in the cilia proteins, Pkd1 and Pkd2, alter skeletal defects following changes in mechanical stress [110,112].

CRANIAL NC DEVELOPMENT AND EVOLUTION

Since their inception in common ancestors of the gnathostome lineage, jaws have evolved a variety of shapes and functions, while keeping a basic bauplan, suggesting a set of underlying features that are common to all jaws. The Hinge and Caps model proposes an archetypal organization of embryonic jaws that would form the basis for the generally conserved gnathostome jaw architecture (Figure 7.5A) [174]. It aims to address if gnathostome jaws all form in a similar manner and hypothesizes a mechanism to coordinate upper and lower jaw development. In this model, the Hinge orchestrates the registry of upper and lower jaws and sets up polarity within the developing jaw (Figure 7.5A). Evidence supporting this model includes Fgf8 as a Hinge-associated source of patterning information and the Cap-associated expression and patterning activity

of Satb2 in mice and chick embryos [140,175]. In addition, the lower jaw to upper jaw homeosis observed in *Dlx5/6* mutant mice demonstrates the existence of a Hinge-centric patterning program [68]. Recent studies in a chondrichthyan have shown that genes with known polarized jaw expression patterns also have conserved Hinge- (*Emx1/2, Nkx3.2*) or Cap-centric (*Dlx3, Prx1, Msx1, Tbx2, Bmp4*) expression patterns (Figure 7.5A) [176]. However, at least some ventralizing signals in the mandible, such as Edn1, seem to be specific to the lower jaw [59,60].

Is the Hinge and Caps model in accordance with our current knowledge of pharyngeal D−V patterning? Numerous experimental studies of D−V patterning have highlighted the asymmetric expression and activity of Edn1 at the ventral/distal tip of the mandibular process, and the equally asymmetric expression of D−V patterning genes, such as Hand2 and Jag1b, in ventral and dorsal cells of the lower jaw, respectively [59,60]. While this ventro-centric patterning appears to contradict the Hinge-centric organization posited by the Hinge and Caps model, these two models reflect parallel aspects of jaw development. One of these aspects is the joint-centric registry of gnathostome upper and lower jaws, while the other is the inherent morphological difference between upper and lower jaws.

Instead of focusing on the genetic commonalities underlying the jaw bauplan, another focus of research has started to identify pathways contributing to morphological variation observed in nature. Studies of Darwin's finches indicate that while Bmp4 and Calmodulin (CaM) signaling independently regulate beak width/depth and length through their activities in the prenasal cartilage primordium, the TgfbIIr/Bcatenin/Dkk3 trio regulates beak length and depth through its activity in the premaxillary bone primordium (Figure 7.5B) [177−179]. While this two-module program of beak development may

(A) Hinge and Caps model

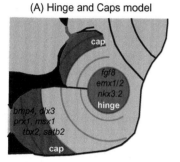

(B) Developmental basis of beak shape in Darwin's finches

Protein	Element patterned		Dimension regulated		
	Prenasal cartilage	Premaxillary bone	Depth	Width	Length
Bmp4	+	- (indirect effect)	+	+	0/-
CaM	+	0/- (indirect effect)	0	0	+
TGFBrII/ Bcat/Dkk3	0	+	+	0	+

(C) Pathways of morphological diversification in cichlids (1) and dogs (2)

(D) Evolution of neural crest – mesoderm boundary in the osteichthyan skull vault

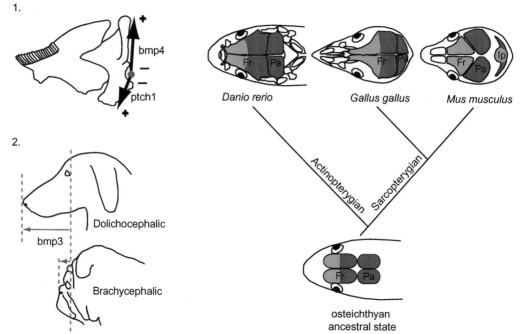

FIGURE 7.5 **Comparative developmental studies of genetic basis for skull evolution.** (A) Hinge and Caps model for upper and lower jaws. Anterior view of left side of an embryonic mouse face showing genes expressed in the hinge (red) and caps (blue) in the MdP and MxP. (B) Table summarizing control of beak morphology in Darwin's finches (redrawn after Mallarino et al. 2010). (C) Quantitative trait linkage (QTL) analysis of the jaw closure apparatus in cichlid fish (1) and facial length in dogs (2) (1. redrawn after Albertson et al. [181]; 2. redrawn after Schoenebeck et al. [173]). (D) Shifts in the boundary between NC- (blue) and mesoderm- (red) derived membrane bones of the skull vault in zebrafish, chick, and mouse suggest an ancestral boundary within the frontal bone of an ancestral osteichthyan that gave rise to both ray-finned fishes (actinopterygians) and lobe-finned fishes (sarcopterygians) which gave rise to tetrapods (redrawn after Chai and Maxson [183] and Cubbage and Mabee [22]).

explain how depth and width can evolve independently, it does not directly identify loci where genomic sequence differences contribute to phenotypic differences. In contrast, across breed genome wide association studies (ABGWAS) conducted in dogs and quantitative trait locus analysis (QTLA) conducted in cichlid fish have started to meet this challenge. Interestingly, two such studies have identified sequence polymorphisms associated with Bmp sequences: different *bmp4* alleles regulate the length of the coronoid process in the cichlid jaw [181], and *bmp3* variants are causative of morphological differences between long- and short-headed dogs (Figure 7.5C) [173]. Another QTLA study found that allelic variants of the Hh-receptor Ptch1 contribute to differences in retroarticular process length—a character state modulating the tradeoff between force and velocity during lower jaw depression in cichlids (Figure 7.5C) [182].

How have changes in the lineal origins of different bones influenced evolutionary variation in the skull? As discussed above, recent advances in lineage mapping of the NC versus mesoderm contributions to the head skeleton in mice, chicken, and zebrafish give a novel perspective on skull evolution [19,40,41,46]. The contrasting derivation of the avian- and murine-frontals casts doubt over their homology such that renaming the avian frontal as "frontoparietal" has been proposed [184]. This dual NC/mesoderm derivation of the frontal bone has been recently demonstrated in zebrafish using Cre-mediated methods to fluorescently label NC (Figure 7.5D) [19,46], suggesting that the ancestral condition in osteichthyans is a frontoparietal bone with dual NC/mesoderm derivation. A novel perspective given by this new reconstruction is that the ancestral NC−mesoderm boundary did not correspond to a cranial suture. Consequently, the registry between the mammalian coronal suture and NC−mesoderm boundary is evolutionarily derived within the

tetrapod lineage and appears restricted to mammals (Figure 7.5D), further suggesting that the mammalian coronal suture may be an evolutionary innovation that perhaps allowed increased brain size. No avian sutures are in a suitable position to form long-term growth centers contributing growth in the fronto-occipital plane, equivalent to the function of the mammalian coronal suture [41].

Following the paradigm of homology inferred from NC-derivation, mammalian skulls across more than 300 extinct and extant taxa have been examined, leading to the proposal that the dual NC/mesoderm origin of the interparietal can be explained as an evolutionary consequence of the fusion between the NC-derived postparietals and the mesoderm-derived tabulars of more basal lineages [185]. The authors conclude that the interparietal consists of four elements, which have variably fused through evolution, providing a developmental basis to the persistent trend of simplification via fusions and losses of bony elements shown by the premammalian synapsid fossil record [186].

CONCLUSIONS AND PERSPECTIVES

Skeletogenic NCC start out life similar to other pluripotent NC progenitors but rapidly enter unique environments, where they encounter signals that establish a modular organization quite distinct from the axial and appendicular skeleton. This includes NC versus mesodermal origins of skeletal elements, segmentation into pharyngeal arches rather than somites, D−V domains within each pharyngeal segment, midline migrations and interactions that form the cranial base and palate at the extreme anterior end of the skull, and NC versus mesodermal domains of the skull vault.

Here we have tried to highlight recent genetic studies. These include long-term fate-mapping studies in fish and mice, conditional loss-of-function studies of a huge number of factors in the NC using *Wnt1:CRE* in mice, and manipulations of epithelial signals. Modularity provides a framework for understanding how genetic and environmental influences alter EM development. Modularity also provides a bridge between evolution and development, as many craniofacial differences between species clearly arise through module-specific changes in early embryonic development.

These studies have set the stage for more molecular and cellular approaches, which will ultimately reveal the underlying mechanisms and whether or not they act in such modular fashion. Going forward, a major challenge in the field will be translating early patterning events in embryos (e.g., A−P and D−V specification of NCC) into precise spatiotemporal control of craniofacial skeletal growth, survival, morphogenesis, and differentiation. How do cranial cartilages and bones form with the appropriate size and shape? Conditional mutants in mice have begun to tease apart the various tissue-specific requirements for craniofacial genes, but this is just the tip of the iceberg. The generation of transgenic lines in zebrafish, which allows imaging of cell shape changes and their underlying molecular dynamics in skeletal precursors, promises to contribute a lot to our understanding. Efforts to look outside of these model systems, at craniofacial variation in other species, are also generating new ideas as to modularity and constraints on the development of the vertebrate head. Finally, with all of this new information, it seems that the time has come to begin to assemble the wealth of new genetic information into computational models incorporating spatial and temporal dynamics of interacting craniofacial tissues to build a more comprehensive picture of the integrated GRN underlying the craniofacial skeleton.

References

[1] Knight RD, Schilling TF. Cranial neural crest and development of the head skeleton. Adv Exp Med Biol 2006;589:120−33.

[2] Platt JB. Ectodermic origin of the cartilages of the head. Anat Anz 1893;506−9.

[3] Landacre FL. The fate of the neural crest in the head of the urodeles. J Comp Neurol 1921;1−43.

[4] Stone LS. Experiments showing the role of migrating neural crest in the formation of head skeleton and loose connective tissue in *Rana palustris*. Roux's Arch 1929;40−77.

[5] Blentic A, Tandon P, Payton S, et al. The emergence of ectomesenchyme. Dev Dyn 2008;237(3):592−601.

[6] Knight RD, Javidan Y, Zhang TL, Nelson S, Schilling TF. AP2-dependent signals from the ectoderm regulate craniofacial development in the zebrafish embryo. Development 2005;132(13):3127−38.

[7] Wang WD, Melville DB, Montero-Balaguer M, Hatzopoulos AK, Knapik EW. Tfap2a and Foxd3 regulate early steps in the development of the neural crest progenitor population. Dev Biol 2011;360(1):173−85.

[8] Cox SG, Kim H, Garnett AT, Medeiros DM, An W, Crump JG. An essential role of variant histone H3.3 for ectomesenchyme potential of the cranial neural crest. PLoS Genet 2012;8(9):16.

[9] Dupin E, Calloni GW, Le Douarin NM. The cephalic neural crest of amniote vertebrates is composed of a large majority of precursors endowed with neural, melanocytic, chondrogenic and osteogenic potentialities. Cell Cycle 2010;9(2):238−49.

[10] Ishii M, Arias AC, Liu L, Chen YB, Bronner ME, Maxson RE. A stable cranial neural crest cell line from mouse. Stem Cell Dev 2012;21(17):3069−80.

[11] Hochgreb-Hägele T, Bronner ME. A novel FoxD3 gene trap line reveals neural crest precursor movement and a role for FoxD3 in their specification. Dev Biol 2013;374(1):1−11.

[12] Das A, Crump JG. Bmps and id2a act upstream of Twist1 to restrict ectomesenchyme potential of the cranial neural crest. PLoS Genet 2012;8(5):e1002710.

[13] Mundell NA, Labosky PA. Neural crest stem cell multipotency requires Foxd3 to maintain neural potential and repress mesenchymal fates. Development 2011;138(4):641−52.

[14] Breau MA, Pietri T, Stemmler MP, Thiery JP, Weston JA. A nonneural epithelial domain of embryonic cranial neural folds gives rise to ectomesenchyme. Proc Natl Acad Sci USA 2008;105(22):7750−5.

[15] Janvier P. Early vertebrates. Oxford, England: Oxford University Press; 1996.

[16] Gilbert SF, Bender G, Betters E, Yin M, Cebra-Thomas JA. The contribution of neural crest cells to the nuchal bone and plastron of the turtle shell. Integr Comp Biol 2007;47(3):401−8.

[17] McGonnell IM, Graham A. Trunk neural crest has skeletogenic potential. Curr Biol 2002;12(9):767−71.

[18] Smith M, Hickman A, Amanze D, Lumsden A, Thorogood P. Trunk neural crest origin of caudal fin mesenchyme in the zebrafish Brachydanio rerio. Proc R Soc B Biol Sci 1994;256(1346):137−45.

[19] Kague E, Gallagher M, Burke S, Parsons M, Franz-Odendaal T, Fisher S. Skeletogenic fate of zebrafish cranial and trunk neural crest. PLoS One 2012;7(11): e47394.

[20] Noden DM. Craniofacial development: new views on old problems. Anat Rec 1984;208:1−13.

[21] Gross JB, Hanken J. Review of fate-mapping studies of osteogenic cranial neural crest in vertebrates. Dev Biol 2008;317:389−400.

[22] Cubbage CC, Mabee PM. Development of the cranium and paired fins in the zebrafish Danio rerio (Ostariophysi, Cyprinidae). J Morphol 1996;229:121−60.

[23] Lee RTH, Thiery JP, Carney TJ. Dermal fin rays and scales derive from mesoderm, not neural crest. Curr Biol 2013;23(9):R336−7.

[24] Sobkow L, Epperlein HH, Herklotz S, Straube WL, Tanaka EM. A germline GFP transgenic axolotl and its use to track cell fate: dual origin of the fin mesenchyme during development and the fate of blood cells during regeneration. Dev Biol 2006;290 (2):386−97.

[25] Hunt P, Krumlauf R. Deciphering the Hox code—clues to patterning branchial regions of the head. Cell 1991;66(6):1075−8.

[26] Schilling TF, Knight RD. Origins of anteroposterior patterning and Hox gene regulation during chordate evolution. Philos Trans R Soc Lond B Biol Sci 2001;356 (1414):1599−613.

[27] Rijli FM, Mark M, Lakkaraju S, Dierich A, Dollé P, Chambon P. A homeotic transformation is generated in the rostral branchial region of the head by disruption of Hoxa-2, which acts as a selector gene. Cell 1993;75(7):1333−49.

[28] Gendron-Maguire M, Mallo M, Zhang M, Gridley T. Hoxa-2 mutant mice exhibit homeotic transformation of skeletal elements derived from cranial neural crest. Cell 1993;75(7):1317−31.

[29] Hunter MP, Prince VE. Zebrafish Hox paralogue group 2 genes function redundantly as selector genes to pattern the second pharyngeal arch. Dev Biol 2002;247(2):367−89.

[30] Baltzinger M, Ori M, Pasqualetti M, Nardi I, Rijli FM. Hoxa2 knockdown in Xenopus results in hyoid to mandibular homeosis. Dev Dyn 2005;234(4):858−67.

[31] Pasqualetti M, Ori M, Nardi I, Rijli FM. Ectopic Hoxa2 induction after neural crest migration results in homeosis of jaw elements in Xenopus. Development 2000;127(24):5367−78.

[32] Grammatopoulos GA, Bell E, Toole L, Lumsden A, Tucker AS. Homeotic transformation of branchial arch identity after Hoxa2 overexpression. Development 2000;127(24):5355−65.

[33] Santagati F, Minoux M, Ren SY, Rijli FM. Temporal requirement of Hoxa2 in cranial neural crest skeletal morphogenesis. Development 2005;132(22):4927−36.

[34] Crump JG, Swartz ME, Eberhart JK, Kimmel CB. Moz-dependent Hox expression controls segment-specific fate maps of skeletal precursors in the face. Development 2006;133(14):2661−9.

[35] Miller CT, Maves L, Kimmel CB. moz regulates Hox expression and pharyngeal segmental identity in zebrafish. Development 2004;131(10):2443−61.

[36] Laue K, Daujat S, Crump JG, et al. The multidomain protein Brpf1 binds histones and is required for Hox gene expression and segmental identity. Development 2008;135(11):1935−46.

[37] Haberland M, Mokalled MH, Montgomery RL, Olson EN. Epigenetic control of skull morphogenesis by histone deacetylase 8. Gene Dev 2009;23 (14):1625−30.

[38] Minoux M, Antonarakis GS, Kmita M, Duboule D, Rijli FM. Rostral and caudal pharyngeal arches share a common neural crest ground pattern. Development 2009;136(4):637−45.

[39] Noden DM. The role of the neural crest in patterning of avian cranial skeletal, connective and muscle tissues. Dev Biol 1983;96(1):144−65.

[40] Evans DJR, Noden DM. Spatial relations between avian craniofacial neural crest and paraxial mesoderm cells. Dev Dyn 2006;235(5):1310−25.

[41] Jiang XB, Iseki S, Maxson RE, Sucov HM, Morriss-Kay GM. Tissue origins and interactions in the mammalian skull vault. Dev Biol 2002;241(1):106−16.

[42] Matsuoka T, Ahlberg PE, Kessaris N, et al. Neural crest origins of the neck and shoulder. Nature 2005;436(7049):347−55.

[43] Yoshida T, Vivatbutsiri P, Morriss-Kay G, Saga Y, Iseki S. Cell lineage in mammalian craniofacial mesenchyme. Mech Dev 2008;125(9−10):797−808.

[44] Noden DM. The role of neural crest in patterning of avian cranial skeletal, connective and muscle tissues. Dev Biol 1983;96:144−65.

[45] Thompson H, Ohazama A, Sharpe PT, Tucker AS. The origin of the stapes and relationship to the otic

capsule and oval window. Dev Dyn 2012;241 (9):1396–404.

[46] Mongera A, Singh AP, Levesque MP, Chen YY, Konstantinidis P, Nüsslein-Volhard C. Genetic lineage labeling in zebrafish uncovers novel neural crest contributions to the head, including gill pillar cells. Development 2013;140(4):916–25.

[47] Simon C, Lickert H, Götz M, Dimou L. Sox10-iCreERT2: a mouse line to inducibly trace the neural crest and oligodendrocyte lineage. Genesis 2012;50 (6):506–15.

[48] Freyer L, Aggarwal V, Morrow BE. Dual embryonic origin of the mammalian otic vesicle forming the inner ear. Development 2011;138(24):5403–14.

[49] Wang S-K, Komatsu Y, Mishina Y. Potential contribution of neural crest cells to dental enamel formation. Biochem Biophys Res Commun 2011;415(1):114–9.

[50] Richter CA, Amin S, Linden J, Dixon J, Dixon MJ, Tucker AS. Defects in middle ear cavitation cause conductive hearing loss in the Tcof1 mutant mouse. Hum Mol Genet 2010;19(8):1551–60.

[51] Thompson H, Tucker AS. Dual origin of the epithelium of the mammalian middle ear. Science 2013;339 (6126):1453–6.

[52] Couly G, Creuzet S, Bennaceur S, Vincent C, Le Douarin NM. Interactions between Hox-negative cephalic neural crest cells and the foregut endoderm in patterning the facial skeleton in the vertebrate head. Development 2002;129(4):1061–73.

[53] David NB, Saint-Etienne L, Tsang M, Schilling TF, Rosa FM. Requirement for endoderm and FGF3 in ventral head skeleton formation. Development 2002;129(19):4457–68.

[54] Jheon AH, Schneider RA. The cells that fill the bill: neural crest and the evolution of craniofacial development. J Dent Res 2009;88(1):12–21.

[55] Ozeki H, Kurihara Y, Tonami K, Watatani S, Kurihara H. Endothelin-1 regulates the dorsoventral branchial arch patterning in mice. Mech Dev 2004;121 (4):387–95.

[56] Rieder MJ, Green GE, Park SS, et al. A human homeotic transformation resulting from mutations in PLCB4 and GNAI3 causes auriculocondylar syndrome. Am J Hum Genet 2012;90(5):907–14.

[57] Graham A, Richardson J. Developmental and evolutionary origins of the pharyngeal apparatus. EvoDevo 2012;3(1):24.

[58] Clouthier DE, Hosoda K, Richardson JA, et al. Cranial and cardiac neural crest defects in endothelin-A receptor-deficient mice. Development 1998;125(5):813–24.

[59] Clouthier DE, Garcia E, Schilling TF. Regulation of facial morphogenesis by endothelin signaling: insights from mice and fish. Am J Med Genet A 2010;152A (12):2962–73.

[60] Medeiros DM, Crump JG. New perspectives on pharyngeal dorsoventral patterning in development and evolution of the vertebrate jaw. Dev Biol 2012;371 (2):121–35.

[61] Miller CT, Swartz ME, Khuu PA, Walker MB, Eberhart JK, Kimmel CB. mef2ca is required in cranial neural crest to effect Endothelin1 signaling in zebrafish. Dev Biol 2007;308(1):144–57.

[62] Walker MB, Miller CT, Swartz ME, Eberhart JK, Kimmel CB. phospholipase C, beta 3 is required for Endothelin1 regulation of pharyngeal arch patterning in zebrafish. Dev Biol 2007;304(1):194–207.

[63] Kimmel CB, Ullmann B, Walker M, Miller CT, Crump JG. Endothelin 1-mediated regulation of pharyngeal bone development in zebrafish. Development 2003;130(7):1339–51.

[64] Nair S, Li W, Cornell R, Schilling TF. Requirements for Endothelin type-A receptors and Endothelin-1 signaling in the facial ectoderm for the patterning of skeletogenic neural crest cells in zebrafish. Development 2007;134(2):335–45.

[65] Tavares AL, Garcia EL, Kuhn K, Woods CM, Williams T, Clouthier DE. Ectodermal-derived Endothelin1 is required for patterning the distal and intermediate domains of the mouse mandibular arch. Dev Biol 2012;371(1):47–56.

[66] Sato T, Kurihara Y, Asai R, et al. An endothelin-1 switch specifies maxillomandibular identity. Proc Natl Acad Sci USA 2008;105(48):18806–11.

[67] Ruest LB, Clouthier DE. Elucidating timing and function of endothelin-A receptor signaling during craniofacial development using neural crest cell-specific gene deletion and receptor antagonism. Dev Biol 2009;328(1):94–108.

[68] Depew MJ, Lufkin T, Rubenstein JLR. Specification of jaw subdivisions by Dix genes. Science 2002;298(5592):381–5.

[69] Verzi MP, Agarwal P, Brown C, McCulley DJ, Schwarz JJ, Black BL. The transcription factor MEF2C is required for craniofacial development. Dev Cell 2007;12(4):645–52.

[70] Funato N, Chapman SL, McKee MD, et al. Hand2 controls osteoblast differentiation in the branchial arch by inhibiting DNA binding of Runx2. Development 2009;136(4):615–25.

[71] Barron F, Woods C, Kuhn K, Bishop J, Howard MJ, Clouthier DE. Downregulation of Dlx5 and Dlx6 expression by Hand2 is essential for initiation of tongue morphogenesis. Development 2011;138(11):2249–59.

[72] Barbosa AC, Funato N, Chapman S, et al. Hand transcription factors cooperatively regulate development

of the distal midline mesenchyme. Dev Biol 2007;310 (1):154–68.

[73] Miller CT, Yelon D, Stainier DY, Kimmel CB. Two endothelin 1 effectors, hand2 and bapx1, pattern ventral pharyngeal cartilage and the jaw joint. Development 2003;130(7):1353–65.

[74] Talbot JC, Johnson SL, Kimmel CB. hand2 and Dlx genes specify dorsal, intermediate and ventral domains within zebrafish pharyngeal arches. Development 2010;137(15):2506–16.

[75] Zuniga E, Rippen M, Alexander C, Schilling TF, Crump JG. Gremlin 2 regulates distinct roles of BMP and Endothelin 1 signaling in dorsoventral patterning of the facial skeleton. Development 2011;138 (23):5147–56.

[76] Sperber SM, Dawid IB. barx1 is necessary for ectomesenchyme proliferation and osteochondroprogenitor condensation in the zebrafish pharyngeal arches. Dev Biol 2008;321(1):101–10.

[77] Nichols JTPL, Moens CB, Kimmel CB. barx1 represses joints and promotes cartilage in the craniofacial skeleton 2013;140(13):2765–75.

[78] Alexander C, Zuniga E, Blitz IL, et al. Combinatorial roles for BMPs and Endothelin 1 in patterning the dorsal-ventral axis of the craniofacial skeleton. Development 2011;138(23):5135–46.

[79] Swartz ME, Sheehan-Rooney K, Dixon MJ, Eberhart JK. Examination of a palatogenic gene program in zebrafish. Dev Dyn 2011;240(9):2204–20.

[80] Mani P, Jarrell A, Myers J, Atit R. Visualizing canonical Wnt signaling during mouse craniofacial development. Dev Dyn 2010;239(1):354–63.

[81] Bonilla-Claudio M, Wang J, Bai Y, Klysik E, Selever J, Martin JF. Bmp signaling regulates a dose-dependent transcriptional program to control facial skeletal development. Development 2012;139(4):709–19.

[82] Zuniga E, Stellabotte F, Crump JG. Jagged-Notch signaling ensures dorsal skeletal identity in the vertebrate face. Development 2010;137(11):1843–52.

[83] Humphreys R, Zheng W, Prince LS, et al. Cranial neural crest ablation of Jagged1 recapitulates the craniofacial phenotype of Alagille syndrome patients. Hum Mol Genet 2012;21(6):1374–83.

[84] Mead TJ, Yutzey KE. Notch pathway regulation of neural crest cell development in vivo. Dev Dyn 2012;241(2):376–89.

[85] Kawakami M, Umeda M, Nakagata N, Takeo T, Yamamura K. Novel migrating mouse neural crest cell assay system utilizing P0-Cre/EGFP fluorescent time-lapse imaging. BMC Dev Biol 2011;11:17.

[86] Reisoli E, De Lucchini S, Nardi I, Ori M. Serotonin 2B receptor signaling is required for craniofacial morphogenesis and jaw joint formation in Xenopus. Development 2010;137(17):2927–37.

[87] Muenke M, Beachy PA. Genetics of ventral forebrain development and holoprosencephaly. Curr Opin Genet Dev 2000;10(3):262–9.

[88] Balk K, Biesecker LG. The clinical atlas of Greig cephalopolysyndactyly syndrome. Am J Med Genet A 2008;146A(5):548–57.

[89] Dennis JF, Kurosaka H, Iulianella A, et al. Mutations in Hedgehog acyltransferase (Hhat) perturb Hedgehog signaling, resulting in severe acrania-holoprosencephaly-agnathia craniofacial defects. PLoS Genet 2012;8(10):e1002927.

[90] Sarmah B, Wente SR. Inositol hexakisphosphate kinase-2 acts as an effector of the vertebrate Hedgehog pathway. Proc Natl Acad Sci USA 2010;107(46):19921–6.

[91] Wada N, Nohno T, Kuratani S. Dual origins of the prechordal cranium in the chicken embryo. Dev Biol 2011;356(2):529–40.

[92] Yen HY, Ting MC, Maxson RE. Jagged1 functions downstream of Twist1 in the specification of the coronal suture and the formation of a boundary between osteogenic and non-osteogenic cells. Dev Biol 2010;347 (2):258–70.

[93] Hu D, Marcucio RS, Helms JA. A zone of frontonasal ectoderm regulates patterning and growth in the face. Development 2003;130(9):1749–58.

[94] Marcucio RS, Cordero DR, Hu D, Helms JA. Molecular interactions coordinating the development of the forebrain and face. Dev Biol 2005;284 (1):48–61.

[95] Foppiano S, Hu D, Marcucio RS. Signaling by bone morphogenetic proteins directs formation of an ectodermal signaling center that regulates craniofacial development. Dev Biol 2007;312(1):103–14.

[96] Chong HJ, Young NM, Hu D, et al. Signaling by SHH rescues facial defects following blockade in the brain. Dev Dyn 2012;241(2):247–56.

[97] Wada N, Javidan Y, Nelson S, Carney TJ, Kelsh RN, Schilling TF. Hedgehog signaling is required for cranial neural crest morphogenesis and chondrogenesis at the midline in the zebrafish skull. Development 2005;132(17):3977–88.

[98] Eberhart JK, Swartz ME, Crump JG, Kimmel CB. Early Hedgehog signaling from neural to oral epithelium organizes anterior craniofacial development. Development 2006;133(6):1069–77.

[99] Dougherty M, Kamel G, Shubinets V, Hickey G, Grimaldi M, Liao EC. Embryonic fate map of first pharyngeal arch structures in the sox10: kaede zebrafish transgenic model. J Craniofac Surg 2012;23 (5):1333–7.

[100] Javidan Y, Schilling TF. Development of cartilage and bone. Zebrafish: 2nd Edition Methods Cell Biol 2004;76:415–36.

[101] Schwend T, Loucks EJ, Ahlgren SC. Visualization of Gli activity in craniofacial tissues of hedgehog-pathway reporter transgenic zebrafish. PLoS One 2010;5(12):e14396.

[102] Lan Y, Jiang R. Sonic hedgehog signaling regulates reciprocal epithelial–mesenchymal interactions controlling palatal outgrowth. Development 2009;136 (8):1387–96.

[103] Gitton Y, Benouaiche L, Vincent C, et al. Dlx5 and Dlx6 expression in the anterior neural fold is essential for patterning the dorsal nasal capsule. Development 2011;138(5):897–903.

[104] Han J, Mayo J, Xu X, et al. Indirect modulation of Shh signaling by Dlx5 affects the oral-nasal patterning of palate and rescues cleft palate in Msx1-null mice. Development 2009;136(24):4225–33.

[105] Xiong W, He F, Morikawa Y, et al. Hand2 is required in the epithelium for palatogenesis in mice. Dev Biol 2009;330(1):131–41.

[106] Baek J-A, Lan Y, Liu H, Maltby KM, Mishina Y, Jiang R. Bmpr1a signaling plays critical roles in palatal shelf growth and palatal bone formation. Dev Biol 2011;350(2):520–31.

[107] Brugmann SA, Allen NC, James AW, Mekonnen Z, Madan E, Helms JA. A primary cilia-dependent etiology for midline facial disorders. Hum Mol Genet 2010;19(8):1577–92.

[108] Zaghloul NA, Brugmann SA. The emerging face of primary cilia. Genesis 2011;49(4):231–46.

[109] Corbit KC, Aanstad P, Singla V, Norman AR, Stainier DYR, Reiter JF. Vertebrate smoothened functions at the primary cilium. Nature 2005;437 (7061):1018–21.

[110] Khonsari RH, Seppala M, Pradel A, et al. The bucco-hypophyseal canal is an ancestral vertebrate trait maintained by modulation in sonic hedgehog signaling. BMC Biol 2013;11:27.

[111] Tobin JL, Di Franco M, Eichers E, et al. Inhibition of neural crest migration underlies craniofacial dysmorphology and Hirschsprung's disease in Bardet-Biedl syndrome. Proc Natl Acad Sci USA 2008;105 (18):6714–9.

[112] Hou B, Kolpakova-Hart E, Fukai N, Wu K, Olsen BR. The polycystic kidney disease 1 (Pkd1) gene is required for the responses of osteochondroprogenitor cells to midpalatal suture expansion in mice. Bone 2009;44(6):1121–33.

[113] Lin C, Fisher AV, Yin Y, et al. The inductive role of Wnt-β-Catenin signaling in the formation of oral apparatus. Dev Biol 2011;356(1):40–50.

[114] Song Z, Liu C, Iwata J, et al. Mice with tak1 deficiency in neural crest lineage exhibit cleft palate associated with abnormal tongue development. J Biol Chem 2013;288(15):10440–50.

[115] Cobourne MT, Sharpe PT. Making up the numbers: the molecular control of mammalian dental formula. Semin Cell Dev Biol 2010;21(3):314–24.

[116] Tucker AS, Matthews KL, Sharpe PT. Transformation of tooth type induced by inhibition of BMP signaling. Science 1998;282(5391):1136–8.

[117] Jia S, Zhou J, Gao Y, et al. Roles of Bmp4 during tooth morphogenesis and sequential tooth formation. Development 2013;140(2):423–32.

[118] Li L, Lin M, Wang Y, Cserjesi P, Chen Z, Chen Y. BmprIa is required in mesenchymal tissue and has limited redundant function with BmprIb in tooth and palate development. Dev Biol 2011;349 (2):451–61.

[119] Duverger O, Zah A, Isaac J, et al. Neural crest deletion of Dlx3 leads to major dentin defects through down-regulation of Dspp. J Biol Chem 2012;287 (15):12230–40.

[120] Duverger O, Isaac J, Zah A, et al. In vivo impact of Dlx3 conditional inactivation in neural crest-derived craniofacial bones. J Cell Physiol 2013;228 (3):654–64.

[121] Ting MC, Wu NL, Roybal PG, et al. EphA4 as an effector of Twist1 in the guidance of osteogenic precursor cells during calvarial bone growth and in craniosynostosis. Development 2009;136 (5):855–64.

[122] Merrill AE, Bochukova EG, Brugger SM, et al. Cell mixing at a neural crest-mesoderm boundary and deficient ephrin-Eph signaling in the pathogenesis of craniosynostosis. Hum Mol Genet 2006;15 (8):1319–28.

[123] Roybal PG, Wu NL, Sun J, Ting MC, Schafer CA, Maxson RE. Inactivation of Msx1 and Msx2 in neural crest reveals an unexpected role in suppressing heterotopic bone formation in the head. Dev Biol 2010;343(1–2):28–39.

[124] Satokata I, Ma L, Ohshima H, et al. Msx2 deficiency in mice causes pleiotropic defects in bone growth and ectodermal organ formation. Nat Genet 2000;24 (4):391–5.

[125] Han J, Ishii M, Bringas P, Maas RL, Maxson RE, Chai Y. Concerted action of Msx1 and Msx2 in regulating cranial neural crest cell differentiation during frontal bone development. Mech Dev 2007;124(9–10):729–45.

[126] Komatsu Y, Yu PB, Kamiya N, et al. Augmentation of Smad-dependent BMP signaling in neural crest cells causes craniosynostosis in mice. J Bone Miner Res 2013;28(6):1422–33.

[127] Morriss-Kay GM, Wilkie AOM. Growth of the normal skull vault and its alteration in craniosynostosis: insights from human genetics and experimental studies. J Anat 2005;207(5):637−53.

[128] Jerome LA, Papaioannou VE. DiGeorge syndrome phenotype in mice mutant for the T-box gene, Tbx1. Nat Genet 2001;27(3):286−91.

[129] Piotrowski T, Ahn DG, Schilling TF, et al. The zebrafish Van Gogh mutation disrupts tbx1, which is involved in the DiGeorge deletion syndrome in humans. Development 2003;130(20):5043−52.

[130] Hu TH, Yamagishi H, Maeda J, McAnally J, Yamagishi C, Srivastava D. Tbx1 regulates fibroblast growth factors in the anterior heart field through a reinforcing autoregulatory loop involving forkhead transcription factors. Development 2004;131 (21):5491−502.

[131] Moon AM, Guris DL, Seo JH, et al. Crkl deficiency disrupts Fgf8 signaling in a mouse model of 22q11 deletion syndromes. Dev Cell 2006;10(1):71−80.

[132] Arnold JS, Werling U, Braunstein EM, et al. Inactivation of Tbx1 in the pharyngeal endoderm results in 22q11DS malformations. Development 2006;133(5):977−87.

[133] Bachiller D, Klingensmith J, Shneyder N, et al. The role of chordin/Bmp signals in mammalian pharyngeal development and DiGeorge syndrome. Development 2003;130(15):3567−78.

[134] Garg V, Yamagishi C, Hu TH, Kathiriya IS, Yamagishi H, Srivastava D. Tbx1 a DiGeorge syndrome candidate gene, is regulated by Sonic hedgehog during pharyngeal arch development. Dev Biol 2001;235(1):62−73.

[135] Zhang LF, Zhong T, Wang YX, Jiang Q, Song HY, Gui YH. TBX1, a DiGeorge syndrome candidate gene, is inhibited by retinoic acid. Int J Dev Biol 2006;50(1):55−61.

[136] Brito JM, Teillet MA, Le Douarin NM. Induction of mirror-image supernumerary jaws in chicken mandibular mesenchyme by Sonic Hedgehog-producing cells. Development 2008;135(13):2311−9.

[137] Balczerski B, Matsutani M, Castillo P, Osborne N, Stainier DY, Crump JG. Analysis of sphingosine-1-phosphate signaling mutants reveals endodermal requirements for the growth but not dorsoventral patterning of jaw skeletal precursors. Dev Biol 2012;362(2):230−41.

[138] Aggarwal VS, Carpenter C, Freyer L, Liao J, Petti M, Morrow BE. Mesodermal Tbx1 is required for patterning the proximal mandible in mice. Dev Biol 2010;344(2):669−81.

[139] Frank DU, Fotheringham LK, Brewer JA, et al. An Fgf8 mouse mutant phenocopies human 22q11 deletion syndrome. Development 2002;129(19): 4591−603.

[140] Trumpp A, Depew MJ, Rubenstein JLR, Bishop JM, Martin GR. Cre-mediated gene inactivation demonstrates that FGF8 is required for cell survival and patterning of the first branchial arch. Genes Dev 1999;13(23):3136−48.

[141] Ahlgren SC, Thakur V, Bronner-Fraser M. Sonic hedgehog rescues cranial neural crest from cell death induced by ethanol exposure. Proc Natl Acad Sci USA 2002;99(16):10476−81.

[142] Jeong JH, Mao JH, Tenzen T, Kottmann AH, McMahon AP. Hedgehog signaling in the neural crest cells regulates the patterning and growth of facial primordia. Genes Dev 2004;18(8):937−51.

[143] Smoak IW, Byrd NA, Abu-Issa R, et al. Sonic hedgehog is required for cardiac outflow tract and neural crest cell development. Dev Biol 2005;283 (2):357−72.

[144] Mesbah K, Rana MS, Francou A, et al. Identification of a Tbx1/Tbx2/Tbx3 genetic pathway governing pharyngeal and arterial pole morphogenesis. Hum Mol Genet 2012;21(6):1217−29.

[145] Singh N, Gupta M, Trivedi CM, Singh MK, Li L, Epstein JA. Murine craniofacial development requires Hdac3-mediated repression of Msx gene expression. Dev Biol 2013;377(2):333−44.

[146] Huang T, Liu Y, Huang M, Zhao X, Cheng L. Wnt1-cre-mediated conditional loss of Dicer results in malformation of the midbrain and cerebellum and failure of neural crest and dopaminergic differentiation in mice. J Mol Cell Biol 2010;2(3):152−63.

[147] Zehir A, Hua LL, Maska EL, Morikawa Y, Cserjesi P. Dicer is required for survival of differentiating neural crest cells. Dev Biol 2010;340(2):459−67.

[148] Nie X, Wang Q, Jiao K. Dicer activity in neural crest cells is essential for craniofacial organogenesis and pharyngeal arch artery morphogenesis. Mech Dev 2011;128(3−4):200−7.

[149] Newbern J, Zhong J, Wickramasinghe RS, et al. Mouse and human phenotypes indicate a critical conserved role for ERK2 signaling in neural crest development. Proc Natl Acad Sci USA 2008;105 (44):17115−20.

[150] Nakamura T, Gulick J, Colbert MC, Robbins J. Protein tyrosine phosphatase activity in the neural crest is essential for normal heart and skull development. Proc Natl Acad Sci USA 2009;106(27): 11270−5.

[151] Johnson CW, Hernandez-Lagunas L, Feng W, Melvin VS, Williams T, Artinger KB. Vgll2a is required for neural crest cell survival during zebrafish craniofacial development. Dev Biol 2011;357(1):269−81.

[152] Ito Y, Yeo JY, Chytil A, et al. Conditional inactivation of Tgfbr2 in cranial neural crest causes cleft palate and calvaria defects. Development 2003;130(21):5269–80.

[153] Dudas M, Kim J, Li W-Y, et al. Epithelial and ectomesenchymal role of the type I TGF-beta receptor ALK5 during facial morphogenesis and palatal fusion. Dev Biol 2006;296(2):298–314.

[154] Ko SO, Chung IH, Xu X, et al. Smad4 is required to regulate the fate of cranial neural crest cells. Dev Biol 2007;312(1):435–47.

[155] Nie X, Deng C-x, Wang Q, Jiao K. Disruption of Smad4 in neural crest cells leads to mid-gestation death with pharyngeal arch, craniofacial and cardiac defects. Dev Biol 2008;316(2):417–30.

[156] Iwata J, Hacia JG, Suzuki A, Sanchez-Lara PA, Urata M, Chai Y. Modulation of noncanonical TGF-β signaling prevents cleft palate in Tgfbr2 mutant mice. J Clin Invest 2012;122(3):873–85.

[157] Iwata J, Suzuki A, Pelikan RC, et al. Smad4-Irf6 genetic interaction and TGFβ-mediated IRF6 signaling cascade are crucial for palatal fusion in mice. Development 2013;140(6):1220–30.

[158] Yang X, Kilgallen S, Andreeva V, Spicer DB, Pinz I, Friesel R. Conditional expression of Spry1 in neural crest causes craniofacial and cardiac defects. BMC Dev Biol 2010;10:48. Available from: http://dx.doi.org/10.1186/1471-213X-10-48.

[159] Xu X, Han J, Ito Y, Bringas Jr. P, Urata MM, Chai Y. Cell autonomous requirement for Tgfbr2 in the disappearance of medial edge epithelium during palatal fusion. Dev Biol 2006;297(1):238–48.

[160] Fuchs S, Herzog D, Sumara G, et al. Stage-specific control of neural crest stem cell proliferation by the small Rho GTPases Cdc42 and Rac1. Cell Stem Cell 2009;4(3):236–47.

[161] Sakai D, Trainor PA. Treacher Collins syndrome: unmasking the role of Tcof1/treacle. Int J Biochem Cell Biol 2009;41(6):1229–32.

[162] Jones NC, Lynn ML, Gaudenz K, et al. Prevention of the neurocristopathy Treacher Collins syndrome through inhibition of p53 function. Nat Med 2008;14(2):125–33.

[163] Smith FHD, Young NM, Lainoff AJ, Jamniczky HA, Maltepe E, Hallgrimsson B, et al. The effect of hypoxia on facial shape variation and disease phenotypes in chicken embryos. Dis Model Mech 2013.

[164] Li J, Shi Y, Sun J, Zhang Y, Mao B. *Xenopus* reduced folate carrier regulates neural crest development epigenetically. PLoS One 2011;6:11.

[165] Gray JD, Nakouzi G, Slowinska-Castaldo B, et al. Functional interactions between the LRP6 WNT co-receptor and folate supplementation. Hum Mol Genet 2010;19(23):4560–72.

[166] Wang G, Bieberich E. Prenatal alcohol exposure triggers ceramide-induced apoptosis in neural crest-derived tissues concurrent with defective cranial development. Cell Death Dis 2010;1:e46. Available from: http://dx.doi.org/1038/cddis.2010.22.

[167] Dong J, Sulik KK, Chen S-y. Nrf2-mediated transcriptional induction of antioxidant response in mouse embryos exposed to ethanol *in vivo*: implications for the prevention of fetal alcohol spectrum disorders. Antioxid Redox Signal 2008;10(12):2023–33.

[168] Saito H, Yamamura K, Suzuki N. Reduced bone morphogenetic protein receptor type 1A signaling in neural-crest-derived cells causes facial dysmorphism. Dis Model Mech 2012;5(6):948–55.

[169] Tang S, Snider P, Firulli AB, Conway SJ. Trigenic neural crest-restricted Smad7 over-expression results in congenital craniofacial and cardiovascular defects. Dev Biol 2010;344(1):233–47.

[170] Milunsky JM, Maher TA, Zhao GP, et al. TFAP2A mutations result in branchio-oculo-facial syndrome (82, pg 1171, 2008). Am J Hum Genet 2009;84(2):301.

[171] Pai AC. Developmental genetics of a lethal mutation muscular dysgenesis (MDG) in mouse 1. Genetic analysis and gross morphology. Dev Biol 1965;11(1):82–92.

[172] Shwartz Y, Farkas Z, Stern T, Aszodi A, Zelzer E. Muscle contraction controls skeletal morphogenesis through regulation of chondrocyte convergent extension. Dev Biol 2012;370(1):154–63.

[173] Schoenebeck JJ, Hutchinson SA, Byers A, et al. Variation of BMP3 contributes to dog breed skull diversity. PLoS Genet 2012;8:8.

[174] Depew MJ, Simpson CA. 21(st) century neontology and the comparative development of the vertebrate skull. Dev Dyn 2006;235(5):1256–91.

[175] Fish JL, Villmoare B, Köbernick K, et al. Satb2, modularity, and the evolvability of the vertebrate jaw. Evol Dev 2011;13(6):549–64.

[176] Compagnucci C, Debiais-Thibaud M, Coolen M, et al. Pattern and polarity in the development and evolution of the gnathostome jaw: both conservation and heterotopy in the branchial arches of the shark, *Scyliorhinus canicula*. Dev Biol 2013;377(2):428–48.

[177] Abzhanov A, Kuo WP, Hartmann C, Grant BR, Grant PR, Tabin CJ. The calmodulin pathway and evolution of elongated beak morphology in Darwin's finches. Nature 2006;442(7102):563–7.

[178] Abzhanov A, Protas M, Grant BR, Grant PR, Tabin CJ. Bmp4 and morphological variation of beaks in Darwin's finches. Science 2004;305(5689):1462–5.

[179] Mallarino R, Grant PR, Grant BR, Herrel A, Kuo WP, Abzhanov A. Two developmental modules establish

3D beak-shape variation in Darwin's finches. Proc Natl Acad Sci USA 2011;108(10):4057−62.

[180] Albertson RC, Kocher TD. Assessing morphological differences in an adaptive trait: a landmark-based morphometric approach. J Exp Zool 2001;289:385−403.

[181] Albertson RC, Streelman JT, Kocher TD, Yelick PC. Integration and evolution of the cichlid mandible: the molecular basis of alternate feeding strategies. Proc Natl Acad Sci USA 2005;102(45):16287−92.

[182] Roberts RB, Hu Y, Albertson RC, Kocher TD. Craniofacial divergence and ongoing adaptation via the hedgehog pathway. Proc Natl Acad Sci USA 2011;108(32):13194−9.

[183] Chai Y, Maxson RE. Recent advances in craniofacial morphogenesis. Dev Dyn 2006;235:2353−75.

[184] Noden DM, Schneider RA. Neural crest cells and the community of plan for craniofacial development: historical debates and current perspectives. Adv Exp Med Biol 2006;589:1−23.

[185] Koyabu D, Maier W, Sanchez-Villagra MR. Paleontological and developmental evidence resolve the homology and dual embryonic origin of a mammalian skull bone, the interparietal. Proc Natl Acad Sci USA 2012;109(35):14075−80.

[186] Sidor CA. Simplification as a trend in synapsid cranial evolution. Evolution 2001;55(7):1419−42.

Neural Crest Cell and Placode Interactions in Cranial PNS Development

Stephen J. Fleenor and Jo Begbie

Department of Physiology, Anatomy and Genetics, University of Oxford, South Parks Road, Oxford OX1 3QX, UK

OUTLINE

THE CRANIAL PERIPHERAL NERVOUS SYSTEM

The cranial peripheral nervous system (PNS) consists of the cranial nerves and their associated sensory and parasympathetic ganglia (Figure 8.1). The head also contains the paired special sense organs, such as the inner ear and olfactory sensory epithelium. Unlike the PNS in the trunk, which is derived

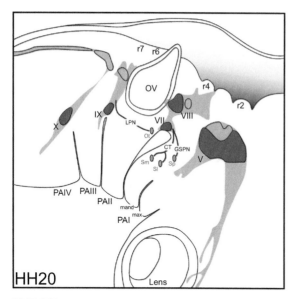

FIGURE 8.1 **Early development of the cranial PNS.**
Cartoon of the hindbrain region of a Hamburger–Hamilton
(HH) stage 20 chick embryo depicting cranial nerves V, VII,
VIII, IX, and X and the PAs. Neurons emigrating from the
epibranchial, trigeminal, and otic placodes coalesce to form
the distal aspects of the sensory ganglia (blue). Cranial
NCC form the proximal aspects (orange), as well as ectomesenchymal cells, Schwann cells, melanocytes, and neurons
of the parasympathetic ganglia (green). The parasympathetic ganglia are innervated by visceromotor nerve fibers
associated with cranial nerves IX and VII that emanate from
the hindbrain: LPN, lesser petrosal nerve; CT, chorda tym-
pani; and GSPN, greater superficial petrosal nerve. While
the submandibular (Sm) and sublingual (Sl) ganglia are
localized to the mandibular aspect (mand) of PAI, the sphenopalatine (Sp) ganglion is localized to the maxillary (max)
aspect. Ot, otic ganglion; OV, otic vesicle; r2–7, respective
rhombomeres of the hindbrain.

exclusively from NCC, the cranial PNS has a
dual embryonic origin arising from both NCC
and ectodermal placodes. A dual embryonic
origin for the cranial PNS had been described
in many studies across a diverse group of vertebrates from the late 1800s onward. However
the introduction of quail-chick transplants
really allowed the careful tracing of derivatives to produce more detailed fate maps, such

as that carried out for the cranial peripheral
ganglia [1].

The cranial sensory ganglia (CSG) are intricately connected with the cranial nerves (CNs).
For example, the trigeminal ganglion associates
with CNV, the geniculate ganglion with the facial
nerve (CNVII) and the vestibuloacoustic ganglion
with CNVIII. The petrosal and superior ganglia
form part of the glossopharyngeal nerve (CNIX),
while the nodose and jugular ganglia associate
with the vagal nerve (X) (see Figure 8.1).

The distal ganglia of CNVII, IX, and X are collectively known as the epibranchial ganglia for
their position at the top of the branchial or pharyngeal arches (PAs). Fate-mapping studies
show that the sensory neurons of the proximal
ganglia are derived from NCC, while those of
the distal ganglia are derived from the trigeminal and epibranchial placodes (see Figures 8.1
and 8.2) [1–3]. NCC can generate sensory neurons in the distal ganglia, but they mainly produce the non-neuronal cells, Schwann cells, and
satellite glia associated with the ganglia [1,4].

The facial CNVII and glossopharyngeal
CNIX also have cranial parasympathetic ganglia that associate with them (see Figure 8.1).
Sphenopalatine, sublingual, and submandibular ganglia are on CNVII, and the otic ganglion
on CNIX. All of the parasympathetic neurons
and associated non-neuronal cells are NCC-
derived, but an essential interaction with the
placode-derived sensory neurons [5,6] will be
discussed later.

The cranial PNS receives some of its sensory
information from the paired sense organs. While
the retina of the eye is really an extension of the
diencephalon, the lens of the eye is formed from
an ectodermal placode. Similarly, the olfactory
sensory epithelium responsible for smell and the
inner ear responsible for hearing and balance are
derived from ectodermal placodes. However, an
interdependent relationship exists between NCC
and placodes in the generation of these structures, as will be discussed later.

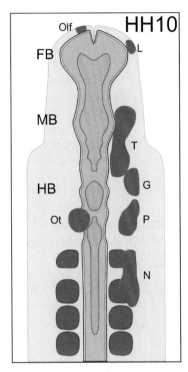

FIGURE 8.2 Cranial NCC/placode progenitor domains.
Fate map of a HH stage 10 chick embryo showing progenitor domains of cranial NCC (orange) and placodes (blue). Though the placode progenitor domains are bilateral, only one side is shown. FB, forebrain; MB, midbrain; and HB, hindbrain divisions of the neural tube. Olf, olfactory; L, lens; Ot, otic; T, trigeminal; G, geniculate; P, petrosal; and N, nodose progenitor domains. Rostral somites are dark gray. *Based on Ref. [1].*

ECTODERMAL PLACODES

Ectodermal placodes are the embryonic source of a large proportion of the cranial sensory PNS [7–9]. Morphologically placodes are first visible as transient ectodermal thickenings at stereotypical locations in the vertebrate head (see Figures 8.2 and 8.3) [1]. While the ectodermal placodes are frequently grouped together, the characteristics and behaviors of the ectoderm can be quite distinct for the individual placodes. For example, while the majority of placodes are indeed large thickened ectoderm, the trigeminal

ophthalmic placode forms from numerous small punctate loci across the placodal area [10,11]. The thickened ectoderm of the otic and lens placodes will involute to form vesicles that pinch off from the surface ectoderm, while the ectoderm of the olfactory placode will simply involute and the epibranchial placodes remain straight and superficial [9]. Similarly a large number but not all of the placodes will produce migratory cells that delaminate from the ectoderm, while the lateral line placodes, comprising a series specific to aquatic vertebrates, themselves migrate as primordia along the body axis [7–9,12]. Furthermore, the derivatives they generate will vary from complex structures such as the inner ear to simple sensory neurons alone or the lens of the eye. So while it is in some ways informative to consider what they have in common, it is also important not to think of all ectodermal placodes as the same thing but to consider what it is that gives them their specific identity.

The most anterior placode, the olfactory placode, gives rise to the sensory and respiratory epithelia of the nasal cavity, including olfactory sensory neurons, sustentacular supporting cells, and mucus-producing glands in addition to the GnRH neurons that migrate to populate the CNS [13,14]. It was previously thought that the olfactory placode generated the glial olfactory ensheathing cells, but recent lineage labeling analysis has demonstrated a NCC origin [15,16]. It has also recently been shown that NCC can intercalate into the olfactory placode [15,17]; however, these cells are incorporated into the general olfactory derivatives rather than having an organizing function, and therefore will not be considered as an NCC/placode interaction.

The otic placode, sitting alongside the hindbrain, generates the complex epithelial structure of the inner ear: the semicircular canals of the vestibular system and the cochlea or cochlear duct of the acoustic system, as well as the mechanosensory hair cells and the sensory

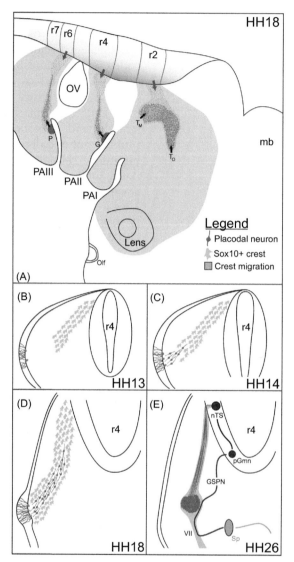

FIGURE 8.3 NCC interactions with placodal derivatives. (A) Cartoon of the hindbrain region of a HH stage 18 chick embryo depicting migratory streams of cranial NCC (orange and light brown) and neural derivatives of the cranial sensory placodes (blue). Black arrows show direction of migration from the placodes (blue thickenings). Brown arrows show direction of NCC migration from the dorsal aspects of rhombomeres 2, 4, and 6/7 (r2–7, respectively). While NCC migratory paths invade the PAs and head (light brown), the Sox10+ subpopulation (orange) migrates specifically to the placodes, forming corridors to guide placodal neural migration and hindbrain innervation. PAIV, with associated nodose placode and neuroblasts guided by r6 Sox10+ NCC, is not shown. OV, otic vesicle; Olf, olfactory placode; mb, midbrain. (B–E) Development of sensory ganglia within the Sox10+ crest cell corridors. At the onset of placodal neurogenesis (B), Sox10+ cells (orange) migrate toward the placode but have not yet made contact. As neurons (blue) begin to emanate toward the hindbrain (C), the Sox10+ stream extends to funnel them internally. By the peak of placodal neurogenesis (D), the Sox10+ corridor has extended to the placode and completely encompasses the migratory stream. Several days later in development, the placode ceases neurogenesis (E) and morphologically resembles the rest of the surface ectoderm.

neurons that innervate them [18]. As in the olfactory placode, recent studies have shown that NCC (besides those giving the pigment cells of the stria vascularis) and even neuro-epithelial cells can intercalate into the otic placodal epithelium, but the cells are suggested to simply incorporate rather than have any functional interaction [19].

A number of placodes will generate only one cell type, such as those producing the sensory neurons of the CSG [7,20]. The trigeminal placode that gives rise to the somatosensory neurons of the trigeminal ganglion can be split in two, with the ophthalmic placode sitting alongside the midbrain/rostral hindbrain and the maxillomandibular placode lying more ventral. The epibranchial placodes crown the PAs and give rise to the viscerosensory neurons of the epibranchial ganglia. Additionally, the lateral line series of placodes of anamniotes generate the mechanoreceptive neuromasts and associated sensory neurons of the lateral line system [21].

The Pre-Placodal Region

While ectodermal placodes were first identified morphologically as thickenings of ectoderm and fate-mapped to the stereotyped locations shown in Figure 8.2, the discovery of earlier molecular markers such as Six1 and Eya1, along with more recent fate maps in chick, zebrafish, and *Xenopus*, has shown that they originate within a horse shoe-shaped domain around the anterior neural plate [9,22,23]. The fate-mapping experiments show a degree of intermingling between cells for the different placodes and suggest that this domain is competent to generate any of the placodes, leading to it being called the pan-placodal region (PPR) [24,25]. Further

evidence for this competence comes from experiments carried out in amphibians, which show that following rotation of the PPR, the more caudal cranial ectoderm can contribute to the rostral olfactory placode when transplanted early [26].

An interesting question is therefore this: How are the specific placodes positioned at their stereotypical locations? One answer to this lies in cell sorting, which has been suggested to occur in the chick to segregate olfactory and lens placode precursors, which are initially intermingled [24]. However, recent fate maps and time-lapse analysis in *Xenopus* suggest that no large-scale cell movements occur, but that the PPR segregates by some other mechanism [27]. This other mechanism could lie in the response of the overlying ectoderm to distinct cues, as experiments studying the inductive cues for individual placodes have shown that they come from tissues adjacent to the presumptive placode along the anterior–posterior axis: for example, BMP7 from the pharyngeal endoderm induces the epibranchial placodes, while Wnt and FGF from the neural tube induces the ophthalmic trigeminal placode, and FGF from the mesoderm followed by FGF and Wnt from the neural tube induces the otic placode [8,28–31]. This has given rise to a two-step model for placode development, as *in vitro* experiments show that naive ectoderm needs to be made competent by first grafting it into the PPR before it can respond to otic placode-inducing cues; the localization of the second, specific cues would position the placode [32]. Recent studies have suggested a further mechanism for segregation and positioning of individual placodes, implicating NCC and placode interactions, as discussed below.

At this stage, the cranial sensory ganglion has innervated the nucleus of the solitary tract (nTS) in the hindbrain. Parasympathetic ganglia (green; shown is the sphenopalatine ganglion [Sp]), formed from NCC that migrate rostrally into PAI from the r4 stream (see Figure 8.1), are innervated by nerve fibers that associate with the cranial nerves (shown is the greater superficial petrosal nerve [GSPN]). pGmn, pre-ganglionic visceromotor neuron.

INTERACTIONS

NCC Interactions with the Pre-Placodal Region

The lens of the eye develops from the lens placode, which is found in the anterior of the embryo lateral to the olfactory placode (see Figure 8.2). Classically, it is considered that interactions between the optic vesicle and surface ectoderm are required for lens development. However, using molecular markers of early stages of lens development, it is becoming clear that while the optic vesicle is important for later stages, early development of the lens placode can occur independently [12]. Indeed, experiments carried out in chick show that explants taken at the otic level of the early PPR become lens in the absence of other signals [33–35]. Ablation and co-culture experiments have demonstrated that this lens forming potential is actively suppressed in non-lens ectoderm by NCC that migrate beneath a large part of the cranial ectoderm [33]. Expression analysis showed that NCC express a number of TGFβ ligands and that ectopic expression of constitutively active Smad3 in presumptive lens ectoderm repressed lens formation. However, TGFβ signaling alone is not sufficient to have this effect; it also induces expression of Wnt2b in the overlying ectoderm that actives the canonical Wnt signaling pathway to suppress lens potential [36]. As the NCC cannot physically migrate past the optic vesicle, this leaves the overlying ectoderm free to become lens. It's suggested that this interaction between the NCC and the PPR thus aligns the optic vesicle and lens with each other.

While the characterization of NCC-mediated lens repression has so far only been demonstrated in the chick, it is interesting to note that lens specification in mouse occurs only after the optic vesicle makes contact with pre-lens ectoderm and releases BMP4 [37]. Prior to this, mesenchymal cells (possibly of NCC origin) underly the pre-lens ectoderm. Thus, it is possible that, in addition to requiring optic vesicle-derived BMP signaling, lens specification in mouse depends on the physical clearing out of repressive NCC by the expanding optic vesicle.

Another example of NCC cell interaction with the PPR has come out in a recent study focusing on the segmentation of the epibranchial series of placodes in *Xenopus laevis* [38]. Here, an initially uniform field of Eya1-positive ectoderm is segmented by the underlying NCC cell streams where the two cell populations are engaged in "chase-and-run" behavior. The "chase" is carried out by the NCC, which express CXCR4 and are attracted to Sdf1 expressed by the placodal cells. On contact with the NCC, transient adhesion complexes are established that lead to an N-cadherin dependent downregulation of focal adhesions in placodal cells and an increase in their directional migration causing the placodes to "run away" from the NCC [38]. Evidence from the frog and zebrafish suggests that in the absence of NCC interactions with the placode, the epibranchial placodes are not segregated [38]. However, we have shown in the chick, and others have shown in the mouse, that following ablation of NCC, the epibranchial placodes were segregated and positioned normally in the embryo, and that the problems lay in the migration of their derivatives [5,28,39,40]. Thus, the "chase-and-run" behavior described in this study may be an aquatic-specific phenomenon.

NCC Interactions with Placodal Derivatives

Cranial Sensory Ganglia

As introduced previously, the neurons of the distal portion of the CSG are derived from the trigeminal (ophthalmic and maxillomandibular) and epibranchial (geniculate, petrosal, and nodose) neurogenic placodes (see Figure 8.1)

[1–3]. With the exception of the trigeminal ophthalmic placode, which generates post-mitotic neurons, these placodes produce neuroblasts that delaminate and actively migrate toward the hindbrain [3,10,11,41]. These sensory neuroblasts coalesce to form ganglia and send their projections internally to connect with the CNS. Importantly, the distinct CSG innervate the hindbrain at specific axial levels, coordinating their sensory input with the correct hindbrain region. This raises the question of how the migration of neuroblasts born at a distance from the CNS is coordinated to choreograph the correct construction and integration of the sensory PNS. Studies in a number of species have shown that the elegant solution is that the neuroblasts interact with the NCC that migrate out from the CNS into the periphery.

Our work in chick has shown that the pattern of sensory neuroblast migration mirrors the hindbrain cranial NCC migratory streams [39]. Furthermore, NCC ablation either mechanically or molecularly shows that the NCC migratory streams are required for the formation of CSG connected to the correct location in the developing hindbrain [39,42]. However, this requirement is more about axial positioning than target innervation, since ablation of the CNS after NCC emigration resulted in normal ganglion formation and projection to the correct axial level of the hindbrain, but continued extension of projections past their hindbrain targets [39]. These manipulations affect not only the epibranchial ganglia that were the main focus of our study, but also the vestibuloacoustic ganglion and the trigeminal ganglion, where the ophthalmic and maxillomandibular lobes became separated [39,42].

In support of this finding in chick, mouse mutants with defects in cranial NCC are often accompanied by defective CSG [5,6,43,44]. Interestingly, given that the NCC appear to provide registration between hindbrain and the placodal derivatives, defects have been described in a number of Hox gene mutant mice (HoxA3 and Hoxa1/b1 double mutants), as well as being apparent in chick following ectopic HoxB1 expression [40,45,46]. Indeed, a recent study has shown that in the absence of r4-derived NCC, r2-derived NCC can guide the geniculate placodal neuroblasts, but that the ganglion is ectopically located and fused to the trigeminal ganglion, the usual target of r2-derived NCC [44].

A number of mouse mutants have misrouted NCC cell migration. Best studied of these with respect to CSG development are the mice defective for class 3 semaphorin/neuropilin (SEMA3/NRP) signaling [47,48]. In the absence of NRP signaling, either through knockout of the receptor NRP 1 or 2, or ligand SEMA3A or 3F, NCC can invade territory adjacent to rhomobomere 3 (r3) that is usually NCC-free [47,48]. In single mutants, this results in an ectopic Sox10-positive NCC bridge between the trigeminal and hyoid NCC streams, and marker analysis shows that this is associated with ectopic sensory neurons of placodal origin [48]. Double mutants are more severe with fusion of trigeminal, vestibuloacoustic, and geniculate ganglia [48]. Similar fusions are seen in Krox20 mouse mutants where r3 is eliminated, and in chick r3 replacements where the NCC streams are no longer separated [49,50].

Still, while these mouse mutants highlight the important interaction between the NCC and the placodal derivatives, they do not give an insight into the mechanisms by which it occurs. With this in mind, the transgenic mouse line where beta-catenin is inactivated in NCC, using Wnt1-Cre is of particular interest [51]. Here, the NCC migrate normally into the periphery but then undergo apoptosis; correspondingly, the CSG fail to connect correctly to the hindbrain, suggesting that the NCC themselves are required rather than simply altering the environment as they migrate.

Analysis of a panel of zebrafish mutants designed to determine which subpopulation of cranial NCC is important in CSG formation also shows defects in CSG formation associated with defects in cranial NCC [52]. Surprisingly, this analysis shows that the chondrogenic or ectomesenchymal subpopulation that fills the branchial arch is more important than the neuroglial subpopulation. Interestingly, the study also shows a correlation with disruption of specific CSG formation and defects in the associated ceratobranchial skeletal element, thus suggesting integration with the patterning of the ventral branchial arch region as well as dorsally with the hindbrain [52].

In summary, failure of correct hindbrain NCC cell streaming results in a failure of the placodal-derived CSG to form correctly and connect accurately to their target. This is of relevance for human disease, as there are human syndromes that affect NCC migration. Where analysis has been done on mouse models of these, there are some effects on the cranial PNS development. For example, in mice mutant for Tcof1, the gene responsible for Treacher-Collins syndrome, analysis showed that there were reduced numbers of migrating NCC and the associated sensory ganglia were hypoplastic [76]. Similar effects were seen in the chick when NCC migration was reduced by interfering with expression levels of Mid1, the gene responsible for X-linked Opitz syndrome [53]. Furthermore, Tbx1 mutant mice, used as a model for DiGeorge syndrome (DGS; del22q11), which show defects in cranial NCC migration also have defects in cranial PNS patterning [54,55].

Mechanisms Underlying NCC and Placodal Neuroblast Interaction in CSG Development: Molecular

The molecular basis for the interaction between NCC and placodal neuroblasts in the forming CSG has been best studied in the chick trigeminal ganglion. These studies demonstrate an important role for NCC in the aggregation of the placode neurons to form a ganglion, and implicate both Slit/Robo and Wnt signaling.

Expression analysis in chick shows that Slit1 is expressed by the NCC, while its receptor Robo2 is expressed by trigeminal placodal neurons [56,57]. Perturbing Slit/Robo signaling in chick using a dominant negative Robo2 or RNAi-mediated knockdown of either Slit1 or Robo2 results in defects in trigeminal ganglion formation that phenocopy NCC ablation with diffuse ganglia, which fail to connect accurately to the hindbrain [57]. Further work suggests that the coalescence of the trigeminal neurons is mediated by a change in the subcellular localization of N-cadherin, with increased localization at intercellular junctions in response to NCC-derived Slit1 activating Robo2 signaling in the placodal neurons [56]. Similarly, studies in zebrafish have shown fragmented CSG in the absence of N-cadherin [58,59].

Wnt signaling is known to be important in the differentiation of the trigeminal ophthalmic placode [29,30] and has been implicated in trigeminal ganglion formation by studies addressing the role of Wise, a secreted Wnt signal modulator in the chick [60]. Wise is co-expressed with Wnt6 in the surface ectoderm overlying the trajectory of the trigeminal ophthalmic ganglion. Ectopic expression of either Wise or Wnt6 led to the generation of ectopic ganglia, and data shows that Wise-expressing cells attract NCC and cluster. Conversely, following morpholino knockdown of Wise in the surface ectoderm, trigeminal neurons associate with fewer NCC [60]. This non-cell-autonomous effect demonstrates a reciprocal interaction between the placodes and NCC.

Finally, it is important to note that NCC express fibronectin, an extracellular matrix protein that serves as a highly permissible substrate for placode-derived neuronal migration *in vitro* [61] (see below). However, whether this is due simply to cell adhesion or participates in a broader cell communication mechanism has not been demonstrated. By and large, considering how important the interactions between NCC and placodes are, it is surprising that so little is known about the ligands and receptors governing NCC–placode interactions. Many candidate molecules are currently being, and remain to be, tested.

Mechanisms Underlying NCC and Placodal Neuroblast Interaction in CSG Development: Cellular

To better understand the interaction between NCC and migrating placodal neuroblasts in the forming CSG, we have analyzed the cellular relationship between the two populations. 3D reconstructions of chick and mouse embryos show that the NCC are localized to the periphery of the migrating placodal neuroblasts, delineating a pathway between the placodal epithelium and the hindbrain (see Figure 8.3) [61]. Our *in vitro* analysis of placodal cell migration on NCC compared with fibronectin or mesoderm demonstrate that the NCC are not providing an active guidance cue, but rather that mesoderm is a less favorable substrate for migration. The cranial NCC organized into 3D structures when migrating from the dorsal neural tube (diencephalon to caudal rhombomere levels) in culture of the neural tube showing that this is an intrinsic property of the NCC. In the embryo however, our analysis shows that the extension of the NCC corridor to the placode requires the presence of placodal neuroblasts (see Figure 8.3) [61]. This shows the importance of reciprocal interactions between NCC and placode for the organization of the CSG and their afferent innervation of the hindbrain.

As discussed previously, mouse mutants with defects in NCC patterning frequently display aberrant CSG patterning, but there are studies that show that genetic NCC ablation has less of an effect [5,44,48,62]. We propose that the NCC corridor provides a physical corridor for inherently migratory neuroblasts to traverse a non-neural territory, rather than

providing an active guidance mechanism. In this scenario, neuroblasts could still migrate in the absence of NCC, even if less efficiently, while a misplaced corridor caused by a NCC migration defect would have more effect on CSG patterning. However, it should be noted that in the chick the neuroblasts did not migrate significantly internally in the absence of NCC and this may reflect species differences.

The interactions between NCC and placodal derivatives could reflect a broader developmental mechanism relating to interactions between neurons and glia. Indeed, studies in the developing lateral line system have shown that the glial cells are attracted by the placodally derived neurons, but that the placodal axons require the glial cells for proper organization and fasciculation [63]. Studies carried out in *Drosophila* show a reciprocal co-dependence of glial cells and PNS neurons for glial differentiation and sensory axon path finding into the CNS, respectively [64,65]. In addition, the idea of physical constraint by glial cells has been suggested in peripheral nerve repair: following nerve damage, Schwann cells de-differentiate and form "bridges" to guide regenerating axons [66,67].

Cranial Parasympathetic Ganglia

Recent work has shown that interactions between NCC and placodal derivatives play a wider role than organizing the CSG and are required for orchestrating the development of cranial visceral circuits in the PNS (see Figures 8.1 and 8.3) [5,6,44]. Fate-mapping studies have determined the rhombomeric origin of the NCC that generate the different parasympathetic ganglia in the head. The sphenopalatine and submandibular ganglia associated with pharyngeal arch 1 (PA1) actually come from r4 NCC normally destined for PA2 (see Figure 8.1) [6]. These r4 NCC migrate anteriorly through the placode-derived geniculate ganglion, and the absence of this sensory ganglion in neurogenin 2 (neurog2) mutant mice leads to a loss of these parasympathetic ganglia [6]. Similarly, the otic ganglion is generated by r6 NCC that migrate anteriorly through the placode-derived petrosal ganglion and fails to form in the neurog2 mutant mice [44,68]. In a complementary study, a panel of mutant mice lines were analyzed to assess the contribution of different neuronal components of the visceral circuit, which again demonstrated the reciprocal interaction between the placode-derived sensory ganglion and NCC in the formation of parasympathetic ganglia (see Figure 8.3) [5]. This study also demonstrated that the placode-derived sensory ganglia are required for the extension of the visceromotor axons from the hindbrain to their target in the parasympathetic ganglion, thus completing the circuit [5].

EVOLUTIONARY SIGNIFICANCE

The intricate co-interaction of cranial NCC and placodal derivatives raises intriguing questions about their co-evolution. Indeed, they both emerge as striking evolutionary novelties of the vertebrate head, and there is no known ancestor that has one population but not the other. The cephalochordate amphioxus produces peripheral sensory neurons from a nonneural ectodermal field that expresses Tlx [69,70], a marker of placode-derived CSG [71]. Tunicates, which are phylogenetically more closely related to vertebrates than cephalochordates [72], produce sensory neurons derived from a placode-like ectodermal field [73], but also produce non-neural migratory pigment cells that resemble NCC-derived melanocytes [74]. This suggests distinct evolutionary origins of NCC and placodes, with placode-like nonneural ectoderm serving as the ancestral source of peripheral sensory neurons.

However, while these rudimentary structures exist in tunicates and cephalochordates, a key distinction between vertebrates and the rest of Chordata is the organization of sensory neurons into ganglia. Undoubtedly, this organization

allows for robust higher-order processing essential for advanced vertebrate behaviors such as predation. While placodes are responsible for contributing the majority of cranial sensory neurons, they are not sufficient to create a functional PNS network; as they are born in the periphery, they depend on NCC interactions to properly pattern them with the hindbrain. In addition, because cranial sensory placodes generate committed neuroblasts rather than multipotent stem cells (as NCC are), placode-derived neurons depend on NCC for glial ensheathment. Thus, while the emergence of CSG in the vertebrate lineage probably required the evolution of neurogenic placodes, it ultimately depended on the co-evolution of NCC to organize and ensheath the placodal neurons.

CONCLUSIONS AND PERSPECTIVES

Ultimately, the coordinated development of the cranial PNS with the CNS is enabled by the reciprocal cellular and molecular interactions of NCC and placodes. The regularly segmented pattern of NCC emigration from the hindbrain out to the PAs organizes the axial location and innervation of the CSG with the appropriate targets in the CNS. Reciprocally, specified epibranchial placodes guide the extension outward of the NCC corridors that constrain the migration inward of the epibranchial placode-derived neuroblasts. Further, the sensory ganglia derived from the epibranchial placodes serve as guideposts for the formation of parasympathetic ganglia from NCC and visceromotor innervation from the hindbrain. This shows the role for the interactions in registering peripheral and central components of the nervous system. However, the interactions between the NCC and the pre-placodal domain suggest a broader role in coordinating development with the periphery, with the alignment of the presumptive lens with the optic cup through exclusion of NCC migration. Our overall understanding of the interactions between NCC and placodes is really in its infancy, particularly at the molecular level, and we will be greatly advanced by the identification and testing of candidate ligands and receptors. Recent advances in cell sorting has allowed for the precise isolation of cell types [75], and transcriptomic analysis between differentiation states or between control and ablated conditions also promises to shed much light on the molecular interactions between NCC and placodes. To this end, it will be exciting to see the progress we make over the next few years in understanding in detail how interactions between NCC and placode shape crucial steps of PNS development.

Acknowledgments

Thanks go to many colleagues for stimulating discussions, and to all members of the Begbie lab for their hard work. Work in the Begbie lab has been funded by the Anatomical Society, BBSRC and John Fell Fund OUP. SJF supported by Wellcome studentship 092920/Z/10/Z.

References

[1] D'Amico-Martel A, Noden DM. Contributions of placodal and neural crest cells to avian cranial peripheral ganglia. Am J Anat 1983;166(4):445−68.

[2] Thompson H, Blentic A, Watson S, Begbie J, Graham A. The formation of the superior and jugular ganglia: insights into the generation of sensory neurons by the neural crest. Dev Dyn 2010;239(2):439−45.

[3] Blentic A, Chambers D, Skinner A, Begbie J, Graham A. The formation of the cranial ganglia by placodally-derived sensory neuronal precursors. Mol Cell Neurosci 2011;46(2):452−9.

[4] Harlow DE, Yang H, Williams T, Barlow LA. Epibranchial placode-derived neurons produce BDNF required for early sensory neuron development. Dev Dyn 2011;240(2):309−23.

[5] Coppola E, Rallu M, Richard J, et al. Epibranchial ganglia orchestrate the development of the cranial neurogenic crest. Proc Natl Acad Sci USA 2010;107(5):2066−71.

[6] Takano-Maruyama M, Chen Y, Gaufo GO. Placodal sensory ganglia coordinate the formation of the cranial visceral motor pathway. Dev Dyn 2010;239(4):1155−61.

[7] Baker CV, Bronner-Fraser M. Vertebrate cranial placodes I. Embryonic induction. Dev Biol 2001;232 (1):1−61.

[8] Ladher RK, O'Neill P, Begbie J. From shared lineage to distinct functions: the development of the inner ear and epibranchial placodes. Development 2010;137 (11):1777−85.

[9] Schlosser G. Making senses development of vertebrate cranial placodes. Int Rev Cell Mol Biol 2010;283: 129−234.

[10] Begbie J, Ballivet M, Graham A. Early steps in the production of sensory neurons by the neurogenic placodes. Mol Cell Neurosci 2002;21(3):502−11.

[11] McCabe KL, Sechrist JW, Bronner-Fraser M. Birth of ophthalmic trigeminal neurons initiates early in the placodal ectoderm. J Comp Neurol 2009;514(2):161−73.

[12] Gunhaga L. The lens: a classical model of embryonic induction providing new insights into cell determination in early development. Philos Trans R Soc Lond B Biol Sci 2011;366(1568):1193−203.

[13] Balmer CW, LaMantia AS. Noses and neurons: induction, morphogenesis, and neuronal differentiation in the peripheral olfactory pathway. Dev Dyn 2005;234 (3):464−81.

[14] Whitlock KE. Developing a sense of scents: plasticity in olfactory placode formation. Brain Res Bull 2008;75 (2−4):340−7.

[15] Forni PE, Taylor-Burds C, Melvin VS, Williams T, Wray S. Neural crest and ectodermal cells intermix in the nasal placode to give rise to GnRH-1 neurons, sensory neurons, and olfactory ensheathing cells. J Neurosci 2011;31(18):6915−27.

[16] Barraud P, Seferiadis AA, Tyson LD, et al. Neural crest origin of olfactory ensheathing glia. Proc Natl Acad Sci USA 2010;107(49):21040−5.

[17] Saxena A, Peng BN, Bronner ME. Sox10-dependent neural crest origin of olfactory microvillous neurons in zebrafish. eLife 2013;2:e00336.

[18] Groves AK, Fekete DM. Shaping sound in space: the regulation of inner ear patterning. Development 2012;139(2):245−57.

[19] Freyer L, Aggarwal V, Morrow BE. Dual embryonic origin of the mammalian otic vesicle forming the inner ear. Development 2011;138(24):5403−14.

[20] Begbie J. Induction and patterning of neural crest and ectodermal placodes and their derivatives. In: Rubenstein JLR, Rakic P, editors. Comprehensive developmental neuroscience: patterning and cell type specification in the developing CNS and PNS. Amsterdam: Academic Press; 2013. p. 239−58.

[21] Ghysen A, Dambly-Chaudiere C. The lateral line microcosmos. Gen Dev 2007;21(17):2118—30.

[22] Streit A. The preplacodal region: an ectodermal domain with multipotential progenitors that contribute to sense organs and cranial sensory ganglia. Int J Dev Biol 2007;51(6—7):447—61.

[23] Grocott T, Tambalo M, Streit A. The peripheral sensory nervous system in the vertebrate head: a gene regulatory perspective. Dev Biol 2012;370(1):3—23.

[24] Bhattacharyya S, Bailey AP, Bronner-Fraser M, Streit A. Segregation of lens and olfactory precursors from a common territory: cell sorting and reciprocity of Dlx5 and Pax6 expression. Dev Biol 2004;271(2):403—14.

[25] Streit A. Early development of the cranial sensory nervous system: from a common field to individual placodes. Dev Biol 2004;276(1):1—15.

[26] Jacobson AG. The determination and positioning of the nose, lens and ear. III. Effects of reversing the antero-posterior axis of epidermis, neural plate and neural fold. J Exp Zool 1963;154:293—303.

[27] Pieper M, Eagleson GW, Wosniok W, Schlosser G. Origin and segregation of cranial placodes in *Xenopus laevis*. Dev Biol 2011;360(2):257—75.

[28] Begbie J, Brunet JF, Rubenstein JL, Graham A. Induction of the epibranchial placodes. Development 1999;126(5):895—902.

[29] Canning CA, Lee L, Luo SX, Graham A, Jones CM. Neural tube derived Wnt signals cooperate with FGF signaling in the formation and differentiation of the trigeminal placodes. Neural Dev 2008;3:35.

[30] Lassiter RN, Dude CM, Reynolds SB, Winters NI, Baker CV, Stark MR. Canonical Wnt signaling is required for ophthalmic trigeminal placode cell fate determination and maintenance. Dev Biol 2007;308 (2):392—406.

[31] Holzschuh J, Wada N, Wada C, et al. Requirements for endoderm and BMP signaling in sensory neurogenesis in zebrafish. Development 2005;132(16):3731—42.

[32] Martin K, Groves AK. Competence of cranial ectoderm to respond to FGF signaling suggests a two-step model of otic placode induction. Development 2006;133(5):877—87.

[33] Bailey AP, Bhattacharyya S, Bronner-Fraser M, Streit A. Lens specification is the ground state of all sensory placodes, from which FGF promotes olfactory identity. Dev Cell 2006;11(4):505—17.

[34] Sjodal M, Edlund T, Gunhaga L. Time of exposure to BMP signals plays a key role in the specification of the olfactory and lens placodes *ex vivo*. Dev Cell 2007;13(1):141—9.

[35] Sullivan CH, Braunstein L, Hazard-Leonards RM, et al. A re-examination of lens induction in chicken embryos: *in vitro* studies of early tissue interactions. Int J Dev Biol 2004;48(8—9):771—82.

[36] Grocott T, Johnson S, Bailey AP, Streit A. Neural crest cells organize the eye via TGF-beta and canonical Wnt signalling. Nat Commun 2011;2:265.

[37] Furuta Y, Hogan BL. BMP4 is essential for lens induction in the mouse embryo. Gen Dev 1998;12(23):3764—75.

[38] Theveneau E, Steventon B, Scarpa E, et al. Chase-and-run between adjacent cell populations promotes directional collective migration. Nat Cell Biol 2013;15 (7):763—72.

[39] Begbie J, Graham A. Integration between the epibranchial placodes and the hindbrain. Science 2001;294 (5542):595—8.

[40] Gavalas A, Studer M, Lumsden A, Rijli FM, Krumlauf R, Chambon P. Hoxa1 and Hoxb1 synergize in patterning the hindbrain, cranial nerves and second pharyngeal arch. Development 1998;125(6):1123—36.

[41] Graham A, Blentic A, Duque S, Begbie J. Delamination of cells from neurogenic placodes does not involve an epithelial-to-mesenchymal transition. Development 2007;134(23):4141—5.

[42] Osborne NJ, Begbie J, Chilton JK, Schmidt H, Eickholt BJ. Semaphorin/neuropilin signaling influences the positioning of migratory neural crest cells within the hindbrain region of the chick. Dev Dyn 2005;232 (4):939—49.

[43] Graham A, Begbie J, McGonnell I. Significance of the cranial neural crest. Dev Dyn 2004;229(1):5—13.

[44] Chen Y, Takano-Maruyama M, Gaufo GO. Plasticity of neural crest-placode interaction in the developing visceral nervous system. Dev Dyn 2011;240(8): 1880—8.

[45] Bell E, Wingate RJ, Lumsden A. Homeotic transformation of rhombomere identity after localized Hoxb1 misexpression. Science 1999;284(5423):2168—71.

[46] Watari N, Kameda Y, Takeichi M, Chisaka O. Hoxa3 regulates integration of glossopharyngeal nerve precursor cells. Dev Biol 2001;240(1):15—31.

[47] Gammill LS, Gonzalez C, Bronner-Fraser M. Neuropilin 2/semaphorin 3F signaling is essential for cranial neural crest migration and trigeminal ganglion condensation. Dev Neurobiol 2007;67(1):47—56.

[48] Schwarz Q, Vieira JM, Howard B, Eickholt BJ, Ruhrberg C. Neuropilin 1 and 2 control cranial gangliogenesis and axon guidance through neural crest cells. Development 2008;135(9):1605—13.

[49] Kuratani SC, Eichele G. Rhombomere transplantation repatterns the segmental organization of cranial nerves and reveals cell-autonomous expression of a homeodomain protein. Development 1993;117(1):105—17.

[50] Schneider-Maunoury S, Topilko P, Seitandou T, et al. Disruption of Krox-20 results in alteration of rhombomeres 3 and 5 in the developing hindbrain. Cell 1993;75(6):1199—214.

[51] Brault V, Moore R, Kutsch S, et al. Inactivation of the beta-catenin gene by Wnt1-Cre-mediated deletion results in dramatic brain malformation and failure of craniofacial development. Development 2001;128 (8):1253–64.

[52] Culbertson MD, Lewis ZR, Nechiporuk AV. Chondrogenic and gliogenic subpopulations of neural crest play distinct roles during the assembly of epibranchial ganglia. PloS One 2011;6(9):e24443.

[53] Latta EJ, Golding JP. Regulation of PP2A activity by Mid1 controls cranial neural crest speed and gangliogenesis. Mech Dev 2012;128(11–12):560–76.

[54] Calmont A, Thapar N, Scambler PJ, Burns AJ. Absence of the vagus nerve in the stomach of Tbx1-/- mutant mice. Neurogastroenterol Motil 2011;23(2):125–30.

[55] Vitelli F, Morishima M, Taddei I, Lindsay EA, Baldini A. Tbx1 mutation causes multiple cardiovascular defects and disrupts neural crest and cranial nerve migratory pathways. Hum Mol Genet 2002;11(8):915–22.

[56] Shiau CE, Bronner-Fraser M. N-cadherin acts in concert with Slit1-Robo2 signaling in regulating aggregation of placode-derived cranial sensory neurons. Development 2009;136(24):4155–64.

[57] Shiau CE, Lwigale PY, Das RM, Wilson SA, Bronner-Fraser M. Robo2-Slit1 dependent cell-cell interactions mediate assembly of the trigeminal ganglion. Nat Neurosci 2008;11(3):269–76.

[58] Kerstetter AE, Azodi E, Marrs JA, Liu Q. Cadherin-2 function in the cranial ganglia and lateral line system of developing zebrafish. Dev Dyn 2004;230(1):137–43.

[59] Knaut H, Blader P, Strahle U, Schier AF. Assembly of trigeminal sensory ganglia by chemokine signaling. Neuron 2005;47(5):653–66.

[60] Shigetani Y, Howard S, Guidato S, Furushima K, Abe T, Itasaki N. Wise promotes coalescence of cells of neural crest and placode origins in the trigeminal region during head development. Dev Biol 2008;319 (2):346–58.

[61] Freter S, Fleenor SJ, Freter R, Liu KJ, Begbie J. Cranial neural crest cells form corridors prefiguring sensory neuroblast migration. Development 2013;140:3595–3600.

[62] Golding JP, Trainor P, Krumlauf R, Gassmann M. Defects in pathfinding by cranial neural crest cells in mice lacking the neuregulin receptor ErbB4. Nat Cell Biol 2000;2(2):103–9.

[63] Gilmour DT, Maischein HM, Nusslein-Volhard C. Migration and function of a glial subtype in the vertebrate peripheral nervous system. Neuron 2002;34 (4):577–88.

[64] Sepp KJ, Auld VJ. Reciprocal interactions between neurons and glia are required for Drosophila peripheral nervous system development. J Neurosci 2003;23 (23):8221–30.

[65] Sepp KJ, Schulte J, Auld VJ. Peripheral glia direct axon guidance across the CNS/PNS transition zone. Dev Biol 2001;238(1):47–63.

[66] Nguyen QT, Sanes JR, Lichtman JW. Pre-existing pathways promote precise projection patterns. Nat Neurosci 2002;5(9):861–7.

[67] Parrinello S, Napoli I, Ribeiro S, et al. EphB signaling directs peripheral nerve regeneration through Sox2-dependent Schwann cell sorting. Cell 2010;143 (1):145–55.

[68] Takano-Maruyama M, Chen Y, Gaufo GO. Differential contribution of Neurog1 and Neurog2 on the formation of cranial ganglia along the anterior-posterior axis. Dev Dyn 2012;241(2):229–41.

[69] Lu TM, Luo YJ, Yu JK. BMP and Delta/Notch signaling control the development of amphioxus epidermal sensory neurons: insights into the evolution of the peripheral sensory system. Development 2012;139 (11):2020–30.

[70] Kaltenbach SL, Yu JK, Holland ND. The origin and migration of the earliest-developing sensory neurons in the peripheral nervous system of amphioxus. Evol Dev 2009;11(2):142–51.

[71] Logan C, Wingate RJ, McKay IJ, Lumsden A. Tlx-1 and Tlx-3 homeobox gene expression in cranial sensory ganglia and hindbrain of the chick embryo: markers of patterned connectivity. J Neurosci 1998;18 (14):5389–402.

[72] Delsuc F, Brinkmann H, Chourrout D, Philippe H. Tunicates and not cephalochordates are the closest living relatives of vertebrates. Nature 2006;439(7079):965–8.

[73] Mazet F, Hutt JA, Milloz J, Millard J, Graham A, Shimeld SM. Molecular evidence from Ciona intestinalis for the evolutionary origin of vertebrate sensory placodes. Dev Biol 2005;282(2):494–508.

[74] Abitua PB, Wagner E, Navarrete IA, Levine M. Identification of a rudimentary neural crest in a non-vertebrate chordate. Nature 2012;492(7427):104–7.

[75] Pan Y, Ouyang Z, Wong WH, Baker JC. A new FACS approach isolates hESC derived endoderm using transcription factors. PLoS One 2011;6(3):e17536.

[76] Dixon J, Jones NC, Sandell LL, et al. Tcof1/Treacle is required for neural crest cell formation and proliferation deficiencies that cause craniofacial abnormalities. Proc Natl Acad Sci USA 2006;103(36):13403–8.

Neural Crest Cells in Ear Development

Lisa Sandell

Birth Defects Center—MCCB, School of Dentistry, University of Louisville, Louisville KY

OUTLINE

THE EAR

The ear is the vertebrate sense organ of hearing and balance. Neural crest cells (NCC), notable for their many and varied contributions to other parts of the vertebrate head and body, likewise contribute multiple cell types and tissues to the ear and are essential for its development. Ear function is important in all vertebrates because it is required for equilibrium, enabling motion and balance. In terrestrial animals the ear is additionally adapted for detecting sound in air. In social animals such as humans, it plays a critical role in communication and can be important for learning and social interactions. Thus, understanding the development of the ear, and the contribution of NCC in formation of this important sense organ, is relevant both for basic science knowledge and human clinical concern.

In higher vertebrates the ear is composed of three distinct interconnected compartments, the inner ear, middle ear, and outer ear, each of which requires developmental contribution from NCC (Figure 9.1). The inner ear is the sensory component that detects motion,

Neural Crest Cells.
DOI: http://dx.doi.org/10.1016/B978-0-12-401730-6.00010-7

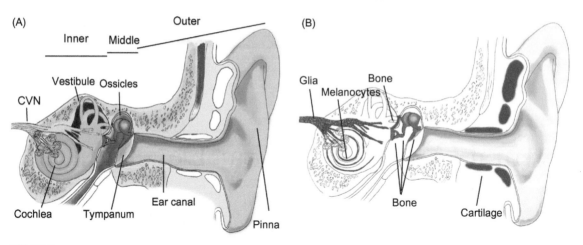

FIGURE 9.1   **Schematic of human ear illustrating the inner, middle, and outer ear compartments and NCC contribution to each part.** (A) The mammalian inner ear consists of the spiral cochlea and semicircular canals of the vestibular system embedded within the squamosal bone of the skull. The cochleovestibular nerve connects the sensory apparatus of the inner ear to the brain. The middle ear is an air-filled compartment containing three small ossicles bridging the gap between the inner ear and the tympanum or ear drum. The outer ear is an anatomical funnel that channels sound waves from the external world to the tympanum. It consists of the ear canal and the external pinna. (B) NCC contribute to each of the compartments of the ear. In the inner ear, NCC give rise to glial cells of the cochlea-vestibular nerve, the melanocyte intermediate cells of the stria vascularis, and a small portion of the bony otic capsule. In the middle ear, NCC give rise to the ossicles and are responsible for generating the air space surrounding the ossicles. In the outer ear, NCC give rise to the cartilages and fibrous tissues of the ear canal and pinna. CVN, cochleovestibular nerve.

direction, gravity, and sound. It is an elaborately shaped, liquid-filled chamber imbedded within the cranium that contains mechanosensory receptors and their associated sensory nerves. Adjacent to the inner ear is the middle ear, which consists of the tympanic membrane and a small bone or series of bones called ossicles spanning an air-filled chamber. It functions to transmit and convert sound-wave energy from the external air environment to vibrations of the liquid within the inner ear. The outer ear is an anatomical funnel that focuses and filters sound waves from the exterior world to the tympanic membrane inside the head. It consists of the ear canal and, in mammals, an external pinna. Neural crest cells are important for the development of each of the three compartments of the ear, contributing a numerically small but functionally essential

set of cells to the inner ear, and being a predominant progenitor cell type giving rise to tissues of the middle ear and outer ear.

Evolutionarily, the three compartments of the ear arose sequentially. Fish, including bony fish and cartilaginous fish, have inner ear structures only, lacking the middle ear and outer ear. Amphibians have middle ear structures in addition to inner ears, but lack outer ear components. Reptiles, birds, and mammals have outer ears in addition to inner and middle ears, with mammals being the only group to have an external pinna. It is likely that the middle ear arose evolutionarily as an adaptation for improved detection of sound in an air environment, with the outer ear being a further progression for improved collection of sound waves in air and spatial localization of sound source.

TIMING AND POSITION OF OTIC NCC EMIGRATION

The middle and outer ear develop from fusion and remodeling of the first and second pharyngeal arches, which are filled with NCC, while the inner ear arises from the otic vesicle located initially directly posterior to the dorsal portion of the second arch NCC stream. Study of middle ear ossicle development predates the identification of NCC. The formation of the middle ear ossicles from the first and second pharyngeal arches was described initially by Carl Reichert in his seminal comparative anatomical study of pig, bird, and frog nearly two centuries ago [1]. The outer ear pinna and ear canal also derive from the first and second pharyngeal arches and the cleft between them, an ontogeny defined by morphological analysis of a variety of vertebrate embryos, including human [2,3].

The first and second pharyngeal arches that are developmentally remodeled to form the middle and outer ear are filled with NCC that migrate from defined axial levels of the midbrain and hindbrain. The organization and origin of NCC migrating to the arches were defined based on classic lineage tracing experiments, initially in amphibian and subsequently avian and cultured mouse embryos [4–9]. In chick and mouse, the mesenchyme of the first pharyngeal arch (mandibular arch) is populated by NCC that emerge from the posterior midbrain and from the anterior hindbrain at the level of rhombomere 1 and 2 (R1 and R2). First arch NCC contributing to ear development are primarily those located in the posterior portion of the arch, originating from R1–R2. The second pharyngeal arch (hyoid arch) is populated by NCC that emigrate from the neural tube primarily at the level of R4, with a very minor contribution from R3 and R5. Thus, the middle ear and outer ear, which are known to develop by remodeling of the

first and second pharyngeal arches, derive from NCC that fill those arches—namely, NCC originating from a region of the neural tube spanning from the posterior midbrain to R4.

It is well accepted that the middle and outer ear develop from the first and second pharyngeal arches and that NCC populate those arches. Less noted is the contribution of NCC to the inner ear, which is often described as arising entirely from the otic vesicle. The otic vesicle forms directly adjacent to the proximal portion of the second arch NCC stream and NCC contribute several cell types to the developing inner ear. Lineage tracing studies of avian embryos demonstrate that glial support Schwann cells of the cochleovestibular nerve are derived from NCC emigrating from the rostral myelencephalon [10], presumably the region corresponding to R4. Quail-chick chimeric analysis indicates a portion of the bony capsule that surrounds the membranous labyrinth of the inner ear is derived from NCC, and transgenic lineage tracing in mice demonstrates that that these NCC originate from R4 [7,11]. With respect to inner ear NCC derivatives, arguably the most important is a population of melanocytes of the stria vascularis in the cochlea, which are essential for inner ear function. The axial origin of the inner ear melanocyte progenitor NCC has yet to be defined.

The timing of emigration of otic NCC has been determined by lineage tracing studies primarily in chick and mouse. In chick, tracing of NCC derivatives by timed radiolabeling or by quail-chick chimeric transplants indicates that the first and second arch NCC streams emerge from the neural tube between stages 9 and 11 (7 somites to 13 somites) [12]. In mouse, NCC emerge from the neural tube to enter the first arch stream as early as the 1–2 somite stage and continue through the 13 somite stage, that is, from embryonic day 8.0 (E8.0) through E9.0 [5,13]. This interval corresponds to Carnegie stage 9–10, or week 3–4 of human embryo development (clinical gestational age 5−6 weeks from last ovulation). The stages of emergence and early migration of otic NCC precede morphological formation of the first and second pharyngeal arches, and also precede morphological formation of the otic placodes.

The timing and migratory route of the NCC melanocyte progenitors populating the cochlear stria vascularis are not known. The migration characteristics have, however, been established for NCC melanoblast progenitors of the dermis of the trunk. In chick and mouse, some of the NCC melanoblast progenitors destined to form skin melanocytes of the body migrate along a dorsolateral pathway [14]. Other trunk dermis melanocytes arise from Schwann cell NCC progenitors, which migrate along a more ventral route [15]. The dorsolateral migrating trunk melanoblasts emerge approximately 24 h later than the corresponding ventrally migrating NCC that give rise to peripheral nerves [16]. In the head region of the mouse, dorsal migrating dermal melanoblast NCC are observed at the 25−26 somite stage [17]. Based on the migration characteristics of dermal trunk melanocytes, it may be speculated that the melanocytes of the inner ear may arise from NCC migrating with the cochleovestibular nerve Schwann progenitors, or may arrive in a later wave of migration of melanoblast progenitor NCC following a more dorsal route, or both.

In addition to lineage tracing analysis, the timing of otic NCC development has also been inferred from the temporal interval when these cells are sensitive to perturbation. Exposure of mouse or primate embryos to retinoic acid disrupts formation or migration of first and second arch NCC populations and causes malformations of the middle ear bones [18,19]. Administration of this teratogen at precisely timed intervals relative to the onset of pregnancy defines a discreet temporal window when middle ear development is sensitive to disruption. From these experiments it can be inferred that an interval within mouse embryonic day 8, specifically, E8 plus 4−5 h [18], up to E8.5 [20] is the critical period when the formation, migration, or development of NCC destined for the middle ear may be disrupted. This temporal window corresponds with the gestational time window of 2−5 weeks when excessive retinoids cause teratogenic ear phenotypes in humans [21].

NCC DERIVATIVES IN THE INNER EAR

Within the inner ear, the contribution of NCC is relatively limited compared to the variety of cell types derived from the otic placode. The otic placode gives rise to all the epithelia and sensory tissues of the inner ear and neurons of the cochleovestibular nerve [22−24]. The otic placode forms directly posterior to the second pharyngeal arch NCC migratory stream. It invaginates or cavitates to become the otic vesicle. This simple epithelial sac subsequently undergoes dramatic morphogenic remodeling to form the complex interconnected structures of the inner ear: the vestibular system for sensing motion and gravity, and the cochlea for sensing sound. The cochlea and vestibular system, collectively known as the membranous labyrinth, are lined with patches of specialized mechano-sensory units, which are connected to the brain via the cochleovestibular nerve (cranial nerve VIII). The membranous labyrinth is suspended within a fluid-filled bony case known as the otic capsule, which becomes integrated with the squamosal portion of the temporal bone of the skull around the time of birth. Of all the varied cell types within these structures, three are derived from NCC: glial Schwann cells of the cochleovestibular nerve, melanocytes of the cochlea, and a portion of the otic capsule. Transgenic lineage tracing in mice also indicates the presence of some NCC-derived cells within the sensory epithelium of the inner ear [25], although this observation has yet to be confirmed by alternate methodologies.

NCC contribution as progenitors of glial Schwann cells of the cochleovestibular nerve and satellite glia of the cochleovestibular ganglion were demonstrated by quail-chick chimera transplantation [10]. The distribution of NCC glial cells within the cochleovestibular nerve is extensive and has been confirmed by molecular and genetic analysis in mouse [26]. Inside the otic capsule, NCC-derived glia are closely associated with neurites, axons and neuronal cell bodies of the cochlear spiral ganglion and the vestibular ganglion of Scarpa. Outside the otic capsule NCC give rise to satellite glial cells of the cochleovestibular ganglion and axons that connect the ganglion to the hindbrain. In addition to the glial cells, NCC may also give rise to a small number of neuronal cells specifically within the vestibular ganglion [10,26].

The otic capsule is dense lamellar bone that surrounds the membranous labyrinth. NCC produce some otic capsule bone as demonstrated first by quail-chick chimera analysis [27] and subsequently by transgenic lineage tracing in mice [11]. The majority of the otic capsule is derived from other mesenchyme, with NCC contribution being limited to a small region of the outer surface of the pars canalicularis that surrounds the semicircular canals of the vestibular system [11,28,29].

Perhaps the most critical NCC derivatives within the inner ear are the population of melanocytes located within the stria vascularis of the cochlea. The identity and essential role of these melanocyte cells has been revealed through multiple lines of evidence over a long history. In 1851 Corti [30] described that pigmented cells were present within a vascular strip, or stria vascularis, within the cochlea of various mammals. Functionally, a correlation between pigmentation and hearing has long been noted, being remarked upon even by

Darwin his 1859 *Origin of the Species* [31]. That pigment cells, or melanocytes, within the skin were derivatives of NCC became clear years later by chimeric transplantation and *in vitro* analysis of amphibian and chick embryos and by mouse tissue explants [32–34]. That the pigmented cells within the stria vascularis could be characterized as melanocytes was demonstrated by electron microscopy [35]. Finally, the essential role of NCC-derived melanocytes in inner ear function was elucidated by analysis of mice with mutations causing defects in pigmentation and deafness, as well as by characterization of corresponding neurocristopathy syndromes in humans (Table 9.1).

Neurocristopathies are conditions that arise from abnormal development of NCC [63,64]. Waardenburg syndrome (WS) subtypes 1–4 are human neurocristopathies characterized by defects in pigmentation and variable sensorineural hearing loss ranging from profound deafness at birth to progressive loss of hearing after speech [65,66]. Vestibular defects may also be present in WS patients [67,68], although the incidence of vertigo is controversial. The different subtypes of WS are characterized on the basis of involvement of additional organs and tissues, corresponding in many cases to different genetic causes. WS1 is estimated at a prevalence of 1/42,000, accounting for 1–3% of congenital deafness [66]. Mutations responsible for WS have been identified in humans, and analogous mutations causing deafness and coat spotting phenotypes have been characterized in mice. While the mouse mutations are excellent models for the corresponding human syndromes, concordance of phenotypes is not complete. Notably, the mouse mutant *splotch*, in which WS gene *Pax3* is mutated, does not exhibit auditory defects [47].

Analysis of mice with cochlear defects and spotting phenotypes enabled the discovery that the essential function of NCC-derived melanocytes within the stria vascularis is the maintenance of the specialized endolymphatic fluid that fills the cochlear duct [69]. The

TABLE 9.1   Human Neurocristopathies of Inner Ear Melanocytes and Corresponding Mouse Models

Gene	Human Syndrome	Mouse Mutation
Mitf	WS2A (MIM 193510) Sensorineural hearing loss [36] Albinism, ocular, with sensorineural deafness (MIM 103470), sensorineural deafness, vestibular hypofunction [37] Tietz (MIM 103500) Profound sensorineural deafness [38]	*Microphthalmia* [39,40] Abnormal morphology of cochlea and saccule [41]
Pax3	WS1 (MIM: 193500), Variable sensorineural deafness [42–44] WS3 (MIM: 148820) Variable sensorineural deafness [45] CDH (MIM:122880) [46] Sensorineural hearing loss	*splotch* No auditory defects in heterozygous mutants [47,48]
Snai2	WS2D (MIM: 608890) sensorineural hearing loss [49]	*Snai2* targeted mutation Hyperactivity, circling (suggests vestibular defect) [49,50]
Sox10	WS2E (MIM: 611584) sensorineural hearing loss [51]	*dominant megacolon* No ear defect reported [52,53]
Edn3	WS4B (MIM: 613265) may include sensorineural hearing loss [54]	*lethal spotting*, no ear defect reported [55]
Ednrb	WS4A (MIM: 277580) may include sensorineural hearing loss [56]	*piebald, piebald-lethal*, deafness, loss of cochlear melanocytes [57,58]
Kit	Piebald Trait (MIM: 172800), deafness in some cases [59]	*viable dominant spotting*, sensorineural deafness [60–62]

endolymphatic fluid is enriched in potassium ions and carries a positive potential, which is essential for signal transduction of the mechano-sensory units within the cochlea. Deficiency or absence of NCC-derived melanocytes results in loss of ionic potential [69] and causes age-dependent degeneration of the stria vascularis [70,71]. A secondary consequence of the disrupted endolymph is degeneration of the cochlea and the saccule, collapse of the membranous labyrinth, loss of hair cells, and degeneration of the neurons of the spiral ganglia, a phenomenon observed in mouse models and human patients [72,73]. Although strial melanocytes, like other melanocytes, produce the pigment melanin, melanin production is not their essential function within the cochlea.

Presence of melanocytes is sufficient for hearing function, even if the melanocytes are deficient in production of melanin, such as those found in albino mice and humans. Although melanin is not required for auditory function, it may aid in long-term hearing protection [74] and survival of cells within the stria vascularis during aging and following noise injury [75].

The mutations causing WS in humans or deafness and spotting in mice may be tied to a single molecular network that includes the transcription factor MITF [76–78]. MITF regulates NCC differentiation into melanocytes [79,80] and is also required for survival of NCC melanocyte progenitors [81,82]. *Pax3* is required for induction of NCC [83] and acts synergistically with *Sox10* to regulate expression of *Mitf*. EDN3 signaling of through EDNRB stimulates survival and proliferation of early NCC and NCC committed to melanocyte differentiation [84]. Collectively, these genes function to regulate the induction, survival, and differentiation of NCC melanocyte progenitors involved in skin pigmentation and maintenance of the inner ear endolymphatic fluid.

NCC IN MIDDLE EAR AND OUTER EAR DEVELOPMENT

Just as NCC contribute critical cell types during formation of the sensory inner ear, likewise NCC are essential for development of the middle and outer ear. The two nonsensory parts of the ear are distinct from each other functionally and anatomically, but are closely linked in terms of embryonic development. Both are formed from the first and second pharyngeal arches, which are populated with NCC, and by growth and remodeling of the cleft and pouch between them [85] (Figure 9.2A–G).

The middle ear is an air-filled cavity embedded within the squamosal portion of the temporal bone of the cranium. It is present in mammals, birds, reptiles, and some amphibians. The middle ear cavity is contiguous with the Eustachian tube, which connects to the back of the nasal cavity and allows for air pressure equalization. The exterior surface of the middle ear cavity is the tympanic membrane, the interface between the middle and outer ear. Attached to the tympanic membrane and spanning the middle ear cavity are small bone ossicles. In reptiles and birds there is a single ossicle known as the columella. In mammals there is a chain of three ossicles, the

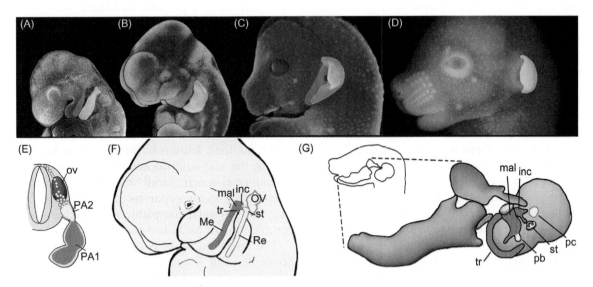

FIGURE 9.2 **Contribution of first and second pharyngeal arch NCC to outer and middle ear.** (A–D) Remodeling of the first and second pharyngeal arches to form outer ear in mouse embryos from E9.5 to E17.5. Distribution of otic NCC represented schematically. (A) E9.5, (B) E11.5, (C) E14.5, (D) E17.5. (E–G) Schematic representation of otic NCC contribution of middle ear bones and cartilages. (E) E9.5, transverse section anterior to otic vesicle, (F) E11.5, lateral view, (G) E17.5 lateral view. R1 + R2 NCC indicated in pink, R4 NCC indicated in yellow. (inc) incus, (mal) malleus, (Me) Meckel's cartilage, (OV) otic vesicle, (PA1) pharyngeal arch 1, (PA2) pharyngeal arch 2, (pb) processus brevis, (pc) pars canalicularis of otic capsule, (Re) Reichert's cartilage, (st) stapes, (tr) tympanic ring.

malleus, incus, and stapes. One end of the ossicle or ossicle chain is inserted into the tympanic membrane, and the other into an opening in the otic capsule, thus forming a physical bridge between the outer and inner ear. In some mammals a bony covering called the auditory bulla encases the middle ear cavity along with the inner ear capsule.

The function of the middle ear is to amplify the relatively weak signal of airborne sound vibrations by relaying the motion of the tympanic membrane to a tiny opening in the inner ear. Because liquids, such as endolymphatic fluid, are much denser than air, most of the energy of air sound waves would be deflected if received by the fluid-filled inner ear directly. The middle ear serves to transfer the motion of the air sound wave striking the relatively large surface area of the tympanic membrane to the relatively small oval window of the inner ear, thereby effectively concentrating the signal and converting sound waves into mechanical vibrations. Malformation of the ossicles, or fixation with respect to each other and their surroundings, inhibits this transfer and results in hearing loss.

The outer ear, which is present only in amniotes, is an anatomical funnel. In reptiles and birds this consists of an ear canal, also known as the external auditory meatus. In mammals only there is a pinna or auricle, the part of the ear visible outside the head. The structure of the external pinna is an elastic cartilage support that is contiguous with a cartilage sleeve lining the outer ear canal. The function of the outer ear is to allow entry of sound waves from the external environment to the tympanic membrane within the head. In mammals, the external pinna functions additionally to filter incoming sound waves so that the location of a sound source may be identified.

The developmental formation of the middle and outer ear have been a topic of study for a number of years, yet the important

contribution of NCC to these nonsensory structures has been elucidated only relatively recently. The pharyngeal arch origins of the nonsensory parts of the ear were defined initially by classical embryology [1–3]. That these arches are filled with NCC that give rise to skeletal elements was demonstrated subsequently by Horstradius in the 1940s in amphibians [86], an experimental verification of earlier observations of Platt [87,88].

For the middle ear, direct lineage tracing demonstrating that the middle ear ossicles derive from NCC has been performed in chick, mouse, and frog. For chick, NCC contribution to the middle ear columella has been an issue of some controversy. Initial quail-chick chimeric analysis suggested that the columella was derived entirely from NCC [27]. Later refinements revealed that the source of columella NCC progenitors was restricted to NCC emanating from R4 [4,9,85,89–91] but differed with respect to a possible additional contribution from mesoderm.

Mouse transgenic lineage tracing analysis demonstrated that NCC contribute to the three middle ear ossicles in mammals [92]. The distinct contributions of NCC of the second arch stream were determined using an R4-specific lineage trace system in mouse [11]. R4 NCC derivatives were observed in a small projection of the malleus, known as the processus brevis, and in the stapedial footplate. Comparison of mesodermal-specific and NCC-specific lineage tracing transgenic reporters in mice indicates that the stapedial footplate is derived from a combination of NCC and mesoderm [29].

The contribution of NCC to the middle ear bones of adult *Xenopus* is largely in agreement with that observed for chick and mouse, with the exception that the equivalent of the stapedial footplate contains NCC derivatives from the branchial arch posterior to the hyoid arch in addition to those of the second arch [93]. Similar discrepancies between amphibians and amniotes occurs with respect to the axial origin of NCC contributing to facial skull bones, which in amniotes are derived from forebrain and midbrain NCC but in frog are derived from the mandibular and hyoid arch [93]. The different axial origins of otic and skull NCC in frogs relative to amniote species may be related to the fact that frogs undergo metamorphosis.

The ossicles of the middle ear are endochondral bones that form from cartilage templates. The malleus and the incus, which are unique to mammals, form as cartilage condensations initially at the proximal end of the first arch cartilage, known as Meckel's cartilage (Figure 9.2F). The primordial malleus and incus form as two projections perpendicular to the long axis of the first arch cartilage [94]. In an analogous fashion, the stapes or columella cartilage condenses initially at the proximal end of the second arch cartilage, which in mammals is known as Reichart's cartilage.

An essential part of middle ear morphology is the space where tissue is absent—the air-filled tympanic cavity. Initially, the middle ear ossicles form as cartilage condensations within a solid mesenchyme. This mesenchyme disappears in a process known as cavitation or pneumatization, leaving the ossicles suspended in an air space contiguous with the Eustachian tube [95–97]. The cavitation process, which begins during gestation and is completed postnatally, is essential to free the ossicles so that they may move and thereby mechanically transmit the motion of the tympanic membrane to the oval window. The disappearing mesenchyme, like the condensing ossicle cartilages that remain, is of NCC origin [98]. So too is the bony covering, the auditory bulla, that encases the middle and inner ear in some mammals [98]. *Tcof1* mutant mice, in which survival of cranial NCC is impaired [99], have reduced growth of mesenchyme-filled bulla and exhibit a defect in middle ear cavitation [98]. The result suggests that adequate NCC survival and proliferation are needed for sufficient growth of the mesenchyme-filled bulla, and that a size threshold is necessary for middle ear cavitation to occur.

The tympanic membrane, the interface between the middle and outer ear, depends also upon NCC for development, although NCC are not the direct progenitors of the membrane itself. Instead, NCC give rise to the skeletal support bones to which the membrane attaches. In mammals the membrane forms in conjunction with the tympanic ring, a transient C-shaped skeletal element that anchors the edges of the tympanic membrane within the developing cranium [100]. The tympanic ring is a dermal bone that grows by ossification of NCC from the first arch, around the first arch cleft, and into the second arch territory. The ectoderm of the proximal first arch cleft becomes attached to the extending ring and, and, as the arches grow the ectoderm is pulled inward to form the tympanic membrane. Genetic or environmental perturbation of tympanic ring development disrupts formation of the tympanic membrane in a corresponding fashion. Mouse embryos have duplications of the tympanic ring by virtue of mutation of *Hoxa2* [101,102], and have corresponding duplications of the tympanic membrane [100].

NCC also play a critical role in development of the outer ear. The canal and pinna of the mammalian outer ear are formed by growth and remodeling of the first and second pharyngeal arches [2,3], which are filled with mesenchyme of NCC origin [103]. Transgenic lineage tracing demonstrates NCC presence within the developing pinna [104]. The cavity of the outer ear canal is formed by remodeling and deepening of the cleft between the first and second pharyngeal arches. In human, the mature canal is lined with a sleeve of cartilages which form initially as isolated bar-like condensations attached to the second arch cartilage along the inferior aspect of the canal [105].

NEUROCRISTOPATHIES OF THE MIDDLE AND OUTER EAR

Because middle ear and outer ear tissues are derived from NCC, abnormalities in NCC development can produce congenital defects in either of the two nonsensory compartments of the ear (Figure 9.3). Any malformation that blocks air sound waves from reaching the tympanic membrane or prevents the middle ear ossicles from transmitting the motion mechanically to the inner ear can result in conductive hearing loss. Malformations of the external ear may also affect personal physical appearance and can have important social and psychological ramifications.

Neurocristopathies of the middle ear include abnormalities of the shape, position, or articulation of the ossicles. Such defects can cause discontinuity of the ossicle chain or displacement of the stapes relative to the oval window. Neurocristopathy may also impede the process of cavitation leaving the middle ear cavity incompletely cleared, resulting in ossicle fixation. In addition to hearing impairment owing to ossicle fixation, incomplete clearance of the middle ear cavity can lead to a condition known as "residual mesenchyme in adults" [106]. This presence of uncleared residual tissue within the middle ear, which is of NCC origin, is a contributing factor in otitis media, a condition that imposes considerable economic burden world-wide and is a leading cause of preventable hearing loss [107].

Neurocristopathies of the outer ear may cause malformation of the external auricle, the ear canal, or both. Malformations of the external auricle include microtia, anotia, and auricular tags or pits. Microtia (MIM: 600674) is underdevelopment of the external auricle, categorized in four classes, ranging from mild structural malformation (class I) to complete absence of the auricle (class IV, also known as anotia). Collectively, the four classes of microtia are observed at an estimated prevalence of 2/10,000 [108]. Congenital aural atresia (MIM: 607842) is malformation of the ear canal such that the ear canal is incomplete or absent. Stenosis of the ear canal refers to underdevelopment or dysplasia such that the canal is narrowed or blocked. Auricular tags and pits, small fleshy nodules or sinuses, respectively, are minor malformations that may represent defects in fusion and closure of the embryonic pharyngeal arches.

Microtia and congenital aural atresia occur primarily as isolated anomalies without other defects, but may also present as part of a syndrome [109–111]. Examples of first and second pharyngeal arch syndromes include Treacher-Collins syndrome (MIM: 154500), Hemifacial microsomia (MIM: 164210), Auriculocondylar syndrome 1 and 2 (MIM: 602483 and 614669),

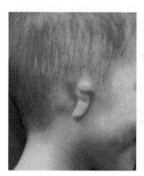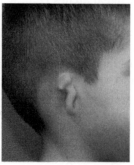

FIGURE 9.3   **Neurocristopathies of the outer and middle ear include microtia and congenital aural atresia.** Two patients with class III microtia and congenital aural atresia. In each case there is complete absence of the ear canal and a small, malformed external auricle with lobule. Appearance of such external auricles may be repaired by reconstructive surgery. *Source: Photographs courtesy of Russel H. Griffiths, MD, with permission from parents.*

and Velocardialfacial syndrome (MIM: 192430).

Microtia and congenital aural atresia may occur bilaterally or on one side only, with the right side being more commonly affected than the left [112]. When the middle and outer ear are coincidently involved and one side is more affected than the other, the severity of middle and outer ear malformations are coincident on the same side [113,114]. When the middle and outer ear are both involved, structural malformations of the inner ear may also be observed [113,114].

Patients with middle and outer ear malformations may be aided by surgical reconstruction. Surgical repair and reconstruction of the outer ear can normalize appearance of the external auricle. In some cases of aural atresia, hearing may be improved by canaloplasty if the middle ear and inner ear are functional, although such surgeries are complex and may risk potential damage to the facial nerve. If aural atresia or middle ear defect is severe but the inner ear remains functional, a bone anchored hearing aid can enable hearing through direct bone conduction of sound vibrations.

GENETIC REGULATION OF NCC MIDDLE AND OUTER EAR DEVELOPMENT

In contrast to inner ear neurocristopathies, which arise from melanocyte defects, neurocristopathies of the outer ear and middle ear are caused by development or patterning defects of bone and cartilage. Whereas most inner ear neurocristopathies are caused by disruptions of a single network, abnormalities of the middle and outer ear may arise from more diverse causes [115]. Development of the nonsensory parts of the ear is sensitive to mutations in extracellular signaling pathways, intracellular signaling cascades, DNA damage repair, and patterning. Nonsensory ear malformations are observed for mutations that disrupt NCC induction, survival, and differentiation into bone and cartilage derivatives. Many mutations affecting development of the middle and outer ear affect patterning the first and second pharyngeal arches, both at early stages prior to NCC migration and at later stages when NCC have reached their destinations within the arches.

Mutations affecting middle and outer ear development have been reviewed extensively elsewhere [85,116]. Listed below are a selected subset of genes and pathways that highlight the role of NCC in formation of the nonsensory vertebrate ear. Mouse conditional tissue-specific mutations have revealed many of the factors that are required within the NCC directly.

Tfap2a: TFAP2A is a transcription factor important for NCC induction, survival, and differentiation. Its function in otic development was revealed in large part by analysis of zebrafish mutants [117–119]. Mutation of *Tfap2a* in mouse, including a NCC-specific knockout, results in ossicle abnormalities [120–123]. Importantly, mutations of *TFAP2A* are associated with the human syndrome branchio-oculofacial syndrome (MIM: 113620) [124], which is characterized by defects of the outer and middle ear and conductive hearing loss.

Tcof1: The *Tcof1* gene product, Treacle, is a phosphoprotein associated with nucleoli [125], centrosomes and kinetochores [126]. *Tcof1* is required for survival of cranial NCC [99]. Mice heterozygous for a *Tcof1* mutation exhibit severe conductive hearing loss owing to a defect in cavitation of the middle ear and defects in growth of the auditory bulla [98]. Mutations of *TCOF1* are associated with Treacher-Collins syndrome (MIM: 154500) in human [127,128], the phenotype of which includes conductive hearing loss owing to reduction of outer ear structures and middle ear defects.

Hox anterior–posterior patterning: The *Hox* family of genes encode transcription factors that are necessary for patterning the anterior–posterior axis of the developing embryo. Targeted mouse mutations of *Hoxa2* demonstrate that this gene is required for formation of the external ear, as deletion results in absence of pinna [101,102]. Temporal and NCC-specific knock out of *Hoxa2* demonstrate that the requirement for *Hoxa2* occurs after the cells have migrated and filled the arches [129]. Mutation of *HOXA2* has been identified as the causative lesion responsible for some cases of human microtia [130]. *Hoxa1* and *Hoxb1* are together required for pinna formation, likely owing to their function in patterning the hindbrain and pharyngeal arches rather than a direct effect on NCC in the pharyngeal arches [131].

Endothelin signaling: Endothelins are secreted peptide ligands that function through G-protein coupled receptors. EDN1 and its cognate receptor EDNRA mediate a critical signal for establishing the ventral/distal identity of the first pharyngeal arch [132]. Targeted knock out of *Ednra* in mouse causes malformation of the malleus and incus, mislocalization of the stapes, loss of the tympanic ring, and absence of ear canal [104,133]. Mutation of other components of the signaling pathway, including defects in *Edn1* or *Ece1*, which encodes an enzyme necessary for activation of EDN1, produce similar phenotypes [134,135]. Temporal and NCC-specific knock out of *Ednra* reveals that endothelin signaling is required in NCC within a very narrow time window of development, likely being needed for patterning NCC shortly after arrival in the arches [136].

BMP signaling: Bone morphogenetic proteins (BMPs), secreted growth factors of the TGFβ superfamily that signal through their cognate receptors, are important for bone and cartilage formation, embryo patterning, and induction of NCC [137]. Mutation of *Bmp5* [138] results in reduction of pinna size likely owing to a defect in growth of the NCC-derived outer ear cartilage. Disruption of the gene for the BMP type 1 receptor *Acvr1* specifically within NCC results in a subtle defect in the shape of the malleus [139].

PLCB4 and GNAI3: Specific missense mutations in PLCB4 and GNAI3 have been shown to be responsible for Auriculocondylar syndrome type 2 and 1, respectively (MIM: 614669 and 602483) [140,141]. The syndromes are characterized by distinctive malformations of the external auricle, defects that may possibly result from dominant negative or gain of function disruption of endothelin signaling. *PLCB4* encodes phospholipase C beta 4, and *GNAI3* encodes a G protein, both of which are known to be important for transduction of many extracellular signals. Mutation of the orthologous *Plcb3* in zebrafish disrupts Endothelin 1 regulation of pharyngeal arch patterning [142].

HMX1 is a homeodomain transcription factor expressed in otic NCC [143]. Mutation causes defects in the pinna in mouse [144] and rat [145] and has been shown to be the causative mutation in Oculoauricular syndrome (MIM:612109) in human [146].

Conditional mutations within NCC: Mouse conditional mutations within NCC reveal the nature of genes needed directly within NCC and also highlight the importance of NCC in formation of the middle and outer ear. As mentioned above, these include *Tfap2a, Hoxa2, Ednra,* and *Acvr1.* Other genes producing middle and outer ear phenotypes when conditionally inactivated in NCC include *Sox9* [147], *Myc* [148], *Fgfr1* [149], *Ext1* [150], *Ptpn11* [151], *Efnb1* [152], and *Smo* [153].

NCC AND EVOLUTION OF THE EAR

The dramatic variety of vertebrate craniofacial forms is thought to be due in large part to NCC, and evolution of the ear is a prime example of this influence. As proposed by Gans and Northcutt [154,155], the radiation of vertebrates was made possible by the acquisition of specialized sensory organs, all derived developmentally from combined contribution of epidermal neurogenic placodes and NCC. The ear, which exhibits exceptional morphological variation across vertebrates, is one such sense organ.

Compelling evidence suggests that vertebrate NCC are evolutionarily derived from pigmented cells present in an ancestral chordate [156]. The extant urochordate *Ciona intestinalis* has melanocyte cells that arise at the neural plate border. These cells express many hallmark genes of NCC, and become migratory when subjected to ectopic expression of a single gene *Twist*. Interestingly, these pigmented cells give rise to a light-sensing receptor, the ocellus, and a gravity-sensing receptor, a single-celled otolith present within the head region of tunicates [157]. While the single-celled otolith is clearly very distinct from the vestibular system of vertebrate inner ear, it is worth noting that the link between NCC-like pigment cells and spatial-sensory function may have roots deep in evolution.

It seems likely that the primitive ear functioned primarily as an organ to sense motion and balance. However, basic auditory sensation may also be a very ancient feature of vertebrates [158]. The vestibular-auditory inner ear organ may be evolutionarily related to the lateral line system, a system of sensory units present on the body surface in fish and amphibians [159–161]. Both are mechano-sensory systems comprised of placodally derived neurons supported by or surrounded by NCC-derived glial cells and bone.

Whereas the role of NCC in evolution of the inner ear may be theoretical and difficult to visualize, the role of NCC in variation of the middle and outer ear is clear and obvious. The importance of NCC in evolution of the middle and outer ear can be appreciated by considering the differences in middle ear ossicles between mammals and non-mammalian jawed vertebrates. Reptiles, amphibians, and birds have a single ossicle known as the columella or stapes. Mammals, on the other hand, have three ossicles, the malleus, incus, and stapes, with the stapes being homologous to the single ossicle in other vertebrates. Thus, relative to more ancestral vertebrate forms, the mammalian middle ear acquired two new ossicles in addition to the stapes [162].

Based on comparative anatomy, Reichert [1] proposed in 1837 that the two mammalian-specific ossicles, the malleus and incus, are homologous to the quadrate and articular, the two bones responsible for articulation of the jaw joint in non-mammalian species. Reichert's initial insight, that the novel mammalian ossicles are evolutionary derivatives of ancestral jaw joint bones, has been supported by a substantial body of data from multiple species, the fossil record, and by molecular analysis. In non-mammalian jawed vertebrates, the jaw joint forms between the articular and quadrate (or palatoquadrate) bones, which are derivatives of the first arch NCC that form at the proximal end of Meckel's cartilage during embryogenesis [4,9]. Molecularly, the joint between them is marked by expression of *Bapx1* [163,164]. In these species, first arch NCC give rise to multiple bones within the mandible including the angular and the dentary. The body of the columella, along with a process of the articular bone (the retroarticular process), derive from second pharyngeal arch NCC (Figure 9.4).

In mammals the articular and quadrate bones, which form the jaw joint in non-mammalian jawed vertebrates, are absent, and the jaw joint forms between the squamosal bone of the skull and the dentary bone of the mandible. Lacking the articular and quadrate bones of the jaw, mammals have, instead, the articulated malleus and incus of the middle ear. There are many parallels between the articular

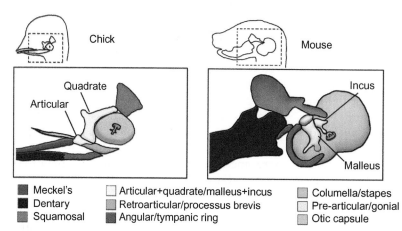

FIGURE 9.4 **A schematic representation of bones and cartilages of the middle ear and jaws in avian and mammalian species.** The mammalian malleus and incus are homologous to the articular and quadrate of the chick. The mammalian stapes is homologous to the single columella of birds. The processus brevis of the mammalian malleus is homologous to the retroarticular process of avian species. *Source: Adapted from Ref. [162].*

and quadrate bones and the malleus and incus. Each articulated pair forms at the proximal end of Meckel's cartilage during embryogenesis [85], and, in each case, the joint between is molecularly marked by expression of *Bapx1* [164]. The evolutionary derivation of the malleus and the incus from the articular and quadrate is further supported by the pharyngeal arch origins of portions of the bones. Lineage tracing using an R4-specific transgenic reporter in mouse reveals that the processus brevis of the malleus and the footplate of the stapes are derived from second arch NCC, an ontogeny that supports that the small process of the malleus is homologous to the retroarticular process, and the stapes to the columella [11]. The tympanic ring, a first arch NCC derivative, is homologous to the angular bone of avian species.

Evolutionary variation of NCC-derived ear elements is not limited to differences between non-mammalian and mammalian vertebrates. Striking evolutionary variations occur also between different orders, families, and genre as well. For example, whereas the air-filled middle ear cavity and external pinna are evolutionary adaptations for improved hearing of airborne sound, the ears of marine mammals evolved for function in an aquatic environment. Cetacean ears utilize the NCC-derived

mandible and mandibular fat pad to sense sound via bone conduction [165] for matching impedance of sound waves in water. Also in marine mammals, the external pinna is reduced or absent, an adaptation of obvious hydrodynamic advantage.

CONCLUDING REMARKS

NCC are important for many aspects of ear development, producing glial cells and melanocytes in the sensory inner ear, bony ossicles of the middle ear, and cartilages of the outer ear canal and pinna. In addition to producing multiple tissue types, NCC are crucial for the morphological development of the empty spaces of the ear in terrestrial vertebrates: the external ear canal and the air cavity surrounding the middle ear ossicles, each of which are critical for conduction of sound signal to the inner ear. NCC contribution in ear formation is clearly diverse and extensive. Knowledge of the developmental biology of otic NCC has great potential to elucidate aspects of evolutionary variation in ear morphology and to reveal mechanisms underlying a variety of congenital ear abnormalities in humans. In spite of the obvious importance of NCC in development of the ear, many aspects of otic NCC biology remain obscure.

Of obvious benefit would be increased understanding of otic NCC regulation during growth and remodeling of the first and second pharyngeal arches into the external ear canal and pinna. What signals and interactions direct morphological growth and remodeling? What genetic and environmental perturbations disrupt the process during embryogenesis? Can stem cells with NCC characteristics be induced to grow cartilage for surgical repair of congenital defects of the external ear?

Because the external ear is found only in mammals, questions about the role of NCC in external ear development must be addressed with a mammalian system. Generation of new mouse models of human microtia, and study of existing models, are each likely to provide additional insights into NCC development during morphogenesis of the external ear canal and pinna. New transgenic lineage tracing reporter strains marking specific otic progenitor cell populations would also be useful. Existing transgenic reporter mice that label all NCC, or those that label R4 NCC exclusively, have been invaluable for understanding NCC contribution in the ear. Lineage trace reporter strains that specifically mark other populations of NCC, such as first arch NCC or specific mesodermal populations, will no doubt also yield new insights about how the external ear canal and pinna are formed.

Another topic that demands attention is the process of clearing of the middle ear cavity. Incomplete cavitation results in persistence of NCC mesenchyme within the middle ear, which can cause conductive hearing loss and is strongly linked with otitis media. What genetic or environmental factors impede complete cavitation? Does normal clearance occur by apoptosis, phagocytosis, by redistribution of cells within an expanding space, or by a combination of these mechanisms? Mouse mutants with defects in these distinct processes may shed light on the mechanisms

involved. Long-term lineage tracing analysis of middle ear mesenchyme in chick or frog or other organisms may also reveal mechanisms of the cavitation process.

Equally important is a deeper understanding of NCC within the developing inner ear. The melanocytes of the stria vascularis are essential for inner ear function yet little is known about the developmental characteristics of these cells. Do the strial melanocytes arise from late-migrating melanoblast-specific NCC progenitors, or from the population of NCC that are progenitors of Schwann cells of the cochleovestibular nerve? By what migratory routes do melanocytes take up position in the developing stria? Do NCC-derived Schwann cells or Schwann-associated progenitors of the cochleovestibular nerve have the potential to differentiate into melanocytes? Answers to these questions may inform approaches to promote regeneration or to develop stem cell therapies for the treatment of sensorineural hearing defects or age related hearing loss.

From a basic science perspective, the study of otic NCC may serve as a paradigm for understanding the overall role of NCC in evolution of vertebrate forms. The ancestral predecessor cells of NCC may have produced melanocytes and generated gravity-sensing otoliths. The advent of NCC, a migratory cell type capable of generating multiple tissue fates, is thought to have played a key role in the radiation of vertebrates. These properties of NCC also undoubtedly contributed to evolution of the ear from a probable simple motion detector to a complex organ capable of sensing equilibrium and sound and, ultimately, an organ of communication.

References

[1] Reichert KB. Ueber die Visceralbogen der Wirbelthiere im Allgemeinen und deren Metamorphose bei den Säugethieren und Vögeln. Archiv für Anatomie, Physiologie und wissenschaftliche Medicin 1837; Sittenfeld.

[2] His W. Anatomie menschlicher Embryonen. Vogel; 1880.

[3] Streeter GL. Development of the auricle in the human embryo. Contrib Embryol 1922;69:111.

[4] Kontges G, Lumsden A. Rhombencephalic neural crest segmentation is preserved throughout craniofacial ontogeny. Development 1996;122(10):3229–42.

[5] Serbedzija GN, Bronner-Fraser M, Fraser SE. Vital dye analysis of cranial neural crest cell migration in the mouse embryo. Development 1992;116:297–307.

[6] Stone LS. Further experiments on the extirpation and transplantation of mesectoderm in *Amblystoma punctatum*. J Exp Zool 1926;44(1):95–131.

[7] Le Lievre CS, Le Douarin NM. Mesenchymal derivatives of the neural crest: analysis of chimaeric quail and chick embryos. J Embryol Exp Morphol 1975;34:125–54.

[8] Lumsden A, Sprawson N, Graham A. Segmental origin and migration of neural crest cells in the hindbrain region of the chick embryo. Development 1991;113(4):1281–91.

[9] Couly GF, Coltey PM, Le Douarin NM. The triple origin of skull in higher vertebrates: a study in quail-chick chimeras. Development 1993;117(2):409–29.

[10] D'Amico-Martel A, Noden D. Contributions of placodal and neural crest cells to avian cranial peripheral ganglia. Am J Anat 1983;166:445–68.

[11] O'Gorman S. Second branchial arch lineages of the middle ear of wild-type and Hoxa2 mutant mice. Dev Dyn 2005;234(1):124–31.

[12] Tosney K. The segregation and early migration of cranial neural crest cells in the avian embryo. Dev Biol 1982;89:13–24.

[13] Chan WY, Tam PPL. A morphological and experimental study of the mesencephalic neural crest cells in the mouse embryo using wheat-germ agglutinin gold conjugate as the cell marker. Development 1988;102(2):427–42.

[14] Weston JA. A radioautographic analysis of the migration and localization of trunk neural crest cells in the chick. Dev Biol 1963;6(3):279–310.

[15] Adameyko I, Lallemend F, Aquino JB, Pereira JA, Topilko P, Müller T, et al. Schwann cell precursors from nerve innervation are a cellular origin of melanocytes in skin. Cell 2009;139(2):366–79.

[16] Erickson CA. From the crest to the periphery: control of pigment cell migration and lineage segregation. Pigment Cell Res 1993;6(5):336–47.

[17] Nakayama A, Nguyen M-TT, Chen CC, Opdecamp K, Hodgkinson CA, Arnheiter H. Mutations in microphthalmia, the mouse homolog of the human deafness gene MITF, affect neuroepithelial and neural crest-derived melanocytes differently. Mech Dev 1998;70(1–2):155–66.

[18] Mallo M. Retinoic acid disturbs mouse middle ear development in a stage-dependent fashion. Dev Biol 1997;184(1):175–86.

[19] Wei X, Makori N, Peterson PE, Hummler H, Hendrickx AG. Pathogenesis of retinoic acid-induced ear malformations in a primate model. Teratology 1999;60(2):83–92.

[20] Vieux-Rochas M, Coen L, Sato T, Kurihara Y, Gitton Y, Barbieri O, et al. Molecular dynamics of retinoic acid-induced craniofacial malformations: implications for the origin of Gnathostome Jaws. PLoS ONE 2007;2(6):e510.

[21] Rosa FW, Wilk AL, Kelsey FO. Vitamin A congeners. Teratology 1986;33(3):355–64.

[22] Barald KF, Kelley MW. From placode to polarization: new tunes in inner ear development. Development 2004;131(17):4119–30.

[23] Giraldez F, Fritzsch B. The molecular biology of ear development -"Twenty years are nothing". Int J Dev Biol 2007;51(6/7):429.

[24] Ohyama T, Groves AK. Generation of Pax2-Cre mice by modification of a Pax2 bacterial artificial chromosome. Genesis 2004;38(4):195–9.

[25] Freyer L, Aggarwal V, Morrow BE. Dual embryonic origin of the mammalian otic vesicle forming the inner ear. Development 2011;138(24):5403–14.

[26] Breuskin I, Bodson M, Thelen N, Thiry M, Borgs L, Nguyen L, et al. Glial but not neuronal development in the cochleo-vestibular ganglion requires Sox10. J Neurochem 2010;114(6):1827–39.

[27] Le Lievre CS. Participation of neural crest-derived cells in the genesis of the skull in birds. J Embryol Exp Morphol 1978;47:17–37.

[28] McPhee JR, Van De Water TR. Epithelial—mesenchymal tissue interactions guiding otic capsule formation: the role of the otocyst. J Embryol Exp Morphol 1986;97(1):1–24.

[29] Thompson H, Ohazama A, Sharpe PT, Tucker AS. The origin of the stapes and relationship to the otic capsule and oval window. Dev Dyn 2012;241(9):1396–404.

[30] Corti A. Recherches sur l'organe de l'ouie des mammiferes 1851.

[31] Darwin C. On the origin of species by means of natural selection. Murray; 1859.

[32] Dorris F. The production of pigment *in vitro* by chick neural crest. Wilhelm Roux' Arch Entwickl 1938;138(3–4):323–34.

[33] DuShane GP. An experimental study of the origin of pigment cells in amphibia. J Exp Zool 1935;72(1):1–31.

[34] Rawles ME. Origin of pigment cells from the neural crest in the mouse embryo. Physiol Zool 1947;248–66.

[35] Hilding D, Ginzberg RD. Pigmentation of the stria vascularis the contribution of neural crest melanocytes. Acta Otolaryngol 1977;84(1-6):24—37.

[36] Giebel LB, Spritz RA. Mutation of the KIT (mast/stem cell growth factor receptor) protooncogene in human piebaldism. Proc Natl Acad Sci USA 1991; 88(19):8696—9.

[37] Cable J, Barkway C, Steel KP. Characteristics of stria vascularis melanocytes of viable dominant spotting (WvWv) mouse mutants. Hear Res 1992; 64(1):6—20.

[38] Geissler EN, Ryan MA, Housman DE. The dominant-white spotting (W) locus of the mouse encodes the c-kit proto-oncogene. Cell 1988;55(1):185—92.

[39] Ruan H-B, Zhang N, Gao X. Identification of a novel point mutation of mouse proto-oncogene c-kit through N-ethyl-N-nitrosourea mutagenesis. Genetics 2005;169(2):819—31.

[40] Bolande RP. The neurocristopathies: a unifying concept of disease arising in neural crest maldevelopment. Hum Pathol 1974;5(4):409—29.

[41] Jones MC. The neurocristopathies: reinterpretation based upon the mechanism of abnormal morphogenesis. Cleft Palate Craniofac J 1990;27(2):136—40.

[42] Pingault V, Ente D, Dastot-Le Moal F, Goossens M, Marlin S, Bondurand N. Review and update of mutations causing Waardenburg syndrome. Hum Mutat 2010;31(4):391—406.

[43] Read AP, Newton VE. Waardenburg syndrome. J Med Genet 1997;34(8):656—65.

[44] Black FO, Pesznecker SC, Allen K, Gianna CA. Vestibular phenotype for Waardenburg syndrome? Otol Neurotol 2001;22(2):188—94.

[45] Kaneaster SK, Saunders JE. Congenital vestibular disorders. In: Weber PC, editor. Vertigo and disequilibrium a practical guide to diagnosis and management. New York, NY: Thieme Medical Publishers; 2008. p. 119—24.

[46] Steel KP, Barkway C. Another role for melanocytes: their importance for normal stria vascularis development in the mammalian inner ear. Development 1989;107(3):453—63.

[47] Tassabehji M, Read AP, Newton VE, Harris R, Balling R, Gruss P, et al. Waardenburg's syndrome patients have mutations in the human homologue of the Pax-3 paired box gene. Nature 1992;355(6361):635—6.

[48] Coppens AG, Salmon I, Heizmann CW, Kiss R, Poncelet L. Postnatal maturation of the dog stria vascularis—an immunohistochemical study. Anat Rec A Discov Mol Cell Evol Biol 2003;270A(1):82—92.

[49] Steel KP, Bock GR. Hereditary inner-ear abnormalities in animals: relationships with human abnormalities. Arch Otolaryngol Head Neck Surg 1983;109(1):22.

[50] Nadol Jr JB, Merchant SN. Histopathology and molecular genetics of hearing loss in the human. Int J Pediatr Otorhinolaryngol 2001;61(1):1—15.

[51] Merchant SN, McKenna MJ, Baldwin CT, Milunsky A, Nadol JB. Otopathology in a case of type I Waardenburg's syndrome. Ann Otol Rhinol Laryngol 2001;110(9):875—82.

[52] Murillo-Cuesta S, Contreras J, Zurita E, Cediel R, Cantero M, Varela-Nieto I, et al. Melanin precursors prevent premature age-related and noise-induced hearing loss in albino mice. Pigment Cell Melanoma Res 2010;23(1):72—83.

[53] Ohlemiller KK, Rybak Rice ME, Lett JM, Gagnon PM. Absence of strial melanin coincides with age-associated marginal cell loss and endocochlear potential decline. Hear Res 2009;249(1—2):1—14.

[54] Price ER, Fisher DE. Sensorineural deafness and pigmentation genes: melanocytes and the Mitf transcriptional network. Neuron 2001;30(1):15—18.

[55] Asher JH, Friedman TB. Mouse and hamster mutants as models for Waardenburg syndromes in humans. J Med Genet 1990;27(10):618—26.

[56] Tachibana M, Kobayashi Y, Matsushima Y. Mouse models for four types of Waardenburg syndrome. Pigment Cell Res 2003;16(5):448—54.

[57] Yasumoto K, Yokoyama K, Shibata K, Tomita Y, Shibahara S. Microphthalmia-associated transcription factor as a regulator for melanocyte-specific transcription of the human tyrosinase gene. Mol Cell Biol 1994;14(12):8058—70.

[58] Hou L, Arnheiter H, Pavan WJ. Interspecies difference in the regulation of melanocyte development by SOX10 and MITF. Proc Natl Acad Sci USA 2006;103 (24):9081—5.

[59] Opdecamp K, Nakayama A, Nguyen MT, Hodgkinson CA, Pavan WJ, Arnheiter H. Melanocyte development in vivo and in neural crest cell cultures: crucial dependence on the Mitf basic-helix-loop-helix-zipper transcription factor. Development 1997;124(12):2377—86.

[60] Hornyak TJ, Hayes DJ, Chiu L-Y, Ziff EB. Transcription factors in melanocyte development: distinct roles for Pax-3 and Mitf. Mech Dev 2001;101(1—2):47—59.

[61] Sato T, Sasai N, Sasai Y. Neural crest determination by co-activation of Pax3 and Zic1 genes in Xenopus ectoderm. Development 2005;132(10):2355—63.

[62] Saldana-Caboverde A, Kos L. Roles of endothelin signaling in melanocyte development and melanoma. Pigment Cell Melanoma Res 2010;23(2):160—70.

[63] Tassabehji M, Newton VE, Read AP. Waardenburg syndrome type 2 caused by mutations in the human microphthalmia (MITF) gene. Nat Genet 1994;8(3):251—5.

[64] Morell R, Spritz RA, Ho L, Pierpont J, Guo W, Friedman TB, et al. Apparent digenic inheritance of

Waardenburg syndrome type 2 (WS2) and Autosomal Recessive Ocular Albinism (AROA). Hum Mol Genet 1997;6(5):659–64.

[65] Amiel J, Watkin PM, Tassabehji M, Read AP, Winter RM. Mutation of the MITF gene in albinism-deafness syndrome (Tietz syndrome). Clin Dysmorphol 1998;7(1):17.

[66] Hodgkinson CA, Moore KJ, Nakayama A, Steingrímsson E, Copeland NG, Jenkins NA, et al. Mutations at the mouse microphthalmia locus are associated with defects in a gene encoding a novel basic-helix-loop-helix-zipper protein. Cell 1993; 74(2):395–404.

[67] Hughes MJ, Lingrel JB, Krakowsky JM, Anderson KP. A helix-loop-helix transcription factor-like gene is located at the mi locus. J Biol Chem 1993; 268(28):20687–90.

[68] Deol MS. The relationship between abnormalities of pigmentation and of the inner ear. Proc R Soc Lond B Biol Sci 1970;175(1039):201–17.

[69] Baldwin CT, Hoth CF, Amos JA, da-Silva EO, Milunsky A. An exonic mutation in the HuP2 paired domain gene causes Waardenburg's syndrome. Nature 1992;355(6361):637–8.

[70] Morell R, Friedman TB, Moeljopawiro S, Hartono S, Asher JH. A frameshift mutation in the HuP2 paired domain of the probable human homolog of murine Pax-3 is responsible for Waardenburg syndrome type 1 in an Indonesian family. Hum Mol Genet 1992;1(4):243–7.

[71] Hoth C, Milunsky A, Lipsky N, Sheffer R, Clarren S, Baldwin C. Mutations in the paired domain of the human PAX3 gene cause Klein-Waardenburg syndrome (WS-III) as well as Waardenburg syndrome type I (WS-I). Am J Hum Genet 1993;52(3):455.

[72] Asher JH, Sommer A, Morell R, Friedman TB. Missense mutation in the paired domain of PAX3 causes craniofacial-deafness-hand syndrome. Hum Mutat 1996;7(1):30–5.

[73] Steel KP, Smith RJH. Normal hearing in Splotch (Sp/+), the mouse homologue of Waardenburg syndrome type 1. Nat Genet 1992;2(1):75–9.

[74] Epstein DJ, Vekemans M, Gros P. Splotch (Sp2H), a mutation affecting development of the mouse neural tube, shows a deletion within the paired homeodomain of Pax-3. Cell 1991;67(4):767–74.

[75] Sánchez-Martín M, Rodríguez-García A, Pérez-Losada J, Sagrera A, Read AP, Sánchez-García I. SLUG (SNAI2) deletions in patients with Waardenburg disease. Hum Mol Genet 2002;11(25):3231–6.

[76] Pérez-Losada J, Sánchez-Martı M, Rodrı A, Sánchez ML, Orfao A, Flores T, et al. Zinc-finger transcription factor Slug contributes to the function of the stem cell factor c-kit signaling pathway. Blood 2002; 100(4):1274–86.

[77] Pingault V, Bondurand N, Kuhlbrodt K, Goerich DE, Prehu MO, Puliti A, et al. SOX10 mutations in patients with Waardenburg–Hirschsprung disease. Nat Genet 1998;18(2):171–3.

[78] Herbarth B, Pingault V, Bondurand N, Kuhlbrodt K, Hermans-Borgmeyer I, Puliti A, et al. Mutation of the Sry-related Sox10 gene in dominant megacolon, a mouse model for human Hirschsprung disease. Proc Natl Acad Sci USA 1998;95 (9):5161–5.

[79] Southard-Smith EM, Kos L, Pavan WJ. SOX10 mutation disrupts neural crest development in Dom Hirschsprung mouse model. Nat Genet 1998; 18(1):60–4.

[80] Edery P, Attie T, Amiel J, Pelet A, Eng C, Hofstra RM, et al. Mutation of the endothelin-3 gene in the Waardenburg–Hirschsprung disease (Shah-Waardenburg syndrome). Nat Genet 1996;12 (4):442–4.

[81] Baynash AG, Hosoda K, Giaid A, Richardson JA, Emoto N, Hammer RE, et al. Interaction of endothelin-3 with endothelin-B receptor is essential for development of epidermal melanocytes and enteric neurons. Cell 1994;79(7):1277–85.

[82] Attié T, Till M, Pelet A, Amiel J, Edery P, Boutrand L, et al. Mutation of the endothelin-receptor B gene in Waardenburg–Hirschsprung disease. Hum Mol Genet 1995;4(12):2407–9.

[83] Matsushima Y, Shinkai Y, Kobayashi Y, Sakamoto M, Kunieda T, Tachibana M. A mouse model of Waardenburg syndrome type 4 with a new spontaneous mutation of the endothelin-B receptor gene. Mamm Genome 2002;13(1):30–5.

[84] Hosoda K, Hammer RE, Richardson JA, Baynash AG, Cheung JC, Giaid A, et al. Targeted and natural (piebald-lethal) mutations of endothelin-B receptor gene produce megacolon associated with spotted coat color in mice. Cell 1994;79(7):1267–76.

[85] Mallo M. Formation of the outer and middle ear, molecular mechanisms. Curr Top Dev Biol 2003;85–113.

[86] Horstadius S, Sellman S. Experimentelle untersuchungen uber die determination des knorpeligen Kopfskelettes bei Urodelen. Nov Act Reg Soc Scient Ups Ser 1946;13:1–170.

[87] Platt JB. Ectodermal origin of the cartilages of the head. Anat Anz 1893;8:506–9.

[88] Platt JB. The development of the cartilaginous skull and of the branchial and hypoglossal musculature in *Necturus*. Morphol Jb 1897;25:377–464.

[89] Noden DM. Patterns and organization of craniofacial skeletogenic and myogenic mesenchyme: a perspective. Prog Clin Biol Res 1982;101:167–203.

[90] Noden D. Craniofacial development; new views on old problems. Anat Rec 1984;208:1–13.

[91] Chapman SC. Can you hear me now? Understanding vertebrate middle ear development. Front Biosci 2011;16:1675.

[92] Abe M, Ruest LB, Clouthier DE. Fate of cranial neural crest cells during craniofacial development in endothelin-A receptor-deficient mice. Int J Dev Biol 2007;51(2):97.

[93] Gross JB, Hanken J. Segmentation of the vertebrate skull: neural-crest derivation of adult cartilages in the clawed frog, Xenopus laevis. Integr Comp Biol 2008;48(5):681–96.

[94] Hall BK, Miyake T. All for one and one for all: condensations and the initiation of skeletal development. BioEssays 2000;22(2):138–47.

[95] Jaskoll TF, Maderson PFA. A histological study of the development of the avian middle ear and tympanum. Anat Rec 1978;190(2):177–99.

[96] Palva T, Ramsay H. Fate of the mesenchyme in the process of pneumatization. Otol Neurotol 2002;23(2):192–9.

[97] Palva T, Pääkkö P, Ramsay H, Chrobok V, Simáková E. Apoptosis and regression of embryonic mesenchyme in the development of the middle ear spaces. Acta Otolaryngol 2003;123(2):209.

[98] Richter CA, Amin S, Linden J, Dixon J, Dixon MJ, Tucker AS. Defects in middle ear cavitation cause conductive hearing loss in the Tcof1 mutant mouse. Hum Mol Genet 2010;19(8):1551–60.

[99] Dixon J, Jones NC, Sandell LL, Jayasinghe SM, Crane J, Rey JP, et al. Tcof1/Treacle is required for neural crest cell formation and proliferation deficiencies that cause craniofacial abnormalities. Proc Natl Acad Sci USA 2006;103:13403–8.

[100] Mallo M, Gridley T. Development of the mammalian ear: coordinate regulation of formation of the tympanic ring and the external acoustic meatus. Development 1996;122(1):173–9.

[101] Gendron-Maguire M, Mallo M, Zhang M, Gridley T. Hoxa-2 mutant mice exhibit homeotic transformation of skeletal elements derived from cranial neural crest. Cell 1993;75:1317–31.

[102] Rijli FM, Mark M, Lakkaraju S, Dierich A, Dolle P, Chambon P. A homeotic transformation is generated in the rostral branchial region of the head by disruption of Hoxa-2, which acts as a selector gene. Cell 1993;75:1333–49.

[103] Chai Y, Jiang X, Ito Y, Bringas Jr. P, Han J, Rowitch DH, et al. Fate of the mammalian cranial neural crest during tooth and mandibular morphogenesis. Development 2000;127(8):1671–9.

[104] Clouthier DE, Hosoda K, Richardson JA, Williams SC, Yanagisawa H, Kuwaki T, et al. Cranial and cardiac neural crest defects in endothelin-A receptor-deficient mice. Development 1998;125(5):813–24.

[105] Ikari Y, Katori Y, Ohtsuka A, Rodríguez-Vázquez JF, Abe H, Kawase T, Murakami G, Abe S-i. Fetal development and variations in the cartilages surrounding the human external acoustic meatus. Annals of Anatomy - Anatomischer Anzeiger 2013;195(2): 128–136.

[106] Jaisinghani VJ, Paparella MM, Schachern PA, Schneider DS, Le CT. Residual mesenchyme persisting into adulthood. Am J Otolaryngol 1999;20(6):363–70.

[107] Monasta L, Ronfani L, Marchetti F, Montico M, Vecchi Brumatti L, Bavcar A, et al. Burden of disease caused by otitis media: systematic review and global estimates. PLoS ONE 2012;7(4):e36226.

[108] Luquetti DV, Leoncini E, Mastroiacovo P. Microtia-anotia: a global review of prevalence rates. Birth Defects Res A Clin Mol Teratol 2011;91(9):813–22.

[109] Passos-Bueno MR, Ornelas CC, Fanganiello RD. Syndromes of the first and second pharyngeal arches: a review. Am J Med Genet A 2009;149A (8):1853–9.

[110] Johnson JM, Moonis G, Green GE, Carmody R, Burbank HN. Syndromes of the first and second branchial arches, part 2: syndromes. Am J Neuroradiol 2011;32(2):230–7.

[111] Johnson JM, Moonis G, Green GE, Carmody R, Burbank HN. Syndromes of the first and second branchial arches, part 1: embryology and characteristic defects. Am J Neuroradiol 2011;32(1):14–19.

[112] Kösling S, Omenzetter M, Bartel-Friedrich S. Congenital malformations of the external and middle ear. Eur J Radiol 2009;69(2):269–79.

[113] Mayer TE, Brueckmann H, Siegert R, Witt A, Weerda H. High-resolution CT of the temporal bone in dysplasia of the auricle and external auditory canal. Am J Neuroradiol 1997;18(1):53–65.

[114] Vrabec JT, Lin JW. Inner ear anomalies in congenital aural atresia. Otol Neurotol 2010;31(9):1421.

[115] Luquetti DV, Heike CL, Hing AV, Cunningham ML, Cox TC. Microtia: epidemiology and genetics. Am J Med Genet A 2011.

[116] Fekete DM. Development of the vertebrate ear: insights from knockouts and mutants. Trends Neurosci 1999;22(6):263–9.

[117] Knight RD, Nair S, Nelson SS, Afshar A, Javidan Y, Geisler R, et al. Lockjaw encodes a zebrafish tfap2a required for early neural crest development. Development 2003;130(23):5755–68.

[118] Li W, Cornell RA. Redundant activities of Tfap2a and Tfap2c are required for neural crest induction and development of other non-neural ectoderm derivatives in zebrafish embryos. Dev Biol 2007; 304(1):338–54.

[119] Barrallo-Gimeno A, Holzschuh J, Driever W, Knapik EW. Neural crest survival and differentiation in zebrafish depends on mont blanc/tfap2a gene function. Development 2004;131:1463–77.

[120] Ahituv N, Erven A, Fuchs H, Guy K, Ashery–Padan R, Williams T, et al. An ENU-induced mutation in AP-2α leads to middle ear and ocular defects in Doarad mice. Mamm Genome 2004;15(6):424–32.

[121] Zhang J, Hagopian-Donaldson S, Serbedzija G, Elsemore J, Plehn-Dujowich D, McMahon AP, et al. Neural tube, skeletal and body wall defects in mice lacking transcription factor AP-2. Nature 1996; 381(6579):238–41.

[122] Brewer S, Feng W, Huang J, Sullivan S, Williams T. Wnt1-Cre-mediated deletion of AP-2alpha causes multiple neural crest-related defects. Dev Biol 2004;267:135–52.

[123] Nottoli T, Hagopian-Donaldson S, Zhang J, Perkins A, Williams T. AP-2-null cells disrupt morphogenesis of the eye, face, and limbs in chimeric mice. Proc Natl Acad Sci USA 1998;95:13714–19.

[124] Milunsky JM, Maher TA, Zhao G, Roberts AE, Stalker HJ, Zori RT, et al. TFAP2A mutations result in branchio-oculo-facial syndrome. Am J Hum Genet 2008;82(5):1171–7.

[125] Valdez BC, Henning D, So RB, Dixon J, Dixon MJ. The Treacher Collins syndrome (TCOF1) gene product is involved in ribosomal DNA gene transcription by interacting with upstream binding factor. Proc Natl Acad Sci USA 2004;101 (29):10709–14.

[126] Sakai D, Dixon J, Dixon MJ, Trainor PA. Mammalian neurogenesis requires Treacle-Plk1 for precise control of spindle orientation, mitotic progression, and maintenance of neural progenitor cells. PLoS Genet 2012;8(3):e1002566.

[127] Gladwin AJ, Dixon J, Loftus SK, Edwards S, Wasmuth JJ, Hennekam RC, et al. Treacher Collins syndrome may result from insertions, deletions or splicing mutations, which introduce a termination codon into the gene. Hum Mol Genet 1996;5(10):1533–8.

[128] The Treacher Collins Syndrome Collaborative G., Dixon J, Edwards SJ, Gladwin AJ, Dixon MJ, Loftus SK, et al. Positional cloning of a gene involved in the pathogenesis of Treacher Collins syndrome. Nat Genet 1996;12(2):130–6.

[129] Santagati F, Minoux M, Ren SY, Rijli FM. Temporal requirement of Hoxa2 in cranial neural crest skeletal morphogenesis. Development 2005;132(22):4927–36.

[130] Alasti F, Sadeghi A, Sanati MH, Farhadi M, Stollar E, Somers T, et al. A Mutation in HOXA2 is responsible for autosomal-recessive microtia in an Iranian family. Am J Hum Genet 2008;82(4):982–91.

[131] Gavalas A, Studer M, Lumsden A, Rijli FM, Krumlauf R, Chambon P. Hoxa1 and Hoxb1 synergize in patterning the hindbrain, cranial nerves and second pharyngeal arch. Development 1998; 125(6):1123–36.

[132] Clouthier DE, Garcia E, Schilling TF. Regulation of facial morphogenesis by endothelin signaling: insights from mice and fish. Am J Med Genet A 2010;152A(12):2962–73.

[133] Sato T, Kawamura Y, Asai R, Amano T, Uchijima Y, Dettlaff-Swiercz DA, et al. Recombinase-mediated cassette exchange reveals the selective use of Gq/ G11-dependent and -independent endothelin 1/ endothelin type A receptor signaling in pharyngeal arch development. Development 2008;135(4):755–65.

[134] Yanagisawa H, Yanagisawa M, Kapur RP, Richardson JA, Williams SC, Clouthier DE, et al. Dual genetic pathways of endothelin-mediated intercellular signaling revealed by targeted disruption of endothelin converting enzyme-1 gene. Development 1998;125(5):825–36.

[135] Kurihara Y, Kurihara H, Suzuki H, Kodama T, Maemura K, Nagai R, et al. Elevated blood pressure and craniofacial abnormalities in mice deficient in endothelin-1. Nature 1994;368(6473):703–10.

[136] Ruest L-B, Clouthier DE. Elucidating timing and function of endothelin-A receptor signaling during craniofacial development using neural crest cell-specific gene deletion and receptor antagonism. Dev Biol 2009;328(1):94–108.

[137] Kishigami S, Mishina Y. BMP signaling and early embryonic patterning. Cytokine Growth Factor Rev 2005;16(3):265–78.

[138] Kingsley DM, Bland AE, Grubber JM, Marker PC, Russell LB, Copeland NG, et al. The mouse short ear skeletal morphogenesis locus is associated with defects in a bone morphogenetic member of the TGFβ superfamily. Cell 1992;71(3):399–410.

[139] Dudas M, Sridurongrit S, Nagy A, Okazaki K, Kaartinen V. Craniofacial defects in mice lacking BMP type I receptor Alk2 in neural crest cells. Mech Dev 2004;121:173–82.

[140] Gordon CT, Vuillot A, Marlin S, Gerkes E, Henderson A, AlKindy A, et al. Heterogeneity of mutational mechanisms and modes of inheritance in auriculocondylar syndrome. J Med Genet 2013;50(3):174–86.

[141] Rieder MJ, Green GE, Park SS, Stamper BD, Gordon CT, Johnson JM, et al. A human homeotic transformation resulting from mutations in PLCB4 and GNAI3 causes auriculocondylar syndrome. Am J Hum Genet 2012;90(5):907–14.

[142] Walker MB, Miller CT, Swartz ME, Eberhart JK, Kimmel CB. Phospholipase C, beta 3 is required for

Endothelin1 regulation of pharyngeal arch patterning in zebrafish. Dev Biol 2007;304(1):194–207.

[143] Yoshiura K-I, Leysens NJ, Reiter RS, Murray JC. Cloning, characterization, and mapping of the mouse homeobox GeneHmx1. Genomics 1998;50(1):61–8.

[144] Munroe R, Prabhu V, Acland G, Johnson K, Harris B, O'Brien T, et al. Mouse H6 Homeobox 1 (Hmx1) mutations cause cranial abnormalities and reduced body mass. BMC Dev Biol 2009;9(1):27.

[145] Quina LA, Kuramoto T, Luquetti DV, Cox TC, Serikawa T, Turner EE. Deletion of a conserved regulatory element required for Hmx1 expression in craniofacial mesenchyme in the dumbo rat: a newly identified cause of congenital ear malformation. Dis Model Mech 2012;5(6):812–22.

[146] Schorderet DF, Nichini O, Boisset G, Polok B, Tiab L, Mayeur H, et al. Mutation in the human homeobox gene NKX5-3 causes an oculo-auricular syndrome. Am J Hum Genet 2008;82(5):1178–84.

[147] Mori-Akiyama Y, Akiyama H, Rowitch DH, de Crombrugghe B. Sox9 is required for determination of the chondrogenic cell lineage in the cranial neural crest. Proc Natl Acad Sci USA 2003;100 (16):9360–5.

[148] Wei K, Chen J, Akrami K, Galbraith GC, Lopez IA, Chen F. Neural crest cell deficiency of c-myc causes skull and hearing defects. Genesis 2007;45(6):382–90.

[149] Trokovic N, Trokovic R, Mai P, Partanen J. Fgfr1 regulates patterning of the pharyngeal region. Genes Dev 2003;17(1):141–53.

[150] Iwao K, Inatani M, Matsumoto Y, Ogata-Iwao M, Takihara Y, Irie F, et al. Heparan sulfate deficiency leads to Peters anomaly in mice by disturbing neural crest TGF-β2 signaling. J Clin Invest 2009;119(7):1997–2008.

[151] Nakamura T, Gulick J, Colbert MC, Robbins J. Protein tyrosine phosphatase activity in the neural crest is essential for normal heart and skull development. Proc Natl Acad Sci USA 2009;106(27):11270–5.

[152] Davy A, Aubin J, Soriano P. Ephrin-B1 forward and reverse signaling are required during mouse development. Genes Dev 2004;18:572–83.

[153] Jeong J, Mao J, Tenzen T, Kottmann AH, McMahon AP. Hedgehog signaling in the neural crest cells regulates the patterning and growth of facial primordia. Genes Dev 2004;18(8):937–51.

[154] Northcutt RG, Gans C. The genesis of neural crest and epidermal placodes: a reinterpretation of vertebrate origins. Q Rev Biol 1983;58(1):1–28.

[155] Gans C, Northcutt R. Neural crest and the origin of vertebrates: a new head. Science 1983;220: 268–74.

[156] Abitua PB, Wagner E, Navarrete IA, Levine M. Identification of a rudimentary neural crest in a non-vertebrate chordate. Nature 2012;492(7427): 104–7.

[157] Nishida H, Satoh N. Cell lineage analysis in ascidian embryos by intracellular injection of a tracer enzyme: I. Up to the eight-cell stage. Dev Biol 1983; 99(2):382–94.

[158] Corwin JT, Northcutt RG. Auditory centers in the elasmobranch brain stem: deoxyglucose autoradiography and evoked potential recording. Brain Res 1982;236(2):261–73.

[159] Kalmijn AJ. Functional evolution of lateral line and inner ear sensory systems. In: Coombs S, Görner P, Münz H, editors. The mechanosensory lateral line. New York, NY: Springer; 1989. p. 187–215.

[160] Baker CVH, O'Neill P, McCole RB. Lateral line, otic and epibranchial placodes: developmental and evolutionary links? J Exp Zool B Mol Dev Evol 2008;310B(4):370–83.

[161] Streit A. Origin of the vertebrate inner ear: evolution and induction of the otic placode. J Anat 2001; 199(1–2):99–103.

[162] Anthwal N, Joshi L, Tucker AS. Evolution of the mammalian middle ear and jaw: adaptations and novel structures. J Anat 2013;222(1):147–60.

[163] Miller CT, Yelon D, Stainier DYR, Kimmel CB. Two endothelin 1 effectors, hand2 and bapx1, pattern ventral pharyngeal cartilage and the jaw joint. Development 2003;130(7):1353–65.

[164] Tucker AS, Watson RP, Lettice LA, Yamada G, Hill RE. Bapx1 regulates patterning in the middle ear: altered regulatory role in the transition from the proximal jaw during vertebrate evolution. Development 2004;131(6):1235–45.

[165] Nummela S, Thewissen JGM, Bajpai S, Hussain T, Kumar K. Sound transmission in archaic and modern whales: anatomical adaptations for underwater hearing. Anat Rec 2007;290(6):716–33.

Neural Crest Cells in Ocular Development

Sinu Jasrapuria-Agrawal and Peter Y. Lwigale

Department of Biochemistry and Cell Biology, Rice University, Houston, TX, USA

INTRODUCTION

Vertebrate eye development is a complex process that involves intricate interactions between the surface ectoderm, neuroectoderm, and mesenchymal cells of neural crest (NC) origin. Neural crest cells (NCC) that contribute to the ocular anlage migrate a relatively long distance from the neural tube into the presumptive eye region, also known as the periocular region. Cell tracking experiments using DiI [1],

interspecies neural crest transplantation [2–5], or transgenesis [6] have mapped the NCC migrating from the region between the rostral diencephalon and the metencephalon to the periocular region (Figure 10.1A and B). NCC from the rostral diencephalic regions migrate between the ectoderm and the optic vesicles and later combine with NCC from the mesencephalic and metencephalic regions to form the periocular NCC (Figure 10.1C). Although NCC are generated along the axis of

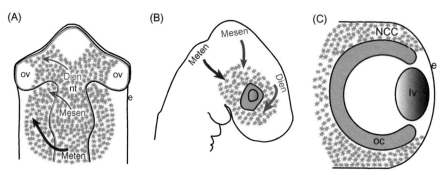

FIGURE 10.1 **Origin and migration of cranial NCC into the periocular region during eye development in chick.**
(A) At the 12-somite stage [7], NCC (green) originating from the neural tube region between the rostral diencephalon and metencephalon migrate laterally during the optic vesicle stage of eye development. The arrows represent populations from the diencephalic (red), mesencephalic (purple), and metencephalic (blue) regions. (B) At embryonic day 2, NCC from the three regions coalesce in the periocular region but avoid the invaginating optic vesicle and presumptive lens region. (C) Representation of a cross section through a rudimentary eye at E3 showing the lens vesicle and optic cup surrounded by NCC and ectoderm. ov, optic vesicle; nt, neural tube; e, ectoderm; oc, optic cup; lv, lens vesicle; NCC, neural crest cells in the periocular region.

the developing embryo, their ability to give rise to ocular structures diminishes towards the caudal regions [8]. The ability of cranial NCC to generate normal ocular structure has been linked to the lack of Hox gene function in the presumptive ocular neural crest territory, since ectopic expression of *Hoxa2*, *-a3*, or *-b4* affects the formation of ocular structures [9,10].

Upon the formation of the rudimentary eye, the surface ectoderm gives rise to the lens vesicle and overlying ectoderm, and the neuroectoderm gives rise to the optic cup, whereas the periocular mesenchyme (consisting of the NCC and cranial mesoderm) fills in the space between the optic cup and ectoderm (Figure 10.1C). The rudimentary eye functions as the basic structure for eye development, which undergoes subsequent development to form the anterior segment (cornea, iris, lens, ciliary body, and trabecular meshwork), eyelids, retina, and the ocular blood vessels and muscles. With the exception of the lens and retina, NCC either give rise to or contribute to most of the ocular tissues [3,4]. As the NCC differentiate into ocular tissues, nerves originating from the trigeminal and ciliary ganglia, as well as the

oculomotor nerve, project into the developing eye and provide sensory, sympathetic, and parasympathetic innervation. Because the trigeminal and ciliary ganglia are comprised of neural crest-derived and ectodermal placode-derived neurons [11,12], a significant portion of ocular innervation can be considered an indirect contribution of NCC to the eye.

During their migration into the periocular region, NCC express markers such as the neuronal carbohydrate epitope HNK-1 and the transcription factor Sox10. However, these markers are downregulated in the periocular region [2,13] as the NCC take on new identities that are crucial to their differentiation into various ocular tissues. The behavior of NCC as they migrate from the neural tube to the periocular region and their subsequent differentiation into ocular tissues is conserved between chick, mouse, and human (Figure 10.2) [14–17]. However, some differences exist between chick and mouse cornea development. During chick cornea development, two waves of NCC migration form, respectively, the endothelium and stroma of the cornea [16]. In mouse, these layers form

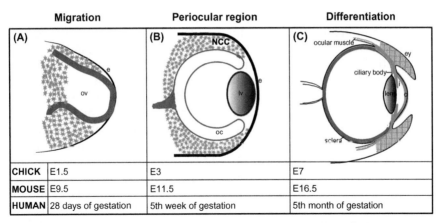

FIGURE 10.2 **Cranial neural crest behavior during eye development and differentiation into ocular tissue is conserved between chick, mouse, and human.** Cranial neural crest behavior during eye development (A, B) and differentiation into ocular tissues (C) is conserved between chick, mouse and human. The relative times during development at which NCC (green) migrate into the presumptive ocular region, contribute to the mesenchyme of the rudimentary eye, and differentiate into various ocular tissues are indicated. ov, optic vesicle; e, ectoderm; oc, optic cup; lv, lens vesicle; NCC, neural crest cells in periocular region; ey, eyelids; c, cornea; i, iris.

from a single mass of NCC that migrate between the lens and ectoderm and eventually differentiate into cornea endothelium and stroma [14]. NCC are undifferentiated as they migrate and occupy the ocular region. Morphogenic and inductive processes that form the lens vesicle and optic cup localize NCC in the periocular region and provide migration and differentiation cues that instruct these cells to generate various ocular tissues. In this chapter, we summarize studies that have contributed to our current knowledge of the cellular and molecular mechanisms underlying neural crest contribution to the eye, and the ocular malformation associated with NCC.

NEURAL CREST AND EARLY EYE DEVELOPMENT

NCC that migrate into the presumptive eye region prior to the formation of the lens and optic cup play a critical role in organizing the eye during early development. This was evident from co-culture studies when NCC abolished expression of the lens-specific gene δ-crystallin by presumptive lens ectoderm [18]. These results were further confirmed *in vivo* when cranial neural tube ablation in chick resulted in the formation of ectopic lenses, which stained positive for *Pax6* and δ-crystallin in the absence of neural crest. Subsequently, it was recognized that NCC suppress lens formation via transforming growth factor beta (TGF-β) signaling, which activates *Smad3* and *Wnt* signaling to repress *Pax6* in the non-lens-forming cranial ectoderm [19]. Thus, the intrinsic lens potential of the cranial ectoderm is restricted by NCC, which permit lens formation only in the NC-free zone at the interface between the optic vesicle and surface ectoderm. Formation of the lens in the NC-free zone is crucial for its subsequent alignment with the optic cup and critical for normal vision. Likewise, the developing eye is required for proper migration of NCC. Genetic ablation of the optic vesicles in *chokh/rx3* eyeless zebrafish mutants prevents anterior migration of cranial NCC, resulting in defects in the neurocranium and orbits [20]. In addition, platelet-derived growth factor (PDGF) secreted by the optic vesicle regulates neural crest migration via their expression of platelet-derived growth factor receptor (PDGFR) that is modulated by miR140 [21]. Signaling between periocular mesenchyme and the optic cup is also involved in the formation of the retinal pigmented epithelium (RPE). When optic vesicle explants were cultured in the absence of the periocular mesenchyme, the RPE-specific gene *Mitf* and other markers were not expressed, and instead the neural retina-specific gene *Chx10* was ectopically expressed [22]. However, the RPE forms normally when mesenchyme-free optic vesicles are cultured in the presence of the TGF-β family member activin A [22]. Together, these studies indicate that signaling between the neural crest and optic vesicles is crucial for the early stages of eye development. Not only are these early interactions indispensible for ocular development but also essential for craniofacial development, since abnormalities in eye and craniofacial development are closely associated [23].

SIGNALING BETWEEN NCC AND THE OPTIC CUP

NCC continue to surround the optic vesicle as it undergoes morphogenesis to form the optic cup but they avoid the anterior region where the lens vesicle detaches from the surface ectoderm (Figure 10.1C). Despite the loss of typical NCC markers such as Sox10 and HNK-1 [2,13] and their transition from mesenchymal to fibroblastic morphology, [24]. NCC remain relatively undifferentiated while they receive signals from the surrounding ocular tissues that determine their future identity. Signaling between the NCC and the optic cup is crucial for the formation of the anterior

segment and posterior structures such as the ocular muscles. The morphogen retinoic acid (RA) is a major signaling component of the optic cup, lens, and surface ectoderm [25]. Paracrine RA signaling via retinoic acid receptors (RAR-$\alpha/\beta/\gamma$) expressed by the NCC regulates the expression of genes that are involved in directing NCC migration, proliferation, and differentiation. RA signaling impairs apoptosis of NCC and controls the expression of the transcription factors *Pitx2*, *Foxc1*, *Eya2* [25–27] that are crucial for the proper patterning and development of the anterior segment. In humans, null mutations in *Pitx2* and *Foxc1* cause Axenfeld–Rieger's syndrome, a condition that includes dysgenesis of the anterior segment characterized by thickening of the corneal stroma, absence of the anterior chamber, and malformation of the iris [28,29]. Similar defects were observed in knockout mice lacking *Pitx2* or *Foxc1* [10,26]. Furthermore, *Pitx2* acts upstream of *Dkk2* to regulate canonical *Wnt* signaling [30], which feeds back into *Pitx2* and maintains its expression [31]. Several *Wnt* genes and receptors are expressed in the anterior eye during development [22,32–34], but the function of *Wnt* signaling in the NCC is not clear. In culture, Wnt/β-catenin signaling promotes neural crest differentiation into melanoblasts, and their proliferation [35,36]. But in mice, the constitutive activation of β-catenin (*Ctnnb1*) in NCC promotes sensory neurogenesis in all cell lineages, including melanocytes [37]. In contrast, activation of β-catenin in zebrafish promotes formation of melanocytes but represses neural cell lineages [38]. This inconsistency is due to the fact that in mouse embryos, Wnt/β-catenin signaling controls sensory fate acquisition prior to formation of melanocyte lineage [39]. Since NCC do not differentiate into melanocytes or neurons in the ocular region, it is unlikely that Wnt/β-catenin signaling plays a similar role, and function remains a topic for future study.

SIGNALING BETWEEN NCC AND THE LENS

Lens ablation experiments have shown that the lens plays a crucial role during the formation of the cornea and development of the anterior segment [40–44]. These studies show that in the absence of the lens, NCC precociously migrate into the cornea-forming region, but fail to form the cornea endothelium and stroma, suggesting that the lens is involved in neural crest migration and/or differentiation. During development of the anterior segment, NCC migrate between the lens and overlaying ectoderm to form the cornea, while those that remain in the periocular region form the ciliary body, iris, trabecular meshwork, and also contribute to the pericytes of the ocular blood vessels and to the ocular muscles. Several studies have shown that TGF signaling from the lens is involved in NCC migration and subsequent development of the anterior segment. Disruption of TGF-β1 [45,46] or TGF-β2 [47–49] in mice via loss of function or overexpression leads to defective ocular development. Reduction in size of the cornea stroma [48] and absence of the corneal endothelium [49] suggest that lens-derived TGF signaling augments NCC migration. Not only do lens-derived signals augment NCC migration. The lens expresses *Semaphorin3A* (*Sema3A*), a secreted chemorepellent [50–52] that forms a barrier to ocular cells that express its receptor *Neuropilin1* (*Npn1*). In chick, *Sema3A* is steadily expressed by the lens during ocular development whereas, expression of the *Npn1* is downregulated by NCC that migrate between the ectoderm and lens to form the cornea endothelium and stroma, but maintained in the non-migratory NCC that contribute to the periocular tissues [13]. Therefore, the lens also plays a critical role in directing NCC migration and localizing them in regions where they receive different signals to differentiate into various ocular tissues.

In addition to guiding NCC migration, signaling by the lens is required during both the early and late stages of cornea development. Shortly after its formation, the lens vesicle induces the overlying ectoderm to secrete extracellular matrix (primary stroma), which acts as a substratum for neural crest migration during cornea development [16,53]. The primary stroma is rich in extracellular molecules such as collagen, fibronectin, laminin, and hyaluronan, which create a conducive environment for NCC migration [54–56]. In chick, a first wave of NCC migrates on the primary stroma adjacent to the lens and forms the inner most layer of the cornea (cornea endothelium). This is followed by a second wave of NCC migration into the primary stroma, which differentiates into stromal keratocytes. The keratocytes synthesize the matrix of the mature cornea that consists of collagens and proteoglycans that are vital to corneal transparency [57]. Together, the neural crest-derived cornea endothelium and stroma comprise 90% of the cornea. Signals from the lens are required for NCC differentiation into cornea endothelium and keratocytes. The cornea endothelium and keratocytes do not form in the absence of the lens [40,41]. Lack of differentiation of neural crest into corneal cells was also observed in mice after genetically ablating the lens by targeted deletion of the αA-crystallin gene [58]. Similarly, cavefish in which the lens degenerates shortly after the formation of the rudimentary eye show massive migration of periocular mesenchyme into the ocular region, but tissues of the anterior segment do not form. However, normal eye development is restored when surface fish lens is transplanted into the cavefish at the rudimentary eye stage [59]. The surface fish to cavefish lens transplantation studies suggest that signaling from the lens is required after the rudimentary eye stage for differentiation of NCC into tissues of the anterior segment.

SIGNALING BETWEEN NCC AND ECTODERM

Periocular NCC also contribute to the mesenchyme of the eyelids, and in birds to skeletogenic condensations known as the scleral ossicles [3,4,60]. The formation of the eyelids and scleral ossicles is initiated by interactions between the NCC and periocular ectoderm. At the initiation of eyelid formation, the forkhead transcription factors foxc1, foxc2, and foxl2 are expressed in the neural crest mesenchyme of the presumptive eyelid buds [10,61]. Unlike foxc1 and foxc2, which are also expressed in the presumptive eyelid ectoderm, expression of foxl2 is restricted to the NCC. Eyelids do not form in foxl2-null mice, [62] and overexpression of Notch signaling via ectopic expression of the Notch1 intracellular domain in NCC downregulates foxl2 also resulting in the eyes-open at birth (EOB) phenotype [63]. Also the foxc1 and foxc2 knockout mice exhibit the EOB phenotype [10]. Furthermore, the EOB phenotype is observed in the absence of other genes expressed by NCC including Fgf10, bone morphogenetic protein 4 (BMP4), and Dkk2 [64]. Signaling between the NCC-derived Fgf10 and the Fgfr2 receptor in the ectoderm stimulates epithelial migration [65], promotes the expression of BMP4 in the NCC, and suppresses Wnt signaling via sfrp1 to maintain foxc2 [64]. Furthermore, it is possible that Pitx2 augmentation of Dkk2 in the NCC plays a similar role since Pitx2-null mice display the EOB phenotype [30]. These studies show that signaling between the NCC and periocular ectoderm are critical for the induction and elongation of eyelids during development.

The scleral ossicles form a ring around the developing eye in avians, reptilians, and pisces [66]. Development of scleral ossicles in avians is initiated by transient thickening of the ectoderm in the periocular region of the conjunctiva, which result in the formation of papillae that correspond to subsequent condensation of the subjacent NCC [67,68]. Ablation of papillae prior to the condensation of the NCC [69] or placement of a barrier between the ectoderm and NCC [70] prevents the formation of the scleral ossicles. Together, these studies suggest that epithelial—mesenchymal interactions via a diffusible factor are required for the formation of scleral ossicles. In an effort to determine the nature of the signals involved in the epithelial—mesenchymal interactions, a study by Franz-Odendaal [71] showed that shh is expressed in the papillae prior to NCC condensation, maintains its signal to the epithelium in a positive feed-back mechanism, and that inhibition of shh signaling in the papillae with cyclopamine prevented ossicle development. In a follow-up publication, Duench and Franz-Odendaal [72] identified the expression of Indian hedgehog (ihh) in the papillae during induction, and that of patched1 (ptc1), a receptor for shh, in the epithelium and mesenchyme during ossicle development. Inhibition of BMP via bead implantation adjacent to the papillae reduced the expression of ihh, suggesting that these pathways interact during development of the scleral ossicles [72]. Given that BMP and Hh interact during skeletal development, [72–74] and that ihh is involved in the development of neural crest-derived bones, [75] the two pathways may be involved in NCC differentiation into scleral ossicles.

OTHER CONTRIBUTION OF NCC TO OCULAR TISSUES

The mesodermal component of the periocular mesenchyme gives rise to the ocular blood vessels and muscles [76]. NCC contribute to these tissues as pericytes and alpha smooth muscle-positive cells to the ocular blood vessels, iris stroma, and eyelids [4,77]. NCC that contribute to the ocular blood vessels and iris stroma belong to the *Npn1*-positive population that resides adjacent to the optic cup [13], probably in response to vascular endothelial growth factor (VEGF) signaling from the optic cup. Pericytes (also known as vascular smooth muscle cells or mural cells) are contractile cells that wrap around the surface of the vascular tube. The signals involved in pericyte differentiation may originate from the ocular blood vessels since only the adjacent NCC form pericytes. NCC are recruited to the developing vasculature via PDGF and EGF signaling. Co-culture of vascular endothelial cells with pericytes showed that pericyte migration, proliferation, and recruitment to the endothelial tube are dependent on the presence of endothelial cells and under the influence of PDGF and EGF signaling [78]. This study also showed that abrogation of both PDGF and EGF signaling in quail embryos prevented pericyte recruitment to developing vasculature resulting in increased vessel width and vascular hemorrhage due to decreased deposition of basement membrane. In chick, *PDGF* and *PDGFRβ* are expressed in the periocular region where ocular blood vessels form [79], suggesting that signaling between NCC and periocular vascular mesoderm plays a role during development of ocular blood vessels.

NCC also contribute to myogenic structures of the eye, including the anterior surface of the presumptive iris, the trabecular meshwork and ciliary body of the iridocorneal angle, and stroma of the eyelids. All these tissues stain positive for the muscle marker α-SMA shortly after their formation [80,81]. Similarly, these tissues express members of the TGF-β superfamily such as *Activin*, *BMP4/7* that are involved in myogenesis [45,80–82]. Knockout mice lacking either BMP4 or BMP7 show diminished ocular development, but due to early embryonic lethality they cannot be studied for events that occur during late ocular development [83].

Contribution to the connective tissues of the extraocular muscles occurs in the medial-posterior regions of the eye [3,84]. During early eye development, the presumptive extraocular muscle mesoderm is intermingled with NCC that form the sclera. Experiments in which the trunk paraxial mesoderm was grafted into the cranial region showed that the grafted mesoderm could form extraocular muscles indicating that NCC direct myogenesis in this region [84]. Also, neural crest ablation experiments suggest that interactions between NCC and mesoderm are required for proper extraocular muscle development. Genetic ablation of cranial NCC in zebrafish using morpholinos against *Sox10* or *FoxD3* results in malformation of several tissues including the extraocular muscles [85]. Its also been suggested that RA signaling in the NCC is required for normal positioning of the extraocular muscles [26]. Although *Pitx2* is downstream of RA signaling, its expression by NCC is not required for proper development of extraocular muscles [27].

INDIRECT NEURAL CREST CONTRIBUTION TO THE EYE VIA TRIGEMINAL AND CILIARY NERVES

Nerves that innervate the ocular tissues (except the retina) originate from the trigeminal and ciliary ganglia. Both ganglia are derived from neural crest and ectodermal placode cells [11,12]. The NCC that contribute to the trigeminal and ciliary ganglia migrate from the neural tube region encompassing the mesencephalon and metencephalon [11,86]. Although this region overlaps with the origin of the NCC, the trigeminal NCC cease to migrate shortly after they exit from the neural tube and intermix with the trigeminal placode that ingress from the overlying ectoderm [87,88]. NCC that give rise to ciliary neurons originate from the mesencephalic region of the neural tube, migrate ventral-laterally, and aggregate close to the ophthalmic branch of the trigeminal ganglion and oculomotor nerve located in the posterior eye region [12,89,90]. The NCC within these two ganglia differentiate into neurons and glia, but the placode cells form only neurons.

The trigeminal ganglion provides sensory innervation to the cornea, iris, extraocular muscles, and eyelids [86,91–93]. Sensory nerves that innervate the face and the eye project from the ophthalmic and maxillary branches of the trigeminal ganglion [84,94]. As observed during NCC migration, Sema3A/Npn1 signaling regulates trigeminal sensory nerve projection into the developing cornea [52,95]. Other studies have shown that Robo-Slit signaling is also involved in guiding trigeminal nerves during cornea innervation [96]. Of particular interest to neural crest contribution to the eye is the selective innervation of the cornea by neural crest-derived trigeminal nerves with no contribution from the placode-derived nerves. Experiments using quail-chick chimeras and tissue ablation showed that despite the dual composition of the trigeminal ganglion by both neural crest- and placode-derived cells, only the neural crest-derived nerves innervate the cornea [86]. Sensory nerves in the cornea secrete neural transmitters including acetylcholine, substance P, and calcitonin gene-related peptide, which stimulate proliferation of cells in the cornea epithelium [97,98]. Loss of sensory innervation of the cornea results in a vision-threatening clinical condition known as neurotrophic keratopathy, characterized by reduced corneal sensation, corneal epithelial defects, and perforation of the stroma [97,99].

The ciliary ganglion provides parasympathetic innervation to the iris, ciliary muscle, extraocular muscles, and eyelids [100,101]. Ciliary innervation of the iris regulates the amount of light entering the eye by controlling pupil dilation and constriction. Innervation of the ciliary muscle plays a role in visual acuity by regulating lens accommodation. Damage to the ciliary ganglion due to disease or trauma causes a neurological condition known as tonic pupil (Adie's syndrome), whereby the pupil does not respond to light and lens accommodation is slow [102,103].

NEURAL CREST RELATED OCULAR DEFECTS

Anterior segment dysgenesis (ASD) refers to anomalies in the structure of the mature anterior segment of the eye caused by developmental disorders associated with the NCC, lens, and optic cup. Clinical screening of patients with ocular defects and genetic analysis of the developing eye have lead to the identification of several genes (mainly transcription factors) that act as molecular cues during the patterning of NCC. ASD affects multiple ocular tissues, making classification of disorders difficult due to overlap in clinical features. To avoid the use of different clinical terms to describe the same condition [104], features of malformation affecting each ocular tissue in the anterior segment have been characterized. Most of the abnormalities due to ASD are associated with elevated intraocular pressure, increased incidence of glaucoma, and loss of corneal transparency. They fall into a broad category of panocular diseases described in Table 10.1. Different gene disorders cause overlapping phenotypes, suggesting that ASD genes act via similar pathways and could be upstream or downstream target of each other. For example, *Pitx2* binds *Foxc1* and represses the activation of putative *Foxc1* target genes [108]. Similarly, *FGF19* is regulated by *Foxc1*, and in zebrafish reduced *FGF19* activity leads to ASD [109].

Since NCC receive signals from other ocular tissues, genes associated with development of

TABLE 10.1 Ocular Defects Associated with NCC

Defects	Clinical Features	Associated Genes
ANTERIOR SEGMENT DYSGENESIS (ASD)		
Aniridia	Corneal opacity, absence of iris, ciliary body may be hypoplastic. Increased corneal thickness in some cases.	PAX6 [70,97]
Anterior Segment Mesenchymal Dysgenesis	Iris adhesions to the cornea and corneal opacity. Corneas are dysplastic (abnormal migration or function of NCC). Lens opacities. Elevated intraocular pressure is usually not present.	PITX3 [97], FOXE3 [105], (TGF-β signaling) [40–44], MAF [106]
Axenfeld-Rieger Syndrome	Polycoria (more than 1 pupil), abnormal iris, thickening of corneal stroma, and absence of anterior chamber.	BMP signaling [70], PITX2 [30], FOXC1 [2], PAX6 [70,97], FOXE3[2]
Cataract	Clouding of the lens.	PAX6 [70,97], MAF [106], PITX3 [106]
Coloboma	Missing pieces from ocular tissues (iris, ciliary body, lens), appear as gap/notches. Associated with microphthalmia.	PAX6 [70,97]
Congenital Primary Aphakia	Absence of lens, iris, ciliary body, trabecular meshwork.	FOXE3 [2], PITX2 [30]
Iris hypoplasia/ Iridogoniodysgenesis (IRID1)	Iris hypoplasia, Abnormal angle iris strands to trabecular meshwork, cornea synechiae.	PITX2 [30], FOXC1 [2]
Microphthalmia/ Anophthalmia	Eyeballs are abnormally small. Associated with coloboma, cataract and microcornea.	MAF [106], PAX6 [70–97]
Peters Anomaly	Corneal opacity (with iris/lens adhesion), Pupil (polycoria), abnormal iris, cornea synechiae.	PITX2 [30], PAX6 [70–97], FOXC1[2], CYP1B1 [107]
Primary congenital glaucoma (PCG)	Abnormal iris, cornea (synechiae).	CYP1B1 [107]
OTHER DISEASE		
Adie's syndrome [92,93]	Pupil does not respond to light and lens accommodation is slow, due to defects in the ciliary ganglion.	ND
Blepharophimosis, ptosis and epicanthus inversus syndrome (BPES)	Eyelid defects, poor eyelid closure, absence of levator smooth muscles. Palpebral fissures (distance between upper and lower eyelids) is small.	FOXL2 (Notch signaling) [58]
Eyelids open at birth (EOB)	Poor eyelid closure.	FOXL2 (Notch signaling) [58] Dkk2 (Wnt signaling) [30,60] Fgf10, Fgf2 (BMP signaling) [60]
Neurotrophic keratitis [87,89]	Reduced corneal sensation, corneal epithelial defects, neovascularization, poor corneal healing, perforation/melting of the stroma due to the impairment of trigeminal ganglion.	ND

ND, Not determined.

the lens or optic cup may cause ASD through regulation of ASD-related transcription factors. Mutations affecting TGF-β signaling in the lens cause malformations similar to ASD [28,40,47]. In addition, the lens-specific genes *MAF* and *Pitx3* cause ASD and induce cataract formation. Loss of *Pax6*, which is required for lens development, affects neural crest migration and differentiation, resulting in ASD [110]. Other neural crest-related ocular defects affect specific tissues. For example, aberrant expression of *Notch1*, which is crucial for proper expression of *Foxl2* in the periocular mesenchyme causes reduced eyelid size or blepharophimosis syndrome [63]. Similar defects in *Foxl2* in mouse cause EOB [63]. Malformation of the Trigeminal ganglion leads to absence of corneal nerves resulting in corneal anesthesis and melting of the stroma associated with neurotrophic keratitis [111]. Similarly, malformation or loss of the ciliary nerves affects the iris and accommodation of the lens and causes Adie's syndrome [102,103].

CONCLUSIONS AND PERSPECTIVES

Since the early classical experiments revealed the contribution of neural crest mesenchyme to ocular tissues, substantial progress has been made in elucidating the mechanisms involved in the generation of multiple ocular tissues for these multipotent cells. The fact that misregulation and mutation of genes causes ocular defects in different animal models provides further evidence that the timing of neural crest migration and interaction with the developing lens, retina, and ectoderm are crucial for proper differentiation into various ocular tissues. So far, the generation of various transgenic animal models, screening for modifier genes, characterization of regulatory elements, in conjunction with the identification of key players such as the RA, TGF-β, BMP, and

Wnt signaling pathways, and transcription factors such as Pax6, Pitx3, and Foxc, have all contributed to our understanding of NCC development. However, questions remain about mechanisms that control NCC differentiation into specific ocular tissues such as the cornea and stroma of the iris. Future studies will benefit from newly available transgenic lines and data analysis by ChIP-Seq, RNA-Seq, and Mass Spec. Increasing the list of target genes and specific binding partners will provide better characterization of the genetic and transcriptional regulatory networks crucial for NCC differentiation, which in turn will increase our understanding of eye development and provide valuable insight for therapies aimed at treating ocular disorders and diseases.

NCC-derived cornea endothelium and stromal keratocytes are quiescent in the adult, but damage and disease render them dysfunctional. With increasing knowledge, there is enormous potential in using stem cells to repair neural crest-derived ocular tissues. The derivation of keratocytes from human embryonic stem cells (hESCs) provides a valuable tool for studying human cornea development and associated disease [112]. Generation of cornea endothelial cells and keratocytes from neural crest or pluripotent stem cells has tremendous implications for tissue-engineering and cell-based therapies for treatment of ocular diseases. The good news is that neural crest function in ocular development is conserved across species despite differences in how their eyes form. Therefore, future studies utilizing a combination of classical techniques and new approaches including genomic and cell biological studies that integrate the cell-intrinsic molecular mechanisms and interactions can be applied to the popular model organisms (chick, mouse, *Xenopus*, and zebrafish) for developmental biology research.

Finally, the evolution of the eye is a subject of significant interest. These studies address similarities and diversions between eyes ranging from simple photoreceptors that sense light to complex multicellular organs of vision. In this chapter, we have discussed the roles of NCC in vertebrate eye development. But the role of NCC in the evolution of the eye remains an open subject for investigation. Interestingly, some invertebrates such as the squid form complex eyes that arc homologous to vertebrates despite the absence of NCC. Therefore, some interesting questions arise: How does the squid form complex ocular structure such as the cornea, and iris, in the absence of NCC? Do squid have cells that perform signaling roles similar to those of NCC during vertebrate ocular development? These questions will increase our understanding of whether the molecular mechanisms and cellular interactions of NCC are a vertebrate invention or if they already existed in non-neural crest cells of the invertebrate cranial mesenchyme.

References

[1] Trainor PA, Tam PPL. Cranial paraxial mesoderm and neural crest cells of the mouse embryo—codistribution in the craniofacial mesenchyme but distinct segregation in branchial arches. Development 1995;121(8):2569–82.

[2] Mears AJ, Jordan T, Mirzayans F, Dubois S, Kume T, Parlee M, et al. Mutations of the forkhead/winged-helix gene, FKHL7, in patients with Axenfeld–Rieger anomaly. Am J Hum Genet 1998;63(5):1316–28.

[3] Johnston MC, Noden DM, Hazelton RD, Coulombre JL, Coulombre AJ. Origins of avian ocular and periocular tissues. Exp Eye Res 1979;29(1):27–43.

[4] Creuzet S, Vincent C, Couly G. Neural crest derivatives in ocular and periocular structures. Int J Dev Biol 2005;49(2–3):161–71.

[5] Lwigale PY, Cressy PA, Bronner-Fraser M. Corneal keratocytes retain neural crest progenitor cell properties. Dev Biol 2005;288(1):284–93.

[6] Nishimura DY, Swiderski RE, Alward WL, Searby CC, Patil SR, Bennet SR, et al. The forkhead transcription factor gene FKHL7 is responsible for glaucoma phenotypes which map to 6p25. Nat Genet 1998;19(2):140–7.

[7] Hamburger V, Hamilton HL. A series of normal stages in the development of the chick embryo. J Morphol 1951;88(1):49–92.

[8] Nishimura DY, Searby CC, Alward WL, Walton D, Craig JE, Mackey DA, et al. A spectrum of FOXC1 mutations suggests gene dosage as a mechanism for developmental defects of the anterior chamber of the eye. Am J Hum Genet 2001;68(2):364–72.

[9] Smith RS, Zabaleta A, Kume T, Savinova OV, Kidson SH, Martin JE, et al. Haploinsufficiency of the transcription factors FOXC1 and FOXC2 results in aberrant ocular development. Hum Mol Genet 2000;9(7):1021–32.

[10] Kume T, Deng KY, Winfrey V, Gould DB, Walter MA, Hogan BL. The forkhead/winged helix gene Mf1 is disrupted in the pleiotropic mouse mutation congenital hydrocephalus. Cell 1998;93(6):985–96.

[11] D'Amico-Martel A, Noden DM. Contributions of placodal and neural crest cells to avian cranial peripheral ganglia. Am J Anat 1983;166(4):445–68.

[12] Lee VM, Sechrist JW, Luetolf S, Bronner-Fraser M. Both neural crest and placode contribute to the ciliary ganglion and oculomotor nerve. Dev Biol 2003;263(2):176–90.

[13] Lwigale PY, Bronner-Fraser M. Semaphorin3A/neuropilin-1 signaling acts as a molecular switch regulating neural crest migration during cornea development. Dev Biol 2009;336(2):257–65.

[14] Pei YF, Rhodin JA. The prenatal development of the mouse eye. Anat Rec 1970;168(1):105–25.

[15] Hoar RM. Embryology of the eye. Environ Health Perspect 1982;44:31–4.

[16] Hay ED, Revel JP. Fine structure of the developing avian cornea. Monogr Dev Biol 1969;1:1–144.

[17] Sowden JC. Molecular and developmental mechanisms of anterior segment dysgenesis. Eye 2007;21(10):1310–8.

[18] Bailey AP, Bhattacharyya S, Bronner-Fraser M, Streit A. Lens specification is the ground state of all sensory placodes, from which FGF promotes olfactory identity. Dev Cell 2006;11(4):505–17.

[19] Grocott T, Johnson S, Bailey AP, Streit A. Neural crest cells organize the eye via TGF-beta and canonical Wnt signalling. Nat Commun 2011;2:265.

[20] Langenberg T, Kahana A, Wszalek JA, Halloran MC. The eye organizes neural crest cell migration. Dev Dyn 2008;237(6):1645–52.

[21] Eberhart JK, He X, Swartz ME, Yan YL, Song H, Boling TC, et al. MicroRNA Mirn140 modulates Pdgf signaling during palatogenesis. Nat Genet 2008;40(3):290–8.

[22] Fuhrmann S, Levine EM, Reh TA. Extraocular mesenchyme patterns the optic vesicle during early eye development in the embryonic chick. Development 2000;127(21):4599–609.

[23] Kish PE, Bohnsack B, Gallina D, Kasprick DS, Kahana A. The eye as an organizer of craniofacial development. Genesis 2011;49(4):222–30.

[24] Bard JB, Hay ED. The behavior of fibroblasts from the developing avian cornea. Morphology and movement *in situ* and *in vitro*. J Cell Biol 1975;67 (2Pt.1):400−18.

[25] Matt N, Dupe V, Garnier JM, Dennefeld C, Chambon P, Mark M, et al. Retinoic acid-dependent eye morphogenesis is orchestrated by neural crest cells. Development 2005;132(21):4789−800.

[26] Matt N, Ghyselinck NB, Pellerin I, Dupe V. Impairing retinoic acid signalling in the neural crest cells is sufficient to alter entire eye morphogenesis. Dev Biol 2008;320(1):140−8.

[27] Evans AL, Gage PJ. Expression of the homeobox gene Pitx2 in neural crest is required for optic stalk and ocular anterior segment development. Hum Mol Genet 2005;14(22):3347−59.

[28] Chang TC, Summers CG, Schimmenti LA, Grajewski AL. Axenfeld−Rieger syndrome: new perspectives. Br J Ophthalmol 2012;96(3):318−22.

[29] Reis LM, Tyler RC, Volkmann Kloss BA, Schilter KF, Levin AV, Lowry RB, et al. PITX2 and FOXC1 spectrum of mutations in ocular syndromes. Eur J Hum Genet 2012;20(12):1224−33.

[30] Gage PJ, Qian M, Wu D, Rosenberg KI. The canonical Wnt signaling antagonist DKK2 is an essential effector of PITX2 function during normal eye development. Dev Biol 2008;317(1):310−24.

[31] Zacharias AL, Gage PJ. Canonical Wnt/beta-Catenin signaling is required for maintenance but not activation of Pitx2 expression in neural crest during eye development. Dev Dyn 2010;239(12):3215−25.

[32] Fokina VM, Frolova EI. Expression patterns of Wnt genes during development of an anterior part of the chicken eye. Dev Dyn 2006;235(2):496−505.

[33] Jin EJ, Burrus LW, Erickson CA. The expression patterns of Wnts and their antagonists during avian eye development. Mech Dev 2002;116(1−2):173−6.

[34] Fuhrmann S. Wnt signaling in eye organogenesis. Organogenesis 2008;4(2):60−7.

[35] Dunn KJ, Williams BO, Li Y, Pavan WJ. Neural crest-directed gene transfer demonstrates Wnt1 role in melanocyte expansion and differentiation during mouse development. Proc Natl Acad Sci USA 2000;97:10050−5.

[36] Jin EJ, Erickson CA, Takada S, Burrus LW. Wnt and BMP signaling govern lineage segregation of melanocytes in the avian embryo. Dev Biol 2001;233 (1):22−37.

[37] Lee HY, Kleber M, Hari L, Brault V, Suter U, Taketo MM, et al. Instructive role of Wnt/beta-catenin in sensory fate specification in neural crest stem cells. Science 2004;303(5660):1020−3.

[38] Dorsky RI, Moon RT, Raible DW. Control of neural crest cell fate by the Wnt signalling pathway. Nature 1998;396(6709):370−3.

[39] Hari L, Miescher I, Shakhova O, Suter U, Chin L, Taketo M, et al. Temporal control of neural crest lineage generation by Wnt/beta-catenin signaling. Development 2012;139(12):2107−17.

[40] Beebe DC, Coats JM. The lens organizes the anterior segment: specification of neural crest cell differentiation in the avian eye. Dev Biol 2000;220 (2):424−31.

[41] Zak NB, Linsenmayer TF. Analysis of corneal development with monoclonal antibodies. I. Differentiation in isolated corneas. Dev Biol 1985;108(2):443−54.

[42] Zinn KM. Changes in corneal ultrastructure resulting from early lens removal in the developing chick embryo. Invest Ophthalmol 1970;9(3):165−82.

[43] Genis-Galvez JM. Role of the lens in the morphogenesis of the iris and cornea. Nature 1966;210 (5032):209−10.

[44] Genis-Galvez JM, Santos-Gutierrez L, Rios-Gonzalez A. Causal factors in corneal development: an experimental analysis in the chick embryo. Exp Eye Res 1967;6(1):48−56.

[45] Flugel-Koch C, Ohlmann A, Piatigorsky J, Tamm ER. Disruption of anterior segment development by TGF-beta1 overexpression in the eyes of transgenic mice. Dev Dyn 2002;225(2):111−25.

[46] Reneker LW, Silversides DW, Xu L, Overbeek PA. Formation of corneal endothelium is essential for anterior segment development—a transgenic mouse model of anterior segment dysgenesis. Development 2000;127(3):533−42.

[47] Ittner LM, Wurdak H, Schwerdtfeger K, Kunz T, Ille F, Leveen P, et al. Compound developmental eye disorders following inactivation of TGFbeta signaling in neural-crest stem cells. J Biol 2005;4(3):11.

[48] Sanford LP, Ormsby I, Gittenberger-de Groot AC, Sariola H, Friedman R, Boivin GP, et al. TGFbeta2 knockout mice have multiple developmental defects that are non-overlapping with other TGFbeta knockout phenotypes. Development 1997;124(13):2659−70.

[49] Saika S, Liu CY, Azhar M, Sanford LP, Doetschman T, Gendron RL, et al. TGFbeta2 in corneal morphogenesis during mouse embryonic development. Dev Biol 2001;240(2):419−32.

[50] Chilton JK, Guthrie S. Cranial expression of class 3 secreted semaphorins and their neuropilin receptors. Dev Dyn 2003;228(4):726−33.

[51] Giger RJ, Wolfer DP, De Wit GM, Verhaagen J. Anatomy of rat semaphorin III/collapsin-1 mRNA expression and relationship to developing nerve tracts

during neuroembryogenesis. J Comp Neurol 1996;375 (3):378—92.

[52] Lwigale PY, Bronner-Fraser M. Lens-derived Semaphorin3A regulates sensory innervation of the cornea. Dev Biol 2007;306(2):750—9.

[53] Hay ED. Development of the vertebrate cornea. Int Rev Cytol 1980;63:263—322.

[54] Coulombre AJ, Coulombre JL. Lens development. I. Role of the lens in eye growth. J Exp Zool 1964;156:39—47.

[55] Toole BP, Trelstad RL. Hyaluronate production and removal during corneal development in the chick. Dev Biol 1971;26(1):28—35.

[56] Fitch JM, Birk DE, Linsenmayer C, Linsenmayer TF. Stromal assemblies containing collagen types IV and VI and fibronectin in the developing embryonic avian cornea. Dev Biol 1991;144(2):379—91.

[57] Hassell JR, Birk DE. The molecular basis of corneal transparency. Exp Eye Res 2010;91(3):326—35.

[58] Kaur S, Key B, Stock J, McNeish JD, Akeson R, Potter SS. Targeted ablation of alpha-crystallin-synthesizing cells produces lens-deficient eyes in transgenic mice. Development 1989;105(3):613—9.

[59] Yamamoto Y, Jeffery WR. Central role for the lens in cave fish eye degeneration. Science 2000;289 (5479):631—3.

[60] Fyfe DM, Hall BK. The origin of the ectomesenchymal condensations which precede the development of the bony scleral ossicles in the eyes of embryonic chicks. J Embryol Exp Morphol 1983;73:69—86.

[61] Crisponi L, Deiana M, Loi A, Chiappe F, Uda M, Amati P, et al. The putative forkhead transcription factor FOXL2 is mutated in blepharophimosis/ptosis/epicanthus inversus syndrome. Nat Genet 2001;27 (2):159—66.

[62] Uda M, Ottolenghi C, Crisponi L, Garcia JE, Deiana M, Kimber W, et al. Foxl2 disruption causes mouse ovarian failure by pervasive blockage of follicle development. Hum Mol Genet 2004;13 (11):1171—81.

[63] Zhang Y, Kao WW, Pelosi E, Schlessinger D, Liu CY. Notch gain of function in mouse periocular mesenchyme downregulates FoxL2 and impairs eyelid levator muscle formation, leading to congenital blepharophimosis. J Cell Sci 2011;124(Pt 15):2561—72.

[64] Huang J, Dattilo LK, Rajagopal R, Liu Y, Kaartinen V, Mishina Y, et al. FGF-regulated BMP signaling is required for eyelid closure and to specify conjunctival epithelial cell fate. Development 2009;136 (10):1741—50.

[65] Tao H, Shimizu M, Kusumoto R, Ono K, Noji S, Ohuchi H. A dual role of FGF10 in proliferation and

coordinated migration of epithelial leading edge cells during mouse eyelid development. Development 2005;132(14):3217—30.

[66] Watanabe K, Bruder SP, Caplan AI. Transient expression of type II collagen and tissue mobilization during development of the scleral ossicle, a membranous bone, in the chick embryo. Dev Dyn 1994;200 (3):212—26.

[67] Murray PD. The development of the conjunctival papillae and of the scleral bones in the embryo chick. J Anat 1943;77(Pt 3):225—40.2.

[68] Coulombre AJ, Coulombre JL. The skeleton of the eye. I. Conjunctival papillae and scleral ossicles. Dev Biol 1962;5:382—401.

[69] Coulombre AJ, Coulombre JL. The skeleton of the eye. II. Overlap of the scleral ossicles of the domestic fowl. Dev Biol 1973;33(2):257—67.

[70] Pinto CB, Hall BK. Toward an understanding of the epithelial requirement for osteogenesis in scleral mesenchyme of the embryonic chick. J Exp Zool 1991;259 (1):92—108.

[71] Franz-Odendaal TA. Toward understanding the development of scleral ossicles in the chicken, Gallus gallus. Dev Dyn 2008;237(11):3240—51.

[72] Duench K, Franz-Odendaal TA. BMP and Hedgehog signaling during the development of scleral ossicles. Dev Biol 2012;365(1):251—8.

[73] Bitgood MJ, McMahon AP. Hedgehog and Bmp genes are coexpressed at many diverse sites of cell—cell interaction in the mouse embryo. Dev Biol 1995;172 (1):126—38.

[74] Kim HJ, Rice DP, Kettunen PJ, Thesleff I. FGF-, BMP- and Shh-mediated signalling pathways in the regulation of cranial suture morphogenesis and calvarial bone development. Development 1998;125 (7):1241—51.

[75] Abzhanov A, Rodda SJ, McMahon AP, Tabin CJ. Regulation of skeletogenic differentiation in cranial dermal bone. Development 2007;134 (17):3133—44.

[76] Couly GF, Coltey PM, Le Douarin NM. The developmental fate of the cephalic mesoderm in quail-chick chimeras. Development 1992;114(1):1—15.

[77] Etchevers HC, Vincent C, Le Douarin NM, Couly GF. The cephalic neural crest provides pericytes and smooth muscle cells to all blood vessels of the face and forebrain. Development 2001;128(7):1059—68.

[78] Stratman AN, Schwindt AE, Malotte KM, Davis GE. Endothelial-derived PDGF-BB and HB-EGF coordinately regulate pericyte recruitment during vasculogenic tube assembly and stabilization. Blood 2010;116 (22):4720—30.

[79] Kwiatkowski S, Munjaal RP, Lee T, Lwigale PY. Expression of pro- and anti-angiogenic factors during the formation of the periocular vasculature and development of the avian cornea. Dev Dyn 2013;242 (6):738–51.

[80] Jensen AM. Potential roles for BMP and Pax genes in the development of iris smooth muscle. Dev Dyn 2005;232(2):385–92.

[81] Link BA, Nishi R. Development of the avian iris and ciliary body: the role of activin and follistatin in coordination of the smooth-to-striated muscle transition. Dev Biol 1998;199(2):226–34.

[82] Zhao S, Chen Q, Hung FC, Overbeek PA. BMP signaling is required for development of the ciliary body. Development 2002;129(19):4435–42.

[83] Jena N, Martin-Seisdedos C, McCue P, Croce CM. BMP7 null mutation in mice: developmental defects in skeleton, kidney, and eye. Exp Cell Res 1997;230 (1):28–37.

[84] Noden DM. Patterning of avian craniofacial muscles. Dev Biol 1986;116(2):347–56.

[85] Bohnsack BL, Gallina D, Thompson H, Kasprick DS, Lucarelli MJ, Dootz G, et al. Development of extraocular muscles requires early signals from periocular neural crest and the developing eye. Arch Ophthalmol 2011;129(8):1030–41.

[86] Lwigale PY. Embryonic origin of avian corneal sensory nerves. Dev Biol 2001;239(2):323–37.

[87] Stark MR, Sechrist J, Bronner-Fraser M, Marcelle C. Neural tube-ectoderm interactions are required for trigeminal placode formation. Development 1997;124 (21):4287–95.

[88] Shiau CE, Lwigale PY, Das RM, Wilson SA, Bronner-Fraser M. Robo2-Slit1 dependent cell–cell interactions mediate assembly of the trigeminal ganglion. Nat Neurosci 2008;11(3):269–76.

[89] Narayanan CH, Narayanan Y. Determination of the embryonic origin of the mesencephalic nucleus of the trigeminal nerve in birds. J Embryol Exp Morphol 1978;43:85–105.

[90] D'Amico-Martel A. Temporal patterns of neurogenesis in avian cranial sensory and autonomic ganglia. Am J Anat 1982;163(4):351–72.

[91] Fackelmann K, Nouriani A, Horn AK, Buttner-Ennever JA. Histochemical characterisation of trigeminal neurons that innervate monkey extraocular muscles. Prog Brain Res 2008;171:17–20.

[92] Kessler JA, Bell WO, Black IB. Interactions between the sympathetic and sensory innervation of the iris. J Neurosci 1983;3(6):1301–7.

[93] Nakamura A, Hayakawa T, Kuwahara S, Maeda S, Tanaka K, Seki M, et al. Morphological and immuno-histochemical characterization of the trigeminal ganglion neurons innervating the cornea and upper eyelid of the rat. J Chem Neuroanat 2007; 34(3-4):95–101.

[94] Marfurt CF. The somatotopic organization of the cat trigeminal ganglion as determined by the horseradish peroxidase technique. Anat Rec 1981;201(1):105–18.

[95] Kubilus JK, Linsenmayer TF. Developmental guidance of embryonic corneal innervation: roles of Semaphorin3A and Slit2. Dev Biol 2010;344(1):172–84.

[96] Schwend T, Lwigale PY, Conrad GW. Nerve repulsion by the lens and cornea during cornea innervation is dependent on Robo-Slit signaling and diminishes with neuron age. Dev Biol 2012;363 (1):115–27.

[97] Cavanagh HD, Colley AM. The molecular basis of neurotrophic keratitis. Acta Ophthalmol Suppl 1989;192:115–34.

[98] Mishima S. The effects of the denervation and the stimulation of the sympathetic and trigeminal nerve on the mitotic rate of the corneal epithelium in the rabbit. Jpn J Ophthalmol 1957;1:65–73.

[99] Bonini S, Rama P, Olzi D, Lambiase A. Neurotrophic keratitis. Eye 2003;17(8):989–95.

[100] Jaeger RJ, Benevento LA. A horseradish peroxidase study of the innervation of the internal structures of the eye. Evidence for a direct pathway. Invest Ophthalmol Vis Sci 1980;19(6):575–83.

[101] Neuhuber W, Schrodl F. Autonomic control of the eye and the iris. Auton Neurosci 2011;165(1):67–79.

[102] Thompson HS. Adie's syndrome: some new observations. Trans Am Ophthalmol Soc 1977;75:587–626.

[103] Perkin GD. Neuro-ophthalmological syndromes for neurologists. J Neurol Neurosurg Psychiatry 2004;75 (Suppl 4):20–3.

[104] Idrees F, Vaideanu D, Fraser SG, Sowden JC, Khaw PT. A review of anterior segment dysgeneses. Sur Ophthalmol 2006;51(3):213–31.

[105] Semina EV, Brownell I, Mintz-Hittner HA, Murray JC, Jamrich M. Mutations in the human forkhead transcription factor FOXE3 associated with anterior segment ocular dysgenesis and cataracts. Hum Mol Genet 2001;10(3):231–6.

[106] Semina EV, Murray JC, Reiter R, Hrstka RF, Graw J. Deletion in the promoter region and altered expression of Pitx3 homeobox gene in aphakia mice. Hum Mol Genet 2000;9(11):1575–85.

[107] Sitorus R, Ardjo SM, Lorenz B, Preising M. CYP1B1 gene analysis in primary congenital glaucoma in Indonesian and European patients. J Med Genet 2003;40(1):e9.

[108] Berry FB, Lines MA, Oas JM, Footz T, Underhill DA, Gage PJ, et al. Functional interactions between FOXC1 and PITX2 underlie the sensitivity to FOXC1 gene dose in Axenfeld–Rieger syndrome and anterior segment dysgenesis. Hum Mol Genet 2006;15 (6):905–19.

[109] Tamimi Y, Skarie JM, Footz T, Berry FB, Link BA, Walter MA. FGF19 is a target for FOXC1 regulation in ciliary body-derived cells. Hum Mol Genet 2006;15(21):3229–40.

[110] Thaung C, West K, Clark BJ, McKie L, Morgan JE, Arnold K, et al. Novel ENU-induced eye mutations in the mouse: models for human eye disease. Hum Mol Genet 2002;11(7):755–67.

[111] Muller LJ, Vrensen GF, Pels L, Cardozo BN, Willekens B. Architecture of human corneal nerves. Invest Ophthalmol Vis Sci 1997;38(5):985–94.

[112] Chan AA, Hertsenberg AJ, Funderburgh ML, Mann MM, Du Y, Davoli KA, et al. Differentiation of human embryonic stem cells into cells with corneal keratocyte phenotype. PloS One 2013;8(2): e56831.

The Cardiac Neural Crest and Their Role in Development and Disease

Joshua W. Vincentz and Anthony B. Firulli

Riley Heart Research Center, Herman B Wells Center for Pediatric Research Division of Pediatrics Cardiology, Departments of Anatomy, Biochemistry, and Medical and Molecular Genetics, Indiana University Medical School, 1044 W. Walnut St. R4302E, Indianapolis, IN 46202-5225, USA

OUTLINE

Neural Crest Cells.
DOI: http://dx.doi.org/10.1016/B978-0-12-401730-6.00012-0

205

INTRODUCTION

The human heart originates as a simple, tubular structure that ultimately assumes considerable complexity as it develops into a fully functional organ. Much of this complexity comes from the integration of a number of extra-cardiac cell types, cells that must migrate to the heart from outside of the field of cells that forms this initial heart tube. The cardiac neural crest cells (cNCC) are a key population of extra-cardiac cells that invades the arterial pole of the nascent heart and contributes to the cardiac outflow tract (OFT). Reflective of their important roles in heart formation, defects in the specification, migration, and differentiation of the cNCC are thought to cause a number of congenital heart defects (CHDs) within humans. CHDs are the most frequently occurring birth defect, and they are observed at a frequency as high as 1% of all live births [1]. In many cases, these defects require immediate surgical interventions. Indeed, although surgical correction is life saving, these young CHD patients are at higher risk for developing adult-related cardiac diseases due to the pathological remodeling that results from suboptimal cardiac function (see also [2]).

Like all NCC, the cNCC are generated at the dorsal lip of the neural tube within a specific rostrocaudal domain just below that of the cranial NCC and just above that of the vagal NCC between the otic placode and somite 3 [3,4] (Figure 11.1). This NCC-fated neuroepithelium first undergoes a cell shape transition, known as an epithelial-to-mesenchymal transition (EMT). The now mesenchymal (or rather, ecto-mesenchymal) NCC can now migrate from the leading edge of the forming neural tube ventrally toward the already forming heart [3,5–8]. When these cells arrive in the pharyngeal arches (PA), they contribute to the formation of the pharyngeal arch arteries (PAA). The PAA undergo a non-symmetrical remodeling, which results in the

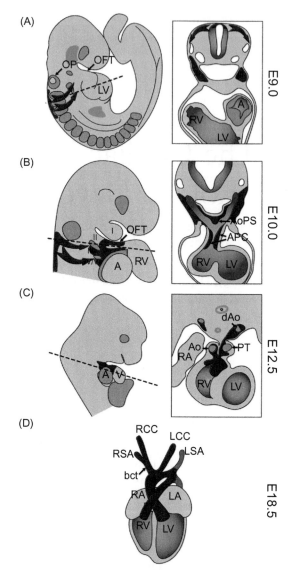

FIGURE 11.1   cNCC development in mice. (A) Cardiac NCC (blue) specify at the dorsal lip of the neural tube between the otic placode (OP) and somite 3. Cardiac NCC then undergo EMT, emigrating as waves from the neural tube toward the caudal pharyngeal arches (arrows). (B) The first wave of cNCC continue their migration, moving farthest into the OFT and forming the aorticopulmonary septum (AoPS) and the aorticopulmonary cushions (APC). Subsequent migratory waves arrive in the pharyngeal arches where they form the PAAs. (C) These cells subsequently remodel to form the septum between the aorta

formation of the great arteries: the dorsal aorta, the brachiocephalic root of the internal carotid, and the pulmonary arteries. The cNCC differentiate into the vascular smooth muscle of these vessels and form the aortic and pulmonary valves. This chapter will review what is known about the basic processes of the cNCC, the human diseases that result from malfunction of the cNCC, and the animal models that have generated insight into these processes. Ultimately, this chapter will provide a broad overview of the signaling molecules and transcriptional regulators that transform this migratory cell population into both a functional connection between the heart and blood vessels, and a barrier that divides the pulmonary and systemic circulatory systems.

NCC SPECIFICATION

Like all NCC, cNCC specify from the neuroepithelium at the dorsal lip of the forming neural tube (Figure 11.1). The genesis of the cNCC involves cell−cell communications between the nonneural surface ectoderm and the underlying dorsal lip of the neural plate [9], whereby both overlying ectoderm and underlying neural plate cells adopt an NCC fate [10−12]. Inducers of this NCC specification include the signaling molecules bone morphogenic proteins (BMPs), fibroblast growth factors (FGFs), Notch, retinoic acid (RA), and Wnts [13−18]. These signaling factors subsequently orchestrate a transcriptional response

within the now-fated cNCC that includes the upregulation of key transcription factors such as the homeodomain proteins Dlx3 and Dlx5 [19], Msx1 and Msx2 [20], the Zn-finger Zic proteins [21,22], and the paired box homeodomain factors Pax3 and Pax7 [23−27]. These factors collectively define the NCC boarder via a number of additional transcriptional regulators, whose expression "marks" the now-specified NCC. These include the helix-span-helix factor TFAP2α [19,28], the winged helix protein FoxD3 [29], the UAS-bHLH protein c-Myc [30], and the HMG-box factors Sox9 and Sox10 [31−33]. These factors allow the NCC to become competent to undergo EMT. Finally, expression of the in the Snail family Zn-finger factors (Snai1 and Snai2) [34] and the bHLH factor Twist1 marks the NCC undergoing EMT, although it should be noted that, in mammals, Twist1 also marks non-NCC mesenchyme [35].

The consequences of this transcriptional cascade are the generation of migratory cells. This transition involves the regulation of a number of cell adhesion molecules. The epithelial-expressed E-cadherin is downregulated within the NCC via the actions of Snail (Snai1) and, likely, Twist1 [36−38]. Expression of Sox9, the Snail-related factor Slug (Snai2), and FoxD3 allow for EMT competence, initiation of EMT, and the expression of migratory cell adhesion proteins, respectively [39]. In addition to E-cadherin, N-cadherin, Ncam, and Cad6b expression is reduced within the NCC, while expression of Cad7, Cad11, and α4β1 and α5β1 integrins is upregulated [40].

(Ao) and pulmonary trunk (PT), the membranous septum between the left and right ventricles (LV and RV) and the smooth muscle of the aortic arch, including the dorsal aorta (dAo). (D) Cardiac NCC ultimately contribute to the majority of smooth muscle in the aortic arch, including the right subclavian artery (RSA), the left and right common carotid arteries (LCC and RCC), the ductus arteriosus (da), and the Ao and PT.

MIGRATION AND MORPHOGENESIS OF THE CNCC

After assuming competence to migrate, the cNCC move ventrolaterally through pharyngeal arches 3, 4, and 6, and subsequently proliferate (Figure 11.1). From there, a subpopulation of cNCC moves into the arterial pole of the heart, termed the cardiac OFT. The remaining cNCC contribute to the third, fourth, and sixth PAAs. The migrating cells that enter the cardiac OFT form both the septum that divides pulmonary and systemic circulation, and, presumably, a subset of the parasympathetic cardiac ganglia [3]. Additionally, migratory cNCC move into the forming esophagus and parasympathetic enteric plexus of the forming mid- and hindgut [7]. Finally, the migrating cNCC contribute to the thymus and parathyroid glands, which form from the endoderm of the caudal pharynx and migrating cNCC from pharyngeal arches 3−6 [2,41]. Given the complex morphogenetic program undertaken by the cNCC, we can see that the phenotypic abnormalities associated with cNCC dysfunction can result from improper genesis, migration, remodeling, proliferation, survival, and/or differentiation. Cardiac NCC migrate in waves of temporally distinct populations. The first wave of cells moves directly into the cardiac OFT and contributes to the aorticopulmonary septum—the most distal structure formed. Waves of cNCC subsequently populate the pharyngeal arches. These cells will form the PAAs [42]. To be migratory, NCC must undergo EMT, whereby they adopt a flattened shape with both filopodia and lamellipodia. The intermediate filament protein vimentin marks NCC that have undergone EMT [40].

Also important in NCC migration is the production of secreted glycoproteins that influence cell−cell interactions among the NCC and with other cell types, collectively known as extracellular matrix (ECM). Although ECM changes dynamically and is thus difficult to analyze, one NCC-secreted factor, Tenascin C, has been shown to be essential for NCC migration. Reducing Tenascin C expression blocks the initial emigration of NCC from the neural tube [43]. Fibronectin is also important for NCC migration; α4β1 and α5β1 integrins are expressed in NCC and allow for cellular movement on fibronectin, but not on other ECM molecules, such as laminin, vitronectin, or type 1 collagen. Blocking access to α5β1 integrins in cranial NCC inhibits their migration on fibronectin [44]. Hand in hand with the ECM, NCC express and secrete proteases to allow for breakdowns of cell−cell interactions and ECM. In frog, NCC overexpression of the metalloprotease ADAM13 enhances the invasion of NCC within surrounding tissues, whereas expression of a dominant negative (DN) form of the protein inhibits NCC migration [45]. Similar findings are observed with the protease regulating protein Timp2. Timp2 is expressed within migrating cNCC and its overexpression promotes NCC migration [46] (whereas its reduced expression inhibits NCC migration).

Understanding the migration process in 3D within the developing embryo relies heavily upon cell lineage analysis, which, in the avian system, employs both cNCC dye injection and chick-quail chimeras. Both techniques enable NCC migration paths to be traced from the point of dye injection or cell implantation to their final destinations. In the now-classic studies from the Kirby lab using these technologies, it was shown that ablation of cNCC within the neural tube prior to their migration results in a severe OFT abnormality, observed in some humans with CHD, termed persistent truncus arteriosis (PTA; Figure 11.2). In both NCC-ablated chicks and human patients with PTA, the aorta and pulmonary trunk fail to septate, thus opening the circuits of oxygenated and nonoxygenated circulation [3,41,47]. Grafting quail NCC into the region that contained the now-ablated chick cNCC showed that, indeed, the cNCC were the contributing source of the aorticopulmonary septum of the avian OFT.

Determining if mammals conserved this pattern of cNCC distribution proved more difficult due simply to the impracticality of manipulating a developing embryo *in utero*

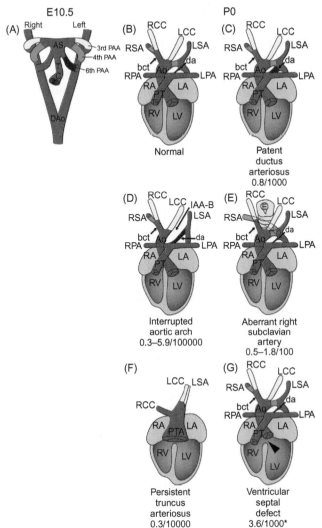

FIGURE 11.2 **Structural organization of the symmetric PAA populated and remodeled by the cNCC.** (A) Schematic of the PAA at E10.5. The first and second PAA (not shown) are bilaterally reabsorbed. Both right and left third PAA (yellow) contribute to the formation of the common carotid arteries (LCC and RCC). The right fourth PAA (green) contributes to the right subclavian artery (RSA), whereas the left fourth PAA remodels to form the medial aortic arch. The right sixth PAA almost completely regresses, contributing only to the base of the right pulmonary artery, and the left sixth PAA (blue) forms the left pulmonary artery and the ductus arteriosus (da), which closes after birth. (B) Illustration of a phenotypically normal postnatal (P0) heart showing the fully remodeled great vessels. Aorta (Ao), pulmonary trunk (PT), left and right pulmonary arteries (LPA and RPA), carotids (LCC and RCC), and subclavian arteries (LSA and RSA) conform to a particular pattern. (C) Failure of the da to close postnatally results in a condition termed patent ductus arteriosus (PDA). (D, E) Dysfunctional fourth PAA development can lead to an absence of the medial aortic arch, a condition known as (D) interrupted aortic arch type B (IAA-B) or (E) aberrant origin of the right subclavian artery. In this instance, the RSA connects to the descending aorta near the origin of the LSA and projects behind the esophagus (E), a condition known as retroesophageal RSA. (F) Failure of the Ao and PT to septate into distinct vessels results in a condition termed PTA. (G) Failure of the distal-most cNCC to septate the left and right ventricles (LV and RV) causes a condition termed VSD (arrowhead). The frequency of depicted OFT defects in the human population is denoted [198,199]. * The frequency of all VSDs, not only those thought to be associated with NCC dysfunction, is denoted.

compared to one self-contained within an egg (*in ovo*). Some comparative evidence was available through the study of a mouse mutant family termed *Splotch*, which harbored a series of different mutations in the gene encoding the transcription factor Pax3 [48−51]. Pax3 is expressed within the dorsal neural tube in both NCC-forming and non-NCC-forming ectoderm. *PAX3* mutation in humans is associated with Waardenburg syndrome, symptoms of which include pigmentation and hearing defects [52,53]. Splotch mice, like the cNCC-ablated chicken embryos, presented with PTA and PAA patterning defects [54,55]. Pax3 is not required for NCC specification or migration, as NCC do migrate in loss-of-function mice. It is the requirement for Pax3 in NCC expansion that limits the number of cNCC that ultimately invade the pharyngeal arches and OFT [55−57]. Thus, mechanistically, this absence of an aorticopulmonary septum can be attributed to decreased proliferation of cNCC that limits their numbers [58,59], which in turn limits expansion of the progenitor pool that then indirectly limits the number of normally migrating NCC [60,61].

The breakthrough in tracking the migration of NCC directly within the mouse came with the advent of lineage marking tools that, when introduced into mice via transgenics, provided the ability to follow the NCC throughout their migration. The first lineage marker was isolated from the cell surface protein-coding gene *Connexin 43* (*Cx43*). An upstream fragment of the *Cx43* transcriptional promoter used to drive *β-galactosidase* expression allowed for enzymatic staining of tissues that expressed this enzyme, thus allowing one to follow the cNCC expressing the *Cx43*-transgene independent of direct embryo manipulations [62]. Later, with the development of Cre-recombinase technologies, cells within any lineage could be permanently marked by the excision of an inhibitory transcriptional and translational "stop" sequence flanked by bacterial *loxP* elements placed 5' of the transcriptional start site of the *β-galactosidase* reporter gene integrated into the ubiquitously active *Rosa26* locus [63,64]. Additional gene promoters from the *Wnt1*, *Pax3*, and *P0* genes allowed development of Cre-expressing lines that mark nearly all NCC (including the cNCC) and all of their progeny [8,49,65]. Through careful analysis of these genetically modified mouse embryos, the migration paths and final destinations of the cNCC were determined in detail (Figure 11.1). Given that the NCC (and their progeny) become permanently marked by activated *Rosa26* β-galactosidase activity, their final contribution to the heart is possible even though the promoter elements driving Cre expression are downregulated within these cells. We know from these studies that, similar to the chick, the cardiac contribution from the cNCC includes the smooth muscle surrounding the ascending aorta and aortic arch, the proximal carotid and coronary arteries, and both the aortic and pulmonary valves. Perhaps more importantly, these proven NCC recombination tools also allow for a refined analysis of the role individual genes play within NCC populations in mice.

The chemotactic signaling cues that guide the NCC to their final destinations within the heart have parallels with the mechanisms that instruct neurons to pattern correctly [66]. Ephrins are cell surface ligands for Eph receptors, which are also membrane bound [67]. Through cell-to-cell contact, these ligand/receptor complexes coordinate cell movement throughout the developing embryo. *EphrinB2* (*EfnB2*) is expressed within the mesoderm of the head and body, whereas one of EfnB2's known receptors, EphA4, is expressed within the NCC. Ligand−receptor binding results in an auto-phosphorylation of the receptor, which in turn triggers a signaling cascade that ultimately modulates gene expression within the NCC. Conversely, evidence indicates that EfnB2 is also phosphorylated on its carboxy-terminal tail and thereby signals within the cells surrounding the NCC [68,69]. This bi-directional communication suggests that feedback signaling to the surrounding mesoderm through which the NCC are moving fine-tunes the amount of cell adhesion between the migrating NCC and their extracellular environment, conveying directionality to NCC movement. There are multiple Ephrins and Eph receptors expressed throughout the developing embryo, and it is likely that additional family members play roles in NCC migration.

Semaphorin ligands and their receptors and co-receptors, neuropilins and plexins, respectively, are another axon guidance signaling family found to be important for guiding NCC to their prescribed destinations [70−72]. Cardiac NCC express *PlexinA2*, while the surrounding mesoderm expresses and secretes semaphorin 3C (Sema3C) [70]. Ablation of

Sema3C in mice results in PTA, demonstrating the non-cell-autonomous role of the surrounding mesenchyme for cNCC guidance [73]. Gene expression analysis shows that, within the cNCC occupying the aorticopulmonary cushions (which are two spiral-shaped cushions within the cardiac OFT), *PlexinA2* mRNA is reduced in these mutants. Interestingly, in cranial NCC, loss of either Sema3F or Neuropilin2 function alters NCC migration into the first and second pharyngeal arches [74]. The combinatorial power observed in such ligand/receptor pairs allows for highly directed migratory programs, which, when disrupted from either the ligand or receptor component of the signaling circuit, will disrupt normal NCC migration. Within the chick, NCC are repelled form sources of Sema3A, Sema3F, and Sema6, showing the importance of non-cell-autonomous effectors of NCC path finding [72,75]. Circumstantial but supportive evidence of the importance of ligand/receptor phosphorylation and their downstream signaling cascades for NCC migration can be seen in both gain-of-function and loss-of-function models of the anti-phosphatase Paladin [76], modulation of protein phosphatase 2A activity [77], expression of a DN form of the serine/threonine kinase Pak1 [78], and the modulation of Rho/Rac signaling [79–81]. Further understanding of the complexities of intracellular signaling will be required to better link events occurring on the cell surface with those in the nucleus of both NCC and the surrounding tissues.

Cell adhesion molecules such as connexins and cadherins integrate the aforementioned cell-signaling information being transmitted to the migrating cNCC and surrounding mesoderm. Both Cx43 loss-of-function and gain-of-function analyses reveal cardiac defects within the OFT [82,83]; however, an NCC autonomous role for Cx43 is unlikely, as Wnt1-Cre specific deletion of *Cx43* within the NCC does not result in any obvious cardiac phenotypes

[84]. Both loss-of-function and gain-of-function of N-cadherin within the NCC disrupts cell migration patterns. Cx43 and N-cadherin co-localize within migrating NCC, suggesting that they form a key complex that allows for interactions at the cell surface, as well as intercellular interactions with the actin cytoskeleton. N-cadherin also interacts with P120 catenin, another factor important for cell migration. Interestingly, the cellular distribution of P120 catenin is altered in both Cx43 and N-cadherin depleted NCC [85]. The tight junction scaffolding protein Cingulin1 also plays a role in regulating NCC migration [86]. Knockdown of Cingulin1 increases the migratory NCC domain within the neural tube and overexpression is observed to augment migration, expanding the premigratory NCC population associated with decreased expression of RhoA. Collectively these observations highlight the importance that gap- and tight-junction complexes hold in the migration of the NCC.

Another example of altered NCC migration resulting from faulty signaling is the loss-of-function mutation in the neuregulin receptor *ErbB4*. In *ErbB4* null embryos, NCC moving into the first and second pharyngeal arches cross paths, resulting in a fusion of forming ganglia. ErbB4 expression within the hindbrain acts to repel NCC migration, excluding their contribution to regions of the hindbrain [87–90]. Taken together, these observations clearly show that numerous intercellular and intracellular communication mechanisms integrate to coordinate NCC migration.

Similar to *ErbB4* null embryos, deletion of the bHLH transcription factor *Twist1* results in an NCC stream-mixing phenotype [91]. *Twist1* null embryos die by E11.5 and exhibit severe craniofacial and cardiac defects [92,93]. *Twist1* is expressed in both the NCC and the sounding mesenchymal tissue, making identification of cell autonomous vs. non-cell-autonomous contributions to the NCC migratory process complex [93,94]. From our efforts, we know that NCC migration into the OFT is diminished in *Twist1* systemic null embryos and that there is a cell-autonomous cell adhesion defect in the cNCC occupying the OFT [93]. These aggregations of cNCC form clumps or nodules that are marked exclusively by the expression of the related bHLH factors Hand1 and Hand2. Cardiac NCC aggregates do not display normal cNCC gene expression, suggesting that these cNCC have mis-migrated or, more interestingly, may have adopted an alternate cell fate [93]. We now know that the latter mechanism holds true and the details of this are covered in the sections below.

INVASION OF THE CNCC INTO THE PHARYNGEAL ARCHES AND OFT CUSHIONS

As mentioned above, the migrating cNCC first populate the caudal pharyngeal arches 3, 4, and 6, where they then proliferate (Figure 11.1). Each PAA will play a unique role in the formation of the fully remodeled aortic arch. The migrating cells must therefore respond to refined signals that guide them specifically to their intended destination [40]. In cranial NCC, VEGF expression within pharyngeal arch 2 is a powerful attractor of NCC; however, cNCC are not affected by this signal. The Kirby lab reports that Fgf8 expression within the pharyngeal endoderm and ectoderm guides cNCC to the caudal arches and maintains viability of the cells within the lumen of pharyngeal arch 4 [40]. The ligand/receptor signaling circuit of Slit and Roundabout (Robo) also guides cells to pharyngeal arch 4. Decreased amounts of Slit cause cNCC destined for arch 4 to move to arches 3 or 6. Without the support of the cNCC, arch 4 degrades, resulting in a clinical condition called interrupted aortic arch (Figure 11.2) [95]. Ephrin/Eph signaling is also implicated in arch invasion. Studies performed

in frog revealed a DN form of the ephrin receptor, EphA2, causes cNCC destined for arch 3 to instead move to arch 4. However, analysis of *EphA2* mouse mutants does not recapitulate this observation, suggesting that evolutionary variations in the chemotactic signals are likely to occur [95].

REMODELING OF THE PHARYNGEAL ARCH ARTERIES

The initial orientation of the PAAs is shown in Figure 11.2. The right and left first and second PAA regress in humans. Both left and right sides of the third PAAs remodel to form the internal carotid arteries. The right fourth PAA contributes to the right subclavian artery, whereas the left fourth PAA remodels to contribute to the aortic arch at the region between the left carotid artery and the ductus arteriosus. The right sixth PAA contributes to the most proximal portion of the right pulmonary artery and its more distal domains regress. The sixth PAA remodels to form the left pulmonary artery and the ductus arteriosus, which shunts pulmonary circulation to the aorta during gestation and is lost a few days after birth.

The ultimate differentiated fate of the cNCC that form the arch arteries is vascular smooth muscle that acts to support the vessel integrity under hemodynamic load. In addition to migration, modulating the number of cNCC in the arches through proliferation and maintenance of cell viability as they arrive collectively dictates the morphogenic process. A reduction in NCC number at any location could result from mismigration, low proliferation, high cell death, or combinations of each mechanism. When interrogating a phenotype in which NCC number is reduced, consideration of cell growth and death along the migration path is required.

Interestingly, the remodeling fourth and sixth PAA exhibit asymmetric structural outcomes. Thus, the signaling circuits that control remodeling intersect with those that govern the left-right asymmetry programs. One signaling circuit that plays roles in arch artery remodeling is that of platelet-derived growth factor (Pdgf) through the Pdgf receptors (Pdgfr) α and β. Loss-of-function studies of a systemic *Pdgfrα* mouse mutant revealed significant cardiac OFT defects [96]. In chimera analysis in which mutant *Pdgfrα* embryonic stem (ES) cells were injected into wild-type host blastocysts, mutant ES cells clearly failed to contribute to the pharyngeal arches in resultant chimeric embryos [96]. In an NCC-specific knockout of *Pdgfrα*, NCC migration was found to be unaffected and NCC differentiation into vascular smooth muscle appeared normal; however, the third, fourth, and sixth PAA appeared either dilated or with a significantly reduced luminal diameter than control littermates, although phenotypic penetrance was incomplete. As both *Pdgfrα* and *Pdgfrβ* could be contributing to arch remodeling, an NCC-specific double knockout was generated [97]. In this study, doubly null embryos exhibited the same arch defects observed in the *Pdgfrα* conditional null, but at full penetrance. This shows partial compensation from Pdgfrβ in the absence of Pdgfrα. Additionally, NCC migration into the OFT and aortic sac was now observed to be affected showing overlapping functions of these receptors in cNCC remodeling.

Endothelins and their receptors also play a role in PAA remodeling. These ligand/receptor partners play roles in vasoconstriction and raising blood pressure. During development, family members of this signaling pathway are expressed within the NCC. Loss-of-function mouse models for the endothelin receptor ET_A [98], and the endothelin-converting enzyme *Ece-1* [99] result in interrupted aortic arch and loss of the right subclavian artery. Arch formation in both mouse models was initially found to be normal; however, defective arch remodeling occurs in arches 3, 4, and 6, whereby

arches that should regress persist, and vice versa [100]. ET_A chimera studies suggest a cell autonomous role for endothelin signaling, as mutant cells are excluded from incorporation into the pharyngeal arches, forming arch arteries, and the OFT [101]. Similar to Pdgf receptor studies, gene redundancy and overlapping function within this pathway are in play. Gene deletion of *Ece-2* results in no detectable cardiac phenotypes; however, *Ece-1/Ece-2* double knockouts exhibit more severe OFT defects than either single *Ece-1* mutant present [102].

A specific role for the homeobox transcription factor HoxA3 can be associated with the development of the third PAA. *HoxA3* homozygous null mutant mice exhibit bilateral third PAA degeneration by embryonic day 11.5, resulting in the malformation of the carotid arteries [103].

Of the all the PAAs, the sixth arch is most sensitive to left/right positioning, as the right sixth arch is almost completely regressed. The Tgfβ superfamily ligand Nodal is asymmetrically expressed within the left side of the lateral mesoderm and establishes sidedness [104,105]. Downstream of Nodal signaling lies the homeobox transcription factor Pitx2. The Pitx2 isoform *Pitx2c* is expressed in a left-sided pattern within the pharyngeal arch mesoderm and the underling mesoderm of the second heart field [106]. Conditional *Pitx2c* mutants exhibit arch remodeling defects including randomization of the direction of the aortic arch, but strangely, *Pitx2c*-expressing cells are not observed to contribute to the fourth or sixth arches [107]. Data shows that *Pitx2c* expression within the second heart field mesoderm alters OFT morphology non-cell-autonomously such that there is an uneven distribution of blood supply, thereby maintaining Pdgfα and Vegf2 signaling on the left side and thus supporting progression of the left sixth arch artery and cutting off the right sixth arch from blood flow and promoting its regression.

SEPTATION OF THE OFT

The cNCC that leave in the first wave of migration are the cells that move into the distal, middle, and proximal OFT, where they will form the septum between the aorta and the pulmonary trunk and contribute to the aortic and pulmonary valves. The distal cells populate the region termed the aortic sac, which sits between the fourth and sixth PAA. The cNCC expand, remodel, and separate the fourth arch (forming the aortic arch) from the left sixth PAA (forming the pulmonary arteries). The mid-level OFT cNCC fork into two spiral-shaped cushions (termed aorticopulmonary cushions, or, alternatively, the conotruncal cushions or the OFT cushions) that are connected at their distal base and anchored to the walls of the aortic sac. Collectively, the distal-medial condensation of cNCC is termed the aorticopulmonary septation complex [108,109]. Septation occurs when distal cells move into the medial OFT, replacing the cushions with a spiral-shaped distal-medial septum. Finally, the proximal OFT closes following a distal to proximal progression in a "zipper-like" manner, and myocardial cells from the proximal domains of the heart invade the proximal OFT, fully separating the pulmonary and systemic circulations.

When insufficient numbers of cNCC are present due to perturbed migration or increased cell death, clinical conditions such as PTA, DORV, and interrupted aortic arch manifest or occur. A number of mouse models that present with these phenotypes are discussed below.

VALVULOGENESIS IN THE OFT

The cNCC that move farthest into the OFT contribute the aorticopulmonary cushions, which ultimately give rise to the aortic and pulmonary valves (also called the semilunar valves). These thin leaflets prevent blood that has exited the heart from reentering the ventricular chambers. The atrioventricular valves each have three cusps composed of stratified compartments of different ECM molecules [110]. Although the atrioventricular cushions receive abundant contributions of cNCC, cNCC contribution to the mature semilunar valve leaflets is not substantial, indicating that these cells are replaced during OFT remodeling [8]. Studies in mutant mice indicate that instructive signals from the cNCC—among them, Notch—are important to regulate second heart field contribution to the valves, and maturation of the aorticopulmonary cushion mesenchyme [111].

Failure of proper valve remodeling underlies bicuspid aortic valve disease (BAVD), an extremely common CHD [1] in which the aortic valves contain two, typically unequally sized cusps. Unlike the other CHD discussed in this chapter, BAVD rarely causes clinical complications during childhood. Rather, patients with BAVD typically require clinical intervention, sometimes entailing aortic valve replacement, only after midlife. Indeed, the probability of aortic valve replacement is 1−1.5% for patients from 1 to 19 years of age, but 27−30% for patients aged from 60 to 79 years [112]. Both *Pax3* and *Fgf8* mutant mice provide animal models of BAVD [111,113].

NCC DIFFERENTIATION

The cNCC that contribute to the PAA and the cells that migrate into the OFT to form the septum between systemic and pulmonary circulations ultimately differentiate into vascular smooth muscle cells. The majority of both engineered phenotypic abnormalities and cNCC-derived CHDs are associated with reduced number of cNCC, through problems of migration or viability, as well as a few defects in actual cNCC differentiation, have been identified. Smooth myogenesis is arguably the most complex muscle differentiation program to understand given that, in contrast to skeletal and cardiac myogenesis, its differentiation is reversible [114–116]. At the core of the smooth muscle program is the cis-element termed a CArG box that binds to the MADS box serum response factor (SRF). The myocardin family of SAP domain cofactors interacts with SRF to drive the smooth muscle program in NCC and non-NCC-derived smooth muscle [114–118]. The related myocardin-related transcription factor B (MrtfB) is expressed with the cNCC that populate the PAA and loss-of-function mice die after birth from PAA defects [119]. Normal numbers of cNCC are present in the OFTs of *MrtfB* mutants, indicating normal migration and survival, but interrogation of smooth muscle gene expression showed robust reduction in smooth muscle α-actin, demonstrating defective differentiation as the mechanism of these defects. Gain-of-function studies expressing a DN Notch signaling inhibitor, DN-Maml, within the cNCC produced a similar differentiation phenotype associated with the reduction of the Notch responsive bHLH factors Hey1, 2, and 3 in addition to smooth muscle α-actin, and Sm22α [120].

In our recent findings, we know that Twist1 also plays a role in modulation of differentiation of the cNCC [121]. In addition to the migration phenotypes described in *Twist1* null embryos [93] the adhesion defect described is the result of post-migration *trans-differentiation* of cNCC from smooth muscle to sympathetic-like neurons. Analysis of OFT cNCC within an NCC-restricted *Twist1* conditional knockout using both a pre-migration NCC-specific Cre (*Wnt1-Cre*) as well as a post-migration NCC-specific Cre (*Hand1^{Cre}* [122]) allows for basic timing control for when *Twist1* is deleted from the OFT cNCC [121]. Results of these studies confirm that the adhesion defects observed in the systemic null *Twist1* mice are NCC autonomous and that the phenotype appears when Twist1 is deleted after the cNCC are already in the OFT. These data suggest that the abnormal cNCC require Twist1 to control cell specification or differentiation after the cells arrive. The aggregated cNCC express a marker gene profile consistent with that of a sympathetic neuron and the aggregations resemble sympathetic ganglia [121]. Mechanistically, Twist1 is required to repress the neuronal program within the cNCC. Twist1 accomplishes this through direct repression of a master regulator of sympathetic neurogenesis, Phox2b, through direct DNA binding to the *Phox2b* promoter as well as protein–protein interactions with Phox2b and another *trans-activator* of *Phox2b*, Sox10. Protein interactions are dependent on a carboxy-terminal sequence termed the Twist-box [123]. Indeed, mice that express a *Twist1* mutant allele (the *Charlie Chaplin Twist1* allele) also develop ectopic neuron aggregations within the cardiac OFT completely independent of the OFT structural abnormalities observed in *Twist1* ablation models [93,121]. The Twist1 mutant OFT defects include PTA and DORV with associated ventricular septal defects (VSD) as well as retroesophageal right subclavian artery. These structural defects become significantly less penetrant when *Twist1* is ablated post-NCC migration [121]. The neuronal cell fate conversion is completely dependent on Phox2b, as removal of *Phox2b* on either a *Twist1 Charlie Chaplin* or *null* background rescues the gain-of-function cell-fate conversion phenotype. In additional gain-of-function analysis, ectopic expression of Twist1 within all of the NCC results in a robust reduction of neurons within the endogenous sympathetic ganglia. Together, these data show that in addition to roles for Twist1 as a modulator of NCC migration, Twist1 acts cell-autonomously as a neuronal cell fate repressor within a distinct subpopulation of cNCC that maintain the ability to adopt a neuronal fate. This cNCC subpopulation is marked by the expression of the Twist-family factors *Hand1* and *Hand2*; however, their role in the cell fate conversion is negligible, as embryos that are triple null for *Twist1*, *Hand1*, and *Hand2*, although severely compromised in viability, still exhibit ectopic OFT neuronal aggregations [121]. This finding is important; it indicates that the neurogenic program that is activated within the cNCC is non-canonical given endogenous sympathetic neurons are dependent on the functions of Hand2 [124,125] (Figure 11.3).

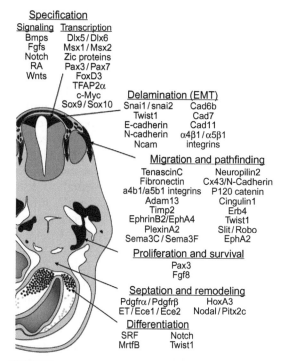

Specification

Signaling	Transcription
Bmps	Dlx5 / Dlx6
Fgfs	Msx1 / Msx2
Notch	Zic proteins
RA	Pax3 / Pax7
Wnts	FoxD3
	TFAP2α
	c-Myc
Sox9 / Sox10	

Delamination (EMT)

Snai1 / snai2	Cad6b
Twist1	Cad7
E-cadherin	Cad11
N-cadherin	α4β1 / α5β1
Ncam	integrins

Migration and pathfinding

TenascinC	Neuropilin2
Fibronectin	Cx43/N-Cadherin
a4b1/a5b1 integrins	P120 catenin
Adam13	Cingulin1
Timp2	Erb4
EphrinB2/EphA4	Twist1
PlexinA2	Slit/Robo
Sema3C / Sema3F	EphA2

Proliferation and survival

Pax3
Fgf8

Septation and remodeling

Pdgfrα / Pdgfrβ	HoxA3
ET/Ece1/Ece2	Nodal/Pitx2c

Differentiation

SRF	Notch
MrtfB	Twist1

FIGURE 11.3 **Summary of the molecular mechanisms governing cNCC development.** The development of cNCC, along with the molecular mediators of their maturation, from their specification at the boarder of the neural tube, their delamination, via EMT, from the dorsal neural tube, their migration to the pharyngeal arches, their proliferation and survival upon invasion of the pharyngeal arches, and their ultimate differentiation to remodel and septate the OFT are depicted.

EVOLUTION OF NEURAL CREST CELL CONTRIBUTION TO THE HEART

Cardiac NCC must differentiate into smooth muscle and connective tissue to completely septate the aorta and pulmonary arteries and the two ventricular chambers in both birds and mammals. However, this complete septation does not occur in single-ventricled vertebrates, such as fish, amphibians, and (with the exception of crocodilians) reptiles. Nonetheless, cNCC do form in zebrafish and the frog, *Xenopus laevis*, and although the role of these cells in the development of the fish or amphibian heart is comparatively poorly understood, the variation in cNCC plasticity and migration behavior becomes evident when we examine them in these evolutionary contexts (Figure 11.4).

In zebrafish, cNCC originate from a broader axial level along the neural tube, extending rostral to the otic placode, a region corresponding to cranial NCC in birds and mammals [126]. Fate-mapping studies reveal cNCC contributions not only to the pharyngeal arches and the OFT but to all segments of the zebrafish heart, including the ventricle, atrioventricular canal, and atrium. Intriguingly, marker analyses suggest that, unlike birds and mammals, after migrating deep into the heart, zebrafish cNCC can form myocardium. Thus, zebrafish embryos draw cNCC from a more broadly specified progenitor pool, which then migrates more extensively than those of birds and mammals and ultimately differentiates into cell types that amniote NCC (at least those which have been studied) do not [126,127].

X. laevis also has a single ventricular chamber; however, the *Xenopus* OFT nonetheless features a spiral septum that directs blood flow to the pulmocutaneous and systemic arteries [128,129]. This septum is not composed of cNCCs, but rather derivatives of the second

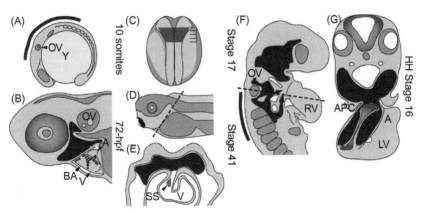

FIGURE 11.4 **Comparative analysis of cNCC contributions to the fish, frog, and chick heart.** (A) In the zebrafish at the 10 somite stage, the cNCC originate between rhombomere 1 and somite 6 (blue line). (B) By 72 h postfertilization (hpf), the cNCC have migrated to all cardiac segments (BA, bulbus arteriosus; V, ventricle; A, atrium). (C) In *Xenopus*, the cNCC originate between somites 1–4 (dorsal view shown at stage 17). (D) A section of a stage 41 embryo (dotted line) shows (E) the cNCC (blue) have migrated to the pharyngeal arches but not to the heart. Rather, the spiral septum (SS) within the OFT is derived from second heart field cells (red). (F) In the chick, the cNCC originate between the otic vesicle (OV) and somite 3 (blue line). A section through the OFT (dotted line) shows that (G) the cardiac neural crest cells (blue) have migrated to the pharyngeal arches and the aorticopulmonary cushions (APC), but not to the other chambers of the heart.

heart field [130]. Indeed, *Xenopus* cNCC cease migration after arrival in the pharyngeal arches and are thus dispensable for OFT septation. Although *Xenopus* cNCC do generate the aortic sac and aortic arch arteries, rather than invading the aorticopulmonary cushions, they instead maintain a sharp boundary with the myocardial lineage of the OFT.

Zebrafish and *Xenopus* therefore provide contrasting model systems to chick and mouse cNCC development and may therefore provide useful tools with which to investigate cNCC migration to the pharyngeal arches and aorticopulmonary cushions. In chick and mouse, the cNCCs migrate specifically to the pharyngeal arches and aorticopulmonary cushions [3,8], whereas in zebrafish these cells are dispersed throughout the heart, and in *Xenopus* they are entirely excluded from the heart proper. It is therefore of interest to examine the molecular regulators of cNCC migration in these organisms, as they may yield insight into human CHDs in which cNCC migration is disrupted. As mentioned previously, cNCC

migration in the mouse is regulated by ephrins, Ephs, semaphorins, plexins/neuropilins, and cell adhesion and ECM protein expression [12,40,131]. Indeed, in zebrafish cNCC migration may be regulated by semaphorin 3D and neuropilin 1A [132]. Although a handful of known migratory factors, such as Eph/ephrins [95,133] and cadherins [133] regulate NCC migration in *Xenopus*, cNCC migration, specifically, has not been well studied in this model system, and it is unknown why these cells fail to invade the aorticopulmonary cushions.

As previously mentioned, zebrafish cNCCs differentiate into cardiomyocytes, whereas chick and mouse cNCCs differentiate into smooth muscle and connective tissue. This suggests that cNCCs may display developmental plasticity across evolution. Therefore, studying cNCC differentiation in multiple organisms, especially those featuring uniquely specialized OFT structures, may provide insight into the mechanisms controlling cell fate. For example, in crocodilians, connective tissue nodules termed "cog-teeth" project into the subpulmonary OFT, where they evidently provide a unique means of regulating blood pressure resistance [134]. These "cog-teeth" are affected by vagal innervation and controlled through muscular contraction. Assessing cNCC contribution to these structures might illuminate alternative mechanisms governing cNCC potency during development and disease.

MOLECULAR REGULATORS OF CNCC DEVELOPMENT AND LINKS TO CONGENITAL HEART DISEASE

As described above, specification, migration and differentiation are distinct processes that involve integration of cell—cell communication, proliferation, and survival programs to generate a sufficient number of cNCC for normal cardiogenesis to proceed. Equally important is the participation of non-NCC derived tissues through which cNCC move and communicate. Ultimately, both cNCC autonomous and non-cell-autonomous changes in gene expression in response to the extracellular signaling coordinates the progression of heart formation. In this section, we focus upon the factors that influence in the formation of the OFT and that have links to CHDs encountered in neonates.

22q11 Deletion Syndrome

22q11 deletion syndrome, also commonly referred to as DiGeorge syndrome, is the most common human chromosomal microdeletion syndrome and is caused by varying deletions within the *q11* band of chromosome 22. A number of congenital defects associated with NCC dysfunction are typical of this syndrome, including defects of varying penetrance within the palate, thymus and parathyroid glands,

kidney, and inner ear. This syndrome also displays variable cardiac OFT defects, including tetralogy of Fallot, interrupted aortic arch, PTA, and associated VSDs [135,136]. These phenotypes are sometimes referred to as CATCH-22 (for Cardiac defects, Abnormal facial appearance, Thymic hypoplasia, Cleft palate, Hypocalcaemia, with chromosome 22 deletion). Candidate gene analysis within the DiGeorge syndrome-deleted region determined that loss of the gene encoding the T-box transcription factor Tbx1 causes this disease [137—140]. Mice haploinsufficient for *Tbx1* show abnormal development of the fourth PAA, whereas homozygous null *Tbx1* mutants present the more severe cardiac and gland phenotypes observed in DiGeorge patients [137]. Interestingly, although Tbx1 disruption engenders NCC defects, Tbx1 is not expressed within the cNCC. Tbx1 is instead expressed within the mesoderm of the second heart field and in the pharyngeal endoderm and its functional role relative to NCC appears to be maintaining NCC cell proliferation non-cell-autonomously [140,141]. This effect is, in part, related to the expression of Fgf8. Replacement of Fgf8 expression within mutant embryos with reduced Tbx1 expression partially rescues OFT phenotypes [142]. Within the second heart field, Tbx1 also regulates expression of a second Fgf, Fgf10 [143], as well as another regulator of heart development, the MADS box transcription factor Mef2c [144]. Continued study of the Tbx1 downstream targets within the second heart field will further add to our understanding of the importance of the second heart field in the expansion of the cNCC and patterning and septation of the OFT.

Alagille Syndrome

Alagille syndrome is an autosomal dominant disease that is linked to gene mutations in either the Notch ligand NOTCH2 [145] or the Notch receptor Jagged1 (JAG1) [146,147]. Like DiGeorge syndrome, there are a number of phenotypes that occur with varying penetrance within the spectrum of this disorder. Cardiac phenotypes that implicate the cNCC include tetralogy of Fallot, VSDs, pulmonary hypoplasia, and pulmonary valve stenosis. Notch proteins are a family of cell surface ligands that interact with membrane-bound receptors in adjoining cells to effectively convey signaling to both cells. As previously mentioned, these factors are crucial to communication between the cNCC and other cell types within the developing OFT and pharyngeal arches [18,148,149]. Expression of Jag1 in the endothelium activates Notch1 within both the NCC-derived and non-NCC-derived vascular smooth muscle, propagating smooth muscle differentiation. Inhibition of Notch within the cNCC results in reduced vascular smooth muscle differentiation and, consequently, in defects within the forming PAA. Mechanistically, it appears that activated Notch within the cNCC directly regulates Jag1 expression in the differentiating smooth muscle, thus allowing these cells to respond directly to Notch. If Notch is not activated or inhibited in the smooth muscle cells, this upregulated source of Jag1 is blocked.

Noonan Syndrome

Noonan syndrome phenotypes include short stature and both craniofacial and cardiac abnormalities. Penetrance of cardiac defects is high (80–90%) and includes both NCC-associated [150] and non-NCC-associated [151] phenotypes. VSDs, pulmonary valve stenosis, tetralogy of Fallot, and aortic coarctation, in which the aorta narrows proximal to the ductus arteriosus, are among the phenotypes associated with function of the cNCC. Noonan syndrome is associated with several gene mutations, but nearly 50% of all Noonan

syndrome patients have mutations in *PTPN11*, which encodes a tyrosine protein phosphatase called SHP-2. Shp-2 has been shown to play a role in the formation of the aortic and pulmonary valves. Mutations in SHP-2 are also associated with LEOPARD syndrome, which has phenotypic similarities to Noonan syndrome but involves different *PTPN11* mutations. SHP-2 phosphatase activity is increased in Noonan syndrome but is inhibited in LEOPARD syndrome, suggesting that maintaining a critical level of SHP-2 enzymatic activity within the cNCC is essential for OFT and valve development [150]. 10–15% of Noonan syndrome patients have mutations in SOS1, which encodes a guanine nucleotide exchange factor. A SOS1 mouse model reveals cardiac defects associated with activation of Ras/MAP Kinase signaling [152].

4q Deletion Syndrome

4q deletion syndrome is a rare disorder that presents with abnormalities of the hands and feet, face, and, in nearly 60% of cases, the heart. These cardiac defects include tetralogy of Fallot, aortic and pulmonary valve stenosis, VSD, PTA, and patent ductus arteriosus (PDA) [153–156]. Several genes that play direct roles in cardiogenesis are located within the deleted loci. These include the metalloprotease Tolid-like 1 (TLL1), the oxidoreductase 15-hydroxyprostaglandin dehydrogenase (HPGD; which degrades prostaglandins), and the bHLH transcription factor *HAND2* [153]. What is confusing regarding this syndrome is the lack of strong correlation between any one gene within the deleted locus and the resultant phenotypes observed in patients, suggesting a multi-gene causality. For example, some patients that present with cardiac defects that have lost *HAND2*, but, conversely, other patients that have lost *HAND2* exhibit no cardiac defects. Mouse loss-of-function models show that some of the cardiac defects observed in 4q deletion syndrome appear in these gene targeted mice.

Loss-of-function in *Tll1* in mice is lethal between E13.5 and E16.5 and leads to DORV with large VSD, in addition to cardiac defects not directly associated with cNCC [157]. Tll1 is an important regulator of Bmp-1 signaling and likely modulates other ligands within the Tgfβ superfamily. *HPGD* null mice die as neonates within 2 days of birth. These mice exhibit PDA, a condition where the sixth PAA derived ductus fails to close after pulmonary function is established, but no other cardiac defects [158]. Although this recapitulation of PDA is implicating, the lack of patent ductus in many 4q deletion patients in which *HPGD* is deleted is confounding.

HAND2 is an extremely attractive candidate gene for 4q deletion syndrome given its established roles in both limb and heart development [94,159–161]. Systemic *Hand2* deletion results in embryonic death at E9.5, yielding little insight on the impact on cNCC [162]. Conditional deletion of *Hand2* using Wnt1-Cre results in mice that die at E12.5 due to defects in sympathetic neurogenesis [125,163]. Cardiac phenotypes include DORV, interrupted aortic arch, and VSDs. Additionally, cNCC-specific *Hand2* deletion phenotypes are sensitive to gene dosage of the related factor *Hand1*. When *Hand2* is deleted in cells expressing *Hand1*, mice develop the more severe PTA with 50% penetrance [164]. This observation could well explain the apparent randomness to *Hand2* association with congenital defects. In a haploinsufficient state, *Hand2* is sensitive to the levels of expression of other bHLH proteins like Hand1. For example, in limb morphogenesis, polydactyly occurring due to Twist1 haploinsufficiency can be fully rescued by concurrently making *Hand2* haploinsufficient [160]. It is likely that additional genetic mutations outside the 4q deletion may influence the impact of *Hand2* haploinsufficiency via a similar mechanism. Finally, *Hand2* also plays critical roles within the second heart field. Deletion of *Hand2* using a range of temporally and spatially distinct second heart field Cre mouse lines reveals a spectrum OFT phenotypes that include tricuspid atresia, and varying degrees of hypoplastic right heart [165]. Given the importance of the second heart field in the migration and expansion of the cNCC Hand2's role within these cardiac progenitor cells are equally likely to contribute to OFT defects in 4q deletion syndrome.

DIETARY FACTORS INFLUENCING CARDIAC NEURAL CREST CELL DEVELOPMENT

In addition to genetic factors, maternal diet can affect cNCC development [166]. For example, folate plays important roles in one-carbon metabolism, and, consequently, DNA methylation and cell proliferation [167]. Folate deficiency is primarily associated with neural tube defects [168,169]. Nonetheless, folate deficiencies are also associated with increased risk of CHDs. A mutant mouse model in which folic acid cellular uptake has been compromised displays abnormal apoptosis in the OFT and the forming interventricular septum. These phenotypes correlate with reduced *Pax3* expression in the presumptive migrating NCC [170]. Although the direct effects of folate upon cNCC gene expression and cell proliferation remain unclear, it is likely that folic acid supplementation during early pregnancy may reduce the risk of CHDs [171].

Fetal Alcohol Syndrome

Fetal alcohol syndrome features neuronal, craniofacial, and cardiac defects that can likely be attributed to disrupted development of the NCC and their derivatives. In mice, embryonic alcohol exposure during the developmental stages in which the cNCC invade the pharyngeal arches phenocopies cardiovascular aspects of 22q11 deletion syndrome, including membranous VSD and aortic arch patterning defects [172]. Indeed, epidemiological studies have linked alcohol consumption by pregnant women periconception and an increased risk of OFT defects in their offspring [173].

There are multiple potential mechanism(s) by which ethanol exposure might inhibit cNCC development. Impaired cell migration [174–176] and/or increased apoptosis [177,178] have all been proposed to contribute to NCC dysfunction in models of ethanol toxicity. Further evidence indicates that ethanol exposure influences NCC gene expression [179] and may disrupt signaling mediated through another factor critical for cNCC development, RA.

Retinoic Acid

RA, more commonly known as vitamin A, is teratogenic in excessive doses during pregnancy [180,181]. For example, a RA derivative and analog used to treat acne, isotretinoin (more commonly known as Accutane), causes RA embryopathy in babies whose mothers were prescribed the drug during pregnancy. In addition to craniofacial, thymic, and neuronal phenotypes, RA embryopathy patients display CHDs such as transposition of the great vessels, tetralogy of Fallot, double outlet right ventricle, truncus arteriosus, VSD, interrupted aortic arch type B, retroesophageal right subclavian artery, and aortic arch hypoplasia [180,182].

RA deficiency, similarly, leads to congenital abnormalities, including cNCC-associated OFT defects such as PTA, VSD, and aortic arch artery defects [183–185]. Mutation in RA receptors, or of factors necessary to synthesize endogenous RA, leads to OFT defects reminiscent of 22q11 deletion syndrome [186–189]. Studies in *Xenopus* suggest that ethanol inhibits RA biosynthesis, thereby diminishing RA signaling to levels below those critical for normal development [190]. Further studies in chick support these findings, indicating that RA supplementation can ameliorate the effects of ethanol toxicity, suggesting that prolonged ethanol exposure interferes with RA signaling [176].

Genetic studies in mice have shed light upon the mechanisms by which RA signaling affects NCC development. Mice mutant for two RA receptors, RARα and RARβ, display completely penetrant PTA [191]. However, when these two RARs are knocked out specifically in NCC, aorticopulmonary septation proceeds normally, indicating that RA signaling influences NCC development indirectly, through other cell types in the OFT [192].

The second heart field progenitor cells, which produce the myocardium of the OFT, are particularly sensitive to RA deficiency [193]. As mentioned with respect to 22q11 deletion syndrome, the second heart field and its derivatives regulate maturation of the cNCC non-cell-autonomously through cell-signaling cues. The secreted signaling factor TGFβ2 is upregulated in embryonic hearts when RA signaling is disrupted [194], whereas expression of a T-Box transcription factor related to Tbx1, *Tbx2*, is downregulated in the RA-treated OFT myocardium [195]. RA-associated OFT defects can be partially rescued when TGFβ2 levels are correspondingly reduced [194]. Thus, it has been proposed that RA signaling inhibits TGFβ signaling via Tbx2, and that excessive TGFβ2 activity may disrupt NCC patterning and/or aorticopulmonary cushion development, as well as development of the OFT myocardium, thus causing CHDs [195].

Thus, like 22q11 deletion syndrome, the OFT defects ensuing from aberrant retinoic signaling likely arise due to abnormal tissue–tissue interactions between the NCC and derivatives of the second heart field. Indeed, diminished RA levels appear to accelerate the recovery from arterial growth delay characteristic of $Tbx1^{+/-}$ mutant mice, indicating that RA levels may influence the penetrance and severity of the aortic arch patterning defects seen in 22q11 patients [196].

CONCLUSIONS AND PERSPECTIVES

As this chapter has illustrated, cNCC are integral to proper cardiac development and function and, given the relative frequency of CHDs, have a substantial impact upon human health. Improvements in surgical intervention made since the mid-twentieth century have drastically increased the likelihood that a child born with a severe CHD will survive their first year of life; however, the long-term prognosis for these patients remains less certain. Ongoing research has made considerable advances toward expanded genetic testing and prenatal care with the ultimate goal of diminishing the risk of and improving clinical outcomes for CHDs associated with NCC dysfunction.

cNCC dysfunction is not only responsible for a significant proportion of CHDs but also is likely to contribute to adult cardiac diseases, such as aortic valve disease. In addition to integrating a broadly disparate body of knowledge into a comprehensive picture of cNCC development, future cNCC studies face the challenge of understanding the mechanisms by which the disruption of developmental processes impact health into adulthood. A lingering question is this: After the OFT and aortic arch has formed, do cNCC remain in the heart and, if so, can they be modulated to perform restorative functions to alleviate damage and disease? Like all NCC, the cNCC display a remarkable developmental plasticity that has

been integral to the evolution of the four-chambered heart and may hold significant therapeutic potential as advances are made in stem cell biology. Although controversial, studies have suggested that neural crest stem cells may have therapeutic potential following myocardial infarction [197], and further investigations into such clinical endeavors will certainly expand our understanding of this fascinating cell population born of the neural ectoderm.

Acknowledgments

We would like to thank the scientists who have worked to collectively produce our current understanding of the cardiac neural crest. We thank the Riley Heart Research Center for providing an outstanding scientific environment. Infrastructural support at the Herman B Wells Center is partially supported by the Riley Children's Foundation and Division of Pediatric Cardiology. Grant support for this work was provided by NIH 1R0AR061392-01 and support from the American Heart Association. The Authors declare that there are no conflicts of interests associated with this work.

References

[1] Hoffman JI, Kaplan S, Liberthson RR. Prevalence of congential heart disease. Am Heart J 2004;147:425–39.
[2] Chin AJ, Saint-Jeannet JP, Lo CW. How insights from cardiovascular developmental biology have impacted the care of infants and children with congenital heart disease. Mech Dev 2012;129:75–97.
[3] Kirby ML, Gale TF, Stewart DE. Neural crest cells contribute to normal aorticopulmonary septation. Science 1983;220(4601):1059–61.
[4] Kirby ML. Plasticity and predetermination of mesencephalic and trunk neural crest transplanted into the region of the cardiac neural crest. Dev Biol 1989;134(2):402–12.
[5] Gittenberger-de Groot AC, Blom NM, Aoyama N, Sucov H, Wenink AC, Poelmann RE. The role of neural crest and epicardium-derived cells in conduction system formation. Novartis Found Symp 2003;250:125–34 discussion 34–41.
[6] Bronner-Fraser M. Origins and developmental potential of the neural crest. Exp Cell Res 1995;218(2):405–17.

[7] Kirby ML, Waldo KL. Neural crest and cardiovascular patterning. Circ Res 1995;77(2):211–5.

[8] Jiang X, Rowitch DH, Soriano P, McMahon AP, Sucov HM. Fate of the mammalian cardiac neural crest. Development 2000;127(8):1607–16.

[9] Stoller JZ, Epstein JA. Cardiac neural crest. Sem Cell Dev Biol 2005;16(6):704–15.

[10] Selleck MA, Bronner-Fraser M. The genesis of avian neural crest cells: a classic embryonic induction. Proc Natl Acad Sci USA 1996;93(18):9352–7.

[11] Inman KE, Ezin M, Bronner-Fraser M, Trainor PA. Role of cardiac neural crest in morphogenesis of the heart and great vessels. In: Rosenthal N, Harvery RP, editors. Heart development and regeneration. 1st ed. London: Academic Press; 2010. p. 417–62.

[12] Bronner ME. Formation and migration of neural crest cells in the vertebrate embryo. Histochem Cell Biol 2012;138(2):179–86.

[13] Heeg-Truesdell E, LaBonne C. A slug, a fox, a pair of sox: transcriptional responses to neural crest inducing signals. Birth Defects Res C Embryo Today 2004;72 (2):124–39.

[14] Basch ML, Garcia-Castro MI, Bronner-Fraser M. Molecular mechanisms of neural crest induction. Birth Defects Res C Embryo Today 2004;72(2):109–23.

[15] Basch ML, Bronner-Fraser M. Neural crest inducing signals. Adv Exp Med Biol 2006;589:24–31.

[16] Steventon B, Carmona-Fontaine C, Mayor R. Genetic network during neural crest induction: from cell specification to cell survival. Sem Cell Dev Biol 2005;16 (6):647–54.

[17] High F, Epstein JA. Signalling pathways regulating cardiac neural crest migration and differentiation. Novartis Found Symp 2007;283:152–61 discussion 61–64, 238–241.

[18] Mead TJ, Yutzey KE. Notch pathway regulation of neural crest cell development in vivo. Dev Dyn 2012;241(2):376–89.

[19] Luo T, Matsuo-Takasaki M, Sargent TD. Distinct roles for Distal-less genes Dlx3 and Dlx5 in regulating ectodermal development in Xenopus. Mol Reprod Dev 2001;60(3):331–7.

[20] Ishii M, Han J, Yen HY, Sucov HM, Chai Y, Maxson Jr RE. Combined deficiencies of Msx1 and Msx2 cause impaired patterning and survival of the cranial neural crest. Development 2005;132(22):4937–50.

[21] Nakata K, Nagai T, Aruga J, Mikoshiba K. Xenopus Zic family and its role in neural and neural crest development. Mech Dev 1998;75(1–2):43–51.

[22] Merzdorf CS. Emerging roles for zic genes in early development. Dev Dyn 2007;236(4):922–40.

[23] Lacosta AM, Muniesa P, Ruberte J, Sarasa M, Dominguez L. Novel expression patterns of Pax3/Pax7 in early trunk neural crest and its melanocyte and non-melanocyte lineages in amniote embryos. Pigment Cell Res 2005;18(4):243–51.

[24] Otto A, Schmidt C, Patel K. Pax3 and Pax7 expression and regulation in the avian embryo. Anat Embryol 2006;211(4):293–310.

[25] Lacosta AM, Canudas J, Gonzalez C, Muniesa P, Sarasa M, Dominguez L. Pax7 identifies neural crest, chromatophore lineages and pigment stem cells during zebrafish development. Int J Dev Biol 2007;51 (4):327–31.

[26] Minchin JE, Hughes SM. Sequential actions of Pax3 and Pax7 drive xanthophore development in zebrafish neural crest. Dev Biol 2008;317(2):508–22.

[27] Maczkowiak F, Mateos S, Wang E, Roche D, Harland R, Monsoro-Burq AH. The Pax3 and Pax7 paralogs cooperate in neural and neural crest patterning using distinct molecular mechanisms, in Xenopus laevis embryos. Dev Biol 2010;340(2):381–96.

[28] Williams T, Tjian R. Characterization of a dimerization motif in AP-2 and its function in heterologous DNA-binding proteins. Science 1991;251(4997): 1067–71.

[29] Kos R, Reedy MV, Johnson RL, Erickson CA. The winged-helix transcription factor FoxD3 is important for establishing the neural crest lineage and repressing melanogenesis in avian embryos. Development 2001;128(8):1467–79.

[30] Bellmeyer A, Krase J, Lindgren J, LaBonne C. The protooncogene c-myc is an essential regulator of neural crest formation in Xenopus. Dev Cell 2003;4(6):827–39.

[31] Hong CS, Saint-Jeannet JP. Sox proteins and neural crest development. Sem Cell Dev Biol 2005;16 (6):694–703.

[32] Sakai D, Suzuki T, Osumi N, Wakamatsu Y. Cooperative action of Sox9, Snail2 and PKA signaling in early neural crest development. Development 2006;133(7):1323–33.

[33] Wahlbuhl M, Reiprich S, Vogl MR, Bosl MR, Wegner M. Transcription factor Sox10 orchestrates activity of a neural crest-specific enhancer in the vicinity of its gene. Nucl Acids Res 2012;40(1):88–101.

[34] Taneyhill LA, Coles EG, Bronner-Fraser M. Snail2 directly represses cadherin6B during epithelial-to-mesenchymal transitions of the neural crest. Development 2007;134(8):1481–90.

[35] O'Rourke MP, Tam PP. Twist functions in mouse development. Int J Dev Biol 2002;46(4):401–13.

[36] Akitaya T, Bronner-Fraser M. Expression of cell adhesion molecules during initiation and cessation of neural crest cell migration. Dev Dyn 1992;194(1):12–20.

[37] Inoue T, Chisaka O, Matsunami H, Takeichi M. Cadherin-6 expression transiently delineates specific

rhombomeres, other neural tube subdivisions, and neural crest subpopulations in mouse embryos. Dev Biol 1997;183(2):183—94.

[38] Yuen HF, Chua CW, Chan YP, Wong YC, Wang X, Chan KW. Significance of TWIST and E-cadherin expression in the metastatic progression of prostatic cancer. Histopathology 2007;50(5):648—58.

[39] Cheung M, Chaboissier MC, Mynett A, Hirst E, Schedl A, Briscoe J. The transcriptional control of trunk neural crest induction, survival, and delamination. Dev Cell 2005;8(2):179—92.

[40] Kirby ML, Hutson MR. Factors controlling cardiac neural crest cell migration. Cell Adh Migr 2010;4 (4):609—21.

[41] Bockman DE, Kirby ML. Dependence of thymus development on derivatives of the neural crest. Science 1984;223(4635):498—500.

[42] Boot MJ, Gittenberger-De Groot AC, Van Iperen L, Hierck BP, Poelmann RE. Spatiotemporally separated cardiac neural crest subpopulations that target the outflow tract septum and pharyngeal arch arteries. Anat Rec Discov Mol Cell Evol Biol 2003;275 (1):1009—18.

[43] Tucker RP. Abnormal neural crest cell migration after the in vivo knockdown of tenascin-C expression with morpholino antisense oligonucleotides. Dev Dyn 2001;222(1):115—9.

[44] Alfandari D, Cousin H, Gaultier A, Hoffstrom BG, DeSimone DW. Integrin alpha5beta1 supports the migration of Xenopus cranial neural crest on fibronectin. Dev Biol 2003;260(2):449—64.

[45] Alfandari D, Cousin H, Gaultier A, et al. Xenopus ADAM 13 is a metalloprotease required for cranial neural crest-cell migration. Curr Biol 2001;11(12):918—30.

[46] Cantemir V, Cai DH, Reedy MV, Brauer PR. Tissue inhibitor of metalloproteinase-2 (TIMP-2) expression during cardiac neural crest cell migration and its role in proMMP-2 activation. Dev Dyn 2004;231(4):709—19.

[47] Kirby ML, Bockman DE. Neural crest and normal development: a new perspective. Anat Rec 1984;209 (1):1—6.

[48] Dickie MM. New splotch alleles in the mouse. J Hered 1964;55:97—101.

[49] Li J, Chen F, Epstein JA. Neural crest expression of Cre recombinase directed by the proximal Pax3 promoter in transgenic mice. Genesis 2000;26(2):162—4.

[50] Barber TD, Barber MC, Cloutier TE, Friedman TB. PAX3 gene structure, alternative splicing and evolution. Gene 1999;237(2):311—9.

[51] Conway SJ, Henderson DJ, Kirby ML, Anderson RH, Copp AJ. Development of a lethal congenital heart defect in the splotch (Pax3) mutant mouse. Cardiovasc Res 1997;36(2):163—73.

[52] Foy C, Newton V, Wellesley D, Harris R, Read AP. Assignment of the locus for Waardenburg syndrome type I to human chromosome 2q37 and possible homology to the Splotch mouse. Am J Hum Genet 1990;46(6):1017—23.

[53] Tassabehji M, Newton VE, Leverton K, et al. PAX3 gene structure and mutations: close analogies between Waardenburg syndrome and the Splotch mouse. Hum Mol Genet 1994;3(7):1069—74.

[54] Franz T. Persistent truncus arteriosus in the Splotch mutant mouse. Anat Embryol 1989;180(5):457—64.

[55] Epstein JA, Li J, Lang D, et al. Migration of cardiac neural crest cells in splotch embryos. Development 2000;127(9):1869—78.

[56] Conway SJ, Henderson DJ, Copp AJ. Pax3 is required for cardiac neural crest migration in the mouse: evidence from the splotch (Sp2H) mutant. Development 1997;124(2):505—14.

[57] Chan WY, Cheung CS, Yung KM, Copp AJ. Cardiac neural crest of the mouse embryo: axial level of origin, migratory pathway and cell autonomy of the splotch (Sp2H) mutant effect. Development 2004;131 (14):3367—79.

[58] Wilson DB. Proliferation in the neural tube of the splotch (Sp) mutant mouse. J Comp Neurol 1974;154 (3):249—55.

[59] Conway SJ, Bundy J, Chen J, Dickman E, Rogers R, Will BM. Decreased neural crest stem cell expansion is responsible for the conotruncal heart defects within the splotch (Sp(2H))/Pax3 mouse mutant. Cardiovasc Res 2000;47(2):314—28.

[60] Henderson DJ, Ybot-Gonzalez P, Copp AJ. Overexpression of the chondroitin sulphate proteoglycan versican is associated with defective neural crest migration in the Pax3 mutant mouse (splotch). Mech Dev 1997;69(1—2):39—51.

[61] Serbedzija GN, McMahon AP. Analysis of neural crest cell migration in Splotch mice using a neural crest-specific LacZ reporter. Dev Biol 1997;185(2):139—47.

[62] Lo CW, Cohen MF, Huang GY, et al. Cx43 gap junction gene expression and gap junctional communication in mouse neural crest cells. Dev Genet 1997;20 (2):119—32.

[63] Zambrowicz BP, Imamoto A, Fiering S, Herzenberg LA, Kerr WG, Soriano P. Disruption of overlapping transcripts in the ROSA beta geo 26 gene trap strain leads to widespread expression of beta-galactosidase in mouse embryos and hematopoietic cells. Proc Natl Acad Sci USA 1997;94(8):3789—94.

[64] Soriano P. Generalized lacZ expression with the ROSA26 Cre reporter strain. Nat Genet 1999;21(1):70—1.

[65] Yamauchi Y, Abe K, Mantani A, et al. A novel transgenic technique that allows specific marking of the

neural crest cell lineage in mice. Dev Biol 1999;212 (1):191–203.

[66] Krull CE. Neural crest cells and motor axons in avians: common and distinct migratory molecules. Cell Adh Migr 2010;4(4):631–4.

[67] Nakamoto M. Eph receptors and ephrins. Int J Biochem Cell Biol 2000;32(1):7–12.

[68] Davy A, Aubin J, Soriano P. Ephrin-B1 forward and reverse signaling are required during mouse development. Genes Dev 2004;18(5):572–83.

[69] Mellott DO, Burke RD. Divergent roles for Eph and ephrin in avian cranial neural crest. BMC Dev Biol 2008;8:56.

[70] Brown CB, Feiner L, Lu MM, et al. PlexinA2 and semaphorin signaling during cardiac neural crest development. Development 2001;128(16):3071–80.

[71] Gitler AD, Brown CB, Kochilas L, Li J, Epstein JA. Neural crest migration and mouse models of congenital heart disease. Cold Spring Harb Symp Quant Biol 2002;67:57–62.

[72] Toyofuku T, Yoshida J, Sugimoto T, et al. Repulsive and attractive semaphorins cooperate to direct the navigation of cardiac neural crest cells. Dev Biol 2008;321(1):251–62.

[73] Feiner L, Webber AL, Brown CB, et al. Targeted disruption of semaphorin 3C leads to persistent truncus arteriosus and aortic arch interruption. Development 2001;128(16):3061–70.

[74] Gammill LS, Gonzalez C, Bronner-Fraser M. Neuropilin 2/semaphorin 3F signaling is essential for cranial neural crest migration and trigeminal ganglion condensation. Dev Neurobiol 2007;67(1):47–56.

[75] Eickholt BJ, Mackenzie SL, Graham A, Walsh FS, Doherty P. Evidence for collapsin-1 functioning in the control of neural crest migration in both trunk and hindbrain regions. Development 1999;126(10): 2181–9.

[76] Roffers-Agarwal J, Hutt KJ, Gammill LS. Paladin is an antiphosphatase that regulates neural crest cell formation and migration. Dev Biol 2012;371(2):180–90.

[77] Latta EJ, Golding JP. Regulation of PP2A activity by Mid1 controls cranial neural crest speed and gangliogenesis. Mech Dev 2012;128(11–12):560–76.

[78] Bisson N, Wedlich D, Moss T. The p21-activated kinase Pak1 regulates induction and migration of the neural crest in *Xenopus*. Cell Cycle 2012;11(7):1316–24.

[79] Faure S, Fort P. Atypical RhoV and RhoU GTPases control development of the neural crest. Small GTPases 2011;2(6):310–3.

[80] Shoval I, Kalcheim C. Antagonistic activities of Rho and Rac GTPases underlie the transition from neural crest delamination to migration. Dev Dyn 2012;241 (7):1155–68.

[81] Clay MR, Halloran MC. Regulation of cell adhesions and motility during initiation of neural crest migration. Curr Opin Neurobiol 2011;21(1):17–22.

[82] Xu X, Francis R, Wei CJ, Linask KL, Lo CW. Connexin 43-mediated modulation of polarized cell movement and the directional migration of cardiac neural crest cells. Development 2006;133(18):3629–39.

[83] Huang GY, Wessels A, Smith BR, Linask KK, Ewart JL, Lo CW. Alteration in connexin 43 gap junction gene dosage impairs conotruncal heart development. Dev Biol 1998;198(1):32–44.

[84] Kretz M, Eckardt D, Kruger O, et al. Normal embryonic development and cardiac morphogenesis in mice with Wnt1-Cre-mediated deletion of connexin43. Genesis 2006;44(6):269–76.

[85] Xu X, Li WE, Huang GY, et al. N-cadherin and Cx43alpha1 gap junctions modulates mouse neural crest cell motility via distinct pathways. Cell Commun Adh 2001;8(4–6):321–4.

[86] Wu CY, Taneyhill LA. Annexin a6 modulates chick cranial neural crest cell emigration. PloS One 2012;7 (9):e44903.

[87] Farlie PG, Kerr R, Thomas P, et al. A paraxial exclusion zone creates patterned cranial neural crest cell outgrowth adjacent to rhombomeres 3 and 5. Dev Biol 1999;213(1):70–84.

[88] Golding JP, Trainor P, Krumlauf R, Gassmann M. Defects in pathfinding by cranial neural crest cells in mice lacking the neuregulin receptor ErbB4. Nat Cell Biol 2000;2(2):103–9.

[89] Golding JP, Dixon M, Gassmann M. Cues from neuroepithelium and surface ectoderm maintain neural crest-free regions within cranial mesenchyme of the developing chick. Development 2002;129(5):1095–105.

[90] Golding JP, Sobieszczuk D, Dixon M, et al. Roles of erbB4, rhombomere-specific, and rhombomere-independent cues in maintaining neural crest-free zones in the embryonic head. Dev Biol 2004;266(2):361–72.

[91] Soo K, O'Rourke MP, Khoo P-L, et al. Twist function is required for the morphogenesis of the cephalic neural tube and the differentiation of the cranial neural crest cells in the mouse embryo. Dev Biol 2002;247 (2):251–70.

[92] Chen ZF, Behringer RR. Twist is required in head mesenchyme for cranial neural tube morphogenesis. Genes Dev 1995;9:686–99.

[93] Vincentz JW, Barnes RM, Rodgers R, Firulli BA, Conway SJ, Firulli AB. An absence of Twist1 results in aberrant cardiac neural crest morphogenesis. Dev Biol 2008;320:131–9.

[94] Barnes RM, Firulli ABA. Twist of insight, the role of twist-family bHLH factors in development. Int J Dev Biol 2009;53(7):909–24.

[95] Smith A, Robinson V, Patel K, Wilkinson DG. The EphA4 and EphB1 receptor tyrosine kinases and ephrin-B2 ligand regulate targeted migration of branchial neural crest cells. Curr Biol 1997;7 (8):561–70.

[96] Tallquist MD, Soriano P. Cell autonomous requirement for PDGFRalpha in populations of cranial and cardiac neural crest cells. Development 2003;130 (3):507–18.

[97] Richarte AM, Mead HB, Tallquist MD. Cooperation between the PDGF receptors in cardiac neural crest cell migration. Dev Biol 2007;306(2):785–96.

[98] Clouthier DE, Hosoda K, Richardson JA, et al. Cranial and cardiac neural crest defects in endothelin-A receptor-deficient mice. Development 1998;125(5):813–24.

[99] Yanagisawa H, Yanagisawa M, Kapur RP, et al. Dual genetic pathways of endothelin-mediated intercellular signaling revealed by targeted disruption of endothelin converting enzyme-1 gene. Development 1998;125(5):825–36.

[100] Yanagisawa H, Hammer RE, Richardson JA, Williams SC, Clouthier DE, Yanagisawa M. Role of Endothelin-1/Endothelin-A receptor-mediated signaling pathway in the aortic arch patterning in mice. J Clin Invest 1998;102(1):22–33.

[101] Clouthier DE, Williams SC, Hammer RE, Richardson JA, Yanagisawa M. Cell-autonomous and nonautonomous actions of endothelin-A receptor signaling in craniofacial and cardiovascular development. Dev Biol 2003;261(2):506–19.

[102] Yanagisawa H, Hammer RE, Richardson JA, et al. Disruption of ECE-1 and ECE-2 reveals a role for endothelin-converting enzyme-2 in murine cardiac development. J Clin Invest 2000;105(10):1373–82.

[103] Kameda Y. Hoxa3 and signaling molecules involved in aortic arch patterning and remodeling. Cell Tissue Res 2009;336(2):165–78.

[104] Lohr JL, Danos MC, Yost HJ. Left-right asymmetry of a nodal-related gene is regulated by dorsoanterior midline structures during *Xenopus* development. Development 1997;124(8):1465–72.

[105] Wagner MK, Yost HJ. Left-right development: the roles of nodal cilia. Curr Biol 2000;10(4):R149–51.

[106] Liu C, Liu W, Palie J, Lu MF, Brown NA, Martin JF. Pitx2c patterns anterior myocardium and aortic arch vessels and is required for local cell movement into atrioventricular cushions. Development 2002;129 (21):5081–91.

[107] Yashiro K, Shiratori H, Hamada H. Haemodynamics determined by a genetic programme govern asymmetric development of the aortic arch. Nature 2007;450(7167):285–8.

[108] Waldo K, Zdanowicz M, Burch J, et al. A novel role for cardiac neural crest in heart development. J Clin Invest 1999;103(11):1499–507.

[109] Waldo K, Miyagawa-Tomita S, Kumiski D, Kirby ML. Cardiac neural crest cells provide new insight into septation of the cardiac outflow tract: aortic sac to ventricular septal closure. Dev Biol 1998;196 (2):129–44.

[110] Combs MD, Yutzey KE. Heart valve development: regulatory networks in development and disease. Circ Res 2009;105(5):408–21.

[111] Jain R, Engleka KA, Rentschler SL, et al. Cardiac neural crest orchestrates remodeling and functional maturation of mouse semilunar valves. J Clin Invest 2011;121(1):422–30.

[112] Sabet HY, Edwards WD, Tazelaar HD, Daly RC. Congenitally bicuspid aortic valves: a surgical pathology study of 542 cases (1991 through 1996) and a literature review of 2,715 additional cases. Mayo Clin Proc 1999;74(1):14–26.

[113] Macatee TL, Hammond BP, Arenkiel BR, Francis L, Frank DU, Moon AM. Ablation of specific expression domains reveals discrete functions of ectoderm- and endoderm-derived FGF8 during cardiovascular and pharyngeal development. Development 2003;130 (25):6361–74.

[114] Hirschi KK, Majesky MW. Smooth muscle stem cells. Anat Rec Discov Mol Cell Evol Biol 2004;276 (1):22–33.

[115] Miano JM. Serum response factor: toggling between disparate programs of gene expression. J Mol Cell Card 2003;35(6):577–93.

[116] Wang Z, Wang DZ, Pipes GC, Olson EN. Myocardin is a master regulator of smooth muscle gene expression. Proc Natl Acad Sci USA 2003;100(12):7129–34.

[117] Chen J, Kitchen CM, Streb JW, Miano JM. Myocardin: a component of a molecular switch for smooth muscle differentiation. J Mol Cell Cardiol 2002;34(10):1345–56.

[118] Firulli AB. Another hat for myocardin. J Mol Cell Cardiol 2002;34(10):1293–6.

[119] Li J, Zhu X, Chen M, et al. Myocardin-related transcription factor B is required in cardiac neural crest for smooth muscle differentiation and cardiovascular development. Proc Natl Acad Sci USA 2005;102 (25):8916–21.

[120] High FA, Zhang M, Proweller A, et al. An essential role for Notch in neural crest during cardiovascular development and smooth muscle differentiation. J Clin Invest 2007;117(2):353–63.

[121] Vincentz JW, Firulli BA, Lin A, Spicer DB, Howard MJ, Firulli AB. Twist1 controls a cell specification switch governing cell fate decisions within the

cardiac neural crest. PLoS Genet 2013;9(3): e10034051—e100340514.

[122] Barnes RM, Firulli B, Conway SJ, Vincentz JW, Firulli AB. Analysis of the Hand1 cell lineage reveals novel contributions to cardiovascular, neural crest, extra-embryonic, and lateral mesoderm derivatives. Dev Dyn 2010;239:3086—97.

[123] Biakel P, Kern B, Yang X, et al. A twist code determines the onset of osteoblast differentiation. Dev Cell 2004;6:423—35.

[124] Xu H, Firulli AB, Wu X, Zhang X, Howard MJ. HAND2 synergistically enhances transcription of dopamine-B-hydroxylase in the presence of Phox2a. Dev Biol 2003;262:183—93.

[125] Hendershot TJ, Liu H, Clouthier DE, et al. Conditional deletion of Hand2 reveals critical functions in neurogenesis and cell type-specific gene expression for development of neural crest-derived noradrenergic sympathetic ganglion neurons. Dev Biol 2008;319(2):179—91.

[126] Sato M, Yost HJ. Cardiac neural crest contributes to cardiomyogenesis in zebrafish. Dev Biol 2003;257 (1):127—39.

[127] Li YX, Zdanowicz M, Young L, Kumiski D, Leatherbury L, Kirby ML. Cardiac neural crest in zebrafish embryos contributes to myocardial cell lineage and early heart function. Dev Dyn 2003;226 (3):540—50.

[128] De Graaf AR. Investigations into the distributions of blood in the heart and aortic arches of *Xenopus laevis* (DAUD). Exp Biol 1957;34(2):143—74.

[129] Farmer CG. Evolution of the vertebrate cardio-pulmonary system. Annu Rev Physiol 1999;61: 573—92.

[130] Lee YH, Saint-Jeannet JP. Cardiac neural crest is dispensable for outflow tract septation in *Xenopus*. Development 2011;138(10):2025—34.

[131] McKeown SJ, Wallace AS, Anderson RB. Expression and function of cell adhesion molecules during neural crest migration. Dev Biol 2013;373(2):244—57.

[132] Sato M, Tsai HJ, Yost HJ. Semaphorin3D regulates invasion of cardiac neural crest cells into the primary heart field. Dev Biol 2006;298(1):12—21.

[133] Helbling PM, Tran CT, Brandli AW. Requirement for EphA receptor signaling in the segregation of *Xenopus* third and fourth arch neural crest cells. Mech Dev 1998;78(1—2):63—79.

[134] Axelsson M. The crocodilian heart; more controlled than we thought? Exp Physiol 2001;86(6):785—9.

[135] Espstein JA. Developing models of DiGeorge syndrome. Trends Genet 2001;17(10):13—7.

[136] Baldini A. The 22q11.2 deletion syndrome: a gene dosage perspective. Sci World J 2006;6:1881—7.

[137] Lindsay EA, Vitelli F, Su H, et al. Tbx1 haploinsufficieny in the DiGeorge syndrome region causes aortic arch defects in mice. Nature 2001;410(6824):97—101.

[138] Merscher S, Funke B, Epstein JA, et al. TBX1 is responsible for cardiovascular defects in velo-cardio-facial/DiGeorge syndrome. Cell 2001;104(4):619—29.

[139] Vitelli F, Morishima M, Taddei I, Lindsay EA, Baldini A. Tbx1 mutation causes multiple cardiovascular defects and disrupts neural crest and cranial nerve migratory pathways. Hum Mol Genet 2002;11 (8).915—22.

[140] Xu H, Morishima M, Wylie JN, et al. Tbx1 has a dual role in the morphogenesis of the cardiac outflow tract. Development 2004;131(13):3217—27.

[141] Zhang Z, Cerrato F, Xu H, et al. Tbx1 expression in pharyngeal epithelia is necessary for pharyngeal arch artery development. Development 2005;132 (23):5307—15.

[142] Vitelli F, Lania G, Huynh T, Baldini A. Partial rescue of the Tbx1 mutant heart phenotype by Fgf8: genetic evidence of impaired tissue response to Fgf8. J Mol Cell Cardiol 2010;49(5):836—40.

[143] Watanabe Y, Zaffran S, Kuroiwa A, et al. Fibroblast growth factor 10 gene regulation in the second heart field by Tbx1, Nkx2-5, and Islet1 reveals a genetic switch for down-regulation in the myocardium. Proc Natl Acad Sci USA 2012;109(45):18273—80.

[144] Pane LS, Zhang Z, Ferrentino R, Huynh T, Cutillo L, Baldini A. Tbx1 is a negative modulator of Mef2c. Hum Mol Genet 2012;21(11):2485—96.

[145] McDaniell R, Warthen DM, Sanchez-Lara PA, et al. NOTCH2 mutations cause Alagille syndrome, a heterogeneous disorder of the notch signaling pathway. Am J Hum Genet 2006;79(1):169—73.

[146] Oda T, Elkahloun AG, Pike BL, et al. Mutations in the human Jagged1 gene are responsible for Alagille syndrome. Nat Genet 1997;16(3):235—42.

[147] Li L, Krantz ID, Deng Y, et al. Alagille syndrome is caused by mutations in human Jagged1, which encodes a ligand for Notch1. Nat Genet 1997;16 (3):243—51.

[148] Jain R, Rentschler S, Epstein JA. Notch and cardiac outflow tract development. Ann N Y Acad Sci 2010;1188:184—90.

[149] Manderfield LJ, High FA, Engleka KA, et al. Notch activation of Jagged1 contributes to the assembly of the arterial wall. Circulation 2012;125(2):314—23.

[150] Stewart RA, Sanda T, Widlund HR, et al. Phosphatase-dependent and -independent functions of Shp2 in neural crest cells underlie LEOPARD syndrome pathogenesis. Dev Cell 2010;18(5):750—62.

[151] Araki T, Chan G, Newbigging S, Morikawa L, Bronson RT, Neel BG. Noonan syndrome cardiac

defects are caused by PTPN11 acting in endocardium to enhance endocardial-mesenchymal transformation. Proc Natl Acad Sci USA 2009;106(12):4736–41.

[152] Chen P-C, Wakimoto H, Conner D, et al. Activation of multiple signaling pathways causes developmental defects in mice with a Noonan syndrome-associated Sos1 mutation. J Clin Invest 2010;120 (12):4353–65.

[153] Xu W, Ahmad A, Dagenais S, Iyer RK, Innis JW. Chromosome 4q deletion syndrome: narrowing the cardiovascular critical region to 4q32.2-q34.3. Am J Med Genet A 2012;158A(3):635–40.

[154] Taub PJ, Wolfeld M, Cohen-Pfeffer J, Mehta L. Mandibular distraction in the setting of chromosome 4q deletion. J Plast Reconstr Aesthet Surg 2012;65(4): e95–8.

[155] Strehle EM, Yu L, Rosenfeld JA, et al. Genotype–phenotype analysis of 4q deletion syndrome: proposal of a critical region. Am J Med Genet A 2012;158A(9):2139–51.

[156] Markiewicz MR, Verschueren D, Assael LA. Chromosome 4q deletion syndrome: craniofacial characteristics associated with monosomy of the long arm of chromosome 4q. Cleft Palate Craniofac J 2010;47(5):518–22.

[157] Clark TG, Conway SJ, Scott IC, et al. The mammalian Tolloid-like 1 gene, Tll1, is necessary for normal septation and positioning of the heart. Development 1999;126(12):2631–42.

[158] Coggins KG, Latour A, Nguyen MS, Audoly L, Coffman TM, Koller BH. Metabolism of PGE2 by prostaglandin dehydrogenase is essential for remodeling the ductus arteriosus. Nat Med 2002;8 (2):91–2.

[159] Vincentz JW, Barnes RM, Firulli AB. Hand factors as regulators of cardiac morphogenesis and implications for congenital heart defects. Birth Defects Res Clin Mol Teratol 2011;91(6):485–94.

[160] Firulli BA, Krawchuk D, Centonze VE, et al. Altered Twist1 and Hand2 dimerization is associated with Saethre-Chotzen syndrome and limb abnormalities. Nat Genet 2005;37(4):373–81.

[161] Firulli ABA. HANDful of questions: the molecular biology of the HAND-subclass of basic helix-loop-helix transcription factors. Gene 2003;312C:27–40.

[162] Srivastava D, Thomas T, Lin Q, Kirby ML, Brown D, Olson EN. Regulation of cardiac mesodermal and neural crest development by the bHLH transcription factor, dHAND. Nat Genet 1997;16(2):154–60.

[163] Hendershot TJ, Liu H, Sarkar AA, et al. Expression of Hand2 is sufficient for neurogenesis and cell type-specific gene expression in the enteric nervous system. Dev Dyn 2007;236(1):93–105.

[164] Barnes RM, Firulli BA, VanDusen NJ, et al. Hand2 loss-of-function in Hand1-expressing cells reveals distinct roles in epicardial and coronary vessel development. Circ Res 2011;108:940–9.

[165] Takatoshi T, Jun M, Hiroyuki Y, Black BL, Olson EN, Srivastava D. Hand2 is required for second heart field development during cariogenesis weinstein cardiovasulat conference abstract 2008;Abstract 161.

[166] Keyte A, Hutson MR. The neural crest in cardiac congenital anomalies. Differentiation 2012;84 (1):25–40.

[167] Rosenquist TH. Folate, homocysteine and the cardiac neural crest. Dev Dyn 2013;242(3):201–18.

[168] Hibbard ED. The figlu-excretion test and defective folic-acid metabolism in pregnancy. Lancet 1964;2 (7370):1146–9.

[169] Hibbard E, Smithells RW. Folic acid metabolism and human embryopathy. Lancet 1965;285(7398):1254.

[170] Tang LS, Wlodarczyk BJ, Santillano DR, Miranda RC, Finnell RH. Developmental consequences of abnormal folate transport during murine heart morphogenesis. Birth Defects R A Clin Mol Teratol 2004;70 (7):449–58.

[171] Linask KK. The heart-placenta axis in the first month of pregnancy: induction and prevention of cardiovascular birth defects. J Pregnancy 2013;2013:320413.

[172] Daft PA, Johnston MC, Sulik KK. Abnormal heart and great vessel development following acute ethanol exposure in mice. Teratology 1986;33(1):93–104.

[173] Carmichael SL, Shaw GM, Yang W, Lammer EJ. Maternal periconceptional alcohol consumption and risk for conotruncal heart defects. Birth Defects Res A Clin Mol Teratol 2003;67(10):875–8.

[174] Hassler JA, Moran DJ. Effects of ethanol on the cytoskeleton of migrating and differentiating neural crest cells: possible role in teratogenesis. J Craniofac Genet Dev Biol Suppl 1986;2:129–36.

[175] Nyquist-Battie C, Freter M. Cardiac mitochondrial abnormalities in a mouse model of the fetal alcohol syndrome. Alcohol Clin Exp Res 1988;12(2):264–7.

[176] Satiroglu-Tufan NL, Tufan AC. Amelioration of ethanol-induced growth retardation by all-trans-retinoic acid and alpha-tocopherol in shell-less culture of the chick embryo. Rep Toxicol 2004;18 (3):407–12.

[177] Dunty Jr WC, Chen SY, Zucker RM, Dehart DB, Sulik KK. Selective vulnerability of embryonic cell populations to ethanol-induced apoptosis: implications for alcohol-related birth defects and neurodevelopmental disorder. Alcohol Clin Exp Res 2001;25 (10):1523–35.

[178] Flentke GR, Garic A, Amberger E, Hernandez M, Smith SM. Calcium-mediated repression of beta-

catenin and its transcriptional signaling mediates neural crest cell death in an avian model of fetal alcohol syndrome. Birth Defects Res A Clin Mol Teratol 2011;91(7):591−602.

[179] Wentzel P, Eriksson UJ. Altered gene expression in neural crest cells exposed to ethanol in vitro. Brain Res 2009;1305(Suppl):S50−60.

[180] Lammer EJ, Chen DT, Hoar RM, et al. Retinoic acid embryopathy. N Engl J Med 1985;313(14):837−41.

[181] Mulder GB, Manley N, Grant J, et al. Effects of excess vitamin A on development of cranial neural crest-derived structures: a neonatal and embryologic study. Teratology 2000;62(4):214−26.

[182] Coberly S, Lammer E, Alashari M. Retinoic acid embryopathy: case report and review of literature. Pediatr Pathol Lab Med 1996;16(5):823−36.

[183] Pan J, Baker KM. Retinoic acid and the heart. Vitam Horm 2007;75:257−83.

[184] Wilson JG, Roth CB, Warkany J. An analysis of the syndrome of malformations induced by maternal vitamin A deficiency. Effects of restoration of vitamin A at various times during gestation. Am J Anat 1953;92(2):189−217.

[185] Wilson JG, Warkany J. Cardiac and aortic arch anomalies in the offspring of vitamin A deficient rats correlated with similar human anomalies. Pediatrics 1950;5(4):708−25.

[186] Kawaguchi R, Yu J, Honda J, et al. A membrane receptor for retinol binding protein mediates cellular uptake of vitamin A. Science 2007;315(5813):820−5.

[187] Niederreither K, Vermot J, Messaddeq N, Schuhbaur B, Chambon P, Dolle P. Embryonic retinoic acid synthesis is essential for heart morphogenesis in the mouse. Development 2001;128(7):1019−31.

[188] Pasutto F, Sticht H, Hammersen G, et al. Mutations in STRA6 cause a broad spectrum of malformations including anophthalmia, congenital heart defects, diaphragmatic hernia, alveolar capillary dysplasia, lung hypoplasia, and mental retardation. Am J Hum Genet 2007;80(3):550−60.

[189] Ghyselinck NB, Wendling O, Messaddeq N, et al. Contribution of retinoic acid receptor beta isoforms to the formation of the conotruncal septum of the embryonic heart. Dev Biol 1998;198(2):303−18.

[190] Yelin R, Schyr RB, Kot H, et al. Ethanol exposure affects gene expression in the embryonic organizer and reduces retinoic acid levels. Dev Biol 2005;279 (1):193−204.

[191] Lee RY, Luo J, Evans RM, Giguere V, Sucov HM. Compartment-selective sensitivity of cardiovascular morphogenesis to combinations of retinoic acid receptor gene mutations. Circ Res 1997;80(6):757−64.

[192] Jiang X, Choudhary B, Merki E, Chien KR, Maxson RE, Sucov HM. Normal fate and altered function of the cardiac neural crest cell lineage in retinoic acid receptor mutant embryos. Mech Dev 2002;117 (1−2):115−22.

[193] Ryckebusch L, Wang Z, Bertrand N, et al. Retinoic acid deficiency alters second heart field formation. Proc Natl Acad Sci USA 2008;105(8):2913−8.

[194] Li P, Pashmforoush M, Sucov HM. Retinoic acid regulates differentiation of the secondary heart field and TGFbeta-mediated outflow tract septation. Dev Cell 2010;18(3):480−5.

[195] Sakabe M, Kokubo H, Nakajima Y, Saga Y. Ectopic retinoic acid signaling affects outflow tract cushion development through suppression of the myocardial Tbx2-Tgf2 pathway. Development 2012;139(2): 385−95.

[196] Ryckebusch L, Bertrand N, Mesbah K, et al. Decreased levels of embryonic retinoic acid synthesis accelerate recovery from arterial growth delay in a mouse model of DiGeorge syndrome. Circ Res 2010;106(4):686−94.

[197] Tamura Y, Matsumura K, Sano M, et al. Neural crest-derived stem cells migrate and differentiate into cardiomyocytes after myocardial infarction. Arterioscler Thromb Vasc Biol 2011;31(3):582−9.

[198] Lin AE, Belmont J, Malik S. Cardiorespiratory organs: heart. 2nd ed. New York NY: Oxford University Press; 2006.

[199] Hoffman JI, Kaplan S. The incidence of congenital heart disease. J Am Coll Cardiol 2002;39(12): 1890−900.

Neural Crest Cells in Enteric Nervous System Development and Disease

Amanda J. Barlow

Department of Surgery, University of Wisconsin

INTRODUCTION

The gastrointestinal (GI) tract is an endodermally derived structure that extends from the mouth to the anus. The GI tract is divided along its length into three segments: the foregut, midgut, and hindgut. The foregut is composed of the esophagus, stomach, and duodenum.

Neural Crest Cells.
DOI: http://dx.doi.org/10.1016/B978-0-12-401730-6.00013-2

The midgut encompasses the small intestine and length of the colon that extends up to the splenic flexure, while the remainder of the colon constitutes the hindgut. The GI tract functions to transport, absorb, digest, and excrete food and liquids taken in by the mouth. These functions are regulated by the nervous system that is present along the entire length of the GI tract, termed the enteric nervous system (ENS). The ENS monitors and responds to the state of the lumen by controlling luminal secretions and blood flow and regulating the muscular activity that controls peristalsis.

The ENS is derived from neural crest cells (NCC) that emigrate the neural tube and transit into and then along the length of the gut wall. The NCC that constitute the ENS have the most extensive migratory journey of all of the NCC populations within the embryo due to the vast length of the gut. The discovery of the NCC origin of the ENS was first reported by Yntema and Hammond [1]. They performed elegant ablations of the neural tube in avian embryos and noted the absence of enteric ganglia along the GI tract [1]. The precise contribution of NCC from different axial levels of the embryo was further elucidated using isotopic and isochronic neural tube transplants between quail and chick embryos [2–4]. In mice, NCC emigrate from the vagal region of the neural tube, adjacent to somites 1–7, at embryonic day (E) 8.5, from where they travel toward and then into the foregut (Figure 12.1A) [5]. Over the next 4–5 days, these enteric NCC (ENCC) migrate caudally to colonize the entire length of the gut wall forming the majority of the ENS (Figure 12.1A–D) [5]. A second population of NCC from the sacral region of the embryo, caudal to the 24th pair of somites, enters the hindgut later (E13.5–E14 in mice) and migrates in a rostral direction toward the small intestine, constituting around 13% of the ENS within the distal small intestine and colon [6–8]. Sacral NCC enter the hindgut upon arrival of the vagal NCC, but their migration into the bowel wall is independent of vagal NCC since they are present within the colon following vagal NCC ablation [7,9]. In humans, vagal NCC migration toward and then into the gut begins after the fourth gestational week and takes 3 weeks to complete [10–12]. To date, a contribution of sacral NCC to ENS development has not been determined in humans.

Time lapse imaging of fluorescently labeled ENCC has shown that the rostro-caudal colonization of the midgut occurs as a wave by vagal ENCC that are interconnected within chains [13–16]. The ENCC migration wavefront advances along the gut at speeds of 50–80 μm/h, although their trajectories are complex and unpredictable [13,15]. Once vagal ENCC reach the cecum, their migration is halted for several hours, and ENCC chains begin to breakdown, producing more isolated cells that are more exploratory within the local microenvironment [15]. It has been shown using a photoconvertible protein, humanized KikumeGR driven by endothelin receptor B regulatory sequences, that the vagal ENCC that have advanced along the midgut and through the cecum will only colonize the proximal portion of the hindgut [17]. The majority of the hindgut is eventually colonized by vagal ENCC that traverse the mesentery between the intestines, termed transmesenteric NCC (tmNCC) (Figure 12.1E) [17]. These tmNCC exit the midgut along its length between E10.5 and E11.5 as individual cells and cross the mesentery at speeds of 48 μm/h. These tmNCC accumulate along the mesenteric border of the hindgut before invading the mesenchyme between E11.5 and E12.5. Once they enter the hindgut, they coalesce with other vagal ENCC that have migrated through the cecum and then they advance along the hindgut (Figure 12.1E) [17]. In contrast, sacral ENCC travel more slowly along the hindgut (18 μm/h), often individually in close association with extrinsic nerve fibers [8,18].

These different NCC populations integrate to form the ENS, a complex portion of the

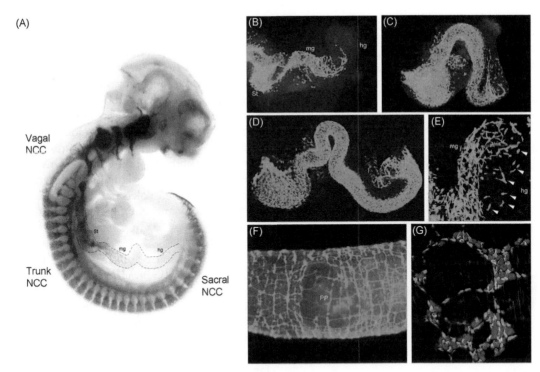

FIGURE 12.1 **NCC and ENS development.** A. *Sox10 in situ* hybridization showing NCC within the embryo at E10.25. Gut is outlined with black dashed line. B-D. Shows the progressive colonization of the gut by NCC labeled with yellow fluorescent protein, YFP from $R26R^{stopYFP}$;Wnt1^{Cre} transgenic mice at E10.25 (B), E11.5 (C) and E12.5 (D). (E). Shows tmNCC (white arrowheads) transiting from the midgut through the mesentery at E10.5. (F). Small intestinal myenteric plexus at P21 stretched across a Peyer's patch (PP). (G). Interconnected neurons labeled with Hu (red) and glia stained with GFAP (green) in small intestinal ganglia at P21. St, stomach; mg, midgut; hg, hindgut; PP, Peyer's patch.

peripheral and autonomic nervous systems. ENCC eventually differentiate into an extensive network of neurons and glial cells that organize into two distinct ganglionic networks (Figure 12.1). The outer myenteric (Auerbach's) plexus is located between the circular and longitudinal smooth muscle layers and extends along the length of the gut, while the inner submucosal (Meissner's) plexus is restricted to the small and large intestines. These plexuses coordinately control gut motility, fluid exchange across the mucosal surface, blood flow, and gut hormonal secretion [19].

Failure of ENCC to complete colonization of the entire length of the gut wall results in the most common neurocristopathy in humans, termed Hirschsprung disease (HSCR). This is a congenital disorder, affecting 1:5000 live births, with a male to female bias of 4:1. HSCR is characterized by the absence of enteric neurons (aganglionosis) along variable lengths of the distal bowel [20]. Patients diagnosed with the disease suffer with intestinal obstruction caused by tonic contraction of the muscle in the aganglionic regions of the colon [19]. Our understanding of the etiology and pathogenesis of HSCR has been greatly enhanced by the use of genetic mutations in rodents and neural tube ablation/grafting experiments performed in avian embryos (Table 12.1) [21–23].

TABLE 12.1   Human and Mouse Genes Associated with ENS Defects and HSCR

Gene	Human Etiology	Homozygous Phenotype in Mice
RET	Non-syndromic HSCR/MEN2A	Total intestinal AG
GDNF	Non-syndromic HSCR	Total intestinal AG
GFRα1	ND	Total intestinal AG
NTN	Non-syndromic HSCR	Defective number of submucosal neurons and excitatory nerve fiber density
EDNRB	Non-syndromic HSCR/Waardenburg Syndrome Type 4 (WS)	Distal colonic AG
EDN3	Non-syndromic HSCR/Waardenburg Syndrome Type 4 (WS)	Distal colonic AG
ECE-1	Non-syndromic HSCR	Distal colonic AG
SOX10	Syndromic and Non-syndromic HSCR/WS	Total intestinal AG
PHOX2B	Congenital central hypoventilation syndrome (CCSH) with HSCR	Total intestinal AG
ZEB2	Mowat-Wilson Syndrome (MSW) with HSCR	Colonic and partial small intestinal AG
PAX3	ND	Total intestinal AG
ASCL1	ND	Esophageal AG
HAND2	ND	Disorganized ganglia and loss of VIP and nNOS neurons in conditional KO
KIA1279 (KBP)	Goldberg–Shprintzen syndrome with HSCR	ND
L1CAM	X-linked hydrocephalus with HSCR	Delayed migration of ENCC along the intestine
TCF4	Pitt–Hopkins syndrome with HSCR	ND
NRG1	Non-syndromic HSCR	ND
β1-integrin	ND	Distal colonic AG

MEN2A, multiple endocrine neoplasia type IIA; AG, aganglionosis; ND, not determined.

Formation of a complete ENS requires the coordinated survival, migration, proliferation, and eventual differentiation of ENCC. These processes also need to be integrated with the extensive growth and morphological changes of the gut tube. Numerous molecules involved in regulating ENCC colonization of the gut have been identified and their expression is restricted either to ENCC or the gut microenvironment, or is present in both cell populations. This chapter focuses on the main signaling pathways and molecules and their role(s) in ENS development and the etiology and pathogenesis of HSCR.

GDNF, GFRα1, AND RET SIGNALING

Normal signaling through the receptor tyrosine kinase gene, RET, is essential for ENS development, as mutations in RET account for 7–35% of sporadic cases and approximately 50% of familial cases of HSCR (Table 12.1) [24,25]. In addition, non-coding mutations of RET affect the susceptibility to HSCR [26–28]. Ret signaling is propagated by dimerization through the formation of a complex between one of the four neurotrophic factor ligands, such as glial cell line-derived neurotrophic factor (GDNF) and its glysosylphosphatidylinositol (GPI)-anchored co-receptor, GFRα1. Expression of *Ret* and *Gfrα1* is activated in ENCC upon entry into the foregut, whereas *Gdnf* is detected in gut mesenchyme prior to ENCC arrival in the foregut [29]. In contrast to the colonic aganglionosis observed in HSCR patients with *RET* mutations, loss of *Gdnf*, *Gfrα1*, or *Ret* in mice causes a lack of enteric ganglia throughout the small and large intestines and most of the stomach [30].

RET exists in two main isoforms, RET9 and RET51, that are generated by alternative splicing. These isoforms are identical except for a carboxy terminal intracellular sequence that follows tyrosine 1062 (Y1062), a multidocking site for mitogen-activated protein kinase (MAPK)

and phosphatidylinositol 3-kinase (PI3K) [31]. Mice expressing single Ret isoforms exhibit either complete NCC colonization of the gut ($Ret^{9/9}$) or phenocopy the colonic aganglionosis observed in HSCR patients ($Ret^{51/51}$), thereby providing a comparable mouse model to study HSCR development [32]. However, a different strain of $Ret^{9/9}$ mice has been observed to exhibit an absence of ENS in the distal bowel [33]. The disparity between these results may reflect technical differences in either the methods employed to generate the transgenic mice or genetic background, but these results collectively support the important role played by Ret in ENCC colonization of the gut and formation of a functional ENS.

NEUTURIN, GFRα2, AND RET SIGNALING

Neuturin (NTN) is another member of the GDNF ligand family and activates RET signaling by binding to GFRα2. Ntn and Gfrα2 knockout mice exhibit an ENS along the entire length of the gut but demonstrate gut motility defects due to reduced density of nerve fibers projecting from cholinergic neurons (Table 12.1) [34,35]. Rare HSCR patients have been identified with NTN mutations, although a concomitant RET mutation was also present, which is indicative of the synergistic interactions of components of the same signaling pathway (Table 12.1) [36].

ENDOTHELIN 3 (EDN3)/ ENDOTHELIN RECEPTOR B (EDNRB) SIGNALING

A key role in ENS development has been attributed to the endothelin B signaling pathway, as mutations in the G-protein coupled receptor, EDNRB, its ligand endothelin 3 (EDN3), and endothelin-converting enzyme ECE-1 account for approximately 5% of the HSCR cases in humans (Table 12.1) [20]. Ednrb is expressed in both ENCC and some regions of the gut mesenchyme [37,38], while Edn3 expression is restricted to the gut mesenchyme, with the highest levels detected in the cecum [38,39]. EDNRB signaling is also required for development of melanocytes (another NCC derivative), and mutations in EDNRB and EDN3 are associated with Waardenburg syndrome (WS), which manifests as colonic aganglionosis, pigmentation abnormalities, and deafness (Table 12.1) [20,40]. Mice carrying mutations in Ednrb and Edn3 exhibit similar phenotypes to those in humans showing delayed gut colonization during embryogenesis resulting in variable lengths of colonic aganglionosis and pigmentation defects (Table 12.1) [37,38,41−43]. Therefore, endothelin receptor signaling regulates many aspects of ENS development.

Transcription Factors

ENS development is also regulated by a network of transcription factors, including the SRY-related high mobility group (HMG) box transcription factor, SOX10. SOX10 mutations constitute less than 5% of HSCR cases and are also associated with WS [20,40,44,45]. Sox10 expression is initiated in vagal NCC as they migrate from the neural tube toward the foregut and is responsible for the activation of Ednrb and Ret in ENCC once they enter the gut (Figure 12.1) [46−48]. In the absence of Sox10, mice die at birth and demonstrate a failure of NCC migration into the foregut, while Sox10 heterozygous mice exhibit aganglionosis and pigmentation defects characteristic of WS type 4 (Table 12.1) [43,44,49,50].

Mutations in the paired-like homeodomain transcription factor 2b (PHOX2B) have been identified in HSCR patients often in association with congenital central hypoventilation syndrome (CCHS) (Table 12.1) [51]. Phox2b is expressed prior to the entry of vagal NCC into the gut, and its mutation in mice causes a loss of

NCC beyond the foregut due to extensive apoptotic NCC death [52]. Similar to Sox10, Phox2b regulates the expression of *Ret*, the loss of which may underlie the pathogenesis of HSCR [53].

Although the zinc finger E-box homeodomain transcription factor 2, ZEB2 (ZFHX1B/SIP1) has not previously been associated with isolated HSCR, ZEB2 may in fact be a susceptibility gene for this disorder (Table 12.1) [25]. This is because nonsense mutations in *ZEB2* have been identified in patients with Mowat−Wilson syndrome (MWS), who display distinct craniofacial defects, mental retardation, and sometimes HSCR [54]. *Zeb2* has a similar pattern of expression as *Sox10* in vagal NCC as they migrate toward the foregut in humans and mice [50,55−57]. *Zeb2* is present in all enteric progenitor cells along the gut as well as some stomach mesenchymal cells. As development proceeds, its expression is downregulated in neuronal cells but is maintained within glia [50,57]. *Zeb2* mutant mice die at approximately E9.5 with a variety of cardiovascular and neural defects caused by the absence of NCC formation (Table 12.1) [58]. Conditional deletion of *Zeb2* in NCC results in craniofacial, cardiac, pigment, and peripheral nervous system deficiencies, including partial small intestinal and total colonic aganglionosis (Table 12.1) [57]. *Zeb2* plays an essential role in the epithelial-to-mesenchymal transition (EMT) required for delamination of NCC from the neural tube [58]. The mechanisms by which *Zeb2* controls ENS development beyond NCC EMT remain to be determined.

KIAA1279

KIAA1279 encodes a kinesin superfamily 1 (Kif1)-binding protein termed KBP. Homozygous mutations in KIAA1279 have been described in Goldberg−Shprintzen megacolon syndrome (GOSHS), which is characterized by HSCR, microcephaly, craniofacial defects, short stature, and learning difficulties (Table 12.1)

[26,59]. KBP has been shown to regulate axonal growth and maintenance through its interactions with SCG10 [60], actin filaments, and microtubules [61]. Therefore, the HSCR phenotype in KIAA1279 mutations could result from changes in cytoskeletal proteins, which then causes defective neuronal development [61].

L1CAM

L1CAM belongs to the immunoglobulin gene superfamily of neural cell adhesion molecules. It is expressed in most ENCC along the gut in avians and mice [62,63]. Mutations in L1CAM have been detected in HSCR patients together with X-linked hydrocephalus, although it is present in only 3% of cases (Table 12.1) [64,65]. Loss of *L1cam* in mice results in a delayed colonization of the gut by ENCC, but these mice exhibit a complete functional ENS at birth [62].

TCF4

TCF4 is a basic helix-loop-helix transcription factor expressed widely throughout the mammalian developing central nervous system (CNS) and mediates the Wnt-β-catenin pathway [66]. Heterozygous missense mutations in TCF4 have been identified in patients with Pitt−Hopkins syndrome (PHS), which is defined by mental retardation, wide mouth, hyperventilation, and in rare cases, HSCR (Table 12.1) [67].

Neuregulins

Mutations and deletions in neuregulin 1 (*NRG1*) and 3 (*NRG3*), the ligands for *ErbB* receptor tyrosine kinases have been identified in HSCR patients (Table 12.1) [68−70]. *NRG1* is expressed in the intestinal mucosa and enteric ganglia of mice and humans [71,72], while *ErbB2 and ErbB3* receptors are expressed in vagal NCC as they enter the foregut and later in adult intestinal epithelium [73,74]. It has been proposed that the HSCR phenotype resulting from *NRG1* mutations could be mediated by misregulation of *Sox10* [70].

SIGNALING PATHWAY INTERACTIONS IN ENS DEVELOPMENT AND HSCR

ENS formation requires the coordination and integration of numerous different signaling pathways. HSCR patients have been identified with multiple defects demonstrating that genetic interactions occur between RET and EDNRB, RET and PHOX2B, and RET and Bardet-Biedl syndrome proteins [20,66]. Genetic studies in mice have confirmed and expanded these data to show interactions between *Ret* and *Ednrb*, *Ret* and *Edn3*, *Sox10* and *Ednrb*, *Sox10* and *Edn3*, *Sox10* and *Zeb2*, *L1cam* and *Ednrb*, *L1cam* and *Edn3*, and *Notch* and *Hedgehog* [75,76]. Some of these interactions may be direct, as has been described for *Sox10* and *Ednrb* [48], which activate common downstream intracellular pathway components or synergistic/antagonistic regulation of similar developmental processes, such as proliferation or migration [38,77]. As genetic tools improve and expand, our understanding of the cellular and molecular mechanisms that control ENCC colonization of the gut will continue to provide insight into the etiology and pathogenesis of HSCR.

GENES INVOLVED IN ENS DEVELOPMENT, NOT CURRENTLY ASSOCIATED WITH HUMAN DISEASE

The absence of enteric neurons is easily determined, whereas subtle changes in number, constitution, and organization of ganglia that could affect gut motility are more difficult to discern. Detailed examination of conventional and conditional knockout mice has identified a variety of genes involved in ENS development not yet implicated in human GI disorders. During colonization of the gut, ENCC contact each other and the local microenvironment, the dynamics of which must be tightly controlled to allow their advancement along the gut. Intra- and inter-cellular adhesion between ENCC and their microenvironment, respectively, is regulated spatially and temporally along the gut wall by extracellular matrix molecules (ECM). This was demonstrated by conditional deletion of β1-integrin, the major adhesive receptor for ECM in NCC, which led to increased aggregation of ENCC as they migrated through the cecum and along the colon, resulting in disorganized ganglia and incomplete colonization of the gut (Table 12.1) [78,79].

Disorganized ganglia have been reported in mice containing neural-specific deletion of the basic helix-loop-helix transcription factor *Hand2* (Table 12.1) [80]. *Hand2* is expressed within ENCC as they enter the foregut and is required for neuronal differentiation, neurotransmitter expression and patterning of enteric ganglia [80–82]. *Hand2* conditional knockout mice are completely colonized by ENCC, but the disorganization of enteric ganglia results in intestinal obstruction and subsequent death similar to that described in HSCR (Table 12.1) [80]. These defects could be caused by altered cell adhesion, possibly from changes in expression of the neural cell adhesion molecule, *N-CAM*, a direct target of Hand2 [83].

Modulation of bone morphogenetic protein (BMP) signaling in avians and rodents inhibits ENCC migration along the intestines and results in the failure or reduction of enteric ganglion formation [84,85]. *BMP* expression is restricted to the gut mesenchyme, but it regulates ENCC movement and aggregation by modulating polysialylation of the neural cell adhesion molecule, N-CAM [85,86].

REGULATION OF THE NCC POPULATION THAT MIGRATES TOWARD AND THEN INTO THE FOREGUT

The formation of an ENS along the entire gut tube during embryonic development requires the completion of two key steps: first, the arrival and then entry of sufficient numbers of NCC into the foregut, and second, the survival, migration, proliferation, and then differentiation of ENCC within the gut wall. The production and then maintenance of a sufficient migratory population of NCC that travels toward and then into the foregut is the first key process that ensures complete formation of an ENS along the entire gut. Reducing this migratory NCC population using both neural tube ablation experiments in avian embryos and genetic manipulation in mice has clearly demonstrated this importance [23]. Incomplete colonization of the gut by ENCC was observed when differing lengths of the vagal neural tube were ablated in chick embryos [1,9,87]. Further examination showed, despite a normal contribution of NCC from 7 somite lengths of vagal neural tube, the entire length of the gut could in fact be populated by NCC derived from a single somite length [88]. This revealed that a critical NCC number is required to form a complete ENS and that individual populations of vagal NCC from varying axial positions have differential proliferative capacities that affect their ability to colonize the gut [88].

The size of the vagal NCC population that migrates toward and then into the foregut can also be altered by regulating cell death. This has been demonstrated by inhibition of caspase-dependent cell death in vagal NCC as they migrate toward the foregut in avian embryos. Increased numbers of ENCC were detected in the foreguts of these embryos, demonstrating a role for cell survival in the control of the early migratory vagal NCC population [89].

In mice, expression of Sox10 is essential in regulating the size of the NCC progenitor cell population that migrates toward and into the foregut. Of the genes involved in ENS development and HSCR, Sox10 is one of the earliest to be expressed in NCC shortly after they delaminate from the neural tube [90] (Figure 12.1A). Loss of Sox10 caused either by spontaneous (dominant megacolon, $Sox10^{Dom}$) or targeted mutation ($Sox10^{LacZ}$) in mice results in intestinal aganglionosis (Table 12.1) [91–94]. The primary role of Sox10 in NCC is to sustain their survival during the migration to the foregut, since apoptotic cell death eliminates or reduces this population when Sox10 expression is lost or decreased, respectively [43,46,95]. Sox10 is also required to maintain the multipotency of progenitor cells [96–98] and controls the spatial and temporal differentiation of ENS progenitors through interaction/regulation of other key HSCR-related genes, Ednrb, Ret, and Phox2b [47,48,53,97,99]. The closely related protein Sox8 also co-operates with Sox10 to maintain survival of vagal NCC as shown by increased apoptosis of these cells and greater severity and penetrance of intestinal aganglionosis observed in mice containing both Sox8 and Sox10 mutations [95].

Apoptotic cell death underlies the reduction in vagal NCC observed in mice with mutations in the transcription factor Pax3 [100]. Expression of Pax3 is detected in the dorsal neural tube and early migrating NCC, but it is downregulated prior to their arrival at target tissues [101,102]. Pax3 mutant mice have complete intestinal aganglionosis caused by p53-dependent cell death in both the neural tube and migrating vagal NCC (Table 12.1) [47,100,103]. Similar to Sox10, Pax3 also plays a role in the maintenance and proliferation of NCC [104], and deficiencies in these processes contribute to the smaller population of vagal NCC in Pax3 mutants [100]. The ENS defects that occur in the absence of Pax3 could be enhanced by the co-operation with Sox10 to regulate the expression of RET, the main HSCR-associated gene [47,53,99].

The size of the vagal NCC population was shown to be limited in Tcof1 haploinsufficient mice [100,105]. Tcof1 is a nucleolar phosphoprotein that is widely expressed throughout the embryo, and its loss results in p53-dependent cell death in the neural tube [106]. As a consequence, there is a reduction in the delamination of vagal NCC that causes a delay in their migration into and then along the length of the gut. Haploinsufficiency of *Tcof1* alone is unable to cause obvious ENS defects at birth, but it can sensitize *Pax3* heterozygous mice to colonic aganglionosis [100,105].

Conditional deletion of the winged helix forkhead transcription factor, *Foxd3*, in NCC results in complete absence of ENS development [107]. Like *Pax3*, expression of *Foxd3* is present in the neural tube and in migrating NCC [108,109]. Extensive apoptotic cell death eliminates the NCC prior to their arrival at the foregut [107].

The inability of NCC to delaminate from the neural tube underlies the lack of ENS development seen in *Zeb2* mutants [58]. *Zeb2* expression in the neural tube is necessary for the downregulation of E-cadherin that is essential for the EMT that controls NCC delamination from the neural tube. *Zeb2* mutants die early during embryogenesis, which precluded the ability to study the role of Zeb2 in NCC colonization of the gut (Table 12.1) [58]. Whereas, conditional NCC deletion of *Zeb2* results in craniofacial, cardiac, pigment, and peripheral nervous system deficiencies, including partial small intestinal and total colonic aganglionosis [57]. The mechanisms by which *Zeb2* controls ENS development have not currently been determined.

REGULATION OF ENCC SURVIVAL, MIGRATION, PROLIFERATION, AND DIFFERENTIATION WITHIN THE GUT WALL

Following the arrival and entry of sufficient numbers of NCC into the foregut, these cells then proceed in their colonization of the length of the gut. In order to ensure that ENCC migrate to the end of the colon, they need to survive, migrate, proliferate, and eventually differentiate within the gut microenvironment. All of these processes need to be coordinated within the ENCC themselves and the surrounding gut microenvironment.

ENCC Survival Within the Gut Wall

ENCC survival is the initial key step in ensuring that the gut wall continues to be colonized once a sufficient number of NCC has entered the foregut. The expression of *Gdnf*, *Gfrα1*, *Ret*, *Sox10*, *Phox2b*, and *Pax3* all are essential for this process, as the loss of any one of these factors results in either complete aganglionosis of the gut or small and large intestines (Table 12.1; reviewed in [21]). A role in survival of ENCC has also been attributed to the basic helix-loop-helix transcription factor *Ascl1* (*Mash1*), since the absence of *Ascl1* results in the lack of neurons in the esophagus [110].

These studies demonstrate that cell survival plays a key role early in the process of ENS development. However, the absence of observable apoptosis during later migration within normal ENS development [34] coupled with normal enteric neuron numbers in mice mutant for the pro-apoptotic genes *Bax* and *Bid* [34], suggested cell survival was not a key factor during late ENS development. It was not until *Gfrα1* was conditionally deleted during late embryonic stages that extensive widespread cell death was detected in colonic neurons, leading to aganglionosis typical of HSCR [111]. In agreement with previous studies, no signs of apoptosis were apparent and cell death was both caspase and *Bax* independent [111]. Reduction of *Ret* expression or conditional ablation of *Ret* at late developmental stages after completion of migration within the colon also caused colonic aganglionosis due to widespread cell death [33]. These data define a role for cell death in the etiology of HSCR. However, the mechanism that regulates ENCC survival within the intestines during normal ENS formation remains to be determined.

ENCC Migration Along the Gut Wall

If ENCC survival is achieved and maintained, their migration along the gut wall is then mediated by NC cell–cell and cell–environment interactions that are regulated by many different factors. Gdnf is chemoattractive for ENCC *in vitro* [29,112]. *In vivo* in E9 embryos, *Gdnf* is expressed in the mesenchyme of the pharyngeal pouches of the branchial arches along the migratory route of NCC to the foregut, which may facilitate their entry into the foregut [29]. At this stage, high levels of *Gdnf* are also present within the mesenchyme of the stomach and cecum preceding entry of NCC into the foregut. The rostral–caudal directional migration of ENCC along the gut wall is thought to be controlled by the spatial and temporal change in *Gdnf* expression that occurs during development [29,113]. *Gdnf* is initially expressed at high levels in the stomach and lower levels along the gut, whereas later in development expression is increased in the cecum relative to more proximal regions of the gut [29,113]. The migration of tmENCC across the mesentery between the small intestine and colon between E10.5 and E11.5 is also dependent upon the activity of Gdnf-Ret signaling since this process fails to occur when this pathway is altered [17]. The high level of *Gdnf* present in the cecum facilitates the migration of ENCC into this region, but it has been proposed that the activity of Gdnf needs to be regulated in order for ENCC to advance through the cecum into the proximal colon. Endothelin 3 has been suggested as a molecule that modulates the activity of Gdnf in the cecum, since it has a similar pattern of expression in the cecum and is able to inhibit the chemoattractive properties of Gdnf in avian and mouse gut tissue *in vitro* [38,77,113,114]. The directionality of ENCC migration may also be regulated by endothelin receptor B signaling since antagonism of Ednrb with a specific chemical inhibitor or NCC-specific knockout of *Ednrb* (*Ednrb^flex/flex*) results in altered trajectories and reduced speed of movement of ENCC within the colon [16]. Loss of Ednrb activity appears to alter NC cell–cell contacts, which causes a disruption of the ENCC chains that have been demonstrated to regulate the speed of movement along the gut [15,16].

The maintenance of ENCC chains during their migration along the gut has been demonstrated to be dependent upon the expression of the cell adhesion molecule *L1cam*. ENCC express *L1cam* as they colonize the gut and its absence causes a migration delay during early development, although the entire length of the gut is eventually colonized [62]. Perturbation of L1cam *in vitro* led to partial breakdown of ENCC chains at the migration wavefront, producing more isolated ENCC that migrated more slowly along the gut causing colonic aganglionosis [62].

ENCC migration through the cecum is also dependent upon the activity of β1-integrin, the major adhesive receptor for ECM [78,79]. Cell adhesion between ENCC themselves and with their local microenvironment is tightly regulated by ECM. The cecum contains high levels of the ECM, tenascin C, and fibronectin that are able to inhibit or promote ENCC migration *in vitro*, respectively [79]. β1-integrin expression in the cecum has been proposed to control cell adhesion and consequently ENCC migration by modulating the activity of these ECM, since conditional deletion of β1-integrin in NCC results in the incomplete colonization of the length of the gut as a consequence of increased aggregation of ENCC (Table 12.1) [79]. The balance between enteric NC cell–cell and cell–ECM interactions is also mediated by the cell adhesion molecule N-cadherin. Mice containing double knockout of β1-integrin and N-cadherin show more severe aganglionosis of the colon, with altered speed and directionality of movement of ENCC along the gut [115]. Interestingly, these mice do not display the aggregated ganglia observed in the β1-integrin mutants. Therefore, NC cell—cell and cell—ECM interactions are coordinately regulated by the activity of these two molecules [115], confirming previously predicted mathematical modeling results showing that ganglion formation is regulated by differential adhesion of neurons and non-neuronal ENCC [116].

The capacity of ENCC to advance along the gut appears to be controlled by both intrinsic and extrinsic factors since vagal ENCC travel in a rostral to caudal direction along the length of the gut; however, they are equally capable of migrating from the caudal to rostral end of the small intestine *in vitro* in the absence of endogenous ENCC [117]. The extrinsic regulation of ENCC invasion into the colonic microenvironment is considered to be controlled temporally, such that the microenvironment from E13.5 to E14.5 becomes less permissive to ENCC [16,118]. Indeed, colonic aganglionosis observed in *Edn3* and *Ednrb* mutant mice is thought to be caused by increased laminin expression within the mesenchyme that restricts the ability of ENCC to enter and then migrate along the colon [16,118—121]. However, the permissiveness of the gut mesenchyme to ENCC invasion extends beyond E14.5 in mice haploinsufficient for *Tcof1* [105]. These mice show a delay in ENCC migration along the gut until E14.5, yet the entire length of the gut is colonized by E18.5 [105]. The ability of ENCC to advance along the length of the colon may be dependent upon both the number and developmental time at which ENCC enter the colonic microenvironment. This has been demonstrated in mice with combined loss of *Pax3* and *Tcof1* that display colonic aganglionosis at E18.5. *Tcof1* heterozygous and *Pax3;Tcof1* double heterozygous animals should contain similar colonic microenvironments that are equally capable of being colonized by ENCC, and therefore the severe reduction in ENCC numbers must delay their time of entry into the colon which then impedes their migration along the colon, causing aganglionosis [100].

Movement of ENCC within the local gut microenvironment requires dynamic remodeling of their cytoskeleton. Rho GTPases may regulate the actin cytoskeletal changes required for ENCC to respond chemotactically to Gdnf. Indeed, inhibition of the activity of Rac1/Cdc42 and Rho signaling delayed the migration of ENCC along the colon in explant cultures *in vitro*, presumably by reducing the ENCC protrusions and cell—cell contacts at the migration wavefront [122]. Activation of Rac1-GEF may be regulated by Ret, since mice heterozygous for *Ret* were more sensitive to the inhibitors [122]. This could be achieved by the phosphorylation of Ret at serine 696, which has been shown to be essential for Rac1-GEF activation *in vitro* [123]. In support of this, abnormal ENCC colonization of the gut was observed in mice with mutation of Ret at serine 696 [124]. In contrast to a role in NCC migration, Fuchs and colleagues have suggested that *Cdc42* and *Rac1* maintain the proliferation of ENCC. Conditional deletion of *Cdc42* and *Rac1* in mice does not affect the ability of NCC to enter and then migrate along the gut wall initially but reduces the mitotic capacity of these cells, thereby causing aganglionosis of the colon [125].

ENCC Proliferation Within the Gut Wall

The ENCC migration wavefront within the gut has received a lot of attention from researchers, presumably because there are fewer ENCC at this position making it easier to study their molecular and cellular processes within the context of the gut as a whole. ENCC proliferation at the migration wavefront has been proposed to drive the advancement of ENCC along the gut. Mathematical modeling and grafting experiments using avian gut tissue have shown that cells within the first 300 μm of the migration wavefront serve as a proliferative zone that drives the movement of ENCC into previously uncolonized regions of the gut. ENCC in more proximal portions of the gut are essentially non-proliferative, and therefore the gut is colonized by a method of frontal expansion [126]. However, analysis of ENCC proliferation within different regions of the gut both at the migration wavefront and in more proximal regions at E11.5–12.5 in mice has not identified any differential rates of proliferation within ENCC [105,127,128]. It is only later at E13.5 that increased proliferation at the ENCC migration wavefront is detected compared to more proximal regions of the mouse gut [105]. This may reflect ENCC differentiation that is occurring behind the migration wavefront at this stage. The apparent differences between the data from avian and mouse experiments may also be attributed to the fact that the avian guts were cultured *in vitro*, where very little growth of the gut tissue occurs. Since gut tissue is growing extensively *in vivo* during its colonization by ENCC, high rates of ENCC proliferation are required in all regions of the gut during the early stages of development to produce sufficient enteric neurons along the length of the gut tube.

ENCC proliferation has been shown to be regulated by mesenchymal expression of *Gdnf*, such that the availability of GDNF determines the number of enteric neurons generated [34,129]. Experiments have demonstrated that when more GDNF is available, the density of enteric neurons is increased, whereas reducing GDNF availability results in fewer ENCC due to a lower rate of proliferation [34,129]. *Gdnf* haploinsufficiency has been shown to have a greater effect on reducing ENCC numbers at the migration wavefront compared to more proximal regions of the gut at E11.5 and E12.5, causing a migration delay in ENCC along the gut at these stages [130].

In order for ENCC to continue their migration along the gut, a sufficient pool of proliferative ENS progenitor cells needs to be maintained within the ENCC population. Hence, colonic aganglionosis in endothelin signaling mutants has been attributed in part to a

role for Edn3-EdnrB in the maintenance of undifferentiated and uncommitted ENS progenitor cells. Reduced numbers of proliferative progenitors and premature neuronal differentiation is observed at the migration wavefront in the absence of *Edn3* [38,98]. *In vitro*, Edn3 is able to reversibly inhibit the differentiation of ENCC into neurons or glial cells [98]. Interestingly, the capability of Edn3 to control the proliferative uncommitted state of cells by maintaining the expression of Sox10 was exploited to facilitate the isolation and propagation of enteric neural crest progenitor cells from mouse embryonic stem cells (ESCs) [131]. These data demonstrate that Edn3 maintains multipotency of a range of progenitor cells that extend beyond ENCC.

Continued colonization of the gut at late developmental stages was also shown to be regulated by increased ENCC proliferation at the migration wavefront in *Tcof1* haploinsufficient mice [105]. Despite a reduction in the initial number of NCC that migrate toward and then into the foregut, the colon eventually becomes colonized by ENCC as the levels of proliferation and differentiation are balanced at the migration wavefront, thereby maintaining a sufficient number of progenitor cells within the ENCC population capable of continuing their migration along the gut wall [105]. These data suggest that the density of ENCC within the gut wall may be intrinsically "sensed" and then controlled by the modulation of ENCC proliferation and differentiation. This "sensing" mechanism appears to be dependent upon a critical number of ENCC, since the gut is capable of being completely colonized when the initial migratory NCC population is reduced to around 60% of the normal numbers, but more severe reductions cannot be compensated.

This intrinsic regulation of ENCC density in the gut has also been demonstrated in mice where *Foxd3* was specifically deleted in ENCC once they had migrated into the small intestine with an *Ednrb*-specific element driving Cre expression, *Ednrb-iCre*. The loss of enteric progenitor cells expressing *Ednrb-iCre* in these mice was compensated by increased proliferation and glial differentiation of the *Ednrb-iCre* independent population enabling complete colonization of the gut [132]. In addition, enteric neuronal density was maintained in mice with loss of *Hand2* expression in nestin-expressing neural precursors by compensatory differentiation of non-nestin-expressing neural precursor cells [80]. While it is clear that the ENCC density is being tightly controlled, the exact molecular basis of this mechanism has not yet been elucidated.

ENCC Differentiation Within the Gut Wall

Neuronal differentiation of ENCC begins early during mouse embryonic development while ENCC are migrating along the small intestine and continues into postnatal life [133,134]. Several factors expressed within the gut mesenchyme have been shown to induce differentiation of ENCC into neurons, including Gdnf, neurotrophin 3, and BMPs, while Edn3 is inhibitory to neuronal differentiation (reviewed in [21]). The process of neuronal differentiation is also regulated by a cascade of expression of several transcription factors. Initially, all vagal NCC express *Sox10* as they travel toward and then into the gut, whereupon they begin expression of *Phox2b* and *Ascl1* [135]. Since Sox10 maintains the multipotency of ENS progenitor cells, its expression needs to be downregulated in neuronal precursors, which is achieved by the action of Phox2b and Ascl1 protein products [97]. If expression of *Sox10* is continued or enhanced, then neuronal differentiation is impaired [98]. A reduction in *Sox10* expression, as observed in the absence of Notch signaling, causes premature neuronal differentiation of ENCC [136]. *Hand2* expression is also necessary for neuronal differentiation, since the reduction or conditional

deletion of *Hand2* impairs enteric neuron differentiation [80–82,137].

The mature ENS is composed of ganglia that are comprised of phenotypically distinct neuronal subtypes, including interneurons, sensory neurons, and inhibitory and excitatory motor neurons. Each subtype of neurons displays a distinct cell morphology, axonal projection, electrophysiology, and neurotransmitter expression [19]. Unlike the CNS, the expression of pan neuronal markers is not restricted to postmitotic neurons, and ENS neurons are still capable of proliferating while they are undergoing differentiation [128,133]. Neuronal precursors eventually exit the cell cycle and begin expression of distinct neurochemical markers during ENS development [138–140]. The expression of calbindin, neuronal nitric oxide synthase (nNOS), and neuropeptide Y can be detected as early as E11.5 while ENCC are still migrating along the small intestine [139]. The precise molecular regulation of enteric neuron subtype development has not been defined, and the loss of specific neuronal subtypes has not been attributed to a single genetic mutation. However, the production of sufficient numbers of nNOS neurons requires a distinct level of Ret signaling [129,141,142]. *Hand2* is also required for the development of nNOS, vasoactive intestinal peptide (VIP), and calretinin-containing neurons [80,137]. Whereas the development of serotonergic neurons is dependent upon *Ascl1*, calcitonin gene-related peptide (CGRP) neurons require the tyrosine kinase TrkC and its ligand neurotrophin 3 for their development [110,143]. The ENS contains a relatively even distribution of neurons of particular subtypes within ganglia along its length, although the establishment of this organization and patterning remains to be determined.

Enteric glial cells are integrated with neurons within ganglia and function within the gut to regulate motility, mucosal secretion, and host defense (Figure 12.1) [144]. Early expression of the glial marker, brain-specific fatty acid binding protein (B-FABP), has been detected in ENCC within the foregut and midgut at E11.5. However, differentiation of glial cells occurs much later than neurogenesis with the activation of S100b expression at E14.5 [145]. Glial cell differentiation requires Sox10, Notch, glial growth factor 2, BMPs, and ErbB3 but can be inhibited by the transcriptional co-factor HIPK2 in postnatal life [21]. Genetic fate mapping experiments revealed that neurons could be generated from glial cells at postnatal stages *in vitro* and *in vivo*, but this only occurred in response to tissue damage or injury and not during normal ENS development [134,146].

HSCR and Cell Transplantation Therapies

It is clear from all of the research that coordination of many different molecular and cellular processes are needed for the formation of a functioning ENS along the gut. The absence of enteric ganglia from variable lengths of the colon that underlies HSCR is most often confined to the rectosigmoid colon and is termed "short-segment HSCR," which commonly presents as a failure to pass meconium within the first 2 days in newborns. Rare cases of "long-segment HSCR" also occur that affect considerable lengths of the colon and can even extend into the small intestine [20]. Aganglionosis in HSCR patients causes tonic contraction of the muscle resulting in intestinal obstruction and dysmotility. HSCR is typically treated by surgical resection of the aganglionic bowel and performing a "pull-through" of ganglionated bowel [147]. Unfortunately, surgical approaches to treatment are often inadequate, as many patients continue to suffer long-term complications, including fecal incontinence, constipation, and enterocolitis [148]. Alternative therapeutic treatments in the form of neural stem cell replenishments are being considered in an effort to improve ENS function and possibly

overcome the complications associated with surgical removal of the aganglionic bowel. A number of researchers across the world are investigating the potential of ENS stem or progenitor cell differentiation into enteric ganglia that could be used to repopulate the affected bowel in HSCR. A variety of different tissues have been investigated as potential sources of these stem/progenitor cells, including the CNS, ESCs, induced pluripotent stem cells (iPS), and the gut [149,150].

Neural stem cells derived from the CNS of embryonic mice were utilized in transplantation experiments into the pylorus of a model of gastroparesis, a condition that reduces the ability of contents to be emptied from the stomach. New functional NOS neurons were generated within 1 week of the transplantation, and this was associated with improved gastric function; unfortunately, the survival period in the host was very limited [151]. Transplantation of CNS-derived neural stem cells *in vivo* in a rat model of HSCR where enteric neurons had been chemically destroyed successfully enabled the production and long-term survival of new NOS and choline acetyltransferase (ChAT) neurons. More importantly, these rats showed functional restoration of the rectoanal inhibitory reflex [152]. Improvement of colonic motility has also been demonstrated following *in vivo* transplantation of neuroepithelial stem cells into a similar rat HSCR model. These cells were shown to give rise to NOS and ChAT neurons within the colon [153].

Human and mouse ESC are capable of migrating and generating neurons and/or glial cells *in vitro* in cultured embryonic mouse guts [131,154,155]. However, transplantation of mouse ESC into the pylorus of mice *in vivo* did not result in neuronal or glial differentiation 1 week post-transplantation [155]. The use of ESC in transplantation experiments needs to be carefully evaluated from ethical and biological perspectives since the pluripotency of these cells means they have the potential to form teratomas [156]. The efficacy of any cell transplantation experiments also relies on their ability to overcome immunological rejection [157]. Therefore, stem cells derived directly from patients may be the safest and most effective source for transplantation therapies. iPS offer real potential as a source of stem cells for ENS transplantation since the issue of transplantation rejection should be overcome and there is access to unlimited quantities of these cells. These cells can be generated from mature cells by the addition of four transcription factors, Oct4, Sox2, Klf4, and c-Myc, that revert these mature cells into a more primitive pluripotent state. Human and mouse iPS cells have been demonstrated to differentiate into intestinal tissue *in vitro* in the form of gut organoids [158,159]. However, their ability to generate neurons and glia upon transplantation into the gut microenvironment remains to be tested.

The ability to repopulate aganglionic gut and functionally restore ENS activity will probably be best achieved from stem cells obtained directly from gut tissue. This possibility started to become a reality with the isolation and propagation of multipotent progenitor cells from both embryonic and postnatal mouse gut tissue. These cells could even be collected from the ganglionic portion of a HSCR mouse model, really offering great potential for the beginnings of developing therapeutic treatments for HSCR patients [160−164]. These progenitor cells were capable of surviving, migrating, proliferating, and also differentiating when transplanted into aganglionic gut, all of the processes required for ENS development [160−164]. These studies were taken further when ENS progenitor cells were obtained from human postnatal bowel, including from HSCR patients [164−168]. These human cells isolated from HSCR patients restored contractility when grafted into embryonic aganglionic gut [164]. While these results are very encouraging, there are still many challenges ahead to achieve stem cell replacement therapies for patients with HSCR and other gut motility disorders. These include selecting the safest source of the progenitor cell population to be transplanted, determining the most efficient and effective route of delivery of these cells into the gut, investigating whether these progenitor cells need to be manipulated *in vitro* prior to their delivery into host gut tissue, and ensuring the survival of these cells in postnatal aganglionic gut tissue. In addition, the gut microenvironment into which the progenitor cells are injected will also affect the success of any stem cell transplantation strategies as has been demonstrated by the reduced capability of ENCC to migrate within *Ret-* and *Ednrb*-deficient gut tissue [16,169].

HSCR: Beyond Colonic Aganglionosis

Since dysmotility and enterocolitis can still often occur in HSCR patients after surgical removal of the aganglionic portion of the bowel [148], GI problems in these patients may extend beyond the aganglionic region. It is possible that there is altered neuronal circuitry within the "normal" ganglionated bowel of HSCR patients that compromises gut motility. In agreement with this idea, a 50% reduction in myenteric neurons in the bowel proximal to the aganglionic portion and lack of colonic migrating motor complexes was identified in mice mutant for *Edn3* [170]. These reductions in neuronal density were associated with increases in the proportion of nNOS neurons [170,171]. Large reductions in myenteric neuronal density were also found in *Ednrb* mutant and heterozygous animals [172], while the balance of inhibitory and excitatory neurotransmitters was altered in the ganglionated portion of mice containing NCC-conditional deletion of *Ednrb* [173]. These authors revealed an inverse relationship between neuronal density and expression of nNOS and VIP in the colons of these mice. Taken together, these changes, if also present in HSCR patients, could contribute to the postoperative complications that they experience.

CONCLUSIONS AND PERSPECTIVES

Our understanding of ENS development and function has been achieved by extensive basic research using animal models and clinical investigations of human HSCR patients. As a result, the vast majority of cellular and molecular processes that control gut colonization by ENCC have been determined. The challenges that remain as we move forward include analysis of how the composition of individual ganglia are determined and how these integrate to propagate signals along the length of the gut. In addition, the fact that up to 40% of treated HSCR patients experience recurrent enterocolitis suggests that defects in HSCR extend beyond the aganglionic bowel and supports a link between the ENS and innate host immunity. A genetic basis for this relationship was demonstrated in mice where the key HSCR-associated gene, Ret, was shown to be involved in the development of Peyer's patches (PP), the primary inductive sites for host immune defense in the GI tract (Figure 12.1). PP primordia were absent in Ret knockout mice at late developmental stages and reduced in number in $Ret^{51/51}$ animals [174]. Analysis of NCC-conditional Ednrb ($Ednrb^{flex/flex}$) knockout mice showed, not a loss or reduction in PP number, but rather a decrease in PP size compared to heterozygous littermates. These defects appear to be related to altered development of the spleen, the primary source of mature B-lymphocytes for PP. Indeed, the spleen in Ednrb-deficient mice and rats is reduced in size and contains an abnormal architecture compared to heterozygotes or wild types [175–177]. Examinations of HSCR patients have identified many different defects in the immune system, including increased T-lymphocyte and natural killer cells in the bowel wall of HSCR patients that develop enterocolitis, reductions in the number of IgA and IgM containing plasma cells in the lamina propria of the resected bowel in enterocolitis patients, and reduced luminal levels of IgA and IgM [148,178].

Together, these data demonstrate that the ENS plays a vital role in controlling mucosal immunity and that the defects in HSCR can extend beyond the aganglionic portion of the colon even into the ganglionated small intestine. Integration of the ENS and immune systems is probably established during embryonic development, therefore knowledge of the mechanisms that inter-connect these two systems is essential when considering therapeutic treatments for HSCR patients.

Acknowledgments

I would like to thank Professor Vassilis Pachnis for permission to include Figure 12.1A–F generated at the NIMR, UK; Chris Erickson, UW, Madison for kindly providing Figure 12.1F, -G; and Professor Miles Epstein for insightful comments on the chapter.

References

[1] Yntema CL, Hammond WS. The origin of intrinsic ganglia of trunk viscera from vagal neural crest in the chick embryo. J Comp Neurol 1954;101:515–41.

[2] Le Douarin NM, Teillet MA. The migration of neural crest cells to the wall of the digestive tract in avian embryo. J Embryol Exp Morphol 1973;30(1):31–48.

[3] Burns AJ, Douarin NM. The sacral neural crest contributes neurons and glia to the post-umbilical gut: spatiotemporal analysis of the development of the enteric nervous system. Development 1998;125(21):4335–47.

[4] Burns AJ, Le Douarin NM. Enteric nervous system development: analysis of the selective developmental potentialities of vagal and sacral neural crest cells using quail-chick chimeras. The Anat Rec 2001;262(1):16–28.

[5] Kapur RP, Yost C, Palmiter RD. A transgenic model for studying development of the enteric nervous system in normal and aganglionic mice. Development 1992;116(1):167–75.

[6] Serbedzija GN, Burgan S, Fraser SE, Bronner-Fraser M. Vital dye labelling demonstrates a sacral neural crest contribution to the enteric nervous system of chick and mouse embryos. Development 1991;111(4):857–66.

[7] Kapur RP. Colonization of the murine hindgut by sacral crest-derived neural precursors: experimental support for an evolutionarily conserved model. Dev Biol 2000;227(1):146–55.

[8] Wang X, Chan AK, Sham MH, Burns AJ, Chan WY. Analysis of the sacral neural crest cell contribution to the hindgut enteric nervous system in the mouse embryo. Gastroenterology 2011;141(3):992–1002, e1001–1006.

[9] Burns AJ, Champeval D, Le Douarin NM. Sacral neural crest cells colonise aganglionic hindgut in vivo but fail to compensate for lack of enteric ganglia. Dev Biol 2000;219(1):30–43.

[10] Newgreen D, Young HM. Enteric nervous system: development and developmental disturbances—part 2. Pediatr Dev Pathol Off J Soc Pediatr Pathol Paediatr Pathol Soc 2002;5(4):329–49.

[11] Wallace AS, Burns AJ. Development of the enteric nervous system, smooth muscle and interstitial cells of Cajal in the human gastrointestinal tract. Cell Tissue Res 2005;319(3):367–82.

[12] Durbec PL, Larsson-Blomberg LB, Schuchardt A, Costantini F, Pachnis V. Common origin and developmental dependence on c-ret of subsets of enteric and sympathetic neuroblasts. Development 1996;122 (1):349–58.

[13] Young HM, Bergner AJ, Anderson RB, Enomoto H, Milbrandt J, Newgreen DF, et al. Dynamics of neural crest-derived cell migration in the embryonic mouse gut. Dev Biol 2004;270(2):455–73.

[14] Druckenbrod NR, Epstein ML. The pattern of neural crest advance in the cecum and colon. Dev Biol 2005;287(1):125–33.

[15] Druckenbrod NR, Epstein ML. Behavior of enteric neural crest-derived cells varies with respect to the migratory wavefront. Dev Dyn 2007;236(1):84–92.

[16] Druckenbrod NR, Epstein ML. Age-dependent changes in the gut environment restrict the invasion of the hindgut by enteric neural progenitors. Development 2009;136(18):3195–203.

[17] Nishiyama C, Uesaka T, Manabe T, Yonekura Y, Nagasawa T, Newgreen DF, et al. Trans-mesenteric neuralcrest cells are the principal source of the colonic enteric nervous system. Nat Neurosci 2012;15 (9):1211–8.

[18] Erickson CS, Zaitoun I, Haberman KM, Gosain A, Druckenbrod NR, Epstein ML. Sacral neural crest-derived cells enter the aganglionic colon of Ednrb-/- mice along extrinsic nerve fibers. J Comp Neurol 2012;520(3):620–32.

[19] Furness JB. The enteric nervous system and neurogastroenterology. Nat Rev Gastroenterol Hepatol 2012;9 (5):286–94.

[20] Amiel J, Sproat-Emison E, Garcia-Barcelo M, Lantieri F, Burzynski G, Borrego S, et al. Hirschsprung disease, associated syndromes and genetics: a review. J Med Genet 2008;45(1):1–14.

[21] Obermayr F, Hotta R, Enomoto H, Young HM. Development and developmental disorders of the enteric nervous system. Nat Rev Gastroenterol Hepatol 2012;10(1):43–57.

[22] Laranjeira C, Pachnis V. Enteric nervous system development: recent progress and future challenges. Auton Neurosci 2009;151(1):61–9.

[23] Sasselli V, Pachnis V, Burns AJ. The enteric nervous system. Dev Biol 2012;366(1):64–73.

[24] Panza E, Knowles CH, Graziano C, Thapar N, Burns AJ, Seri M, et al. Genetics of human enteric neuropathies. Prog Neurobiol 2012;96(2):176–89.

[25] Pan ZW, Li JC. Advances in molecular genetics of Hirschsprung's disease. Anat Rec (Hoboken) 2012;295 (10):1628–38.

[26] Brooks AS, Bertoli-Avella AM, Burzynski GM, Breedveld GJ, Osinga J, Boven LG, et al. Homozygous nonsense mutations in KIAA1279 are associated with malformations of the central and enteric nervous systems. Am J Hum Genet 2005;77(1):120–6.

[27] Emison ES, Garcia-Barcelo M, Grice EA, Lantieri F, Amiel J, Burzynski G, et al. Differential contributions of rare and common, coding and noncoding Ret mutations to multifactorial Hirschsprung disease liability. Am J Hum Genet 2010;87(1):60–74.

[28] Griseri P, Lantieri F, Puppo F, Bachetti T, Di Duca M, Ravazzolo R, et al. A common variant located in the 3′UTR of the RET gene is associated with protection from Hirschsprung disease. Hum Mutation 2007;28 (2):168–76.

[29] Natarajan D, Marcos-Gutierrez C, Pachnis V, de Graaff E. Requirement of signalling by receptor tyrosine kinase RET for the directed migration of enteric nervous system progenitor cells during mammalian embryogenesis. Development 2002;129 (22):5151–60.

[30] Schuchardt A, D'Agati V, Larsson-Blomberg L, Costantini F, Pachnis V. Defects in the kidney and enteric nervous system of mice lacking the tyrosine kinase receptor Ret. Nature 1994;367(6461):380–3.

[31] Takahashi M. The GDNF/RET signaling pathway and human diseases. Cytokine Growth Factor Rev 2001;12 (4):361–73.

[32] de Graaff E, Srinivas S, Kilkenny C, D'Agati V, Mankoo BS, Costantini F, et al. Differential activities of the RET tyrosine kinase receptor isoforms during mammalian embryogenesis. Genes Dev 2001;15 (18):2433–44.

II. NEURAL CREST CELL DIFFERENTIATION AND DISEASE

[33] Uesaka T, Nagashimada M, Yonemura S, Enomoto H. Diminished Ret expression compromises neuronal survival in the colon and causes intestinal aganglionosis in mice. J Clin Invest 2008;118(5):1890–8.

[34] Gianino S, Grider JR, Cresswell J, Enomoto H, Heuckeroth RO. GDNF availability determines enteric neuron number by controlling precursor proliferation. Development 2003;130(10):2187–98.

[35] Rossi J, Herzig KH, Võikar V, Hiltunen PH, Segerstråle M, Airaksinen MS. Alimentary tract innervation deficits and dysfunction in mice lacking GDNF family receptor alpha2. J Clin Invest 2003;112 (5):707–16.

[36] Doray B, Salomon R, Amiel J, Pelet A, Touraine R, Billaud M, et al. Mutation of the RET ligand, neurturin, supports multigenic inheritance in Hirschsprung disease. Hum Mol Genet 1998;7(9):1449–52.

[37] Lee HO, Levorse JM, Shin MK. The endothelin receptor-B is required for the migration of neural crest-derived melanocyte and enteric neuron precursors. Dev Biol 2003;259(1):162–75.

[38] Barlow A, de Graaff E, Pachnis V. Enteric nervous system progenitors are coordinately controlled by the G protein-coupled receptor EDNRB and the receptor tyrosine kinase RET. Neuron 2003;40(5):905–16.

[39] Leibl MA, Ota T, Woodward MN, Kenny SE, Lloyd DA, Vaillant CR, et al. Expression of endothelin 3 by mesenchymal cells of embryonic mouse caecum. Gut 1999;44(2):246–52.

[40] Pingault V, Ente D, Dastot-Le Moal F, Goossens M, Marlin S, Bondurand N. Review and update of mutations causing Waardenburg syndrome. Hum Mutation 2010;31(4):391–406.

[41] Hosoda K, Hammer RE, Richardson JA, Baynash AG, Cheung JC, Giaid A, et al. Targeted and natural (piebald-lethal) mutations of endothelin-B receptor gene produce megacolon associated with spotted coat color in mice. Cell 1994;79(7):1267–76.

[42] Baynash AG, Hosoda K, Giaid A, Richardson JA, Emoto N, Hammer RE, et al. Interaction of endothelin-3 with endothelin-B receptor is essential for development of epidermal melanocytes and enteric neurons. Cell 1994;79(7):1277–85.

[43] Stanchina L, Baral V, Robert F, Pingault V, Lemort N, Pachnis V, et al. Interactions between Sox10, Edn3 and Ednrb during enteric nervous system and melanocyte development. Dev Biol 2006;295(1):232–49.

[44] Bondurand N, Dastot-Le Moal F, Stanchina L, Collot N, Baral V, Marlin S, et al. Deletions at the SOX10 gene locus cause Waardenburg syndrome types 2 and 4. Am J Hum Genet 2007;81(6):1169–85.

[45] Chaoui A, Watanabe Y, Touraine R, Baral V, Goossens M, Pingault V, et al. Identification and functional analysis of SOX10 missense mutations in different subtypes of Waardenburg syndrome. Hum Mutat 2011;32(12):1436–49.

[46] Kapur RP. Early death of neural crest cells is responsible for total enteric aganglionosis in Sox10(Dom)/Sox10(Dom) mouse embryos. Pediatr Dev Pathol 1999;2(6):559–69.

[47] Lang D, Chen F, Milewski R, Li J, Lu MM, Epstein JA. Pax3 is required for enteric ganglia formation and functions with Sox10 to modulate expression of c-ret. J Clin Invest 2000;106(8):963–71.

[48] Zhu L, Lee HO, Jordan CS, Cantrell VA, Southard-Smith EM, Shin MK. Spatiotemporal regulation of endothelin receptor-B by SOX10 in neural crest-derived enteric neuron precursors. Nat Genet 2004;36 (7):732–7.

[49] Cantrell VA, Owens SE, Chandler RL, Airey DC, Bradley KM, Smith JR, et al. Interactions between Sox10 and EdnrB modulate penetrance and severity of aganglionosis in the Sox10Dom mouse model of Hirschsprung disease. Hum Mol Genet 2004;13 (19):2289–301.

[50] Stanchina L, Van de Putte T, Goossens M, Huylebroeck D, Bondurand N. Genetic interaction between Sox10 and Zfhx1b during enteric nervous system development. Dev Biol 2010;341(2):416–28.

[51] Amiel J, Laudier B, Attié-Bitach T, Trang H, de Pontual L, Gener B, et al. Polyalanine expansion and frameshift mutations of the paired-like homeobox gene PHOX2B in congenital central hypoventilation syndrome. Nat Genet 2003;33(4):459–61.

[52] Pattyn A, Morin X, Cremer H, Goridis C, Brunet JF. The homeobox gene Phox2b is essential for the development of autonomic neural crest derivatives. Nature 1999;399(6734):366–70.

[53] Leon TY, Ngan ES, Poon HC, So MT, Lui VC, Tam PK, et al. Transcriptional regulation of RET by Nkx2-1, Phox2b, Sox10, and Pax3. J Pediatr Surg 2009;44 (10):1904–12.

[54] Wakamatsu N, Yamada Y, Yamada K, Ono T, Nomura N, Taniguchi H, et al. Mutations in SIP1, encoding Smad interacting protein-1, cause a form of Hirschsprung disease. Nat Genet 2001;27(4):369–70.

[55] Espinosa-Parrilla Y, Amiel J, Augé J, Encha-Razavi F, Munnich A, Lyonnet S, et al. Expression of the SMADIP1 gene during early human development. Mech Dev 2002;114(1–2):187–91.

[56] Bassez G, Camand OJ, Cacheux V, Kobetz A, Dastot-Le Moal F, Marchant D, et al. Pleiotropic and diverse expression of ZFHX1B gene transcripts during mouse and human development supports the various clinical manifestations of the "Mowat-Wilson" syndrome. Neurobiol Dis 2004;15(2):240–50.

[57] Van de Putte T, Maruhashi M, Francis A, Nelles L, Kondoh H, Huylebroeck D, et al. Neural crest-specific removal of Zfhx1b in mouse leads to a wide range of neurocristopathies reminiscent of Mowat—Wilson syndrome. Hum Mol Genet 2007;16(12):1423—36.

[58] Van de Putte T, et al. Mice lacking ZFHX1B, the gene that codes for Smad-interacting protein-1, reveal a role for multiple neural crest cell defects in the etiology of Hirschsprung disease-mental retardation syndrome. Am J Hum Genet 2003;72 (2):465—70.

[59] Goldberg RB, Shprintzen RJ. Hirschsprung megacolon and cleft palate in two sibs. J Craniofacial Genet Dev Biol 1981;1(2):185—9.

[60] Lyons DA, Naylor SG, Mercurio S, Dominguez C, Talbot WS. KBP is essential for axonal structure, outgrowth and maintenance in zebrafish, providing insight into the cellular basis of Goldberg—Shprintzen syndrome. Development 2008;135(3):599—608.

[61] Drévillon L, Megarbane A, Demeer B, Matar C, Benit P, Briand-Suleau A, et al. KBP-cytoskeleton interactions underlie developmental anomalies in Goldberg—Shprintzen syndrome. Hum Mol Genet 2013;22 (12):2387—99.

[62] Anderson RB, Turner KN, Nikonenko AG, Hemperly J, Schachner M, Young HM. The cell adhesion molecule l1 is required for chain migration of neural crest cells in the developing mouse gut. Gastroenterology 2006;130(4):1221—32.

[63] Nagy N, Burns AJ, Goldstein AM. Immunophenotypic characterization of enteric neural crest cells in the developing avian colorectum. Dev Dyn 2012;241 (5):842—51.

[64] Nakakimura S, Sasaki F, Okada T, Arisue A, Cho K, Yoshino M, et al. Hirschsprung's disease, acrocallosal syndrome, and congenital hydrocephalus: report of 2 patients and literature review. J Pediatr Surg 2008;43 (5):E13—7.

[65] Jackson SR, Guner YS, Woo R, Randolph LM, Ford H, Shin CE. L1CAM mutation in association with X-linked hydrocephalus and Hirschsprung's disease. Pediatr Surg Int 2009;25(9):823—5.

[66] de Pontual L, Zaghloul NA, Thomas S, Davis EE, McGaughey DM, Dollfus H, et al. Epistasis between RET and BBS mutations modulates enteric innervation and causes syndromic Hirschsprung disease. Proc Natl Acad Sci USA 2009;106(33):13921—6.

[67] Amiel J, Rio M, de Pontual L, Redon R, Malan V, Boddaert N, et al. Mutations in TCF4, encoding a class I basic helix-loop-helix transcription factor, are responsible for Pitt—Hopkins syndrome, a severe epileptic encephalopathy associated with autonomic dysfunction. Am J Hum Genet 2007;80(5):988—93.

[68] Tang CS, Cheng G, So MT, Yip BH, Miao XP, Wong EH, et al. Genome-wide copy number analysis uncovers a new HSCR gene: NRG3. PLoS Genet 2012;8(5):e1002687.

[69] Tang CS, Ngan ES, Tang WK, So MT, Cheng G, Miao XP, et al. Mutations in the NRG1 gene are associated with Hirschsprung disease. Hum Genet 2012;131(1):67—76.

[70] Luzón-Toro B, Torroglosa A, Núñez-Torres R, Enguix-Riego MV, Fernández RM, de Agustín JC, et al. Comprehensive analysis of NRG1 common and rare variants in Hirschsprung patients. PloS One 2012;7(5): e36524.

[71] Britsch S. The neuregulin-I/ErbB signaling system in development and disease. Adv Anat Embryol Cell Biol 2007;190:1—65.

[72] Prigent SA, Lemoine NR, Hughes CM, Plowman GD, Selden C, Gullick WJ. Expression of the c-erbB-3 protein in normal human adult and fetal tissues. Oncogene 1992;7(7):1273—8.

[73] Meyer D, Birchmeier C. Distinct isoforms of neuregulin are expressed in mesenchymal and neuronal cells during mouse development. Proc Natl Acad Sci USA 1994;91(3):1064—8.

[74] Orr-Urtreger A, Trakhtenbrot L, Ben-Levy R, Wen D, Rechavi G, Lonai P, et al. Neural expression and chromosomal mapping of Neu differentiation factor to 8p12-p21. Proc Natl Acad Sci USA 1993;90 (5):1867—71.

[75] Wallace AS, Anderson RB. Genetic interactions and modifier genes in Hirschsprung's disease. World J Gastroenterol WJG 2011;17(45):4937—44.

[76] Ngan ES, Garcia-Barceló MM, Yip BH, Poon HC, Lau ST, Kwok CK, et al. Hedgehog/Notch-induced premature gliogenesis represents a new disease mechanism for Hirschsprung disease in mice and humans. J Clin Invest 2011;121(9):3467—78.

[77] Kruger GM, Mosher JT, Tsai YH, Yeager KJ, Iwashita T, Gariepy CE, et al. Temporally distinct requirements for endothelin receptor B in the generation and migration of gut neural crest stem cells. Neuron 2003;40(5):917—29.

[78] Breau MA, Pietri T, Eder O, Blanche M, Brakebusch C, Fässler R, et al. Lack of beta1 integrins in enteric neural crest cells leads to a Hirschsprung-like phenotype. Development 2006;133(9):1725—34.

[79] Breau MA, Dahmani A, Broders-Bondon F, Thiery JP, Dufour S. Beta1 integrins are required for the invasion of the caecum and proximal hindgut by enteric neural crest cells. Development 2009;136(16):2791—801.

[80] Lei J, Howard MJ. Targeted deletion of Hand2 in enteric neural precursor cells affects its functions in neurogenesis, neurotransmitter specification and gangliogenesis, causing functional aganglionosis. Development 2011;138(21):4789—800.

[81] Hendershot TJ, Liu H, Sarkar AA, Giovannucci DR, Clouthier DE, Abe M, et al. Expression of Hand2 is sufficient for neurogenesis and cell type-specific gene expression in the enteric nervous system. Dev Dyn 2007;236(1):93−105.

[82] D'Autreaux F, Morikawa Y, Cserjesi P, Gershon MD. Hand2 is necessary for terminal differentiation of enteric neurons from crest-derived precursors but not for their migration into the gut or for formation of glia. Development 2007;134(12):2237−49.

[83] Holler KL, Hendershot TJ, Troy SE, Vincentz JW, Firulli AB, Howard MJ. Targeted deletion of Hand2 in cardiac neural crest-derived cells influences cardiac gene expression and outflow tract development. Dev Biol 2010;341(1):291−304.

[84] Goldstein AM, Brewer KC, Doyle AM, Nagy N, Roberts DJ. BMP signaling is necessary for neural crest cell migration and ganglion formation in the enteric nervous system. Mech Dev 2005;122(6):821−33.

[85] Fu M, Vohra BP, Wind D, Heuckeroth RO. BMP signaling regulates murine enteric nervous system precursor migration, neurite fasciculation, and patterning via altered Ncam1 polysialic acid addition. Dev Biol 2006;299(1):137−50.

[86] Faure C, Chalazonitis A, Rhéaume C, Bouchard G, Sampathkumar SG, Yarema KJ, et al. Gangliogenesis in the enteric nervous system: roles of the polysialylation of the neural cell adhesion molecule and its regulation by bone morphogenetic protein-4. Dev Dyn 2007;236(1):44−59.

[87] Peters-van der Sanden MJ, et al. Ablation of various regions within the avian vagal neural crest has differential effects on ganglion formation in the fore-, mid- and hindgut. Dev Dynam 1993;196(3):183−94.

[88] Barlow AJ, Wallace AS, Thapar N, Burns AJ. Critical numbers of neural crest cells are required in the pathways from the neural tube to the foregut to ensure complete enteric nervous system formation. Development 2008;135(9):1681−91.

[89] Wallace AS, Barlow AJ, Navaratne L, Delalande JM, Tauszig-Delamasure S, Corset V, et al. Inhibition of cell death results in hyperganglionosis: implications for enteric nervous system development. Neurogastroenterol Motil 2009;21(7):e749−68.

[90] Kuhlbrodt K, Schmidt C, Sock E, Pingault V, Bondurand N, Goossens M, et al. Functional analysis of Sox10 mutations found in human Waardenburg−Hirschsprung patients. J Biol Chem 1998;273(36):23033−8.

[91] Britsch S, Goerich DE, Riethmacher D, Peirano RI, Rossner M, Nave KA, et al. The transcription factor Sox10 is a key regulator of peripheral glial development. Genes Dev 2001;15(1):66−78.

[92] Herbarth B, Pingault V, Bondurand N, Kuhlbrodt K, Hermans-Borgmeyer I, Puliti A, et al. Mutation of the Sry-related Sox10 gene in dominant megacolon, a mouse model for human Hirschsprung disease. Proc Natl Acad Sci USA 1998;95(9):5161−5.

[93] Lane PW, Liu HM. Association of megacolon with a new dominant spotting gene (Dom) in the mouse. J Heredity 1984;75(6):435−9.

[94] Southard-Smith EM, Kos L, Pavan WJ. Sox10 mutation disrupts neural crest development in Dom Hirschsprung mouse model. Nat Genet 1998;18 (1):60−4.

[95] Maka M, Stolt CC, Wegner M. Identification of Sox8 as a modifier gene in a mouse model of Hirschsprung disease reveals underlying molecular defect. Dev Biol 2005;277(1):155−69.

[96] Paratore C, Eichenberger C, Suter U, Sommer L. Sox10 haploinsufficiency affects maintenance of progenitor cells in a mouse model of Hirschsprung disease. Hum Mol Genet 2002;11(24):3075−85.

[97] Kim J, Lo L, Dormand E, Anderson DJ. SOX10 maintains multipotency and inhibits neuronal differentiation of neural crest stem cells. Neuron 2003;38 (1):17−31.

[98] Bondurand N, Natarajan D, Barlow A, Thapar N, Pachnis V. Maintenance of mammalian enteric nervous system progenitors by SOX10 and endothelin 3 signalling. Development 2006;133 (10):2075−86.

[99] Lang D, Epstein JA. Sox10 and Pax3 physically interact to mediate activation of a conserved c-RET enhancer. Hum Mol Genet 2003;12(8):937−45.

[100] Barlow AJ, Dixon J, Dixon M, Trainor PA. Tcof1 acts as a modifier of Pax3 during enteric nervous system development and in the pathogenesis of colonic aganglionosis. Hum Mol Genet 2013.

[101] Goulding MD, Chalepakis G, Deutsch U, Erselius JR, Gruss P. Pax-3, a novel murine DNA binding protein expressed during early neurogenesis. EMBO J 1991;10(5):1135−47.

[102] Epstein JA, Li J, Lang D, Chen F, Brown CB, Jin F, et al. Migration of cardiac neural crest cells in Splotch embryos. Development 2000;127:1869−78.

[103] Pani L, Horal M, Loeken MR. Rescue of neural tube defects in Pax-3-deficient embryos by p53 loss of function: implications for Pax-3-dependent development and tumorigenesis. Genes Dev 2002;16 (6):676−80.

[104] Conway SJ, Bundy J, Chen J, Dickman E, Rogers R, Will BM. Decreased neural crest stem cell expansion is responsible for the conotruncal heart defects within the splotch (Sp(2H))/Pax3 mouse mutant. Cardiovas Res 2000;47(2):314−28.

[105] Barlow AJ, Dixon J, Dixon MJ, Trainor PA. Balancing neural crest cell intrinsic processes with those of the microenvironment in Tcof1 haploinsufficient mice enables complete enteric nervous system formation. Hum Mol Genet 2012;21(8):1782–93.

[106] Jones NC, Lynn ML, Gaudenz K, Sakai D, Aoto K, Rey JP, et al. Prevention of the neurocristopathy treacher collins syndrome through inhibition of p53 function. Nat Med 2008;14(2):125–33.

[107] Teng L, Mundell NA, Frist AY, Wang Q, Labosky PA. Requirement for Foxd3 in the maintenance of neural crest progenitors. Development 2008;135(9):1615–24.

[108] Dottori M, Gross MK, Labosky P, Goulding M. The winged-helix transcription factor Foxd3 suppresses interneuron differentiation and promotes neural crest cell fate. Development 2001;128(21):4127–38.

[109] Labosky PA, Kaestner KH. The winged helix transcription factor Hfh2 is expressed in neural crest and spinal cord during mouse development. Mech Dev 1998;76(1–2):185–90.

[110] Blaugrund E, Pham TD, Tennyson VM, Lo L, Sommer L, Anderson DJ, et al. Distinct subpopulations of enteric neuronal progenitors defined by time of development, sympathoadrenal lineage markers and Mash-1-dependence. Development 1996;122(1):309–20.

[111] Uesaka T, Jain S, Yonemura S, Uchiyama Y, Milbrandt J, Enomoto H. Conditional ablation of GFRalpha1 in postmigratory enteric neurons triggers unconventional neuronal death in the colon and causes a Hirschsprung's disease phenotype. Development 2007;134(11):2171–81.

[112] Young HM, Hearn CJ, Farlie PG, Canty AJ, Thomas PQ, Newgreen DF. GDNF is a chemoattractant for enteric neural cells. Dev Biol 2001;229(2):503–16.

[113] Burns AJ, Thapar N. Advances in ontogeny of the enteric nervous system. Neurogastroenterol Motil 2006;18(10):876–87.

[114] Nagy N, Goldstein AM. Endothelin-3 regulates neural crest cell proliferation and differentiation in the hindgut enteric nervous system. Dev Biol 2006;293(1):203–17.

[115] Broders-Bondon F, Paul-Gilloteaux P, Carlier C, Radice GL, Dufour S. N-cadherin and beta1-integrins cooperate during the development of the enteric nervous system. Dev Biol 2012;364(2):178–91.

[116] Hackett-Jones EJ, Landman KA, Newgreen DF, Zhang D. On the role of differential adhesion in gangliogenesis in the enteric nervous system. J Theor Biol 2011;287:148–59.

[117] Anderson RB, Bergner AJ, Taniguchi M, Fujisawa H, Forrai A, Robb L, et al. Effects of different regions of the developing gut on the migration of enteric neural crest-derived cells: a role for Sema3A, but not Sema3F. Dev Biol 2007;305(1):287–99.

[118] Hotta R, Anderson RB, Kobayashi K, Newgreen DF, Young HM. Effects of tissue age, presence of neurones and endothelin-3 on the ability of enteric neurone precursors to colonize recipient gut: implications for cell-based therapies. Neurogastroenterol Motil 2010;22(3):331–e386.

[119] Wu JJ, Chen JX, Rothman TP, Gershon MD. Inhibition of in vitro enteric neuronal development by endothelin-3: mediation by endothelin B receptors. Development 1999;126(6):1161–73.

[120] Rothman TP, Chen J, Howard MJ, Costantini F, Schuchardt A, Pachnis V, et al. Increased expression of laminin-1 and collagen (IV) subunits in the aganglionic bowel of ls/ls, but not c-ret -/- mice. Dev Biol 1996;178(2):498–513.

[121] Tennyson VM, Pham TD, Rothman TP, Gershon MD. Abnormalities of smooth muscle, basal laminae, and nerves in the aganglionic segments of the bowel of lethal spotted mutant mice. Anatomical Rec 1986;215(3):267–81.

[122] Stewart AL, Young HM, Popoff M, Anderson RB. Effects of pharmacological inhibition of small GTPases on axon extension and migration of enteric neural crest-derived cells. Dev Biol 2007;307(1):92–104.

[123] Fukuda T, Kiuchi K, Takahashi M. Novel mechanism of regulation of Rac activity and lamellipodia formation by RET tyrosine kinase. J Biol Chem 2002;277(21):19114–21.

[124] Asai N, Fukuda T, Wu Z, Enomoto A, Pachnis V, Takahashi M, et al. Targeted mutation of serine 697 in the Ret tyrosine kinase causes migration defect of enteric neural crest cells. Development 2006;133(22):4507–16.

[125] Fuchs S, Herzog D, Sumara G, Büchmann-Møller S, Civenni G, Wu X, et al. Stage-specific control of neural crest stem cell proliferation by the small rho GTPases Cdc42 and Rac1. Cell Stem Cell 2009;4(3):236–47.

[126] Simpson MJ, Zhang DC, Mariani M, Landman KA, Newgreen DF. Cell proliferation drives neural crest cell invasion of the intestine. Dev Biol 2007;302(2):553–68.

[127] Walters LC, Cantrell VA, Weller KP, Mosher JT, Southard-Smith EM. Genetic background impacts developmental potential of enteric neural crest-derived progenitors in the Sox10Dom model of Hirschsprung disease. Hum Mol Genet 2010;19(22):4353–72.

[128] Young HM, Turner KN, Bergner AJ. The location and phenotype of proliferating neural-crest-derived

cells in the developing mouse gut. Cell Tissue Res 2005;320(1):1–9.

[129] Wang H, Hughes I, Planer W, Parsadanian A, Grider JR, Vohra BP, et al. The timing and location of glial cell line-derived neurotrophic factor expression determine enteric nervous system structure and function. J Neurosci 2010;30(4):1523–38.

[130] Flynn B, Bergner AJ, Turner KN, Young HM, Anderson RB. Effect of Gdnf haploinsufficiency on rate of migration and number of enteric neural crest-derived cells. Dev Dyn 2007;236(1):134–41.

[131] Kawaguchi J, Nichols J, Gierl MS, Faial T, Smith A. Isolation and propagation of enteric neural crest progenitor cells from mouse embryonic stem cells and embryos. Development 2010;137(5):693–704.

[132] Mundell NA, Plank JL, LeGrone AW, Frist AY, Zhu L, Shin MK, et al. Enteric nervous system specific deletion of Foxd3 disrupts glial cell differentiation and activates compensatory enteric progenitors. Dev Biol 2012;363(2):373–87.

[133] Baetge G, Gershon MD. Transient catecholaminergic (TC) cells in the vagus nerves and bowel of fetal mice: relationship to the development of enteric neurons. Dev Biol 1989;132(1):189–211.

[134] Laranjeira C, Sandgren K, Kessaris N, Richardson W, Potocnik A, Vanden Berghe P, et al. Glial cells in the mouse enteric nervous system can undergo neurogenesis in response to injury. J Clin Invest 2011;121(9):3412–24.

[135] Anderson RB, Stewart AL, Young HM. Phenotypes of neural-crest-derived cells in vagal and sacral pathways. Cell Tissue Res 2006;323(1):11–25.

[136] Okamura Y, Saga Y. Notch signaling is required for the maintenance of enteric neural crest progenitors. Development 2008;135(21):3555–65.

[137] D'Autréaux F, Margolis KG, Roberts J, Stevanovic K, Mawe G, Li Z, et al. Expression level of Hand2 affects specification of enteric neurons and gastrointestinal function in mice. Gastroenterol 2011;141(2):576–87 e571–6.

[138] Chalazonitis A, Pham TD, Li Z, Roman D, Guha U, Gomes W, et al. Bone morphogenetic protein regulation of enteric neuronal phenotypic diversity: relationship to timing of cell cycle exit. J Comp Neurol 2008;509(5):474–92.

[139] Pham TD, Gershon MD, Rothman TP. Time of origin of neurons in the murine enteric nervous system: sequence in relation to phenotype. J Comp Neurol 1991;314(4):789–98.

[140] Hao MM, Young HM. Development of enteric neuron diversity. J Cell Mol Med 2009;13(7):1193–210.

[141] Uesaka T, Enomoto H. Neural precursor death is central to the pathogenesis of intestinal aganglionosis in Ret hypomorphic mice. J Neurosci 2010;30(15):5211–8.

[142] Yan H, Bergner AJ, Enomoto H, Milbrandt J, Newgreen DF, Young HM. Neural cells in the esophagus respond to glial cell line-derived neurotrophic factor and neurturin, and are RET-dependent. Dev Biol 2004;272(1):118–33.

[143] Chalazonitis A, Pham TD, Rothman TP, DiStefano PS, Bothwell M, Blair-Flynn J, et al. Neurotrophin-3 is required for the survival-differentiation of subsets of developing enteric neurons. J Neurosci 2001;21(15):5620–36.

[144] Gulbransen BD, Sharkey KA. Novel functional roles for enteric glia in the gastrointestinal tract. Nat Rev Gastroenterol Hepatol 2012;9(11):625–32.

[145] Young HM, Bergner AJ, Muller T. Acquisition of neuronal and glial markers by neural crest-derived cells in the mouse intestine. J Comp Neurol 2003;456(1):1–11.

[146] Joseph NM, He S, Quintana E, Kim YG, Núñez G, Morrison SJ. Enteric glia are multipotent in culture but primarily form glia in the adult rodent gut. J Clin Invest 2011;121(9):3398–411.

[147] Kenny SE, Tam PK, Garcia-Barcelo M. Hirschsprung's disease. Semin Pediatr Surg 2010;19(3):194–200.

[148] Austin KM. The pathogenesis of Hirschsprung's disease-associated enterocolitis. Semin Pediatr Surg 2012;21(4):319–27.

[149] Hotta R, Natarajan D, Burns AJ, Thapar N. Stem cells for GI motility disorders. Curr Opin Pharmacol 2011;11(6):617–23.

[150] Wilkinson DJ, Edgar DH, Kenny SE. Future therapies for Hirschsprung's disease. Semin Pediatr Surg 2012;21(4):364–70.

[151] Micci MA, Kahrig KM, Simmons RS, Sarna SK, Espejo-Navarro MR, Pasricha PJ. Neural stem cell transplantation in the stomach rescues gastric function in neuronal nitric oxide synthase-deficient mice. Gastroenterology 2005;129(6):1817–24.

[152] Dong YL, Liu W, Gao YM, Wu RD, Zhang YH, Wang HF, et al. Neural stem cell transplantation rescues rectum function in the aganglionic rat. Transplant Proc 2008;40(10):3646–52.

[153] Liu W, Wu RD, Dong YL, Gao YM. Neuroepithelial stem cells differentiate into neuronal phenotypes and improve intestinal motility recovery after transplantation in the aganglionic colon of the rat. Neurogastroenterol Motil 2007;19(12):1001–9.

[154] Hotta R, Pepdjonovic L, Anderson RB, Zhang D, Bergner AJ, Leung J, et al. Small-molecule induction of neural crest-like cells derived from human neural progenitors. Stem Cell 2009;27(12):2896–905.

[155] Sasselli V, Micci MA, Kahrig KM, Pasricha PJ. Evaluation of ES-derived neural progenitors as a potential source for cell replacement therapy in the gut. BMC Gastroenterol 2012;12:81.

[156] Ben-David U, Benvenisty N. The tumorigenicity of human embryonic and induced pluripotent stem cells. Nat Rev Cancer 2011;11(4):268–77.

[157] Kulkarni S, Becker L, Pasricha PJ. Stem cell transplantation in neurodegenerative disorders of the gastrointestinal tract: future or fiction? Gut 2012;61(4):613–21.

[158] Spence JR, Mayhew CN, Rankin SA, Kuhar MF, Vallance JE, Tolle K, et al. Directed differentiation of human pluripotent stem cells into intestinal tissue in vitro. Nature 2011;470(7332):105–9.

[159] Ueda T, Yamada T, Hokuto D, Koyama F, Kasuda S, Kanehiro H, et al. Generation of functional gut-like organ from mouse induced pluripotent stem cells. Biochem Biophys Res Commun 2010;391(1):38–42.

[160] Natarajan D, Grigoriou M, Marcos-Gutierrez CV, Atkins C, Pachnis V. Multipotential progenitors of the mammalian enteric nervous system capable of colonising aganglionic bowel in organ culture. Development 1999;126(1):157–68.

[161] Sidebotham EL, Kenny SE, Lloyd DA, Vaillant CR, Edgar DH. Location of stem cells for the enteric nervous system. Pediatr Surg Int 2002;18(7):581–5.

[162] Bixby S, Kruger GM, Mosher JT, Joseph NM, Morrison SJ. Cell-intrinsic differences between stem cells from different regions of the peripheral nervous system regulate the generation of neural diversity. Neuron 2002;35:643–56.

[163] Bondurand N, Natarajan D, Thapar N, Atkins C, Pachnis V. Neuron and glia generating progenitors of the mammalian enteric nervous system isolated from foetal and postnatal gut cultures. Development 2003;130(25):6387–400.

[164] Lindley RM, Hawcutt DB, Connell MG, Almond SL, Vannucchi MG, Faussone-Pellegrini MS, et al. Human and mouse enteric nervous system neurosphere transplants regulate the function of aganglionic embryonic distal colon. Gastroenterology 2008;135(1):205–216, e206.

[165] Lindley RM, Hawcutt DB, Connell MG, Edgar DH, Kenny SE. Properties of secondary and tertiary human enteric nervous system neurospheres. J Pediatr Surg 2009;44(6):1249–55 discussion 1255–1246.

[166] Almond S, Lindley RM, Kenny SE, Connell MG, Edgar DH. Characterisation and transplantation of enteric nervous system progenitor cells. Gut 2007;56(4):489–96.

[167] Rauch U, Hansgen A, Hagl C, Holland-Cunz S, Schafer KH. Isolation and cultivation of neuronal precursor cells from the developing human enteric nervous system as a tool for cell therapy in dysganglionosis. Int J Colorectal Dis 2006;21(6):554–9.

[168] Metzger M, Caldwell C, Barlow AJ, Burns AJ, Thapar N. Enteric nervous system stem cells derived from human gut mucosa for the treatment of aganglionic gut disorders. Gastroenterology 2009;136(7):2214–25 e2211–2213.

[169] Bogni S, Trainor P, Natarajan D, Krumlauf R, Pachnis V. Non-cell-autonomous effects of Ret deletion in early enteric neurogenesis. Development 2008;135(18):3007–11.

[170] Roberts RR, Bornstein JC, Bergner AJ, Young HM. Disturbances of colonic motility in mouse models of Hirschsprung's disease. Am J Physiol Gastrointest Liver Physiol 2008;294(4):G996–1008.

[171] Sandgren K, Larsson LT, Ekblad E. Widespread changes in neurotransmitter expression and number of enteric neurons and interstitial cells of Cajal in lethal spotted mice: an explanation for persisting dysmotility after operation for Hirschsprung's disease? Dig Dis Sci 2002;47(5):1049–64.

[172] Ro S, Hwang SJ, Muto M, Jewett WK, Spencer NJ. Anatomic modifications in the enteric nervous system of piebald mice and physiological consequences to colonic motor activity. Am J Physiol Gastrointest Liver Physiol 2006;290(4):G710–8.

[173] Zaitoun I, Erickson CS, Barlow AJ, Klein TR, Heneghan AF, Pierre JF, et al. Altered neuronal density and neurotransmitter expression in the ganglionated region of Ednrb null mice: implications for Hirschsprung's disease. Neurogastroenterol Motil 2013.

[174] Veiga-Fernandes H, Coles MC, Foster KE, Patel A, Williams A, Natarajan D, et al. Tyrosine kinase receptor RET is a key regulator of Peyer's patch organogenesis. Nature 2007;446(7135):547–51.

[175] Cheng Z, Wang X, Dhall D, Zhao L, Bresee C, Doherty TM, et al. Splenic lymphopenia in the endothelin receptor B-null mouse: implications for Hirschsprung associated enterocolitis. Pediatr Surg Int 2011;27(2):145–50.

[176] Dang R, Sasaki N, Nishino T, Nakanishi M, Torigoe D, Agui T. Lymphopenia in Ednrb-deficient rat was strongly modified by genetic background. Biomed Res 2012;33(4):249–53.

[177] Gosain A, Pierre J, Heneghan A, Barlow A, Erickson C, Epstein M, et al. Impaired cellular immunity in the murine neural crest-specific conditional knockout of endothelin receptor-B model of Hirschsprung's disease; 2013.

[178] Frykman PK, Short SS. Hirschsprung-associated enterocolitis: prevention and therapy. Semin Pediatr Surg 2012;21(4):328–35.

Neural Crest Cells and Peripheral Nervous System Development

Andrew Prendergast and David W. Raible

Department of Biological Structure, University of Washington, Seattle, WA, USA

OUTLINE

Neural Crest Cells.
DOI: http://dx.doi.org/10.1016/B978-0-12-401730-6.00014-4

OVERVIEW

The majority of the peripheral nervous system (PNS) originates from the so-called trunk neural crest. During neurulation, trunk neural crest cells (NCC) are generated within the dorsal-most aspect of the neural plate or neural tube. It must then colonize more ventral aspects of the body by migrating great distances and subsequently undergoing overt differentiation. The timing of these events relative to other major developmental milestones is outlined in Figure 13.1 for mouse, chicken, and zebrafish embryos. Here, we will primarily discuss neurogenesis in the sympathetic chain ganglia or paravertebral ganglia as well as in the dorsal root or spinal sensory ganglia. The anatomy of these structures and some of their projections is illustrated in Figure 13.2. Neural crest contribution to cranial ganglia is discussed in Chapter 9; contribution to the enteric nervous system is discussed in this chapter.

DEFINITION OF TRUNK NEURAL CREST

Trunk neural crest is variably defined, but it can generally be understood to encompass all neural crest posterior to the hindbrain. Nonetheless, fate-mapping studies performed with different techniques in different model organisms have generated an array of boundary definitions and have led investigators to propose additional subdivisions. The most prevalent of these is the compartmentalization of the trunk neural crest into vagal, trunk, and

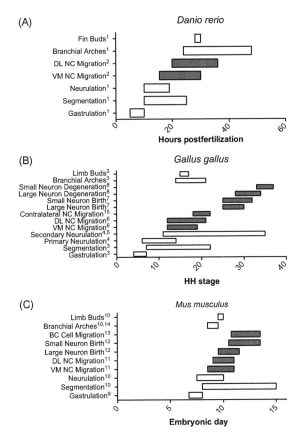

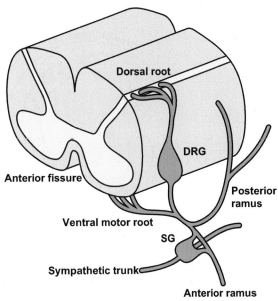

FIGURE 13.2 **Some of the PNS derivatives of the trunk neural crest.** A single segment of spinal cord is diagramed along with one hemisegment of adjacent PNS tissue. Based on images from [16] and modeled as mammalian tissue. Although the DRG and sympathetic ganglia are derived entirely from neural crest, PNS nerves will also contain neuroectodermally derived fibers (e.g., the ventral motor root).

FIGURE 13.1 **Timing of developmental milestones in zebrafish, chick, and mouse embryos with a focus on DRG development.** Important events in DRG development are highlighted in red and the timing of each event is expressed in terms of each organism's most commonly expressed staging system (rather than in absolute time). Developmental milestones are presented for (A) the zebrafish (*Danio rerio*), (B) the chicken (*Gallus gallus*), and (C) the mouse (*Mus musculus*). The following works are cited as support for the intervals displayed in the figure: (1) [1], (2) [2], (3) [3], (4) [4], (5) [5], (6) [6], (7) [7], (8) [8], (9) [9], (10) [10], (11) [11], (12) [12], (13) [13], (14) [14], (15) [15].

sacral neural crest. The rough segment boundaries for these compartments are outlined in Figure 13.3 for zebrafish and chick embryos [17–24]. These compartments reflect distinct developmental outcomes: in amniotes and mammals, vagal and sacral neural crest both contribute to the enteric nervous system

[17,19,25], while in fish, the vagal neural crest is the only known source for the enteric nervous system [11,22,25–28].

DIVERGENT NEURAL CREST MIGRATORY PATHWAYS REFLECT FATE RESTRICTION OF THE TRUNK NEURAL CREST

Trunk neural crest is specified at the lateral edges of the neuroectoderm during neurulation. Following neurulation, these cells delaminate and migrate to colonize their ultimate targets along the migratory pathways illustrated in Figure 13.4. Trunk neural crest migrates along two generally recognized pathways—a ventromedial migration that occurs first, and a

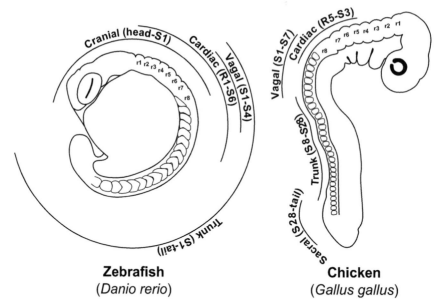

FIGURE 13.3 **Regionalization of neural crest.** A schematized 18SS zebrafish embryo is displayed at left and an HH18 chick embryo is displayed at right. *Source: Based on images from [1] and [3], respectively.*

dorsolateral pathway, into which NCC enter with some delay (the timing of these events are summarized in Figure 13.1). However, some investigators refer to other migratory patterns, including intersomitic migration and separate ventral migratory streams in both chick and mouse embryos [11,22,26–28].

In all species considered here, the ventromedial migratory stream gives rise to sympathetic neurons, sensory neurons of the dorsal root ganglia (DRG), and the glia associated with these neurons [11,22,26,29]. As a result, the trunk PNS across vertebrate taxa is probably derived solely from cells that enter this migratory route. However, the nature of this migration differs somewhat from species to species. In chick and in mouse embryos, this stream passes in part through the sclerotome (a medial subdivision of the somite) [2,11,26,29–32], with fewer cells invading the sclerotome in mice than in chick. In teleost embryos, ventromedially migrating cells travel exclusively between the neural tube and somite [2,26,29–32]. Ventromedial

migration is segmented such that one distinct stream of migrating neural crest is present in each somite. In mouse and chick embryos, this migration is restricted to the anterior half of each somite [2,26,29,31]; in the zebrafish, this migration is instead restricted to the middle of each somite [2,31,33,34].

NEURAL CREST FATE RESTRICTION BEGINS LONG BEFORE THE OBSERVATION OF OVERT DIFFERENTIATION MARKERS

The neural crest is essentially homogenous with respect to marker expression both before and during its early phase of migration. Does this reflect a general homogeneity of developmental potential? Some evidence suggests that NCC exhibit considerable fate plasticity throughout their migration. Labeled premigratory NCC in chick embryos yield

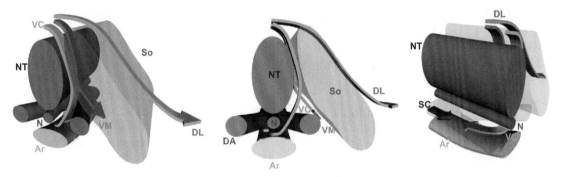

FIGURE 13.4 Migratory pathways of trunk neural crest. Trunk neural crest initially migrates along a ventromedial (VM) pathway. The exact route this pathway takes depends on the model organism—in mice and chicken embryos, these cells infiltrate the sclerotome, while in fish, they travel between the neural tube and somite. This migratory stream gives rise to the autonomic lineage as well as the sensory neurons and glia of the DRG. A slightly delayed migratory stream travels between the somite and the overlying epidermis; this dorsolateral (DL) pathway gives rise to pigment cells. Also shown is the migratory pathway of the vagal neural crest (VC), which is exposed to inductive axial signals and then colonizes the developing gut.

mixed derivatives when examined later [33,34], although some fate bias is apparent, as neuron-only clones still occur with great frequency [33,35]. Lineage-labeling experiments performed *in vivo* in zebrafish [18,20−24,35,36] and chick embryos [17−24,36,37] suggest fate restriction occurs early, possibly prior to the initiation of migration in most cases. Experiments in which cultured neural crest is ablated prior to differentiation selectively depletes particular colony types, verifying this early fate commitment [17,19,37].

However, other work does suggest that neural crest is generally multipotent. So-called neural crest stem cells isolated from E10.5 rat neural tube on the basis of low-affinity NGF receptor expression divide and generate neurons, glia, and other cell types, even when passaged twice as single cell clones [18,20−24,38]. Neural crest stem cells isolated from DRG, gut, and sciatic nerve also continue to demonstrate plasticity in culture [39−41] and can be influenced instructively to differentiate along specific lineages by exposure to distinct growth factors (reviewed in [42]).

Heterochronic transplants—i.e., experiments in which cells are transplanted between two different developmental stages—suggest that neural crest fate restriction generally cannot be overcome by returning those cells to a less differentiated environment [25,43−45], though some transplant experiments suggest lineage restriction may not be consistent over developmental time [22,46,47]. Experiments in quail [11,48−52] or mouse [22,53] clonal cultures generally support these experiments, as clones predominantly give rise to a restricted set of cell types and not the full complement of trunk neural crest derivatives. However, migratory quail NCC harvested from explants and then transplanted into neural tube, suggest this commitment is reversible in some cases, perhaps as a consequence of exposure to culture conditions [54]. It should be noted that the relationship between plasticity in culture and after transplantation is complex, as NCC with stem cell properties *in vitro* show restricted contribution to neural crest-derived structures *in vivo* [40,55−57]. Taken together, the data suggest that some NCC exhibit some commitment and this commitment may occur surprisingly early in development, prior even to neural crest migration.

FACTORS THAT SPECIFY THE NEURAL CREST CONTINUE TO MODIFY NEURAL CREST FATE PROLIFERATION, SURVIVAL, AND FATE CHOICE

Many signaling factors involved in the initial specification of the neural crest from ectoderm have been identified, among them bone morphogenetic proteins (BMPs), fibroblast growth factors (FGFs), retinoic acid (RA), and Wnts (reviewed recently in [11,26,29,58] and elsewhere in this volume). As in many tissues, these signals are reused later in development to influence other aspects of tissue morphogenesis. Consider BMPs, which contribute to neural crest specification through a morphogen gradient/threshold and/or cooperative signaling mechanism in *Xenopus* and zebrafish (reviewed recently in [59] and elsewhere in this volume). Following this initial function, BMPs regulate neural crest delamination and survival. BMPs later play a pivotal role in the specification of sympathetic neurons, and this will be discussed in detail below.

Wnt signaling is known to influence the specification [2,30–32,60–63], delamination [26,29,64–67], and migration [2,31,68,69] of NCC (reviewed more extensively in [33,34,70]), but it also influences fate choice in trunk neural crest. Wnt1 expression is apparent in the dorsal neural tube at E8.5 in mouse embryos prior to neural crest migration [33,71], but it also appears to play a later role in neural crest fate choice. Conditional inactivation of Wnt signaling in mouse NCC results in loss of melanocyte and sensory neuron lineages [35,72]. Hyperactivation of Wnt signaling using a constitutively active beta-catenin expressed in mouse neural crest induces ectopic sensory neuron formation at the expense of other cell types [73]. However, zebrafish NCC injected with beta-catenin mRNA assume a pigment cell fate at the expense of neurons and glia [74].

Similarly, treatment of chick neural crest explants with Wnt3a-conditioned media results in a preponderance of melanocytes at the expense of neurons and glia [75]. It is likely that the timing of Wnt signaling regulates the potential disparate roles of this signaling pathway in cell lineage specification, as temporal disruption of Wnt signaling results in distinct deficits [76,77].

RA, a derivative of vitamin A/retinol, appears to initially regulate whether cells develop as anterior neural plate or neural crest in *Xenopus* [78], but RA continues to influence neural crest development in several model systems. RA emanating from ventral neural tube at later developmental timepoints is positioned appropriately to influence neural crest to develop as neurons [79], possibly via enhanced precursor proliferation. Both isoforms of cellular retinoic acid binding protein (CRABP), an intracellular molecule that binds and transports RA, and at least one of the retinoic acid receptors (RXRs), RXRα, are expressed in both neural crest and DRG [80–87]. Experiments to determine the effect of RA on cultured quail [88] and mouse [89] NCC suggest RA is both mitogenic and influences cells to develop as sympathetic neurons.

THE SURVIVAL AND MULTIPOTENCY OF TRUNK NEURAL CREST IS MAINTAINED BY SEVERAL TRANSCRIPTION FACTORS EXPRESSED DURING MIGRATION

Sox10 Promotes Neural Crest Survival, Supports Glial Development, and Maintains Neural Crest Multipotency

The Sox (SRY-related, HMG box-containing) family of transcription factors contains several members expressed in neural crest (the SoxE subfamily): Sox8, Sox9, and Sox10 (discovered in [90] and reviewed in [91]). Of these, Sox10 is notable among the SoxE family for its continued expression throughout migration and glial differentiation [92]. Sox10 expression is initiated in the neural crest as early as E8.5 in mice [93,94], HH 7/8 in chick [95], St 12 in *Xenopus* [96], and 5SS in fish [97]. Its expression is maintained in Schwann cells and satellite glia of the DRG, though in crest-derived sensory neurons, Sox10 is downregulated concurrent with differentiation [93–95,98,99].

Perturbation of Sox10 function (due to haploinsufficiency) in humans causes Waardenburg–Hirschsprung disease [100,101]. This disorder is characterized by phenotypes consistent with aberrant neural crest development: pigment disorders, deafness, and dysfunction of the gastrointestinal tract due to incomplete enteric nervous system formation. Mouse Sox10 mutants have been isolated both by targeted inactivation [94,98,102,103] and serendipitous discovery [104–106]. These mutants exhibit pigment deficits and absence of peripheral glia, and they are generally unviable. Surprisingly, DRG in Sox10 mutant mice initially develop normally, but these DRG degenerate over several days, probably because they lack glial support [94,102]. Other reports suggest that undifferentiated progenitors in the DRG die by apoptosis [98]. Work in zebrafish suggests migrating NCC require Sox10 for survival [97,106,107]. These results can be reconciled by a model in which Sox10 promotes the survival of a subset of neural crest directly and thereby also indirectly promotes the survival of DRG neurons.

In addition to ensuring the survival of neural crest, Sox10 also appears to maintain neural crest in an undifferentiated, proliferative state. Constitutive activation of Sox10 in cultured neural crest stem cells confers resistance to the differentiating factors BMP2 and TGFβ [99]. Electroporation of Sox10 into chick neural tube generates Sox10 + migrating cells that also express the NCC marker HNK-1 but fail to express other markers of differentiation [92]. A limited body of evidence also suggests

that Sox10 acts as a top-level, positive regulator of transcription factors in the gene networks leading to both sympathetic ganglia differentiation (Figure 13.5A) and DRG differentiation (Figure 13.5B). Mouse embryos carrying truncation mutations in *sox10* fail to express Ascl1, a top-tier transcription factor that is indicative of sympathetic neuron precursors. Likewise, in culture, Sox10 expression appears to maintain Ascl1 expression in the face of TGFβ challenge, a treatment that usually extinguishes neurogenic potential [99]. Overexpression of Sox10 in zebrafish embryos induces the expression of *neurog1*, a hierarchically similar transcription factor in the sensory neurogenic cascade [107].

ErbB/NEUREGULIN SIGNALING SUPPORTS NEURAL CREST MIGRATION AND GLIAL DEVELOPMENT

The ErbB family of tyrosine kinase receptors contains four members in mammals and are receptors of varying affinity for neuregulin (Nrg) (reviewed in [146]). Neuregulin itself comes in multiple isoforms, each of which is generated by alternative splicing from four mammalian loci; of these, Nrg1 appears to be most important to neural crest development [147–152]. Signaling through ErbB receptors involves the formation of receptor dimers; these can be homodimers or heterodimers (reviewed in [146]). Although ErbB3/B2 and ErbB4/B2 heterodimers are believed to be the functional Neuregulin receptor complexes, ErbB4/2 function seems to be confined to cardiac development [153–156]. While ErbB2 is not itself a high-affinity receptor for neuregulin, it remains of interest as its heterodimerization with ErbB3 increases ErbB3's affinity for neuregulin [157]. ErbB3 is expressed in mouse DRG as early as E10 [158]. The expression of neuregulin splice variants is tissue-specific,

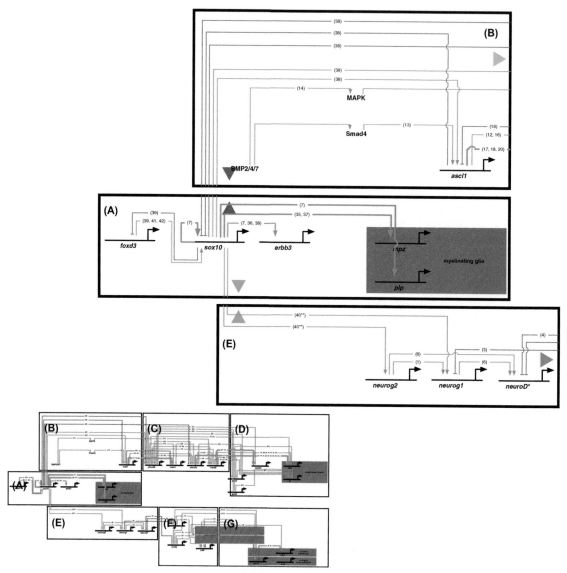

FIGURE 13.5 **Gene regulatory network leading to DRG differentiation, gliogenesis, and autonomic nervous system development.** We summarize diverse sources of evidence in this figure, and it is possible that any given linkage may or may not be direct, since there is a relative paucity of studies that establish direct binding between transcription factors and their targets. We have cited each linkage and described the type of study done in support, so readers can draw their own conclusions as to the strength of the link. Thicker lines indicate direct binding studies. (A) The early/gliogenic component of the PNS transcriptional network. Sox10 directly activates the transcription of myelin-specific genes, but also activates top-tier autonomic and sensory transcription factors. (B) Initial signaling steps that specifies the neurons of the sympathetic ganglia. BMP2/4/7 signaling activates a transcriptional cascade leading to DβH and TH transcription via Ascl1 and Phox2b. (C) Second-tier transcription factors in autonomic specification. Note the extensive cross-regulation in this module. (D) Later events in autonomic specification. Gata and Hand transcription factors are the last to be expressed; DβH and TH indicate overt autonomic neuronal differentiation. (E) Initial steps in the gene regulatory network that

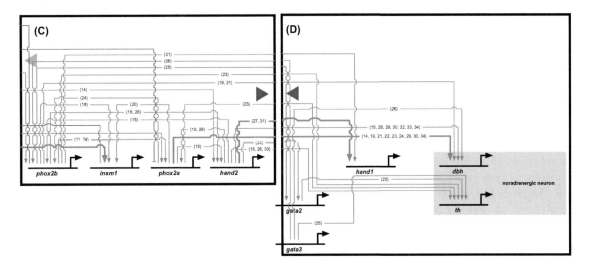

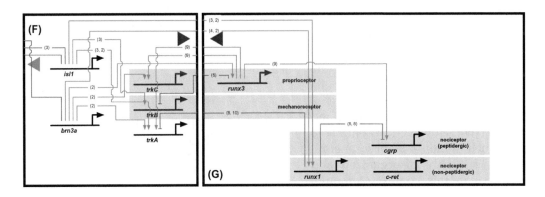

FIGURE 13.5   (Continued).

specifies the neurons of the DRG. An unknown combination of signals initiates the expression of Neurogenins, which cooperate to activate downstream neurogenic factors. (F) Second-tier transcription factors in sensory specification. In contrast to the sympathetic gene regulatory network, the program for specification in DRG is more hierarchical and generates a diversity of cell types through the restricted expression of survival mechanisms and cell-type-specific transcription factors. (G) Later events in sensory specification. Runx transcription factors refine Trk receptor expression to help define particular neuronal subtypes. * There are several *neuroD* genes, each repressed by Isl1 and Brn3a to a different extent; here they are condensed to a single unit for clarity. ** This result was obtained in zebrafish, which has only one *neurog* gene—it is therefore difficult to extrapolate how this regulatory interaction would appear in mammals, which have two *neurog* genes. The following works are cited as support for the interactions displayed in the figure: (1) [108], (2) [109], (3) [110], (4) [111], (5) [112], (6) [113], (7) [114], (8) [115], (9) [116], (10) [117], (11) [118], (12) [119], (13) [120], (14) [121], (15) [122], (16) [123], (17) [124], (18) [125], (19) [126], (20) [127], (21) [128], (22) [129], (23) [130], (24) [131], (25) [132], (26) [133], (27) [134], (28) [135], (29) [136], (30) [137], (31) [138], (32) [139], (33) [140], (34) [141], (35) [142], (36) [94], (37) [143], (38) [99], (39) [92], (40) [107], (41) [144], (42) [145].

with Nrg1 type III expression initiating in the DRG at E10 and remaining strong thereafter [158]. ErbB2/3 and Nrg1 type III are therefore the most important ErbB-related factors implicated in neural crest development.

Neuregulin/ErbB signaling appears to serve multiple functions in neural crest development. Though early studies found no evidence of DRG hypotrophy following Nrg inactivation [155,156], later studies do observe this defect [159], probably because these neurons degenerate between E10.5 and E14.5 in the absence of neuregulin and the effect must therefore be observed at later timepoints. Zebrafish exhibit somewhat different Nrg requirements—they posses three known Nrg genes; of these, both Nrg1 and Nrg2a appear to be important for normal neural crest development [160].

Targeted deletion of ErbB2, ErbB3, or the Nrg1 locus causes neural crest to pile up in the DRG anlage with an inability to migrate more ventrally such that sympathetic ganglia fail to develop [159,160]. Migratory phenotypes are also observed in zebrafish ErbB2 and ErbB3b mutants [160]. However, in contrast to the results obtained in mouse embryos, neural crest migration is highly disorganized and perhaps excessive in this context. Although the sensory neurons of the DRG initially form normally in mouse mutants of both ErbB3 [161] and ErbB2 [162,163], these neurons are rapidly lost due to apoptosis. In the comparable zebrafish alleles, DRG do not appear to form at all, though apoptosis is not excluded as a mechanism for this phenotype [160]. Viewed in conjunction with the divergent observations pertaining to neural crest migration and neuregulin dependence, these results suggest ErbB/Nrg signaling may have somewhat different effects in zebrafish than in mammals. However, in either model, ErbB/Nrg signaling is probably involved in neural crest migration and is certainly essential to the development of DRG neurons.

The most consistent phenotype across all alleles of ErbB studied is a failure of peripheral glial development, an outcome that occurs in every studied mutation of ErbB2, ErbB3, or neuregulin [155,156,158,159,161–164]. Nrg1 type III appears to be specifically required for the formation of these glia, as replacing the endogenous locus with an Nrg cassette in which the Ig domain has been disrupted (yielding an Nrg type III analog that is not susceptible to alternative splicing) is sufficient to rescue loss of Nrg1 function with respect to the glial defect [156,158]. ErbB signaling appears to be necessary for several steps in glial development, including fate acquisition [165], proliferation of glial progenitors [164], and migration to target [164]. Whether the defects observed in DRG neurons following the disruption of ErbB/Nrg signaling are secondary to the glial phenotypes remains an interesting question, particularly in mammals where the defect appears to be apoptosis of neurons.

NEURAL CREST ARE INSTRUCTED TO DEVELOP AS AUTONOMIC NEURONS BY BMP SIGNALS EMANATING FROM THE DORSAL AORTA

BMPs Directly Promote Neuronal Fates

In mice, neural crest derivatives appear to differentiate generally in the order with which they terminate their migration [11]. Under this model, autonomic derivatives are the first to form, followed by DRG (which are situated more dorsally). Melanocytes, derived from the later dorsolateral migratory stream, are the last to develop. It should be noted that some experiments in chick embryos contradict this order of differentiation, finding instead that DRG neurons and sympathetic ganglia differentiate more or less simultaneously [6,166]; but other experiments in chick do support the ventral-to-dorsal differentiation sequence model [36]. We will discuss the formation of the PNS derivatives of the neural crest in the order with which they are classically believed to form, beginning with the development of the sympathetic ganglia and following with the development of the DRG.

The autonomic nervous system encompasses the parasympathetic and sympathetic systems of peripheral nerves and ganglia. Parasympathetic nerves project from the central nervous system, synapsing on ganglia located close to or within target organs [22]. Sympathetic neurons are generally situated as a segmented strip of bilaterally symmetric ganglia lying ventral to and flanking the spinal cord. These are known as the sympathetic chain ganglia or the paravertebral ganglia. These ganglia also receive CNS enervation from preganglionic fibers; the postganglionic fibers then project long distances to enervate target organs [22].

A large body of evidence now suggests that exposure to BMPs during dorsoventral neural crest migration specifies the autonomic lineage. However, the role of BMPs in this process appears to be quite complicated. It is clear that several BMPs are expressed by the dorsal aorta, including BMP2 [167,168], BMP4 [169,170], and BMP7 [170]. All three BMPs are capable of increasing the incidence of either neuronal morphologies or autonomic markers such as tyrosine hydroxylase (TH) in culture [39,121,167,170−174] and these effects are comparable to those observed when neural crest is co-cultured with dorsal aorta explants [170]. Viral expression of a constitutively active BMPR1A receptor in neural crest exerts similar effects [172] and administration of the BMP antagonist Noggin [119] or a dominant negative BMP receptor [127] blocks autonomic specification both *in vitro* and *in vivo*.

By what mechanism does BMP treatment increase the density of presumptive autonomic neurons? It does not appear to accelerate the proliferation of autonomic precursors, as BrdU + neurons are not more prevalent following BMP treatment [121,171]; furthermore, BMPs neither induce apoptosis directly, nor promote the survival of a committed subset of NCC [39,171,175]. After implantation with Noggin-soaked agarose beads, chick embryos exhibit colonization of the sympathogenic anlage (as assessed by expression of Sox10), but the expression of markers of noradrenergic fate is not initiated [119,122]. A nearly identical result is observed when BMPR1A is conditionally deleted from neural crest [120]. These results suggest that although neural crest survives and migrates appropriately, the program of sympathetic neuron differentiation is never induced in the absence of BMP. BMP therefore appears to directly instruct neural crest to adopt autonomic fates.

The conditions under which this instruction can occur appear to be somewhat narrow, however, and this has led to confusing results. Neural crest cultures respond to BMP concentrations in a non-monotonic fashion, with lower doses of BMP2 and BMP4 strongly promoting autonomic differentiation. However, this effect falls off dramatically at higher doses [170,171]. Conversely, BMP7 is less potent than BMP2 or BMP4, and exhibits a monotonic relationship between concentration and autonomic induction [170]. There also may be a narrow temporal window in which neural crest is responsive to BMPs, as cells treated after some time *in vitro* or isolated from older embryos are less responsive to attempts to induce autonomic differentiation [170−173,175,176]. Consequently, neural crest seems to have a very narrow susceptibility to BMP-mediated autonomic specification.

Several investigators have attempted to further characterize the BMP-activated signal cascades affecting sympathetic nervous system development. BMP receptor activation leads to phosphorylation of the R-Smads 1, 5, and 8, which are typically shuttled to the nucleus by Smad4; this is known as canonical BMP signaling (recently reviewed in [177]). Smad4 conditional knockout mouse embryos exhibit normal sympathetic ganglia initially, but by E16.5, ganglion size is reduced as a consequence of reduced proliferation [120]. Surprisingly, R-Smads translocate to the nucleus in sympathetic ganglia even after Smad4 deletion, suggesting that another, as of yet undescribed mechanism may mediate other aspects of BMP-induced sympathetic neurogenesis. Non-canonical BMP (i.e., non-Smad-mediated) signaling is also implicated in noradrenergic differentiation via MAPK and PKA-dependent mechanisms [121,174]. We can conclude, perhaps unsurprisingly, that BMPs influence on neural crest is mediated by multiple, partially redundant arms of the BMP signaling pathway.

BMPs Trigger Changes in Trophic Factor Dependence

In contradiction with the above evidence, some investigators observe only transient increases in TH + cell density [178] or extensive apoptosis following BMP administration [179,180]. These findings may reflect differences in experimental approach, or the narrowness of the window in which BMPs can promote autonomic fates. However, it may also be that they reflect the changes in neurotrophic factor dependence that BMP enforces as a consequence of its pro-autonomic function. During the development of the vertebrate nervous system, far more neurons are generated than can be retained in the adult organism. In one of the most widely appreciated studies in all of developmental biology, Levi-Montalcini, Hamburger, and investigators following their work established that whether neurons live or die is predicated on their acquisition of limited growth factors called neurotrophins (reviewed in [181,182]). A variety of neurotrophins and

their cognate receptors have now been identified, including the high-affinity neurotrophin receptors TrkA, TrkB, and TrkC. As we shall see, these neurotrophin receptors are essential components of all aspects of PNS development and serve as some of the earliest indicators of neuronal subtype in both autonomic and sensory lineages.

One consequence of adopting a neuronal fate is an acquired dependence on the neurotrophic factors that bind the Trk receptors. NGF, BDNF, and NT-3 are primarily (but not exclusively) ligands for the TrkA, TrkB, and TrkC receptors, respectively [183—190]. Genetic inactivation of the neurotrophins and their cognate receptors rapidly revealed that sympathetic ganglia rely on neurotrophic signaling throughout their development. TrkA and NGF knockout mice both exhibit reductions in the size of sympathetic ganglia [191,192]; in the case of NGF−/− mice, sympathetic ganglia cannot be identified in dissection by P14 [192]. TrkC and NT-3 knockout mice exhibit comparable phenotypes that can be duplicated using NT-3 antiserum injection [193—196]. Signaling through TrkA and TrkC therefore appear to be critical to the maintenance of the autonomic nervous system.

Surprisingly, cells in the sympathetic ganglia exhibit varying reliance on these neurotrophic factors at different developmental timepoints. Initially, cultured sympathetic neurons survive only in response to NT-3, but cultures derived from postnatal ganglia exhibit a significantly stronger survival response to NGF than NT-3 [197,198]. This switch in neurotrophin requirements coincides with the expected changes in Trk receptor expression, with TrkC expression declining over developmental time and being replaced in sympathetic ganglia by TrkA [197,198]. How is this transition initiated? BMP appears to induce TrkC expression either directly or indirectly and suppress TrkA expression [199]. Since BMP is only transiently available in the sympathetic anlage, it is likely that BMP confers dependence on NT-3/TrkC signaling for trophic maintenance. This dependence is later transferred to NGF/TrkA, perhaps as a consequence of cessation of BMP exposure.

THE TRANSCRIPTIONAL CASCADE LEADING TO AUTONOMIC NERVOUS SYSTEM DIFFERENTIATION IS EXTENSIVELY CROSS-REGULATORY

Until this point, we have restricted the discussion of peripheral neurogenesis primarily to signaling molecules and survival factors. However, as in other tissues, these factors inevitably cooperate to activate a transcriptional cascade that in turn establishes the functional and morphological changes that are coincident with differentiation. For the remainder of this chapter, we will discuss the transcriptional networks responsible for the generation of the autonomic and sensory lineages.

Autonomic Differentiation Begins with the Expression of Ascl1

Achaete-scute homolog 1 (Ascl1) is a proneural, bHLH transcription factor and the first indicator of neuronal differentiation in the autonomic lineage; it is in fact excluded from all other neural crest derivatives [200,201]. A great deal of evidence suggests that Ascl1 expression is a direct outcome of BMP signaling. The onset of Ascl1 expression coincides with the first exposure of neural crest to BMPs at the dorsal aorta, occurring at 10.5 in mouse [200] and HH15 in chick embryos [202]. The domain of BMP and Ascl1 expressions overlap spatially in the sympathetic anlage and cultured neural crest stem cells express Ascl1

following BMP treatment [167,169]. Abrogation of BMP signaling using Noggin-soaked beads implanted in the path of migrating neural crest [119] or by knocking out Smad4 in neural crest [120] sharply reduces Ascl1 expression. Ascl1 expression also persists when sympathogenic neural crest is deprived of axial signals, a property other genes in the sympathetic program do not share [203]. Ascl1 appears to occur earlier than any other marker of the autonomic lineage and is dependent on BMP signaling.

Early studies of Ascl1 function appear to place it as a master proneural gene for the specification of the sympathetic lineage. Consistent with this hypothesis, neural crest explants from Ascl1−/− mice fail to form neurons in culture [204] and forced expression of Ascl1 in neural crest stem cells [123] or in *Xenopus* embryos [127] is sufficient to induce neuronal markers. However, studies of targeted inactivation of Ascl1 in an *in vivo* context yield results inconsistent with this model. Although many markers of sympathetic neurogenesis (such as Phox2a, TH, and DβH) are absent at early stages of autonomic development, others (such as c-Ret, or the neurofilament proteins) are present [204], and thoracic sympathetic ganglia express the full complement of sympathetic markers within a few days [205−207]. One implication of these results is that Ascl1 is one of a set of at least partially redundant proneural factors and is merely the first to appear. Therefore, in its absence, sympathetic ganglia either fail to form or are hypertrophic due to delayed neuronal differentiation or altered proliferation of precursors.

Expression of the Phox2 Transcription Factors is Essential for the Continued Development of the Sympathetic Ganglia

Rather than being a master regulatory gene, Ascl1 is an early and significant component in a transcriptional network that utilizes extensive cross-regulation and functional overlap rather than one in which an orderly sequence of gene expression leads to differentiation. The paired homeobox Phox2 genes are the next transcription factors expressed in the sympathetic anlage, and their relationship to Ascl1 and other sympathogenic genes typifies this complexity. Phox2b is expressed in sympathetic ganglia at the earliest stages of ganglion formation and precedes that of Phox2a or later noradrenergic markers [118,176,208]. Targeted inactivation of Phox2b does not impair Ascl1 expression and Ascl1 deletion does not affect Phox2b expression, so initiation of either gene is clearly independent of initiation of the other [118,207]. Despite this apparent redundancy, the consequences of Phox2b loss-of-function are dire—although neural crest migrates appropriately, colonizes the sympathetic anlage, and expresses Ascl1, by E13.5, all prospective sympathetic neurons are lost due to apoptosis and the sympathetic ganglia are consequently not maintained [118]. Despite its later expression, Phox2b appears to be hierarchically equivalent to Ascl1.

By contrast, Phox2a is the first transcription factor in the autonomic cascade that seems to depend on prior factors. Phox2a was discovered first, though it is expressed slightly later than Phox2b, being initiated at E10.5 in mouse embryos and HH18 in chicken embryos [203,209,210]. Phox2a expression appears to be entirely dependent on Phox2b-mediated activation [118], although its expression can be induced in culture via forced expression of Ascl1 [123]. Multiple *in vitro* studies have established that Phox2a directly binds and activates the promoters for DβH and TH [126,128,129], and forced expression of Phox2a in culture or *in vivo* confirms these results [130,131], though these results may depend on the type of cell line employed and whether other signaling factors such as cAMP are

present in sufficient quantity [123]. Taken together, these results show that Phox2a is initiated by first-tier factors and contributes directly to sympathetic differentiation. This behavior will be observed in many of the transcription factors discussed subsequently.

Insm1 Is Activated Downstream of Phox2b and Promotes Proliferation in Nascent Sympathetic Ganglia

Insm1 is a zinc-finger transcription factor expressed as early as E9.5 in the sympathetic anlage, making it one of the earliest-expressed genes implicated in sympathetic neuron development [125]. Targeted deletion of Insm1 impairs neither Ascl1 nor Phox2b expression, suggesting it is situated downstream of both genes or functions in a parallel or different pathway [125]. In Insm1 knockout mice, the expression of later sympathetic markers are reduced but eventually restored, reflecting the extensive redundancy of the sympathogenic gene regulatory network [125]. Insm1 loss-of-function does permanently reduce the size of sympathetic ganglia, and this is coincident with a reduction in proliferating cells [125]. Furthermore, in the absence of Insm1, Ascl1 expression is upregulated; this suggests an important role for Insm1 in terminating Ascl1 expression [125]. Insm1 therefore represents a second-tier factor that regulates proliferation and shuts off earlier factors in the network; the latter is a function that is surprisingly rare in the autonomic cascade.

Hand 1 and 2 Mediate Specific Aspects of Noradrenergic Differentiation

The Hand transcription factors are bHLH transcription factors that appear to mediate the adoption of a restricted set of noradrenergic characteristics. Hand2 is expressed first of the two Hand factors, coming on at E10.5 in mouse embryos in sympathetic ganglion precursors [211,212]; both genes are expressed in chick embryos by HH17 [213]. Both Hand1 and 2 appear to predominantly lie downstream of the other factors discussed in the transcriptional network leading to sympathetic neuron differentiation, but as with most other genes in this network, evidence exists to suggest they also regulate upstream elements potentially as components of feedback loops. It is difficult to place Hand2 in the hierarchy of the sympathogenic transcriptional regulatory network due to contradictory evidence. Nonetheless, all studies agree that expression of TH and DβH are lost following Hand2 deletion [134,135,139,140] and expanded following Hand2 overexpression [122,213]. Furthermore there is additional evidence to show that Hand2 directly binds the DβH promoter and activates it [136–138].

It is somewhat less clear whether any of the other transcription factors in the network are susceptible to manipulation of Hand2 levels, but the weight of the evidence suggests Hand2 expression functions later in the hierarchy of sympathetic neuron differentiation [122,134,135,139,140,211], with perhaps only Hand1 being wholly dependent on Hand2 function [134]. Hand1 in fact may itself be wholly dispensable for sympathetic neuron formation, as knockouts appear to have normal TH and DβH expression. Investigations as to the mechanism by which Hand2 contributes to sympathetic ganglion development are also murky. Some investigators find reduced proliferation as a consequence of forced Hand expression in cultured NCC [213], suggesting Hand promotes cell cycle exit, but another group working in Hand2 knockout mice observes a reduction in sympathetic progenitors [135]. It is difficult to reconcile these results, and so the role of the Hand genes in sympathetic

neuron development therefore remains a fertile field of inquiry.

Gata Transcription Factors Are Situated Near the End of the Sympathogenic Transcriptional Cascade

Although transcription factor expression in the sympathetic lineage is hardly sequential, the Gata factors appear to be expressed the latest of the genes discussed here. The Gata family of genes contain six members in mammals, all of which are zinc-finger transcription factors [214]. Two of these, Gata2 and Gata3, exhibit specific expression in the sympathetic ganglia and appear to influence their development. Gata2 expression in the sympathetic lineage begins at HH18 [203] appearing slightly later than that of Phox2b, Hand2, and Phox2a [132]. Gata2 expression is dependent on BMP signaling, and abrogation of Gata2 expression using dominant negative constructs reduces TH staining and causes hypertrophy in sympathetic ganglia [132]. Surprisingly, the same authors find that Gata2 overexpression does not itself lead to the expression of any other sympathetic marker, but does induce expression of other pan-neuronal markers [132]. Paradoxically, the function of Gata3 in sympathetic ganglia appears to vary depending on the model organism employed. Gata3 expression is initiated in sympathetic ganglia at E11.5 in mouse embryos [215] but apparently is never expressed in the sympathetic ganglia of chick embryos [132] and therefore may only be relevant to mammalian sympathetic ganglia development. Targeted deletion of Gata3 in mouse embryos leads to reductions in sympathetic ganglia volume and TH expression [133,216]. Taken together, the Gata factors appear to be positioned last in the gene regulatory sequence, being primarily responsible for activating the enzymes in the catecholamine biosynthetic pathway or other neuronal differentiation genes.

Lessons Learned from the Autonomic Transcription Factor Cascade

What can we glean from this detailed investigation of the autonomic gene regulatory network? The network is initiated by the expression of the top-tier factors Ascl1 and Phox2b in the sympathetic anlage, and this initiation appears to be wholly dependent on BMP signaling emanating from the dorsal aorta. The expression of second-tier factors (defined as those dependent on expression of Ascl1 and Phox2b) follows.

These later factors (Phox2a, Insm1, and Hand2) extensively activate one another and promote the continued expression of the top-tier factors. At the same time, they promote the expression of noradrenergic markers. The autonomic cascade is notable for its lack of negative regulation and its relative egalitarianism among factors. This may reflect the relative simplicity of the sympathetic lineage, which does not exhibit the cell-type diversity observed in the DRG. As we shall see, that network is far more hierarchical and less compressed in developmental time. This might also be compensated for in part by the prevalence of cross-regulation amongst the players involved.

DRG FORMATION OCCURS IN OVERLAPPING WAVES OF NEUROGENESIS THAT CAN BE FOLLOWED USING THE Trk RECEPTOR FAMILY AS MARKERS OF CELL FATE

As in the autonomic lineage, the sensory neurons of the DRG must acquire neurotrophic factors to survive. In the autonomic lineage, Trk receptor dependence is more or less equivalent across all cells but varies with developmental time. By contrast, Trk receptor requirements in DRG appear to vary primarily with cell subtype. The sensory neurons of the DRG convey several sensory modalities from the periphery to the central nervous system; these are proprioception, mechanoreception, thermoreception, and nociception. Each sensation is conveyed by distinct neuronal subtypes (reviewed in [217]) and consequently, the undifferentiated neural crest must undergo fate decisions to give rise to this neuronal diversity. The expression of each individual neurotrophin receptor is associated with a different sensory modality carried by DRG. Trk expression is complicated by the fact that DRG neurons initially exhibit somewhat broad expression of these receptors, which is then segregated as development progresses.

DRG Neurogenesis Is Traditionally Understood to Occur in Distinct Waves

In chick and mouse embryos, large-diameter, mechanoreceptive and proprioceptive neurons are the first to be born [12,33]. These cells express the neurotrophin receptors TrkB and TrkC [218–221]. Proprioceptive neurons later downregulate TrkB, whereas mechanoreceptive neurons express various combinations of TrkB, TrkC, and Ret before downregulating TrkC [112,222]. Mechanoreceptors are therefore often referred to as TrkB + and proprioceptors as TrkC +; this is an oversimplification, however, as some coexpression of Trk receptors clearly persists postnatally [223,224]. In a slightly delayed but overlapping phase of neurogenesis, small-diameter, nociceptive neurons are born [12,33]. These cells initially express TrkA, although this expression will be later downregulated in the subset of cells that differentiate as non-peptidergic nociceptors [115,117,225,226] (and see [227] for a study of TRP/Trk receptor coexpression).

What is the source of this second wave of neurogenesis? Between E10.75 and E13.5 in mouse embryos, Krox20/Egr2+ boundary cap cells located at the dorsal root proliferate and infiltrate the DRG [13]. Ultimately, these cells will contribute to the Schwann glia of the dorsal root, satellite glia adjacent to TrkB/C+ large-diameter neurons, and a subset of the TrkA+ nociceptive neuronal population [13]. In chick embryos an additional wave of contralateral neural crest migration into the nascent DRG between HH18–22 has been observed [15]. As with boundary cap cells, these late-infiltrating cells primarily differentiate as TrkA+ nociceptive neurons [15]. Cells in the peripheral layer of the chick DRG, present as early as HH21 (this layer is itself partially derived from the contralaterally-migrating cells already described) divide, infiltrate the DRG core, and differentiate primarily as TrkA+ neurons [166]. Therefore, this second wave of neurogenesis is probably derived from multiple spatially distinct but lineally related progenitor populations. It is perhaps inappropriate, then, to consider it a single "wave" except insofar as TrkA+ neurons tend to arise later than TrkB/C+ neurons.

Ablation of TrkC Signaling Causes the Loss of Large-Diameter Neurons

Genetic ablation or other perturbation of a given Trk receptor or its prevalent ligand generally leads to the loss of the neurons conveying the sensory modality with which its expression is correlated, suggesting that each Trk receptor supports the survival of the submodality with which it is associated. TrkC, the high-affinity receptor for NT-3, is the first Trk receptor expressed in DRG, potentially appearing in some migrating neural crest even prior to overt neuronal differentiation [224,228–231] although this claim is disputed [232]. Mice in which the TrkC receptor has been genetically ablated [233] or disrupted using anti-TrkC antibodies [234]

exhibit significant loss of DRG neurons—about 20% following receptor knockout and about 50% following antibody treatment. In both cases, this loss falls most heavily on large-diameter proprioceptive neurons. Genetic ablation of the TrkC ligand NT-3 leads to a much more severe defect with investigators observing between 50% and 80% fewer neurons in mutants than in wild-type mice [193,194,223,235–238]. Clearly, TrkC/NT-3 signaling is essential for the retention of large-diameter neurons.

Yet although proprioceptors are obviously lost following this manipulation [233,239], the magnitude of neuronal loss is incompatible with a model where only postmitotic TrkC+ proprioceptors are dependent on NT-3. This discrepancy can be explained if NT-3 is required to support a cycling progenitor population, and indeed a role for NT-3 in maintenance of actively cycling DRG neuronal progenitors has been observed *in vivo* [238,240] and in various culture models [197,241]. However, other studies find no evidence that Trk+ cells divide or that progenitors are depleted following NT-3 inactivation [232,235]. It is likely that NT-3 supports the maintenance of other neuronal subtypes via weaker interactions with the TrkA and TrkB receptors. Indeed, apoptotic TrkB neurons are observed following NT-3 loss-of-function [232], and double knockouts of BDNF and NT-3 do not have an additive effect on neuron loss, suggesting overlap in their neurotrophin requirements [223]. NT-3 may also directly promote or accelerate the adoption of neural or glial fates by undifferentiated neural crest [242,243]. Thus, the requirement for NT-3 is broader than at TrkC+ neurons alone.

Loss of TrkB Function Causes Loss of a Subset of DRG Neurons

Interpreting experiments concerning the dependence of nascent DRG on TrkB/BDNF is somewhat more straightforward than for TrkC/ NT-3. TrkB knockout mice do not respond to painful stimuli and exhibit approximately 30% fewer DRG neurons than wild-type controls [244]; knocking out BDNF has a comparable effect on DRG neuron density [245,246]. As is the case for NT-3, BDNF may additionally act as a mitogenic factor for neuronal precursors [240] or as a maturation factor [242]. TrkB/ BDNF signaling therefore appears to be somewhat more tightly cell-type restricted than is observed for NT-3.

Removal of TrkA and its Ligand Cause Loss of Most Small-Diameter Neurons

As TrkA is ultimately the most widely expressed of the Trk receptors, and since it supports the maintenance of nociceptive neurons (themselves the most prevalent of the DRG neurons), it is perhaps unsurprising that knocking out TrkA leads to massive losses in DRG volume. Following TrkA loss of function, mice DRG are reduced 70–90% of wild-type size [191,247,248]. The distribution of surviving neurons is increased in cell diameter and surviving embryos do not respond to a variety of painful stimuli, suggesting nociceptors are specifically lost. Reducing or ablating NGF function using antibodies administered *in utero* [249–252] or through targeted gene inactivation [192] parallel the results of TrkA loss of function almost exactly. As with TrkB/BDNF, TrkA signaling appears to correlate with a single cell type—in this case, the small-diameter, nociceptive neurons.

Trk Receptor Studies Are Complicated by the Fact That Trk Receptor Expression is Not Exclusive

To what extent DRG neuron progenitors are committed to their fates as they express multiple Trk receptor subtypes is somewhat unclear. TrkB+/TrkC+ cells are frequently observed at early stages of DRG development, though their expression eventually resolves into two largely distinct populations by E14.5 in mouse embryos [112,232]. TrkA expression initially overlaps with that of TrkB or TrkC expression at E11.5 in mice [110], and most TrkA+ or TrkB+ cells in chick embryos also initially express TrkC [224], though this coexpression disappears rapidly. And though the expression of each of the Trk receptors correlates strongly with the neuronal diameter in concordance with the sensory modality it purportedly represents (TrkC+/proprioceptors being larger than TrkB+/mechanoreceptors being larger than TrkA+/nociceptors [219,220,253]), it must be noted that there is considerable overlap in the size distributions of TrkA-, TrkB-, and TrkC-expressing cells. This may mean that neuronal precursors are merely biased toward one subtype or another, rather than being exclusively restricted to the TrkB/ C+ or TrkA+ waves as they are traditionally conceived. Alternatively, and arguably more plausibly, cells may commit to one fate or another long before Trk expression resolves, perhaps because there is a general requirement for other neurotrophins earlier in their development. A closer dissection of the developmental potential of Trk-expressing cells therefore remains a productive field of inquiry.

Zebrafish embryos present an intriguing alternative strategy for the development of the DRG. Zebrafish DRG consist initially of only one or two crest-derived neurons, but the population of DRG neurons increases gradually over developmental time, apparently supported by local progenitors that express the neural crest/glial marker Sox10 [254]. It should be noted that the temporal dynamics of neuronal subtype generation have not yet been tested in the zebrafish owing to the relatively poor quality of Trk receptor markers in fish relative to other models.

THE DRG TRANSCRIPTIONAL CASCADE IS HIERARCHICAL AND SORTS CELLS INTO SUBTYPES

The Neurogenins are the Top-Tier Transcription Factors in the DRG Transcriptional Cascade

The expression of the bHLH transcription factors Neurog1 and Neurog2 in neural crest is correlated with the two major waves of DRG neurogenesis. In mice and chick embryos, Neurog2 is expressed in neural crest just before Neurog1 [108,113,255], and Neurog1 knockout mice fail to develop nociceptive neurons but retain proprioceptive and mechanoreceptive neurons [108]. However, Neurog2 is not itself a master proneural gene—Neurog2 knockout mice express Neurog1 with a lag of 12—18 h and ultimately produce DRG (albeit delayed in their production of proprioceptive and mechanoreceptive neurons); it is only when both Neurog1 and Neurog2 are knocked out that DRG are fully lost [108]. It is apparent then, that Neurog1 and Neurog2 exhibit some functional overlap and that Neurog2 activates the early expression of Neurog1. Neurog1 itself seems to be required for the production of nociceptive neurons but is apparently competent to direct the development of proprioceptive and mechanoreceptive neurons as well. Neurog genes therefore appear to serve the role of Ascl1/Phox2b in the DRG cascade— that of the top-tier proneural factor.

Overexpression experiments involving neurogenins have outcomes consistent with the loss-of-function data. Forced expression of Neurogenin mRNA greatly increases neurogenesis in *Xenopus*, zebrafish, and chick embryos. These neurons express sensory markers, and their preponderance comes at the expense of glia derived from the same lineages [113,256—258]. Cultured cortical progenitors respond similarly following Neurog1 viral infection [259]. Taken together, these results indicate that Neurogenins not only promote a sensory neuron fate but also actively suppress the adoption of glial fates.

The fate commitment imposed by Neurogenin expression is fairly strong, albeit not absolute. Zebrafish retain only a single Neurogenin gene; knocking it down results in the loss of cranial sensory ganglia, DRG, and all sensory hair cell innervation [260]. In the case of zebrafish DRG, cells deficient in Neurogenin1 adopt a glial fate instead [261]. In mouse, Neurog2+ cells, while capable of adopting autonomic fates, differentiate as sensory neurons far more frequently than the progenitors from which they are derived [262]. Neurog2 is in fact capable of compensating for Ascl1 loss of function in specifying sympathetic ganglia [263]. Challenging neural tube explants with BMP2, a factor that induces an autonomic fate from undifferentiated crest cells, does not eliminate the expression of Neurog1, Neurog2, or NeuroD in nascent sensory neurons [264]. Due to their powerful proneural influence, Neurog transcription factors appear to be good candidates for top-tier sensory fate determinants.

In the sympathetic lineage, it is clear that BMPs act as a local instructive signal, imparting a neuronal fate to undifferentiated neural crest by initiating Ascl1 expression. Finding a similar signal in the initiation of Neurog expression has been challenging; however, studies in zebrafish suggest that neural crest migration itself might be somehow linked to the onset of Neurog1 expression. Embryos lacking Erbb3 or Erbb2 receptors fail to express Neurog1 in the DRG anlage despite their normal neural crest formation and early migration [160]. Erbb mutants exhibit increasingly disorganized neural crest migration over time, with cells failing to colonize the DRG anlage. Similar results are obtained in observations of Sonic hedgehog (Shh) mutants. Consistent with this, Neurog1 is not expressed in Hedgehog receptor Smoothened mutants, which also exhibit disorganized neural crest migration [265]. Neurog1 expression is also lost following inactivation of the gene Reck, which encodes an inhibitor of metalloproteinases [266]. In culture systems, cells become hypermigratory following Reck inhibition; presumably as a consequence of elevated metalloproteinase activity increasing extracellular matrix permeability [267]. In zebrafish lacking Reck expression, NCC are also hypermigratory [266]. These results might indicate that migratory behavior itself can act as a key determinant of neuronal fate, perhaps by exposing cells to other, as of yet unknown proneural signals. However, other data from zebrafish suggests that neural crest fate is set early, long before migration through the DRG anlage [35]. It therefore remains unclear exactly how Erbb signaling, Shh signaling, or Reck-mediated inhibition of MMPs contribute to DRG specification. Examination of the roles of these factors in other species is also needed.

THE SUBSEQUENT EXPRESSION OF OTHER TRANSCRIPTION FACTORS IS ESSENTIAL TO DRG DEVELOPMENT AND MAINTENANCE

Brn3a and Islet1 Are Initiated After the Neurog Factors and Cooperate to Regulate Many Targets

Following expression of the Neurogenin genes, the activity of a host of other proneural transcription factors is initiated in the nascent DRG, forming a complex hierarchical cascade of gene expression (illustrated in Figure 13.5). Brn3a is a class IV Pou domain transcription factor expressed from E10.5 onward in mouse DRG, just as sensory neurons exit the cell cycle [268–270]. Elimination of Brn3a alone has no effect on DRG neuronal density prenatally; however, expression of the neuronal markers Brn3b and Brn3c is reduced and expression of

all Trk receptors is reduced prenatally as DRG neurons gradually die [271,272]. Brn3a directly activates the transcription of many later markers of sensory neuron fate, including Runx1, Runx3, and Drg11 [111]. Coincident with its expression at the end of sensory neuron precursor division, Brn3a also represses the proneural bHLH transcription factor NeuroD1 [111]. It therefore seems to act as a transitional factor between initial specification and differentiation.

Expression of Brn3a coincides roughly with the onset of the LIM homeodomain transcription factor Islet1, another essential factor in the development of motor neurons [273]. When conditionally deleted from neural crest in mouse embryos using a Wnt1-Cre/loxP-Isl1 transgenic strategy, DRG and sympathetic ganglia exhibit apoptotic degeneration between E11.5 and E14.5. Some investigators have shown that this neuronal loss falls primarily on TrkB- and TrkA-expressing neurons [110], while others observe a deficit in TrkC-expressing neurons [270]. Isl1, like Brn3a, appears to activate later neuronal markers while inhibiting the expression of proneural genes that precede its expression—Runx1 expression is reduced in Isl1 knockout DRG, but NeuroD4 and Neurog1 exhibit increased expression [110].

Although neither is essential for initial DRG formation, both Brn3a and Isl1 appear to be critical for the maintenance of DRG neurons as development progresses. Brn3a and Isl1 are co-localized in E12.5 mouse DRG [109]. However, Brn3a and Isl1 do not appear to regulate each other's expression. Brn3a knockouts retain Isl1 expression [109,272] and Isl1 knockouts retain Brn3a expression [109,110]. Both genes appear to regulate a number of transcripts in common. However, the effect of knocking out both Brn3a and Isl1 on a given transcript is less than the effects of either single knockout added together, suggesting extensive redundancy and cooperativity in their function [109].

Runx Transcription Factors Further Refine the Dorsal Root Ganglion Sensory Neuron Population

Although the Neurogenins exhibit limited specificity with respect to the segregation of DRG neuronal subtypes, the second-tier transcription factors Brn3a and Isl1 do not. We might then expect other transcription factors to promote differential expression of Trk receptors and their concomitant commitment to proprioceptive, mechanoreceptive, and nociceptive fates. The Runx transcription factors appear to serve this function to a limited extent.

Runx3 is the first of these transcription factors to be expressed in DRG, beginning at E10.5 in mouse embryos [274]. Runx3 generally exhibits strong correlation with TrkC + prospective proprioceptors but is excluded from essentially all TrkB + cells even if those cells also express TrkC [110,112]. This expression pattern seems to reflect a direct suppression of TrkB transcription by Runx3. Expressing Runx3 under the control of the Isl1 promoter (making it essentially panneuronal in the nascent DRG) strongly suppresses TrkB cell-autonomously and Runx3$^{-/-}$ embryos exhibit an expanded TrkB+ population as well as greater coexpression of TrkB and TrkC [112,275]. These results are supported by experiments in a neuroblastoma cell line that expresses TrkB in response to RA treatment. Transfection of these cells with Runx3 prior to RA administration occludes TrkB expression and furthermore Runx3 binds directly to TrkB regulatory elements [275]. In addition to suppressing TrkB expression, Runx3 is essential for the maintenance of TrkC+ neurons. TrkC expression is initiated in Runx3$^{-/-}$ mutant mouse DRG (although it is far weaker than in wt embryos), but this expression is rapidly lost [274,275]. Runx3 therefore influences cells to assume a proprioceptive fate by inhibiting TrkB expression and promoting TrkC expression.

Runx1 expression is slightly delayed relative to Runx3, appearing at approximately E12.5 in mouse DRG; its expression is strongly correlated with that of TrkA [110,112,274]. Runx1 loss-of-function only marginally affects the absolute number of neurons present in mouse DRG with both increases and decreases in neuronal volume, proliferation, and cell death being reported depending on the developmental timepoint observed [115,276]. Runx1 does appear to unambiguously influence the expression of calcitonin gene related peptide (CGRP), a marker of peptidergic nociceptive neurons. Runx1 expression is excluded from CGRP + cells, and loss of Runx1 function leads to an excess of CGRP neurons [112,115]. Runx1 therefore diverts nociceptors toward a non-peptidergic fate at the expense of peptidergic cells.

CONCLUSIONS AND PERSPECTIVES: COMMON FEATURES IN AUTONOMIC AND SENSORY NEURONAL DEVELOPMENT

When viewed in totality, the developmental processes of the autonomic and sensory lineages have much in common. An initially heterogeneous population consisting of committed and multipotent progenitors migrates in phases and along different routes to reach niches for their continued development. Signals emanating from sources located along the route of migration (the spinal cord in the case of the DRG and the dorsal aorta in the case of the autonomic lineage) further influence the development of this heterogeneous cell population.

The expression of bHLH neurogenic transcription factors is the earliest indicator of neuronal differentiation in both the DRG and autonomic lineages; DRG sensory neuron fate is indicated by expression of the Neurogenins, whereas autonomic neurons express Ascl1. These transcription factors then initiate a complex cascade of second-tier transcription factors—Brn3a and Isl1 in the case of DRG neurons and the Phox2 factors in autonomic cells. Although both networks exhibit sequential expression of transcription factors, both are characterized by a great degree of redundancy and cross-regulation. The DRG will ultimately produce multiple neuronal types carrying different sensory modalities; despite this, most transcription factors implicated in their development do not appear to exhibit cell-type specificity (with the notable exception of the Runx transcription factors and perhaps Drg11 [277,278]). In sympathetic ganglia, the gene regulatory network is even more homogenous in its expression and redundant in its function. Each successive factor is expressed in essentially all neurons, and nearly every factor appears to exert positive regulatory effects on most other factors in the network.

Finally, neuronal survival is regulated by the expression and maintenance of a series of neurotrophic Trk receptors. The profile of Trk receptors changes over time in both contexts. In DRG sensory neurons, an initially broad pattern of Trk receptor expression is whittled down such that each Trk receptor defines a particular sensory modality. In contrast, in the autonomic lineage, there appears to be primarily a chronological sequence of Trk receptor expression, such that most cells are dependent on the same neurotrophins for brief developmental intervals.

References

[1] Kimmel CB, Ballard WW, Kimmel SR, Ullmann B, Schilling TF. Stages of embryonic development of the zebrafish. Dev Dyn 1995;203:253–310.

[2] Raible DW, Wood A, Hodsdon W, Henion PD, Weston JA, Eisen JS. Segregation and early dispersal of neural crest cells in the embryonic zebrafish. Dev Dyn 1992;195:29–42.

[3] Hamburger V, Hamilton HL. A series of normal stages in the development of the chick embryo. 1951. Dev Dyn 1992;195:231–72.

[4] Schoenwolf GC, Smith JL. Mechanisms of neurulation: traditional viewpoint and recent advances. Development 1990;109:243–70.

[5] Yang H-J, Wang K-C, Chi JG, et al. Cytokinetics of secondary neurulation in chick embryos: Hamburger and Hamilton stages 16–45. Childs Nerv Syst 2006;22:567–71.

[6] Serbedzija GN, Bronner-Fraser M, Fraser SE. A vital dye analysis of the timing and pathways of avian trunk neural crest cell migration. Development 1989;106:809–16.

[7] Carr VM, Simpson SB. Proliferative and degenerative events in the early development of chick dorsal root ganglia. II. Responses to altered peripheral fields. J Comp Neurol 1978;182:741–55.

[8] Hamburger V, Brunso-Bechtold JK, Yip JW. Neuronal death in the spinal ganglia of the chick embryo and its reduction by nerve growth factor. J Neurosci 1981;1:60–71.

[9] Downs KM, Davies T. Staging of gastrulating mouse embryos by morphological landmarks in the dissecting microscope. Development 1993;118:1255–66.

[10] Theiler K. The house mouse: atlas of embryonic development. New York: Springer-Verlag; 2007.

[11] Serbedzija GN, Fraser SE, Bronner-Fraser M. Pathways of trunk neural crest cell migration in the mouse embryo as revealed by vital dye labelling. Development 1990;108:605–12.

[12] Lawson SN, Biscoe TJ. Development of mouse dorsal root ganglia: an autoradiographic and quantitative study. J Neurocytol 1979;8:265–74.

[13] Maro GS, Vermeren M, Voiculescu O, et al. Neural crest boundary cap cells constitute a source of neuronal and glial cells of the PNS. Nat Neurosci 2004;7:930–8.

[14] Tamarin A, Boyde A. Facial and visceral arch development in the mouse embryo: a study by scanning electron microscopy. J Anat 1977;124:563–80.

[15] George L, Chaverra M, Todd V, Lansford R, Lefcort F. Nociceptive sensory neurons derive from contralaterally migrating, fate-restricted neural crest cells. Nat Neurosci 2007;10:1287–93.

[16] Goldstein B. Anatomy of the peripheral nervous system. Phys Med Rehabil Clin N Am 2001;12:207–36.

[17] Serbedzija GN, Burgan S, Fraser SE, Bronner-Fraser M. Vital dye labelling demonstrates a sacral neural crest contribution to the enteric nervous system of chick and mouse embryos. Development 1991;111:857–66.

[18] Vaglia JL, Hall BK. Regulation of neural crest cell populations: occurrence, distribution and underlying mechanisms. Int J Dev Biol 1999;43:95–110.

[19] Pomeranz HD, Gershon MD. Colonization of the avian hindgut by cells derived from the sacral neural crest. Dev Biol 1990;137:378–94.

[20] Schilling TF, Kimmel CB. Segment and cell type lineage restrictions during pharyngeal arch development in the zebrafish embryo. Development 1994;120:483–94.

[21] Martinsen BJ. Reference guide to the stages of chick heart embryology. Dev Dyn 2005;233:1217–37.

[22] Le Douarin NM. The neural crest. Cambridge: Cambridge University Press; 1982.

[23] Sato M, Yost HJ. Cardiac neural crest contributes to cardiomyogenesis in zebrafish. Dev Biol 2003;257: 127–39.

[24] Shepherd I, Eisen J. Development of the zebrafish enteric nervous system. Methods Cell Biol 2011; 101:143–60.

[25] Shepherd IT, Pietsch I, Elworthy S, Kelsh RN, Raible DW. Roles for GFRalpha1 receptors in zebrafish enteric nervous system development. Development 2004;131:241–9.

[26] Bronner-Fraser M. Analysis of the early stages of trunk neural crest migration in avian embryos using monoclonal antibody HNK-1. Dev Biol 1986;115: 44–55.

[27] Schwarz Q, Maden CH, Vieira JM, Ruhrberg C. Neuropilin 1 signaling guides neural crest cells to coordinate pathway choice with cell specification. Proc Natl Acad Sci USA 2009;106:6164–9.

[28] Schwarz Q, Maden CH, Davidson K, Ruhrberg C. Neuropilin-mediated neural crest cell guidance is essential to organise sensory neurons into segmented dorsal root ganglia. Development 2009;136:1785–9.

[29] Rickmann M, Fawcett JW, Keynes RJ. The migration of neural crest cells and the growth of motor axons through the rostral half of the chick somite. J Embryol Exp Morphol 1985;90:437–55.

[30] Lamers CH, Rombout JW, Timmermans LP. An experimental study on neural crest migration in Barbus conchonius (Cyprinidae, Teleostei), with special reference to the origin of the enteroendocrine cells. J Embryol Exp Morphol 1981;62:309–23.

[31] Eisen JS, Weston JA. Development of the neural crest in the zebrafish. Dev Biol 1993;159:50–9.

[32] Morin-Kensicki EM, Eisen JS. Sclerotome development and peripheral nervous system segmentation in embryonic zebrafish. Development 1997;124:159–67.

[33] Frank E, Sanes JR. Lineage of neurons and glia in chick dorsal root ganglia: analysis in vivo with a recombinant retrovirus. Development 1991;111: 895–908.

[34] Bronner-Fraser M, Fraser SE. Cell lineage analysis reveals multipotency of some avian neural crest cells. Nature 1988;335:161–4.

[35] Raible DW, Eisen JS. Restriction of neural crest cell fate in the trunk of the embryonic zebrafish. Development 1994;120:495–503.

[36] Krispin S, Nitzan E, Kassem Y, Kalcheim C. Evidence for a dynamic spatiotemporal fate map and early fate restrictions of premigratory avian neural crest. Development 2010;137:585–95.

[37] Sieber-Blum M, Sieber F. Heterogeneity among early quail neural crest cells. Brain Res 1984;316:241–6.

[38] Stemple DL, Anderson DJ. Isolation of a stem cell for neurons and glia from the mammalian neural crest. Cell 1992;71:973–85.

[39] Morrison SJ, White PM, Zock C, Anderson DJ. Prospective identification, isolation by flow cytometry, and in vivo self-renewal of multipotent mammalian neural crest stem cells. Cell 1999;96:737–49.

[40] Kruger GM, Mosher JT, Bixby S, Joseph N, Iwashita T, Morrison SJ. Neural crest stem cells persist in the adult gut but undergo changes in self-renewal, neuronal subtype potential, and factor responsiveness. Neuron 2002;35:657–69.

[41] Hagedorn L, Suter U, Sommer L. P0 and PMP22 mark a multipotent neural crest-derived cell type that displays community effects in response to TGF-beta family factors. Development 1999;126:3781–94.

[42] Sommer L. Growth factors regulating neural crest cell fate decisions. Adv Exp Med Biol 2006;589:197–205.

[43] Le Lievre CS, Schweizer GG, Ziller CM, Le Douarin NM. Restrictions of developmental capabilities in neural crest cell derivatives as tested by in vivo transplantation experiments. Dev Biol 1980;77:362–78.

[44] Schweizer G, Ayer-Le Lièvre C, Le Douarin NM. Restrictions of developmental capacities in the dorsal root ganglia during the course of development. Cell Differ 1983;13:191–200.

[45] Bronner-Fraser M, Sieber-Blum M, Cohen AM. Clonal analysis of the avian neural crest: migration and maturation of mixed neural crest clones injected into host chicken embryos. J Comp Neurol 1980;193:423–34.

[46] Baker CV, Bronner-Fraser M, Le Douarin NM, Teillet MA. Early- and late-migrating cranial neural crest cell populations have equivalent developmental potential in vivo. Development 1997;124:3077–87.

[47] Weston JA, Butler SL. Temporal factors affecting localization of neural crest cells in the chicken embryo. Dev Biol 1966;14:246–66.

[48] Henion PD, Weston JA. Timing and pattern of cell fate restrictions in the neural crest lineage. Development 1997;124:4351–9.

[49] Bronner-Fraser M. An antibody to a receptor for fibronectin and laminin perturbs cranial neural crest development in vivo. Dev Biol 1986;117:528–36.

[50] Sieber-Blum M, Cohen AM. Clonal analysis of quail neural crest cells: they are pluripotent and differentiate in vitro in the absence of noncrest cells. Dev Biol 1980;80:96–106.

[51] Baroffio A, Dupin E, Le Douarin NM. Clone-forming ability and differentiation potential of migratory neural crest cells. Proc Natl Acad Sci USA 1988;85: 5325–9.

[52] Dupin E, Baroffio A, Dulac C, Cameron-Curry P, Le Douarin NM. Schwann-cell differentiation in clonal cultures of the neural crest, as evidenced by the anti-Schwann cell myelin protein monoclonal antibody. Proc Natl Acad Sci USA 1990;87:1119–23.

[53] Ito K, Sieber-Blum M. Pluripotent and developmentally restricted neural-crest-derived cells in posterior visceral arches. Dev Biol 1993;156:191–200.

[54] Ruffins S, Artinger KB, Bronner-Fraser M. Early migrating neural crest cells can form ventral neural tube derivatives when challenged by transplantation. Dev Biol 1998;203:295–304.

[55] White PM, Morrison SJ, Orimoto K, Kubu CJ, Verdi JM, Anderson DJ. Neural crest stem cells undergo cell-intrinsic developmental changes in sensitivity to instructive differentiation signals. Neuron 2001;29:57–71.

[56] Bixby S, Kruger GM, Mosher JT, Joseph NM, Morrison SJ. Cell-intrinsic differences between stem cells from different regions of the peripheral nervous system regulate the generation of neural diversity. Neuron 2002;35:643–56.

[57] Joseph NM, He S, Quintana E, Kim Y-G, Núñez G, Morrison SJ. Enteric glia are multipotent in culture but primarily form glia in the adult rodent gut. J Clin Invest 2011;121:3398–411.

[58] Milet C, Monsoro-Burq AH. Neural crest induction at the neural plate border in vertebrates. Dev Biol 2012;366:22–33.

[59] Stuhlmiller TJ, García-Castro MI. Current perspectives of the signaling pathways directing neural crest induction. Cell Mol Life Sci 2012;69:3715–37.

[60] García-Castro MI, Marcelle C, Bronner-Fraser M. Ectodermal Wnt function as a neural crest inducer. Science 2002;297:848–51.

[61] Saint-Jeannet JP, He X, Varmus HE, Dawid IB. Regulation of dorsal fate in the neuraxis by Wnt-1 and Wnt-3a. Proc Natl Acad Sci USA 1997;94:13713–8.

[62] Chang C, Hemmati-Brivanlou A. Neural crest induction by Xwnt7B in *Xenopus*. Dev Biol 1998;194:129–34.

[63] LaBonne C, Bronner-Fraser M. Neural crest induction in *Xenopus*: evidence for a two-signal model. Development 1998;125:2403–14.

[64] Burstyn-Cohen T, Stanleigh J, Sela-Donenfeld D, Kalcheim C. Canonical Wnt activity regulates trunk neural crest delamination linking BMP/noggin signaling with G1/S transition. Development 2004;131:5327–39.

[65] De Calisto J, Araya C, Marchant L, Riaz CF, Mayor R. Essential role of non-canonical Wnt signalling in neural crest migration. Development 2005;132:2587–97.

[66] Shoval I, Ludwig A, Kalcheim C. Antagonistic roles of full-length N-cadherin and its soluble BMP cleavage product in neural crest delamination. Development 2006;134:491–501.

[67] Chalpe AJ, Prasad M, Henke AJ, Paulson A. Regulation of cadherin expression in the chicken neural crest by the Wnt/β-catenin signaling pathway. Cell Adh Migr 2010;4:431–8.

[68] Matthews HK, Marchant L, Carmona-Fontaine C, et al. Directional migration of neural crest cells in vivo is regulated by Syndecan-4/Rac1 and non-canonical Wnt signaling/RhoA. Development 2008;135:1771–80.

[69] Banerjee S, Gordon L, Donn TM, et al. A novel role for MuSK and non-canonical Wnt signaling during segmental neural crest cell migration. Development 2011;138:3287–96.

[70] Klymkowsky M, Cortez Rossi C, Artinger KB. Mechanisms driving neural crest induction and migration in the zebrafish and *Xenopus laevis*. Cell Adh Migr 2010;4:595–608.

[71] Parr BA, Shea MJ, Vassileva G, McMahon AP. Mouse Wnt genes exhibit discrete domains of expression in the early embryonic CNS and limb buds. Development 1993;119:247–61.

[72] Hari L, Brault V, Kléber M, et al. Lineage-specific requirements of beta-catenin in neural crest development. J Cell Biol 2002;159:867–80.

[73] Lee H-Y, Kléber M, Hari L, et al. Instructive role of Wnt/beta-catenin in sensory fate specification in neural crest stem cells. Science 2004;303:1020–3.

[74] Dorsky RI, Moon RT, Raible DW. Control of neural crest cell fate by the Wnt signalling pathway. Nature 1998;396:370–3.

[75] Jin E-J, Erickson CA, Takada S, Burrus LW. Wnt and BMP signaling govern lineage segregation of melanocytes in the avian embryo. Dev Biol 2001;233:22–37.

[76] Lewis JL, Bonner J, Modrell M, et al. Reiterated Wnt signaling during zebrafish neural crest development. Development 2004;131: 1299–308.

[77] Hari L, Miescher I, Shakhova O, et al. Temporal control of neural crest lineage generation by Wnt/-catenin signaling. Development 2012;139:2107–17.

[78] Villanueva S, Glavic A, Ruiz P, Mayor R. Posteriorization by FGF, Wnt, and retinoic acid is required for neural crest induction. Dev Biol 2002;241:289–301.

[79] Wagner M, Han B, Jessell TM. Regional differences in retinoid release from embryonic neural tissue detected by an in vitro reporter assay. Development 1992;116: 55–66.

[80] Maden M, Ong DE, Summerbell D, Chytil F, Hirst EA. Cellular retinoic acid-binding protein and the role of retinoic acid in the development of the chick embryo. Dev Biol 1989;135:124–32.

[81] Dencker L, Annerwall E, Busch C, Eriksson U. Localization of specific retinoid-binding sites and expression of cellular retinoic-acid-binding protein (CRABP) in the early mouse embryo. Development 1990;110:343–52.

[82] Maden M, Hunt P, Eriksson U, Kuroiwa A, Krumlauf R, Summerbell D. Retinoic acid-binding protein, rhombomeres and the neural crest. Development 1991;111:35–43.

[83] Maden M, Ong DE, Chytil F. Retinoid-binding protein distribution in the developing mammalian nervous system. Development 1990;109:75–80.

[84] Vaessen MJ, Meijers JH, Bootsma D, Van Kessel AG. The cellular retinoic-acid-binding protein is expressed in tissues associated with retinoic-acid-induced malformations. Development 1990,110.371–8.

[85] Ruberte E, Dolle P, Chambon P, Morriss-Kay G. Retinoic acid receptors and cellular retinoid binding proteins. II. Their differential pattern of transcription during early morphogenesis in mouse embryos. Development 1991;111:45–60.

[86] Ruberte E, Friederich V, Morriss-Kay G, Chambon P. Differential distribution patterns of CRABP I and CRABP II transcripts during mouse embryogenesis. Development 1992;115:973–87.

[87] Rowe A, Eager NS, Brickell PM. A member of the RXR nuclear receptor family is expressed in neural-crest-derived cells of the developing chick peripheral nervous system. Development 1991;111:771–8.

[88] Henion PD, Weston JA. Retinoic acid selectively promotes the survival and proliferation of neurogenic precursors in cultured neural crest cell populations. Dev Biol 1994;161:243–50.

[89] Ito K, Morita T. Role of retinoic acid in mouse neural crest cell development *in vitro*. Dev Dyn 1995;204:211–8.

[90] Wright EM, Snopek B, Koopman P. Seven new members of the Sox gene family expressed during mouse development. Nucl Acids Res 1993;21:744.

[91] Haldin CE, Labonne C. SoxE factors as multifunctional neural crest regulatory factors. Int J Biochem Cell Biol 2010;42:441–4.

[92] McKeown SJ, Lee VM, Bronner-Fraser M, Newgreen DF, Farlie PG. Sox10 overexpression induces neural crest-like cells from all dorsoventral levels of the neural tube but inhibits differentiation. Dev Dyn 2005;233:430–44.

[93] Kuhlbrodt K, Herbarth B, Sock E, Hermans-Borgmeyer I, Wegner M. Sox10, a novel transcriptional modulator in glial cells. J Neurosci 1998;18:237–50.

[94] Britsch S, Goerich DE, Riethmacher D, et al. The transcription factor Sox10 is a key regulator of peripheral glial development. Genes Dev 2001;15:66–78.

[95] Cheng Y, Cheung M, Abu-Elmagd MM, Orme A, Scotting PJ. Chick sox10, a transcription factor expressed in both early neural crest cells and central nervous system. Brain Res Dev Brain Res 2000;121:233–41.

[96] Honoré SM, Aybar MJ, Mayor R. Sox10 is required for the early development of the prospective neural crest in *Xenopus* embryos. Dev Biol 2003;260:79–96.

[97] Dutton KA, Pauliny A, Lopes SS, et al. Zebrafish colourless encodes sox10 and specifies non-ectomesenchymal neural crest fates. Development 2001;128:4113–25.

[98] Paratore C, Goerich DE, Suter U, Wegner M, Sommer L. Survival and glial fate acquisition of neural crest cells are regulated by an interplay between the transcription factor Sox10 and extrinsic combinatorial signaling. Development 2001;128:3949–61.

[99] Kim J, Lo L, Dormand E, Anderson DJ. SOX10 maintains multipotency and inhibits neuronal differentiation of neural crest stem cells. Neuron 2003;38:17–31.

[100] Pingault V, Bondurand N, Kuhlbrodt K, et al. SOX10 mutations in patients with Waardenburg–Hirschsprung disease. Nat Genet 1998;18:171–3.

[101] Kuhlbrodt K, Schmidt C, Sock E, et al. Functional analysis of Sox10 mutations found in human Waardenburg–Hirschsprung patients. J Biol Chem 1998;273:23033–8.

[102] Sonnenberg-Riethmacher E, Miehe M, Stolt CC, Goerich DE, Wegner M, Riethmacher D. Development and degeneration of dorsal root ganglia in the absence of the HMG-domain transcription factor Sox10. Mech Dev 2001;109:253–65.

[103] Schreiner S, Cossais F, Fischer K, et al. Hypomorphic Sox10 alleles reveal novel protein functions and unravel developmental differences in glial lineages. Development 2007;134:3271–81.

[104] Southard-Smith EM, Kos L, Pavan WJ. Sox10 mutation disrupts neural crest development in Dom Hirschsprung mouse model. Nat Genet 1998;18:60–4.

[105] Herbarth B, Pingault V, Bondurand N, et al. Mutation of the Sry-related Sox10 gene in Dominant megacolon, a mouse model for human Hirschsprung disease. Proc Natl Acad Sci USA 1998;95:5161–5.

[106] Kelsh RN, Eisen JS. The zebrafish colourless gene regulates development of non-ectomesenchymal neural crest derivatives. Development 2000;127:515–25.

[107] Carney TJ, Dutton KA, Greenhill E, et al. A direct role for Sox10 in specification of neural crest-derived sensory neurons. Development 2006;133:4619–30.

[108] Ma Q, Fode C, Guillemot F, Anderson DJ. Neurogenin1 and neurogenin2 control two distinct waves of neurogenesis in developing dorsal root ganglia. Genes Dev 1999;13:1717–28.

[109] Dykes IM, Tempest L, Lee SI, Turner EE. Brn3a and Islet1 Act epistatically to regulate the gene expression program of sensory differentiation. J Neurosci 2011;31:9789–99.

[110] Sun Y, Dykes IM, Liang X, Eng SR, Evans SM, Turner EE. A central role for Islet1 in sensory neuron development linking sensory and spinal gene regulatory programs. Nat Neurosci 2008;11:1283—93.

[111] Eng SR, Dykes IM, Lanier J, Fedtsova N, Turner EE. POU-domain factor Brn3a regulates both distinct and common programs of gene expression in the spinal and trigeminal sensory ganglia. Neural Dev 2007;2:3.

[112] Kramer I, Sigrist M, de Nooij JC, Taniuchi I, Jessell TM, Arber S. A role for runx transcription factor signaling in dorsal root ganglion sensory neuron diversification. Neuron 2006;49:379—93.

[113] Perez SE, Rebelo S, Anderson DJ. Early specification of sensory neuron fate revealed by expression and function of neurogenins in the chick embryo. Development 1999;126:1715—28.

[114] Lee KE, Nam S, Cho E-A, et al. Identification of direct regulatory targets of the transcription factor Sox10 based on function and conservation. BMC Genom 2008;9:408.

[115] Yoshikawa M, Senzaki K, Yokomizo T, Takahashi S, Ozaki S, Shiga T. Runx1 selectively regulates cell fate specification and axonal projections of dorsal root ganglion neurons. Dev Biol 2007;303:663—74.

[116] Nakamura S, Senzaki K, Yoshikawa M, et al. Dynamic regulation of the expression of neurotrophin receptors by Runx3. Development 2008;135:1703—11.

[117] Chen C-L, Broom DC, Liu Y, et al. Runx1 determines nociceptive sensory neuron phenotype and is required for thermal and neuropathic pain. Neuron 2006;49:365—77.

[118] Pattyn A, Morin X, Cremer H, Goridis C, Brunet JF. The homeobox gene Phox2b is essential for the development of autonomic neural crest derivatives. Nature 1999;399:366—70.

[119] Schneider C, Wicht H, Enderich J, Wegner M, Rohrer H. Bone morphogenetic proteins are required in vivo for the generation of sympathetic neurons. Neuron 1999;24:861—70.

[120] Morikawa Y, Zehir A, Maska E, et al. BMP signaling regulates sympathetic nervous system development through Smad4-dependent and -independent pathways. Development 2009;136:3575—84.

[121] Liu H, Margiotta JF, Howard MJ. BMP4 supports noradrenergic differentiation by a PKA-dependent mechanism. Dev Biol 2005;286:521—36.

[122] Howard MJ, Stanke M, Schneider C, Wu X, Rohrer H. The transcription factor dHAND is a downstream effector of BMPs in sympathetic neuron specification. Development 2000;127:4073—81.

[123] Lo L, Tiveron MC, Anderson DJ. MASH1 activates expression of the paired homeodomain transcription factor Phox2a, and couples pan-neuronal and subtype-specific components of autonomic neuronal identity. Development 1998;125:609—20.

[124] Castro DS, Martynoga B, Parras C, et al. A novel function of the proneural factor Ascl1 in progenitor proliferation identified by genome-wide characterization of its targets. Genes Dev 2011;25:930—45.

[125] Wildner H, Gierl MS, Strehle M, Pla P, Birchmeier C. Insm1 (IA-1) is a crucial component of the transcriptional network that controls differentiation of the sympathoadrenal lineage. Development 2008;135:473—81.

[126] Yang C, Kim HS, Seo H, Kim KS. Identification and characterization of potential cis-regulatory elements governing transcriptional activation of the rat tyrosine hydroxylase gene. J Neurochem 1998;71:1358—68.

[127] Parlier D, Ariza A, Christulia F, et al. Xenopus zinc finger transcription factor IA1 (Insm1) expression marks anteroventral noradrenergic neuron progenitors in Xenopus embryos. Dev Dyn 2008;237:2147—57.

[128] Kim HS, Seo H, Yang C, Brunet JF, Kim KS. Noradrenergic-specific transcription of the dopamine beta-hydroxylase gene requires synergy of multiple cis-acting elements including at least two Phox2a-binding sites. J Neurosci 1998;18:8247—60.

[129] Zellmer E, Zhang Z, Greco D, Rhodes J, Cassel S, Lewis EJ. A homeodomain protein selectively expressed in noradrenergic tissue regulates transcription of neurotransmitter biosynthetic genes. J Neurosci 1995;15:8109—20.

[130] Lo L, Morin X, Brunet JF, Anderson DJ. Specification of neurotransmitter identity by Phox2 proteins in neural crest stem cells. Neuron 1999;22:693—705.

[131] Stanke M, Junghans D, Geissen M, Goridis C, Ernsberger U, Rohrer H. The Phox2 homeodomain proteins are sufficient to promote the development of sympathetic neurons. Development 1999;126:4087—94.

[132] Tsarovina K, Pattyn A, Stubbusch J, et al. Essential role of Gata transcription factors in sympathetic neuron development. Development 2004;131:4775—86.

[133] Moriguchi T, Takako N, Hamada M, et al. Gata3 participates in a complex transcriptional feedback network to regulate sympathoadrenal differentiation. Development 2006;133:3871—81.

[134] Morikawa Y, D'Autréaux F, Gershon MD, Cserjesi P. Hand2 determines the noradrenergic phenotype in the mouse sympathetic nervous system. Dev Biol 2007;307:114—26.

[135] Hendershot TJ, Liu H, Clouthier DE, et al. Conditional deletion of Hand2 reveals critical functions in neurogenesis and cell type-specific gene expression for development of neural crest-derived noradrenergic sympathetic ganglion neurons. Dev Biol 2008;319:179—91.

[136] Rychlik JL, Gerbasi V, Lewis EJ. The interaction between dHAND and Arix at the dopamine β-hydroxylase promoter region is independent of direct dHAND binding to DNA. J Biol Chem 2003;278:49652–60.

[137] Rychlik JL, Hsieh M, Eiden LE, Lewis EJ. Phox2 and dHAND transcription factors select shared and unique target genes in the noradrenergic cell type. J Mol Neurosci 2005;27:281–92.

[138] Vincentz JW, VanDusen NJ, Fleming AB, et al. A Phox2- and Hand2-Dependent Hand1 cis-Regulatory element reveals a unique gene dosage requirement for Hand2 during sympathetic neurogenesis. J Neurosci 2012;32:2110–20.

[139] Schmidt M, Lin S, Pape M, et al. The bHLH transcription factor Hand2 is essential for the maintenance of noradrenergic properties in differentiated sympathetic neurons. Dev Biol 2009;329: 191–200.

[140] Lucas ME, Müller F, Rüdiger R, Henion PD, Rohrer H. The bHLH transcription factor hand2 is essential for noradrenergic differentiation of sympathetic neurons. Development 2006;133:4015–24.

[141] Xu H, Firulli AB, Zhang X, Howard MJ. HAND2 synergistically enhances transcription of dopamine-β-hydroxylase in the presence of Phox2a. Dev Biol 2003;262:183–93.

[142] Peirano RI, Goerich DE, Riethmacher D, Wegner M. Protein zero gene expression is regulated by the glial transcription factor Sox10. Mol Cell Biol 2000;20: 3198–209.

[143] Peirano RI, Wegner M. The glial transcription factor Sox10 binds to DNA both as monomer and dimer with different functional consequences. Nucl Acids Res 2000;28:3047–55.

[144] Teng L, Mundell NA, Frist AY, Wang Q, Labosky PA. Requirement for Foxd3 in the maintenance of neural crest progenitors. Development 2008;135: 1615–24.

[145] Mundell NA, Labosky PA. Neural crest stem cell multipotency requires Foxd3 to maintain neural potential and repress mesenchymal fates. Development 2011;138:641–52.

[146] Citri A, Yarden Y. EGF–ERBB signalling: toward the systems level. Nat Rev Mol Cell Biol 2006;7:505–16.

[147] Marchionni MA, Goodearl AD, Chen MS, et al. Glial growth factors are alternatively spliced erbB2 ligands expressed in the nervous system. Nature 1993;362: 312–8.

[148] Zhang D, Sliwkowski MX, Mark M, et al. Neuregulin-3 (NRG3): a novel neural tissue-enriched protein that binds and activates ErbB4. Proc Natl Acad Sci USA 1997;94:9562–7.

[149] Harari D, Tzahar E, Romano J, et al. Neuregulin-4: a novel growth factor that acts through the ErbB-4 receptor tyrosine kinase. Oncogene 1999;18:2681–9.

[150] Higashiyama S, Horikawa M, Yamada K, et al. A novel brain-derived member of the epidermal growth factor family that interacts with ErbB3 and ErbB4. J Biochem 1997;122:675–80.

[151] Busfield SJ, Michnick DA, Chickering TW, et al. Characterization of a neuregulin-related gene, Don-1, that is highly expressed in restricted regions of the cerebellum and hippocampus. Mol Cell Biol 1997;17:4007–14.

[152] Steinthorsdottir V, Stefansson H, Ghosh S, et al. Multiple novel transcription initiation sites for NRG1. Gene 2004;342:97–105.

[153] Lee KF, Simon H, Chen H, Bates B, Hung MC, Hauser C. Requirement for neuregulin receptor erbB2 in neural and cardiac development. Nature 1995;378:394–8.

[154] Gassmann M, Casagranda F, Orioli D, et al. Aberrant neural and cardiac development in mice lacking the ErbB4 neuregulin receptor. Nature 1995;378:390–4.

[155] Meyer D, Birchmeier C. Multiple essential functions of neuregulin in development. Nature 1995;378:386–90.

[156] Kramer R, Bucay N, Kane DJ, Martin LE, Tarpley JE, Theill LE. Neuregulins with an Ig-like domain are essential for mouse myocardial and neuronal development. Proc Natl Acad Sci USA 1996;93:4833–8.

[157] Sliwkowski MX, Schaefer G, Akita RW, et al. Coexpression of erbB2 and erbB3 proteins reconstitutes a high affinity receptor for heregulin. J Biol Chem 1994;269:14661–5.

[158] Meyer D, Yamaai T, Garratt A, et al. Isoform-specific expression and function of neuregulin. Development 1997;124:3575–86.

[159] Britsch S, Li L, Kirchhoff S, et al. The ErbB2 and ErbB3 receptors and their ligand, neuregulin-1, are essential for development of the sympathetic nervous system. Genes Dev 1998;12:1825–36.

[160] Honjo Y, Kniss J, Eisen JS. Neuregulin-mediated ErbB3 signaling is required for formation of zebrafish dorsal root ganglion neurons. Development 2008;135:2615–25.

[161] Riethmacher D, Sonnenberg-Riethmacher E, Brinkmann V, Yamaai T, Lewin GR, Birchmeier C. Severe neuropathies in mice with targeted mutations in the ErbB3 receptor. Nature 1997;389:725–30.

[162] Woldeyesus MT, Britsch S, Riethmacher D, et al. Peripheral nervous system defects in erbB2 mutants following genetic rescue of heart development. Genes Dev 1999;13:2538–48.

[163] Morris JK, Lin W, Hauser C, Marchuk Y, Getman D, Lee KF. Rescue of the cardiac defect in ErbB2 mutant

mice reveals essential roles of ErbB2 in peripheral nervous system development. Neuron 1999;23: 273–83.

[164] Lyons DA, Pogoda H-M, Voas MG, et al. erbb3 and erbb2 are essential for Schwann cell migration and myelination in zebrafish. Curr Biol 2005;15:513–24.

[165] Shah NM, Marchionni MA, Isaacs I, Stroobant P, Anderson DJ. Glial growth factor restricts mammalian neural crest stem cells to a glial fate. Cell 1994; 77:349–60.

[166] George L, Kasemeier-Kulesa J, Nelson BR, Koyano-Nakagawa N, Lefcort F. Patterned assembly and neurogenesis in the chick dorsal root ganglion. J Comp Neurol 2010;518:405–22.

[167] Shah NM, Groves AK, Anderson DJ. Alternative neural crest cell fates are instructively promoted by TGFbeta superfamily members. Cell 1996;85:331–43.

[168] Lyons KM, Hogan BL, Robertson EJ. Colocalization of BMP 7 and BMP 2 RNAs suggests that these factors cooperatively mediate tissue interactions during murine development. Mech Dev 1995;50:71–83.

[169] McPherson CE, Varley JE, Maxwell GD. Expression and regulation of Type I BMP receptors during early avian sympathetic ganglion development. Dev Biol 2000;221:220–32.

[170] Reissmann E, Ernsberger U, Francis-West PH, Rueger D, Brickell PM, Rohrer H. Involvement of bone morphogenetic protein-4 and bone morphogenetic protein-7 in the differentiation of the adrenergic phenotype in developing sympathetic neurons. Development 1996;122:2079–88.

[171] Varley JE, Maxwell GD. BMP-2 and BMP-4, but not BMP-6, increase the number of adrenergic cells which develop in quail trunk neural crest cultures. Exp Neurol 1996;140:84–94.

[172] Varley JE, McPherson CE, Zou H, Niswander L, Maxwell GD. Expression of a constitutively active type I BMP receptor using a retroviral vector promotes the development of adrenergic cells in neural crest cultures. Dev Biol 1998;196:107–18.

[173] Varley JE, Wehby RG, Rueger DC, Maxwell GD. Number of adrenergic and islet-1 immunoreactive cells is increased in avian trunk neural crest cultures in the presence of human recombinant osteogenic protein-1. Dev Dyn 1995;203:434–47.

[174] Wu X, Howard MJ. Two signal transduction pathways involved in the catecholaminergic differentiation of avian neural crest-derived cells in vitro. Mol Cell Neurosci 2001;18:394–406.

[175] Lo L, Sommer L, Anderson DJ. MASH1 maintains competence for BMP2-induced neuronal differentiation in post-migratory neural crest cells. Curr Biol 1997;7:440–50.

[176] Ernsberger U, Reissmann E, Mason I, Rohrer H. The expression of dopamine beta-hydroxylase, tyrosine hydroxylase, and Phox2 transcription factors in sympathetic neurons: evidence for common regulation during noradrenergic induction and diverging regulation later in development. Mech Dev 2000;92:169–77.

[177] Mueller TD, Nickel J. Promiscuity and specificity in BMP receptor activation. FEBS Lett 2012;586:1846–59.

[178] Pisano JM, Colón-Hastings F, Birren SJ. Postmigratory enteric and sympathetic neural precursors share common, developmentally regulated, responses to BMP2. Dev Biol 2000;227:1–11.

[179] Song Q, Mehler MF, Kessler JA. Bone morphogenetic proteins induce apoptosis and growth factor dependence of cultured sympathoadrenal progenitor cells. Dev Biol 1998;196:119–27.

[180] Gomes WA, Kessler JA. Msx-2 and p21 mediate the pro-apoptotic but not the anti-proliferative effects of BMP4 on cultured sympathetic neuroblasts. Dev Biol 2001;237:212–21.

[181] Barde YA. Trophic factors and neuronal survival. Neuron 1989;2:1525–34.

[182] Levi-Montalcini R. The nerve growth factor 35 years later. Science 1987;237:1154–62.

[183] Squinto SP, Stitt TN, Aldrich TH, et al. trkB encodes a functional receptor for brain-derived neurotrophic factor and neurotrophin-3 but not nerve growth factor. Cell 1991;65:885–93.

[184] Soppet D, Escandon E, Maragos J, et al. The neurotrophic factors brain-derived neurotrophic factor and neurotrophin-3 are ligands for the trkB tyrosine kinase receptor. Cell 1991;65:895–903.

[185] Lamballe F, Klein R, Barbacid M. trkC, a new member of the trk family of tyrosine protein kinases, is a receptor for neurotrophin-3. Cell 1991;66:967–79.

[186] Hosang M, Shooter EM. Molecular characteristics of nerve growth factor receptors on PC12 cells. J Biol Chem 1985;260:655–62.

[187] Radeke MJ, Feinstein SC. Analytical purification of the slow, high affinity NGF receptor: identification of a novel 135 kd polypeptide. Neuron 1991;7:141–50.

[188] Hempstead BL, Martin-Zanca D, Kaplan DR, Parada LF, Chao MV. High-affinity NGF binding requires coexpression of the trk proto-oncogene and the low-affinity NGF receptor. Nature 1991;350:678–83.

[189] Kaplan DR, Hempstead BL, Martin-Zanca D, Chao MV, Parada LF. The trk proto-oncogene product: a signal transducing receptor for nerve growth factor. Science 1991;252:554–8.

[190] Kaplan DR, Martin-Zanca D, Parada LF. Tyrosine phosphorylation and tyrosine kinase activity of the trk proto-oncogene product induced by NGF. Nature 1991;350:158–60.

[191] Smeyne RJ, Klein R, Schnapp A, et al. Severe sensory and sympathetic neuropathies in mice carrying a disrupted Trk/NGF receptor gene. Nature 1994;368: 246–9.

[192] Crowley C, Spencer SD, Nishimura MC, et al. Mice lacking nerve growth factor display perinatal loss of sensory and sympathetic neurons yet develop basal forebrain cholinergic neurons. Cell 1994;76:1001–11.

[193] Ernfors P, Lee KF, Kucera J, Jaenisch R. Lack of neurotrophin-3 leads to deficiencies in the peripheral nervous system and loss of limb proprioceptive afferents. Cell 1994;77:503–12.

[194] Fariñas I, Jones KR, Backus C, Wang XY, Reichardt LF. Severe sensory and sympathetic deficits in mice lacking neurotrophin-3. Nature 1994;369:658–61.

[195] ElShamy WM, Linnarsson S, Lee KF, Jaenisch R, Ernfors P. Prenatal and postnatal requirements of NT-3 for sympathetic neuroblast survival and innervation of specific targets. Development 1996;122:491–500.

[196] Zhou XF, Rush R. Sympathetic neurons in neonatal rats require endogenous neurotrophin-3 for survival. J Neurosci 1995;15:6521–30.

[197] DiCicco-Bloom E, Friedman WJ, Black IB. NT-3 stimulates sympathetic neuroblast proliferation by promoting precursor survival. Neuron 1993;11: 1101–11.

[198] Birren SJ, Lo L, Anderson DJ. Sympathetic neuroblasts undergo a developmental switch in trophic dependence. Development 1993;119:597–610.

[199] Kobayashi M, Fujii M, Kurihara K, Matsuoka I. Bone morphogenetic protein-2 and retinoic acid induce neurotrophin-3 responsiveness in developing rat sympathetic neurons. Brain Res Mol Brain Res 1998;53:206–17.

[200] Lo LC, Johnson JE, Wuenschell CW, Saito T, Anderson DJ. Mammalian achaete-scute homolog 1 is transiently expressed by spatially restricted subsets of early neuroepithelial and neural crest cells. Genes Dev 1991;5:1524–37.

[201] Johnson JE, Birren SJ, Anderson DJ. Two rat homologues of Drosophila achaete-scute specifically expressed in neuronal precursors. Nature 1990;346:858–61.

[202] Ernsberger U, Patzke H, Tissier-Seta JP, Reh T, Goridis C, Rohrer H. The expression of tyrosine hydroxylase and the transcription factors cPhox-2 and Cash-1: evidence for distinct inductive steps in the differentiation of chick sympathetic precursor cells. Mech Dev 1995;52:125–36.

[203] Groves AK, George KM, Tissier-Seta JP, Engel JD, Brunet JF, Anderson DJ. Differential regulation of transcription factor gene expression and phenotypic markers in developing sympathetic neurons. Development 1995;121:887–901.

[204] Sommer L, Shah N, Rao M, Anderson DJ. The cellular function of MASH1 in autonomic neurogenesis. Neuron 1995;15:1245–58.

[205] Guillemot F, Lo LC, Johnson JE, Auerbach A, Anderson DJ, Joyner AL. Mammalian achaete-scute homolog 1 is required for the early development of olfactory and autonomic neurons. Cell 1993;75:463–76.

[206] Hirsch MR, Tiveron MC, Guillemot F, Brunet JF, Goridis C. Control of noradrenergic differentiation and Phox2a expression by MASH1 in the central and peripheral nervous system. Development 1998;125: 599–608.

[207] Pattyn A, Guillemot F, Brunet J-F. Delays in neuronal differentiation in Mash1/Ascl1 mutants. Dev Biol 2006;295:67–75.

[208] Pattyn A, Morin X, Cremer H, Goridis C, Brunet JF. Expression and interactions of the two closely related homeobox genes Phox2a and Phox2b during neurogenesis. Development 1997;124:4065–75.

[209] Valarché I, Tissier-Seta JP, Hirsch MR, Martinez S, Goridis C, Brunet JF. The mouse homeodomain protein Phox2 regulates Ncam promoter activity in concert with Cux/CDP and is a putative determinant of neurotransmitter phenotype. Development 1993;119: 881–96.

[210] Tiveron MC, Hirsch MR, Brunet JF. The expression pattern of the transcription factor Phox2 delineates synaptic pathways of the autonomic nervous system. J Neurosci 1996;16:7649–60.

[211] Morikawa Y, Dai Y-S, Hao J, Bonin C, Hwang S, Cserjesi P. The basic helix-loop-helix factor Hand2 regulates autonomic nervous system development. Dev Dyn 2005;234:613–21.

[212] Cserjesi P, Brown D, Lyons GE, Olson EN. Expression of the novel basic helix-loop-helix gene eHAND in neural crest derivatives and extraembryonic membranes during mouse development. Dev Biol 1995;170:664–78.

[213] Howard M, Foster DN, Cserjesi P. Expression of HAND gene products may be sufficient for the differentiation of avian neural crest-derived cells into catecholaminergic neurons in culture. Dev Biol 1999;215:62–77.

[214] Yamamoto M, Ko LJ, Leonard MW, Beug H, Orkin SH, Engel JD. Activity and tissue-specific expression of the transcription factor NF-E1 multigene family. Genes Dev 1990;4:1650–62.

[215] George KM, Leonard MW, Roth ME, Lieuw KH, Kioussis D, Grosveld F, et al. Embryonic expression and cloning of the murine GATA-3 gene. Development 1994; 120: 2673–2686.

[216] Lim KC, Lakshmanan G, Crawford SE, Gu Y, Grosveld F, Engel JD. Gata3 loss leads to embryonic

lethality due to noradrenaline deficiency of the sympathetic nervous system. Nat Genet 2000;25:209–12.

[217] Marmigère F, Ernfors P. Specification and connectivity of neuronal subtypes in the sensory lineage. Nat Rev Neurosci 2007;8:114–27.

[218] Klein R, Parada LF, Coulier F, Barbacid M. trkB, a novel tyrosine protein kinase receptor expressed during mouse neural development. EMBO J 1989;8:3701–9.

[219] Mu X, Silos-Santiago I, Carroll SL, Snider WD. Neurotrophin receptor genes are expressed in distinct patterns in developing dorsal root ganglia. J Neurosci 1993;13:4029–41.

[220] McMahon SB, Armanini MP, Ling LH, Phillips HS. Expression and coexpression of Trk receptors in subpopulations of adult primary sensory neurons projecting to identified peripheral targets. Neuron 1994;12:1161–71.

[221] Kashiba H, Noguchi K, Ueda Y, Senba E. Coexpression of trk family members and low-affinity neurotrophin receptors in rat dorsal root ganglion neurons. Brain Res Mol Brain Res 2003;30:158–64.

[222] Chen AI, de Nooij JC, Jessell TM. Graded activity of transcription factor Runx3 specifies the laminar termination pattern of sensory axons in the developing spinal cord. Neuron 2006;49:395–408.

[223] Liebl DJ, Tessarollo L, Palko ME, Parada LF. Absence of sensory neurons before target innervation in brain-derived neurotrophic factor-, neurotrophin 3-, and TrkC-deficient embryonic mice. J Neurosci 1997;17:9113–21.

[224] Rifkin JT, Todd VJ, Anderson LW, Lefcort F. Dynamic expression of neurotrophin receptors during sensory neuron genesis and differentiation. Dev Biol 2000;227:465–80.

[225] Averill S, McMahon SB, Clary DO, Reichardt LF, Priestley JV. Immunocytochemical localization of trkA receptors in chemically identified subgroups of adult rat sensory neurons. Eur J Neurosci 1995;7:1484–94.

[226] Kashiba H, Ueda Y, Senba E. Coexpression of preprotachykinin-A, alpha-calcitonin gene-related peptide, somatostatin, and neurotrophin receptor family messenger RNAs in rat dorsal root ganglion neurons. Neuroscience 1996;70:179–89.

[227] Kobayashi K, Fukuoka T, Obata K, et al. Distinct expression of TRPM8, TRPA1, and TRPV1 mRNAs in rat primary afferent neurons with aδ/c-fibers and colocalization with trk receptors. J Comp Neurol 2005;493:596–606.

[228] Henion PD, Garner AS, Large TH, Weston JA. trkC-mediated NT-3 signaling is required for the early development of a subpopulation of neurogenic neural crest cells. Dev Biol 1995;172:602–13.

[229] Tessarollo L, Tsoulfas P, Martin-Zanca D, et al. trkC, a receptor for neurotrophin-3, is widely expressed in the developing nervous system and in non-neuronal tissues. Development 1993;118:463–75.

[230] Pinco O, Carmeli C, Rosenthal A, Kalcheim C. Neurotrophin-3 affects proliferation and differentiation of distinct neural crest cells and is present in the early neural tube of avian embryos. J Neurobiol 1993;24:1626–41.

[231] Kahane N, Kalcheim C. Expression of trkC receptor mRNA during development of the avian nervous system. J Neurobiol 1994;25:571–84.

[232] Fariñas I, Wilkinson GA, Backus C, Reichardt LF, Patapoutian A. Characterization of neurotrophin and Trk receptor functions in developing sensory ganglia: direct NT-3 activation of TrkB neurons in vivo. Neuron 1998;21:325–34.

[233] Klein R, Silos-Santiago I, Smeyne RJ, et al. Disruption of the neurotrophin-3 receptor gene trkC eliminates la muscle afferents and results in abnormal movements. Nature 1994;368:249–51.

[234] Lefcort F, Clary DO, Rusoff AC, Reichardt LF. Inhibition of the NT-3 receptor TrkC, early in chick embryogenesis, results in severe reductions in multiple neuronal subpopulations in the dorsal root ganglia. J Neurosci 1996;16:3704–13.

[235] Fariñas I, Yoshida CK, Backus C, Reichardt LF. Lack of neurotrophin-3 results in death of spinal sensory neurons and premature differentiation of their precursors. Neuron 1996;17:1065–78.

[236] Tojo H, Kaisho Y, Nakata M, et al. Targeted disruption of the neurotrophin-3 gene with lacZ induces loss of trkC-positive neurons in sensory ganglia but not in spinal cords. Brain Res 1995;669:163–75.

[237] Tessarollo L, Vogel KS, Palko ME, Reid SW, Parada LF. Targeted mutation in the neurotrophin-3 gene results in loss of muscle sensory neurons. Proc Natl Acad Sci USA 1994;91:11844–8.

[238] ElShamy WM, Ernfors P. A local action of neurotrophin-3 prevents the death of proliferating sensory neuron precursor cells. Neuron 1996;16: 963–72.

[239] Kucera J, Fan G, Jaenisch R, Linnarsson S, Ernfors P. Dependence of developing group Ia afferents on neurotrophin-3. J Comp Neurol 1995;363:307–20.

[240] Memberg SP, Hall AK. Proliferation, differentiation, and survival of rat sensory neuron precursors in vitro require specific trophic factors. Mol Cell Neurosci 1995;6:323–35.

[241] Kalcheim C, Carmeli C, Rosenthal A. Neurotrophin 3 is a mitogen for cultured neural crest cells. Proc Natl Acad Sci USA 2004;89:1661–5.

[242] Wright EM, Vogel KS, Davies AM. Neurotrophic factors promote the maturation of developing sensory

neurons before they become dependent on these factors for survival. Neuron 1992;9:139–50.

[243] Chalazonitis A, Rothman TP, Chen J, Lamballe F, Barbacid M, Gershon MD. Neurotrophin-3 induces neural crest-derived cells from fetal rat gut to develop *in vitro* as neurons or glia. J Neurosci 1994;14:6571–84.

[244] Klein R, Smeyne RJ, Wurst W, et al. Targeted disruption of the trkB neurotrophin receptor gene results in nervous system lesions and neonatal death. Cell 1993;75:113–22.

[245] Ernfors P, Lee KF, Jaenisch R. Mice lacking brain-derived neurotrophic factor develop with sensory deficits. Nature 1994;368:147–50.

[246] Jones KR, Fariñas I, Backus C, Reichardt LF. Targeted disruption of the BDNF gene perturbs brain and sensory neuron development but not motor neuron development. Cell 1994;76:989–99.

[247] Silos-Santiago I, Molliver DC, Ozaki S, et al. Non-TrkA-expressing small DRG neurons are lost in TrkA deficient mice. J Neurosci 1995;15:5929–42.

[248] Minichiello L, Piehl F, Vazquez E, et al. Differential effects of combined trk receptor mutations on dorsal root ganglion and inner ear sensory neurons. Development 1995;121:4067–75.

[249] Carroll SL, Silos-Santiago I, Frese SE, Ruit KG, Milbrandt J, Snider WD. Dorsal root ganglion neurons expressing trk are selectively sensitive to NGF deprivation in utero. Neuron 1992;9:779–88.

[250] Goedert M, Otten U, Hunt SP, et al. Biochemical and anatomical effects of antibodies against nerve growth factor on developing rat sensory ganglia. Proc Natl Acad Sci USA 1984;81:1580–4.

[251] Ruit KG, Elliott JL, Osborne PA, Yan Q, Snider WD. Selective dependence of mammalian dorsal root ganglion neurons on nerve growth factor during embryonic development. Neuron 1992;8:573–87.

[252] Johnson EM, Gorin PD, Brandeis LD, Pearson J. Dorsal root ganglion neurons are destroyed by exposure *in utero* to maternal antibody to nerve growth factor. Science 1980;210:916–8.

[253] Wright DE, Snider WD. Neurotrophin receptor mRNA expression defines distinct populations of neurons in rat dorsal root ganglia. J Comp Neurol 1995;351:329–38.

[254] McGraw HF, Snelson CD, Prendergast A, Suli A, Raible DW. Postembryonic neuronal addition in zebrafish dorsal root ganglia is regulated by Notch signaling. Neural Dev 2012;7:23.

[255] Sommer L, Ma Q, Anderson DJ. Neurogenins, a novel family of atonal-related bHLH transcription factors, are putative mammalian neuronal determination genes that reveal progenitor cell heterogeneity

in the developing CNS and PNS. Mol Cell Neurosci 1996;8:221–41.

[256] Blader P, Fischer N, Gradwohl G, Guillemot F, Strähle U. The activity of neurogenin1 is controlled by local cues in the zebrafish embryo. Development 1997;124:4557–69.

[257] Ma Q, Kintner C, Anderson DJ. Identification of neurogenin, a vertebrate neuronal determination gene. Cell 1996;87:43–52.

[258] Perron M, Opdecamp K, Butler K, Harris WA, Bellefroid EJ. X-ngnr-1 and Xath3 promote ectopic expression of sensory neuron markers in the neurula ectoderm and have distinct inducing properties in the retina. Proc Natl Acad Sci USA 1999;96:14996–5001.

[259] Sun Y, Nadal-Vicens M, Misono S, et al. Neurogenin promotes neurogenesis and inhibits glial differentiation by independent mechanisms. Cell 2001;104:365–76.

[260] Andermann P, Ungos J, Raible DW. Neurogenin1 defines zebrafish cranial sensory ganglia precursors. Dev Biol 2002;251:45–58.

[261] McGraw HF, Nechiporuk A, Raible DW. Zebrafish dorsal root ganglia neural precursor cells adopt a glial fate in the absence of neurogenin1. J Neurosci 2008;28:12558–69.

[262] Zirlinger M, Lo L, McMahon J, McMahon AP, Anderson DJ. Transient expression of the bHLH factor neurogenin-2 marks a subpopulation of neural crest cells biased for a sensory but not a neuronal fate. Proc Natl Acad Sci USA 2002;99:8084–9.

[263] Parras CM. Divergent functions of the proneural genes Mash1 and Ngn2 in the specification of neuronal subtype identity. Genes Dev 2002;16:324–38.

[264] Greenwood AL, Turner EE, Anderson DJ. Identification of dividing, determined sensory neuron precursors in the mammalian neural crest. Development 1999;126:3545–59.

[265] Ungos JM, Karlstrom RO, Raible DW. Hedgehog signaling is directly required for the development of zebrafish dorsal root ganglia neurons. Development 2003;130:5351–62.

[266] Prendergast A, Linbo TH, Swarts T, et al. The metalloproteinase inhibitor Reck is essential for zebrafish DRG development. Development 2012;139: 1141–52.

[267] Morioka Y, Monypenny J, Matsuzaki T, et al. The membrane-anchored metalloproteinase regulator RECK stabilizes focal adhesions and anterior-posterior polarity in fibroblasts. Oncogene 2009;28:1454–64.

[268] Fedtsova NG, Turner EE. Brn-3.0 expression identifies early post-mitotic CNS neurons and sensory neural precursors. Mech Dev 1995;53:291–304.

[269] Eng SR, Gratwick K, Rhee JM, Fedtsova N, Gan L, Turner EE. Defects in sensory axon growth precede

neuronal death in Brn3a-deficient mice. J Neurosci 2001;21:541—9.

[270] Zou M, Li S, Klein WH, Xiang M. Brn3a/Pou4f1 regulates dorsal root ganglion sensory neuron specification and axonal projection into the spinal cord. Dev Biol 2012;364:114—27.

[271] Xiang M, Gan L, Zhou L, Klein WH, Nathans J. Targeted deletion of the mouse POU domain gene Brn-3a causes selective loss of neurons in the brainstem and trigeminal ganglion, uncoordinated limb movement, and impaired suckling. Proc Natl Acad Sci USA 1996;93:11950—5.

[272] McEvilly RJ, Erkman L, Luo L, Sawchenko PE, Ryan AF, Rosenfeld MG. Requirement for Brn-3.0 in differentiation and survival of sensory and motor neurons. Nature 1996;384:574—7.

[273] Pfaff SL, Mendelsohn M, Stewart CL, Edlund T, Jessell TM. Requirement for LIM homeobox gene Isl1 in motor neuron generation reveals a motor neuron-dependent step in interneuron differentiation. Cell 1996;84:309—20.

[274] Levanon D, Bettoun D, Harris-Cerruti C, et al. The Runx3 transcription factor regulates development and survival of TrkC dorsal root ganglia neurons. EMBO J 2002;21:3454—63.

[275] Inoue K-I, Ito K, Osato M, Lee B, Bae S-C, Ito Y. The transcription factor Runx3 represses the neurotrophin receptor TrkB during lineage commitment of dorsal root ganglion neurons. J Biol Chem 2007;282: 24175—84.

[276] Kobayashi A, Senzaki K, Ozaki S, Yoshikawa M, Shiga T. Runx1 promotes neuronal differentiation in dorsal root ganglion. Mol Cell Neurosci 2012;49:23—31.

[277] Saito T, Greenwood A, Sun Q, Anderson DJ. Identification by differential RT-PCR of a novel paired homeodomain protein specifically expressed in sensory neurons and a subset of their CNS targets. Mol Cell Neurosci 1995;6:280—92.

[278] Rebelo S, Reguenga C, Osório L, Pereira C, Lopes C, Lima D. DRG11 immunohistochemical expression during embryonic development in the mouse. Dev Dyn 2007;236:2653—60.

Neural Crest Cells and Pigmentation

Alberto Lapedriza, Kleio Petratou and Robert N. Kelsh

Department of Biology and Biochemistry and Centre for Regenerative Medicine,
University of Bath, Bath, UK

OUTLINE

INTRODUCTION

Pigmentation is a characteristic widespread among the chordates, where its diverse functions include aiding visual acuity, protection against ultraviolet radiation, camouflage, and attraction of a mate. Most pigments are accumulated by specialized cells, termed pigment cells, derived early in embryonic development either directly from the neural plate (e.g.,

pigmented retinal epithelium (PRE) and the pineal organ pigment cells) [1] or from the neural crest [2,3]. Neural crest-derived pigment cells (or chromatophores) are ancestrally very diverse [3] with fish, reptiles, and amphibians having up to at least seven pigment cell types, including melanocytes (usually black or brown), iridophores (reflective silver, blue, or gold), leucophores (white), xanthophores/erythrophores (yellow/red), cyanophores (blue),

Neural Crest Cells.
DOI: http://dx.doi.org/10.1016/B978-0-12-401730-6.00015-6

and irido-erythrophores (violet); Table 14.1. Mammals and birds have secondarily lost most of this diversity, retaining only melano-cytes (and iridophores in birds) [7]. Different pigment cells are characterized by their syn-thesis or dietary accumulation of specific clas-ses of pigments, usually in specialized pigment organelles (Table 14.1). Pigment cells are prominently associated with the skin, residing in the dermis and/or epidermis, but are also found in many internal locations—for example, around the viscera.

Melanocytes are characterized by the pro-duction of melanin, a high molecular weight polymer derived from tyrosine, within membrane-bound organelles known as mela-nosomes. In most species, melanosomes remain exclusively within the melanocytes, but often they can be moved around within the cell along microtubules, a process with a key role in the rapid color changes so characteristic of fish and some reptiles. In contrast, in the skin, as well as in growing hair and feathers, of amniotes (birds and mammals), melano-somes are exported from melanocytes to adja-cent keratinocytes; skin color and tanning

reactions are mostly due to this process of epi-dermal melanosome transfer [8]. To facilitate this, mammalian melanocytes tend to be highly dendritic cells; although their cell body rests on the basal lamina, their dendrites proj-ect into different layers of the epidermis where they are intimately associated with the kerati-nocytes of the skin [9].

Melanocytes produce two kinds of melanin: black to brown eumelanin and yellow to red pheomelanin, although the latter is thought to be restricted to mammals and birds [10]. Eumelanin acts as a photoprotective anti-oxi-dant, whereas pheomelanin is phototoxic and a pro-oxidant [11,12]. In mammals, natural mela-nin pigments consist of a mixture of both eumelanin and pheomelanin in varying propor-tions, with the overall color reflecting the bias in ratios [13]. In humans, black hair contains almost exclusively eumelanins, whereas in red hair pheomelanins predominate. Commonly, mammalian hairs have a banded pattern, com-bining regions (often the tip and the base) with eumelanin and an internal band of pheomela-nin, the so-called agouti pattern. In these cases, the extent of the pheomelanin band determines

TABLE 14.1   Characteristics of Major Pigment Cell Types

Cell Type	Organelle	Pigment	Color	Species
Melanocytes	Melanosome	Eumelanin or Pheomelanin	Black-brown or yellow	Fish, amphibians, reptiles, birds, mammals
Iridophores	Reflective platelets (iridosome)	Crystallized guanine (plus some pteridines)	Iridescent, blue or silver	Fish, amphibians, reptiles
Xanthophores/ Erythrophores	Pterinosomes and carotenoid vesicles	Pteridines (sepiapterines, drosopterines and others) and carotenoids	Yellow-red-orange	Fish, amphibians, reptiles
Leucophores	Reflective platelets	Crystallized purines (guanine)	Reflective, white	Fish
Cyanophores	Unknown	Uncharacterized cyan biochrome	Cyan	Fish
Erythro-iridophores	Reflecting platelets and carotenoid vesicles	Crystallized purines and carotenoids	Violet	Fish

Source: Based on [4–6].

the overall color impression given. For example, in many cats, the black stripes or spots consist of hairs containing predominantly eumelanin, with only a minimal pheomelanin band, whereas the hairs in the background yellow or orange areas contain an extensive pheomelanin band [14].

Melanocyte genetics is perhaps better characterized than that of any other cell type [15]. In contrast, other pigment cells have been relatively neglected, despite fascination with the often stunning results of their presence: the breathtaking diversity of tropical fish coloration, and the kaleidoscopic pigmentation of chameleons, one minute camouflage and the next expressing aggression to a rival, depends largely upon other pigment cell types. Xanthophores are usually highly dendritic cells that appear yellow under white light. They contain granules of pteridine pigment, synthesized from guanosine triphosphate through complex and incompletely defined biochemical pathways [16]. Erythrophores are a closely related cell type and, although appearing red or orange, are sometimes difficult to distinguish from xanthophores, as their appearance depends on the quantity and type of pteridine derivatives present within their granules [17,18]. Both these cell types also may contain carotenoid granules: membrane-bound organelles containing dietary carotenoids [19]. Iridophores, formerly known also as guanophores (appearing blue, silver, or gold), and leucophores (white or cream) are non-dendritic cells that contain organelles termed reflective platelets, composed of crystallized guanine. The distinctive appearance of these two cell types is primarily due to the organization of the reflecting platelets; in leucophores they are found in all orientations, whereas in iridophores the spacing and orientation of the platelets is carefully orchestrated to generate the iridescence by a thin-layer interference mechanism [20]. Cyanophores were identified in mandarin fish as blue pigment cells, owing to a so-far-uncharacterized cyan biochrome within their granules [4]. Finally, irido-erythrophores are a recently discovered pigment cell type containing both reflecting platelets and vesicles containing an uncharacterized violet pigment [6].

The prominence of hair, skin, and eye pigmentation has rendered melanocyte genetics of special interest in the fields of human and mouse genetics [15,21]. As the relevant genes have been identified and studied, our understanding of the complex genetic regulation of melanocyte development has deepened. Complementary genetic screens in zebrafish have supplemented that understanding [22–27] and have finally allowed insight into the origin of and genetic mechanisms controlling other pigment cell types. Specification and fate commitment of pigment cell types from the NC forms an attractive model system for understanding the cellular and genetic basis of stem cell development. Furthermore, numerous human congenital diseases result from disruption of pigment cell development. Further understanding of pigment cell development and the underlying gene regulatory networks (GRNs) will, therefore, not only help clarify the principles of stem cell regeneration and differentiation but will also confer significant medical insight.

In this review, we summarize current understanding of the genetics of pigment cell development, focusing on melanocytes, although briefly considering iridophores and xanthophores. We subdivide the process into five key aspects—fate specification from multipotent precursors, commitment, migration and patterning, proliferation and survival, and differentiation. Although we recognize that development is a continuum, with these different aspects broadly overlapping, we currently lack the tools to portray and explore that continuum. Novel developments in mathematical modeling and systems biology methods for describing GRNs (Figure 14.1) are beginning to change this [29–31], and we anticipate that an integrated view of the genetics will emerge in the next 5 years.

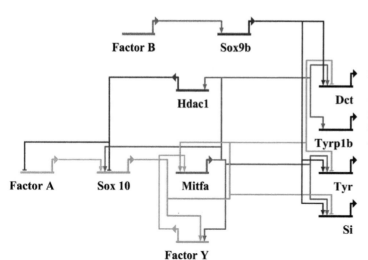

FIGURE 14.1 **GRN of melanophore development in zebrafish.** This GRN describes the behavior of the main genes involved in melanophore development from NCC. Some of the genes are still undefined (e.g., Factor A, comprising several transcription factors) [28], but mathematical modeling has predicted their depicted role in the network. For gene symbols, see text. *Source: Modified from [29].*

TIMING OF PIGMENT CELL DEVELOPMENT

Melanoblasts, the precursors of melanocytes, originate from the neuroepithelium (e.g., PRE) or the neural crest in a manner largely conserved across vertebrate species. Although it still remains unclear how early in neural crest development fate specification occurs, current evidence favors the view that for many cells this is very early, around the time of delamination, since neural crest cells (NCC) begin to express markers characteristic of specific differentiated cell types (but note that other cells are specified later from a postmigratory Schwann cell precursor source) [32,33]. For example, markers such as microphthalmia-associated transcription factor (*Mitf*) and dopachrome tautomerase (*Dct*) (see below) are expressed at these early stages, and cells migrate along pathways that are fate specific [34]. Melanoblasts in chick and mouse migrate exclusively along the dorsolateral pathway (Figure 14.2A) [35]. In zebrafish, the migratory pattern is more complex, with melanoblasts (and indeed differentiating melanocytes) migrating along the dorsolateral and medial pathways, whereas iridophores and xanthophores use the medial

and dorsolateral pathways, respectively (Figure 14.2B) [34]. Importantly, pathway-specific cell migration, at least of melanocytes, requires specification to a melanocyte fate and expression of melanocyte-specific guidance molecules [36]. Recent work in chick highlights the very early segregation, prior to delamination, of melanogenic NCC from their neurogenic counterparts, and emphasizes the roles of mutually exclusive maintenance of *Mitf* and *Foxd3* expression respectively in these two populations (Figure 14.2A) [37].

Pigment Cell Progenitors

The long-standing model of pigment cell origins is implicitly based upon a progressive fate restriction model of neural crest development and postulates the existence of a partially restricted progenitor common to all pigment cell types (i.e., a chromatoblast) [38]. The model was based largely on the shared origin of pigmentary organelles from the endoplasmic reticulum and the presence of pigment cells containing mixtures of organelle types in both pathological and physiological situations [38]. The concept has been widely accepted, but with mouse genetics in the spotlight for many years,

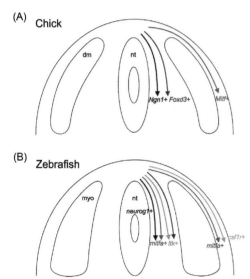

(A) Chick

(B) Zebrafish

FIGURE 14.2 **Migratory pathways of NCC derivatives.** The schematic shows transverse sections of (A) chick and (B) zebrafish embryonic trunks and outlines the migratory pathways followed by distinct NC derivatives, together with representative key marker genes. Red arrows represent *Mitf/mitfa*-positive melanocyte precursors. In zebrafish, the green and blue arrows represent iridophore and xanthophore precursors, respectively. The black arrows represent *Neurogenin1* positive neural progenitors, which primarily generate both neuronal and glial components of the dorsal root ganglia. The purple arrows represent one or more further glial progenitor types, which in chick are known to express *Foxd3* and give rise to SCs. In zebrafish, other neural progenitors remain poorly characterized but express *sox10* and give rise to SCs and sympathetic ganglial components. dm, dermomyotome; myo, myoseptum; nt, neural tube.

it has remained untested. The recent surge of interest in zebrafish genetics allows the question to be addressed directly, although published data remain inconclusive at this point.

Intriguingly, large-scale mutagenesis screens have identified numerous loci that affect all three pigment cell types occurring in zebrafish embryos [22,23,25]. These loci, however, tend to be pleiotropic, and thus it is likely that gene activity is not restricted to the chromatophore lineages. For example, *colorless/sox10* (Figure 14.3E and F) and *mindbomb* (formerly known as *whitetail*) display severe reductions in all pigment cell types when mutated, but clearly have much broader roles in the neural crest [39–41]. Clonal analysis studies of premigratory trunk NCC in wild-type zebrafish embryos show that individual clones expressing these genes may contribute to more than one pigment cell fate, but many clones present with a mix of both pigment and neural derivatives, although the naturally small clone sizes make data interpretation challenging [40,42]. Interestingly, clonal analysis in *sox10* mutants indicated that all pigment cell progenitors, but not neural progenitors, remained stuck in a premigratory

position [40]. Further studies by our group identified *ltk* as a marker of these restricted cells, leading to the hypothesis that *ltk* identifies an early stage multipotent pigment cell precursor, consistent with the chromatoblast hypothesis [43]. However, the potency of these cells has yet to be demonstrated.

The phenotype of the *mitfa* mutant in zebrafish combines absence of melanocytes with increased iridophores and is most readily interpreted as resulting from redirection of fate choice of multipotent precursors, although this remains to be formally tested [44,45]. A recent study relied on co-visualization of *mitfa:gfp*-expressing cells and different chromatophore lineage markers to suggest the existence of a bipotent melano-iridophore precursor [46]. A similar proposal for a class of adult pigment cell precursors has also been made [47].

The idea that there may be multiple types of pigment cell precursors is reinforced by recent work looking at melanocyte origins in mammals and birds. Mitf appears to be specific to the melanocyte lineage in these groups, and melanocytes have long been thought to originate only from migrating Mitf-expressing cells

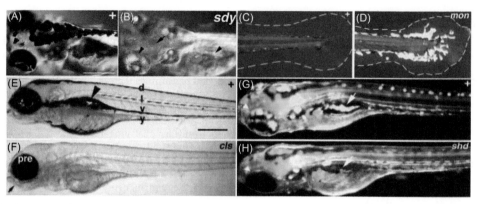

FIGURE 14.3 **Zebrafish pigmentation mutants.** (A and B) Wild-type zebrafish compared with *sandy/tyrosinase* (*sdy/tyr*) homozygous mutants at 6 days postfertilization (dpf). Melanin production is affected, although melanophores survive. Arrows on the mutant embryo point to the positions normally occupied by melanophores. (C and D) Wild-type embryos are compared with *moonshine*/tripartite motif-containing 33 (*mon/trim-33*) homozygous mutants at 6 dpf. *mon* embryos display larger numbers of iridophores. (E and F) Wild-type zebrafish compared with *colourless*/sry-related HMG box10 (*cls/sox10*) homozygous mutants at 6 dpf. Chromatophores (melanophores, xanthophores, iridophores) are absent in *cls* mutants. (G and H) Wild-type zebrafish compared with *shady/leukocyte tyrosine kinase* (*shd/ltk*) homozygous mutants at 6 dpf. *shd* mutants have reduced iridophore numbers. *Source: Modified from [23].*

[48]. However, recent studies incorporating carefully timed ablation experiments identified an unexpected second origin for mammalian and avian embryonic melanocytes from bipotent melanoglial progenitors on the spinal nerves [32,33]. Genetic lineage tracing in mice confirmed these as Schwann cell precursors prior to expression of *Krox20*, although Krox20-expressing cells retained melanocytic potential, which was revealed when contact with nerves was abolished. Thus, here, a close relationship between Schwann cell and melanocyte lineages is implied. In summary, whilst close links between different pigment cells are highly likely, one key unanswered question in the field is whether there is an exclusively chromatoblastic progenitor.

FATE CHOICE/SPECIFICATION

Melanocytes

Certain genes playing a key role in specification of pigment cell fates from the neural crest have been well characterized. At the core of melanocyte fate specification is the transcriptional activator microphthalmia-associated transcription factor (MITF; Mitfa in zebrafish). This conserved transcription factor is a member of the basic helix-loop-helix leucine zipper (bHLH-LZ) family, members of which possess DNA-binding and dimerization domains. MITF has an amino-terminal transactivation domain that triggers transcription and a domain mediating dimerization (HLH-LZ) in the carboxy-terminal region. It is expressed in all melanocytes, throughout their development and maintained in the differentiated cell [44,48–50]. Due to both the absolute requirement for Mitf in melanocyte development and its direct regulation of a large and diverse set of melanocyte genes, it is often referred to as the "master regulator" of melanocyte development [51,52]. In mouse *Mitf* and zebrafish *mitfa* null mutants, all melanocyte markers are absent in embryos and adults, except for transient expression of *Mitf/mitfa* itself [29,44,53]. In humans, *MITF* mutations cause two distinct syndromes: Tietz syndrome (OMIM #103500) and Waardenburg syndrome type 2 A (WS2A; OMIM #193510). Tietz syndrome, also called Tietz albinism-deafness syndrome, is characterized by loss of pigmentation (without affecting the eyes) and complete sensorineural hearing loss [54]. WS2A is characterized by complete absence of pigmentation in patches of skin and hair, especially in the ventral midline and forelock and sensorineural hearing loss [55]. MITF mutations also cause forms of albinism (including ocular albinism) and sensorineural deafness (OMIM #103470). Finally, MITF mutation (amino acid substitution E318K) increases the susceptibility to cutaneous malignant melanoma (OMIM #614456).

MITF expression is driven by several transcription factors, including Sox10, Pax3, CREB, and the Wnt pathway. The protein is required for initial melanoblast specification and melanocyte differentiation, and it has an ongoing role in melanocyte maintenance [44,56]. These multiple roles reflect the extensive range of genes regulated transcriptionally by MITF. The strongest evidence that the requirement for MITF is sustained after melanocytes are fully differentiated comes from studying zebrafish carrying temperature-sensitive mitfa alleles [56]. Inactivation of Mitfa at different stages in embryonic development results in distinct melanocyte phenotypes. When Mitfa is inactivated throughout embryonic development, melanoblast specification fails completely. In contrast, inactivation from 24 hpf onward show a failure of melanization; even after full differentiation, Mitfa inactivation from 72 hpf abolishes melanocyte dendricity and induces apoptosis [56].

Genetic studies in humans, mouse, and zebrafish make clear that SOX10/Sox10, a member of the SOX (Sry-related HMG-box) transcription factor family, has a crucial role in melanocyte development [40,57–61]. Sox10 heterozygous mutant mice display a white belly spot and white extremities, whereas homozygous mutants (Figure 14.4D) lack

essentially all melanoblast marker expression [60]. Furthermore, zebrafish sox10 mutants (also known as colorless (cls)) are missing all three pigment cell types [39]. On close inspection, a few residual specks of melanin are seen in these fish, resulting from Sox9-dependent low-level expression of melanogenic genes, although these have disappeared by 5 days postfertilization (dpf) [29,61].

Interestingly, in both mice and zebrafish, the effects of Sox10 mutations extend to all non-ectomesenchymal (skeletogenic) neural crest fates and to otic development, explaining their embryonic lethality. This is consistent with the fact that human SOX10 mutations have only ever been identified in heterozygous patients and display complex phenotypes. Most commonly they result in Waardenburg–Shah syndrome, also known as Waardenburg syndrome (WS) type 4C (OMIM #613266), characterized by deafness, pigmentary abnormalities, and Hirschsprung disease (aganglionic megacolon); or WS type 2E manifesting with pigmentary abnormalities and congenital sensorineural hearing loss, but lacking Hirschsprung disease (OMIM #611584) [55]. Other SOX10 mutations are associated with a much more severe neurocristopathy known as PCWH featuring peripheral demyelinating neuropathy, central dysmyelination, WS and Hirschsprung disease (OMIM #609136) [63]. Mutations causing WS and PCWH both include nonsense and frameshift mutations resulting in premature truncations; an elegant comparison of the genotype-phenotype comparison suggested that the key feature determining whether a mutation resulted in the more severe PCWH condition lay in whether or not it resulted in nonsense-mediated decay of the mRNA. Thus, where mRNA was not degraded, truncated proteins were formed and acted in a dominant negative manner, giving the more severe phenotype, rather than the haploinsufficient WS phenotype. Finally, Sox10 has been proposed to

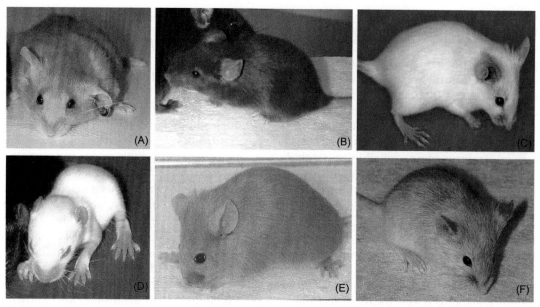

FIGURE 14.4 **Mouse pigmentation mutants.** (A) *Agouti* heterozygous mutant A^y/a (picture by Lluis Montoliu). The dominant allele increases the ratio of pheomelanin to eumelanin. (B) *Dct* homozygous mutant Dct^{slt}/Dct^{slt} (picture by Friedrich Beermann). Melanin levels are decreased. (C) *Mitf* mutant $Mitf^{Mi-b}/Mitf^{Mi-Or}$ (picture by Lynn Lamoreux). Pigmentation is completely absent. (D) *Sox10* mutant $Sox10^{Hry}$ (picture by Andrew Ward). Pigmentation is completely absent. (E) *Oca2* homozygous mutant p/p (picture by Lluis Montoliu). This allele inhibits eumelanin synthesis, leading to a reddish phenotype. (F) *Tyr* homozygous mutant Tyr^{c-m}/Tyr^{c-m} (picture by Lluis Montoliu). Significantly affected melanin production. *Source: Modified from [62].*

promote self-renewal in melanoma cells, whereas its ablation suppresses melanoma progression [64].

Work in mouse and zebrafish indicates that SOX10 is crucial for transcriptional activation of *Mitf/mitfa* from the very earliest stages [40,57,60,61]—that is, for melanocyte fate specification from the neural crest. SOX10 is an early marker of neural crest, and there is some delay before *Mitf* is expressed (e.g., approximately 7 h in zebrafish) [40]. Curran et al. showed that the vast majority of zebrafish NCC transiently turn on *mitfa* [46]. However, only around 50% of them activate expression of melanoblast markers such as *dct*; other cells switch to other fates, including iridophores [46]. Forced expression of *mitfa* in zebrafish *sox10* or *mitfa* mutants, rescues the melanocyte phenotype equally well, strongly suggesting

that the primary role of Sox10 in melanocyte development is to activate *mitfa* expression in melanocyte progenitors (i.e., to specify melanoblasts) [61], although the role in mammals likely extends beyond this (see below).

In mammals, PAX3 has also been shown to have a role in the early specification of melanoblasts. Melanocytes are significantly reduced in heterozygous *Pax3* mutants [65,66]. PAX3 directly activates the *Mitf* promoter [67]. Moreover, SOX10 and PAX3 have been suggested to promote synergistically the activation of the *Mitf* promoter. Hence, PAX3 is proposed to enhance the specificity and effects of SOX10 [68–72]. In postnatal mice, PAX3 has a complex role in driving melanocyte stem cells' fate specification through activation of *Mitf*, while simultaneously repressing ongoing differentiation [73]. Consistent with the mouse data, humans with heterozygous *PAX3* mutations present with Waardenburg syndrome type 1 or 3 (OMIM #193500 and #148820, respectively), characterized by hypopigmented patches, heterochromia irides, dystopia canthorum, and sensorineural deafness (with musculoskeletal abnormalities in WS3). Mutations in *PAX3* can also lead to alveolar rhabdomyosarcoma (OMIM #268220) and craniofacial-deafness-hand syndrome (OMIM #122880).

Acting alongside these intrinsic transcription factors are a variety of extrinsic signals that together drive fate specification of pigment cells from multipotent NCC [74,75]. In the case of melanocytes, the key extrinsic signal is Wnt signaling. Dorsky et al. [76] were the first to show that elevated Wnt signaling biases cranial NCC to adopt a melanocyte fate. Expression studies implicated Wnt1 or Wnt3a as likely ligands mediating these effects *in vivo* [76]. Wnt-induced signaling is mediated by Tcf/Lef transcription factors, direct binding of which was demonstrated on the zebrafish *mitfa* promoter [77]. Similar results obtained using avian and mouse neural crest and human melanocyte cultured cells suggested the conservation of this mechanism, with all studies implicating WNT3A as a likely mediator of melanoblast specification [78–80]. Surprisingly, attempts to identify the Frizzled receptor mediating this Wnt signaling have been unsuccessful to date [81].

Consistent with these studies, genetic manipulation to achieve neural crest-specific loss of Wnt signaling in mouse resulted in the absence of melanocytes, but also sensory neurons, whereas more surprisingly, over-activation of Wnt signaling in neural crest resulted in over-production of sensory neurons, but not melanocytes [82,83]. An elegant dissection of this conundrum showed that the timing of Wnt signaling is critical, with Wnt signaling during neural crest migration crucial to melanocyte specification [84].

Specification of Other Pigment Cell Types

Choosing between alternative fates is a fundamental problem in stem cell biology, tightly connected to the issue of progressive fate restriction—see above. In mouse, many melanocytes seem to be derived from (presumably) bipotent Schwann cell progenitors (SCPs), and hence these cells choose between melanocyte and the alternative Schwann cell fate. Although much remains unknown regarding the molecular basis of this fate decision, current data identifies putatively important interactions between four factors: First, melanocyte formation from SCPs is inhibited by ErbB signaling [33]. Secondly, in mouse and chick, SOX2 represses *Mitf* expression and thus inhibits melanocyte formation. Conversely, MITF represses *Sox2* expression, rendering these mutually repressive interactions important for SCPs' commitment to either Schwann cell fate or melanocyte lineages [33]. Finally, in avian and murine SCPs, loss of *Foxd3* function induces differentiation into melanocytes, suggesting that FOXD3 suppresses a melanocyte fate, while promoting Schwann cell and glial fates [85].

In zebrafish, both iridophores and xanthophores are also formed from NCC, yet current data suggest that the closer relationship is between melanocytes and iridophores. For a long time this was just an inference based upon the reciprocal changes in melanocyte number and iridophore number in *mitfa* mutants [44]. More recently, a molecular basis for this phenotype, strongly implying a bipotential progenitor, has been revealed by work on the transcription factor FoxD3. In *foxd3* mutants iridophores are strongly reduced, but this phenotype is rescued following ablation of both *mitfa* and *foxd3* [45,46,86–90]. Together with the increased *mitfa* expression in *foxd3* mutant embryos, this suggested that direct binding of the *mitfa* promoter by FoxD3 might inhibit *mitfa* expression, a model supported by *in vitro* studies [45]. A similar interaction has been suggested in avian NCC, with FOXD3 repressing melanocyte development via repression of *MITF* [91–93]. At least two other transcription factors are likely to play a major role in iridophore specification. Both *sox10* and *sox9b* mutants lack iridophores [39,94], and in the case of *sox10* at least the defect seems to be in iridophore fate specification [43], although their exact roles need to be determined.

Extrinsic factors are important for iridophore specification too, with signaling via the orphan receptor tyrosine kinase, leukocyte tyrosine kinase (Ltk), crucial for the process [43]. Interestingly, expression of *ltk* is biphasic, and while the late phase expression is clearly in definitive iridophore lineage cells, the early phase expression marks a multipotent progenitor [43]. It appears that Ltk signaling and Wnt signaling perform equivalent roles in specification of iridophores and melanocytes, respectively, and a major challenge is now to understand how multipotent progenitors integrate their responses to these signals to achieve the correct balance of pigment cell types.

Specification of xanthophore fate is less well understood. Zebrafish *sox10* mutants show very severe reductions in xanthophores and in xanthophore markers from the earliest stages, such that a role in xanthophore specification is highly likely [39,40]. Furthermore, Pax3 and Pax7 transcription factors have been strongly implicated in xanthoblast specification and development [95]. Loss of Pax3 results in complete ablation of xanthophores accompanied by an increase in the number of melanophores, suggesting a role for the transcription factor in lineage fate specification. *pax3* mutants lack *pax7a* expression at later stages. Interestingly, embryos lacking only Pax7a present with normal xanthophore numbers but exhibit defects in pigmentation intensity, indicating a role for Pax7a during differentiation. The presence of two *Pax7* orthologs in zebrafish means that functional redundancy may be obscuring the

full role of Pax7 in this cell type [95]. The fact that loss of xanthophores is compensated by expansion of melanophores may suggest the existence of a bipotent melano-xanthoblast precursor, although the timing of the melanocyte increase may simply indicate an indirect effect of the loss of xanthophores [95]. Given the cooperative interactions of SOX10 and PAX3 to induce *Mitf* in mouse [71,72], a possible role for Mitfa in xanthophore development [44] requires further study.

COMMITMENT

Thus far, we have discussed fate specification—the time when cells first show features distinctive of a particular cell fate [42]—but when do pigment cells become committed (i.e., when do they lose their ability to become other cell types)? In general, we do not know the answer to this question *in vivo*. Phenomena such as dedifferentiation and transdifferentiation raise questions over whether cells ever become fully and absolutely committed. Indeed, differentiating melanocytes from avian embryos will dedifferentiate and even revert to a bipotent melanoglial state when placed in culture in the presence of endothelins [96].

A recent examination of a core GRN for zebrafish melanocyte differentiation identified several features of the GRN that might play a role in driving commitment (Figure 14.1) [29]. In particular, given the major role for Sox10 in maintaining multipotency [75,97], loss of *sox10* expression would be expected to contribute to this. Overexpression experiments in the zebrafish embryo, combined with examination of the *sox10* mutant phenotypes, indicate that Mitfa and Sox10 initially establish a positive feedback loop that promotes melanoblast specification; however, subsequently, Mitfa-dependent activation of Hdac1 complex represses *sox10* expression in melanoblasts promoting both differentiation and likely fate commitment [29]. In mammals, the ongoing expression of *Sox10* perhaps indicates that even differentiated melanocytes retain multipotency, but this requires further evaluation.

MIGRATION AND PATTERNING

As described above, many pigment cell progenitors are already fate-specified as they migrate. Mutant phenotypes with white belly spots or decreased ventrally located cells often indicate genes with roles in migration. Recent evidence indicates that fate specification is necessary to allow this patterned migration, at least in the case of melanocytes where the topic has received significant attention in mouse and avian embryos [34].

Three key signaling factors play crucial roles in melanocyte migration. First, endothelins (ET) are ligands for seven pass G-protein-coupled endothelin receptors (Ednr) and have been best studied in the chick where a second Ednrb, EDNRB2, was found to be expressed specifically in early melanoblasts. Furthermore, ET3 binds to EDNRB2 and is expressed lining the lateral pathway [98]. A number of studies suggest that ET3/EDNRB2 signaling is crucial for melanoblast migration through the dorsolateral pathway in chick, mice and *Xenopus* embryos.

Early work demonstrated that mouse and rat mutants for EDNRB present with ventral belly spotting [99,100]. It was further shown that NCC cells express *EdnrB* prior to entering the dorsolateral pathway, while signaling through this receptor appeared to influence mouse embryonic stem cells (ESCs) toward differentiation into melanocyte lineage *in vitro* [101]. Mutations in the human *EDNRB* gene lead to Waardenburg syndrome type 4A, an auditory-pigmentary syndrome characterized by patchy hypopigmentation, congenital sensorineural hearing loss, and Hirschsprung disease (OMIM #277580). Knockdown and

overexpression studies have shown that signaling through EDNRB2 is crucial and sufficient for lateral pathway migration in chick embryos [36]. Although Ednrb2 is absent in zebrafish and mice, recent studies in *Xenopus* embryos further supported a role of the orthologous gene in melanoblast migration [102]. The same study identified the ligand Et3 as expressed at the destination of *ednrb2*-expressing melanoblasts and aiding invasion as well as migration along the ventral pathway. Overall, it appears that EdnrB/Et3 signaling is not specific to the dorsolateral pathway but functions to promote melanoblast migration. Interestingly, Et3 has also been reported to drive proliferation of melanocytes at least in cell culture, suggesting that Ednrb signaling may have multiple functions [103–105]. In zebrafish, *ednrb1a* has a clear role in adult pigment pattern formation [26,106]. However, although *ednrb1a* is expressed in early pigment cells, there are no pigment pattern defects in *ednrb1a* mutant embryos [107].

In birds, in contrast to other vertebrates, melanoblast migration is also stimulated by the expression of *EPHB2*, an ephrin receptor [108]. If *EPHB2* is knocked-down, this can be compensated by overexpression of EDNRB2, suggesting that these two systems act together and are partially redundant.

In mice, the major signaling molecule promoting melanocyte migration has been shown to be KIT. KIT is a type III receptor tyrosine kinase that responds to Stem Cell Factor (SCF, also called Steel factor) binding by inducing PI (3) kinase and MAP kinase cascades. KIT has multiple conserved roles in melanocyte development, including melanoblast survival, proliferation and appropriate dispersal along the dorsolateral migratory pathway. First, *Kit* is expressed from premigratory stages in melanoblast development in mouse and zebrafish [109,110]. Second, *Kit* and *SCF* mutant mice and zebrafish display reduced numbers of melanocytes, with severe deficits in ventral regions [110–112], suggesting reduced migratory capacity of melanocyte precursors. Likewise, human patients heterozygous for dominant mutant *KIT* alleles show hypopigmented patches [113].

Kit signaling was proposed to initiate melanoblast migration, attracting them onto the dorsolateral pathway lined by cells expressing *SCF*. In zebrafish *kit* mutants, melanocytes are specified and differentiate, but they fail to migrate properly and tend to accumulate in premigratory positions (dorsal to the neural tube, and in a cluster posterior to the inner ear) [110,114]. Using a conditional mutant, the sensitive phase for Kit activity in melanocyte migration is restricted to between 1 and 2 dpf [115]. Elegant studies of the phases of KIT requirement during murine melanocyte development used blocking antibody injections to demonstrate the early phase requirement, but also a later one for population of hair follicles [116]. This second role of KIT is proposed to be a chemokinetic effect: KIT increases the melanoblasts' rate of movement and, therefore, their chance of encountering the follicle [117].

It is hypothesized that Kit signaling might regulate *Mitf* expression [118]. Furthermore, the *Kit* promoter contains an E-box motif recognized by MITF [53]. This suggests the possibility of a positive feedback loop between KIT signaling and MITF. However, surprisingly, KIT-negative cells in culture express *Mitf* and several of its targets except for tyrosinase (*Tyr*). This indicates that KIT is neither required for *Mitf* activation nor maintenance, but likely is necessary, at least transiently, for expression of *Tyr* in developing melanoblasts [118].

Finally, stromal cell-derived factor 1 (Sdf1) is a chemokine that appears to have a conserved role in attracting melanocytes to appropriate sites. In zebrafish embryos, *sdf1a* is expressed in a very spatially restricted pattern that controls migration of certain placodal primordia expressing its receptor Cxcr4 [119,120].

Examination of the *choker* mutant and knockdown of *sdf1a* with morpholinos showed that melanocyte patterning in the embryo is partially controlled by the distribution of Sdf1a [121]. *In vitro* studies in mammals have shown a similar chemoattractant role for SDF1 on melanoblasts. The latter express *Cxcr4* and this seems to be necessary *in vivo* for correct positioning of melanoblasts during hair follicle development [122].

Xanthophores

Xanthoblasts are known to migrate along with melanoblasts via the lateral pathway, as well as over the head of the embryos, where they finally occupy the spaces between melanocytes [34]. Colony stimulating factor 1 receptor (*csf1r*, previously known as *fms*) is an early xanthophore marker, detected from 23hpf in migratory NC-derived cells, which co-express the xanthoblast markers *gch* and *xdh* [123]. It is a type III receptor tyrosine kinase closely related to Kit. *csfr1* mutant embryos show a much reduced number of xanthoblasts, and these fail to disperse, suggesting that the receptor may play an important role in xanthoblast migration similar to that of Kit in melanoblast migration.

PROLIFERATION AND SURVIVAL

Given the dramatic increase in cell numbers required for the relatively limited number of NCC to produce the full complement of skin pigment cells, it is clear that proliferation and survival of melanocyte precursors are important. In zebrafish and mammals, survival and proliferation of NC-derived melanoblasts appears to be critically dependent on Kit signaling. In fish, a temperature-sensitive conditional *kit* allele has enabled demonstration that Kit's role in survival begins from around 2 dpf and continues until approximately 4 dpf [115]. A similar transient role has been observed in mammalian melanocytes at around E14.5 [116]. Melanoblast apoptosis in *Kit* mutants results from a failure of the receptor tyrosine kinase to activate the PI(3) kinase and MAP kinase cascades. PI(3) kinase delivers a powerful anti-apoptotic signal via the activation of the kinases PDK1 and AKT [124]. Another target of Kit signaling is *Bcl2*, a known pro-survival gene [125]. Disruption of *Bcl2* function in melanocytes results in their apoptotic loss in mice and in cell culture [125]. Cell-cycle exit, and thus the termination of proliferation, appears to also be MITF-dependent, and results from activation of p21 (WAF1) expression, a cyclin-dependent kinase inhibitor [126].

Little is known of factors regulating survival and, especially, proliferation of other pigment cell types. However, many strong candidates for survival factors are likely included in the collections from mutagenesis screening in zebrafish [23]. These same screens identified numerous genes likely to function in maintenance of melanocytes with many affecting two or more pigment cell types (Figure 14.3) [23]. Identification of the mutant loci is a clear priority for dissecting these and other processes in pigment cell development in zebrafish and is also likely to reveal new players of relevance to mammals.

DIFFERENTIATION

The melanosome is a specialized membrane-bound organelle, which is involved in the synthesis, storage and transport of melanin. Melanosome organellogenesis progresses through four stages of maturation involving multiple enzymatic and structural proteins [127]. Many proteins have been identified as melanosome-specific, including tyrosinase (TYR), tyrosinase-related protein 1 (TRP1), dopachrome tautomerase (DCT), ocular

albinism type 1 protein (OA1), melanoma-associated antigen recognized by T cells (MART-1), vesicle amine transport protein 1 homolog (VAT-1), oculospanin, syntenin, and glycoprotein non-metastatic melanoma protein b (GPNMB) [127]. Their roles are diverse, including melanin synthesis, regulation of ion transport across the melanosome membrane and structural functions, which we will briefly touch upon here. This is a large field and space prevents discussion of other topics here, but recent reviews cover, for example, melanosome biogenesis, PMEL and melanosome fibril formation, Hermansky–Pudlak syndrome, cytoplasmic transport, and identification of novel melanosomal proteins [128–134]. One such protein, Slc24a5, was first identified from its role in the classic zebrafish pigment mutant *golden* and shown to function as a calcium transporter in the melanosome membrane [135]. The role of SLC24A5 was found to be conserved in human pigmentation, underlining the importance of primary pigment cell research in fish [135]. The authors suggested the existence of a group of transmembrane proteins with interdependent functions leading to calcium uptake and pH changes within the organelle, both of which influence melanin synthesis by affecting melanogenic enzyme activity [135,136]. The other transmembrane proteins cooperating with Slc24a5 are the hydrogen transporter, V-ATPase, and a sodium/hydrogen exchanger [137].

Morphology

Postmigratory melanocytes may be highly dendritic cells residing in the basal layer of the epidermis; others lack extensive projections, and are more associated with the dermis [5,138]. In mammals, dendrites are used to transport melanosomes from the cell body ready for secretion and uptake by adjacent keratinocytes. In contrast, in fish, amphibians, and reptiles, melanosomes and pteridine granules are not secreted, but they are often motile, so that dendrite morphology affects coloration and dynamic changes of the skin pigment pattern are observed. Dendrite formation, maintenance and activity are reliant upon a specialized cytoskeleton, especially both peripheral actin filaments and central micro-tubules [139]. Ednrb signaling promotes dendricity in fully differentiated zebrafish melanocytes [107].

In vivo confocal imaging has been employed to study the complex process of dendrite reor-ganization, which was identified as sensitive to hormonal inputs, UV irradiation, and inter-cellular contact [140]. Although actin filaments had been long considered the primary struc-tural component of dendrites, their active role in melanosome transport is now widely accepted. This role may be mediated by actin-binding proteins like myosin Va, melanophilin, RAB27A, and SLP2-A, all shown to transport melanosomes [141–145]. Intriguingly, early results implicated UV irradiation in the elabo-ration of dendrites [146]. Stimulation by beta-endorphin and bone morphogenic protein-4 (BMP-4) has been shown to activate endothe-lin-1, nerve growth factor (NGF), alpha-melanocyte-stimulating hormone (α-MSH), and prostaglandins through cAMP signaling, protein kinase C, and MAP kinase pathways [147,148]. These factors all induce dendrite extension *in vitro* in response to UV-stimuli. The small GTP-binding protein RHO has been implicated in dendrite constriction, whereas RAC1 protein induces the formation of den-drites in melanoma cell lines, suggesting that it might be involved in regulation of melanocyte morphology too [149]. Hence, the balance of RHO and RAC-mediated signaling is believed to play an important role in regulation of melanocyte dendrites [140,150]. Overall, the currently accepted model describes melanophi-lin recruitment to the melanosome membrane by RAB27A, which then allows recruitment of

myosin Va, linking the organelle to the actin cytoskeleton [151].

Studies in zebrafish have demonstrated the conserved effect on the melanophilin ortholog in melanosome transport; *melanophilin* mutants show a melanosome hypo-dispersion pheno-type due to the mutant protein's diminished ability to inhibit microtubule-based dynein-mediated melanosome accumulation in a peri-nuclear position [152].

Melanin Synthesis

Melanin production depends on the cooper-ative action of melanogenic enzymes of the tyrosinase family [153]. These include the rate-limiting oxygenase tyrosinase (TYR), the oxy-genase tyrosinase-related protein 1 (TYRP1), and dopachrome tautomerase TYRP2 (or DCT). The two latter enzymes also function to stabilize TYR. Melanin synthesis begins by a series of TYR-dependent conversions of tyro-sine to DOPAquinone and then to DOPAchrome. For these conversions to take place, the enzyme uses its characteristic binuc-lear copper center [154]. It is of particular interest that at this point the biochemical path-ways diverge to produce either eumelanin or pheomelanin. Insight into these processes has come from a combination of genetics and bio-chemistry, with both mice (Figure 14.4) and humans used as model systems.

The melanocyte master regulator MITF has been shown to be capable of directly activating the expression of these genes through the E-box and M-box binding motifs, in a manner largely conserved across species. Fine tran-scriptional regulation, nevertheless, appears to differ for each of the family members, at least in cultured primary melanoblasts [51,155]. In mouse both MITF and SOX10 appear to play direct roles in melanogenesis through the upregulation of *Tyr* [153,155–157]. Furthermore, distal enhancer elements of *Tyr* were found to be capable of binding SOX10 and Upstream Stimulatory Factor 1 (USF-1). Indeed, USF-1 was found to activate *Tyr* expression upon exposure to UV irradiation [158]. Finally, the MITF-interacting factor TFE3 and other molecules such as glucocorticoids and esters, have been shown to positively affect the expression of *Tyr* [159,160].

Tyrp1 regulation shares much in common with *Tyr*. Thus, the *Tyrp1* promoter also contains M-box and E-box elements recognized by MITF [161–163]. Moreover, there is evidence that the *Tyrp1* promoter is activated by other MITF co-activators—namely USF-1, TFE3, and TFEB [159,162,164] and repressed by the T-box transcription factor TBX2 [165]. Finally, PAX3 has been suggested to compete with TBX2 therefore positively regulating *Tyrp1* [166].

Intriguingly, in both the mouse and zebrafish, *Dct* expression begins significantly before *Tyr* and *Tyrp1*, rendering it one of the very earliest melanoblast markers [114,167,168]. *Dct* expression in different species, including zebrafish, is regulated by MITF [44,53,56,114,163]. There appear to be some interesting differences in the importance of SOX10 for differentiation/maintenance of melanocytes. In zebrafish all evidence converges on the conclusion that the critical role for Sox10 in melanocyte development is in fate specification, via activation of *mitfa* transcription. Crucially, *sox10* expression is minimal beyond approximately 50 hpf, consistent with *in vivo* rescue experiments showing that expression of *mitfa* alone is sufficient to rescue melanocytes in a *sox10* mutant background [29,61]. Interestingly, the evidence here suggests that Sox10 has a partially inhibitory role on melanocyte differentiation, which is removed by an Mitfa-dependent negative feedback loop [29]. In contrast, in mouse and humans expression of many MITF-dependent transcriptional targets, such as *Dct*, is enhanced by SOX10 [169–171]. Most of these studies have been in cultured cells and may not necessarily reflect the *in vivo* roles, with a more

recent study supporting SOX10 dispensability in *Dct* regulation [155]. Consistent with this, a direct test of the ongoing role of SOX10 in mouse melanoblasts freshly *ex vivo* was able to confirm the expected ongoing role only in the case of *Tyr* expression [172]. In adult human melanocytes, the closely related SOX9 is important for continued expression of melanin synthetic enzymes [173].

Intriguingly, PAX3 has been suggested to compete with MITF for the binding site on the *Dct* promoter. When PAX3 is bound, it acts with LEF1 to recruit the GRG4 repressor, which is displaced upon induction of beta-catenin. This elegant system of transcriptional regulation allows PAX3 to aid fate specification in melanocyte precursors, while at the same time preventing terminal differentiation and maintaining sensitivity of the cell to external stimuli [73].

Pigment-Type Switching

The pigment-type switching system (Figure 14.5), which controls whether melanocytes produce black/brown eumelanin or yellow/red pheomelanin, is responsible for many coloration patterns in mammals and birds [174]. The regulator of pigment-type switching is the melanocortin 1 receptor (MC1R), a G-protein-coupled receptor expressed in melanocytes. The molecular mechanisms underlying the switching involves the opposing action of α-melanocyte-stimulating hormone (α-MSH) and agouti signal protein (ASP) [175]. Although in its basal state, MC1R activity is sufficient to produce eumelanin, production significantly increases upon α-MSH stimulation [176]. MC1R signals through both PKA and stimulatory G-protein α subunit (Gαs) activation of adenylyl cyclase, raising levels of cAMP [177]. α-MSH is produced in the skin by keratinocytes and melanocytes, and increased MC1R signaling results in turn in activation of *Mitf*, and then of proteins required for pigment production (e.g., TYR, TYRP1, DCT),

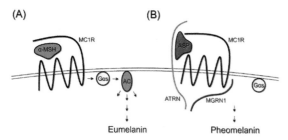

FIGURE 14.5 **Diagram summarizing the pigment cell switching mechanism, which controls the ratio of eumelanin to pheomelanin.** (A) α-melanocyte-stimulating hormone (α-MSH) binding to the melanocortin 1 receptor (MC1R) increases the signaling through the G-protein α subunit (Gαs) to activate adenylyl cyclase (AC), stimulating the production of eumelanin. (B) Agouti signal protein (ASP) competes with α-MSH for binding to MC1R. Attractin (ATRN) serves as an accessory receptor for ASP, and Mahogunin ring finger 1 (MGRN1) binds MC1R, hindering its interaction with the Gαs, and thereby uncoupling the receptor from its downstream effector, adenylyl cyclase. This decreases the eumelanin-to-pheomelanin ratio.

melanosome biogenesis (PMEL117), and transport (RAB27). The result is that melanocytes produce more eumelanin [178].

When MC1R is inactivated by ASP, cAMP levels are decreased and pheomelanin is produced instead. ASP is in mouse secreted by cells of the dermal papilla (a structure at the base of the hair follicle) during the start of the hair growth cycle, and also binds to MC1R [179]. ASP competes with α-MSH for binding to MC1R and also causes a reduction in the rate of new MC1R biosynthesis, thus inhibiting eumelanogenesis and promoting pheomelanogenesis (Figure 14.5B) [180]. Downstream of the receptor, ASP blocks the transcriptional effects of α-MSH and activates genes involved in morphogenesis and cell adhesion promoting a functional dedifferentiation of melanocytes [178].

Two spontaneous mouse mutants identified additional loci required for ASP activity: *mahogany* (*mg*) and *mahoganoid* (*md*) [181]. The *mahogany* locus was identified as the mouse ortholog of the human *attractin* (*ATRN*) gene, a

type I single-pass transmembrane protein [182]. ATRN might serve as an accessory receptor for ASP, which binds to ATRN through a separate region from that which binds MC1R. ATRN may be a part of a noncanonical mechanism of ASP signaling through the MC1R [183]. *Mahoganoid*, encodes a novel RING domain-containing protein [184] termed Mahogunin ring finger 1 (MGRN1), a member of the large class of E3 ubiquitin ligase proteins. MGRN1 binds MC1R, hindering the interaction of MC1R with the Gαs, and thereby uncoupling the receptor from its downstream effector, adenylyl cyclaze [185].

In contrast to the situation observed in mammals, fish melanophores contain only eumelanin and are unable to produce pheomelanin [5]. Melanocortins are involved in melanophore pigmentation in fish, possibly through their binding to Mc1r to increase the levels of cAMP leading to the upregulation of melanogenic genes [186]. Furthermore, knockdown experiments have indicated a role of Mc1r in melanosome dispersal in zebrafish [187]. Recent evidence indicates an unexpected role for Agouti in fish, and dissection of the underlying mechanisms will be particularly fascinating [188].

Iridophore Pigmentation

Iridophores differ from other pigment cell types due to their lack of dendritic processes and of pigment-containing organelles such as melanosomes. Instead, iridescence results from reflecting platelets, which are stacked in such a way as to diffract light [5,189]. Reflective platelets are comprised of crystallized purines: primarily of guanine often combined with varying proportions of adenine, hypoxanthine, and uric acid crystals [190]. Iridophores may or may not present with cellular motility (i.e., movement of organelles within the cell, changing the spacing of their stacking), depending on the model species [191]. When iridophores give rise to the characteristic unchanging silver or white color-ation, the layers of guanine crystals are orga-nized at tightly regulated fixed distances from each other in the cytoplasm. However, in some species platelets are adjusted in response to extracellular cues, such as hormonal inputs, giving rise to alterations in the wavelength of reflected light and, thus, to sometimes stunning changes in pigmentation [191]. Iridophores have also been proposed to contain pteridines, which are responsible for xanthophore pigmen-tation, within their reflective platelets [192]. While currently we know almost nothing about the genetics of the formation and organization of these reflecting platelets the numerous zebrafish mutants affecting iridophore pigmen-tation (Figure 14.3A−D, G, H) [23] are likely to include key genes in each of these processes, the cloning of which could thus illuminate a whole new area of biology.

Xanthophore Pigmentation: The Pteridine Pathway

Little is known currently about xanthophore differentiation, except for some major aspects of pigment synthesis. Xanthophore pigmentation is due to carotenoids derived from the diet and pteridine compounds produced in specialized organelles known as pterinosomes. The pteri-dine biosynthesis pathway is partially character-ized and involves multiple enzymes [193]. Guanine triphosphate (GTP) is initially con-verted into dihydroneopterin triphosphate by GTP cyclohydrolase I (Gch); *gch* expression is one of the very earliest markers of xanthophore lineage development (from approximately 18 hpf in zebrafish) and thus widely used as a specification marker [107,194]. Sequential actions of 6-Pyruvoyl-tetrahydropterin synthase and sepiapterin reductase ultimately yield tetra-hydrobiopterin. Sepiapterin is a derivative of either tetrahydrobiopterin or earlier intermedi-ates and accumulates from 24 hpf. Through a

series of biochemical reactions sepiapterin is converted into 7-oxopterin and eventually into isoxanthopterin or 2,4,7-trioxopteridine through the action of a xanthine oxidase variant (Xod, convertible into xanthine dehydrogenase or Xdh); *xdh* is often also used as an early xantho-phore marker [16,193]. Once again, the zebrafish pigment mutants no doubt include many that could identify key and novel genetic controls on xanthophore differentiation.

CONCLUSIONS AND PERSPECTIVES

Pigmentation plays an important role in multiple processes, ranging from DNA protec-tion from UV light to attraction of a mate. From a human perspective, pigmentation is significant for its association with diverse human disease and with adaptive variants that contribute in no small way to the huge diver-sity characteristic of our species. Even this brief review of our current knowledge in pig-ment cell development makes clear the insight that has come from studying pigmentation in diverse model systems.

Research to date has focused on discovery of melanocyte-specific genes involved in initial specification, survival and proliferation, migra-tion, patterning and differentiation to mature melanin-producing cells, and here mouse and human genetics have been dominant. The recent journey of the zebrafish (and also medaka) into the mainstream of genetic mod-els has allowed for the *in vivo* study of melano-cyte gene function at an unprecedented *in vivo* resolution, neatly complementing the mamma-lian work. These developments also finally open up a new opportunity, to study the genetics of other pigment cells. In particular, the opportunity to study how multipotent stem cells choose between multiple pigment fates offers an exciting and tractable model for understanding multipotency and progressive fate restriction processes *in vivo*.

In this time of systems biology, we are beginning to see the field move away from the traditional linear genetic pathways, to a more realistic appreciation of the complexities of GRNs. This is a major challenge, not least because current tools are far better at identifying the key components and less so at defining their detailed interactions, although developments in bioinformatics, mathematical biology, and systems biology are rapidly changing this. A step change in our understanding will come as geneticists work with bioinformaticians and mathematicians to develop a quantitative model of the GRNs underlying pigment cell development.

References

[1] Sato S, Yamamoto H. Development of pigment cells in the brain of ascidian tadpole larvae: insights into the origins of vertebrate pigment cells. Pigment Cell Res 2001;14:428–36.

[2] Bennett DC, Lamoreux ML. The color loci of mice—a genetic century. Pigment Cell Res 2003;16:333–44.

[3] Kelsh RN. Genetics and evolution of pigment patterns in fish. Pigment Cell Res 2004;17:326–36.

[4] Goda M, Fujii R. Blue chromatophores in two species of callionymid fish. Zool Sci 1995;12:811–3.

[5] Bagnara JT, Matsumoto J. Comparative anatomy and physiology of pigment cells in nonmammalian tissues. In: Nordlund J, Boissy R, Hearing V, King R, Ortonne J-P, editors. The pigmentary system physiology and pathophysiology. Second Ed. Oxford, UK: Blackwell Publishing Ltd; 2007. p. 307–44.

[6] Goda M, Ohata M, Ikoma H, Fujiyoshi Y, Sugimoto M, Fujii R. Integumental reddish-violet coloration owing to novel dichromatic chromatophores in the teleost fish, *Pseudochromis diadema*. Pigment Cell Melanoma Res 2011;24:614–7.

[7] Bagnara JT, Fernandez PJ, Fujii R. On the blue coloration of vertebrates. Pigment Cell Res 2007;20:14–26.

[8] Lin JY, Fisher DE. Melanocyte biology and skin pigmentation. Nature 2007;445:843–50.

[9] Kippenberger S, Bernd A, Bereiter-Hahn J, Ramirez-Bosca A, Kaufmann R. The mechanism of melanocyte dendrite formation: the impact of differentiating keratinocytes. Pigment Cell Res 1998;11:34–7.

[10] Ito S, Wakamatsu K. Quantitative analysis of eumelanin and pheomelanin in humans, mice, and other animals: a comparative review. Pigment Cell Res 2003;16:523–31.

[11] De Leeuw SM, Smit NPM, Van Veldhoven M, et al. Melanin content of cultured human melanocytes and UV-induced cytotoxicity. J Photochem Photobiol B Biol 2001;61:106–13.

[12] Mitra D, Luo X, Morgan A, et al. An ultraviolet-radiation-independent pathway to melanoma carcinogenesis in the red hair/fair skin background. Nature 2012;491:449–53.

[13] Simon JD, Peles D, Wakamatsu K, Ito S. Current challenges in understanding melanogenesis: bridging chemistry, biological control, morphology, and function. Pigment Cell Melanoma Res 2009;22:563–79.

[14] Kaelin CB, Xu X, Hong LZ, et al. Specifying and sustaining pigmentation patterns in domestic and wild cats. Science 2012;337:1536–41.

[15] Lamoreux ML, Delmas V, Larue L, Bennett D. The colors of mice: a model genetic network. Chichester, West Sussex, UK: John Wiley & Sons; 2010.

[16] Ziegler I. The pteridine pathway in zebrafish: regulation and specification during the determination of neural crest cell-fate. Pigment Cell Res 2003;16: 172–82.

[17] Goodrich HB, Hill GA, Arrick MS. The chemical identification of gene-controlled pigments in *Platypoecilus* and *Xiphophorus* and comparisons with other tropical fish. Genetics 1941;26:573–86.

[18] Matsumoto J. Studies on fine structure and cytochemical properties of erythrophores in swordtail, *Xiphophorus helleri*, with special reference to their pigment granules (Pterinosomes). J Cell Biol 1965;27: 493–504.

[19] Ichikawa Y, Ohtani H, Miura I. The erythrophore in the larval and adult dorsal skin of the brown frog, *Rana ornativentris*: its differentiation, migration, and pigmentary organelle formation. Pigment Cell Res 1998;11:345–54.

[20] Morrison RL. A transmission electron microscopic (TEM) method for determining structural colors reflected by lizard iridophores. Pigment Cell Res 1995;8:28–36.

[21] Sturm RA. Molecular genetics of human pigmentation diversity. Hum Mol Genet 2009;18:R9–17.

[22] Henion PD, Raible DW, Beattie CE, Stoesser KL, Weston JA, Eisen JS. Screen for mutations affecting development of zebrafish neural crest. Dev Genet 1996;18:11–7.

[23] Kelsh RN, Brand M, Jiang YJ, et al. Zebrafish pigmentation mutations and the processes of neural crest development. Development 1996;123:369–89.

[24] Odenthal J, Rossnagel K, Haffter P, et al. Mutations affecting xanthophore pigmentation in the zebrafish, *Danio rerio*. Development 1996;123:391−8.

[25] Gaiano N, Amsterdam A, Kawakami K, Allende M, Becker T, Hopkins N. Insertional mutagenesis and rapid cloning of essential genes in zebrafish. Nature 1996;383:829−32.

[26] Johnson SL, Africa D, Walker C, Weston JA. Genetic control of adult pigment stripe development in zebrafish. Dev Biol 1995,167.27−33.

[27] Patton EE, Zon LI. The art and design of genetic screens: zebrafish. Nat Rev Genet 2001;2:956−66.

[28] Dutton JR, Antonellis A, Carney TJ, et al. An evolutionarily conserved intronic region controls the spatiotemporal expression of the transcription factor Sox10. BMC Dev Biol 2008;8:105.

[29] Greenhill ER, Rocco A, Vibert L, Nikaido M, Kelsh RN. An iterative genetic and dynamical modelling approach identifies novel features of the gene regulatory network underlying melanocyte development. PLoS Genet 2011;7:e1002265.

[30] Oliveri P, Davidson EH. Gene regulatory network controlling embryonic specification in the sea urchin. Curr Opin Genet Dev 2004;14:351−60.

[31] Madan Babu M, Teichmann SA. Evolution of transcription factors and the gene regulatory network in *Escherichia coli*. Nucl Acids Res 2003;31:1234−44.

[32] Adameyko I, Lallemend F, Aquino JB, et al. Schwann cell precursors from nerve innervation are a cellular origin of melanocytes in skin. Cell 2009;139:366−79.

[33] Adameyko I, Lallemend F, Furlan A, et al. Sox2 and Mitf cross-regulatory interactions consolidate progenitor and melanocyte lineages in the cranial neural crest. Development 2012;139:397−410.

[34] Kelsh RN, Harris ML, Colanesi S, Erickson CA. Stripes and belly-spots—a review of pigment cell morphogenesis in vertebrates. Semin Cell Dev Biol 2009;20:90−104.

[35] Serbedzija GN, Fraser SE, Bronner-Fraser M. Pathways of trunk neural crest cell migration in the mouse embryo as revealed by vital dye labelling. Development 1990;108:605−12.

[36] Harris ML, Hall R, Erickson CA. Directing pathfinding along the dorsolateral path - the role of EDNRB2 and EphB2 in overcoming inhibition. Development 2008;135:4113−22.

[37] Nitzan E, Krispin S, Pfaltzgraff ER, Klar A, Labosky PA, Kalcheim C. A dynamic code of dorsal neural tube genes regulates the segregation between neurogenic and melanogenic NCC. Development 2013;140: 2269−79.

[38] Bagnara JT, Matsumoto J, Ferris W, et al. Common origin of pigment cells. Science 1979;203:410−5.

[39] Kelsh RN, Eisen JS. The zebrafish colourless gene regulates development of non-ectomesenchymal neural crest derivatives. Development 2000;127: 515−25.

[40] Dutton Ka, Pauliny A, Lopes SS, et al. Zebrafish colourless encodes sox10 and specifies non-ectomesenchymal neural crest fates. Development 2001;128:4113−25.

[41] Cornell RA, Eisen JS. Delta/Notch signaling promotes formation of zebrafish neural crest by repressing Neurogenin 1 function. Development 2002;129: 2639−48.

[42] Raible DW, Eisen JS. Restriction of neural crest cell fate in the trunk of the embryonic zebrafish. Development 1994;120:495−503.

[43] Lopes SS, Yang X, Müller J, et al. Leukocyte tyrosine kinase functions in pigment cell development. PLoS Genet 2008;4:e1000026.

[44] Lister JA, Robertson CP, Lepage T, Johnson SL, Raible DW. Nacre encodes a zebrafish microphthalmia-related protein that regulates neural-crest-derived pigment cell fate. Development 1999;126:3757−67.

[45] Curran K, Raible DW, Lister JA. Foxd3 controls melanophore specification in the zebrafish neural crest by regulation of Mitf. Dev Biol 2009;332:408−17.

[46] Curran K, Lister JA, Kunkel GR, Prendergast A, Parichy DM, Raible DW. Interplay between Foxd3 and Mitf regulates cell fate plasticity in the zebrafish neural crest. Dev Biol 2010;344:107−18.

[47] Budi EH, Patterson LB, Parichy DM. Post-embryonic nerve-associated precursors to adult pigment cells: genetic requirements and dynamics of morphogenesis and differentiation. PLoS Genet 2011;7:16.

[48] Steingrímsson E, Copeland NG, Jenkins NA. Melanocytes and the microphthalmia transcription factor network. Ann Rev Genet 2004;38:365−411.

[49] Hodgkinson CA, Moore KJ, Nakayama A, et al. Mutations at the mouse microphthalmia locus are associated with defects in a gene encoding a novel basic-helix-loop-helix-zipper protein. Cell 1993;74: 395−404.

[50] Tachibana M, Perez-Jurado LA, Nakayama A, et al. Cloning of MITF, the human homolog of the mouse microphthalmia gene and assignment to chromosome 3p14.1−p12.3. Hum Mol Genet 1994;3:553−7.

[51] Cheli Y, Ohanna M, Ballotti R, Bertolotto C. Fifteen-year quest for microphthalmia-associated transcription factor target genes. Pigment Cell Melanoma Res 2010;23:27−40.

[52] Levy C, Khaled M, Fisher DE. MITF: master regulator of melanocyte development and melanoma oncogene. Trends Mol Med 2006;12:406−14.

[53] Opdecamp K, Nakayama A, Nguyen MT, Hodgkinson CA, Pavan WJ, Arnheiter H. Melanocyte

development *in vivo* and in neural crest cell cultures: crucial dependence on the Mitf basic-helix-loop-helix-zipper transcription factor. Development 1997;124: 2377–86.

[54] Izumi K, Kohta T, Kimura Y, et al. Tietz syndrome: unique phenotype specific to mutations of MITF nuclear localization signal. Clin Genet 2008;74:93–5.

[55] Read AP, Newton VE. Waardenburg syndrome. J Med Genet 1997;34:656–65.

[56] Johnson SL, Nguyen AN, Lister JA. mitfa is required at multiple stages of melanocyte differentiation but not to establish the melanocyte stem cell. DeBiol 2011;350:405–13.

[57] Herbarth B, Pingault V, Bondurand N, et al. Mutation of the Sry-related Sox10 gene in dominant megacolon, a mouse model for human Hirschsprung disease. Proc Natl Acad Sci USA 1998;95:5161–5.

[58] Kuhlbrodt K, Schmidt C, Sock E, et al. Functional analysis of Sox10 mutations found in human Waardenburg–Hirschsprung patients. J Biol Chem 1998;273:23033–8.

[59] Pingault V, Bondurand N, Kuhlbrodt K, et al. SOX10 mutations in patients with Waardenburg–Hirschsprung disease. Nat Genet 1998;18:171–3.

[60] Southard-Smith EM, Kos L, Pavan WJ. Sox10 mutation disrupts neural crest development in Dom Hirschsprung mouse model. Nat Genet 1998;18:60–4.

[61] Elworthy S. Transcriptional regulation of mitfa accounts for the sox10 requirement in zebrafish melanophore development. Development 2003;130: 2809–18.

[62] Montoliu L, Oetting W, Bennett D. Color Genes. European Society for Pigment Cell Research 2011. <http://www.espcr.org/micemut> (accessed Oct 2013).

[63] Inoue K, Khajavi M, Ohyama T, et al. Molecular mechanism for distinct neurological phenotypes conveyed by allelic truncating mutations. Nat Genet 2004;36:361–9.

[64] Shakhova O, Sommer L. Testing the cancer stem cell hypothesis in melanoma: The clinics will tell. *Cancer letters* 2012;338:74–81.

[65] Hornyak TJ, Hayes DJ, Chiu LY, Ziff EB. Transcription factors in melanocyte development: distinct roles for Pax-3 and Mitf. Mech Dev 2001;101:47–59.

[66] Guo X-L, Ruan H-B, Li Y, Gao X, Li W. Identification of a novel nonsense mutation on the Pax3 gene in ENU-derived white belly spotting mice and its genetic interaction with c-Kit. Pigment Cell Melanoma Res 2010;23:252–62.

[67] Watanabe A, Takeda K, Ploplis B, Tachibana M. Epistatic relationship between Waardenburg syndrome genes MITF and PAX3. Nat Genet 1998;18: 283–6.

[68] Kamachi Y, Uchikawa M, Kondoh H. Pairing SOX off: with partners in the regulation of embryonic development. Trends Genet 2000;16:182–7.

[69] Wegner M. Secrets to a healthy Sox life: lessons for melanocytes. Pigment Cell Res 2005;18:74–85.

[70] Wissmüller S, Kosian T, Wolf M, Finzsch M, Wegner M. The high-mobility-group domain of Sox proteins interacts with DNA-binding domains of many transcription factors. Nucl Acids Res 2006;34:1735–44.

[71] Potterf SB, Furumura M, Dunn KJ, Arnheiter H, Pavan WJ. Transcription factor hierarchy in Waardenburg syndrome: regulation of MITF expression by SOX10 and PAX3. Hum Genet 2000;107:1–6.

[72] Bondurand N, Pingault V, Goerich DE, et al. Interaction among SOX10, PAX3 and MITF, three genes altered in Waardenburg syndrome. Hum Mol Genet 2000;9:1907–17.

[73] Lang D, Lu MM, Huang L, et al. Pax3 functions at a nodal point in melanocyte stem cell differentiation. Nature 2005;433:884–7.

[74] Betancur P, Bronner-Fraser M, Sauka-Spengler T. Assembling neural crest regulatory circuits into a gene regulatory network. Ann Rev Cell Dev Biol 2010;26:581–603.

[75] Kelsh RN. Sorting out Sox10 functions in neural crest development. BioEssays 2006;28:788–98.

[76] Dorsky RI, Moon RT, Raible DW. Control of neural crest cell fate by the Wnt signalling pathway. Nature 1998;396:370–3.

[77] Dorsky RI, Raible DW, Moon RT. Direct regulation of nacre, a zebrafish MITF homolog required for pigment cell formation, by the Wnt pathway. Genes Dev 2000;14:158–62.

[78] Dunn KJ, Williams BO, Li Y, Pavan WJ. Neural crest-directed gene transfer demonstrates Wnt1 role in melanocyte expansion and differentiation during mouse development. Proc Natl Acad Sci USA 2000;97:10050–5.

[79] Takeda K, Yasumoto K, Takada R, et al. Induction of melanocyte-specific microphthalmia-associated transcription factor by Wnt-3a. J Biol Chem 2000;275: 14013–6.

[80] Jin EJ, Erickson CA, Takada S, Burrus LW. Wnt and BMP signaling govern lineage segregation of melanocytes in the avian embryo. Dev Biol 2001;233: 22–37.

[81] Nikaido M, Law EWP, Kelsh RN. A systematic survey of expression and function of zebrafish frizzled genes. PloS One 2013;8:e54833.

[82] Hari L, Brault V, Kléber M, et al. Lineage-specific requirements of beta-catenin in neural crest development. J Cell Biol 2002;159:867–80.

[83] Lee H-Y, Kléber M, Hari L, et al. Instructive role of Wnt/beta-catenin in sensory fate specification in neural crest stem cells. Science 2004;303:1020–3.

[84] Hari L, Miescher I, Shakhova O, et al. Temporal control of neural crest lineage generation by Wnt/β-catenin signaling. Development 2012;139:2107–17.

[85] Nitzan E, Pfaltzgraff ER, Labosky PA, Kalcheim C. Neural crest and Schwann cell progenitor-derived melanocytes are two spatially segregated populations similarly regulated by Foxd3. *Proceedings of the National Academy of Sciences of the United States of America* 2013;110:12709–14.

[86] Lister JA, Cooper C, Nguyen K, Modrell M, Grant K, Raible DW. Zebrafish Foxd3 is required for development of a subset of neural crest derivatives. Dev Biol 2006;290:92–104.

[87] Montero-Balaguer M, Lang MR, Sachdev SW, et al. The mother superior mutation ablates foxd3 activity in neural crest progenitor cells and depletes neural crest derivatives in zebrafish. Dev Dyn 2006;235:3199–212.

[88] Hochgreb-Hägele T, Bronner ME. A novel FoxD3 gene trap line reveals neural crest precursor movement and a role for FoxD3 in their specification. Dev Biol 2013;374:1–11.

[89] Ignatius MS, Moose HE, El-Hodiri HM, Henion PD. colgate/hdac1 repression of foxd3 expression is required to permit mitfa-dependent melanogenesis. Dev Biol 2008;313:568–83.

[90] Stewart RA, Arduini BL, Berghmans S, et al. Zebrafish foxd3 is selectively required for neural crest specification, migration and survival. Dev Biol 2006;292:174–88.

[91] Harris ML, Erickson CA. Lineage specification in neural crest cell pathfinding. Dev Dyn 2007;236:1–19.

[92] Thomas AJ, Erickson CA. FOXD3 regulates the lineage switch between neural crest-derived glial cells and pigment cells by repressing MITF through a non-canonical mechanism. Development 2009;136:1849–58.

[93] Wan P, Hu Y, He L. Regulation of melanocyte pivotal transcription factor MITF by some other transcription factors. Mol Cell Biochem 2011;354:241–6.

[94] Yan Y-L, Willoughby J, Liu D, et al. A pair of Sox: distinct and overlapping functions of zebrafish sox9 co-orthologs in craniofacial and pectoral fin development. Development 2005;132:1069–83.

[95] Minchin JEN, Hughes SM. Sequential actions of Pax3 and Pax7 drive xanthophore development in zebrafish neural crest. Dev Biol 2008;317:508–22.

[96] Dupin E, Glavieux C, Vaigot P, Le Douarin NM. Endothelin 3 induces the reversion of melanocytes to glia through a neural crest-derived glial-melanocytic progenitor. Proc Natl Acad Sci USA 2000;97:7882–7.

[97] Sommer L. Generation of melanocytes from NCC. Pigment Cell Melanoma Res 2011;24:411–21.

[98] Lecoin L, Sakurai T, Ngo M-T, Abe Y, Yanagisawa M, Le Douarin NM. Cloning and characterization of a novel endothelin receptor subtype in the avian class. Proc Natl Acad Sci USA 1998;95:3024–9.

[99] Hosoda K, Hammer RE, Richardson JA, et al. Targeted and natural (piebald-lethal) mutations of endothelin-B receptor gene produce megacolon associated with spotted coat color in mice. Cell 1994;79:1267–76.

[100] Gariepy CE, Cass DT, Yanagisawa M. Null mutation of endothelin receptor type B gene in spotting lethal rats causes aganglionic megacolon and white coat color. Proc Natl Acad Sci USA 1996;93:867–72.

[101] Pla P, Alberti C, Solov'eva O, Pasdar M, Kunisada T, Larue L. Ednrb2 orients cell migration toward the dorsolateral neural crest pathway and promotes melanocyte differentiation. Pigment Cell Res 2005;18:181–7.

[102] Kawasaki-Nishihara A, Nishihara D, Nakamura H, Yamamoto H. ET3/Ednrb2 signaling is critically involved in regulating melanophore migration in *Xenopus*. Dev Dynam 2011;240:1454–66.

[103] Lahav R, Ziller C, Dupin E, Le Douarin NM. Endothelin 3 promotes neural crest cell proliferation and mediates a vast increase in melanocyte number in culture. Proc Natl Acad Sci USA 1996;93:3892–7.

[104] Reid K, Turnley AM, Maxwell GD, et al. Multiple roles for endothelin in melanocyte development: regulation of progenitor number and stimulation of differentiation. Development 1996;122:3911–9.

[105] Opdecamp K, Kos L, Arnheiter H, Pavan WJ. Endothelin signalling in the development of neural crest-derived melanocytes. Biochem Cell Biol 1998;76:1093–9.

[106] Frohnhöfer HG, Krauss J, Maischein H-M, Nüsslein-Volhard C. Iridophores and their interactions with other chromatophores are required for stripe formation in zebrafish. Development 2013;140:2997–3007.

[107] Parichy DM, Mellgren EM, Rawls JF, Lopes SS, Kelsh RN, Johnson SL. Mutational analysis of endothelin receptor b1 (rose) during neural crest and pigment pattern development in the zebrafish *Danio rerio*. Dev Biol 2000;227:294–306.

[108] Santiago A, Erickson CA. Ephrin-B ligands play a dual role in the control of neural crest cell migration. Development 2002;129:3621–32.

[109] Wehrle-Haller B, Weston JA. Soluble and cell-bound forms of steel factor activity play distinct roles in melanocyte precursor dispersal and survival on the lateral neural crest migration pathway. Development 1995;121:731–42.

[110] Parichy DM, Rawls JF, Pratt SJ, Whitfield TT, Johnson SL. Zebrafish sparse corresponds to an orthologue of c-kit and is required for the morphogenesis of a subpopulation of melanocytes, but is not essential for hematopoiesis or primordial germ cell development. Development 1999;126:3425–36.

[111] Alexeev V, Yoon K. Distinctive role of the cKit receptor tyrosine kinase signaling in mammalian melanocytes. J Invest Dermatol 2006;126:1102–10.

[112] Dooley CM, Mongera A, Walderich B, Nüsslein-Volhard C. On the embryonic origin of adult melanophores: the role of ErbB and Kit signalling in establishing melanophore stem cells in zebrafish. Development 2013;140:1003–13.

[113] Giebel LB, Spritz RA. Mutation of the KIT (mast/stem cell growth factor receptor) protooncogene in human piebaldism. Proc Natl Acad Sci USA 1991;88:8696–9.

[114] Kelsh RN, Schmid B, Eisen JS. Genetic analysis of melanophore development in zebrafish embryos. Dev Biol 2000;225:277–93.

[115] Rawls JF, Johnson SL. Temporal and molecular separation of the kit receptor tyrosine kinase's roles in zebrafish melanocyte migration and survival. Dev Biol 2003;262:152–61.

[116] Nishikawa S, Kusakabe M, Yoshinaga K, et al. In utero manipulation of coat color formation by a monoclonal anti-c-kit antibody: two distinct waves of c-kit-dependency during melanocyte development. EMBO J 1991;10:2111–8.

[117] Jordan SA, Jackson IJ. MGF (KIT ligand) is a chemokinetic factor for melanoblast migration into hair follicles. Dev Biol 2000;225:424–36.

[118] Hou L, Panthier JJ, Arnheiter H. Signaling and transcriptional regulation in the neural crest-derived melanocyte lineage: interactions between KIT and MITF. Development 2000;127:5379–89.

[119] Haas P, Gilmour D. Chemokine signaling mediates self-organizing tissue migration in the zebrafish lateral line. Dev Cell 2006;10:673–80.

[120] Valentin G, Haas P, Gilmour D. The chemokine SDF1a coordinates tissue migration through the spatially restricted activation of Cxcr7 and Cxcr4b. Curr Biol 2007;17:1026–31.

[121] Svetic V, Hollway GE, Elworthy S, et al. Sdf1a patterns zebrafish melanophores and links the somite and melanophore pattern defects in choker mutants. Development 2007;134:1011–22.

[122] Belmadani A, Jung H, Ren D, Miller RJ. The chemokine SDF-1/CXCL12 regulates the migration of melanocyte progenitors in mouse hair follicles. Differentiation 2009;77:395–411.

[123] Parichy DM, Ransom DG, Paw B, Zon LI, Johnson SL. An orthologue of the kit-related gene fms is required for development of neural crest-derived xanthophores and a subpopulation of adult melanocytes in the zebrafish, *Danio rerio*. Development 2000;127:3031–44.

[124] Goding CR. Mitf from neural crest to melanoma: signal transduction and transcription in the melanocyte lineage. Genes Dev 2000;14:1712–28.

[125] McGill GG, Horstmann M, Widlund HR, et al. Bcl2 regulation by the melanocyte master regulator Mitf modulates lineage survival and melanoma cell viability. Cell 2002;109:707–18.

[126] Carreira S, Goodall J, Aksan I, et al. Mitf cooperates with Rb1 and activates p21Cip1 expression to regulate cell cycle progression. Nature 2005;433:764–9.

[127] Hoashi T, Sato S, Yamaguchi Y, Passeron T, Tamaki K, Hearing VJ. Glycoprotein nonmetastatic melanoma protein b, a melanocytic cell marker, is a melanosome-specific and proteolytically released protein. FASEB J 2010;24:1616–29.

[128] Schiaffino MV. Signaling pathways in melanosome biogenesis and pathology. Int J Biochem Cell Biol 2010;42:1094–104.

[129] Palmisano I, Bagnato P, Palmigiano A, et al. The ocular albinism type 1 protein, an intracellular G protein-coupled receptor, regulates melanosome transport in pigment cells. Hum Mol Genet 2008;17:3487–501.

[130] Basrur V, Yang F, Kushimoto T, et al. Proteomic analysis of early melanosomes: identification of novel melanosomal proteins. J Proteome Res 2003;2:69–79.

[131] Giordano F, Bonetti C, Surace EM, Marigo V, Raposo G. The ocular albinism type 1 (OA1) G-protein-coupled receptor functions with MART-1 at early stages of melanogenesis to control melanosome identity and composition. Hum Mol Genet 2009;18:4530–45.

[132] Zhang P, Liu W, Zhu C, et al. Silencing of GPNMB by siRNA inhibits the formation of melanosomes in

melanocytes in a MITF-independent fashion. PLoS One 2012;**7**.

[133] Watt B, Van Niel G, Raposo G, Marks MS. PMEL: a pigment cell-specific model for functional amyloid formation. *Pigment cell & melanoma research* 2013;26:300−15.

[134] Wei A-H, Li W. Hermansky-Pudlak syndrome: pigmentary and non-pigmentary defects and their pathogenesis. Pigment Cell Melanoma Res 2013;26:176−92.

[135] Lamason RL, Mohideen M-APK, Mest JR, et al. SLC24A5, a putative cation exchanger, affects pigmentation in zebrafish and humans. Science 2005;310:1782−6.

[136] Watabe H, Valencia JC, Yasumoto K-I, et al. Regulation of tyrosinase processing and trafficking by organellar pH and by proteasome activity. J Biol Chem 2004;279:7971−81.

[137] Smith DR, Spaulding DT, Glenn HM, Fuller BB. The relationship between Na(+)/H(+) exchanger expression and tyrosinase activity in human melanocytes. Exp Cell Res 2004;298:521−34.

[138] Hirata M, Nakamura K-I, Kondo S. Pigment cell distributions in different tissues of the zebrafish, with special reference to the striped pigment pattern. Dev Dyn 2005;234:293−300.

[139] Lacour JP, Gordon PR, Eller M, Bhawan J, Gilchrest BA. Cytoskeletal events underlying dendrite formation by cultured pigment cells. J Cell Physiol 1992;151:287−99.

[140] Scott G. Rac and rho: the story behind melanocyte dendrite formation. Pigment Cell Res 2002;15:322−30.

[141] Wu X, Bowers B, Rao K, Wei Q, Hammer Ja. Visualization of melanosome dynamics within wild-type and dilute melanocytes suggests a paradigm for myosin V function *in vivo*. J Cell Biol 1998;143:1899−918.

[142] Matesic LE, Yip R, Reuss AE, et al. Mutations in Mlph, encoding a member of the Rab effector family, cause the melanosome transport defects observed in leaden mice. Proc Natl Acad Sci USA 2001;98:10238−43.

[143] Wu XS, Rao K, Zhang H, et al. Identification of an organelle receptor for myosin-Va. Nat Cell Biol 2002;4:271−8.

[144] Wu X, Wang F, Rao K, Sellers JR, Hammer JA. Rab27a is an essential component of melanosome receptor for myosin Va. Mol Biol Cell 2002;13:1735−49.

[145] Fukuda M, Kuroda TS, Mikoshiba K. Slac2-a/melanophilin, the missing link between Rab27 and myosin Va: implications of a tripartite protein complex for melanosome transport. J Biol Chem 2002;277:12432−6.

[146] Jimbow K, Pathak MA, Fitzpatrick TB. Effect of ultraviolet on the distribution pattern of microfilaments and microtubules and on the nucleus in human melanocytes. Yale J Biol Med 1973;46:411−26.

[147] Hara M, Yaar M, Gilchrest BA. Endothelin-1 of keratinocyte origin is a mediator of melanocyte dendricity. J Invest Dermatol 1995;105:744−8.

[148] Hirobe T. Stimulation of dedritogenesis in the epidermal melanocytes of newborn mice by melanocyte-stimulating hormone. J Cell Sci 1978;33:371−83.

[149] Scott GA, Cassidy L. Rac1 mediates dendrite formation in response to melanocyte stimulating hormone and ultraviolet light in a murine melanoma model. J Invest Dermatol 1998;111:243−50.

[150] Threadgill R, Bobb K, Ghosh A. Regulation of dendritic growth and remodeling by Rho, Rac, and Cdc42. Neuron 1997;19:625−34.

[151] Hume AN, Seabra MC. Melanosomes on the move: a model to understand organelle dynamics. Biochem Soc Trans 2011;39:1191−6.

[152] Sheets L, Ransom DG, Mellgren EM, Johnson SL, Schnapp BJ. Zebrafish melanophilin facilitates melanosome dispersion by regulating dynein. Curr Biol 2007;17:1721−34.

[153] Wang N, Hebert DN. Tyrosinase maturation through the mammalian secretory pathway: bringing color to life. Pigment Cell Res 2006;19:3−18.

[154] Hearing VJ, Jiménez M. Mammalian tyrosinase—the critical regulatory control point in melanocyte pigmentation. Int J Biochem 1987;19:1141−7.

[155] Hou L, Arnheiter H, Pavan WJ. Interspecies difference in the regulation of melanocyte development by SOX10 and MITF. Proc Natl Acad Sci USA 2006;103:9081−5.

[156] Yasumoto K, Yokoyama K, Takahashi K, Tomita Y, Shibahara S. Functional analysis of microphthalmia-associated transcription factor in pigment cell-specific transcription of the human tyrosinase family genes. J Biol Chem 1997;272:503−9.

[157] Gaggioli C, Buscà R, Abbe P, Ortonne J-P, Ballotti R. Microphthalmia-associated transcription factor (MITF) is required but is not sufficient to induce the expression of melanogenic genes. Pigment Cell Res 2003;16:374−82.

[158] Galibert MD, Carreira S, Goding CR. The Usf-1 transcription factor is a novel target for the stress-responsive p38 kinase and mediates UV-induced Tyrosinase expression. EMBO J 2001;20:5022−31.

[159] Verastegui C, Bertolotto C, Bille K, Abbe P, Ortonne JP, Ballotti R. TFE3, a transcription factor homologous to microphthalmia, is a potential transcriptional activator of tyrosinase and TyrpI genes. Mol Endocrinol 2000;14:449−56.

[160] Ferguson CA, Kidson SH. Characteristic sequences in the promoter region of the chicken tyrosinase-encoding gene. Gene 1996;169:191−5.

[161] Yasumoto K, Yokoyama K, Shibata K, Tomita Y, Shibahara S. Microphthalmia-associated transcription factor as a regulator for melanocyte-specific transcription of the human tyrosinase gene. Mol Cell Biol 1995;15:1833.

[162] Aksan I, Goding CR. Targeting the microphthalmia basic helix-loop-helix-leucine zipper transcription factor to a subset of E-box elements *in vitro* and *in vivo*. Mol Cell Biol 1998;18:6930−8.

[163] Bertolotto C, Buscà R, Abbe P, et al. Different cis-acting elements are involved in the regulation of TRP1 and TRP2 promoter activities by cyclic AMP: pivotal role of M boxes (GTCATGTGCT) and of microphthalmia. Mol Cell Biol 1998;18:694−702.

[164] Yavuzer U, Goding CR. Melanocyte-specific gene expression: role of repression and identification of a melanocyte-specific factor, MSF. Mol Cell Biol 1994;14:3494−503.

[165] Carreira S, Dexter TJ, Yavuzer U, Easty DJ, Goding CR. Brachyury-related transcription factor Tbx2 and repression of the melanocyte-specific TRP-1 promoter. Mol Cell Biol 1998;18:5099−108.

[166] Galibert MD, Yavuzer U, Dexter TJ, Goding CR. Pax3 and regulation of the melanocyte-specific tyrosinase-related protein-1 promoter. J Biol Chem 1999;274:26894−900.

[167] Steel KP, Davidson DR, Jackson IJ. TRP-2/DT, a new early melanoblast marker, shows that steel growth factor (c-kit ligand) is a survival factor. Development 1992;115:1111−9.

[168] Camp E, Lardelli M. Tyrosinase gene expression in zebrafish embryos. Development 2001;211:150−3.

[169] Potterf SB, Mollaaghababa R, Hou L, et al. Analysis of SOX10 function in neural crest-derived melanocyte development: SOX10-dependent transcriptional control of dopachrome tautomerase. Dev Biol 2001;237:245−57.

[170] Ludwig A, Rehberg S, Wegner M. Melanocyte-specific expression of dopachrome tautomerase is dependent on synergistic gene activation by the Sox10 and Mitf transcription factors. FEBS Lett 2004;556:236−44.

[171] Murisier F, Guichard S, Beermann F. The tyrosinase enhancer is activated by Sox10 and Mitf in mouse melanocytes. Pigment Cell Res 2007;20:173−84.

[172] Hou L, Pavan WJ. Transcriptional and signaling regulation in neural crest stem cell-derived melanocyte development: do all roads lead to Mitf? Cell Res 2008;18:1163−76.

[173] Passeron T, Valencia JC, Namiki T, et al. Upregulation of SOX9 inhibits the growth of human

and mouse melanomas and restores their sensitivity to retinoic acid. J Clin Invest 2009;119:954−63.

[174] Walker WP, Gunn TM. Shades of meaning: the pigment-type switching system as a tool for discovery. Pigment Cell Melanoma Res 2010;23:485−95.

[175] Sakai C, Ollmann M, Kobayashi T, et al. Modulation of murine melanocyte function *in vitro* by agouti signal protein. EMBO J 1997;16:3544−52.

[176] Sánchez-Más J, Hahmann C, Gerritsen I, García-Borrón JC, Jim énez-Cervantes C. Agonist-independent, high constitutive activity of the human melanocortin 1 receptor. Pigment Cell Res 2004;17: 386−95.

[177] García-Borrón JC, Sánchez-Laorden BL, Jiménez-Cervantes C. Melanocortin-1 receptor structure and functional regulation. Pigment Cell Res 2005;18: 393−410.

[178] Le Pape E, Passeron T, Giubellino A, Valencia JC, Wolber R, Hearing VJ. Microarray analysis sheds light on the dedifferentiating role of agouti signal protein in murine melanocytes via the Mc1r. Proc Natl Acad Sci USA 2009;106:1802−7.

[179] Bultman SJ, Michaud EJ, Woychik RP. Molecular characterization of the mouse agouti locus. Cell 1992;71:1195−204.

[180] Rouzaud F, Annereau J-P, Valencia JC, Costin G-E, Hearing VJ. Regulation of melanocortin 1 receptor expression at the mRNA and protein levels by its natural agonist and antagonist. FASEB J 2003;17: 2154−6.

[181] Miller KA, Gunn TM, Carrasquillo MM, Lamoreux ML, Galbraith DB, Barsh GS. Genetic studies of the mouse mutations mahogany and mahoganoid. Genetics 1997;146:1407−15.

[182] Gunn TM, Miller KA, He L, et al. The mouse mahogany locus encodes a transmembrane form of human attractin. Nature 1999;398:152−6.

[183] Hida T, Wakamatsu K, Sviderskaya EV, et al. Agouti protein, mahogunin, and attractin in pheomelanogenesis and melanoblast-like alteration of melanocytes: a cAMP-independent pathway. Pigment Cell Melanoma Res 2009;22:623−34.

[184] He L, Lu X-Y, Jolly AF, et al. Spongiform degeneration in mahoganoid mutant mice. Science 2003;299:710−2.

[185] Pérez-Oliva AB, Olivares C, Jiménez-Cervantes C, García-Borrón JC. Mahogunin ring finger-1 (MGRN1) E3 ubiquitin ligase inhibits signaling from melanocortin receptor by competition with Galphas. J Biol Chem 2009;284:31714−25.

[186] Logan DW, Burn SF, Jackson IJ. Regulation of pigmentation in zebrafish melanophores. Pigment Cell Res 2006;19:206−13.

[187] Richardson J, Lundegaard PR, Reynolds NL, et al. mc1r pathway regulation of zebrafish melanosome dispersion. Zebrafish 2008;5:289—95.

[188] Guillot R, Ceinos RM, Cal R, Rotllant J, Cerdá-Reverter JM. Transient ectopic overexpression of agouti-signalling protein 1 (asip1) induces pigment anomalies in flatfish. PloS One 2012;7:e48526.

[189] Fujii R. Cytophysiology of sh chromatophores. Int Rev Cytol 1993;143:191—255.

[190] Rohrlich ST, Rubin RW. Biochemical characterization of crystals from the dermal iridophores of a chameleon Anolis carolinensis. J Cell Biol 1975;66:635—45.

[191] Fujii R. The regulation of motile activity in fish chromatophores. Pigment Cell Res 2000;13:300—19.

[192] Oliphant LW, Hudon J. Pteridines as reflecting pigments and components of reflecting organelles in vertebrates. Pigment Cell Res 1993;6:205—8.

[193] Ziegler I, McDonald T, Hesslinger C, Pelletier I, Boyle P. Development of the pteridine pathway in the zebrafish, Danio rerio. J Biol Chem 2000;275:18926—32.

[194] Pelletier I, Bally-Cuif L, Ziegler I. Cloning and developmental expression of zebrafish GTP cyclohydrolase I. Mech Dev 2001;109:99—103.

Neural Crest Cells in Vascular Development

Sophie E. Wiszniak and Quenten P. Schwarz

Centre for Cancer Biology, SA Pathology, Adelaide, SA, Australia

OUTLINE

Neural Crest Cells.
DOI: http://dx.doi.org/10.1016/B978-0-12-401730-6.00016-8

313

INTRODUCTION

Neural crest cells (NCC) are a transient population of stem cells that form around the time of neural tube closure during early embryogenesis. NCC are specified from neuroepithelial precursors by inductive signals emanating from the adjacent non-neural ectoderm (as well as from underlying mesoderm—in frogs and fish) that induces these cells to migrate out of the dorsal region of the neural tube into the surrounding tissue. This migration occurs in an anterior to posterior spatiotemporal pattern, such that NCC in the head begin to exit the neural tube first, whereas NCC in the lower trunk exit the neural tube last. After leaving the neural tube, the NCC then migrate to distal regions of the embryo, directed by attractive and repulsive signals. Once these cells arrive at their final destination, they undergo a differentiation program that directs these NCC to become a variety of different cell types. Cranial NCC differentiate into ectomesenchymal derivatives including bone, cartilage, connective tissue, and smooth muscle, as well as sympathetic and sensory neurons. Trunk NCC are more restricted in their developmental potential and differentiate only into sympathetic and sensory neurons, supporting glia, Schwann cells, and melanocytes. The cardiac neural crest also contributes pericytes and smooth muscle cells (SMCs) to the cardiac outflow tract of the heart and is important for ensuring correct septation of the pulmonary circulation from the aorta.

Development of the cardiovascular system intersects with NCC development and differentiation at multiple levels. NCC are directly involved in blood vessel development by contributing components (pericytes and SMCs) that make up vessel walls in a subpopulation of the mature vasculature. NCC also indirectly contribute to correct vessel development and function. The peripheral nervous system is derived primarily from the neural crest and includes all sensory nerves and the autonomic nervous system (ANS), which is subdivided into the sympathetic and parasympathetic divisions. Blood vessels are innervated by sympathetic neurons to control vessel dilation and contraction to regulate blood flow. Patterning of the peripheral nerves and the vasculature is also intimately coordinated. Blood vessels can direct correct axon guidance of peripheral nerves, and concomitantly peripheral nerves can direct proper vascular remodeling. Correct axon and vessel patterning is an essential prerequisite for proper organ and organism function later in development.

The role of the neural crest in contributing to aspects of vascular development, function, and homeostasis will be discussed in this chapter, as well as emerging evidence of the molecular pathways and mechanisms involved in controlling these developmental processes.

BLOOD VESSEL DEVELOPMENT AND STRUCTURE

The Cardiovascular System

All organs and tissues of developing and adult organisms require a source of oxygen and nutrients to maintain correct cell function and survival. The cardiovascular system delivers these requirements to all cells of the body. This system is made up of the heart, which pumps oxygenated blood through an intricate network of vessels to target tissues and cells. The blood circulation is divided into the pulmonary circulation, which pumps deoxygenated blood to the lungs and delivers oxygenated blood back to the heart, where it then passes into the systemic circulatory system, which delivers this oxygenated blood to the rest of the body. Deoxygenated blood from distal tissues is returned to the heart through a network of veins, where it is then directed into the pulmonary circulation to complete the cycle.

Distinction between the pulmonary and systemic circulations is essential, and this is mediated by an exquisite organization of arteries and veins in the heart. Deoxygenated blood is pumped from the right ventricle into the pulmonary artery, which bifurcates and delivers blood to both lungs. Oxygenated blood is delivered back to the heart via the pulmonary vein, and this blood is then pumped from the left ventricle into the aorta, which delivers oxygenated blood into the systemic circulation. The carotid arteries branch off of the aortic arch to deliver blood to the head, the subclavian arteries branch off to deliver blood to the chest and arms, and the descending aorta delivers blood to all parts of the lower body. The main arteries that carry blood away from the heart branch into smaller arterioles, and these further branch into expansive capillary beds that make up most of the mature vasculature in the body. Capillaries are the smallest vessels in the body and facilitate the exchange of oxygen, carbon dioxide, nutrients, hormones, and waste products between target cells and the circulatory system. Capillary beds empty into venules, which then connect to major veins that carry deoxygenated blood back to the heart via the venae cavae.

Blood Vessel Development

During embryogenesis, the blood vasculature forms via two complementary processes, vasculogenesis and angiogenesis. Vasculogenesis refers to the *de novo* formation of vessels from angioblast precursor cells, whereas angiogenesis refers to the generation of new vessels from pre-existing vessels via sprouting [1]. A primordial vessel network is laid down during early embryogenesis by the coalescence and differentiation of angioblasts. Angioblasts are mesoderm-derived endothelial progenitors that are hypothesized to arise from the hemangioblast, a precursor cell

that gives rise to both angioblasts and hematopoetic stem cells [2]. Angioblasts generally acquire an arterial or venous fate and coalesce to form the first embryonic blood vessels, the dorsal aorta and the cardinal vein. Angioblasts also aggregate to form vesicles or blood islands, which then fuse to generate primitive networks of chords; these then undergo further morphogenesis to form endothelial tubes (reviewed in [1,3,4]).

Once this preliminary network is laid down, the mature vasculature develops by an extensive process of angiogenesis and vascular remodeling. In response to angiogenic stimuli (i.e., VEGF-A), endothelial cells change their behavior by loosening cell–cell contacts and become pro-invasive. The newly sprouting vessel is lead by an endothelial "tip cell" that is selected from the population of endothelial cells in the vessel by a molecular process of lateral inhibition. The tip cell extends filopodia in order to sense and respond to chemo-attractive and chemorepulsive signals in the local environment and hence shares many morphological and functional similarities with neuronal growth cones that guide axons. The endothelial cells that trail and support the migrating tip cell are termed "stalk cells" and are important for establishing the vascular lumen. When tip cells make contact with adjacent vessels, they undergo a process of anastomosis, which joins the vessels to form a continuous lumen. Subsequent rounds of angiogenesis, remodeling, specialization, and specification generate the extensive network of arteries, veins and capillaries that constitute the mature vasculature system (reviewed in [1,3]).

Blood Vessel Maturation

Mature blood vessels consist of a tube of endothelial cells that form the inner lining of the vessel wall, surrounded by a layer of mural cells that are embedded within a basement membrane and form the outer abluminal surface (Figure 15.1). Depending on the type and function of the vessel (arterial, venous, or capillary), the composition of mural cells and extracellular matrix (ECM) that make up the vessel wall will differ. During vascular development, nascent vessels undergo a process of blood vessel remodeling and stabilization that involves the recruitment of appropriate mural cells, formation of the basement membrane and ECM in a target organ-specific manner, and innervation of the vessels by the sympathetic nervous system (SNS) to form the final mature vasculature (reviewed in [4,5]).

The mural cell coating of endothelial tubes consists of vascular SMCs (vSMCs) and pericytes. The distinction between pericytes and vSMCs is somewhat ambiguous, and these terms are often used interchangeably, with characterization usually dependent on morphological criteria or the presence of these mural cells on different vessel types [6]. Mural

cell progenitors express PDGFR-β and during active angiogenesis pericytes are recruited to endothelial cell tubes, which express PDGF-B. The number of pericytes recruited in this manner is thought to be dependent on the amount of PDGF-B expressed by the endothelial cells [7]. EphrinB2 is also expressed on mural cells and is essential for maturation of the smallest vessels in the skin, lungs, and intestine [8]. In addition, sphingosine 1-phosphate receptors, angiopoietin and TGFβ, have established roles in mural cell recruitment. The essential requirement of smooth muscle for the function of blood vessels is highlighted by the phenotypes of PDGF-B and PDGFRβ knockout mice. In both examples, endothelial cells fail to recruit mural cells, and embryos subsequently die during development due to hemorrhaging from capillary vessels [9,10].

Different vessels have varying densities of pericyte/vSMC and ECM coverage, which reflects the function of the blood vessel

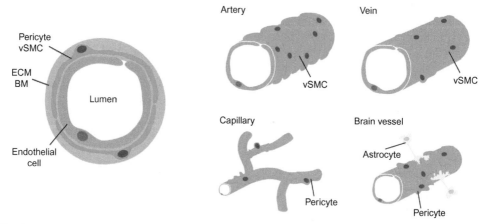

FIGURE 15.1 **Blood vessel structure.** Mature blood vessels consist of a tube of endothelial cells, surrounded by a layer of mural cells (vSMCs and/or pericytes), which are embedded in a basement membrane (BM) made up of ECM components. The composition and coverage of vessels by mural cells depends on the location and function of the vessel. Arteries are covered by vSMCs that form tight concentric rings around the endothelial tube to support the high-pressure blood flow these vessels are exposed to. Veins support low-pressure flow and are covered in a sparse arrangement of vSMCs. Capillaries have a sparse coverage of pericytes that facilitates exchange of gases and nutrients between the blood and surrounding tissue. These pericytes often lie close to vessel branch points, which allows for regulation of flow through capillary branches. Vessels in the brain have an abundant coating of pericytes, as well as coverage by astrocyte end feet, and this extensive coverage helps to protect the brain from potential toxins present in the blood and forms the blood–brain barrier.

(Figure 15.1; reviewed in [4]). Capillaries and small blood vessels have a sparse coverage of pericytes, which facilitates access of the endothelial cells for gas and nutrient exchange with the surrounding tissue. Larger vessels have more specialized mural cell coverage. Arteries are coated with multiple layers of vSMCs that form tight concentric rings around the internal endothelial cells, as well as a layer of fibroblasts embedded in ECM and elastic laminae. This gives arteries the strength to withstand high blood pressure flow, and the vSMC coating contributes to regulation of vessel tone and diameter. Veins generally have an irregular coverage of vSMCs and pericytes and rely on valves to prevent the backflow of blood. Vessels in the brain are further specialized and form a blood–brain barrier to protect the brain from potential toxins that may be present in the blood [5]. Brain vessel endothelial cells form tight junctions and have a high density of pericyte association, as well as numerous contacts with astrocytic foot processes that constitute the blood–brain barrier. Brain pericytes also have pino- and phagocytic activity, thereby cleaning the extracellular fluid as part of the blood–brain barrier [5].

Pericytes and vSMCs are involved in regulating vessel homeostasis by controlling capillary diameter to modulate blood flow. Blood vessels are also innervated by the ANS to regulate these processes. Both of these topics will be discussed in detail later in this chapter.

NEURAL CREST CONTRIBUTION TO BLOOD VESSELS

Neural Crest Contribution to Pharyngeal Arch Artery Development and Patterning

Pharyngeal arch artery (PAA) development has been documented in detail for chick [11] and for mouse [12] and will be discussed below.

The early embryonic vasculature in its most primitive form consists of a basic heart tube and an aortic sac, from which bifurcated ventral aortae project (Figure 15.2). These paired ventral aortae connect with the paired dorsal aortae via a loop at the anterior of the embryo. This loop is called the primitive aortic arch, or the first PAA. Around this stage of early vascular development, the first PAA sends branches ventrally into the head, to form the primitive maxillary artery, and also into the dorsal regions of the forebrain, midbrain, and hindbrain. In the mouse embryo, the primitive aortic arch extends cranio-caudally to form the primitive internal carotid artery, whereas in chick the primitive aortic arch extends in a tight U-shape such that the ventral and dorsal aortae approach and fuse to form the internal carotid artery. As development progresses, the second and third PAAs form one after the other, first appearing as corresponding sprouts from the ventral and dorsal aortae, which eventually connect. At this stage the primitive internal carotid artery also bifurcates to form the primordial cranial and caudal divisions of the circle of Willis. As the fourth PAA appears, the first and second PAAs begin to regress and are transformed into vessels of the mandibular and maxillary process, the primordial external carotid artery, and into capillary beds. The sixth PAA forms soon after this stage, from which the primordium of the pulmonary artery develops. As development proceeds, the aortic arch system becomes bilaterally asymmetrical; in birds this is predominantly right-sided, whereas in mouse the left side is dominant. In mice the aortic sac becomes T-shaped with right and left horns. The dorsal aortae between the third and fourth PAAs regress and the third PAAs remodel to give rise to the right and left common carotid arteries. The right horn of the aortic sac transforms into the right brachiocephalic artery, whereas the left horn together with the left fourth PAA form the definitive aortic arch.

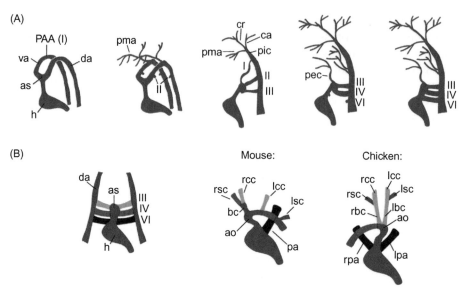

FIGURE 15.2 **Development of the PAAs and remodeling of the great arteries into the heart outflow vessels.** (A) Morphology of the PAAs over a developmental time course. The primitive embryonic vasculature consists of a basic heart tube (h) connected to an aortic sac (as), which bifurcates into paired ventral aortae (va). These extend upward to form the primitive aortic arch, and in mouse this is the first PAA. These connect with the paired dorsal aortae (da). As PAA II forms, the primitive maxillary artery (pma) branches from the PAA. As PAAs II and III form, the primitive internal carotid artery (pic) extends caudally and bifurcates into the cranial (cr) and caudal (ca) divisions of the circle of Willis. The primitive external carotid artery (pec) extends soon after, and PAAs IV and VI form. (B) Ventral view of the bilateral PAAs before remodeling into the heart outflow tract. h, heart; as, aortic sac; da, dorsal aorta. Mouse: Remodeling in the mouse is predominantly left-sided; the third PAAs remodel to give the right and left common carotid arteries (rcc and lcc). The right horn of the aortic sac forms the branchiocephalic artery (bc) off which the right fourth PAA forms the right subclavian artery (rsc). The left horn of the aortic sac and the left fourth PAA form the aorta (ao) from which the left sub-clavian artery (lsc) branches. The right sixth PAA regresses, and the left sixth PAA forms part of the pulmonary artery (pa). Chicken: Remodeling in the chicken is predominantly right-sided. The third PAAs are remodeled into the left and right branchiocephalic arteries. The left fourth PAA regresses and the right forms the aorta. Both sixth PAAs contribute to the left and right pulmonary arteries (lpa and rpa).

The right fourth PAA regresses to form part of the subclavian artery, while the left fourth PAA persists to form part of the aortic arch. The right sixth PAA forms part of the pulmonary arteries and the major capillaries associated with the developing lungs. The left sixth PAA gives rise to the ductus arteriosis that regresses after initiation of breathing post birth. In birds, the third PAAs are remodeled into the left and right brachiocephalic arteries, from which the common carotid arteries branch. The right fourth PAA forms the definitive aorta, while the left regresses. In contrast to mice, in birds both sixth PAAs contribute to formation of the pulmonary arteries.

The cardiac neural crest is essential for the correct development and patterning of the PAAs. Studies in which the cardiac neural crest was mechanically ablated in developing chick embryos have demonstrated that loss of these NCC leads to multiple defects in PAA and subsequent heart development. The disappearance of PAAs 1 and 2 was delayed, and frequently these arteries were enlarged and did not remodel into capillary beds. The development and diameter of PAAs 3 through

6 varied significantly, with some vessels enlarged, and some small or occluded. Bilateral symmetry of the early PAA pattern was disrupted, and the paired arch vessels were often missing. Subsequent heart development was also affected; hearts were often elongated, and the ventricular and atrial chambers formed inappropriately [13,14]. These heart malformations were apparent well before septation of the outflow tract normally occurs, indicating that the neural crest plays an important role in heart development prior to formation of the outflow tract. It is possible that defective PAA development leads to alterations in blood flow that affect correct heart development [15]. These defects are also likely to arise in part from a lack of NCC interactions with the cardiac progenitors of the second heart field. Such NCC interactions have been suggested to induce fibroblast growth factor signaling pathways in the second heart field that control the balance between proliferation and differentiation.

The aforementioned studies also demonstrated that ablation of the cardiac neural crest caused a marked reduction in the amount of mesenchyme occupying the pharyngeal arches; the mesenchyme was not symmetrical and the blood vessel endothelial cells of the PAAs were often in direct contact with the epithelium, rather than being separated by a layer of mesenchyme [14]. This suggested that perhaps the neural crest-derived ectomesenchyme plays an important role during PAA development in either formation of the arteries or stabilization of these vessels. The migrating NCC in pharyngeal arch 3 surround a cord of endothelial precursors, which are subsequently lumenized. In embryos in which the cardiac neural crest was ablated, the third PAA formed and was lumenized, indicating that NCC are not essential for formation of the vessels. However, the vessels became progressively more misshapen, and abnormal ballooning of the PAA occurred at its

connection with the dorsal aorta, suggesting that the neural crest-derived ectomesenchyme may be important for stabilizing the endothelial tube and maintaining vessel integrity as blood flow increases [16].

Ablation of the neural crest in mouse embryos by genetic methods also gives rise to PAA developmental defects reminiscent of those seen in chick embryos, such as the *Splotch* mouse, which contains mutations in the *Pax3* gene [17], and by ablating the neural crest using thymidine kinase [18] or diphtheria toxin [19]. These examples will be discussed in the section below.

Neural Crest Cell Contribution to the Cardiac Outflow Tract

NCC contribution to the cardiac outflow tract is well documented elsewhere (reviewed in [20–22]) and will be touched on only briefly in this chapter. The first evidence for NCC contribution to the cardiovascular system came from pioneering experiments using quail-chick chimeras. Le Lièvre and Le Douarin [23] transplanted quail neural tube and associated neural crest into chick embryos and traced the developmental fate of these grafted NCC in the host embryo. Quail NCC were found in the large arteries that are derived from the PAAs, such as the proximal aorta, pulmonary arteries, branchiocephalic arteries, and the common carotid arteries. The neural crest-derived mesectodermal cells differentiated into smooth muscle and made up the walls of these arteries; they did not contribute to the endothelial lining of the blood vessels, which was entirely of mesodermal origin. These studies were confirmed by Kirby and colleagues [24], who in similar quail-chick chimera experiments demonstrated the presence of NCC in the cardiac outflow tract, in particular in the septum between the aorta and pulmonary trunk. Ablation of the cardiac neural crest disrupted the correct septation of the outflow tract, commonly resulting in conotruncal defects reminiscent of the human conditions persistent truncus arteriosus and double outlet right ventricle [24,25], highlighting the essential role NCC play in correct outflow tract development.

The first evidence for similar contribution of NCC to the outflow tract in mice came from analysis of a transgenic mouse line that expresses *LacZ* under the control of the *connexin 43 (Cx43)* promoter [26]. *Cx43* is highly expressed by NCC as they emigrate from the dorsal neural tube and remains highly expressed in neural crest derivatives such as the outflow tract and cardiac ganglia. Definitive evidence for neural crest contribution to the outflow tract came from lineage tracing studies using Cre/loxP recombination technology. Several transgenic mouse lines have been reported to drive expression of Cre recombinase in the neural crest. These include the *P0*-Cre, *Pax3*-Cre, and *Wnt1*-Cre lines, all of which drive expression of Cre in the dorsal neural tube. When mated to a *LacZ* (R26R) reporter line, these Cre lines allow for lineage tracing of NCC throughout development [27–29]. In all three lines, labeled NCC surrounded the PAAs and contributed to the smooth muscle lining of the aortic arch, common carotid arteries, the branchiocephalic artery, and the right subclavian artery, consistent with contributions seen in chick embryos. One notable difference with that of the chick fate mapping studies is the lack of neural crest contribution to the pulmonary artery. Genetic techniques have been employed to explore whether similar outflow tract abnormalities occur when cardiac neural crest development is disrupted in mammals. The *Splotch* mutant mouse is one such genetic model that has been used to study the effects of cardiac NCC dysfunction. *Splotch* mutant embryos exhibit persistent truncus arteriosus and defects in PAA development [17], which closely resemble the

effects seen in cardiac neural crest-ablated chick embryos. *Splotch* alleles carry mutations in the *Pax3* gene, and while NCC are formed in *Splotch* mutants, the number of NCC migrating in pharyngeal arches 3, 4, and 6 and present in the cardiac outflow tract are significantly reduced [30,31]. This corroborates evidence from the chick models that cardiac NCC are essential for correct outflow tract septation in mammals also.

Further evidence for the requirement of NCC for correct cardiac development in mammals come from studies in mouse embryos in which NCC were ablated using genetic methods. Porras and Brown [18] reported the first study of cardiac NCC ablation in mammals. This study used a *Wnt1*-Cre transgene to drive Cre recombinase expression in NCC, in combination with the PuΔTK selector mouse line, which expresses herpes simplex virus thymidine kinase (TK) after Cre recombination. Upon administration of Ganciclovir (GCK) to the mice, cells expressing TK can no longer undergo DNA synthesis, resulting in cell death. Therefore, this allows for ablation of NCC at specific developmental times and to different extents depending on the dose and timing of GCK administration. Ablation of NCC by the administration of GCK while *in utero* caused severe cardiac malformations, including persistent truncus arteriosus and double outlet right ventricle. Another recent study by Olaopa and colleagues [19] used the *Wnt1*-Cre line to drive expression of activated diphtheria toxin fragment-A. In this case, expression of diphtheria toxin leads to selective death of NCC. This study revealed almost identical results to Porras's; neural crest-ablated embryos exhibited ventricular septal defects and persistent truncus arteriosus. Consistent with the chick ablation studies described earlier, these embryos also demonstrated defects in PAA development, including persistence of the first and second PAAs, and variable remodeling of PAAs 3, 4, and 6 [19].

These and other lineage tracing studies in association with elegant imaging techniques have now defined the contribution of NCC to the PAAs and the cardiac outflow tract. NCC emigrating from hindbrain rhombomeres 5 to somite 3 migrate into the pharyngeal arches and differentiate into vSMCs (tunica media) that coat the PAAs. Interactions between neural crest-derived vSMCs and endothelial cells are essential to control remodeling (stabilization or regression) and maturation of the PAAs. Other cardiac NCC continue to migrate into the primitive outflow tract to participate in the formation of the septum that separates the aortic and pulmonary system. In addition, this population of NCC also give rise to the parasympathetic cardiac ganglia [32].

These ablation experiments in chick and mouse have highlighted the essential requirement of NCC in the formation of the cardiac outflow tract. To achieve this complex remodeling process the development of NCC must be exquisitely controlled at every developmental stage. For example, induction, delamination, migration, endothelial interactions, proliferation, and differentiation must all be coordinated to provide correct numbers, signals, and NCC types for correct PAA remodeling. While some of the molecules controlling these developmental processes have been described (e.g., Pax3 controls neural crest cell numbers, semaphorins are involved in migration, endothelins are involved in PAA remodeling, and Notch is involved in differentiation), there is still much to learn about this process. This list is represented only as an illustration of some of the pathways involved in outflow tract formation and is not meant to be exhaustive. We refer readers to other excellent reviews [33] and book chapters [20] that have discussed our current understanding of the mechanistic control of outflow tract development in detail.

Neural Crest Cell Contribution to Cranial Vessel Smooth Muscle Coating

As discussed above, NCC contribute to the walls of the great arteries (aorta, pulmonary and common carotid) that derive from PAAs 3, 4, and 6 [23]. The contribution of NCC to other more rostral blood vasculature, including vessels derived from PAAs 1 and 2, has also been investigated.

The first evidence for NCC contribution to blood vessels in the brain came from Johnston, who demonstrated, using radioactively labeled NCC in chick embryos, that NCC were present in the meninges of the forebrain [34]. The meninges is a layer of connective tissue that surrounds the brain and encloses a capillary network that provides blood vessels to the neuroectoderm. The blood vasculature in the head is divided into two sectors that are derived from distinct sources; transformation of the first three PAAs form the internal carotid arteries and its derivatives that supply blood to the forebrain and face (branchial sector), whereas blood supply to the midbrain and hindbrain is derived from non-pharyngeal-arch vertebral arteries. These two sectors connect in a vascular structure termed the circle of Willis, which lies immediately ventral to the forebrain—midbrain boundary (Figure 15.3). Etchevers and colleagues [35] performed lineage tracing experiments using quail-chick chimeras to determine the precise contribution of neural crest-derived ectomesenchyme to the blood vessel walls in the head. These studies found that pericytes and SMCs that coated the walls of vessels in the branchial sector were of neural crest origin, whereas vessels of the vertebral sector that supply the midbrain and hindbrain were coated in pericytes of paraxial mesodermal origin. NCC were found in the meninges of the forebrain, as well as the internal carotid arteries. These results were also confirmed by an

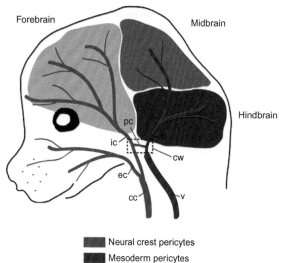

Neural crest pericytes
Mesoderm pericytes

FIGURE 15.3 **Contribution of neural crest to the mural cell coating of vessels in the head.** The mural cell coating of vessels in the branchial sector are derived from the neural crest. These vessels are highlighted in red and supply blood to the forebrain and face. The mural cell coating of vessels that supply the midbrain and hindbrain is of mesodermal origin, and these vessels are highlighted in purple. Vessels of the branchial sector are derived from the common carotid artery (cc), which branches into the internal carotid artery (ic) that supplies the meninges of the forebrain and the external carotid artery (ec) that supplies the face. The blood supply of the midbrain and hindbrain is derived from the vertebral artery (v), which branches into the basilar artery (b). These two blood circulations are joined by the posterior communicating artery (pc) at the circle of Willis (cw).

independent study using quail-chick chimeras [36]. Furthermore, the rostral/caudal origin of NCC in the neural tube determined the positional fate of pericytes, with more distal vessels being coated with pericytes derived from NCC of more rostral origin in the neural tube; for example, meningeal capillaries of the forebrain received NCC from the midbrain—hindbrain boundary, whereas the common carotid artery received cells from rhombomere 5 of the hindbrain [35].

The NCC that form the coating of vessels in the meninges are essential for survival of the forebrain. Mechanical ablation of the neural crest—derived from the posterior forebrain and midbrain causes extensive cell death within the forebrain epithelium [37]. This cell death occurred at a time that coincides with assembly of the perineural vascular network of the forebrain but proceeds the onset of mature vascularization by at least 36 h. Ablation of this region of the neural crest resulted in a loss of ectomesenchymal derivatives (i.e., cells that will form pericytes) in the meninges surrounding the forebrain. The authors propose that these neural crest-derived cells may play a dual role and are essential for secreting trophic

factors necessary for neuroepithelial viability before the mature blood supply has been established, as well as coating the mature vasculature in the meninges.

The first evidence for similar contribution of NCC to mural cell coating of blood vessels in the mouse came from analysis of two mouse LacZ transgenic reporter lines. These lines demonstrated that *Msx* genes (which are highly expressed in NCC [38]) are also expressed in mural cells [39,40]. *Msx2* is expressed in peripheral arteries such as the branchial and femoral arteries in a helicoidal pattern that mimics vSMC organization. *Msx1* is also expressed at a lower level and in a similar pattern in the arteries, but is highly expressed in smaller arterioles and capillaries. Co-localization with markers such as smooth muscle actin confirmed that the *Msx* expressing cells are pericytes/vSMCs [39]. In a subsequent study, *Msx1* and *Msx2* expression was eliminated specifically in vSMCs, and this combined deficiency lead to major defects in blood vessels of the head, including an increase in diameter of the carotid arteries and blood vessel hemorrhaging. This was found to be caused by a decrease in mural cell coverage of the head arteries. The authors propose that the head vSMC defects observed in these mice are a result of loss of a specific NCC subpopulation that depends on functional *Msx* genes to give rise to smooth muscle progenitor cells [40].

Definitive evidence came from explicit NCC lineage tracing studies performed by Yamanishi and colleagues. Using a *P0*-Cre transgenic mouse line to drive expression of an EGFP reporter specifically in NCC, this study reported the presence of NCC in the forebrain [41]. In particular, these NCC began to invade the neuroepithelium coincident with the timing of angiogenesis and migrated in close association with endothelial cells. These cells also expressed the markers PDGFRβ and NG2 and shared a basement membrane with the endothelial cells, confirming their identity as pericytes. Consistent with the chick studies reported above, neural crest-derived pericytes were not found in the midbrain and hindbrain, confirming that the neural crest contribution to central nervous system vasculature is restricted to the forebrain [41].

In contrast to the expanding list of molecules controlling NCC formation of the cardiac outflow tract, there have been few studies as yet exploring the molecular control of neural crest contribution to the vessels of the head. Whether neural crest-derived mural cells exhibit different properties to mesoderm-derived mural cells, and whether this has any role in establishing the blood–brain barrier, remains unresolved.

NCC also contribute to mural cells in the thymus that is derived from the third pharyngeal arch. The first evidence for cells of neural crest origin being present in this organ came from lineage tracing studies using *Wnt1*-Cre R26R transgenic mice. Labeled NCC were found in a circumferential ring around the thymic capsule [29]. Further investigation by two independent studies revealed the precise contribution of NCC to the thymic mesenchyme. Using *Wnt1*-Cre and *Sox10*-Cre transgenic lines

and YFP reporter mice, Foster and colleagues [42] demonstrated the presence of NCC in the adult thymus. Labeled cells formed a three-dimensional network, and associated with structures reminiscent of blood vasculature. These cells also expressed markers of pericytes and SMCs. These results were also confirmed in an independent study by Müller and colleagues [43], who used the *Sox10*-Cre transgenic line and YFP reporter mice to show that NCC in the adult thymus are blood vessel-associated pericytes. Taken together, these studies suggest that NCC migrate into the thymus and differentiate into blood vessel-associated pericytes to support thymic development. Given that these NCC express high levels of vascular endothelial growth factor (VEGF) prior to differentiation into pericytes, they may also promote vessel in-growth to the thymus. In the mature vessel, the neural crest-derived pericytes also play an essential physiological role by promoting egression of mature thymocytes into blood vessels at the corticomedullary junction [44].

The differential embryonic origin of vSMCs in the adult arterial system may have an important impact on cardiovascular disease and the pathogenesis of atherosclerosis. It has been noted that depending on the pathology, different regions of the arterial system can become preferentially calcified with disease progression. In a recent study, Leroux-Berger and colleagues [45] investigated the ability of arteries of different embryonic origins to form calcifications. As discussed earlier, while vSMCs in the aortic arch are derived from the neural crest, vSMCs in the descending aorta are of mesodermal origin. Using *ex vivo* aortic explant cultures, the authors found that aortic arch vessels calcified significantly earlier than descending aorta segments when grown in conditions to induce vascular calcification. A mouse model of spontaneous aortic calcification was also utilized to study differential vessel calcification *in vivo*. This mouse model is deficient in the matrix Gla protein (*Mgp*), a potent inhibitor of calcification. In $Mgp^{-/-}$ mice, legions initiate in the aortic arch and subsequently progress outside the arch region into other arteries, including the dorsal aorta. This ability of vSMCs of NCC origin to form calcifications more readily than those of mesodermal origin perhaps points to the innate ability of NCC to form calcified tissue (cranial NCC form bone and cartilage in the head). This suggests that neural crest-derived vSMCs may maintain a different and perhaps more plastic or "stemlike" gene expression profile, compared to vSMCs of other embryonic origins.

INNERVATION OF THE VASCULATURE BY THE ANS

The ANS

The ANS is part of the peripheral nervous system and controls the function of internal organs in the body. This can include breathing, heart rate, blood flow, salivation, ejaculation, and digestion. The ANS can be subdivided into two distinct components, the parasympathetic nervous system (PSNS) and the SNS (Figure 15.4). The PSNS consists of neurons that have cranial and sacral outflow, and the SNS consists of neurons that have thoracic and lumbar outflow. Postganglionic SNS neurons, on which the pre-ganglionic SNS nerves from the spinal cord terminate, are arranged in a structure termed the sympathetic chain. The sympathetic chains are bilaterally symmetrical and lie on either side, and just ventral to, the vertebral column. Postganglionic SNS neurons extend axons from the sympathetic chain to synapse on their target organs. Pre-ganglionic PSNS nerves extend from the spinal cord to postganglionic neurons that are distinct to the sympathetic chain. Unlike the SNS, the neural crest-derived postganglionic PSNS neurons often sit in close association to their target

organs. All of the neurons of the ANS are developmentally derived from the neural crest.

After being specified in the dorsal neural tube, NCC typically follow one of two distinct migration paths. NCC that exit the neural tube in the developmentally early wave migrate ventrally through the furrows between somites along the intersomitic blood vessels, and then also through the anterior half of the somite. These cells may arrest in the somite tissue adjacent to the neural tube where they will differentiate into the dorsal root ganglia of the sensory nervous system, or they may continue to migrate ventrally and arrest adjacent to the dorsal aorta where they will differentiate into neurons and glia of the SNS. Late-migrating NCC migrate dorsolaterally under the surface ectoderm and differentiate into melanocytes.

Sympathetic NCC precursors that migrate to the dorsal aorta differentiate into sympathetic neurons by the concerted actions of several sympathetic neuron specifying transcription factors including MASH1, PHOX2B, PHOX2A, HAND2, and GATA3 [46]. A recent study by Saito and colleagues [47] found that bone morphogenetic proteins (BMPs) produced by the dorsal aorta induce expression of the chemokine CXCL12 and Neuregulin 1 in the para-aortic region that act as chemoattractants for early migrating sympathoadrenal NCC. BMP signaling was also suggested to promote segregation of the sympathetic and adrenal lineages by specifically activating downstream signaling pathways in only the adrenal precursors. How only a subset of NCC responds to BMP in this model remains to be determined.

Once specified, mature sympathetic neurons become more closely associated and form the ganglia of the sympathetic chain. Neurons also extend axons along the anterior—posterior axis to connect ganglia and form the typical chain-like structure. Some neurons also migrate anteriorly to form the prevertebral

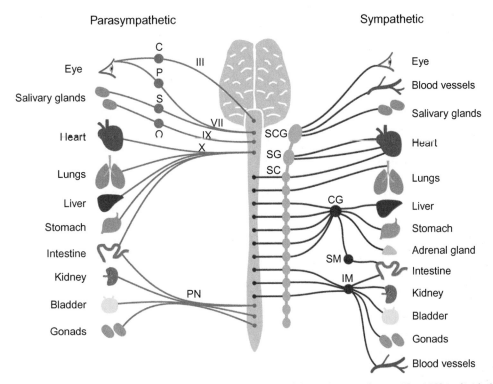

FIGURE 15.4   **The ANS.** All of the neurons of the ANS are derived from the neural crest. The ANS is divided into the sympathetic division and the parasympathetic division. The parasympathetic division (orange) has neurons of cranial outflow, termed the cranial nerves, which target more rostral organs, and also neurons of sacral outflow, termed the pelvic nerves (PN), which target more caudal organs such as the colon, kidney, bladder, and gonads. Cranial nerve III targets the ciliary ganglion (C), which innervates ducts in the eye. Cranial nerve VII also innervates the eye via the pterygopalatine ganglion (P), as well as the salivary glands via the submandibular ganglion (S). Cranial nerve IX also innervates the salivary glands via the otic ganglion (O). Cranial nerve X, also known as the vagus nerve, innervates internal organs such as the heart, lungs, liver, stomach, and intestines. The sympathetic division consists of neurons of thoracic and lumbar outflow. Pre-ganglionic sympathetic nerves (purple) from the spinal cord synapse onto an anatomical structure of ganglia termed the sympathetic chain (SC) (light blue). The bilateral sympathetic chains lie just ventral to the spinal cord. Postganglionic sympathetic neurons (dark blue) then extend axons to their target organs from the sympathetic chain. The sympathetic chain also has specialized cervical ganglia: the superior cervical ganglia (SCG), which send nerves to the head and neck, and the stellate ganglia (SG), which send nerves to the heart. Postganglionic nerves innervate their target organs via peripheral ganglia; the celiac ganglion (CG) innervates the liver and stomach, and the intestines via the superior mesenteric ganglion (SM), and the inferior mesenteric ganglion (IM) innervates caudal organs such as the colon, kidney, bladder, gonads, and distal blood vessels.

ganglia, including the superior cervical ganglia (SCG), which innervate parts of the head and neck, and the stellate ganglia (SG), which innervate the heart.

Both the PSNS and SNS innervate their target organs and vessels via distinct axon paths and act via peripheral ganglia that innervate subsets of distal targets. The major nerve bundles and ganglia are illustrated in Figure 15.4.

Function of Innervation in Blood Vessel Homeostasis

One of the most important aspects of blood vessel homeostasis is the regulation of blood flow and blood pressure, which can be modified by changing blood vessel diameter. Mural cells are at the forefront of regulating blood vessel diameter and do so by controlling blood vessel constriction and dilation. vSMCs and pericytes contain contractile proteins and can constrict around arterioles and capillaries in response to appropriate stimuli [6,48]. Blood vessels also lie in close association with nerve terminals that make contacts with mural cells (and also astrocyte end feet for vessels associated with the blood–brain barrier) [6,48]. Therefore, the role of neural crest in maintenance of blood vessel homeostasis is twofold, with neural crest-derived sympathetic nerves innervating blood vessels, some of which have neural crest-derived vSMC and pericyte coatings.

Studies using electrical stimulation of isolated capillaries have shown that stimulation of pericytes causes constriction; however, stimulating the endothelial walls between pericytes did not constrict the capillary, suggesting that the endothelial cells themselves are not capable of contraction [49]. Contraction of smooth muscle is intimately related to intracellular concentrations of calcium. Contraction is triggered by high calcium levels; this leads to association of calcium ions with calmodulin, which activates myosin light chain kinase (MLCK), which phosphorylates myosin light chain (MLC) and causes its association with alpha-smooth muscle actin (α-SMA) and subsequent cell contraction. Low calcium levels decrease the activity of MLCK, resulting in a decrease in association of MLC with α-SMA, which leads to smooth muscle relaxation [50].

Neurotransmitters can cause alterations of intracellular calcium levels and influence the contractility of pericytes [49]. For example, noradrenaline and ATP are released as neurotransmitters from sympathetic nerves and act on α_1-andrenergic receptors and P2X purinoreceptors, respectively, that are present on blood vessel-associated vSMCs and/or pericytes. Both cause an increase in calcium levels,

resulting in contraction. α_1-Andrenergic receptors are G-protein coupled receptors and stimulation by noradrenaline causes activation of phospholipase C, which catalyses the conversion of phosphatidylinositol-(4,5)-biphosphate (PIP2) into the second messenger phosphoinositol-(1,4,5)-trisphosphate (PIP3), which in turn leads to increases in intracellular calcium upon release of ions from the endoplasmic reticulum. P2X receptors, on the other hand, undergo a conformational change upon binding of ATP, which allows influx of calcium ions into the cell [51,52].

Glutamate, another neurotransmitter, has been shown to dilate blood vessels [49]. Glutamate, via its action on NMDA receptors, has been shown to lead to increased production of nitric oxide (NO) [53]. NO is diffusible and can pass freely though cell membranes of target cells—for example, vSMCs and pericytes. NO activates soluble guanylate cyclase (sGC), which converts GTP to cyclic GMP (cGMP), which acts as a second messenger. cGMP then activates cGMP-dependent protein kinases (cGKs), and activation of cGK triggers a series of downstream reactions that lead to a decrease in intracellular calcium concentration [54]. NO has long been known as a vasodilator and can decrease the contractile tone of pericytes [55,56]. Although there are many other mechanisms through which vessel dilation and constriction are controlled (e.g., prostacyclin and angiotensin), their function within sympathetic neurons is not well established.

Function of Innervation in Cardiac Conduction

In addition to their functions in blood vessel formation, NCC form non-structural components of the cardiovascular system by providing sympathetic and parasympathetic innervation to the heart. Sympathetic nerves from the upper thoracic ganglia and the parasympathetic cardiac nerves coalesce at the cardiac plexus and innervate the heart through the sinoatrial node (SA). Autonomic nerve fibers navigate the remainder of the heart in an epicardial–myocardial gradient. Some of the molecules controlling this innervation are discussed below. Innervation from parasympathetic ganglia and sympathetic ganglia have opposing effects on modifying heart rate and strength of the heartbeat (i.e., sympathetic activity will increase heart rate and contraction strength, while parasympathetic activity has the converse effect) [57].

Mechanisms of Blood Vessel Innervation

It has long been known that sympathetic axon projections, en route to their target organs, travel alongside arterial vessels [58]. However, distinctions must be made between factors important for axon pathfinding along these vessels, factors important for final target innervation and factors involved in synaptic formation on vessel conduits and target organs. Through the recent analysis of genetically engineered mouse mutants, several factors mediating sympathetic projections have been identified and will be discussed below.

Mechanisms of Sympathetic Pathfinding

The first molecule suggested to have a role in sympathetic axon pathfinding was artemin. Artemin is a member of the glial cell-line derived neurotrophic factor family and is expressed in SMCs surrounding the common carotid artery. Sympathetic neurons from the SCG send projections along the internal carotid artery and the external carotid artery to innervate distinct targets and blood vessels. Mouse mutants of either artemin or its receptors Ret or GFRalpha exhibit axonal projection defects from the SCG that lead to aberrant target innervation of the superior tarsus muscle in the anterior region of the face. Although artemin also plays a role in positioning of NCC

prior to axonal extension, strong support of a role in patterning the SNS comes from the finding that ectopic sources of artemin are able to attract SCG nerve fibers *in vivo* [59].

The neurotrophin NT-3 also plays an important role in axonal pathfinding along blood vessels. NT-3 is expressed by many blood vessels during development and is essential to control innervation of several target organs. In mice lacking NT-3, sympathetic axonal extension along blood vessels is considerably reduced. In favor of NT-3 acting as a permissive signal for axonal extension, it has also been found to promote sympathetic axonal growth *in vitro* [60].

Endothelins are a class of vascular-derived factors that are important for sympathetic axon pathfinding, and as opposed to NT-3 and artemin, they are proposed to guide only a subset of neurons from the SCG to innervate their correct targets in the head. Endothelin 3 (Edn3) is preferentially expressed around the third PAA, and subsequently the external carotid artery. The endothelin receptors EdnrA and EdnrB on the other hand are only expressed by a subset of neurons in the SCG. In both *Edn3* and *Ednra* mutant mouse embryos, sympathetic neurons from the SCG fail to extend along the external carotid artery. The authors suggest a model in which sympathetic neurons are predetermined to innervate distinct target tissues, and hence the choice of different vascular routes is mediated by the differential expression of vascular-derived endothelins and their receptors [61].

In addition to the role of vessel-derived attraction (or permissive) signals in sympathetic axon navigation, the class 3 semaphorins, Sema3A and Sema3F, have also been shown to play an essential role in sympathetic axon pathfinding. Semaphorins are secreted axonal guidance molecules that predominantly act as chemorepulsive factors through binding to their cell surface receptors termed neuropilins (Nrp). Mouse mutants of Sema3A and

Sema3F, or their corresponding receptors, Nrp1 and Nrp2, have profound sympathetic axonal guidance defects [62–64]. Indeed, analysis of compound mutants of these ligands or receptors discovered that both signaling pathways cooperate not only to pattern axonal guidance, but also to pattern formation of the sympathetic chains.

Mechanisms of Sympathetic Target Innervation

Nerve growth factor (NGF) is required for sympathetic neuron target innervation. NGF is critically required for sympathetic neuron survival during development [65], which had precluded studies into the role of NGF in axon growth. Therefore, Glebova and Ginty took an elegant approach to study the role of NGF in neuron development in mice by deleting *NGF*, concurrently with deleting *Bax* expression, a pro-apoptotic Bcl-2 family member [66]. This prevented the peripheral nerve death usually observed in *NGF* mutant mice and allowed study of target innervation of these nerves later in development. While proximal projections from the sympathetic ganglia were normal in these mice, growth of axons into peripheral targets such as the heart and salivary glands was severely diminished, indicating that NGF is not required for initial axon growth and guidance, but is required for functional innervation of target organs [66].

Sema3A has also been implicated in regulating correct patterning of sympathetic innervation in the heart. Sympathetic nerves are most highly abundant in the outer epicardial layer of the heart, while the inner endocardial layer is sparsely innervated. These nerves are not involved in the cardiac conduction system, but rather are involved in augmenting cardiac performance, such as heart rate and conduction velocity. Sema3A is a potent chemorepellant for sympathetic axons and is expressed in a gradient in the developing heart, with high expression in the inner endocardial layer and

low expression in the outer layers. In hearts from *Sema3A*$^{-/-}$ mice, sympathetic nerves project aberrantly into the inner layers of the heart, and this results in sinus bradycardia, or slowed heart rate. In contrast, when Sema3A was overexpressed in the heart, sympathetic innervation was reduced, and this caused susceptibility of these mice to ventricular tachycardia, or increased heart rate. This study suggests that Sema3A is expressed in a complementary pattern to sympathetic neuron innervation in the heart, and that Sema3A plays an important role in restricting innervation to the outer epicardial layer of the heart to ensure correct maintenance of cardiac performance and heart rate [67]. In a recent study, Maden and colleagues [64] determined that the Sema3A receptor, Nrp1, is present on sympathetic neurons and is essential for innervation of the heart and dorsal aorta. Intriguingly, another of the Nrp1 ligands, VEGF, is also expressed at the correct time and location to perhaps play a role in cardiac innervation as well [68]. A role for VEGF in this capacity has not yet been examined.

Importantly, in all of the mouse mutants discussed above, sympathetic innervation was still present in many target organs. This suggests that there is likely to be a complex coordination of axonal guidance signals controlling target innervation. In gaining a greater understanding of how the SNS impacts on vascular function, it will be important to determine the full extent of defects in these mutants and to identify novel molecules and signaling pathways controlling sympathetic innervation.

Mechanisms of Synapse Formation on Blood Vessels

To create functional circuits with their target cells, the neural crest-derived sympathetic neurons must position their axons in appropriate locations and then establish a dynamic array of synapses on correct targets. Unlike the typical synapses in the central nervous system, those forming between sympathetic neurons and target cells form large varicosities lacking post-synaptic specializations. VEGF-A and Sema3A are secreted guidance factors that have been implicated in modifying vascular sympathetic innervation and synapse formation. The angiogenic factor VEGF-A promotes axon outgrowth and survival [69], whereas Sema3A promotes growth cone collapse and repulsion. Long and colleagues studied the relative innervation of adult mouse carotid and femoral arteries using the synaptic marker synaptophysin and found that while the femoral artery was densely innervated by sympathetic nerves, the carotid artery was sparsely innervated. VEGF-A was expressed comparably between carotid and femoral arteries, whereas Sema3A expression was higher in the carotid arteries compared to the femoral. Co-culture experiments demonstrated that SCG-derived sympathetic neurons had increased outgrowth toward femoral artery segments compared to carotid artery segments, suggesting Sema3A repels axons from the carotid artery. However, when carotid arteries from mice overexpressing VEGF-A were grown in co-culture with wild-type femoral arteries, SCG neurons had increased outgrowth toward the carotid artery segments compared to the femoral artery segments, suggesting that VEGF-A acts as a chemoattractant that can override Sema3A-mediated repulsion. Both Sema3A and VEGF-A can bind and signal via the Nrp1 receptor, and competitive binding or recruitment of different co-receptors may determine the outcome for axon guidance versus repulsion when exposed to these factors. Taken together, this suggests that gradients of guidance cues work in combination to direct outgrowth and innervation patterns, and in the case of the carotid artery, higher Sema3A expression may blunt the effect of VEGF-A and play a predominant role when both factors are present [70].

NEURAL CREST INVOLVEMENT IN NEUROVASCULAR PATTERNING

Neurovascular Congruence

The co-patterning of the neuronal and vascular systems, or neurovascular congruence, was first recognized at the anatomical level over five centuries ago by the Dutch anatomist Andreas Vesalius [71]. This intimate relationship is evident throughout the body plan and is key to the physiology of both cell types. Blood vessels provide mature neurons with oxygen, nutrients, and survival signals and support neurogenesis through providing a nutrient-rich niche. Concurrently, nerves are required for arterial specification, blood vessel homeostasis, and to control heart rate. The highly ordered overlap of axons and vessels is often controlled by one cell type navigating along the other. Thus, as discussed in previous sections, vessels have been shown to produce signals that attract sympathetic axons (e.g., artemin, NT-3, and Endothelin), and peripheral nerves produce signals that guide vessels (e.g., VEGF-A and CXCL12). The latter of these processes will be discussed in the following section. Not only do blood vessels and nerves follow similar trajectories, they also employ similar guidance cues to navigate through their environment (i.e., VEGF/Nrp1, Slit/ROBO, Eph/Ephrins, Unc/Netrin).

The identification of cell surface receptors such as Nrp1 on both vessels and nerves originally lead to the hypothesis that axons and vessels may align by responding to common cues. At the molecular level, Nrp1 is able to bind two distinct ligands: heparin binding isoforms of VEGF-A (i.e., VEGF164) and members of the class 3 semaphorin family (i.e., Sema3A). A detailed phenotypic analysis of mice lacking Sema3A, VEGF164, and Nrp1 found that although Nrp1 is required for correct neurovascular patterning in the developing limb, neurons and vessels selectively responded to alternative ligands. Even though both ligands are present in the immediate environment encountered by axonal growth cones or endothelial tip cells, vessel growth only required VEGF164 and axonal growth only required Sema3A [72]. The mechanisms controlling the preferential response to VEGF or Sema3A by either vessels or nerves is an ongoing topic of much investigation.

Co-patterning of the vascular and neuronal systems also occurs through interactions of vessels and the NCC precursors of the peripheral nervous system. For example, the first wave of NCC form in the trunk migrates along intersomitic blood vessels toward the dorsal aorta that itself provides instructive signals for the specification of sympathetic neurons and adrenal chromafin cells. Although the mechanisms underpinning this interaction remain elusive, it is tempting to speculate that vessels may provide a permissive or attractive environment to support NCC migration similar to the way they promote axonal navigation.

Blood Vessel Patterning by Peripheral Nerves

In the previous sections we have detailed the role of blood vessels in patterning the sympathetic nerves; however, there is emerging evidence that neural crest-derived peripheral nerves also impact on blood vessel morphogenesis. It has long been known that major nerve bundles and large blood vessels are intimately associated in the periphery of adult tissues. Several recent findings shed light on how this is established during embryogenesis, where the primitive vascular plexus is established prior to axonal innervation. Using genetic studies in mice that lack peripheral sensory axons (neurogenin1/neurogenin2 double KO mice), or Schwann cells (ErbB3 KO mice), Mukouyama and colleagues found that vessel remodeling and arterial specification

was dependent on peripheral nerves. In the absence of nerves, the primitive vascular plexus failed to remodel correctly and arterial differentiation was perturbed. In strong support of the notion that nerves induce these vascular changes, they also found that vessel remodeling and arteriogenesis remained associated with nerves in mice with disorganized axons (Sema3A mutant mice) [73].

The pro-angiogenic factor VEGF-A is expressed in axons and Schwann cells and plays at least some part in this congruence event. Genetic ablation of VEGF-A specifically in peripheral nerves or Schwann cells showed that VEGF-A is required for arteriogenesis; however, other vascular remodeling occurred normally [74]. This finding is consistent with *in vitro* studies in which VEGF-A from axons or Schwann cells induced arterial differentiation of endothelial cells. Finally, it has also been suggested that the chemokine CXCL12 is expressed by Schwann cells to regulate neurovascular congruence in the periphery [75]. Interestingly, the CXCL12 receptor CXCR4 is expressed in only a subset of vessels and may provide a clue to the reason why only a subset of vessels in the environment of nerves acquire an arterial phenotype. Given the large number of proteins co-expressed by nerves and vessels, there will likely be many more factors involved in this congruence.

CONCLUSIONS AND PERSPECTIVES

The interactions of NCC and the cardiovascular system has been recognized at the anatomical level for over five centuries. However, other than the few examples discussed in this chapter, surprisingly little is known about the molecules coordinating their development. Highlighting the essential need to examine this process is the fact that disorders of the outflow tract represent the most common birth defect in humans. A major challenge going forward will be identification of the factors and interactions that promote neural crest contributions to cardiovascular remodeling.

References

[1] Geudens I, Gerhardt H. Coordinating cell behaviour during blood vessel formation. Development 2011;138:4569–83.

[2] Xiong J-W. Molecular and developmental biology of the hemangioblast. Dev Dyn 2008;237:1218–31.

[3] Herbert SP, Stainier DYR. Molecular control of endothelial cell behaviour during blood vessel morphogenesis. Nat Rev Mol Cell Biol 2011;12:551–64.

[4] Jain RK. Molecular regulation of vessel maturation. Nat Med 2003;9:685–93.

[5] Bergers G, Song S. The role of pericytes in blood-vessel formation and maintenance. Neuro-oncology 2005;7:452–64.

[6] Armulik A, Genové G, Betsholtz C. Pericytes: developmental, physiological, and pathological perspectives, problems, and promises. Dev Cell 2011;21:193–215.

[7] Gerhardt H, Betsholtz C. Endothelial–pericyte interactions in angiogenesis. Cell Tissue Res 2003;314:15–23.

[8] Foo SS, Turner CJ, Adams S, et al. Ephrin-B2 controls cell motility and adhesion during blood-vessel-wall assembly. Cell 2006;124:161–73.

[9] Levéen P, Pekny M, Gebre-Medhin S, Swolin B, Larsson E, Betsholtz C. Mice deficient for PDGF B show renal, cardiovascular, and hematological abnormalities. Genes Dev 1994;8:1875–87.

[10] Soriano P. Abnormal kidney development and hematological disorders in PDGF beta-receptor mutant mice. Genes Dev 1994;8:1888–96.

[11] Hiruma T, Hirakow R. Formation of the pharyngeal arch arteries in the chick embryo. Observations of corrosion casts by scanning electron microscopy. Anat Embryol 1995;191:415–23.

[12] Hiruma T, Nakajima Y, Nakamura H. Development of pharyngeal arch arteries in early mouse embryo. J Anat 2002;201:15–29.

[13] Bockman DE, Redmond ME, Waldo K, Davis H, Kirby ML. Effect of neural crest ablation on development of the heart and arch arteries in the chick. Am J Anat 1987;180:332–41.

[14] Bockman DE, Redmond ME, Kirby ML. Alteration of early vascular development after ablation of cranial neural crest. Anat Rec 1989;225:209–17.

[15] Bockman DE, Redmond ME, Kirby ML. Altered development of pharyngeal arch vessels after neural crest ablation. Ann N Y Acad Sci 1990;588:296–304.

[16] Waldo KL, Kumiski D, Kirby ML. Cardiac neural crest is essential for the persistence rather than the formation of an arch artery. Dev Dyn 1996;205:281–92.

[17] Franz T. Persistent truncus arteriosus in the Splotch mutant mouse. Anat Embryol 1989;180:457–64.

[18] Porras D, Brown CB. Temporal-spatial ablation of neural crest in the mouse results in cardiovascular defects. Dev Dyn 2008;237:153–62.

[19] Olaopa M, Zhou H-M, Snider P, et al. Pax3 is essential for normal cardiac neural crest morphogenesis but is not required during migration nor outflow tract septation. Dev Biol 2011;356:308–22.

[20] Brown CB, Baldwin HS. Neural crest contribution to the cardiovascular system. Adv Exp Med Biol 2006;589:134–54.

[21] Stoller JZ, Epstein JA. Cardiac neural crest. Semin Cell Dev Biol 2005;16:704–15.

[22] Hutson MR, Kirby ML. Neural crest and cardiovascular development: a 20-year perspective. Birth Defects Res C Embryo Today 2003;69:2–13.

[23] Le Lièvre CS, Le Douarin NM. Mesenchymal derivatives of the neural crest: analysis of chimeric quail and chick embryos. J Embryol Exp Morphol 1975;34:125–54.

[24] Kirby ML, Gale TF, Stewart DE. Neural crest cells contribute to normal aorticopulmonary septation. Science 1983;220:1059–61.

[25] Kirby ML, Turnage KL, Hays BM. Characterization of conotruncal malformations following ablation of 'cardiac' neural crest. Anat Rec 1985;213:87–93.

[26] Waldo KL, Lo CW, Kirby ML. Connexin 43 expression reflects neural crest patterns during cardiovascular development. Dev Biol 1999;208:307–23.

[27] Yamauchi Y, Abe K, Mantani A, et al. A novel transgenic technique that allows specific marking of the neural crest cell lineage in mice. Dev Biol 1999;212:191–203.

[28] Li J, Chen F, Epstein JA. Neural crest expression of Cre recombinase directed by the proximal Pax3 promoter in transgenic mice. Genesis 2000;26:162–4.

[29] Jiang X, Rowitch DH, Soriano P, McMahon AP, Sucov HM. Fate of the mammalian cardiac neural crest. Development 2000;127:1607–16.

[30] Conway SJ, Henderson DJ, Copp AJ. Pax3 is required for cardiac neural crest migration in the mouse: evidence from the splotch (Sp2H) mutant. Development 1997;124:505–14.

[31] Epstein JA, Li J, Lang D, et al. Migration of cardiac neural crest cells in Splotch embryos. Development 2000;127:1869–78.

[32] Kirby ML, Stewart DE. Neural crest origin of cardiac ganglion cells in the chick embryo: identification and extirpation. Dev Biol 1983;97:433–43.

[33] Kirby ML, Hutson MR. Factors controlling cardiac neural crest cell migration. Cell Adh Migr 2010;4:609–21.

[34] Johnston MC. A radioautographic study of the migration and fate of cranial neural crest cells in the chick embryo. Anat Rec 1966;156:143–55.

[35] Etchevers HC, Vincent C, Le Douarin NM, Couly GF. The cephalic neural crest provides pericytes and smooth muscle cells to all blood vessels of the face and forebrain. Development 2001;128:1059–68.

[36] Korn J, Christ B, Kurz H. Neuroectodermal origin of brain pericytes and vascular smooth muscle cells. J Comp Neurol 2002;442:78–88.

[37] Etchevers HC, Couly G, Vincent C, Le Douarin NM. Anterior cephalic neural crest is required for forebrain viability. Development 1999;126:3533–43.

[38] Bendall AJ, Abate-Shen C. Roles for Msx and Dlx homeoproteins in vertebrate development. Gene 2000;247:17–31.

[39] Goupille O, Saint Cloment C, Lopes M, Montarras D, Robert B. Msx1 and Msx2 are expressed in subpopulations of vascular smooth muscle cells. Dev Dyn 2008;237:2187–94.

[40] Lopes M, Goupille O, Saint Cloment C, Lallemand Y, Cumano A, Robert B. Msx genes define a population of mural cell precursors required for head blood vessel maturation. Development 2011;138:3055–66.

[41] Yamanishi E, Takahashi M, Saga Y, Osumi N. Penetration and differentiation of cephalic neural crest-derived cells in the developing mouse telencephalon. Dev Growth Differ 2012;54:785–800.

[42] Foster K, Sheridan J, Veiga-Fernandes H, Roderick K, Pachnis V, Adams R, et al. Contribution of neural crest-derived cells in the embryonic and adult thymus. J Immunol 2008;180:3183–9.

[43] Müller SM, Stolt CC, Stolt CC, Blum C, Amagai T, Kessaris N, et al. Neural crest origin of perivascular mesenchyme in the adult thymus. J Immunol 2008;180:5344–51.

[44] Zachariah MA, Cyster JG. Neural crest-derived pericytes promote egress of mature thymocytes at the corticomedullary junction. Science 2010;328:1129–35.

[45] Leroux-Berger M, Queguiner I, Maciel TT, Ho A, Relaix F, Kempf H. Pathologic calcification of adult vascular smooth muscle cells differs on their crest or mesodermal embryonic origin. J Bone Miner Res 2011;26:1543–53.

[46] Cane KN, Anderson CR. Generating diversity: mechanisms regulating the differentiation of autonomic neuron phenotypes. Auton Neurosci 2009;151:17–29.

[47] Saito D, Takase Y, Murai H, Takahashi Y. The dorsal aorta initiates a molecular cascade that instructs sympathoadrenal specification. Science 2012;336:1578–81.

[48] Hamilton NB, Attwell D, Hall CN. Pericyte-mediated regulation of capillary diameter: a component of neurovascular coupling in health and disease. Front Neuroenergetics 2010;2(5):1–14.

[49] Peppiatt CM, Howarth C, Mobbs P, Attwell D. Bidirectional control of CNS capillary diameter by pericytes. Nature 2006;443:700–4.

[50] Webb RC. Smooth muscle contraction and relaxation. Adv Physiol Educ 2003;27:201–6.

[51] van Brummelen P, Jie K, van Zwieten PA. Alpha-adrenergic receptors in human blood vessels. Br J Clin Pharmacol 1986;21(Suppl. 1):33S–9S.

[52] Ralevic V. Purines as neurotransmitters and neuromodulators in blood vessels. Curr Vasc Pharmacol 2009;7:3–14.

[53] Garthwaite J. Glutamate, nitric oxide and cell–cell signalling in the nervous system. Trends Neurosci 1991;14:60–7.

[54] Martínez-Ruiz A, Cadenas S, Lamas S. Nitric oxide signaling: classical, less classical, and nonclassical mechanisms. Free Radic Biol Med 2011;51:17–29.

[55] Sakagami K, Kawamura H, Wu DM, Puro DG. Nitric oxide/cGMP-induced inhibition of calcium and chloride currents in retinal pericytes. Microvasc Res 2001;62:196–203.

[56] Haefliger IO, Zschauer A, Anderson DR. Relaxation of retinal pericyte contractile tone through the nitric oxide-cyclic guanosine monophosphate pathway. Invest Ophthalmol Vis Sci 1994;35:991–7.

[57] Hildreth V, Anderson RH, Henderson DJ. Autonomic innervation of the developing heart: origins and function. Clin Anat 2009;22:36–46.

[58] Glebova NO, Ginty DD. Growth and survival signals controlling sympathetic nervous system development. Annu Rev Neurosci 2005;28:191–222.

[59] Honma Y, Araki T, Gianino S, et al. Artemin is a vascular-derived neurotropic factor for developing sympathetic neurons. Neuron 2002;35:267–82.

[60] Kuruvilla R, Zweifel LS, Glebova NO, et al. A neurotrophin signaling cascade coordinates sympathetic neuron development through differential control of TrkA trafficking and retrograde signaling. Cell 2004;118:243–55.

[61] Makita T, Sucov HM, Gariepy CE, Yanagisawa M, Ginty DD. Endothelins are vascular-derived axonal guidance cues for developing sympathetic neurons. Nature 2008;452:759–63.

[62] Kawasaki T, Kitsukawa T, Bekku Y, et al. A requirement for neuropilin-1 in embryonic vessel formation. Development 1999;126:4895–902.

[63] Schwarz Q, Maden CH, Davidson K, Ruhrberg C. Neuropilin-mediated neural crest cell guidance is essential to organise sensory neurons into segmented dorsal root ganglia. Development 2009;136:1785–9.

[64] Maden CH, Gomes J, Schwarz Q, Davidson K, Tinker A, Ruhrberg C. NRP1 and NRP2 cooperate to regulate gangliogenesis, axon guidance and target innervation in the sympathetic nervous system. Dev Biol 2012;369:277–85.

[65] Crowley C, Spencer SD, Nishimura MC, et al. Mice lacking nerve growth factor display perinatal loss of sensory and sympathetic neurons yet develop basal forebrain cholinergic neurons. Cell 1994;76:1001–11.

[66] Glebova NO, Ginty DD. Heterogeneous requirement of NGF for sympathetic target innervation *in vivo*. J Neurosci 2004;24:743–51.

[67] Ieda M, Kanazawa H, Kimura K, et al. Sema3a maintains normal heart rhythm through sympathetic innervation patterning. Nat Med 2007;13:604–12.

[68] Miquerol L, Gertsenstein M, Harpal K, Rossant J, Nagy A. Multiple developmental roles of VEGF suggested by a LacZ-tagged allele. Dev Biol 1999;212:307–22.

[69] Erskine L, Reijntjes S, Pratt T, et al. VEGF signaling through neuropilin 1 guides commissural axon crossing at the optic chiasm. Neuron 2011;70:951–65.

[70] Long JB, Jay SM, Segal SS, Madri JA. VEGF-A and Semaphorin3A: modulators of vascular sympathetic innervation. Dev Biol 2009;334:119–32.

[71] Carmeliet P, Tessier-Lavigne M. Common mechanisms of nerve and blood vessel wiring. Nature 2005;436:193–200.

[72] Vieira JM, Schwarz Q, Ruhrberg C. Selective requirements for NRP1 ligands during neurovascular patterning. Development 2007;134:1833–43.

[73] Mukouyama Y-S, Shin D, Britsch S, Taniguchi M, Anderson DJ. Sensory nerves determine the pattern of arterial differentiation and blood vessel branching in the skin. Cell 2002;109:693–705.

[74] Mukouyama Y-S, Gerber H-P, Ferrara N, Gu C, Anderson DJ. Peripheral nerve-derived VEGF promotes arterial differentiation via neuropilin 1-mediated positive feedback. Development 2005;132:941–52.

[75] Li W, Kohara H, Uchida Y, et al. Peripheral nerve-derived CXCL12 and VEGF-A regulate the patterning of arterial vessel branching in developing limb skin. Dev Cell 2013;24:359–71.

Neural Crest Cells and Cancer: Insights into Tumor Progression

Davalyn R. Powell[a],*, Jenean H. O'Brien[b],*, Heide L. Ford[b] and Kristin Bruk Artinger[a]

[a]Department of Craniofacial Biology, School of Dental Medicine, University of Colorado Anschutz Medical Campus, Aurora, CO, 80045, USA, [b]Department of Pharmacology, School of Medicine, University of Colorado Anschutz Medical Campus, Aurora, CO, 80045, USA

INTRODUCTION

Neural crest cells (NCC) represent an excellent model for investigating developmental

programs usurped by tumor cells. As described in previous chapters, NCC progress through several intricately coordinated steps throughout early embryogenesis. Following specification at the neural plate border, NCC undergo an epithelial-to-mesenchymal transition (EMT),

* These two authors contributed equally.

Neural Crest Cells.
DOI: http://dx.doi.org/10.1016/B978-0-12-401730-6.00017-X

delaminate from the dorsal neural tube, and migrate through various tissues to distant sites where they differentiate to form distinct neural crest derivatives, such as melanocytes and glia. From a gross perspective, this process is mimicked during early stages of metastasis, as cancer cells begin to disseminate from the original tumor location. During this process, epithelial tumor cells can develop mesenchymal characteristics, migrate from their tissue of origin, and spread systemically to metastatic sites. These similarities have inspired many scientists to investigate how neural crest and tumor cell behaviors during embryogenesis and metastasis may be related. At a molecular level, the parallels between neural crest development and tumor progression are extensive. Many of the signaling pathways and transcription factors are shared between the two processes, resulting in the promotion of comparable cellular readouts. In particular, mechanisms driving EMT and cell migration appear to be quite similar and will be explored in the first part of this chapter. The relationship between NCC and tumor cell behavior provides a unique opportunity to study how normal developmental programs may be re-activated to contribute to neoplastic progression.

Further, many of the tissues derived from NCC utilize these same mechanisms to contribute to malignancy in diseases such as melanoma and neuroblastoma. These tumors often exhibit overexpression or misregulation of genes involved in neural crest development, especially genes required for EMT and migration. This re-initiation of developmental programs is often correlated with highly aggressive metastatic cancer and poor patient prognosis. The second part of this chapter will introduce many of the known cancers of neural crest origin and discuss how the molecular reprogramming of differentiated cells to an undifferentiated neural crest-like state contributes to disease progression.

SIMILARITIES BETWEEN NEURAL CREST DELAMINATION/ MIGRATION AND TUMOR PROGRESSION

Metastasis is a multistep process that describes how transformed cells escape from a primary tumor by invading local tissue and circulation, traveling to distant sites, exiting from the bloodstream, proliferating into a secondary lesion, and surviving throughout these new environments. For most solid tumors, the first steps of this progression involve changes that allow tumor cells to detach from an epithelial tissue. As many of these morphological alterations are also observed during normal embryogenesis, in a process called EMT, a role for EMT in tumor progression has been suggested. There are several morphological and functional changes required to progress through an EMT. Epithelial cells are usually organized as stable, polarized cuboidal cells that form close contacts with neighboring epithelial cells, whereas mesenchymal cells are irregular in shape and organization as they are often dynamic motile cells. Events that contribute to changes in shape including disassembly of cell–cell junctions and adhesions, cytoskeletal remodeling, and loss of apicobasal cell polarity must occur relatively early in the EMT process (Figure 16.1). Also, a cell experiencing EMT makes changes to its immediate environment to facilitate motility. These alterations include degrading basement membrane, invading the fibrillar matrix, and building adhesion complexes with new extracellular matrix (ECM). A motile cell must then be able to follow and/or avoid cues directing migration to a distant site and upregulate survival mechanisms to allow for persistence through these changes. Many of the EMT and migratory features utilized by NCC are described in detail by Taneyhill and colleagues in Chapter 3 and by Mayor and associates in Chapter 4. Here, similarities between neural

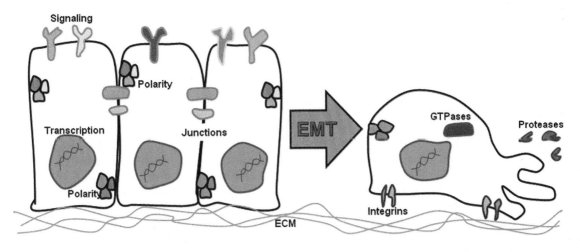

Signaling	Transcription	Junctions	Polarity*	RhoGTPases	ECM, Proteases, Integrins	Migratory Ligands/Receptors
Wnt	Snail1/2 (Slug)	E-cadherin	Scribble	RhoB	Laminin	Ephs
BMP	Twist	N-cadherin	Dlg	Rac	Collagen IV	Ephrins
TGFβ	Sox	Cadherin-6	Lgl	RhoU	MMP-2	Semaphorins
Notch		Cadherin-7	Crb3		MMP-9	Neuropilin
		Cadherin-11	PATJ		ADAM-10	Plexin
		Occludin	aPKC		β1-integrin	SDF-1 (CXCL12)
		ZO-1	Par3			CXCR-4

FIGURE 16.1   **EMT and migration molecules utilized by both neural crest and tumor cells.** As neural crest and transformed cells transition from an epithelial to mesenchymal phenotype and become motile, many of the same molecular programs are engaged. The genes listed have been shown to play a role in EMT and/or migration in both neural crest and tumor cells, *except for the polarity complex genes, which have only been reported in cancer.

crest EMT and migration that have been observed in tumor cells will be explored.

Cell—Cell Junctions

A defining feature of an epithelial tissue is the presence of one or more layers of closely packed cells connected by adherens, tight junctions and gap junctions. The cadherin proteins are important regulators of cell—cell interactions through adherens junctions and are distinctly controlled during neural crest EMT. Many switches in cadherin expression occur in NCC, starting with downregulation of E-cadherin. The loss of E-cadherin is considered a hallmark of EMT in tumor progression [1], as E-cadherin is often referred to as a gatekeeper of epithelial phenotype. E-cadherin has been shown to be important in preventing metastasis, both for its contribution to cell—cell adhesion and its function as a negative regulator of β-catenin signaling [2]. A structural protein located in adherens junctions, β-catenin can function as a transcription factor that stimulates canonical Wnt signaling, important during both normal and cancer development [3]. Overexpression and knockdown studies in numerous cancer models, including pancreatic, breast, bladder, and lung, have demonstrated a role for E-cadherin in inhibition of invasion [4—6]. Further, loss of E-cadherin expression and correlation of this loss to poor prognosis has been observed in patients with several types of cancer, including colon and breast [7,8]. Extensive studies on the role of E-cadherin in tumor progression have been reported; see cited reviews for more details [2,9].

As NCC prepare for migration, N-cadherin and/or cadherin-6 are upregulated. Increased expression of N-cadherin has also been observed in a variety of cancers, including those originating from tissues that embryonically express N-cadherin (pigment cells/melanoma [10]) and those that do not (breast [11] and prostate [12,13]). N-cadherin has been shown to directly promote EMT and motility in many of these tumor models [14,15]. *Cadherin-6* is much less studied in the cancer field, although increased expression is correlated with more advanced stages of renal cell carcinoma [16].

For complete delamination to occur in NCC, N-cadherin must be cleaved and/or downregulated, as *cadherin7* and/or *11*, depending on the species, become the primary cadherins present in migratory NCC [17–19]. Loss of N-cadherin during tumor progression has been observed in a few cancers, including neuroblastoma [13,20]. However, this is one instance where the parallels in cadherin expression between NCC migration and tumor progression depart, as loss of N-cadherin is not typically observed in many cancers. Further, cadherin-7 is reported to be upregulated in malignant melanoma, but may actually inhibit migration [21,22]. The data for *cadherin11* conflict, with overexpression promoting migration in glioblastoma, breast, and prostate cancer [23–25], while other studies suggest silencing and tumor suppressive roles for the molecule [26,27]. As described in developmental settings, disrupting adhesion complexes with either too much or not enough cadherin expression is likely to result in both decreases and increases in cell migration, depending on the context [28,29]. Although migratory mechanisms activated in tumor cells are similar to developmental programs, they are not identical. Since each tumor cell experiences a unique set of genetic mutations and environmental cues, tumor cell migration likely does not require the exact same molecular players and order of events as NCC migration. Overall though, loss of E-cadherin expression and cadherin-switching both play a major role in promoting EMT and migration in neural crest and tumor cells.

A shift in other junctional complexes is also observed as NCC become motile. Tight junctions, primarily located at the apical side of cells in an epithelial layer, are lost as connexin-containing gap junctions are upregulated in migratory NCC [30–32]. Similar trends for junction-associated proteins are observed during tumor progression. Decreases in tight junction proteins occludin and ZO-1 correlate with disease progression in liver, ovarian, endometrial, and breast carcinomas [33–36], where the loss of tight junctions is speculated to be a result of activating EMT and not as a direct driver of the metastatic process. Connexin levels appear to be more context dependent, with both upregulation and downregulation observed during metastatic progression [37,38]. This variation in regulation is speculated to be a result of yet-to-be-determined roles for connexins outside of gap junctions-related functions. Future studies into the functional roles for connexins, both in migratory NCC and tumor promotion, are needed to determine the function in both cell populations. Together, these changes in cell–cell adhesion allow cells to begin to disengage from their neighbors to enable migration.

Cell Polarity

Based on the organization of junctions between epithelial cells, a distinct apical versus basolateral polarity is established. Prior to migration, NCC display apicobasal polarity as part of the neuroepithelium. In addition to junctional complexes, there are several multiprotein complexes designed to maintain epithelial polarity based on localization within the cell, including the *Scribble*, *Crumbs* and *Par*, Crumbs, and Par complexes. Although it is currently unknown exactly how these complexes change during neural crest development, it is anticipated that disassembly and/or rearrangement of these complexes must occur to allow cell migration. Interestingly, there is no data to support that disruption of individual polarity proteins interferes with NCC migration (Grant and Artinger, unpublished). We would speculate that there is a built-in redundancy such that multiple proteins need to be disrupted to see an effect on NCC migration. In alignment with this hypothesis, changes in Scribble, Crumbs, and Par complex assembly have been noted during tumor progression.

With a primary function in restricting basal proteins to maintain a basolateral membrane, the Scribble complex contains *Scribble, Discs large, and Lethal giant larvae* and acts in a mutually antagonistic fashion with the other two complexes. Over 30 years of research has come together to support the idea that the members of the SCRIBBLE complex act as tumor suppressors. All three members of the SCRIBBLE complex are downregulated or mislocalized in a large number of cancers, including cervical, colon, and melanoma, to name a few [39–41]. Several studies have identified the related loss of polarity associated with deconstructed SCRIBBLE complexes as a driver of cancer progression, through decreases in adhesion and migration [42–44]. Further, *Scribble* loss is required for the oncogene Ras to mediate increased invasion *in vitro* and in tumors originating from the eye or antennal imaginal discs in *Drosophila* [45,46].

The CRUMBS complex consists of CRUMBS proteins 1–3 (CRB1–3), protein associated with Lin seven 1 (PALS1), and Pals1-associated tight junction protein (PATJ) and is located in the apical domain of an epithelial cell at the plasma membrane. In kidney epithelial cells, decreased Crb3 protein enhances tumorigenic potential, increases motility, and promotes metastasis [47]. These tumor promotional effects are thought to be the result of disrupting polarity, tight junctions, and contact-inhibited growth of tumor cells, all EMT-related characteristics. In addition, the Crumbs complex PATJ

protein is decreased in cervical cancer cell lines and is necessary to maintain polarity in colon cancer cell lines [48,49].

Also found in the apical region of epithelial cells near tight junctions are the PAR3, PAR6 and atypical protein kinase C (aPKC) proteins, which form the Par complex. Most of the tumor-related studies on the Par complex focus on the tumor promotional roles for aPKC, as observed in ovarian, breast, and non-small cell lung cancer [50–52]. However, Par3 has recently been identified as a potential tumor suppressor in esophageal squamous cell and breast carcinomas, as well as in skin keratoacanthomas [53–55]. Par complex proteins aPKC and Par3 have also been linked to members of the Rho GTPase family, which are important for directing cytoskeletal changes related to motility. Clearly, polarity complexes are important for preventing tumor cell migration, and their loss leads to many prometastatic activities. This suggests that investigation into contributions of polarity complexes in NCC development may be of interest for future research.

Rho GTPases

In order for neural crest or cancer cells to migrate, reorganization of the cytoskeleton must occur to aid in changes from an epithelial to a more mesenchymal cell morphology. In the neural crest, the Rho family of small GTPases is implicated in directing organization of these actin-cytoskeletal structures, ranging from stable structures called stress fibers to more dynamic structures at the leading edge of a migrating cell called lamellopodia. The most recent reports show that inhibition of RhoB/RhoA leads to disassembly of stress fibers and focal adhesions to allow for NCC delamination [56,57]. A similar inhibitory role for RhoB in cancer has been identified. Ectopic expression of RhoB can inhibit tumor cell growth, migration, invasion, and metastasis, specifically of melanoma to the lung [58–60]. In alignment with tumor suppressor activity, RhoB expression decreases as tumors become more aggressive in patients with lung, brain, and head and neck cancers [61–63].

As NCC begin to migrate, upregulation of two Rho GTPases has been observed, Rac (in mouse and chick) and RhoU (in *Xenopus*) [56,64]. Rac signaling also appears to be important in tumor progression, as an upregulation of Rac1 is observed in several cancers, and correlates with increased disease progression specifically in breast, gastric, and testicular cancers [65–67]. Moreover, Rac1 has been shown to promote cancer, often in concert with an oncogenic small GTPase called Ras, as inhibition of Rac in mouse models of lung, colorectal, and skin carcinomas prevented tumor formation [68–70]. RhoU (also known as Wrch1) is a relatively recently identified atypical GTPase, and therefore has not been studied very extensively. *In vitro* studies in transformed mammary epithelial and HeLa cells demonstrate that RhoU overexpression leads to decreased stress fiber formation and increased filopodia (a characteristic motile cell extension), and its inhibition results in decreased motility [71,72]. Thus, RhoU would be anticipated to promote tumor cell migration *in vivo*, similar to its ascribed role in NCC development.

ECM, Proteases, and Integrins

The microenvironment surrounding an epithelial cell provides a barrier function that must be breached if migration is to occur. Immediately surrounding most epithelium is a specialized ECM structure called the basement membrane. This continuous ECM layer is composed primarily of collagen IV and laminin and functions in supporting and separating epithelial cells from surrounding fibrous connective tissue. At the time that NCC begin migrating, there is a breakdown in this normally continuous layer of basement membrane that correlates with an upregulation of several ECM proteases, including matrix metalloproteinases (MMPs) -2, -8, and -9 and a disintegrin and metalloproteinases (ADAMs) -10, and -13, depending on the species [17,73–76]. This perforation in basement membrane allows for migratory NCC to escape into the surrounding fibrous ECM, and a similar degradation of basement membrane is observed in most invasive cancers [77]. Gene programs associated with an increased ability to breakdown basement membrane are upregulated in many tumor cells during neoplastic progression [78] and include MMP-2 and -9 and ADAM-10. These three proteases are upregulated in numerous cancers, including glioblastoma and melanoma [79–81]. Further, upregulation of MMPs/ADAMs in cancer patients contributes to tumor cell aggressiveness and correlates with poor prognosis [82,83]. The list of substrates for these proteases is constantly growing and helps to explain why these enzymes can promote tumor progression [83–85]. Target ECM proteins include basement membrane components, cell-adhesion molecules such as E-cadherin and N-cadherin, growth factors and their receptors, and many cytokines [83,85].

After the basement membrane has been penetrated, motile cells are exposed to a fibrous ECM-rich environment. As the proteases aid in clearing a route for migratory cells to travel through this new substratum, the expression profile for many cell–ECM adhesion molecules, called integrins, changes. Integrin adhesion molecules are heterodimers consisting of α and β subunits that also act as receptors to direct both inside-out (intra- to extracellular) and outside-in (extra- to intracellular) signaling. Migratory NCC upregulate $\alpha1\beta1$-integrin to bind to the ECM ligand laminin, and both $\alpha4\beta1$- and $\alpha5\beta1$-integrins to bind to fibronectin. The common subunit of these protein dimers is $\beta1$-integrin, which has been implicated in several stages of breast tumor progression [86,87], and particularly in advanced stages of tumor recurrence [88]. Overall, the similarities from proteases to integrins suggest that motile cells, whether of neural crest or epithelial tumor cell origin, utilize similar mechanisms to invade new microenvironments.

Migration

Rearrangement of cell junction components, reorganization of cell polarity, activation of GTPases, expression of ECM proteases, and formation of new attachments to the extracellular environment are all in preparation for a critical event for neural crest and tumor cells alike: migration. The migratory process can be engaged by single NCC migrating in chains, or as groups of cells through collective cell migration. The precise mechanism of NCC migration depends on the anterior-posterior position as well as the species. In general, cranial NCC move in a wave, and trunk NCC separate into discrete streams (as reviewed in [89]). A variety of cues from both interacting migratory cells and stationary cells in the surrounding tissue are utilized to direct NCC along designated routes. During neural crest migration, a combination of positive and inhibitory regulators exists, with specific players in each category varying by anterior-posterior position and migratory path taken by the NCC.

There are two major classes of ligands and receptors that primarily act as negative regulators of NCC migration, the ephrins with their Eph receptors, and class 3 semaphorins with their neuropilin/plexin receptors. These signaling molecules regulate a complex array of intracellular pathways and are proposed to prevent migration into specific regions through inducing collapse of cellular protrusions. The Eph tyrosine kinase receptors and plexin-associated tyrosine kinases do this by modulating cytoskeletal dynamics, thereby affecting migration. However, examples where both

categories of ligands/receptors act to promote NCC migration have been reported [90,91]. For example, class 3 semaphorins are primarily reported to repel migrating NCC that populate the trigeminal ganglia, but semaphorin 3C has been shown to attract cardiac NCC [91–93]. In addition, Ephrin-B prevents early migration of NCC along the dorsolateral pathway but promotes melanoblasts migration along this pathway later [90,94]. This context-specific determination of either negative or positive regulation of migration for both ephrins/Ephs and semaphorins aligns with the overall dysregulation of these molecules in tumor models [95,96]. Upregulation of ephrins/Eph receptors correlates with poor prognosis and tumor progression in breast cancer, glioblastoma, and some melanoma studies [97–99]. In contrast, decreased expression is observed in other reports from melanoma, breast, and colon cancer patients, where lower survival rates correlate with loss of EphB receptors [100–102]. Pro-tumor semaphorins have been recognized to promote invasion, metastasis, and EMT in lung, gastric, and ovarian cancer [103–105]. However, there are also reports for loss of semaphorin expression in cancer [96]. As with neural crest migration, this combination of negative and positive effects on outcome appears to be context specific, depending on the specific ligand, the specific receptor, and the tumor type. Since Eph receptors and their ephrin ligands can mediate bidirectional signaling between cells, this complexity predicts the variation in tumor-related functions that these signaling molecules direct. It is interesting to note that *EphB2* receptor and *Semophorin3* negatively regulate NCC migration and, analogously, are downregulated in tumors and have been suggested to act as tumor suppressors [102,106–108].

There are quite a few positive regulators of neural crest migration as well, including the growth factors FGF and PDGF and the chemokine SDF-1 (also known as CXCL12). FGF and

PDGF are upregulated in numerous cancers, where they contribute to EMT and invasion (as reviewed in [109–112]). SDF-1 and its receptor CXCR4 are important for attracting NCC along the anterior migratory stream to populate the craniofacial cartilage and trigeminal ganglia, as well as ventrally to their final destination site in the dorsal root ganglia [113–115]. A very similar function for SDF-1/CXCR4 has been discovered for directing tumor cells to specific metastatic organs, including the lung and bone [116–118], once again indicating the parallel roles for migration-mediating molecules in neural crest and cancer cells.

Signaling and Transcription Factors

NCC delamination and migration are induced by a multitude of signaling pathways and directed through the activation of several transcription factors. The contributions of these signaling molecules to cancer progression are quite extensive and have been excellently reviewed in the following publications. There are many inducers of the EMT program in NCC that have all been implicated in promoting similar functions in tumor models, including members of the Wnt [119,120], BMP [121], TGFβ [122,123], and Notch [124] pathways. These pathways converge through activation of neural crest transcription factors, including *Snail1* and *Snail2* (also known as *Slug*) [125–127], *Twist* [128,129] and *Sox* [130,131] family members, which also direct tumor cell promotion. Many of the transcriptional targets of *Snail1/2* highlight how these proteins can activate EMT, including components of the adherens junctions, tight junctions, and polarity complexes [126]. In addition, in other developing systems, SNAIL proteins can interfere with the cell cycle and provide resistance to apoptosis, thereby allowing migratory embryonic cells as well as cancer cells protection from cell death [132]. It would be interesting to determine if this is also true in NCC

and neural crest–derived cancers. Similar to Snail function, Twist overexpression in tumor cell lines leads to E-cadherin loss, activation of mesenchymal markers, and increased motility [133]. In addition, *Sox10*, a prominent SoxE family member in NCC development, is implicated in pro-migratory tumor cell roles, specifically in melanoma [134,135]. Despite the complexity of these molecular networks, it appears that tumor cells commonly utilize the same mechanisms as neural crest and other developmental programs to activate metastatic programs.

NEURAL CREST-DERIVED CANCERS

While cancer cells in general utilize many of the same cellular and molecular methods to undergo tumor progression and metastasis as NCC use throughout normal embryogenesis, cancers of neural crest origin especially exhibit a re-activation of developmental programs from their neural crest precursors. Transformation of many neural crest derivatives, including melanocytes, glial cells, and adrenal cells, leads to a re-activation of differentiated cells and often an upregulation or misregulation of genes typically involved in the processes of NCC proliferation, survival, migration, and determination (summarized in Table 16.1). Interestingly, these changes in gene expression frequently correlate with highly aggressive tumors and have become predictors for metastatic potential and poor prognoses in patients.

Melanoma

Melanoma results from the transformation of pigment-producing melanocytes into cancerous cells. Although accounting for approximately 4% of all skin cancer incidences, melanoma is by far the most aggressive and

TABLE 16.1   Neural Crest Developmental Genes in Cancers of Neural Crest Derivatives

Gene	Function in Neural Crest Cells (NCC)	Cancers[a]
Sox9/10	NCC specification	Melanoma [134,135,146,147], glioma [188], CCS [215,216]
Snail/Slug	NCC specification, EMT induction	Melanoma [155,156,159], pheochromocytoma [221,222]
Twist1	EMT induction	Melanoma [154], pheochromocytoma [220,222]
N-cadherin	Premigratory cell—cell adhesion	Melanoma [148,149], neuroblastoma [20], pheochromocytoma [220]
Cadherin11	Premigratory cell—cell adhesion	Glioma [23], Ewing's sarcoma [203]
RhoB	Delamination and migration	Glioma [196]
MMPs (1, 2, and 9)	NCC migration, breaking down basement membrane	Melanoma [81,151—154], glioma [80,190—193], pheochromocytoma [220]
Mycn	NCC migration, neuronal differentiation	Neuroblastoma [173—175], pheochromocytoma [219]
PDGF and EGFR	NCC survival and proliferation, differentiation of glia	Glioma [184,188]
Endothelin-3/ EDNRB	NCC survival, enteric NCC proliferation, differentiation, migration	Glioblastoma [197], melanoma [155,156]
Trk receptors (A, B, and C)	Survival and differentiation of neurogenic NCC	Neuroblastoma [172]
Neurotrophin-3	Differentiation of neurons and glia	Neuroblastoma [172]
p75	Marker of multipotent neural crest progenitor cells	Neuroblastoma [172]
S100	Marker of NCC and derivatives	Ectomesenchymoma [209], CCS [215—217]
TCOF1	NCC proliferation	Melanoma [164]
KCTD11	NCC differentiation	Melanoma [164]
MITF	Melanocyte differentiation	Melanoma [139—143], CCS [215-217]

[a]Neural crest-derived cancers in which each gene is known to be misregulated and/or play a role in tumor progression or metastasis.

malignant, causing 80% of all skin cancer deaths [174]. Melanoma incidence has also increased 15-fold over the past 40 years in the United States, making it especially important to understand the mechanisms of the disease [175]. The first stage of melanoma occurrence is the development of a dysplastic nevus that can form from a benign nevus (or mole) or as a new lesion on the skin [174,176]. The development of nevi into melanoma often involves mutations in or misregulation of one or more

hallmark oncogenes including *BRAF*, cyclin checkpoint genes such as *CDKN2A*, *CCND1*, and *INK4A*, and other signaling pathways including MAPK. Interestingly, many of the genes implicated in melanoma progression are genes that are involved in the embryonic development of neural crest-derived melanocytes (the development of melanocytes is discussed in detail by Kelsh and colleagues in Chapter 14). The microphthalmia-associated transcription factor (MITF) is required for melanocyte specification and differentiation during development and for maintenance of melanocyte progenitor cells in adults [169]. MITF is responsible for upregulation of the transcriptional program required for melanin synthesis as well as inducing exit from the cell cycle, allowing melanocytes to differentiate. During melanoma progression, melanocytes no longer exhibit the markers of fully differentiated cells. Melanoma cells often lose expression of *Mitf*-target genes, and this is correlated with poor prognosis in patients [170,171]. However, MITF itself is expressed in nearly all melanomas, and, interestingly, tumors with increased levels of MITF are associated with poor prognosis and resistance to chemotherapy [172]. It is thought that MITF may act as an oncogene in melanoma cells by promoting survival of tumor cells through regulation of anti-apoptotic factor BCL-2 [172,173].

The SoxE family of transcription factors, including *Sox9* and *Sox10*, are early markers of specified NCC and are required for neural crest development. *Sox10* is also important for activating genes required for melanoblast formation [177,178]. Human nevi and melanoma highly express *SOX9* and *SOX10*, and these genes may be negative prognostic markers [136]. In a mouse xenograft model of melanoma, knockdown of *Sox10* in human melanoma cells suppresses proliferation and cell survival and completely abolishes tumor formation *in vivo*, suggesting that *SOX10* plays a crucial role in melanoma progression [137].

Local invasion and metastases of melanomas require many of the same changes in gene expression as occur during normal neural crest EMT. In melanoma, progression from the radial-growth phase to the vertical-growth phase is marked by the loss of E-cadherin and the expression of N-cadherin [148,149]. These changes allow the cells to migrate through the surrounding tissue and also promote cell survival. In addition, a large array of cell-adhesion markers are misregulated in melanoma compared to nevi, including connexin43 [179], the gap junction protein involved in NCC polarity and migration. The transition to metastatic melanoma also involves the upregulation of integrins and MMP2 [152–154], proteins that aid in migration of NCC. Elevated levels of TWIST1, an EMT-inducer, in melanoma have also been shown to upregulate MMP1 and promote metastasis [146]. Another neural crest gene, *Snail1*, a member of the same family of transcription factors and also a key regulator of EMT, is upregulated in invasive melanoma by Endothelin receptor B (EDNRB), which is over-expressed in many human melanoma cell lines [141,142]. EDNRB is required during normal neural crest development for the migration of NCC of the melanocyte and enteric neuron lineages [180,181]. The Snail homolog *Slug* is also required for EMT of NCCs and the initiation of migration. Interestingly, *Slug* has been found to be expressed in human melanoma cells, and in a mouse xenograft model of metastatic melanoma, knockdown of *Slug* produced a significant decrease in metastatic ability [143].

One important aspect of metastatic melanoma is the manner and routes by which melanoma cells migrate within the body. A relatively unique characteristic of melanoma is that tumor cells can migrate along the external surfaces of vascular channels, termed "extravascular migratory metastasis," or EVMM [182], as opposed to the more common intravascular migration. EVMM is likened to neural crest migration, as there is evidence that certain NCC migrate along intersomitic blood vessels [183] or the endothelial cells and vessels of the gut [184]. Recent work to elucidate genes directly involved in human melanoma EVMM has revealed that several of these genes are also involved in neural crest migration, including TCOF1 and KCTD11 [185], suggesting that melanoma cells may upregulate specific neural crest programs when pursuing this migratory path.

Interestingly, several studies have attempted to reintroduce melanoma cells into an embryonic or neural crest-like environment to determine the plasticity of these cells and how the embryonic microenvironment can interact with melanoma cells and influence their invasive behaviors and gene expression [168]. Transplantation of aggressively metastatic human melanoma cells into zebrafish and chick embryos has shown that these cells do not form tumors, but begin to exhibit neural crest-like behaviors through the ability to respond to embryonic signaling cues, migrate, and incorporate into various tissues along the typical route of migrating NCC; with a subset of cells beginning to express some markers of differentiated melanocytes, such as EDNRA, while other cells remain dedifferentiated [186–188]. Highly metastatic melanoma cells are more capable of responding to migratory cues in the embryo and begin to induce expression of neural crest developmental genes, while poorly metastatic cells do not [188]. While these experiments showing changes in gene expression and migration behavior in response to environment are very interesting and demonstrate the developmental potential of melanoma cells, there are some limitations to these types of transplantation experiments. For example, these experiments did not remove the endogenous neural crest population to determine if these cells can respond to embryonic signals in the absence of neighboring migratory NCC. Classic experiments have shown that uncoated latex beads placed into the neural crest migratory pathway will follow along streams of migrating cells [189], suggesting that transplanted objects or cells can be passively carried along the path of migrating cells. Therefore, the melanoma cells may not entirely be actively migrating themselves, thus confounding results that imply that melanoma cells can respond to the neural crest migratory environment. In addition, transplanted melanoma cells have not been

shown to form tumors; however, it is possible that a longer length of time (more than 1–8 days) may be required for these cancer cells to colonize the embryonic microenvironment and form tumors. It should also be noted that the embryonic environment is very different from the adult environment in which these tumors form and that factors such as the immature nature of the developing vasculature in the embryo at the time of transplantation may affect their ability to colonize tumor sites. Thus, while transplantation studies will likely yield important molecular clues as to how melanoma cells take on embryonic characteristics and become more metastatic, the above caveats should be considered when interpreting these results. However, additional experiments have also shown that melanoma cells grown in spheroid cultures under neural crest-like conditions begin to express more neural crest-associated genes and exhibit enhanced migratory and invasive abilities [190,191], further supporting the hypothesis that malignant melanoma cells reinitiate many developmental gene programs during cancer transformation and that these neural crest signatures may contribute to melanoma's highly metastatic capacity.

Neuroblastoma

Neuroblastoma is the most common solid tumor diagnosed in infancy and accounts for 7–10% of all childhood cancer [192]. It is highly heterogeneous and when metastases are present, often in children over 1 year of age, prognosis is very poor [165]. Neuroblastoma is a neuroendocrine tumor derived from neural crest elements of the sympathetic nervous system, most frequently originating in the adrenal glands [165]. One of the hallmark upregulated genes in neuroblastoma cell lines and patient samples is the oncogene *MYCN* [159–161]. High expression of *MYCN* in neuroblastoma patients is associated with highly aggressive disease and poor clinical outcome. Of note, *MYCN* is also important in neural crest development. In chick embryos, *Mycn* is expressed in all early NCC, and during migration is expressed only in cells undergoing neuronal differentiation [193,194].

Other common genetic misregulations that can lead to malignant transformation of neuroblasts are within the Trk family—TrkA, TrkB, and TrkC—of receptor tyrosine kinases that respond to neuronal growth factor ligands nerve growth factor (NGF) and brain-derived neurotrophic factor (BDNF) as well as *neurotrophin-3*. These signaling interactions through Trk are linked to survival and differentiation of specific classes of neurons such as nociceptive, mechanoreceptive, and prorioceptive neurons. Trk receptors are expressed by and required for certain populations of neural crest, specifically neurogenic NCC that populate the dorsal root ganglion [195,196]. The Trk ligand *neurotrophin3* has also been shown to play a role in the differentiation of neurons and glia from NCC in rat [197]. Interestingly, p75, another NGF receptor, is also misregulated in neuroblastoma and is used as a marker of undifferentiated neural crest precursor cells in mouse [198]. Another gene important in the development of neural crest, the cell–cell adhesion gene N-cadherin, is variably expressed in neuroblastoma tumors and cell lines and is downregulated in patients with metastatic disease [20], suggesting that similar to NCC, inhibition of N-cadherin may promote cell migration and possibly survival. These pathways suggest that neuroblastoma cells can reinitiate neural crest developmental genes during the progression of cancer.

Glioma and Glioblastoma

Gliomas account for approximately 30% of all brain and central nervous system tumors and 80% of all malignant brain tumors [199]. Gliomas are cancers of glial cells and are further characterized based on their presumed cell type of origin: astrocytoma, oligodendroglioma, mixed oliogoastrocytoma, and ependymoma [163,200]. Glioblastoma is the highest-grade, most malignant form of astrocytoma, and it is the most common and aggressive type of human brain tumor, with a 5-year survival rate of only 5% [201]. While not all gliomas may be neural crest-derived, since NCC do give rise to certain populations of glial cells within the adult body, some of the mechanisms of glioma progression may be similar to NCC development. Many genetic and chromosomal changes have been implicated in gliomagenesis; some of the most prominent upregulated genes are platelet-derived growth factor (PDGF) and epidermal growth factor receptor (EGFR) [163]. These growth factor pathways, not surprisingly, are thought to have important roles in gliogenesis as well as neural crest proliferation, differentiation, and survival [202,203], suggesting a re-initiation of developmental programs during tumorigenesis. *Sox10*, a neural crest transcription factor, was also found to be upregulated in a mouse model of glioma and is thought to be induced by PDGF signaling [138].

A common characteristic of gliomas, even low-grade tumors, is that they are highly invasive [204]. It is suggested that this is because glioma cells express high levels of MMPs, specifically MMP-2 and -9 in human brain tissue samples [155,156], and that levels of these MMPs increase along with tumor progression [157,158]. MMP-2 and -9, along with MMP-8 are also expressed during neural crest EMT and migration from the neural tube [73,74,76] and have been shown to be required for proper migration of NCC [205,206]. Glioblastomas exhibit a downregulation of *RhoB* [151] and an upregulation of *cadherin11* [23] to promote invasiveness and migration, similar to NCC. Additionally, in glioblastoma stem cells (GSCs), *Endothelin3/EDNRB* signaling has been found to be highly expressed and important in maintaining GSC migration, proliferation, prevention of differentiation, and survival [164]. *Endothelin-3* is important in neural crest survival as well as proliferation, differentiation, and migration of NCC that contribute to enteric nervous system development [207–210]. This suggests that GSC may maintain and utilize some of the same pathways as their neural crest precursors in order to promote disease.

Other Cancers of Neural Crest Origin

Ewing's sarcoma is a rare form of cancer primarily found in children and adolescents. It occurs primarily in the bone and sometimes soft tissue as a poorly differentiated and highly aggressive tumor that metastasizes quickly to the lung, bone marrow, and other tissues [211]. The cell type of origin in Ewing's sarcoma remains controversial. While some believe that it originates in mesenchymal stem cells, others argue that due to the histological and molecular similarities to primitive neuroectodermal cells, Ewing's sarcoma originates in NCC. Ewing's sarcoma is characterized by a chromosomal translocation leading to a fusion of the *EWS* gene and *FLI1*, a member of the *ETS* family of transcription factors, referred to as *EWS-FLI1*. Researchers have found that expression of this translocation in cell culture leads to an upregulation of several neural crest-associated genes that are also upregulated in Ewing's sarcoma, including *Krox20*, *Msx1*, *c-Myc*, *Id2*, *Cadherin11*, and *Runx3* [150], suggesting that Ewing's sarcoma may either originate in NCC or in a multipotent progenitor cell population that is able to express neural crest genes upon expression of the *EWS-FLI1* fusion gene. Interestingly, Ewing's sarcoma cells that highly express the migratory neural crest marker HNK-1 (CD57) are more tumorigenic, possibly due to enhanced adhesion and invasion abilities [212]. Recent studies have shown that driving *EWS-FLI1* in human NCC initiates transformation to a Ewing's sarcoma-like state, suggesting that

NCC could be a cell type of origin for Ewing's sarcoma in humans [213]. It is noted however, that for fully malignant transformation of NCC, other genetic and/or epigenetic events must occur. Regardless of the cellular origin of Ewing's sarcoma, it is clear that these tumor cells are able to turn on neural crest developmental pathways and that this could contribute to cancer progression and metastasis.

Malignant ectomesenchymoma is a rare but aggressive neoplasm composed of both benign neuroectodermal elements and mesenchymal neoplastic elements. The mesenchymal element often consists of a rhabdomyosarcoma (a cancer of the skeletal muscle), while the neuroectodermal elements are most often ganglionic [214,215]. It is termed ectomesenchymoma, as it is believed to arise from remnants of migratory NCC (ectomesenchyme) [216]. While very little is known about these uncommon tumors, recent molecular analysis has confirmed the presence of both differentiated neural and mesenchymal cells within ectomesenchymomal tumors and shown that these cells express many markers of NCC and their derivatives, including vimentin, S100, and neurofilament [166], further demonstrating that these are tumors likely of neural crest origin.

Clear cell sarcoma (CCS) is a rare, aggressive malignancy in adolescents and young adults, characterized by tumors of the soft tissues of the extremities, sometimes referred to as "melanoma of the soft parts" based on evidence of melanocytic differentiation of these cells [217,218]. Due to its similarity to melanoma, as well as various histological and histochemical characteristics, CCS is presumed to arise from NCC, although the exact cellular origin remains to be determined [219,220]. CCS is associated with a chromosomal translocation creating a fusion of Ewing's sarcoma oncogene (*EWS*) and activating transcription factor 1 (*ATF1*) [221]. Genomic profiling of human CCS tumors has shown that they express many neural crest-associated genes, including *SOX10, ERBB3, CTNNA1, S100B,* and *MITF* [139,140]. Recent work has shown that in a mouse model, expression of the *EWS-ATF1* fusion gene induced tumor cells expressing many of these same neural crest genes, especially *S100, Sox10, Mitf,* and that upon lineage-tracing, the cell of origin of these CCS tumors was in fact the neural crest [167]. This supports the hypothesis that CCS in human patients originates from NCC and that maintaining or reinitiating neural crest developmental pathways may contribute to CCS initiation and progression, suggesting potential targets for therapeutic treatments.

Like neuroblastoma, pheochromocytoma is a neural crest-derived tumor of the neuroendocrine system. Pheochromocytoma is a rare tumor, mostly occurring in adults in the medullary adrenal gland chromaffin cells, the cells that synthesize and secrete epinephrine and norepinephrine into the bloodstream. Extra-adrenal pheochromocytomas can also occur and are referred to as paragangliomas. While only about 15% of pheochromocytomas are malignant, once they metastasize, often to the lymph nodes, bone, liver, or lung, there is no curative therapy. Several recent gene expression studies in human pheochromocytoma samples as well as mouse models of the disease have shown misregulation of many genes involved in neural crest development, including the transcription factors *PHOX2B* and *HAND2,* as well as their targets *PHOX2A, GATA2,* and *GATA3* [222], and *MYCN, HAND1,* and *TCFAP2B* [162]. Pheochromocytoma and paraganglioma patients with a mutation in the *SDHB* gene have a higher instance of metastasis [147]. In these tumors, there is an increase in expression levels of neural crest EMT and migration factors *TWIST1, MMP1,* and *MMP2,* as well as a downregulation of the N-cadherin gene *CDH2.* Additional studies have shown that overexpression of the hallmark neural crest EMT gene *SNAIL* correlates with malignancy in pheochromocytoma [144,145].

Due to the heterogeneic nature of tumors and the complex processes of cancer initiation, progression, and metastasis, it is often unclear what the cell of origin for any particular tumor may be. There are certainly other tumors that originate from NCC and neural crest derivatives—for example, Schwann cell neoplasms including neurofibromas, schwannomas, and malignant peripheral nerve sheath tumors [223], as well as tumor types that have yet to be associated with NCC. The emergence of high-throughput genetic screening may help identify even more cancers that exhibit the neural crest transcriptional signature. One characteristic that is common among cancers of neural crest origin is the reacquisition of aspects of embryonic neural crest development, especially the upregulation of pathways involved in neural crest EMT and migration, which may contribute to the highly metastatic nature of many of these cancers.

CONCLUSIONS AND PERSPECTIVES

The parallels observed between normal neural crest development and cancer progression, as well as the re-activation of developmental pathways in tumors of neural crest origin, suggest that NCC are an excellent model for studying tumorigenesis, invasion, and metastasis. Despite some differences described above (cadherin-switching, routes of systemic travel), the mechanisms of EMT and migration utilized by neural crest and tumor cells are remarkably similar. Future studies in both neural crest and cancer fields can benefit from the similarities already observed. Many of the pathways targeted by tumor therapeutics may be of benefit for use in treating neural crest-related disorders or be beneficial in outlining the contributions of these pathways to disease progression. Reciprocally, signaling networks identified in neural crest delamination and migration can provide insight into potential mechanisms that tumor cells utilize during the metastatic process. In both fields, many of the early events necessary to allow a cell to detach from an epithelial tissue and become motile have been identified. Investigation into how these migratory cells travel systemically, find their final destination, and thrive in a new microenvironment is of interest. Further understanding of how reimplementation of embryonic programs can direct metastasis has the potential to reveal exciting new ideas for developmental biologists and cancer researchers alike.

References

[1] Yang J, Weinberg RA. Epithelial–mesenchymal transition: at the crossroads of development and tumor metastasis. Dev Cell 2008;14(6):818–29.

[2] Schmalhofer O, Brabletz S, Brabletz T. E-cadherin, beta-catenin, and ZEB1 in malignant progression of cancer. Cancer Metastasis Rev 2009;28(1–2):151–66.

[3] Valenta T, Hausmann G, Basler K. The many faces and functions of beta-catenin. EMBO J 2012;31 (12):2714–36.

[4] Perl AK, Wilgenbus P, Dahl U, Semb H, Christofori G. A causal role for E-cadherin in the transition from adenoma to carcinoma. Nature 1998;392(6672): 190–3.

[5] Frixen UH, Behrens J, Sachs M, et al. E-cadherin-mediated cell-cell adhesion prevents invasiveness of human carcinoma cells. J Cell Biol 1991;113(1):173–85.

[6] Onder TT, Gupta PB, Mani SA, Yang J, Lander ES, Weinberg RA. Loss of E-cadherin promotes metastasis via multiple downstream transcriptional pathways. Cancer Res 2008;68(10):3645–54.

[7] ElMoneim HM, Zaghloul NM. Expression of E-cadherin, N-cadherin and snail and their correlation with clinicopathological variants: an immunohistochemical study of 132 invasive ductal breast carcinomas in Egypt. Clinics (Sao Paulo) 2011;66(10):1765–71.

[8] Elzagheid A, Buhmeida A, Laato M, El-Faitori O, Syrjanen K, Collan Y, et al. Loss of E-cadherin expression predicts disease recurrence and shorter survival in colorectal carcinoma. APMIS 2012;120(7):539–48.

[9] Jeanes A, Gottardi CJ, Yap AS. Cadherins and cancer: how does cadherin dysfunction promote tumor progression? Oncogene 2008;27(55):6920–9.

[10] Sanders DS, Blessing K, Hassan GA, Bruton R, Marsden JR, Jankowski J. Alterations in cadherin and catenin expression during the biological progression of melanocytic tumours. Mol Pathol 1999;52(3):151–7.

[11] Kovacs A, Dhillon J, Walker RA. Expression of P-cadherin, but not E-cadherin or N-cadherin, relates to pathological and functional differentiation of breast carcinomas. Mol Pathol 2003;56(6):318–22.

[12] Bussemakers MJ, Van Bokhoven A, Tomita K, Jansen CF, Schalken JA. Complex cadherin expression in human prostate cancer cells. Int J Cancer 2000;85 (3):446–50.

[13] Derycke LD, Bracke ME. N-cadherin in the spotlight of cell–cell adhesion, differentiation, embryogenesis, invasion and signalling. Int J Dev Biol 2004;48 (5–6):463–76.

[14] Rezaei M, Friedrich K, Wielockx B, Kuzmanov A, Kettelhake A, Labelle M, et al. Interplay between neural-cadherin and vascular endothelial-cadherin in breast cancer progression. Breast Cancer Res 2012;14 (6):R154.

[15] Monaghan-Benson E, Burridge K. Mutant B-RAF regulates a Rac-dependent cadherin switch in melanoma. Oncogene 2012.

[16] Paul R, Necknig U, Busch R, Ewing CM, Hartung R, Isaacs WB. Cadherin-6: a new prognostic marker for renal cell carcinoma. J Urol 2004;171(1):97–101.

[17] Shoval I, Ludwig A, Kalcheim C. Antagonistic roles of full-length N-cadherin and its soluble BMP cleavage product in neural crest delamination. Development 2007;134(3):491–501.

[18] Nakagawa S, Takeichi M. Neural crest emigration from the neural tube depends on regulated cadherin expression. Development 1998;125(15):2963–71.

[19] Borchers A, David R, Wedlich D. Xenopus cadherin-11 restrains cranial neural crest migration and influences neural crest specification. Development 2001;128 (16):3049–60.

[20] Lammens T, Swerts K, Derycke L, De Craemer A, De Brouwer S, De Preter K, et al. N-cadherin in neuroblastoma disease: expression and clinical significance. PloS One 2012;7(2):e31206.

[21] Moore R, Champeval D, Denat L, Tan SS, Faure F, Julien-Grille S, et al. Involvement of cadherins 7 and 20 in mouse embryogenesis and melanocyte transformation. Oncogene 2004;23(40):6726–35.

[22] Winklmeier A, Contreras-Shannon V, Arndt S, Melle C, Bosserhoff AK. Cadherin-7 interacts with melanoma inhibitory activity protein and negatively modulates melanoma cell migration. Cancer Sci 2009;100(2):261–8.

[23] Kaur H, Phillips-Mason PJ, Burden-Gulley SM, Kerstetter-Fogle AE, Basilion JP, Sloan AE, et al. Cadherin-11, a marker of the mesenchymal phenotype, regulates glioblastoma cell migration and survival in vivo. Mol Cancer Res 2012;10(3):293–304.

[24] Li Y, Guo Z, Chen H, Dong Z, Pan ZK, Ding H, et al. HOXC8-Dependent cadherin 11 expression facilitates breast cancer cell migration through trio and Rac. Genes Cancer 2011;2(9):880—8.

[25] Huang CF, Lira C, Chu K, Bilen MA, Lee YC, Ye X, et al. Cadherin-11 increases migration and invasion of prostate cancer cells and enhances their interaction with osteoblasts. Cancer Res 2010;70(11):4580—9.

[26] Carmona FJ, Villanueva A, Vidal A, Munoz C, Puertas S, Penin RM, et al. Epigenetic disruption of cadherin-11 in human cancer metastasis. J Pathol 2012;228(2):230—40.

[27] Li L, Ying J, Li H, Zhang Y, Shu X, Fan Y, et al. The human cadherin 11 is a pro-apoptotic tumor suppressor modulating cell stemness through Wnt/beta-catenin signaling and silenced in common carcinomas. Oncogene 2012;31(34):3901—12.

[28] Gumbiner BM. Regulation of cadherin-mediated adhesion in morphogenesis. Nat Rev Mol Cell Biol 2005;6(8):622—34.

[29] Taneyhill LA. To adhere or not to adhere: the role of Cadherins in neural crest development. Cell Adh Migr 2008;2(4):223—30.

[30] Fishwick KJ, Neiderer TE, Jhingory S, Bronner ME, Taneyhill LA. The tight junction protein claudin-1 influences cranial neural crest cell emigration. Mech Dev 2012;129(9—12):275—83.

[31] Wu CY, Jhingory S, Taneyhill LA. The tight junction scaffolding protein cingulin regulates neural crest cell migration. Dev Dyn 2011;240(10):2309—23.

[32] Aaku-Saraste E, Hellwig A, Huttner WB. Loss of occludin and functional tight junctions, but not ZO-1, during neural tube closure—remodeling of the neuroepithelium prior to neurogenesis. Dev Biol 1996;180(2):664—79.

[33] Orban E, Szabo E, Lotz G, Kupcsulik P, Paska C, Schaff Z, et al. Different expression of occludin and ZO-1 in primary and metastatic liver tumors. Pathol Oncol Res 2008;14(3):299—306.

[34] Kurrey NK, Amit K, Bapat SA. Snail and Slug are major determinants of ovarian cancer invasiveness at the transcription level. Gynecol Oncol 2005;97(1):155—65.

[35] Tobioka H, Isomura H, Kokai Y, Tokunaga Y, Yamaguchi J, Sawada N. Occludin expression decreases with the progression of human endometrial carcinoma. Hum Pathol 2004;35(2):159—64.

[36] Martin TA, Mansel RE, Jiang WG. Loss of occludin leads to the progression of human breast cancer. Int J Mol Med 2010;26(5):723—34.

[37] Czyz J, Szpak K, Madeja Z. The role of connexins in prostate cancer promotion and progression. Nat Rev Urol 2012;9(5):274—82.

[38] El-Saghir JA, El-Habre ET, El-Sabban ME, Talhouk RS. Connexins: a junctional crossroad to breast cancer. Int J Dev Biol 2011;55(7—9):773—80.

[39] Nakagawa S, Yano T, Nakagawa K, Takizawa S, Suzuki Y, Yasugi T, et al. Analysis of the expression and localisation of a LAP protein, human scribble, in the normal and neoplastic epithelium of uterine cervix. Br J Cancer 2004;90(1):194—9.

[40] Gardiol D, Zacchi A, Petrera F, Stanta G, Banks L. Human discs large and scrib are localized at the same regions in colon mucosa and changes in their expression patterns are correlated with loss of tissue architecture during malignant progression. Int J Cancer 2006;119(6):1285—90.

[41] Kuphal S, Wallner S, Schimanski CC, Bataille F, Hofer P, Strand S, et al. Expression of Hugl-1 is strongly reduced in malignant melanoma. Oncogene 2006;25(1):103—10.

[42] Zhan L, Rosenberg A, Bergami KC, Yu M, Xuan Z, Jaffe AB, et al. Deregulation of scribble promotes mammary tumorigenesis and reveals a role for cell polarity in carcinoma. Cell 2008;135(5):865—78.

[43] Pearson HB, Perez-Mancera PA, Dow LE, Ryan A, Tennstedt P, Bogani D, et al. SCRIB expression is deregulated in human prostate cancer, and its deficiency in mice promotes prostate neoplasia. J Clin Invest 2011;121(11):4257—67.

[44] Schimanski CC, Schmitz G, Kashyap A, Bosserhoff AK, Bataille F, Schafer SC, et al. Reduced expression of Hugl-1, the human homologue of Drosophila tumour suppressor gene lgl, contributes to progression of colorectal cancer. Oncogene 2005;24(19):3100—9.

[45] Dow LE, Elsum IA, King CL, Kinross KM, Richardson HE, Humbert PO. Loss of human Scribble cooperates with H-Ras to promote cell invasion through deregulation of MAPK signalling. Oncogene 2008;27(46):5988—6001.

[46] Wu M, Pastor-Pareja JC, Xu T. Interaction between Ras(V12) and scribbled clones induces tumour growth and invasion. Nature 2010;463(7280):545—8.

[47] Karp CM, Tan TT, Mathew R, Nelson D, Mukherjee C, Degenhardt K, et al. Role of the polarity determinant crumbs in suppressing mammalian epithelial tumor progression. Cancer Res 2008;68(11):4105—15.

[48] Storrs CH, Silverstein SJ. PATJ a tight junction-associated PDZ protein, is a novel degradation target of high-risk human papillomavirus E6 and the alternatively spliced isoform 18 E6. J Virol 2007;81(8):4080—90.

[49] Michel D, Arsanto JP, Massey-Harroche D, Beclin C, Wijnholds J, Le Bivic A. PATJ connects and stabilizes apical and lateral components of tight junctions in human intestinal cells. J Cell Sci 2005;118(Pt 17):4049—57.

[50] Eder AM, Sui X, Rosen DG, Nolden LK, Cheng KW, Lahad JP, et al. Atypical PKCiota contributes to poor prognosis through loss of apical-basal polarity and cyclin E overexpression in ovarian cancer. Proc Natl Acad Sci USA 2005;102(35):12519–24.

[51] Kojima Y, Akimoto K, Nagashima Y, Ishiguro H, Shirai S, Chishima T, et al. The overexpression and altered localization of the atypical protein kinase C lambda/iota in breast cancer correlates with the pathologic type of these tumors. Hum Pathol 2008;39(6):824–31.

[52] Regala RP, Weems C, Jamieson L, Khoor A, Edell ES, Lohse CM, et al. Atypical protein kinase C iota is an oncogene in human non-small cell lung cancer. Cancer Res 2005;65(19):8905–11.

[53] Zen K, Yasui K, Gen Y, Dohi O, Wakabayashi N, Mitsufuji S, et al. Defective expression of polarity protein PAR-3 gene (PARD3) in esophageal squamous cell carcinoma. Oncogene 2009;28(32):2910–8.

[54] Xue B, Krishnamurthy K, Allred DC, Muthuswamy SK. Loss of Par3 promotes breast cancer metastasis by compromising cell–cell cohesion. Nat Cell Biol 2012.

[55] Ellenbroek SI, Iden S, Collard JG. Cell polarity proteins and cancer. Semin Cancer Biol 2012;22(3):208–15.

[56] Shoval I, Kalcheim C. Antagonistic activities of Rho and Rac GTPases underlie the transition from neural crest delamination to migration. Dev Dyn 2012;241(7):1155–68.

[57] Groysman M, Shoval I, Kalcheim C. A negative modulatory role for rho and rho-associated kinase signaling in delamination of neural crest cells. Neural Dev 2008;3:27.

[58] Chen Z, Sun J, Pradines A, Favre G, Adnane J, Sebti SM. Both farnesylated and geranylgeranylated RhoB inhibit malignant transformation and suppress human tumor growth in nude mice. J Biol Chem 2000;275(24):17974–8.

[59] Jiang K, Sun J, Cheng J, Djeu JY, Wei S, Sebti S. Akt mediates Ras downregulation of RhoB, a suppressor of transformation, invasion, and metastasis. Mol Cell Biol 2004;24(12):5565–76.

[60] Bousquet E, Mazieres J, Privat M, Rizzati V, Casanova A, Ledoux A, et al. Loss of RhoB expression promotes migration and invasion of human bronchial cells via activation of AKT1. Cancer Res 2009;69(15):6092–9.

[61] Mazieres J, Antonia T, Daste G, Muro-Cacho C, Berchery D, Tillement V, et al. Loss of RhoB expression in human lung cancer progression. Clin Cancer Res 2004;10(8):2742–50.

[62] Forget MA, Desrosiers RR, Del M, Moumdjian R, Shedid D, Berthelet F, et al. The expression of rho proteins decreases with human brain tumor progression: potential tumor markers. Clin Exp Metastasis 2002;19(1):9–15.

[63] Adnane J, Muro-Cacho C, Mathews L, Sebti SM, Munoz-Antonia T. Suppression of rho B expression in invasive carcinoma from head and neck cancer patients. Clin Cancer Res 2002;8(7):2225–32.

[64] Fort P, Guemar L, Vignal E, Morin N, Notarnicola C, de Santa Barbara P, et al. Activity of the RhoU/Wrch1 GTPase is critical for cranial neural crest cell migration. Dev Biol 2011;350(2):451–63.

[65] Schnelzer A, Prechtel D, Knaus U, Dehne K, Gerhard M, Graeff H, et al. Rac1 in human breast cancer: overexpression, mutation analysis, and characterization of a new isoform, Rac1b. Oncogene 2000;19(26):3013–20.

[66] Pan Y, Bi F, Liu N, Xue Y, Yao X, Zheng Y, et al. Expression of seven main Rho family members in gastric carcinoma. Biochem Biophys Res Commun 2004;315(3):686–91.

[67] Kamai T, Yamanishi T, Shirataki H, Takagi K, Asami H, Ito Y, et al. Overexpression of RhoA, Rac1, and Cdc42 GTPases is associated with progression in testicular cancer. Clin Cancer Res 2004;10(14):4799–805.

[68] Kissil JL, Walmsley MJ, Hanlon L, Haigis KM, Bender Kim CF, Sweet-Cordero A, et al. Requirement for Rac1 in a K-ras induced lung cancer in the mouse. Cancer Res 2007;67(17):8089–94.

[69] Espina C, Cespedes MV, Garcia-Cabezas MA, Gomez del Pulgar MT, Boluda A, Oroz LG, et al. A critical role for Rac1 in tumor progression of human colorectal adenocarcinoma cells. Am J Pathol 2008;172(1):156–66.

[70] Wang Z, Pedersen E, Basse A, Lefever T, Peyrollier K, Kapoor S, et al. Rac1 is crucial for Ras-dependent skin tumor formation by controlling Pak1-Mek-Erk hyperactivation and hyperproliferation in vivo. Oncogene 2010;29(23):3362–73.

[71] Chuang YY, Valster A, Coniglio SJ, Backer JM, Symons M. The atypical Rho family GTPase Wrch-1 regulates focal adhesion formation and cell migration. J Cell Sci 2007;120(Pt 11):1927–34.

[72] Tao W, Pennica D, Xu L, Kalejta RF, Levine AJ. Wrch-1, a novel member of the Rho gene family that is regulated by Wnt-1. Genes Dev 2001;15(14):1796–807.

[73] Cai DH, Vollberg Sr. TM, Hahn-Dantona E, Quigley JP, Brauer PR. MMP-2 expression during early avian cardiac and neural crest morphogenesis. Anat Rec 2000;259(2):168–79.

[74] Monsonego-Ornan E, Kosonovsky J, Bar A, Roth L, Fraggi-Rankis V, Simsa S, et al. Matrix metalloproteinase 9/gelatinase B is required for neural crest cell migration. Dev Biol 2012;364(2):162–77.

[75] Alfandari D, Wolfsberg TG, White JM, DeSimone DW. ADAM 13: a novel ADAM expressed in somitic mesoderm and neural crest cells during *Xenopus laevis* development. Dev Biol 1997;182(2):314–30.

[76] Giambernardi TA, Sakaguchi AY, Gluhak J, Pavlin D, Troyer DA, Das G, et al. Neutrophil collagenase (MMP-8) is expressed during early development in neural crest cells as well as in adult melanoma cells. Matrix Biol 2001;20(8):577–87.

[77] Nair SA, Jagadeeshan S, Indu R, Sudhakaran PR, Pillai MR. How intact is the basement membrane? Role of MMPs. Adv Exp Med Biol 2012;749:215–32.

[78] Hanahan D, Weinberg RA. The hallmarks of cancer. Cell 2000;100(1):57–70.

[79] Roy R, Yang J, Moses MA. Matrix metalloproteinases as novel biomarkers and potential therapeutic targets in human cancer. J Clin Oncol 2009;27 (31):5287–97.

[80] Kast RE, Halatsch ME. Matrix metalloproteinase-2 and −9 in glioblastoma: a trio of old drugs-captopril, disulfiram and nelfinavir-are inhibitors with potential as adjunctive treatments in glioblastoma. Arch Med Res 2012;43(3):243–7.

[81] Hofmann UB, Houben R, Brocker EB, Becker JC. Role of matrix metalloproteinases in melanoma cell invasion. Biochimie 2005;87(3–4):307–14.

[82] Turpeenniemi-Hujanen T. Gelatinases (MMP-2 and -9) and their natural inhibitors as prognostic indicators in solid cancers. Biochimie 2005;87 (3–4):287–97.

[83] Bauvois B. New facets of matrix metalloproteinases MMP-2 and MMP-9 as cell surface transducers: outside-in signaling and relationship to tumor progression. Biochim Biophys Acta 2012;1825(1): 29–36.

[84] Rydlova M, Holubec Jr. L, Ludvikova Jr. M, Kalfert D, Franekova J, Povysil C, et al. Biological activity and clinical implications of the matrix metalloproteinases. Anticancer Res 2008;28(2B):1389–97.

[85] Moss ML, Stoeck A, Yan W, Dempsey PJ. ADAM10 as a target for anti-cancer therapy. Curr Pharm Biotechnol 2008;9(1):2–8.

[86] Weaver VM, Petersen OW, Wang F, Larabell CA, Briand P, Damsky C, et al. Reversion of the malignant phenotype of human breast cells in three-dimensional culture and in vivo by integrin blocking antibodies. J Cell Biol 1997;137(1):231–45.

[87] Imanishi Y, Hu B, Jarzynka MJ, Guo P, Elishaev E, Bar-Joseph I, et al. Angiopoietin-2 stimulates breast cancer metastasis through the alpha(5)beta(1) integrin-mediated pathway. Cancer Res 2007;67 (9):4254–63.

[88] Barkan D, Chambers AF. Beta1-integrin: a potential therapeutic target in the battle against cancer recurrence. Clin Cancer Res 2011;17(23):7219–23.

[89] Theveneau E, Mayor R. Neural crest delamination and migration: from epithelium-to-mesenchyme transition to collective cell migration. Dev Biol 2012;366(1):34–54.

[90] Santiago A, Erickson CA. Ephrin-B ligands play a dual role in the control of neural crest cell migration. Development 2002;129(15):3621–32.

[91] Toyofuku T, Yoshida J, Sugimoto T, Yamamoto M, Makino N, Takamatsu H, et al. Repulsive and attractive semaphorins cooperate to direct the navigation of cardiac neural crest cells. Dev Biol 2008;321 (1):251–62.

[92] Gammill LS, Gonzalez C, Bronner-Fraser M. Neuropilin 2/semaphorin 3F signaling is essential for cranial neural crest migration and trigeminal ganglion condensation. Dev Neurobiol 2007;67 (1):47–56.

[93] Lwigale PY, Bronner-Fraser M. Semaphorin3A/neuropilin-1 signaling acts as a molecular switch regulating neural crest migration during cornea development. Dev Biol 2009;336(2):257–65.

[94] Krull CE, Lansford R, Gale NW, Collazo A, Marcelle C, Yancopoulos GD, et al. Interactions of Eph-related receptors and ligands confer rostrocaudal pattern to trunk neural crest migration. Curr Biol 1997;7 (8):571–80.

[95] Xi HQ, Wu XS, Wei B, Chen L. Eph receptors and ephrins as targets for cancer therapy. J Cell Mol Med 2012;16(12):2894–909.

[96] Rehman M, Tamagnone L. Semaphorins in cancer: biological mechanisms and therapeutic approaches. Semin Cell Dev Biol 2012.

[97] Brantley-Sieders DM, Jiang A, Sarma K, Badu-Nkansah A, Walter DL, Shyr Y, et al. Eph/ephrin profiling in human breast cancer reveals significant associations between expression level and clinical outcome. PLoS One 2011;6(9):e24426.

[98] Nakada M, Niska JA, Miyamori H, McDonough WS, Wu J, Sato H, et al. The phosphorylation of EphB2 receptor regulates migration and invasion of human glioma cells. Cancer Res 2004;64(9):3179–85.

[99] Udayakumar D, Zhang G, Ji Z, Njauw CN, Mroz P, Tsao H. EphA2 is a critical oncogene in melanoma. Oncogene 2011;30(50):4921–9.

[100] Easty DJ, Mitchell PJ, Patel K, Florenes VA, Spritz RA, Bennett DC. Loss of expression of receptor tyrosine kinase family genes PTK7 and SEK in metastatic melanoma. Int J Cancer 1997;71 (6):1061–5.

[101] Batlle E, Bacani J, Begthel H, et al. EphB receptor activity suppresses colorectal cancer progression. Nature 2005;435(7045):1126–30.

[102] Noren NK, Foos G, Hauser CA, Pasquale EB. The EphB4 receptor suppresses breast cancer cell tumorigenicity through an Abl-Crk pathway. Nat Cell Biol 2006;8(8):815–25.

[103] Martin-Satue M, Blanco J. Identification of semaphorin E gene expression in metastatic human lung adenocarcinoma cells by mRNA differential display. J Surg Oncol 1999;72(1):18–23.

[104] Miyato H, Tsuno NH, Kitayama J. Semaphorin 3C is involved in the progression of gastric cancer. Cancer Sci 2012;103(11):1961–6.

[105] Tseng CH, Murray KD, Jou MF, Hsu SM, Cheng HJ, Huang PH. Sema3E/plexin-D1 mediated epithelial-to-mesenchymal transition in ovarian endometrioid cancer. PLoS One 2011;6(4):e19396.

[106] Guo DL, Zhang J, Yuen ST, Tsui WY, Chan AS, Ho C, et al. Reduced expression of EphB2 that parallels invasion and metastasis in colorectal tumours. Carcinogenesis 2006;27(3):454–64.

[107] Lantuejoul S, Constantin B, Drabkin H, Brambilla C, Roche J, Brambilla E. Expression of VEGF, semaphorin SEMA3F, and their common receptors neuropilins NP1 and NP2 in preinvasive bronchial lesions, lung tumours, and cell lines. J Pathol 2003;200 (3):336–47.

[108] Neufeld G, Kessler O. The semaphorins: versatile regulators of tumour progression and tumour angiogenesis. Nat Rev Cancer 2008;8(8):632–45.

[109] Ahmad I, Iwata T, Leung HY. Mechanisms of FGFR-mediated carcinogenesis. Biochim Biophys Acta 2012;1823(4):850–60.

[110] Wesche J, Haglund K, Haugsten EM. Fibroblast growth factors and their receptors in cancer. Biochem J 2011;437(2):199–213.

[111] Liu KW, Hu B, Cheng SY. Platelet-derived growth factor receptor alpha in glioma: a bad seed. Chin J Cancer 2011;30(9):590–602.

[112] Kono SA, Heasley LE, Doebele RC, Camidge DR. Adding to the mix: fibroblast growth factor and platelet-derived growth factor receptor pathways as targets in non-small cell lung cancer. Curr Cancer Drug Targets 2012;12(2):107–23.

[113] Belmadani A, Tran PB, Ren D, Assimacopoulos S, Grove EA, Miller RJ. The chemokine stromal cell-derived factor-1 regulates the migration of sensory neuron progenitors. J Neurosci 2005;25(16): 3995–4003.

[114] Kasemeier-Kulesa JC, McLennan R, Romine MH, Kulesa PM, Lefort F. CXCR4 controls ventral migration of sympathetic precursor cells. J Neurosci 2010;30(39):13078–88.

[115] Olesnicky Killian EC, Birkholz DA, Artinger KB. A role for chemokine signaling in neural crest cell migration and craniofacial development. Dev Biol 2009;333(1):161–72.

[116] Dewan MZ, Ahmed S, Iwasaki Y, Ohba K, Toi M, Yamamoto N. Stromal cell-derived factor-1 and CXCR4 receptor interaction in tumor growth and metastasis of breast cancer. Biomed Pharmacother 2006;60(6):273–6.

[117] Kang Y, Siegel PM, Shu W, Drobnjak M, Kakonen SM, Cordon-Cardo C, et al. A multigenic program mediating breast cancer metastasis to bone. Cancer Cell 2003;3(6):537–49.

[118] Kucia M, Reca R, Miekus K, Wanzeck J, Wojakowski W, Janowska-Wieczorek A, et al. Trafficking of normal stem cells and metastasis of cancer stem cells involve similar mechanisms: pivotal role of the SDF-1-CXCR4 axis. Stem Cell 2005;23(7):879–94.

[119] Polakis P. Wnt signaling in cancer. Cold Spring Harb Perspect Biol 2012;4:5.

[120] Niehrs C, Acebron SP. Mitotic and mitogenic Wnt signalling. EMBO J 2012;31(12):2705–13.

[121] Ye L, Mason MD, Jiang WG. Bone morphogenetic protein and bone metastasis, implication and therapeutic potential. Front Biosci 2011;16:865–97.

[122] Miyazono K, Ehata S, Koinuma D. Tumor-promoting functions of transforming growth factor-beta in progression of cancer. Ups J Med Sci 2012;117(2):143–52.

[123] Heldin CH, Vanlandewijck M, Moustakas A. Regulation of EMT by TGFbeta in cancer. FEBS Lett 2012;586(14):1959–70.

[124] Wang J, Sullenger BA, Rich JN. Notch signaling in cancer stem cells. Adv Exp Med Biol 2012;727:174–85.

[125] Sanchez-Tillo E, Liu Y, de Barrios O, Siles L, Fanlo L, Cuatrecasas M, et al. EMT-activating transcription factors in cancer: beyond EMT and tumor invasiveness. Cell Mol Life Sci CMLS 2012;69(20):3429–56.

[126] Wu Y, Zhou BP. Snail: more than EMT. Cell Adh Migr 2010;4(2):199–203.

[127] Barrallo-Gimeno A, Nieto MA. The Snail genes as inducers of cell movement and survival: implications in development and cancer. Development 2005;132 (14):3151–61.

[128] Yang J, Mani SA, Weinberg RA. Exploring a new twist on tumor metastasis. Cancer Res 2006;66 (9):4549–52.

[129] Qin Q, Xu Y, He T, Qin C, Xu J. Normal and disease-related biological functions of Twist1 and underlying molecular mechanisms. Cell Res 2012;22(1):90–106.

[130] Castillo SD, Sanchez-Cespedes M. The SOX family of genes in cancer development: biological relevance and opportunities for therapy. Expert Opin Ther Targets 2012;16(9):903–19.

[131] Dong C, Wilhelm D, Koopman P. Sox genes and cancer. Cytogenet Genome Res 2004;105 (2–4):442–7.

[132] Vega S, Morales AV, Ocana OH, Valdes F, Fabregat I, Nieto MA. Snail blocks the cell cycle and confers resistance to cell death. Genes Dev 2004;18 (10):1131–43.

[133] Yang J, Mani SA, Donaher JL, Ramaswamy S, Itzykson RA, Come C, et al. Twist, a master regulator of morphogenesis, plays an essential role in tumor metastasis. Cell 2004;117(7):927–39.

[134] Mascarenhas JB, Littlejohn EL, Wolsky RJ, Young KP, Nelson M, Salgia R, et al. PAX3 and SOX10 activate MET receptor expression in melanoma. Pigment Cell Melanoma Res 2010;23(2):225–37.

[135] Seong I, Min HJ, Lee JH, Yeo CY, Kang DM, Oh ES, et al. Sox10 controls migration of B16F10 melanoma cells through multiple regulatory target genes. PLoS One 2012;7(2):e31477.

[136] Bakos RM, Maier T, Besch R, Mestel DS, Ruzicka T, Sturm RA, et al. Nestin and SOX9 and SOX10 transcription factors are coexpressed in melanoma. Exp Dermatol 2010;19(8):e89–94.

[137] Shakhova O, Zingg D, Schaefer SM, Hari L, Civenni G, Blunschi J, et al. Sox10 promotes the formation and maintenance of giant congenital naevi and melanoma. Nat Cell Biol 2012;14(8):882–90.

[138] Ferletta M, Uhrbom L, Olofsson T, Ponten F, Westermark B. Sox10 has a broad expression pattern in gliomas and enhances platelet-derived growth factor-B--induced gliomagenesis. Mol Cancer Res 2007;5 (9):891–7.

[139] Segal NH, Pavlidis P, Noble WS, Antonescu CR, Viale A, Wesley UV, et al. Classification of clear-cell sarcoma as a subtype of melanoma by genomic profiling. J Clin Oncol 2003;21(9):1775–81.

[140] Karamchandani JR, Nielsen TO, van de Rijn M, West RB. Sox10 and S100 in the diagnosis of soft-tissue neoplasms. Appl Immunohistochem Mol Morphol 2012;20(5):445–50.

[141] Ross DT, Scherf U, Eisen MB, et al. Systematic variation in gene expression patterns in human cancer cell lines. Nat Genet 2000;24(3):227–35.

[142] Bittner M, Meltzer P, Chen Y, Jiang Y, Seftor E, Hendrix M, et al. Molecular classification of cutaneous malignant melanoma by gene expression profiling. Nature 2000;406(6795):536–40.

[143] Gupta PB, Kuperwasser C, Brunet JP, Ramaswamy S, Kuo WL, Gray JW, et al. The melanocyte differentiation program predisposes to metastasis after neoplastic transformation. Nat Genet 2005;37 (10):1047–54.

[144] Hayry V, Salmenkivi K, Arola J, Heikkila P, Haglund C, Sariola H. High frequency of SNAIL-expressing cells confirms and predicts metastatic potential of phaeochromocytoma. Endocr Relat Cancer 2009;16 (4):1211–8.

[145] Waldmann J, Slater EP, Langer P, Buchholz M, Ramaswamy A, Walz MK, et al. Expression of the transcription factor snail and its target gene twist are associated with malignancy in pheochromocytomas. Ann Surg Oncol 2009;16(7):1997–2005.

[146] Weiss MB, Abel EV, Mayberry MM, Basile KJ, Berger AC, Aplin AE. TWIST1 Is an ERK1/2 Effector that promotes invasion and regulates MMP-1 expression in human melanoma cells. Cancer Res 2012;72 (24):6382–92.

[147] Loriot C, Burnichon N, Gadessaud N, Vescovo L, Amar L, Libe R, et al. Epithelial to mesenchymal transition is activated in metastatic pheochromocytomas and paragangliomas caused by SDHB gene mutations. J Clin Endocrinol Metab 2012;97(6): E954–62.

[148] Hsu MY, Wheelock MJ, Johnson KR, Herlyn M. Shifts in cadherin profiles between human normal melanocytes and melanomas. J Investig Dermatol Symp Proc 1996;1(2):188–94.

[149] Danen EH, de Vries TJ, Morandini R, Ghanem GG, Ruiter DJ, van Muijen GN. E-cadherin expression in human melanoma. Melanoma Res 1996;6(2):127–31.

[150] Hu-Lieskovan S, Zhang J, Wu L, Shimada H, Schofield DE, Triche TJ. EWS-FLI1 fusion protein up-regulates critical genes in neural crest development and is responsible for the observed phenotype of Ewing's family of tumors. Cancer Res 2005;65(11):4633–44.

[151] Baldwin RM, Parolin DA, Lorimer IA. Regulation of glioblastoma cell invasion by PKC iota and RhoB. Oncogene 2008;27(25):3587–95.

[152] Danen EH, Ten Berge PJ, Van Muijen GN, Van 't Hof-Grootenboer AE, Brocker EB, Ruiter DJ. Emergence of alpha 5 beta 1 fibronectin- and alpha v beta 3 vitronectin-receptor expression in melanocytic tumour progression. Histopathology 1994;24 (3):249–56.

[153] Hofmann UB, Westphal JR, Waas ET, Becker JC, Ruiter DJ, van Muijen GN. Coexpression of integrin alpha(v)beta3 and matrix metalloproteinase-2 (MMP-2) coincides with MMP-2 activation: correlation with melanoma progression. J Invest Dermatol 2000;115 (4):625–32.

[154] Brooks PC, Stromblad S, Sanders LC, et al. Localization of matrix metalloproteinase MMP-2 to

the surface of invasive cells by interaction with integrin alpha v beta 3. Cell 1996;85(5):683−93.

[155] Forsyth PA, Wong H, Laing TD, Rewcastle NB, Morris DG, Muzik H, et al. Gelatinase-A (MMP-2), gelatinase-B (MMP-9) and membrane type matrix metalloproteinase-1 (MT1-MMP) are involved in different aspects of the pathophysiology of malignant gliomas. Br J Cancer 1999;79(11−12):1828−35.

[156] Forsyth PA, Wong H, Laing TD, Rewcastle NB, Morris DG, Muzik H, et al. Localization of gelatinase-A and gelatinase-B mRNA and protein in human gliomas. Neuro-oncology 2000;2(3):145−50.

[157] Rao JS, Steck PA, Mohanam S, Stetler-Stevenson WG, Liotta LA, Sawaya R. Elevated levels of M(r) 92,000 type IV collagenase in human brain tumors. Cancer Res 1993;53(10 Suppl):2208−11.

[158] Sawaya RE, Yamamoto M, Gokaslan ZL, Wang SW, Mohanam S, Fuller GN, et al. Expression and localization of 72 kDa type IV collagenase (MMP-2) in human malignant gliomas in vivo. Clin Exp Metastasis 1996;14(1):35−42.

[159] Brodeur GM, Seeger RC, Schwab M, Varmus HE, Bishop JM. Amplification of N-myc in untreated human neuroblastomas correlates with advanced disease stage. Science 1984;224(4653):1121−4.

[160] Schwab M, Alitalo K, Klempnauer KH, Varmus HE, Bishop JM, Gilbert F, et al. Amplified DNA with limited homology to myc cellular oncogene is shared by human neuroblastoma cell lines and a neuroblastoma tumour. Nature 1983;305(5931):245−8.

[161] Seeger RC, Brodeur GM, Sather H, Dalton A, Siegel SE, Wong KY, et al. Association of multiple copies of the N-myc oncogene with rapid progression of neuroblastomas. N Engl J Med 1985;313(18):1111−6.

[162] Hattori Y, Kanamoto N, Kawano K, Iwakura H, Sone M, Miura M, et al. Molecular characterization of tumors from a transgenic mouse adrenal tumor model: comparison with human pheochromocytoma. Int J Oncol 2010;37(3):695−705.

[163] Maher EA, Furnari FB, Bachoo RM, Rowitch DH, Louis DN, Cavenee WK, et al. Malignant glioma: genetics and biology of a grave matter. Genes Dev 2001;15(11):1311−33.

[164] Liu Y, Ye F, Yamada K, Tso JL, Zhang Y, Nguyen DH, et al. Autocrine endothelin-3/endothelin receptor B signaling maintains cellular and molecular properties of glioblastoma stem cells. Mol Cancer Res 2011;9(12):1668−85.

[165] Brodeur GM. Neuroblastoma: biological insights into a clinical enigma. Nat Rev Cancer 2003;3(3):203−16.

[166] Altenburger DL, Wagner AS, Eslin DE, Pearl GS, Pattisapu JV. A rare case of malignant pediatric ectomesenchymoma arising from the falx cerebri. J Neurosurg Pediatr 2011;7(1):94−7.

[167] Yamada K., Ohno T., Aoki H., Semi K., Watanabe A., Moritake H., et al. EWS/ATF1 expression induces sarcomas from neural crest-derived cells in mice. The Journal of clinical investigation 2013.

[168] Hendrix MJ, Seftor EA, Seftor RE, Kasemeier-Kulesa J, Kulesa PM, Postovit LM. Reprogramming metastatic tumour cells with embryonic microenvironments. Nat Rev Cancer 2007;7(4):246−55.

[169] Levy C, Khaled M, Fisher DE. MITF: master regulator of melanocyte development and melanoma oncogene. Trends Mol Med 2006;12(9):406−14.

[170] Salti GI, Manougian T, Farolan M, Shilkaitis A, Majumdar D, Das Gupta TK. Micropthalmia transcription factor: a new prognostic marker in intermediate-thickness cutaneous malignant melanoma. Cancer Res 2000;60(18):5012−6.

[171] Takeuchi H, Kuo C, Morton DL, Wang HJ, Hoon DS. Expression of differentiation melanoma-associated antigen genes is associated with favorable disease outcome in advanced-stage melanomas. Cancer Res 2003;63(2):441−8.

[172] Garraway LA, Widlund HR, Rubin MA, Getz G, Berger AJ, Ramaswamy S, et al. Integrative genomic analyses identify MITF as a lineage survival oncogene amplified in malignant melanoma. Nature 2005;436(7047):117−22.

[173] McGill GG, Horstmann M, Widlund HR, Du J, Motyckova G, Nishimura EK, et al. Bcl2 regulation by the melanocyte master regulator Mitf modulates lineage survival and melanoma cell viability. Cell 2002;109(6):707−18.

[174] Miller AJ, Mihm Jr. MC. Melanoma. N Engl J Med 2006;355(1):51−65.

[175] Chudnovsky Y, Khavari PA, Adams AE. Melanoma genetics and the development of rational therapeutics. J Clin Invest 2005;115(4):813−24.

[176] Uong A, Zon LI. Melanocytes in development and cancer. J Cell Physiol 2010;222(1):38−41.

[177] Potterf SB, Mollaaghababa R, Hou L, Southard-Smith EM, Hornyak TJ, Arnheiter H, et al. Analysis of SOX10 function in neural crest-derived melanocyte development: SOX10-dependent transcriptional control of dopachrome tautomerase. Dev Biol 2001;237 (2):245−57.

[178] Mollaaghababa R, Pavan WJ. The importance of having your SOX on: role of SOX10 in the development of neural crest-derived melanocytes and glia. Oncogene 2003;22(20):3024−34.

[179] Rezze GG, Fregnani JH, Duprat J, Landman G. Cell adhesion and communication proteins are

differentially expressed in melanoma progression model. Hum Pathol 2011;42(3):409–18.

[180] Lee HO, Levorse JM, Shin MK. The endothelin receptor-B is required for the migration of neural crest-derived melanocyte and enteric neuron precursors. Dev Biol 2003;259(1):162–75.

[181] Shin MK, Levorse JM, Ingram RS, Tilghman SM. The temporal requirement for endothelin receptor-B signalling during neural crest development. Nature 1999;402(6761):496–501.

[182] Lugassy C, Barnhill RL. Angiotropic melanoma and extravascular migratory metastasis: a review. Adv Anatom Pathol 2007;14(3):195–201.

[183] Schwarz Q, Maden CH, Vieira JM, Ruhrberg C. Neuropilin 1 signaling guides neural crest cells to coordinate pathway choice with cell specification. Proc Natl Acad Sci USA 2009;106(15):6164–9.

[184] Nagy N, Mwizerwa O, Yaniv K, Carmel L, Pieretti-Vanmarcke R, Weinstein BM, et al. Endothelial cells promote migration and proliferation of enteric neural crest cells via beta1 integrin signaling. Dev Biol 2009;330(2):263–72.

[185] Lugassy C, Lazar V, Dessen P, van den Oord JJ, Winnepenninckx V, Spatz A, et al. Gene expression profiling of human angiotropic primary melanoma: selection of 15 differentially expressed genes potentially involved in extravascular migratory metastasis. Eur J Cancer 2011;47(8):1267–75.

[186] Kulesa PM, Kasemeier-Kulesa JC, Teddy JM, Margaryan NV, Seftor EA, Seftor RE, et al. Reprogramming metastatic melanoma cells to assume a neural crest cell-like phenotype in an embryonic microenvironment. Proc Natl Acad Sci USA 2006;103(10):3752–7.

[187] Lee LM, Seftor EA, Bonde G, Cornell RA, Hendrix MJ. The fate of human malignant melanoma cells transplanted into zebrafish embryos: assessment of migration and cell division in the absence of tumor formation. Dev Dyn 2005;233(4):1560–70.

[188] Bailey CM, Morrison JA, Kulesa PM. Melanoma revives an embryonic migration program to promote plasticity and invasion. Pigment Cell Melanoma Res 2012;25(5):573–83.

[189] Bronner-Fraser M. Effects of different fragments of the fibronectin molecule on latex bead translocation along neural crest migratory pathways. Dev Biol 1985;108(1):131–45.

[190] Ramgolam K, Lauriol J, Lalou C, Lauden L, Michel L, de la Grange P, et al. Melanoma spheroids grown under neural crest cell conditions are highly plastic migratory/invasive tumor cells endowed with immunomodulator function. PloS One 2011;6(4): e18784.

[191] Ghislin S, Deshayes F, Lauriol J, Middendorp S, Martins I, Al-Daccak R, et al. Plasticity of melanoma cells induced by neural cell crest conditions and three-dimensional growth. Melanoma Res 2012;22 (3):184–94.

[192] Gurney JG, Ross JA, Wall DA, Bleyer WA, Severson RK, Robison LL. Infant cancer in the U.S.: histology-specific incidence and trends, 1973 to 1992. J Pediatr Hematol Oncol 1997;19(5):428–32.

[193] Wakamatsu Y, Watanabe Y, Nakamura H, Kondoh H. Regulation of the neural crest cell fate by N-myc: promotion of ventral migration and neuronal differentiation. Development 1997;124(10):1953–62.

[194] Nelms BL, Labosky PA. Transcriptional control of neural crest development. San Rafael, CA; 2010.

[195] Henion PD, Garner AS, Large TH, Weston JA. trkC-mediated NT-3 signaling is required for the early development of a subpopulation of neurogenic neural crest cells. Dev Biol 1995;172(2):602–13.

[196] Martin-Zanca D, Barbacid M, Parada LF. Expression of the trk proto-oncogene is restricted to the sensory cranial and spinal ganglia of neural crest origin in mouse development. Genes Dev 1990;4(5):683–94.

[197] Chalazonitis A, Rothman TP, Chen J, Lamballe F, Barbacid M, Gershon MD. Neurotrophin-3 induces neural crest-derived cells from fetal rat gut to develop in vitro as neurons or glia. J Neurosci 1994;14(11 Pt 1):6571–84.

[198] Rao MS, Anderson DJ. Immortalization and controlled in vitro differentiation of murine multipotent neural crest stem cells. J Neurobiol 1997;32(7):722–46.

[199] Goodenberger ML, Jenkins RB. Genetics of adult glioma. Cancer Genet 2012;205(12):613–21.

[200] Louis DN, Ohgaki H, Wiestler OD, Cavenee WK, Burger PC, Jouvet A, et al. The 2007 WHO classification of tumours of the central nervous system. Acta Neuropathol 2007;114(2):97–109.

[201] Dolecek TA, Propp JM, Stroup NE, Kruchko C. CBTRUS statistical report: primary brain and central nervous system tumors diagnosed in the United States in 2005–2009. Neuro-oncology 2012;14(Suppl 5):v1–49.

[202] Smith CL, Tallquist MD. PDGF function in diverse neural crest cell populations. Cell Adh Migr 2010;4 (4):561–6.

[203] Garcez RC, Teixeira BL, Schmitt Sdos S, Alvarez-Silva M, Trentin AG. Epidermal growth factor (EGF) promotes the in vitro differentiation of neural crest cells to neurons and melanocytes. Cell Mol Neurobiol 2009;29(8):1087–91.

[204] Rao JS. Molecular mechanisms of glioma invasiveness: the role of proteases. Nat Rev Cancer 2003;3 (7):489–501.

[205] Anderson RB. Matrix metalloproteinase-2 is involved in the migration and network formation of enteric neural crest-derived cells. Int J Dev Biol 2010;54(1):63—9.

[206] Duong TD, Erickson CA. MMP-2 plays an essential role in producing epithelial-mesenchymal transformations in the avian embryo. Dev Dyn 2004;229(1):42—53.

[207] Stone JG, Spirling LI, Richardson MK. The neural crest population responding to endothelin-3 in vitro includes multipotent cells. J Cell Sci 1997;110(Pt 14):1673—82.

[208] Lahav R, Ziller C, Dupin E, Le Douarin NM. Endothelin 3 promotes neural crest cell proliferation and mediates a vast increase in melanocyte number in culture. Proc Natl Acad Sci USA 1996;93(9):3892—7.

[209] Lahav R, Dupin E, Lecoin L, Glavieux C, Champeval D, Ziller C, et al. Endothelin 3 selectively promotes survival and proliferation of neural crest-derived glial and melanocytic precursors in vitro. Proc Natl Acad Sci USA 1998;95(24):14214—9.

[210] Nagy N, Goldstein AM. Endothelin-3 regulates neural crest cell proliferation and differentiation in the hindgut enteric nervous system. Dev Biol 2006;293 (1):203—17.

[211] Lin PP, Wang Y, Lozano G. Mesenchymal stem cells and the origin of Ewing's sarcoma. Sarcoma 2011;.

[212] Wahl J, Bogatyreva L, Boukamp P, Rojewski M, van Valen F, Fiedler J, et al. Ewing's sarcoma cells with CD57-associated increase of tumorigenicity and with neural crest-like differentiation capacity. Int J Cancer 2010;127(6):1295—307.

[213] von Levetzow C, Jiang X, Gwye Y, von Levetzow G, Hung L, Cooper A, et al. Modeling initiation of Ewing sarcoma in human neural crest cells. PloS One 2011;6(4):e19305.

[214] Holimon JL, Rosenblum WI. "Gangliorhabdomyosarcoma": a tumor of ectomesenchyme. Case report. J Neurosurg 1971;34(3):417—22.

[215] Naka A, Matsumoto S, Shirai T, Ito T. Ganglioneuroblastoma associated with malignant mesenchymoma. Cancer 1975;36(3):1050—6.

[216] Freitas AB, Aguiar PH, Miura FK, Yasuda A, Soglia J, Soglia F, et al. Malignant ectomesenchymoma. Case report and review of the literature. Pediatr Neurosur 1999;30(6):320—30.

[217] Mackey SL, Hebel J, Cobb MW. Melanoma of the soft parts (clear cell sarcoma): a case report and review of the literature. J Am Acad Dermatol 1998;38 (5 Pt 2):815—9.

[218] Dim DC, Cooley LD, Miranda RN. Clear cell sarcoma of tendons and aponeuroses: a review. Arch Pathol Lab Med 2007;131(1):152—6.

[219] Mii Y, Miyauchi Y, Hohnoki K, Maruyama H, Tsutsumi M, Dohmae K, et al. Neural crest origin of clear cell sarcoma of tendons and aponeuroses. Ultrastructural and enzyme cytochemical study of human and nude mouse-transplanted tumours. Virchows Archiv Pathol Anat Histopathol 1989;415 (1):51—60.

[220] Kindblom LG, Lodding P, Angervall L. Clear-cell sarcoma of tendons and aponeuroses. An immunohistochemical and electron microscopic analysis indicating neural crest origin. Virchows Archiv Pathol Anat Histopathol 1983;401(1):109—28.

[221] Zucman J, Delattre O, Desmaze C, Epstein AL, Stenman G, Speleman F, et al. EWS and ATF-1 gene fusion induced by t(12;22) translocation in malignant melanoma of soft parts. Nat Genet 1993;4(4):341—5.

[222] Szabo PM, Pinter M, Szabo DR, Zsippai A, Patocs A, Falus A, et al. Integrative analysis of neuroblastoma and pheochromocytoma genomics data. BMC Med Genom 2012;5:48.

[223] Carroll SL. Molecular mechanisms promoting the pathogenesis of Schwann cell neoplasms. Acta Neuropathol 2012;123(3):321—48.

TISSUE ENGINEERING AND REPAIR

Neurocristopathies: The Etiology and Pathogenesis of Disorders Arising from Defects in Neural Crest Cell Development

Kristin E. Noack Watt[a,b] and Paul A. Trainor[a,b]

[a]Stowers Institute for Medical Research, Kansas City, MO 64110, [b]Department of Anatomy and Cell Biology, University of Kansas School of Medicine, Kansas City, KS 66160

OUTLINE

Neural Crest Cells.
DOI: http://dx.doi.org/10.1016/B978-0-12-401730-6.00018-1

INTRODUCTION

Neural Crest Cell Formation

Neural crest cells (NCC) are induced during embryonic development at the border between the neural ectoderm and non-neural ectoderm, which gives rise to the central nervous system and epidermis, respectively. Cell lineage tracing has indicated that both neural ectoderm and non-neural ectoderm give rise to NCC [1], but the vast majority appear to come from the neural ectoderm. Explants of neural ectoderm, however, do not typically give rise to NCC endogenously. The formation of NCC therefore is a multistep process, requiring contact mediated tissue interactions between neural and non-neural ectoderm [1–3] in concert with a complex series of molecular signals including a gradient of bone morphogenetic protein (BMP). Whereas low levels of BMP lead to neural plate formation and high levels of BMP specify epidermis, intermediate levels of BMP are required for NCC induction [4]. However, BMP alone is not sufficient to induce NCC [5]. Additional signals involved in NCC formation include Wnts, fibroblast growth factors (Fgfs), and retinoic acid (RA), but the timing, location, and actual requirement for these signals varies between organisms [4,6–8]. For example, recently it was provocatively proposed that NCC in avian embryos are specified by Pax7 during early gastrulation [9]. Furthermore, this process is driven by Fgf signaling emanating from the non-neural ectoderm [10]. Fgf signaling mediated Pax gene activity also plays a role in NCC formation in Xenopus [11], however the Fgf signal comes from the underlying mesoderm [12]. While these signaling pathways are generally associated with NCC induction in avian and aquatic species, to date an unequivocal role for either BMP, Wnt, Fgf, or RA signaling in mammalian NCC induction has not been observed in loss-of-function or inhibition assays [13]. Rather, these signals

appear to be more important for specifying sublineages within the mammalian NCC. Nonetheless, the combination of BMP, Wnt, Fgf, and RA signals is required at the proper time, in the right place, and in the proper dose to establish competence to form NCC in aquatic and avian species. This period of competence coincides with the onset of expression of neural plate border specifier genes such as Pax3, Pax7, and Tfap2a, followed by more specific NCC specifier genes that include Snail1/2, Sox9/10, and Foxd3. These NCC specifier genes then directly regulate NCC effector genes that govern the delamination, migration, and differentiation of NCC (Figure 17.1). Perturbation of any of these signals may result in a disruption of NCC development leading to the pathogenesis of specific neurocristopathies (outlined in Table 17.1).

Treacher Collins Syndrome (Mandibulofacial Dystosis)

A specific example of a neurocristopathy that results from perturbation of NCC formation is Treacher Collins syndrome (TCS). TCS is a congenital craniofacial disorder characterized by hypoplasia of the mandible and zygomatic complex, cleft palate, downward-sloping palpebral fissures, colobomas (notches) of the lower eyelids, and abnormalities of the external ears and middle ear ossicles that may result in conductive hearing loss (Figure 17.2A and B). However, there is significant variation in phenotypic severity among individuals diagnosed with TCS. Mutations in three different genes are currently known to cause TCS: TCOF1, POLR1C, and POLR1D. Disruption of TCOF1 causes TCS in an autosomal dominant manner [14–17]. Mutations in POLR1C and POLR1D cause TCS in an autosomal recessive and autosomal dominant manner, respectively [18].

TCOF1 encodes the nucleolar phosphoprotein Treacle, which localizes to the nucleolus

Induction	Migration	Neuronal differentiation	Cartilage and bone differentiation

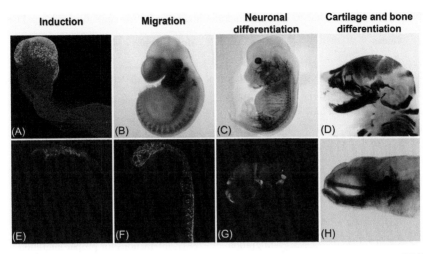

FIGURE 17.1   **Stages of NCC development in mice and zebrafish.** (A) Wnt1-Cre YFP mouse at E8.5 showing induction and early migration of cranial NCC. Green labels NCC, blue is DAPI. (B) E10.5 mouse showing migration of lacZ stained cranial NCC into the frontonasal process and pharyngeal arches, cardiac NCC toward the heart, and trunk NCC into the dorsal root ganglia. (C) NCC differentiation into the peripheral nervous system in an E14.5 mouse. (D) NCC differentiation into bone, stained with alizarin red, and cartilage, stained with alcian blue, of the craniofacial skeleton of an E18.5 mouse. (E) 14 hpf sox10:GFP zebrafish showing induction and early migration of NCC. (F) 25 hpf sox10:GFP zebrafish showing migration of NCC in the head, pharyngeal arches. (G) NCC differentiation into cranial ganglia in a 48 hpf zebrafish stained for HuC. (H) NCC differentiation into cartilage, stained with alcian blue, and bone, stained with alizarin red, in a 5 dpf zebrafish.

[19,20] where it interacts with Upstream Binding Factor (UBF) to stimulate transcription of ribosomal DNA by RNA Polymerase I [21]. POLR1C and POLR1D are subunits of RNA Polymerase I and III and interact to form a dimer [22,23]. RNA polymerases I and III are responsible for transcribing ribosomal RNA (rRNA) genes, which are processed and incorporated into the ribosome. Thus, all three genes implicated in TCS have roles in ribosome biogenesis.

Ribosome biogenesis is important for cell growth and proliferation as ribosomes perform essential functions in translating messenger RNA (mRNA) into polypeptides in the cell. Ribosomes consist of four rRNAs and over 70 ribosomal proteins (RPs), as well as additional factors [24,25]. RNA polymerase I (Pol I) transcribes the 28S, 18S, and 5.8S rRNAs, while RNA polymerase III transcribes the 5S rRNA, and each process occurs in the nucleolus.

A eukaryotic ribosome consists of two subunits—the 40S (small subunit) and the 60S (large subunit)—in which RPs are assembled onto the rRNA in the nucleolus. Once the pre-40S and pre-60S have formed, they are exported to the cytoplasm where the mature ribosome both decodes the mRNA and forms the peptide bonds between amino acids [24]. The process of ribosome biogenesis can be divided into several steps: transcription of ribosomal DNA, covalent modifications to the pre-rRNA, processing of the pre-rRNA to form mature rRNA, assembly of the rRNA and RPs, and transport of the maturing ribosomes to the cytoplasm [24,26] (Figure 17.3). Disruption of any of these steps leads to disorders termed ribosomopathies.

Treacle plays a role in stimulating the transcription of rRNA, and may also have a role in covalent modification of pre-rRNA, specifically in the process of 2'-O-methylation [27].

TABLE 17.1   Neurocristopathies: Defects in the Formation, Migration or Differentiation of Neural Crest Cells

Name	Genes	Role in Neural Crest	Mechanisms
Treacher Collins	*TCOF1, POLR1C, POLR1D*	Formation	NCC survival; deficient ribosome biogenesis
Diamond Blackfan anemia	*RPS17, RPS19, RPS24, RPL5, RPL11, RPL35A*	Formation	Role in ribosome biogenesis
Mandibulofacial dystosis	*EFTUD2*	Formation	Component of the spliceosome complex which produces mature mRNAs; cell proliferation
Miller	*DHODH*	Formation	DHODH is involved in uracil synthesis; cell proliferation
Nager	*SF3B4*	Formation	SF3B4 is a component of pre-mRNA spliceosomal complex; important for cell proliferation
Waardenburg syndrome	*PAX3*	Formation	Melanoblast proliferation
	SNAIL2, EDNRB, EDN3, SOX10	Migration	EMT, interactions for proper migration
Branchio-oculo-facial	*TFAP2A*	Migration	Role in EMT
Mowat−Wilson	*SIP1*	Migration	Role in EMT
Auriculo-condylar	*PLCB4, GNAI3*	Migration	Edn1 and Dlx signaling
DiGeorge	*TBX1*	Migration	Signaling from the endoderm to NCC
Craniofrontonasal dysplasia	*EFNB1*	Migration	Mixing at neural crest-mesoderm border
Familial dysautonomia	*IKBKAP*	Migration	IKBKAP is involved in forming a complex which is important for transcription of proteins that affect the cytoskeleton and cell movement
Piebaldism	*KIT, SNAIL2*	Migration	Role in EMT (Snail2), melanocyte migration and survival (Kit)
Hirschsprung	*RET, GDNF, NTN, EDNRB, EDN3, SOX10, SIP1, L1CAM*	Migration	Signals for NCC guidance during migration; also roles in survival and proliferation
	PHOX2B, HASH1, HAND2	Differentiation	Neuronal differentiation
Congenital central hypoventilation	*PHOX2B* and others (see text)	Differentiation	Neuronal differentiation
Craniosynostosis	*FGFR1, FGFR2, FGFR3, TWIST1, MSX2*	Differentiation	Premature differentiation of osteoblasts
Bamforth-Lazarus	*FOXE1*	Differentiation	Role in chondrogenesis; interacts with MSX1 and TGFβ3
Oculocutaneous albinism	*TYR, OCA2, TRP1, MATP*	Differentiation	Inability to produce melanin
Branchio-oto-renal	*EYA1, SIX1, SIX5*	Formation and Differentiation	Pouch patterning, regulation of placodal neuronal progenitor proliferation and subsequent neuronal differentiation; Six1 has a role in NCC formation

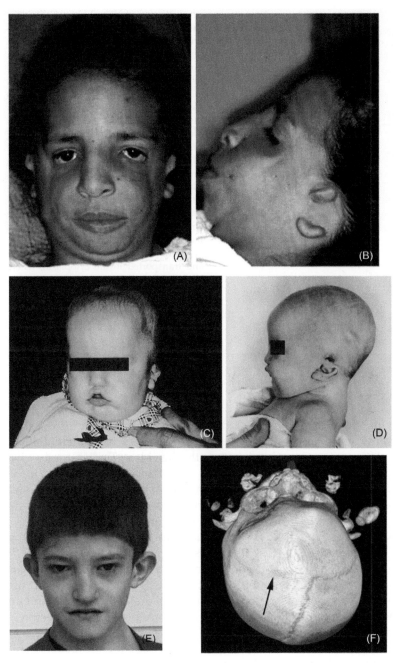

FIGURE 17.2 **Craniofacial phenotypes as a result of perturbation of NCC development.** (A and B) TCS; adapted from [237]. (C and D) BOFS; adapted from [238]. (E) DiGeorge syndrome; adapted from [239]. (F) CS. Arrow points to fused left coronal suture; compare to unfused suture on the right. Adapted from [240].

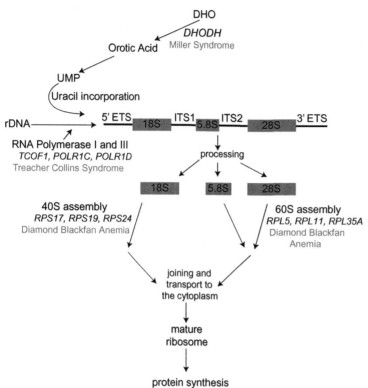

FIGURE 17.3 **The process of ribosome biogenesis and associated neurocristopathies.** Ribosome biogenesis involves the production of rRNA by RNA polymerases I and III, a process disrupted by mutations in *TCOF1*, *POLR1C*, and *POLR1D* in TCS. Production of rRNA also requires uracil, and production of uracil is disrupted by mutations in *DHODH* in Miller syndrome. After the pre-rRNA is synthesized by RNA Polymerases I and III, it generates a transcript that is then processed into the mature rRNA. RPs assemble with the rRNA to form the small (40S) and large (60S) subunits of the ribosome. Mutations in the RP genes of the large and small subunits disrupt ribosome biogenesis and cause DBA. After synthesis of the ribosomal subunits, they are joined and transported to the cytoplasm, where further maturation occurs, and then function to synthesize proteins in the cell.

Consistent with this, NOP56, which is a component of a small nucleolar ribonucleoprotein complex that directs 2'-O-methylation of pre-rRNA, has previously been shown to directly interact with Treacle [27,28]. Furthermore, $Tcof1^{+/-}$ mice exhibit decreased levels of 18S rRNA methylation compared to wild-type mice. The role of *Tcof1* in NCC development has been well studied in mice. During embryonic development, *Tcof1* is expressed in a dynamic manner, with particularly strong expression in the neural ectoderm between embryonic day (E) 8.5–9.5, which corresponds with the spatiotemporal generation of NCC. Additionally, *Tcof1* is expressed strongly at E9.5 [29] in the NCC-derived frontonasal and pharyngeal arch mesenchyme, which presage formation of much of the craniofacial skeleton. These expression patterns imply a role for *Tcof1*

in NCC development, particularly during formation and possibly early differentiation. Consistent with this idea, $Tcof1^{+/-}$ mice display severe craniofacial defects including hypoplastic frontal, nasal, and maxilla bones as well as cleft palate which mimic the severe form of TCS in humans [29]. Thus $Tcof1^{+/-}$ mice serve as an important model for studying the pathogenesis of TCS.

From around E7.5–10.5, $Tcof1^{+/-}$ embryos exhibit elevated levels of apoptosis in the neural ectoderm together with a concomitant decrease in proliferation [29]. Collectively this leads to the diminished production of NCC, which can account for the general cranioskeletal hypoplasia present in $Tcof1^{+/-}$ embryos. Microarray comparisons between wild-type and $Tcof1^{+/-}$ embryos subsequently revealed that the apoptosis present in the neural ectoderm of $Tcof1^{+/-}$ embryos, was p53 dependent [30]. Interestingly, Treacle is known to play important roles in ribosome biogenesis and disruption of ribosome biogenesis causes nucleolar stress activation of p53 [31]. Consistent with this, $Tcof1^{+/-}$ embryos show diminished mature ribosome production. Furthermore, pharmacological inhibition of p53 via *in utero* treatment of pregnant mice with pifithrin-α, or via genetic suppression with $p53^{+/-}$ mice, can reduce neuroepithelial apoptosis, restore the NCC population, and prevent the pathogenesis of TCS in $Tcof1^{+/-}$ mice [30].

Collectively, this suggests that haploinsufficiency of Tcof1 perturbs the process of ribosome biogenesis, leading to diminished mature ribosome generation that is insufficient to meet the proliferative needs of the neural ectoderm. At the same time, perturbed ribosome biogenesis causes nucleolar stress activation of p53 and the apoptotic elimination of progenitor NCC, which manifests as cranioskeletal anomalies in $Tcof1^{+/-}$ mice [30]. Thus by extrapolation, perturbed ribosome biogenesis and the diminished production of NCC underlies the pathogenesis of TCS in humans. TCS can therefore be classified as both a ribosomopathy and neurocristopathy.

POLR1C or *POLR1D* are subunits of RNA polymerases I and III and logically would be presumed to be required in all cells at all times during embryonic and adult life. In agreement with this, yeast strains with mutations in *POLR1C* or *POLR1D* are lethal [22,32]. In contrast, human mutations in *POLR1C* or *POLR1D* result in craniofacial anomalies that are characteristic of TCS, suggesting that in more complex organisms, RNA polymerase I and III transcription may be differentially regulated in distinct tissues [33]. The roles of POLR1C and POLR1D in NCC development are currently not known. However, the phenotypic similarity arising from *TCOF1*, *POLR1C*, and *POLR1D* mutations in humans in association with TCS

is suggestive of similar key roles in ribosome biogenesis. Consistent with this idea, the three key processes in ribosome biogenesis—pre-rRNA transcription, rRNA processing, and ribosomal ribonucleoprotein assembly—all occur in the nucleolus [34]. Therefore, the demonstration of deficient ribosome biogenesis, nucleolar stress activation of p53, and diminished production of NCC in *POLR1C* and *POLR1D* loss-of-function animal models is highly anticipated as part of a unified mechanism for understanding the etiology and pathogenesis of TCS.

Diamond Blackfan Anemia

The three genes involved in TCS are all involved in the process of ribosome biogenesis, which leads us to consider the role of NCC formation in other ribosomopathies such as Diamond Blackfan anemia (DBA). DBA is characterized by anemia, reticulocytopenia, macrocytosis, and a selective decrease or absence of erythroid precursors [35]. However, a range of craniofacial and cardiac defects as well as thumb abnormalities are observed in 40−62% of patients [35]. The craniofacial phenotype of DBA has considerable overlap with TCS and can include cleft palate and microtia. Unlike TCS, which exhibits deficient production of rRNA, DBA is caused by mutations in RPs, including *RPS17*, *RPS19*, *RPS24*, *RPL5*, *RPL11*, and *RPL35A*, (reviewed in [35,36]). However, in about half of patients with DBA, the genetic mutation remains unknown.

Mutations in *RPL5* and *RPL11* are associated with craniofacial and cardiac anomalies more often than mutations in the other RPs linked to DBA [37], which suggests a specific role for RPL5 in NCC development and variable tissue-specific roles for individual RPs. For example, a zebrafish mutant in *rpl11* exhibits metabolic and hematopoietic defects [38], as well as diminished expression of neurogenic markers [39]. The induction of NCC was not investigated in these models; however, given the diminished expression of neurogenic markers in *rpl11* zebrafish morphants, it seems possible that NCC induction could be impaired by the significant apoptosis observed in these embryos. Inhibition of p53 prevented the morphant *rpl11* phenotype [39] and was also successful in preventing the phenotype in mice and zebrafish with mutations in *rps19* [40,41].

Recently, L-leucine treatment has been used to improve the phenotype in mouse and zebrafish models of DBA [42,43]. In Rps19-deficient mice, L-leucine administered in the drinking water improved hematopoiesis and also led to a down-regulation of p53 [43]. When zebrafish embryos injected with *rps14* or *rps19* morpholinos (MO) were developed in media supplemented with 100 mM L-leucine, this treatment was able to ameliorate craniofacial defects [42]. L-leucine supplementation has also been used to treat anemia in some DBA patients [44,45]. Amino acids such as L-leucine are known to stimulate TORC1 (target of rapamycin complex 1, also known as mTOR), which is an important regulator of ribosome biogenesis, cell growth, and cell proliferation. The mechanism by which L-leucine signals to TORC1 was recently elucidated in yeast and mammalian cells. Leucyl-tRNA synthetase (LeuRS), which charges the tRNA with leucine, functions to signal leucine availability to TORC1 [42,46]. TORC1 then acts to stimulate ribosome biogenesis acting on the RNA Polymerases and protein translation. TORC1 will phosphorylate proteins of the RNA Pol I initiation complex to stimulate transcription of rDNA by RNA Pol I [47,48] and also controls phosphorylation of Maf1, which is an important regulator of RNA Pol III [49,50]. Phosphorylation of 4E-BP1 by TORC1 releases it from the translation initiation complex, allowing for translation of RPs [51].

Although the impact of L-leucine supplementation on NCC has not been properly investigated, nutritional supplementation with L-leucine could be a promising treatment option for other neurocristopathies that have an underlying deficiency in ribosome biogenesis as a part of their pathogenesis. The craniofacial anomalies observed in DBA therefore likely have a similar pathogenesis to TCS. Disruption of RPs affects ribosome biogenesis (Figure 17.3), leading to nucleolar stress activation of p53, neuroepithelial cell death and diminished NCC formation and proliferation. Investigation of NCC development in specific animal models of DBA is needed in the future to better understand the potential roles of RPs in NCC and to provide a deeper understanding of the role of NCC in the pathogenesis of DBA.

Miller Syndrome (Genee–Weidemann; Wildervanck–Smith; Postaxial Acrofacial Dystosis)

Miller syndrome is characterized by a combination of craniofacial anomalies including orofacial clefts, micrognathia, malar hypoplasia, cup-shaped ears, and coloboma of the lower eyelid, plus postaxial limb deformities. Whole-exome sequencing identified autosomal recessive mutations in *dihydroorotate dehydrogenase* (*DHODH*) in association with the etiology of Miller syndrome [52]. DHODH is an enzyme involved in *de novo* pyrimidine biosynthesis, catalyzing the oxidation of DHO to orotic acid, which is then converted to uracilmonophosphate. Uracil is one of the bases of RNA, which is integral to ribosome biogenesis (Figure 17.3). DHODH activity in Miller syndrome—associated alleles is diminished, consistent with there being a threshold of activity for DHODH. As with the genes mutated in association with TCS and DBA, it is surprising that mutations in *DHODH* would lead to such a specific phenotype, as uracil and RNA synthesis are global processes. However, *Dhodh in situ* hybridization in E10.5 mouse embryos revealed a spatiotemporal requirement for *Dhodh* in the pharyngeal arches, forelimbs, hindlimbs, and somites during embryonic development [53]. These are the embryonic precursors of the tissues affected in Miller syndrome. Mutations in *DHODH* may result in a rate limiting effect on cell division during specific stages of facial and limb development. While there is not yet a genetic model to study Miller syndrome, interesting results have been obtained by using the DHODH inhibitor, leflunomide. Treatment of zebrafish embryos with leflunomide leads to disruption of NCC development. Specifically, markers of NCC such as *foxd3*, *sox10*, and *dct* were either reduced (*foxd3*) or absent (*sox10*, *dct*), which led to an almost complete loss of melanocyte progenitors, together with reduced numbers of glial cells [54]. Other tissues were not affected which suggests DHODH may play a specific functional role in NCC formation. Further studies are necessary to determine if the formation of NCC is affected by any limit on cell division or if there are alternate functions of DHODH that may impact other steps of NCC development such as survival, migration, or differentiation.

Neural Crest Cell Migration

Progenitor NCC are initially integrated within the neural ectoderm where they are morphologically indistinguishable from the other neuroepithelial cells. However, following the activation of specifier genes, NCC undergo an epithelial-to-mesenchymal transformation (EMT) after which they delaminate from the neuroepithelium. This process involves the breakdown of tight junctions and adherens junctions between progenitor NCC cells in association with considerable changes in their cell adhesion properties through the downregulation of cingulin and alphaN-catenin, and

E-cadherin and N-cadherin, respectively [55]. The process is driven in large part by the Snail family of transcriptional repressors. Snail1 and Snail2 can directly repress the cell adhesion molecule E-cadherin by binding to its promoter, thus facilitating cell migration [56,57]. In contrast to avians, fish, and amphibians, a requirement for the *Snail* genes in the delamination and migration of mammalian NCC has not been demonstrated. Conditional loss-of-function analyses of *Snail1* and *Snail2* either individually or in combination does not inhibit the generation of migrating NCC in mice [58,59]. To date, only mutations in *Zfhx1b* (*SIP1*, *Zeb2*) have been shown to affect mammalian NCC delamination and migration. *Zfhx1b* is a repressor of E-cadherin and in the absence of *Zfhx1b*, E-cadherin mediated cell adhesion persists, preventing EMT [60]. As part of the EMT and delamination process, progenitor NCC concomitantly acquire polarity and the basal lamina of the neural ectoderm breaks down. Collectively, this facilitates the emergence of migrating NCC, which go on to colonize nearly the entire embryo, contributing to a large variety of cell and tissue types.

NCC emerge along nearly the entire anterior—posterior axis and are generally divided into distinct cranial, cardiac, vagal, and trunk populations. The cranial neural crest demonstrates astonishing multipotentiality giving rise to the majority of the bone and cartilage of the head and face, as well as to nerve ganglia, smooth muscle, connective tissue and pigment cells [61–63]. The most striking aspect of cranial NCC migration is the apparent segregation of frontonasal and individual pharyngeal arch populations from one another. These patterns are highly conserved in vertebrate species as disparate as amphibians, teleosts, avians, marsupials, and mammals [63–68] (see also Figure 17.1B and F). Briefly, forebrain and rostral midbrain NCC colonize the frontonasal and periocular regions. Caudal midbrain derived NCC populate the maxillary component of the first pharyngeal arch [62,63]. The hindbrain is transiently partitioned into seven contiguous segments called rhombomeres [69], and rhombomere-derived NCC form discrete segregated migrating streams that populate the first, second, third, and fourth pharyngeal arches [62,63]. The segregation of distinct cranial NCC populations is critical to prevent fusions of the cranial ganglia and skeletal elements and also to limit mixing of NCC with different genetic constitutions [61]. This process is collectively orchestrated by both cell-autonomous signals within NCC as well as non-cell-autonomous signals emanating from the surrounding ectoderm, mesoderm, and endoderm tissues with which the NCC interact during their migration.

Cardiac NCC arise from the neural ectoderm at the level of somites 1–3, migrate laterally, and differentiate into neurons, smooth muscle, and pericytes in the great arteries [70], as well as contributing to the endocardial cushions and septum of the heart [71]. Vagal NCC originate in the neural ectoderm at the levels of somites 1–7 and migrate laterally and ventrally to give rise to the vast array of neurons and glia that constitute the enteric nervous system (ENS) within the gastrointestinal tract. Colonization of the gastrointestinal tract represents the longest migratory path taken by any population of NCC.

Trunk NCC migration follows three distinct pathways—ventral, ventrolateral, and dorsolateral—and like cranial NCC migration, it is influenced by the segmented nature of the embryo. The trunk of the embryo is initially divided into somites. These are bilaterally paired segments of mesoderm that become subdivided dorsally into a dermomyotome and ventrally into a sclerotome that give rise to the reiterated pattern of the axial musculoskeleton. Trunk NCC initially migrate between the somites following the intersomitic blood vessels that may provide a source of chemotactic factors [72]. These NCC will eventually differentiate primarily into neurons and glia of the

sympathetic and parasympathetic ganglia as well as contribute to the adrenal gland [73]. NCC that follow the ventrolateral pathway initially invade the anterior half of the sclerotome. Those NCC that remain in the sclerotome will differentiate into sensory neurons and glia of the dorsal root ganglia, while NCC that pass through the sclerotome will contribute to the sympathetic ganglia [74,75]. The other pathway taken by the NCC is the dorsolateral path, between the epidermis and the dermomyotome [74,76]. These NCC will differentiate into melanocytes and other types of pigmented cells.

Several signaling molecules similarly guide the migration of both cranial and trunk NCC and can be classified as being repulsive (Semaphorin/Neuropilin), attractive (Sdf1b/Cxcr4), or both (Eph/Ephrin). For example, in avians, fish, and mammals, the Neuropilin receptors are expressed in migrating NCC, while the Semaphorin ligands are expressed in adjacent mesoderm. Disrupting Semaphorin/Neuropilin signaling interactions perturbs the segregated pattern of NCC migration leading to fused streams of NCC and consequently fused cranial ganglia and dorsal root ganglia [77–81]. Cxcr4-expressing NCC will differentiate into sensory neurons [82], and interactions with Sdf1 have an important role in directing trunk NCC cells to the dorsal root ganglion [82] and sympathetic ganglion [83]. Cxcr4/Sdf1 signaling also facilitates the migration of cardiac NCC into the heart [84]. Regardless of their attractive or repulsive nature, these receptor ligand interactions illustrate the involvement of bipartite, NCC autonomous and non-cell-autonomous signals in the regulation of NCC migration, the disruption of which can lead to the pathogenesis of specific neurocristopathies.

Branchio-Oculo-Facial Syndrome

Branchio-oculo-facial syndrome (BOFS) is a rare autosomal dominant disorder characterized by facial, ocular, and cutaneous anomalies [62]. Facial anomalies may include hypertelorism, broad nasal tip, upslanted palpebral fissures, cleft lip with or without cleft palate, malformed and prominent pinnae, and hearing loss from inner ear and/or petrous bone anomalies (Figure 17.2C and D). Ocular features may include microphthalmia, anophthalmia, coloboma, and nasolacrimal duct stenosis. The cutaneous defects range from thin skin or a patch of hair to erythematous lesions. The majority of individuals diagnosed with BOFS carry a mutation in transcription factor *TFAP2A* [85]. *TFAP2A* is expressed during multiple stages of NCC development, including formation and migration, and has been shown to have roles in NCC specification [86]. Mice carrying mutations in *Tfap2a* fail to close their neural tube and have numerous craniofacial anomalies associated with deficiencies of NCC-derived craniofacial cartilage and neurons [87,88]. Increased cell death, measured by TUNEL staining, in the neuroepithelial cells as well as migrating NCC may be the underlying reason for the severe craniofacial phenotype in *Tfap2a* mutant mice [88]. Similarly, reduced expression of *foxd3* in zebrafish *low* (*tfap2a*) mutants demonstrates a role for *tfap2a* in the formation of NCC. Cranial cartilage and neuronal defects also suggest a role for *tfap2a* in differentiation. Transplants from wild-type embryos into *low* mutants restored some cartilage; however, transplants from *low* mutants into wild-type embryos failed to migrate and did not form cartilage. This reveals both cell autonomous as well as non-cell-autonomous roles for *tfap2a* in NCC development [89]. *tfap2a* may be required in a cell-autonomous manner for NCC survival, migration, and contribution to the pharyngeal arches [89]. Later non-cell-autonomous roles for *tfap2a* may be important to induce skeletogenesis [90]. Consistent with this idea, another zebrafish *tfap2a* mutant, *mob^{m610}*, exhibits normal NCC induction and initiation of migration, but

subsequently the NCC stop migrating, are unable to differentiate and undergo apoptosis [91]. A subtractive gene expression screen of *Tfap2a* mutant versus wild-type mice indicated that *Tfap2a* may serve to promote proliferation by repressing genes involved in terminal differentiation (e.g., *Klf-4*) [92]. In humans, *TFAP2A* is expressed in migratory (but not premigratory) NCC at Carnegie stage 12 and continues to be expressed in the maxillary mesenchyme and surface ectoderm [93]. This suggests that *TFAP2A* plays a key role in the survival of migrating NCC as well as in their differentiation in the pathogenesis of BOFS.

Piebaldism

Piebaldism is an autosomal dominant condition in which melanocyte development is disrupted, resulting in patches of hair and skin, including areas on the midforehead, abdomen, and extremities, that lack pigmentation. Mutations in *KIT* [94] (mast-stem cell growth factor receptor) and Snail family member *SNAI2* [95] have been reported to cause piebaldism. Kit is a tyrosine kinase receptor for the Kit ligand (Kitl, also called Steel) and mice carrying mutations in *Kit* or *Kitl* show an absence of NCC-derived melanocytes [96–99]. In mouse embryos, *Kit* is expressed in melanocyte precursors while *Kitl* is expressed in the pharyngeal arch mesenchyme and the dorsal dermatome in the trunk [100]. Kitl has two forms, a membrane-bound and secreted form, and the secreted form is thought to serve as a chemotactic factor for Kit [101] while the membrane-bound form may regulate adhesion [102]. In the absence of the Kit receptor or ligand, melanocyte precursors remain in the migration staging area and fail to survive demonstrating the importance of *Kit* and *Kitl* for the survival and migration of melanoblasts [101]. A role for Kit in migration is further supported by the zebrafish mutant *sparse*, which also exhibits defects in pigmentation [103]. Cell transplantation studies between *sparse* mutants and wild-type zebrafish demonstrated that *kit* acts cell autonomously to promote melanocyte migration. These studies also showed a role for *kit* in melanocyte survival [103]. Conversely, expression of a constitutively active form of the Kit receptor resulted in increased migration in cells transplanted into mice along with diminished markers of differentiated melanocytes implying that Kit has an important role in the initiation of migratory behavior [104]. Thus, Kit signaling may have additional roles after initiation of migration in melanocyte maintenance in humans [105].

Snail2 is expressed in premigratory and early migratory NCC in avian [106] and aquatic [107] embryos, where it functions primarily in NCC specification and initiation of EMT. As part of this process, *Snail2* directly represses the cell adhesion molecule Cadherin6, which helps enable migration [108]. *Snail2* mutant mice have a similar pigmentation phenotype to *Kit* and *Kitl* mutants [109]. Analysis of *Kit* and *Kitl* function in *Snail2* mutant mice showed that Kit^{+} cells in *Snail2*$^{-/-}$ mutants migrated very little in response to Kitl suggesting that *Kit* may act upstream of *Snail2* [109]. A possible mechanism entails that melanocyte precursors are unable to undergo proper EMT and migrate, and instead undergo apoptosis, leading to hypopigmentation. Given the importance of Kit in melanocyte migration, it is not surprising that mutations in *SNAI2*, a key player in EMT, have been discovered to result in piebaldism [95]. *SNAI2* mutations are also known to cause Waardenburg syndrome (described later in this chapter), which phenotypically overlaps with piebaldism with respect to pigmentation defects.

Hirschsprung Disease

Hirschsprung disease (HSCR) describes the absence of enteric neurons from varying lengths of the gut due to a failure of enteric NCC to colonize the entire gut. HSCR is associated with mutations in a number of different genes, including *RET, GDNF, NTN, EDNRB, EDN3, KBP, β-1 Integrin, L1Cam, PHOX2B, HASH1, HAND2, ZFHX1B,* and *SOX10* [110]. The majority of the ENS is colonized by the vagal NCC population, which migrates, proliferates, and then differentiates into a number of different types of neurons that populate the entire length of the adult gut. In humans, the enteric NCC colonize the gut during weeks 4−7 [111]. NCC migrate in an anterior to posterior wave within the gut mesenchyme to colonize the gut by E15 in mice [112]. At the same time, the gut undergoes elongation and looping to form different regions of the gastrointestinal tract as well as bring the small intestine into a position adjacent to the colon. A recent study demonstrated that a subpopulation of the enteric NCC (called transmesenteric) will migrate from the small intestine across the mesentery and into the colon where they will give rise to a portion of the ENS in the colon [113].

Several signaling pathways play a role in the proper migration of enteric NCC, including glial cell line-derived neurotrophic factor (GDNF), which binds its co-receptor GDNF family receptor α1 (GFRα1), and activates the RET receptor tyrosine kinase. RET and GFRα1 are expressed in enteric NCC upon entering the gut, while GDNF is expressed in an anterior to posterior gradient in the gut mesenchyme and likely induces NCC to enter the gut and promote their migration [114]. Mutations in *RET* account for approximately half of all familial cases of HSCR and 15−35% of sporadic cases [110]. Ret signaling plays a role in survival, migration, proliferation, and differentiation of NCC [110]. Deletion of *Ret* in the mouse leads to complete intestinal aganglionosis [115], while *ret* knockdown in zebrafish leads to an absence of enteric neurons [116]. Homozygous *Gdnf* [117−119] and *GFRα1* [120] mutations in mice lead to a failure of enteric NCC to colonize the gut beyond the esophagus, which is very similar to the phenotype of *Ret* mutants. Another Ret ligand, neurturin (NTN), has also been implicated in HSCR [121]. The co-receptor for NTN is GFRα2 and, similar to GDNF and GFRα1, mice with mutations in *Ntn* or *GFRα2* exhibit defects in ENS formation [122,123].

Sox10 and *Phox2b* are two transcription factors that stimulate expression of *Ret*, and the Edn3/Ednrb signaling pathway has also been shown to have genetic interactions with both *Sox10* and *Ret*. *Sox10*^{Dom} mutants carry a frameshift mutation that results in a truncated protein. Homozygous mutants have severe NCC defects including a lack of enteric ganglia. In contrast, in *Edn3*$^{ls/ls}$ (*lethal spotting*) mutants, TuJ1-positive cells extend only as far as the stomach. Colonization of the gut is also affected in *Sox10*$^{Dom/+}$; *Edn3*$^{ls/+}$ mice, which display a small population of TuJ1-positive cells in the small intestine. A similar result was seen in *Sox10*$^{Dom/+}$; *Endrb*$^{s/s}$ (*piebald*) mice, which have a decreased number of *Sox10*-positive migrating vagal NCC and delayed colonization of the gut [124].

Of mice with heterozygous mutations in *Ret* and heterozygous mutations in *Edn3*, 35% have an absence of enteric NCC in the intestine at E12.5; if both copies of *Edn3* are disrupted, the percentage increases to 70%. *Ednrb*$^{s/s}$ mice very rarely display megacolon; however, all compound *Ret*$^{+/-}$; *Ednrb*$^{s/s}$ mice exhibit megacolon [125,126]. While the *Ednrb*^{sl} allele (*piebald lethal*) results in the deletion of all coding exons, homozygous *Ednrb*^{s} mutations result in a decrease of around 75% of the transcript [125]. Although the degree of aganglionosis in compound mutants is not as severe as *Ret*$^{-/-}$ mice, it is more severe than *Ednrb*$^{sl/sl}$ [126]. This is consistent with expression analyses that show co-localization of *Ret* and *Ednrb* in a subpopulation of migrating enteric NCC in the gut, further supporting a genetic interaction between these pathways during ENS development [125].

These studies illustrate the complex genetic interactions that occur during enteric NCC migration and the formation of the ENS. Varying the level of expression of these different factors affect the degree of colonization of the gut. This complexity is also demonstrated in part by the variation in human phenotypes, in which some patients will have total intestinal aganglionosis while others may have only aganglionosis in the most distal portion of the colon. Thus, migrating enteric NCC must appropriately balance continued proliferation with differentiation in order to achieve proper formation of the ENS and full colonization of the gut.

HSCR therefore is a complex disorder involving many signaling interactions and reflects the complexity of NCC development. Many of the genes implicated in HSCR have cell-autonomous roles in NCC migration, but there are also contributions from proliferation, survival, and differentiation of NCC in the etiology of this disease. Investigation of HSCR-associated syndromes, such as Mowat–Wilson syndrome and Waardenburg syndrome Type 4 (WS4), has provided additional information on the pathways involved in ENS development and other tissues that are also impacted by the same pathways. The genes involved in Mowat–Wilson syndrome versus WS4 have in common a role in ENS development, but their roles in other tissues may vary. For example, pigmentation defects are not observed in Mowat–Wilson syndrome but are a primary characteristic of WS4. Conversely, while Mowat–Wilson syndrome has distinct craniofacial anomalies, these are not observed in WS4. This raises the possibility of non-cell-autonomous effects such as environmental interactions on NCC as contributing factors in the variation among these syndromes.

Mowat–Wilson Syndrome

Mowat–Wilson syndrome is characterized by a distinct facial phenotype, intellectual deficiency, corpus callosum agenesis, epilepsy, congenital heart defects, Hirschprung disease (HSCR), and genitourinary anomalies. The facial characteristics include a high forehead, prominent chin, an M-shaped upper lip, open mouth, posteriorly rotated ears, and saddle nose with a prominent rounded tip [127]. The phenotype of Mowat–Wilson syndrome is indicative of a perturbation of cranial, cardiac, and vagal NCC development. Mutations in the transcription factor *ZFHX1B* are known to cause Mowat–Wilson syndrome in a dominant manner [127], and both heterozygous deletions and truncating mutations have been described [128,129]. However, the majority of cases are the result of *de novo* mutations.

In human embryos, *ZFHX1B* is expressed in the neural tube, pharyngeal arch mesenchyme, cranial ganglia, dorsal root ganglia, sympathetic ganglia, and the ENS, all consistent with a role in NCC development [130,131]. In mouse embryos, *Zfhx1b* is expressed at E8.5 in the neuroepithelium and then later in the premigratory and migratory cranial and vagal NCC. *Zfhx1b* mutant mice die around E9.5, are severely developmentally delayed, and fail to turn. Defects are visible in homozygous mutant embryos at E8.5 and include an open neural tube and hypoplasia or agenesis of the first pharyngeal arch. Cranial NCC do form, but very few at the cardiac and vagal levels, reflecting the important role of *Zfhx1b* in EMT [60]. *Zfhx1b* is known to downregulate E-cadherin and in *Zfhx1b* homozygous mutants, E-cadherin persists in the neuroepithelium, such that there is no sharp boundary between the neural plate and ectoderm [132,133]. The specific function of *Zfhx1b* in NCC was investigated via conditional deletion with *Wnt1-Cre*. Mutant mice survive to birth and show similar phenotypes to human Mowat–Wilson syndrome. These phenotypes include craniofacial anomalies including hypoplastic mandible and incomplete ossification of the nasal and premaxilla bones, heart defects, sympathetic neuron defects, and HSCR. There was partial colonization of the gut, showing that some of the vagal NCC are able to migrate [134]. The defects observed in *Zfhx1b* mutant mice are indicative of a perturbation in EMT and migration of NCC. However, *Zfhx1b* may also be important in NCC differentiation and this will be a critical avenue to explore in future investigations.

Waardenburg Syndrome

Waardenburg syndrome (WS) is an autosomal dominant disease affecting 1:42,000 births and is characterized by depigmented patches of the skin and hair (partial albinism), blue eyes or heterochromia irides, sensorineural hearing loss, and in some cases spina bifida [135]. WS exhibits considerable phenotypic overlap with piebaldism with respect to pigmentation. There are four types of Waardenburg syndrome (WS1–WS4) that are distinguished on the basis of additional developmental defects. To date, mutations in six distinct genes have been associated with WS. Ninety percent of WS1 and WS3 are due to loss-of-function mutations in *PAX3*, an important regulator of early neural crest specification [136–138]. *MITF* is disrupted in 15% of WS2, while *SNAI2* is perturbed in fewer than 5% of individuals with WS2 [139,140]. Mutations in *EDN3* and *EDNRB* account for a similarly small fraction of WS2 incidence (<5%) but feature more prominently in WS4 (20–30%). Mutations in *SOX10* are responsible for about 15% of WS2 cases and 50% of WS4 cases, respectively [139]. Thus WS2 is the most genetically heterogeneous of the four subtypes of WS, and approximately 70% of patients with type 2 and 15–35% of type 4 patients do not exhibit mutations in any of these six genes.

Pax3 performs numerous functions during embryogenesis in NCC formation, proliferation, and differentiation into glia and melanocytes. In human embryos, *PAX3* is expressed in premigratory NCC in the dorsal neural tube and in the dermomyotome at Carnegie stage 12, the dorsal root ganglion and presumptive melanocytes and musculature at Carnegie stage 15, and in the maxillary process at Carnegie stage 18 [93]. In WS1 and WS3, mutations in *PAX3* likely result in the decreased formation and proliferation of NCC. In mice, *Pax3* is expressed prior to neural tube closure within the dorsal neuroepithelium and in NCC and their derivatives including craniofacial structures, dorsal root ganglia, and somitic mesoderm [141]. The functions of *Pax3* have been investigated using *Splotch* mutant mice that carry mutations in *Pax3*. Mutant mice display neural tube, cardiac, and pigmentation defects, in addition to a curly tail and general developmental delay [142–144]. *Splotch*$^{-/-}$ mice expressing the melanoblast marker *Dct-LacZ*, exhibit greatly reduced numbers of melanoblasts compared to control littermates [145]. *Pax3* mutations also lead to increased neuroepithelial apoptosis suggesting that *Pax3* is required to expand a pool of restricted progenitor cells or melanoblasts that are specified early in development. Consistent with a role in proliferation, *PAX3* has been found to be mutated in the human tumor rhabdomyosarcoma [146,147] and contributes to tumor cell survival in melanoma [148].

Pax3 is known to interact with other genes, such as *Mitf* and *Sox10*, which are implicated in WS2. *Mitf* (microphthalmia transcription factor) is a transcription factor known to play a key role in melanocyte differentiation. Mutations in *Mitf* in mice lead to a variety of phenotypes depending on the nature of the mutation, but all homozygous mutants have some defect in pigmentation [149]. Melanoblasts were not observed entering the dorsolateral pathway in *Mitf* homozygous mutants, suggesting that proper dosage of *Mitf* plays a critical role in the survival of melanoblasts up to the migration staging area of NCC development [145]. *Pax3* and *Mitf* may therefore act in parallel in different aspects of melanocyte development, with Pax3 promoting proliferation of progenitor cells and Mitf promoting survival of committed melanoblasts [145].

Sox10 also plays multiple roles during NCC development, including maintaining multipotency of neural crest stem cells [150], influencing fate decisions of the melanocyte lineage [151–153], and maintaining pluripotency in migrating ENS progenitors [154]. A mouse containing a mutation in *Sox10*, the *Dom* mutant, which has been used as a model for HSCR [155], also has pigmentation defects as seen in WS4 [156]. Homozygous mutants show total aganglionosis and lack expression of NCC-derived melanocyte markers such as *Mitf*, *Dct*, and *Kit*. Sox10 has been shown to activate Mitf expression *in vitro* in cell culture [151,152] and also *in vivo* in zebrafish, illustrating its role in this process [153]. A *sox10* mutant zebrafish, known as *colorless*, also lacks the majority of pigment cells. In these fish, the NCC that would normally adopt nonmesenchymal fates (such as melanocytes) fail to migrate and differentiate and instead undergo apoptosis [157,158]. Similarly, increased apoptosis, was observed in *Dom* mutant mice before NCC entered the gut, and it has been suggested that *Sox10* also plays a cell-autonomous role in enteric NCC [159,160]. Consistent with this, even heterozygous mutants show deficient ENS colonization of the distal colon, which is due to the failure of NCC to migrate and differentiate properly [161]. Thus, *Sox10* is especially important for the survival of migratory melanocyte and enteric NCC precursors. In humans, *SOX10* is expressed in the migrating cells of the ENS, otic vesicles, as well as the cranial, sympathetic, and dorsal root ganglia at Carnegie

Stage 13 [162]. The expression in enteric NCC is especially relevant, as WS4 patients exhibit HSCR.

Sox10 also influences another signaling pathway implicated in WS2 and WS4, the Endothelin3 (*EDN3*)/Endothelin receptor B (*EDNRB*) pathway. *Ednrb* contains Sox10 binding sites and it has been shown that endothelin signaling is necessary for the maintenance and proliferation of Sox10 expressing enteric NCC. Multiple *in vitro* studies have shown roles for endothelin signaling in migration, proliferation, and survival of NCC (reviewed in [163]). Furthermore, *Ednrb* is required between E10−E12.5 in mice, which correlates to the timing of the migration and differentiation of enteric NCC and the migration of melanoblasts. These studies indicated a role for *Ednrb* in the initiation of melanoblast migration and possibly survival [164]. Mouse models carrying mutations in *Ednrb* (*piebald* and *piebald lethal*) or *Edn3* (*lethal spotting*) are mostly white and have megacolon [165−167], reflecting the WS4 phenotype. The white spotting seen in *Ednrb* mutants is dependent on the level of *Ednrb* expression with homozygous mutants nearly completely white and heterozygous mutants displaying white spotting [165]. Megacolon, however, was only observed in homozygous mutants, indicating that the melanocyte and enteric neuron lineages require a different minimum threshold of Ednrb signaling. Ednrb signaling has been shown to have both cell-autonomous and non-cell-autonomous functions during melanocyte development, but the non cell autonomous functions observed *in vitro* have not been demonstrated *in vivo* [168]. Interestingly, the NCC-specific excision of *Ednrb* via *Wnt1*-Cre resulted in the same phenotype as *Ednrb* mutant mice [169]. This demonstrated that Ednrb is important in a cell autonomous manner in migrating NCC [169]. Although white spotting differences may partly distinguish WS2 and WS4 patients, such that those with higher levels of functional EDNRB develop WS2, while individuals with lower levels of EDNRB develop Hirschsprung disease−associated WS4, to date, no correlation has been found between the type of mutation and phenotype [139].

All of the genes implicated in WS have a common role in melanocyte development from NCC and do so through complex interactions among themselves and other signals in the embryo. Cell autonomous signals, such as *Ednrb*, within migrating NCC are especially important. The association of melanocytes with deafness may be associated with a lack of melanoblast-derived strial intermediate cells leading to degeneration of the organ of Corti and resulting in hearing loss [170], or an alternative mechanism involving *Sox10*, which is expressed during inner ear development [171,172]. Further examination of the WS genes—*PAX3*, *MITF*, *SOX10*, *SNAIL2*, *EDN3*, *EDNRB*—and their functions in NCC development will provide a better understanding of the etiology of this disease. A portion of the WS genes—*SOX10*, *EDN3*, and *EDNRB*—are also associated with HSCR.

Metastatic Cancer

EMT facilitates the initiation of NCC delamination and migration during normal embryonic development. As described above, neurocristopathies such as BOFS, piebaldism, HSCR, Mowat−Wilson syndrome, and WS result from disrupted EMT and migration of NCC, leading to hypoplasia of NCC-derived tissues. In contrast, other neurocristopathies such as metastatic cancer arise from reactivation of EMT and migration. As primary tumors develop, cancer cells reactivate EMT, allowing them to disseminate from their original primary tumor location into the circulatory system or other permissive routes, where they spread systemically throughout the body. Once these circulating cells come to a halt, they can re-establish epithelial characteristics via the mesenchymal to epithelial process (MET) and invade new territories. Thus, EMT/MET facilitates colonizing and establishing secondary tumor sites. This is particularly true of melanoma and neuroblastoma where re-establishing the proliferative and migratory characteristics of NCC leads to tumor metastasis.

Melanoma

Melanoma, which arises from the melanocyte lineage, progresses through a variety of stages of development, including formation of dysplastic nevi (moles) followed by a radial growth phase, a vertical growth phase, and metastasis [173]. During the radial growth phase, melanocytes increase proliferation and grow laterally, while during the vertical growth phase the melanoma cells cross the basement membrane and acquire migratory ability, becoming metastatic [173]. During melanoma progression, melanocytes no longer exhibit the markers of fully differentiated cells. Although accounting for approximately 4% of all skin cancer incidences, melanoma is by far the most aggressive and malignant, causing 80% of all skin cancer deaths [174]. The most common mutation associated with melanoma occurs in BRAF, which makes it constitutively active (V600E) [175]. Expression of this allele in mouse fibroblasts leads to increased proliferation and loss of contact inhibition, reflecting endogenous steps in the development of malignant melanoma [176]. Overexpression of the human V600E mutation in BRAF under the control of the mitfa promoter in zebrafish results in development of ectopic melanocytes. Moreover, overexpression in p53 mutant fish leads to melanoma [177].

Mitf, which is considered a master regulator of melanogenesis, is central to several pathways implicated in the development of melanoma [178]. Mitf is essential for proliferation and survival of melanoma cells due to its regulation of genes involved in cell cycle progression (CDK2) and apoptosis (BCL2) [179]. Downstream of BRAF, MAPK/ERK activation will phosphorylate Mitf leading to its activation [178]. Mitf then acts in several ways to direct melanocyte differentiation, cell cycle progression, and apoptosis. Tumors with increased levels of MITF are associated with poor prognosis and resistance to chemotherapy [180]. However, despite the role of Mitf in numerous aspects of melanocyte development, its precise role in melanoma is not entirely clear as Mitf is amplified in some melanomas, while it is lost in others [178].

As with NCC migration and differentiation, numerous factors in the microenvironment surrounding melanoma, the mutations in the melanoma cells, and the interactions between the cells and the environment can impact the ability of melanoma cells to undergo EMT and metastasize.

Neuroblastoma

Neuroblastoma, is the most common extracranial solid tumor diagnosed in infancy, accounting for 7—10% of all childhood cancer [181]. Neuroblastoma is highly heterogeneous and when metastases are present, often in children over 1 year of age, prognosis is very poor [182]. Interestingly, some NB tumors (in infants <1 year) can spontaneously regress. These tumors can receive proper developmental signals to either differentiate properly or undergo apoptosis, demonstrating that reactivation of the proper developmental program can eliminate tumors. This also provides support for the idea that dysregulation of developmental programs results in proliferation and migration associated with metastatic cancers.

NCC contribute to the autonomic and peripheral ganglia of the peripheral nervous system as well as the adrenal gland. Disruptions of NCC development, specifically the sympathoadrenal lineage, are thought to be a major cause of neuroblastoma, making it a neuroendocrine disorder [183]. The sympathoadrenal lineage of the trunk NCC contribute to the sympathetic ganglia and adrenal gland. As NCC migrate toward the dorsal aorta during embryogenesis, they are exposed to BMP signals, which induce the expression of *Mash1* (*mammalian achaete-scute homolog 1*) and also *Phox2b*. These then activate expression of *Phox2a*, which is a positive regulator of noradrenergic traits during neuronal development. Two key enzymes involved in the synthesis of noradrenaline are tyrosine hydroxylase (TH) and dopamine β-hydroxylase (DBH).

An important factor in neuroblastoma is the status of Mycn. Amplification of Mycn occurs in approximately one-third of neuroblastoma tumors, and amplification is a marker of poor prognosis. Mycn is a transcription factor involved in cell cycle progression and stem cell homeostasis. In chick embryos, Mycn is initially expressed in all migrating NCC and expression persists in cells undergoing neuronal differentiation [184,185]. Decreased Mycn is associated with terminal differentiation and exit from the cell cycle, while increased Mycn is associated with proliferation, migration, and activation of the cell cycle in neuroblasts. Mice in which *Mycn* is overexpressed in NCC via the TH promoter develop aggressive neuroblastoma tumors [186]. Mycn interacts with a number of different genes that contribute to proliferation and survival. *Mycn* overexpression leads to overexpression of *Id2*, which is an inhibitor of Rb. This inhibition of Rb allows neuroblastomas to overcome a block to cell cycle progression [187]. In neuroblastoma cell lines, *MDM2* is a direct target of MYCN, and inhibition of *MYCN* leads to a decrease in *MDM2* and a subsequent increase in p53 and apoptosis [188]. However, a more recent study

showed that p53 is also a direct target of MYCN and that increased MYCN is associated with increased p53. The authors suggest that increased *MDM2* expression in the presence of MYCN may also be dependent on the higher expression of p53 [189].

Metastatic cancers such as melanoma and neuroblastoma, together with BOFS, piebaldism, and Mowat−Wilson syndrome, illustrate divergent neurocristopathies that result from perturbation of the delamination and migration of NCC. Understanding the developmental mechanisms that guide these processes may prove beneficial not only for developing potential preventative treatments for congenital neurocristopathies, but also for the treatment of metastatic cancer.

22q11.2 Deletion Syndrome

22q11.2 deletion syndrome, which encompasses DiGeorge syndrome (DGS), is characterized by cardiac defects including interrupted aortic arch, tetralogy of Fallot, as well as hypoplastic thymus, and occasionally mild craniofacial defects, cleft palate, and hypothyroidism (Figure 17.2E). Most patients have a deletion within 22q11.2, and perhaps the most well-studied gene in this region is *TBX1*.

Homozygous mutations of *Tbx1* in mice lead to a characteristic DGS phenotype that includes cardiac outflow tract abnormalities, abnormal facial structures including cleft palate, hypoplastic thymus, and parathyroid glands, and abnormal vertebrae [190]. The cardiac defects associated with mutations in *Tbx1* are likely due to the impaired development of the fourth pharyngeal arch, which encompasses formation of the fourth pharyngeal arch artery (PAA). Studies in *Tbx1*$^{-/-}$ mice have revealed that NCC migration is affected as evidenced by fusion of migratory NCC streams [191], ectopic, misdirected NCC [192], and a reduced number NCC in the fourth pharyngeal arch. Even though reduced numbers of migrating NCC were observed in the hypoplastic fourth pharyngeal arch of $Tbx1^{-/-}$ mice, NCC differentiation was considered normal [193].

$Tbx1$ is a transcription factor expressed in the pharyngeal endoderm, the mesodermal core of the pharyngeal arches, and a region known as the second heart field, which is important for outflow tract development [192,194]. The expression of $Tbx1$ suggests that it has a role in the early remodeling of the PAA [192], and although not expressed by NCC, $Tbx1$ influences the expression of other genes implicated in NCC migration. Loss of $Tbx1$ impacts tissue-specific gene expression altering the environment through which the NCC migrate, and consequently alters the migration of NCC [191,192]. Tbx1 controls the expression of $Gbx2$ at the time of the fourth PAA specification [191], and $Tbx1^{-/-}$ embryos show reduced $Gbx2$ expression, especially in the fourth pharyngeal arch. Ablation of $Gbx2$ in the pharyngeal surface ectoderm (using $Tfap2a$-Cre) disrupts cardiac NCC migration and formation of the fourth PAA. Furthermore, $Slit2$, a Robo ligand, is downregulated in both $Tbx1^{-/-}$ and $Gbx2^{-/-}$ mice. $Robo1$, a receptor expressed on NCC, is also downregulated and its expression pattern is disorganized in mutant mice, suggesting that the migration of cardiac NCC is affected in the $Tbx1^{-/-}$ mice [191].

Cardiac NCC are derived from the region between the otic placode and the third somite and are important for proper heart development, including outflow tract morphogenesis and the innervation of the heart by the parasympathetic nervous system [71]. Cardiac NCC pattern the pharyngeal arches and their derivatives including the arch arteries, and migrate into the cardiac outflow tract where they participate in forming the outflow septum [195]. Disruptions in cardiac NCC migration therefore can lead to cardiovascular as well as other anomalies characteristic of disorders such as 22q11.2 deletion syndrome. Several other signals and pathways interact with Tbx1 including Shh, RA, BMP, and Pitx2, and these interactions help to explain the variation seen in the human phenotype.

In summary, these neurocristopathies demonstrate the complexity of NCC migration, which involves signals that act both cell autonomously and non-cell autonomously and are expressed in the NCC as well as surrounding tissues. 22q11.2 deletion syndrome/DGS is an example of non-cell autonomous factors that affect NCC development in the pathogenesis of a neurocristopathy. Mutations in genes that affect EMT and NCC migration can lead to NCC deficiencies as demonstrated by BOFS, piebaldism, HSCR, Mowat–Wilson, Waardenburg, and 22q11 deletions syndromes. Alternatively, the inappropriate reactivation of NCC-developmental programs and EMT can result in metastatic cancers such as neuroblastoma and melanoma. Although each of these neurocristopathies have some deficiency in NCC migration, a wide range of phenotypes are observed in a variety of tissues including pigment cells, enteric neurons, cardiac cells, and craniofacial cartilage and bone. It is not well understood why one particular subset of NCC may be more affected in one neurocristopathy versus another, however the timing, tissue-specific requirement, and also surrounding environment of the NCC all have a role in the specific phenotypes observed.

Neural Crest Cell Differentiation

The final phase of NCC development is differentiation and in many instances this occurs simultaneously with migration. Migrating NCC comprise a population of multipotential, bipotential, and unipotential progenitor cells [13], which raises the issue as to what signals and mechanisms establish lineage segregation and cell fate selection within the neural crest population to give rise to derivatives as diverse as sensory neurons, pigment cells, cartilage, and bone. Ultimately, NCC differentiation is regulated by a combination of cell autonomous signals intrinsic to NCC balanced with non-cell autonomous signals that are extrinsic. One of the most important differentiation decisions to be made with respect to cranial NCC is whether to initially become neurogenic or non-neurogenic (mesenchymal) [63]. Two SoxE transcription factors, *Sox9* and *Sox10*, are expressed from the outset in migratory NCC and quickly define the initial subdivision of NCC with distinct lineage potential. *Sox9* distinguishes NCC of mesenchymal (bone, cartilage, and connective tissue) potential, while *Sox10* demarcates prospective neurogenic and pigment-derived NCC.

The *Sox10*-positive migrating NCC population goes on to give rise to neurons and glia of the peripheral nervous system (Figure 17.1C) as well as generate melanocytes and similar cells that generate the pigment of the skin. The peripheral nervous system can be subdivided into the sensory nervous system (SNS), which conveys sensory information to the central nervous system from mechanoreceptors, thermoreceptors, nociceptors, and proprioreceptors, and the autonomic nervous system (ANS), which governs involuntary organ control and functions such as breathing and digestion. The ENS, sympathetic nervous system, and parasympathetic nervous system are all subdivisions of the ANS. Autonomic neurogenesis is regulated by BMP signaling, as migrating NCC exposed to exogenous BMP2 preferentially differentiate into sympathetic neurons [196]. The effect of BMP signaling on autonomic neuron development is mediated by *Mash1*, *Phox2a*, and *Phox2b*; each of these genes is upregulated in response to BMPs and can individually promote sympathetic neuron generation [196–199]. Consistent with this, mice homozygous for a mutation in *Mash1* die

shortly after birth due to the impaired development of the sympathetic, parasympathetic, and enteric ganglia [198]. In mice lacking *Phox2b*, autonomic ganglia fail to form properly and degenerate because *Phox2b* is required for the maintenance of *Mash1* expression [200].

Congenital Central Hypoventilation Syndrome

Congenital central hypoventilation syndrome (CCHS) is a rare disorder characterized by shallow breathing (during sleep only or both while awake and asleep), ANS dysregulation, HSCR, and cancer such as neuroblastoma. ANS dysregulation can affect both breathing and heartbeat, such that some patients require a pacemaker. CCHS is inherited in an autosomal dominant fashion, but most patients present with a *de novo* mutation in *PHOX2B*. The causative mutations in *PHOX2B* include frameshift mutations and polyalanine expansions [201]. The severity of the ventilation phenotype has been associated with longer alanine expansions, with shorter expansions resulting in late-onset CCHS (LO-CCHS) [202]. A *Phox2b* polyalanine expansion mouse model was generated with a +7 alanine expansion to mimic the human mutation. Mutants were found to have irregular breathing patterns, be unresponsive to increases in CO_2, and to have deficiencies in neurons that receive input from O_2-sensing receptors, resulting in death soon after birth [203]. Alanine expansion mutations in *Phox2b* result in diminished transactivation of DBH and the expansion region can cause protein misfolding and aggregation [204,205].

In mice, *Phox2b* is expressed in the central nervous system and peripheral nervous system, esophagus, small and large intestine, and undifferentiated neural crest-derived cells in the gut [206,207]. *PHOX2B* is expressed in similar locations in humans, including the autonomic nervous system in the presumptive sympathetic ganglia and enteric ganglia [201]. In the *Phox2b* knockout mouse, NCC migration is perturbed particularly with respect to vagal NCC that enter the foregut at E10.5 but do not migrate further or survive, resulting in agenesis of enteric neurons and the pathogenesis of colonic aganglionosis [206]. Phox2b is required not only for NCC migration but also for maintaining the expression of *Mash1*, which plays important roles in promoting *TH*, *DBH*, and *RET* expression, and neuronal differentiation [206]. Consistent with these roles, homozygous *Phox2b* mutant mice that die around midgestation (shortly after E10.5) can be rescued to E18.5 via treatment with noradrenergic agonists [208].

Mutations in other genes have been reported in CCHS, including *RET*, *GDNF*, *EDN3*, *BDNF*, *ASCL*, *PHOX2A*, *GFRA1*, *BMP2*, and *ECE1* [209]. Many of these genes exhibit known associations with other neurocristopathy disorders such as HSCR, which coincidentally is a component of CCHS. This illustrates the reiterated use of the same signaling pathways in multiple aspects of NCC development and the variety of neurocristopathies that similarly arise from their perturbation.

Oculocutaneous Albinism

Oculocutaneous albinism (OCA) is a form of albinism that may be considered an error of melanocyte differentiation and affects approximately 1 in 20,000 people. Melanocytes are able to form and migrate, but disruptions in the melanin-producing enzymes downstream of Mitf render the melanocytes unable to produce pigment. Tyr, Tyrp1, and Dct are the major enzymes involved in melanin synthesis. There are currently six different types of OCA [210]. OCA1A results from a mutation in *TYR* that has no activity, while OCA1B also results from a mutation in *TYR*, but with decreased activity. OCA1TS is particularly interesting in

that the mutation in *TYR* has made it temperature sensitive such that is inactive above 35°C. Thus, darker hair exists on cooler parts of the body, while in warmer areas hair is white and TYR inactive. OCA2 is associated with the loss of the *P* gene, which encodes the P protein, a melanosome membrane protein that could possibly regulate and process the transport of TYR. OCA3 is caused by mutations in *TRP1*, and individuals remain minimally pigmented. OCA4 is caused by mutations in *MATP*, a protein involved in processing and trafficking of melanosome proteins. Mutations in OCA4 are modeled by the mouse mutant known as under-white (*uw*), of which there are multiple alleles [211].

Craniosynostosis

There are two modes of bone formation—endochondral and intramembranous ossification. The long bones of the skeleton and some of the facial bones are formed by endochondral ossification in which chondrocytes form an initial cartilage template that undergoes hypertrophy and is replaced by osteoblasts. In contrast, the flat bones of the cranium are formed by intramembranous ossification in which an initial condensation of mesenchyme calcifies directly to form the bone. Cranial NCC contribute the majority of cartilage and bone in the head and face (Figure 17.1D and H). The commitment of NCC to differentiate into cartilage is marked by expression of *Sox9* [212], which will activate expression of the type II collagen, (*Col2a1*) [213]. Consistent with this, *Wnt1-Cre* conditional deletion of *Sox9* results in complete absence of NCC-derived cartilage and endochondral bones in the head [214]. Chondrocyte differentiation through *Sox9* is achieved in part by the inhibition of osteoblast-promoting genes such as *β-catenin* [215]. Indeed, expression of a stable form of *β-catenin* inhibits chondrogenesis, mimicking the loss of *Sox9*. Conversely, the conditional deletion of *β-catenin* in chondrocytes mimics overexpression of *Sox9* [216]. Thus, an antagonistic relationship exists between *Sox9* and *β-catenin* in the regulation of cartilage and bone development [214]. Sox9 will also stimulate expression of *Runx2*, an early marker of bone formation. Runx2 is then key in initiating the expression of more mature osteogenic markers such as *osterix, osteocalcin, osteopontin*, and *alkaline phosphatase* as well as type I collagen (*Col1a1*). Mice carrying a mutation in *Runx2* fail to form any bone and have only cartilaginous skeletons [217]. Many signaling pathways and transcription factors have been shown to influence the expression of *Runx2*. The TGFβ, BMP, Hh, and Fgf signaling pathways all influence expression of transcription factors related to osteoblast specification and differentiation. The influence of these pathways on craniofacial bone development, particularly the Fgfs, is evident in the pathogenesis of craniosynostosis (CS) syndromes. The mammalian cranial vault consists of two frontal bones, two parietal bones, and one occipital bone. Whereas the frontal bones are NCC derived, the occipital and parietal bones are mesoderm derived [218,219]. Sutures are NCC-derived mesenchymal junctions between the calvarial bones that maintain separation facilitating birth and postnatal brain and associated skull growth. Underlying the cranium is the dura mater, which is NCC derived and serves as a signaling source during calvaria and suture development [220].

CS is defined by the premature fusion of the cranial sutures and is estimated to occur in 1 out of 2500 births (Figure 17.2F). The majority of CS cases are nonsyndromic, while an estimated 15% are syndromic [221]. Most CS syndromes are associated with mutations in the FGF receptors (*FGFR*) −1, −2, and −3 as well as *TWIST* (Saethre−Chotzen syndrome), *MSX2* (Boston-type CS), and *EFNB1* (craniofrontonasal dysplasia). Crouzon, Jackson−Weiss, Apert, Beare−Stevenson, and Antley−Bixler

syndromes are all associated with mutations in *FGFR2*. Pfeiffer syndrome is associated with mutations in *FGFR1*, and some in *FGFR2*, while Muenke syndrome and a variant of Crouzon syndrome are associated with *FGFR3*. Fgfs are receptor tyrosine kinases and activate signaling cascades through PI3K as well as MAPK. Specificity in Fgf signaling is obtained through tissue-specific expression, alternative splicing, and other modifications. *Fgfr1* is expressed in the calvarial mesenchyme and later in osteoblasts. *Fgfr2* is expressed in regions of osteoprogenitor proliferation at osteogenic fronts while *Fgfr3* is expressed at low levels in the sutural osteogenic front [222,223]. Alternative splicing of *Fgfr2* is tissue specific with *b* forms being expressed in epithelial cells and *c* forms in mesenchymal cells [224].

Studies of *Fgfr2* show that there may be different functions for *Fgfr2* in osteoblasts and suture fusion. Conditional disruption of *Fgfr2* in osteoblast lineages using *Dermo1-Cre* showed that *Fgfr2* is required for osteoblast proliferation but not differentiation or survival [225]. However, most mutations in the *FGFRs* associated with CS are activating mutations, which result in premature bone formation. Osteoblast proliferation, differentiation, apoptosis, and mineralization must be balanced during suture development and disruption of any one of these processes can impact the proper formation of the suture. $Fgfr2^{S252W}$ mutant mice serve as a model for Apert syndrome and display midface hypoplasia as well as coronal suture fusion. Conditional $Fgfr2^{S252W}$ mice have been used to investigate the function of *Fgfr2* in NCC (*Wnt1-Cre*) versus paraxial mesoderm (*Mesp1-Cre*). Expression of $Fgfr2^{S252W}$ in NCC led to midface hypoplasia, but not suture fusion in most mice (10/12) studied. In contrast, conditional expression of $Fgfr2^{S252W}$ in the mesoderm led to coronal suture fusion, which was due to ectopic proliferation and ectopic induction of osteogenesis within the sutural mesenchyme [226].

These studies of the *Fgfr2* Apert mutant mice show that alteration in the mesoderm can elicit CS, but there is also evidence for CS resulting from altered differentiation specifically in NCC. Mice that constitutively express active Bmpr1a (a BMP receptor) in the skull and sutures fail to maintain Smad-dependent BMP signaling in cranial NCC, resulting in CS. Interestingly, quantitative RT-PCR revealed that *Fgfr1, Fgfr2,* and *Fgf2* were upregulated, as was pERK1/2 by Western blot. The mutant phenotype was rescued using a Bmpr1a inhibitor drug, but the same Fgf signaling enhancement was observed in the rescued mutants, which suggests that elevated Fgf signaling alone is not necessarily sufficient to cause CS [227]. Thus, there are multiple complex mechanisms that underpin the etiology and pathogenesis of CS.

Mutations in *TWIST1* are associated with the Saethre–Chotzen form of CS. *Twist1* is a transcription factor that is expressed in early migrating NCC, cranial mesenchyme, pharyngeal arches, the outflow tract of the heart, cranial bones, and sutural mesenchyme [228]. *Twist1* functions in the regulation of EMT and in NCC proliferation and differentiation. Haploinsufficiency of *Twist1* leads to increased *Fgfr2* expression in the suture mesenchyme, suggesting that *Twist1* is upstream of *Fgfr2*. Furthermore, BMP signaling was also enhanced in *Twist1*$^{+/-}$ mice [229]. *Twist1* and the BMP signaling target *Msx2* are co-expressed in the developing calvarial bones and sutures and have similar heterozygous mutant phenotypes, including a foramen in the skull vault. The phenotype is not due to a defect in NCC migration or survival, but rather a defect in NCC differentiation. Mouse embryos heterozygous for mutations in both *Twist* and *Msx2* exhibited a synergistic enhancement of defects in differentiation and proliferation, which resulted in a larger foramen. The diminished differentiation was demonstrated by reduced expression of osteogenic markers *Runx2* and *ALP* [230].

Thus, BMP signaling, which enhances *Msx2* expression, and *Twist1*, which represses Fgf signaling, cooperate in the differentiation and proliferation of the NCC-derived frontal bone during calvarial development. Consistent with this idea, inhibiting Fgfr signaling was able to rescue the suture fusion phenotype in *Twist1*$^{+/-}$ embryos. This suggests that elevated Fgf and/ or BMP signaling may also be associated with pathogenesis of the Saethre–Chotzen form of CS.

In addition to Fgf and BMP signaling, *Twist1* can also influence Eph/ephrin signaling. Deletion of *Twist1* specifically in NCC via *Wnt1-Cre* results in some ectopically located NCC in the presumptive parietal bone, which is normally a mesoderm-derived bone. Postnatal day 0 (P0) mice show fusion of the frontal and parietal bone. No difference in proliferation or apoptosis, as has previously been demonstrated in CS models, was observed but rather the defect was attributed to a disruption of the NCC-mesoderm boundary. In contrast to other studies that show *Msx2* and *Twist1* act in parallel pathways [230], *Msx2* shows a broader expression domain in *Twist1*$^{+/-}$ mice. Consequently, decreased expression of *Msx2* is able to rescue the suture fusion phenotype [231].

EphA4 expression was also diminished in *Twist*$^{+/-}$ mice, suggesting that *EphA4* functions downstream of *Twist1*. Furthermore, *EphA4* expression was similarly restored upon a decrease in *Msx2* [231]. Interestingly, *EphA2* and *EphA4* are expressed in mesoderm and not in NCC, and it is well known that Eph/ephrin signaling plays key roles in boundary formation and establishing the segmental migration of NCC [232]. This implies that maintaining appropriate NCC-mesoderm boundaries may also be important for normal calvarial development and in the pathogenesis of CS. Both the Eph receptors and their ephrin ligands are membrane bound and contain a transmembrane domain, facilitating both forward (in the cell of the receptor) and reverse (in the cell of the ligand) signaling. Thus a disruption in bidirectional signaling could be a contributing factor in the disorganized NCC-mesoderm boundary. Consistent with this idea, *EPHA4* mutations in human patients result in nonsyndromic coronal CS [231] and similarly *EphA4*$^{-/-}$ mutant mice exhibit CS and expanded *ALP* expression. Therefore *Twist1* and *EphA4* appear to function in separating migratory osteogenic cells in the coronal suture [233].

Eph/ephrin signaling is also important in other aspects of cranial skeleton development as mutations in *EFNB1* cause craniofrontonasal syndrome [234,235]. Collectively, these studies demonstrate that altered migration and boundary formation between NCC and mesoderm is another potential mechanism for CS. Thus, CS, which is primarily considered to be a NCC differentiation disorder, can also arise as a defect in NCC migration through disruption of the NCC—mesoderm interface.

CONCLUSIONS AND PERSPECTIVES

Development of a variety of tissues is affected by disruptions of different stages of NCC development. Neurocristopathies that arise due to a disruption of NCC formation include TCS, DBA, and Miller syndromes. Interestingly, an underlying feature of each of these syndromes is a defect in ribosome biogenesis (Figure 17.3) and a disruption in craniofacial development. However, each syndrome remains distinct and reflects the need for a more specific understanding of how each of the identified genes functions in NCC development and survival. Neurocristopathies that arise due to defects in EMT and NCC migration take several forms, including BOFS, HSCR, Mowat—Wilson syndrome, and piebaldism. The aberrant reactivation of EMT can lead to metastatic cancers such as melanoma and neuroblastoma. While a similar step of NCC development is disrupted in these syndromes, very different phenotypes result. Other neurocristopathies arising from disruptions in migration include cell autonomous disorders such as HSCR and WS, and non cell autonomous disorders as in DGS. Neurocristopathies that result from a disruption of NCC differentiation may affect very specific tissues. In the case of CCHS, neuronal differentiation is altered. Pigment differentiation is affected in OCA, and cranial suture development and osteogenesis is disrupted in CS. CS is an interesting example of a neurocristopathy that can arise from either aberrant NCC migration or premature differentiation.

Interestingly, a number of neurocristopathies have overlapping phenotypic characteristics, such as piebaldism, WS, and OCA, but these arise from alterations in different phases of NCC development. Other neurocristopathies such as TCS and DBA arise from deficiencies in the same pathway—namely, ribosome biogenesis—but still have unique characteristics. Knowing the underlying etiology and pathogenesis of individual neurocristopathies is essential for developing potential therapeutic treatments.

One potential therapy under investigation is the use of stem cells such as adipose-derived stem cells for craniofacial regeneration and repair. Adipose-derived stem cells (ASC) have several advantages, including that they are easily harvested from the patient themselves and they are abundant due to multiple donor sites, which yield high numbers of stem cells. ASC can differentiate into both osteogenic and chondrogenic cells *in vitro*, and have been used in clinical studies to improve calvarial defects [236]. It would be interesting to learn whether ASC have applications in other tissues, such as in cardiac repair in DGS or neuronal generation as a treatment for HSCR or CCHS.

Other preventative methods for neurocristopathies that arise due to a deficiency in NCC formation may involve nutritional supplementation and suppression of apoptosis. L-leucine treatment has been used to successfully treat DBA in humans [44,45] and animal models [42,43]. Embryonic zebrafish and mice that model DBA were treated with L-leucine and showed improved craniofacial and hematopoietic development [42,43]. L-leucine stimulates ribosome biogenesis through the mTOR pathway and may be a viable treatment option for other ribosome biogenesis-related neurocristopathies such as TCS. Further research is needed to determine if L-leucine treatment may be able to prevent the early loss of NCC in these models and if the pathogenesis of these syndromes is primarily related to ribosome biogenesis. Even if ribosome biogenesis is a common underlying feature, alternative treatments to L-leucine may be necessary depending on how ribosome biogenesis is disrupted. An additional method for prevention of these phenotypes is inhibition of apoptosis. p53 inhibition has shown to be effective in the improved survival of models of TCS (*Tcof1*), DBA (*Rps19*), and WS (*Pax3*). This could be useful in neurocristopathies that arise due to deficiencies in NCC formation and migration. Again, it will be important to determine the pathogenesis of the specific syndrome in order to understand if inhibiting apoptosis is a viable preventative measure.

The neurocristopathies presented here have in common a disruption of one or more phases of NCC development. However, these syndromes vary widely in the stage of NCC development that is disrupted and the tissue types that are affected, which highlights the complexity of NCC development. Some neurocristopathies that have the same phenotype arise from disruptions in different pathways, while other neurocristopathies arise from disruptions in the same pathway and have different phenotypes. Understanding the etiology and pathogenesis of individual conditions and knowing whether they arise due to defects in NCC formation

migration, and/or differentiation will be instrumental in designing realistic avenues for therapeutic prevention of neurocristopathies.

Acknowledgments

The authors thank Molly Hague, Naomi Tjaden, and Jennifer Dennis for images used in Figure 17.1A, B and C and D, respectively. We also thank Dr Tom Schilling for providing sox10:GFP zebrafish that were used to generate images in Figure 17.1E and F. Research in the Trainor laboratory is supported by the Stowers Institute for Medical Research, and National Institute of Dental and Craniofacial Research (DE 016082).

References

[1] Selleck MA, Bronner-Fraser M. Origins of the avian neural crest: the role of neural plate—epidermal interactions. Development 1995;121(2):525—38.

[2] Rollhauser-ter Horst J. Artificial neural induction in amphibia. I. Sandwich explants. Anat Embryol (Berl) 1977;151:309—16.

[3] Moury JD, Jacobson AG. The origins of neural crest cells in the axolotl. Dev Biol 1990;141(2):243—53.

[4] Mayor R, Aybar M. Induction and development of neural crest in *Xenopus laevis*. Cell Tissue Res 2001;305 (2):203—9.

[5] LaBonne C, Bronner-Fraser M. Neural crest induction in *Xenopus*: evidence for a two-signal model. Development 1998;125(13):2403—14.

[6] Lewis JL, Bonner J, Modrell M, Ragland JW, Moon RT, Dorsky RI, et al. Reiterated Wnt signaling during zebrafish neural crest development. Development 2004;131 (6):1299—308.

[7] Streit A, Berliner AJ, Papanayotou C, Sirulnik A, Stern CD. Initiation of neural induction by FGF signalling before gastrulation. Nature 2000;406(6791):74—8.

[8] Villanueva S, Glavic A, Ruiz P, Mayor R. Posteriorization by FGF, Wnt, and retinoic acid is required for neural crest induction. Dev Biol 2002;241 (2):289—301.

[9] Basch ML, Bronner-Fraser M, García-Castro MI. Specification of the neural crest occurs during gastrulation and requires Pax7. Nature 2006;441(7090): 218—22.

[10] Stuhlmiller TJ, García-Castro MI. FGF/MAPK signaling is required in the gastrula epiblast for avian neural crest induction. Development 2012;139(2):289—300.

[11] Maczkowiak F, Matéos S, Wang E, Roche D, Harland R, Monsoro-Burq AH. The Pax3 and Pax7 paralogs cooperate in neural and neural crest patterning using

distinct molecular mechanisms, in *Xenopus laevis* embryos. Dev Biol 2010;340(2):381−96.

[12] Monsoro-Burq A-H, Fletcher RB, Harland RM. Neural crest induction by paraxial mesoderm in *Xenopus* embryos requires FGF signals. Development 2003;130 (14):3111−24.

[13] Crane JF, Trainor PA. Neural crest stem and progenitor cells. Ann Rev Cell Dev Biol 2006;22(1):267−86.

[14] The Treacher Collins Syndrome Collaborative Group, Dixon J, Edwards SJ, Gladwin AJ, Dixon MJ, Loftus SK, et al. Positional cloning of a gene involved in the pathogenesis of Treacher Collins syndrome. Nat Genet 1996;12(2):130−6.

[15] Gladwin AJ, Dixon J, Loftus SK, Edwards S, Wasmuth JJ, Hennekam RCM, et al. Treacher Collins syndrome may result from insertions, deletions or splicing mutations, which introduce a termination codon into the gene. Hum Mol Genet 1996;5(10):1533−8.

[16] Edwards SJ, Gladwin AJ, Dixon MJ. The mutational spectrum in Treacher Collins syndrome reveals a predominance of mutations that create a premature-termination codon. Am J Med Genet 1997;60:515−24.

[17] Wise CA, Chiang LC, Paznekas WA, Sharma M, Musy MM, Ashley JA, et al. TCOF1 gene encodes a putative nucleolar phosphoprotein that exhibits mutations in Treacher Collins syndrome throughout its coding region. Proc Natl Acad Sci USA 1997;94 (7):3110−5.

[18] Dauwerse JG, Dixon J, Seland S, Ruivenkamp CAL, van Haeringen A, Hoefsloot LH, et al. Mutations in genes encoding subunits of RNA polymerases I and III cause Treacher Collins syndrome. Nat Genet 2011;43(1):20−2.

[19] Marsh KL, Dixon J, Dixon MJ. Mutations in the Treacher Collins syndrome gene lead to mislocalization of the nucleolar protein treacle. Hum Mol Genet 1998;7(11):1795−800.

[20] Winokur ST, Shiang R. The Treacher Collins syndrome (TCOF1) gene product, treacle, is targeted to the nucleolus by signals in its C-Terminus. Hum Mol Genet 1998;7(12):1947−52.

[21] Valdez BC, Henning D, So RB, Dixon J, Dixon MJ. The Treacher Collins syndrome (TCOF1) gene product is involved in ribosomal DNA gene transcription by interacting with upstream binding factor. Proc Natl Acad Sci USA 2004;101(29):10709−14.

[22] Lalo D, Carles C, Sentenac A, Thuriaux P. Interactions between three common subunits of yeast RNA polymerases I and III. Proc Natl Acad Sci USA 1993;90 (12):5524−8.

[23] Larkin RM, Guilfoyle TJ. Reconstitution of yeast and arabidopsis RNA polymerase α-like subunit heterodimers. J Biol Chem 1997;272(19):12824−30.

[24] Lafontaine DLJ, Tollervey D. The function and synthesis of ribosomes. Nat Rev Mol Cell Biol 2001;2 (7):514−20.

[25] Kressler D, Hurt E, Baβler J. Driving ribosome assembly. Biochim Biophys Acta 2010;1803(6):673−83.

[26] Tschochner H, Hurt E. Pre-ribosomes on the road from the nucleolus to the cytoplasm. Trends Cell Biol 2003;13(5):255−63.

[27] Gonzales B, Henning D, So RB, Dixon J, Dixon MJ, Valdez BC. The Treacher Collins syndrome (TCOF1) gene product is involved in pre-rRNA methylation. Hum Mol Genet 2005;14(14):2035−43.

[28] Hayano T, Yanagida M, Yamauchi Y, Shinkawa T, Isobe T, Takahashi N. Proteomic analysis of human Nop56p-associated Pre-ribosomal ribonucleoprotein complexes: possible link between Nop56p and the nucleolar protein treacle responsible for Treacher Collins syndrome. J Biol Chem 2003;278(36): 34309−19.

[29] Dixon J, Jones NC, Sandell LL, Jayasinghe SM, Crane J, Rey J-P, et al. Tcof1/Treacle is required for neural crest cell formation and proliferation deficiencies that cause craniofacial abnormalities. Proc Natl Acad Sci USA 2006;103(36):13403−8.

[30] Jones NC, Lynn ML, Gaudenz K, Sakai D, Aoto K, Rey J-P, et al. Prevention of the neurocristopathy Treacher Collins syndrome through inhibition of p53 function. Nat Med 2008;14(2):125−33.

[31] Rubbi CP, Milner J. Disruption of the nucleolus mediates stabilization of p53 in response to DNA damage and other stresses. EMBO J 2003;22(22):6068−77.

[32] Mann C, Buhler J-M, Treich I, Sentenac A. RPC40, a unique gene for a subunit shared between yeast RNA polymerases A and C. Cell 1987;48(4):627−37.

[33] Tseng H. Cell-type-specific regulation of RNA polymerase I transcription: a new frontier. BioEssays 2006;28(7):719−25.

[34] Boulon S, Westman BJ, Hutten S, Boisvert F-M, Lamond AI. The nucleolus under stress. Mol Cell 2010;40(2):216−27.

[35] Lipton JM, Ellis SR. Diamond-Blackfan anemia: diagnosis, treatment, and molecular pathogenesis. Hematol Oncol Clin North Am 2009;23(2):261−82.

[36] Narla A, Ebert BL. Ribosomopathies: human disorders of ribosome dysfunction. Blood 2010;115(16):3196−205.

[37] Gazda HT, Sheen MR, Vlachos A, Choesmel V, O'Donohue M-F, Schneider H, et al. Ribosomal protein L5 and L11 mutations are associated with cleft palate and abnormal thumbs in Diamond-Blackfan anemia patients. Am J Hum Genet 2008;83(6):769−80.

[38] Danilova N, Sakamoto KM, Lin S. Ribosomal protein L11 mutation in zebrafish leads to haematopoietic and metabolic defects. Br J Haematol 2011;152(2):217−28.

[39] Chakraborty A, Uechi T, Higa S, Torihara H, Kenmochi N. Loss of ribosomal protein L11 affects zebrafish embryonic development through a p53-dependent apoptotic response. PLoS ONE 2009;4(1): e4152.

[40] Jaako P, Flygare J, Olsson K, Quere R, Ehinger M, Henson A, et al. Mice with ribosomal protein S19 deficiency develop bone marrow failure and symptoms like patients with Diamond-Blackfan anemia. Blood 2011;118(23):6087−96.

[41] Danilova N, Sakamoto KM, Lin S. Ribosomal protein S19 deficiency in zebrafish leads to developmental abnormalities and defective erythropoiesis through activation of p53 protein family. Blood 2008;112 (13):5228−37.

[42] Payne EM, Virgilio M, Narla A, Sun H, Levine M, Paw BH, et al. L-leucine improves the anemia and developmental defects associated with Diamond-Blackfan anemia and del(5q) MDS by activating the mTOR pathway. Blood 2012;120(11):2214−24.

[43] Jaako P, Debnath S, Olsson K, Bryder D, Flygare J, Karlsson S. Dietary L-leucine improves the anemia in a mouse model for Diamond-Blackfan anemia. Blood 2012;120(11):2225−8.

[44] Cmejlova J, Dolezalova L, Pospisilova D, Petrtylova K, Petrak J, Cmejla R. Translational efficiency in patients with Diamond-Blackfan anemia. Haematologica 2006;91(11):1456−64.

[45] Pospisilova D, Cmejlova J, Hak J, Adam T, Cmejla R. Successful treatment of a Diamond-Blackfan anemia patient with amino acid leucine. Haematologica 2007;92(5):e66−7.

[46] Bonfils G, Jaquenoud M, Bontron S, Ostrowicz C, Ungermann C, De Virgilio C. Leucyl-tRNA synthetase controls TORC1 via the EGO Complex. Mol Cell 2012;46(1):105−10.

[47] Hannan KM, Brandenburger Y, Jenkins A, Sharkey K, Cavanaugh A, Rothblum L, et al. mTOR-Dependent regulation of ribosomal gene transcription requires S6K1 and is mediated by phosphorylation of the carboxy-terminal activation domain of the nucleolar transcription factor UBF. Mol Cell Biol 2003;23(23):8862−77.

[48] Mayer C, Zhao J, Yuan X, Grummt I. mTOR-dependent activation of the transcription factor TIF-IA links rRNA synthesis to nutrient availability. Genes Dev 2004;18(4):423−34.

[49] Wei Y, Tsang CK, Zheng XFS. Mechanisms of regulation of RNA polymerase III-dependent transcription by TORC1. EMBO J 2009;28(15):2220−30.

[50] Michels AA, Robitaille AM, Buczynski-Ruchonnet D, Hodroj W, Reina JH, Hall MN, et al. mTORC1 directly phosphorylates and regulates human MAF1. Mol Cell Biol 2010;30(15):3749−57.

[51] Powers T, Walter P. Regulation of ribosome biogenesis by the rapamycin-sensitive TOR-signaling pathway in Saccharomyces cerevisiae. Mol Biol Cell 1999;10 (4):987−1000.

[52] Ng SB, Buckingham KJ, Lee C, Bigham AW, Tabor HK, Dent KM, et al. Exome sequencing identifies the cause of a Mendelian disorder. Nat Genet 2010;42 (1):30−5.

[53] Rainger J, Bengani H, Campbell L, Anderson E, Sokhi K, Lam W, et al. Miller (Genée−Wiedemann) syndrome represents a clinically and biochemically distinct subgroup of postaxial acrofacial dysostosis associated with partial deficiency of DHODH. Hum Mol Genet 2012;21(18):3969−83.

[54] White RM, Cech J, Ratanasirintrawoot S, Lin CY, Rahl PB, Burke CJ, et al. DHODH modulates transcriptional elongation in the neural crest and melanoma. Nature 2011;471(7339):518−22.

[55] Taneyhill LA, Schiffmacher AT. Cadherin dynamics during neural crest cell ontogeny. In: Frans van R, editor. Progress in molecular biology and translational science. Academic Press; 2013. p. 291−315 [Chapter 13].

[56] Cano A, Perez-Moreno MA, Rodrigo I, Locascio A, Blanco MJ, del Barrio MG, et al. The transcription factor snail controls epithelial−mesenchymal transitions by repressing E-cadherin expression. Nat Cell Biol 2000;2(2):76−83.

[57] Bolós V, Peinado H, Pérez-Moreno MA, Fraga MF, Esteller M, Cano A. The transcription factor Slug represses E-cadherin expression and induces epithelial to mesenchymal transitions: a comparison with Snail and E47 repressors. J Cell Sci 2003;116 (3):499−511.

[58] Jiang R, Lan Y, Norton CR, Sundberg JP, Gridley T. The Slug gene is not essential for mesoderm or neural crest development in mice. Dev Biol 1998;198 (2):277−85.

[59] Murray SA, Gridley T. Snail family genes are required for left−right asymmetry determination, but not neural crest formation, in mice. Proc Natl Acad Sci USA 2006;103(27):10300−4.

[60] Van de Putte T, Maruhashi M, Francis A, Nelles L, Kondoh H, Huylebroeck D, et al. Mice lacking Zfhx1b, the gene that codes for Smad-interacting protein-1, reveal a role for multiple neural crest cell defects in the etiology of Hirschsprung disease mental retardation syndrome. Am J Hum Genet 2003;72(2):465−70.

[61] Trainor PA, Krumlauf R. Hox genes, neural crest cells and branchial arch patterning. Curr Opin Cell Biol 2001;13(6):698−705.

[62] Fujimoto A, Lipson M, Lacro RV, Shinno NW, Boelter WD, Jones KL, et al. New autosomal dominant

branchio-oculo-facial syndrome. Am J Med Genet 1987;27(4):943−51.

[63] Noden DM, Trainor PA. Relations and interactions between cranial mesoderm and neural crest populations. J Anat 2005;207(5):575−601.

[64] Horigome N, Myojin M, Ueki T, Hirano S, Aizawa S, Kuratani S. Development of cephalic neural crest cells in embryos of *Lampetra japonica*, with special reference to the evolution of the jaw. Dev Biol 1999;207 (2):287−308.

[65] McCauley DW, Bronner-Fraser M. Neural crest contributions to the lamprey head. Development 2003;130 (11):2317−27.

[66] Epperlein H, Meulemans D, Bronner-Fraser M, Steinbeisser H, Selleck MA. Analysis of cranial neural crest migratory pathways in axolotl using cell markers and transplantation. Development 2000;127 (12):2751−61.

[67] Noden DM. An analysis of migratory behavior of avian cephalic neural crest cells. Dev Biol 1975;42 (1):106−30.

[68] Vaglia JL, Smith KK. Early differentiation and migration of cranial neural crest in the opossum, *Monodelphis domextica*. Evol Dev 2003;5(2):121−35.

[69] Vaage S. The segmentation of the primitive neural tube in chick embryos (*Gallus domesticus*). Adv Anat Embryol Cell Biol 1969;41:1−88.

[70] Brown C, Baldwin HS. Neural crest contribution to the cardiovascular system. In: Saint-Jeannet J-P, editor. Neural crest induction and differentiation. US: Springer; 2006. p. 134−54.

[71] Keyte A, Hutson MR. The neural crest in cardiac congenital anomalies. Differentiation 2012;84(1):25−40.

[72] Schwarz Q, Maden CH, Vieira JM, Ruhrberg C. Neuropilin 1 signaling guides neural crest cells to coordinate pathway choice with cell specification. Proc Natl Acad Sci USA 2009;106(15):6164−9.

[73] Kulesa PM, Lefcort F, Kasemeier-Kulesa JC. The migration of autonomic precursor cells in the embryo. Auton Neurosci 2009;151(1):3−9.

[74] Erickson CA, Duong TD, Tosney KW. Descriptive and experimental analysis of the dispersion of neural crest cells along the dorsolateral path and their entry into ectoderm in the chick embryo. Dev Biol 1992;151 (1):251−72.

[75] Teillet M-A, Kalcheim C, Le Douarin NM. Formation of the dorsal root ganglia in the avian embryo: segmental origin and migratory behavior of neural crest progenitor cells. Dev Biol 1987;120(2):329−47.

[76] Serbedzija GN, Fraser SE, Bronner-Fraser M. Pathways of trunk neural crest cell migration in the mouse embryo as revealed by vital dye labelling. Development 1990;108(4):605−12.

[77] Yu H-H, Moens CB. Semaphorin signaling guides cranial neural crest cell migration in zebrafish. Dev Biol 2005;280(2):373−85.

[78] Eickholt BJ, Mackenzie SL, Graham A, Walsh FS, Doherty P. Evidence for collapsin-1 functioning in the control of neural crest migration in both trunk and hindbrain regions. Development 1999;126(10): 2181−9.

[79] Kawasaki T, Bekku Y, Suto F, Kitsukawa T, Taniguchi M, Nagatsu I, et al. Requirement of neuropilin 1-mediated Sema3A signals in patterning of the sympathetic nervous system. Development 2002;129 (3):671−80.

[80] Schwarz Q, Maden CH, Davidson K, Ruhrberg C. Neuropilin-mediated neural crest cell guidance is essential to organise sensory neurons into segmented dorsal root ganglia. Development 2009;136 (11):1785−9.

[81] Maden CH, Gomes J, Schwarz Q, Davidson K, Tinker A, Ruhrberg C. NRP1 and NRP2 cooperate to regulate gangliogenesis, axon guidance and target innervation in the sympathetic nervous system. Dev Biol 2012;369 (2):277−85.

[82] Belmadani A, Tran PB, Ren D, Assimacopoulos S, Grove EA, Miller RJ. The chemokine stromal cell-derived factor-1 regulates the migration of sensory neuron progenitors. J Neurosci 2005;25(16):3995−4003.

[83] Kasemeier-Kulesa JC, McLennan R, Romine MH, Kulesa PM, Lefcort F. CXCR4 controls ventral migration of sympathetic precursor cells. J Neurosci 2010;30 (39):13078−88.

[84] Escot S, Blavet C, Härtle S, Duband J-L, Fournier-Thibault C. Mis-regulation of SDF1-CXCR4 signalling impairs early cardiac neural crest cell migration leading to conotruncal defects. Circ Res 2013;113 (5):505−16.

[85] Milunsky JM, Maher TM, Zhao G, Wang Z, Mulliken JB, Chitayat D, et al. Genotype−phenotype analysis of the branchio-oculo-facial syndrome. Am J Med Genet A 2011;155(1):22−32.

[86] de Crozé N, Maczkowiak F, Monsoro-Burq AH. Reiterative AP2a activity controls sequential steps in the neural crest gene regulatory network. Proc Natl Acad Sci USA 2011;108(1):155−60.

[87] Zhang J, Hagopian-Donaldson S, Serbedzija G, Elsemore J, Plehn-Dujowich D, McMahon AP, et al. Neural tube, skeletal and body wall defects in mice lacking transcription factor AP-2. Nature 1996;381 (6579):238−41.

[88] Schorle H, Meier P, Buchert M, Jaenisch R, Mitchell PJ. Transcription factor AP-2 essential for cranial closure and craniofacial development. Nature 1996;381 (6579):235−8.

[89] Knight RD, Nair S, Nelson SS, Afshar A, Javidan Y, Geisler R, et al. Lockjaw encodes a zebrafish tfap2a required for early neural crest development. Development 2003;130(23):5755–68.

[90] Knight RD, Javidan Y, Zhang T, Nelson S, Schilling TF. AP2-dependent signals from the ectoderm regulate craniofacial development in the zebrafish embryo. Development 2005;132(13):3127–38.

[91] Barrallo-Gimeno A, Holzschuh J, Driever W, Knapik EW. Neural crest survival and differentiation in zebrafish depends on mont blanc/tfap2a gene function. Development 2004;131(7):1463–77.

[92] Pfisterer P, Ehlermann J, Hegen M, Schorle HA. Subtractive gene expression screen suggests a role of transcription factor AP-2α in control of proliferation and differentiation. J Biol Chem 2002;277(8):6637–44.

[93] Betters E, Liu Y, Kjaeldgaard A, Sundström E, García-Castro MI. Analysis of early human neural crest development. Dev Biol 2010;344(2):578–92.

[94] Giebel LB, Spritz RA. Mutation of the KIT (mast/stem cell growth factor receptor) protooncogene in human piebaldism. Proc Natl Acad Sci USA 1991;88 (19):8696–9.

[95] Sánchez-Martín M, Pérez-Losada J, Rodríguez-García A, González-Sánchez B, Korf BR, Kuster W, et al. Deletion of the SLUG (SNAI2) gene results in human piebaldism. Am J Med Genet A 2003;122A(2):125–32.

[96] Chabot B, Stephenson DA, Chapman VM, Besmer P, Bernstein A. The proto-oncogene c-kit encoding a transmembrane tyrosine kinase receptor maps to the mouse W locus. Nature 1988;335(6185):88–9.

[97] Geissler EN, Ryan MA, Housman DE. The dominant-white spotting (W) locus of the mouse encodes the c-kit proto-oncogene. Cell 1988;55 (1):185–92.

[98] Huang E, Nocka K, Beier DR, Chu T-Y, Buck J, Lahm H-W, et al. The hematopoietic growth factor KL is encoded by the Sl locus and is the ligand of the c-kit receptor, the gene product of the W locus. Cell 1990;63(1):225–33.

[99] Zsebo KM, Williams DA, Geissler EN, Broudy VC, Martin FH, Atkins HL, et al. Stem cell factor is encoded at the Sl locus of the mouse and is the ligand for the c-kit tyrosine kinase receptor. Cell 1990;63(1):213–24.

[100] Keshet E, Lyman SD, Williams DE, Anderson DM, Jenkins NA, Copeland NG, et al. Embryonic RNA expression patterns of the *c-kit* receptor and its cognate ligan suggest multiple functional roles in mouse development. EMBO J 1991;10(9):2425–35.

[101] Wehrle-Haller B, Weston JA. Soluble and cell-bound forms of steel factor activity play distinct roles in melanocyte precursor dispersal and survival on the lateral neural crest migration pathway. Development 1995;121(3):731–42.

[102] Tabone-Eglinger S, Wehrle-Haller M, Aebischer N, Jacquier M-C, Wehrle-Haller B. Membrane-bound kit ligand regulates melanocyte adhesion and survival, providing physical interaction with an intraepithelial niche. FASEB J 2012;26(9):3738–53.

[103] Parichy DM, Rawls JF, Pratt SJ, Whitfield TT, Johnson SL. Zebrafish sparse corresponds to an orthologue of c-kit and is required for the morphogenesis of a subpopulation of melanocytes, but is not essential for hematopoiesis or primordial germ cell development. Development 1999;126(15):3425–36.

[104] Alexeev V, Yoon K. Distinctive role of the cKit receptor tyrosine kinase signaling in mammalian melanocytes. J Invest Dermatol 2006;126(5):1102–10.

[105] Grichnik JM, Burch JA, Burchette J, Shea CR. The SCF//KIT pathway plays a critical role in the control of normal human melanocyte homeostasis 1998;111 (2):233–8.

[106] Nieto M, Sargent M, Wilkinson D, Cooke J. Control of cell behavior during vertebrate development by Slug, a zinc finger gene. Science 1994;264(5160):835–9.

[107] Carl TF, Dufton C, Hanken J, Klymkowsky MW. Inhibition of neural crest migration in *Xenopus* using antisense slug RNA. Dev Biol 1999;213(1):101–15.

[108] Taneyhill LA, Coles EG, Bronner-Fraser M. Snail2 directly represses cadherin6B during epithelial-to-mesenchymal transitions of the neural crest. Development 2007;134(8):1481–90.

[109] Pérez-Losada J, Sánchez-Martín M, Rodríguez-García A, Sánchez ML, Orfao A, Flores T, et al. Zinc-finger transcription factor Slug contributes to the function of the stem cell factor c-kit signaling pathway. Blood 2002;100(4):1274–86.

[110] Heanue TA, Pachnis V. Enteric nervous system development and Hirschsprung's disease: advances in genetic and stem cell studies. Nat Rev Neurosci 2007;8(6):466–79.

[111] Fu M, Chi Hang Lui V, Har Sham M, Nga Yin Cheung A, Kwong Hang Tam P. HOXB5 expression is spatially and temporarily regulated in human embryonic gut during neural crest cell colonization and differentiation of enteric neuroblasts. Dev Dyn 2003;228(1):1–10.

[112] Druckenbrod NR, Epstein ML. The pattern of neural crest advance in the cecum and colon. Dev Biol 2005;287(1):125–33.

[113] Nishiyama C, Uesaka T, Manabe T, Yonekura Y, Nagasawa T, Newgreen DF, et al. Trans-mesenteric neural crest cells are the principal source of the colonic enteric nervous system. Nat Neurosci 2012;15 (9):1211–8.

[114] Natarajan D, Marcos-Gutierrez C, Pachnis V, de Graaff E. Requirement of signalling by receptor tyrosine kinase RET for the directed migration of enteric nervous system progenitor cells during mammalian embryogenesis. Development 2002;129(22):5151−60.

[115] Schuchardt A, D'Agati V, Larsson-Blomberg L, Costantini F, Pachnis V. Defects in the kidney and enteric nervous system of mice lacking the tyrosine kinase receptor Ret. Nature 1994;367(6461):380−3.

[116] Shepherd IT, Pietsch J, Elworthy S, Kelsh RN, Raible DW. Roles for GFRα1 receptors in zebrafish enteric nervous system development. Development 2004;131 (1):241−9.

[117] Sanchez MP, Silos-Santiago I, Frisen J, He B, Lira SA, Barbacid M. Renal agenesis and the absence of enteric neurons in mice lacking GDNF. Nature 1996;382(6586):70−3.

[118] Pichel JG, Shen L, Sheng HZ, Granholm A-C, Drago J, Grinberg A, et al. Defects in enteric innervation and kidney development in mice lacking GDNF. Nature 1996;382(6586):73−6.

[119] Moore MW, Klein RD, Farinas I, Sauer H, Armanini M, Phillips H, et al. Renal and neuronal abnormalities in mice lacking GDNF. Nature 1996;382 (6586):76−9.

[120] Cacalano G, Fariñas I, Wang L-C, Hagler K, Forgie A, Moore M, et al. GFRα1 is an essential receptor component for GDNF in the developing nervous system and kidney. Neuron 1998;21(1):53−62.

[121] Doray B, Salomon R, Amiel J, Pelet A, Touraine RL, Billaud M, et al. Mutation of the RET ligand, neurturin, supports multigenic inheritance in Hirschsprung disease. Hum Mol Genet 1998;7(9):1449−52.

[122] Heuckeroth RO, Enomoto H, Grider JR, Golden JP, Hanke JA, Jackman A, et al. Gene targeting reveals a critical role for neurturin in the development and maintenance of enteric, sensory, and parasympathetic neurons. Neuron 1999;22(2):253−63.

[123] Rossi J, Herzig K-H, Voilkar V, Hiltunen PV, Segerstrale M, Airaksinen MS, et al. Alimentary tract innervation deficits and dysfunction in mice lacking GDNF family receptor α2. J Clin Invest 2003;112(5):707−16.

[124] Stanchina L, Van de Putte T, Goossens M, Huylebroeck D, Bondurand N. Genetic interaction between Sox10 and Zfhx1b during enteric nervous system development. Dev Biol 2010;341(2):416−28.

[125] McCallion AS, Stames E, Conlon RA, Chakravarti A. Phenotype variation in two-locus mouse models of Hirschsprung disease: tissue-specific interaction between Ret and Ednrb. Proc Natl Acad Sci USA 2003;100(4):1826−31.

[126] Carrasquillo MM, McCallion AS, Puffenberger EG, Kashuk CS, Nouri N, Chakravarti A. Genome-wide association study and mouse model identify interaction between RET and EDNRB pathways in Hirschsprung disease. Nat Genet 2002;32(2):237−44.

[127] Mowat DR, Wilson MJ, Goossens M. Mowat−Wilson syndrome. J Med Genet 2003;40(5):305−10.

[128] Wakamatsu N, Yamada Y, Yamada K, Ono T, Nomura N, Taniguchi H, et al. Mutations in SIP1, encoding Smad interacting protein-1, cause a form of Hirschsprung disease. Nat Genet 2001;27(4):369−70.

[129] Cacheux V, Dastot-Le Moal F, Kääriäinen H, Bondurand N, Rintala R, Boissier B, et al. Loss-of-function mutations in SIP1 Smad interacting protein 1 result in a syndromic Hirschsprung disease. Hum Mol Genet 2001;10(14):1503−10.

[130] Espinosa-Parrilla Y, Amiel J, Augé J, Encha-Razavi F, Munnich A, Lyonnet S, et al. Expression of the SMADIP1 gene during early human development. Mech Dev 2002;114(1−2):187−91.

[131] Bassez G, Camand OJA, Cacheux V, Kobetz A, Dastot-Le Moal F, Marchant D, et al. Pleiotropic and diverse expression of ZFHX1B gene transcripts during mouse and human development supports the various clinical manifestations of the "Mowat−Wilson" syndrome. Neurobiol Dis 2004;15 (2):240−50.

[132] Comijn J, Berx G, Vermassen P, Verschueren K, van Grunsven L, Bruyneel E, et al. The two-handed E box binding zinc finger protein SIP1 downregulates E-cadherin and induces invasion. Mol Cell 2001;7 (6):1267−78.

[133] van Grunsven LA, Michiels C, Van de Putte T, Nelles L, Wuytens G, Verschueren K, et al. Interaction between Smad-interacting protein-1 and the corepressor c-terminal binding protein is dispensable for transcriptional repression of E-cadherin. J Biol Chem 2003;278(28):26135−45.

[134] Van de Putte T, Francis A, Nelles L, van Grunsven LA, Huylebroeck D. Neural crest-specific removal of Zfhx1b in mouse leads to a wide range of neurocristopathies reminiscent of Mowat−Wilson syndrome. Hum Mol Genet 2007;16(12):1423−36.

[135] Banerjee AK. Waardenburg's syndrome associated with ostium secundum atrial septal defect. J R Soc Med 1986;79(11):677−8.

[136] Baldwin CT, Hoth CF, Amos JA, da-Silva EO, Milunsky A. An exonic mutation in the HuP2 paired domain gene causes Waardenburg's syndrome. Nature 1992;355(6361):637−8.

[137] Tassabehji M, Read AP, Newton VE, Harris R, Balling R, Gruss P, et al. Waardenburg's syndrome patients have mutations in the human homologue of the Pax-3 paired box gene. Nature 1992;355 (6361):635−6.

[138] Hoth CF, Milunsky A, Lipsky N, Sheffer R, Clarren SK, Baldwin CT. Mutations in the paired domain of the human PAX3 gene cause Klein–Waardenburg syndrome (WS-III) as well as Waardenburg syndrome type I (WS-I). Am J Hum Genet 1993;52(3):455–62.

[139] Pingault V, Ente D, Dastot-Le Moal F, Goossens M, Marlin S, Bondurand N. Review and update of mutations causing Waardenburg syndrome. Hum Mutat 2010;31(4):391–406.

[140] Sánchez-Martín M, Rodríguez-García A, Pérez-Losada J, Sagrera A, Read AP, Sánchez-García I. SLUG (SNAI2) deletions in patients with Waardenburg disease. Hum Mol Genet 2002;11(25):3231–6.

[141] Goulding MD, Chalepakis G, Deutsch U, Erselius JR, Gruss P. Pax-3, a novel murine DNA binding protein expressed during early neurogenesis. EMBO J 1991;10(5):1135–47.

[142] Olaopa M, Zhou H-m, Snider P, Wang J, Schwartz RJ, Moon AM, et al. Pax3 is essential for normal cardiac neural crest morphogenesis but is not required during migration nor outflow tract septation. Dev Biol 2011;356(2):308–22.

[143] Nelms BL, Pfaltzgraff ER, Labosky PA. Functional interaction between Foxd3 and Pax3 in cardiac neural crest development. Genesis 2011;49(1):10–23.

[144] Epstein DJ, Vekemans M, Gros P. splotch (Sp2H), a mutation affecting development of the mouse neural tube, shows a deletion within the paired homeodomain of Pax-3. Cell 1991;67(4):767–74.

[145] Hornyak TJ, Hayes DJ, Chiu L-Y, Ziff EB. Transcription factors in melanocyte development: distinct roles for Pax-3 and Mitf. Mech Dev 2001;101(1–2):47–59.

[146] Barr FG, Galili N, Holick J, Biegel JA, Rovera G, Emanuel BS. Rearrangement of the PAX3 paired box gene in the paediatric solid tumour alveolar rhabdomyosarcoma. Nat Genet 1993;3(2):113–7.

[147] Shapiro DN, Sublett JE, Li B, Downing JR, Naeve CW. Fusion of PAX3 to a member of the forkhead family of transcription factors in human alveolar rhabdomyosarcoma. Cancer Res 1993;53(21):5108–12.

[148] Scholl FA, Kamarashev J, Murmann OV, Geertsen R, Dummer R, Schäfer BW. PAX3 is expressed in human melanomas and contributes to tumor cell survival. Cancer Res 2001;61(3):823–6.

[149] Steingrimsson E, Moore KJ, Lamoreux ML, Ferre-D'Amare AR, Burley SK, Sanders Zimring DC, et al. Molecular basis of mouse microphthalmia (mi) mutations helps explain their developmental and phenotypic consequences. Nat Genet 1994;8(3):256–63.

[150] Kelsh RN. Sorting out Sox10 functions in neural crest development. BioEssays 2006;28(8):788–98.

[151] Potterf SB, Furumura M, Dunn KJ, Arnheiter H, Pavan WJ. Transcription factor hierarchy in Waardenburg syndrome: regulation of MITF expression by SOX10 and PAX3. Hum Genet 2000;107(1):1–6.

[152] Bondurand N, Pingault V, Goerich DE, Lemort N, Sock E, Caignec CL, et al. Interaction among SOX10, PAX3 and MITF, three genes altered in Waardenburg syndrome. Hum Mol Genet 2000;9(13):1907–17.

[153] Elworthy S, Lister JA, Carney TJ, Raible DW, Kelsh RN. Transcriptional regulation of mitfa accounts for the sox10 requirement in zebrafish melanophore development. Development 2003;130(12):2809–18.

[154] Bondurand N, Natarajan D, Barlow A, Thapar N, Pachnis V. Maintenance of mammalian enteric nervous system progenitors by SOX10 and endothelin 3 signalling. Development 2006;133(10):2075–86.

[155] Southard-Smith EM, Kos L, Pavan WJ. SOX10 mutation disrupts neural crest development in Dom Hirschsprung mouse model. Nat Genet 1998;18(1):60–4.

[156] Southard-Smith EM, Angrist M, Ellison JS, Agarwala R, Baxevanis AD, Chakravarti A, et al. The Sox10Dom mouse: modeling the genetic variation of Waardenburg-Shah (WS4) syndrome. Genome Res 1999;9(3):215–25.

[157] Dutton KA, Pauliny A, Lopes SS, Elworthy S, Carney TJ, Rauch J, et al. Zebrafish colourless encodes sox10 and specifies non-ectomesenchymal neural crest fates. Development 2001;128(21):4113–25.

[158] Elworthy S, Pinto JP, Pettifer A, Cancela ML, Kelsh RN. Phox2b function in the enteric nervous system is conserved in zebrafish and is sox10-dependent. Mech Dev 2005;122(5):659–69.

[159] Kapur RP. Early death of neural crest cells is responsible for total enteric aganglionosis in Sox10 Dom /Sox10 Dom mouse embryos. Pediatr Dev Pathol 1999;2(6):559–69.

[160] Herbarth B, Pingault V, Bondurand N, Kuhlbrodt K, Hermans-Borgmeyer I, Puliti A, et al. Mutation of the Sry-related Sox10 gene in dominant megacolon, a mouse model for human Hirschsprung disease. Proc Natl Acad Sci USA 1998;95(9):5161–5.

[161] Paratore C, Eichenberger C, Suter U, Sommer L. Sox10 haploinsufficiency affects maintenance of progenitor cells in a mouse model of Hirschsprung disease. Hum Mol Genet 2002;11(24):3075–85.

[162] Touraine RL, Attié-Bitach T, Manceau E, Korsch E, Sarda P, Pingault V, et al. Neurological phenotype in Waardenburg syndrome type 4 correlates with novel

SOX10 truncating mutations and expression in developing brain. Am J Hum Genet 2000;66(5):1496—503.

[163] Saldana-Caboverde A, Kos L. Roles of endothelin signaling in melanocyte development and melanoma. Pigment Cell Melanoma Res 2010;23 (2):160—70.

[164] Shin MK, Levorse JM, Ingram RS, Tilghman SM. The temporal requirement for endothelin receptor-B signalling during neural crest development. Nature 1999;402(6761):496—501.

[165] Hosoda K, Hammer RE, Richardson JA, Baynash AG, Cheung JC, Giaid A, et al. Targeted and natural (piebald-lethal) mutations of endothelin-B receptor gene produce megacolon associated with spotted coat color in mice. Cell 1994;79(7):1267—76.

[166] Baynash AG, Hosoda K, Giaid A, Richardson JA, Emoto N, Hammer RE, et al. Interaction of endothelin-3 with endothelin-B receptor is essential for development of epidermal melanocytes and enteric neurons. Cell 1994;79(7):1277—85.

[167] Pavan WJ, Tilghman SM. Piebald lethal (sl) acts early to disrupt the development of neural crest-derived melanocytes. Proc Natl Acad Sci USA 1994;91(15):7159—63.

[168] Hou L, Pavan WJ, Shin MK, Arnheiter H. Cell-autonomous and cell non-autonomous signaling through endothelin receptor B during melanocyte development. Development 2004;131(14):3239—47.

[169] Druckenbrod NR, Powers PA, Bartley CR, Walker JW, Epstein ML. Targeting of endothelin receptor-B to the neural crest. Genesis 2008;46(8):396—400.

[170] Tachibana M, Kobayashi Y, Matsushima Y. Mouse models for four types of Waardenburg syndrome. Pigment Cell Res 2003;16(5):448—54.

[171] Watanabe K-i, Takeda K, Katori Y, Ikeda K, Oshima T, Yasumoto K-i, et al. Expression of the Sox10 gene during mouse inner ear development. Mol Brain Res 2000;84(1—2):141—5.

[172] Dutton K, Abbas L, Spencer J, Brannon C, Mowbray C, Nikaido M, et al. A zebrafish model for Waardenburg syndrome type IV reveals diverse roles for Sox10 in the otic vesicle. Dis Models Mech 2009;2(1—2):68—83.

[173] Chin L. The genetics of malignant melanoma: lessons from mouse and man. Nat Rev Cancer 2003;3 (8):559—70.

[174] Miller AJ, Mihm Jr. MC. Melanoma. New Engl J Med 2006;355(1):51—65.

[175] Davies H, Bignell GR, Cox C, Stephens P, Edkins S, Clegg S, et al. Mutations of the BRAF gene in human cancer. Nature 2002;417(6892):949—54.

[176] Mercer K, Giblett S, Green S, Lloyd D, DaRocha Dias S, Plumb M, et al. Expression of endogenous oncogenic V600EB-raf induces proliferation and developmental defects in mice and transformation of primary fibroblasts. Cancer Res 2005;65(24):11493—500.

[177] Patton EE, Widlund HR, Kutok JL, Kopani KR, Amatruda JF, Murphey RD, et al. BRAF mutations are sufficient to promote nevi formation and cooperate with p53 in the genesis of melanoma. Curr Biol 2005;15(3):249—54.

[178] Levy C, Khaled M, Fisher DE. MITF: master regulator of melanocyte development and melanoma oncogene. Trends Mol Med 2006;12(9):406—14.

[179] Wellbrock C, Rana S, Paterson H, Pickersgill H, Brummelkamp T, Marais R. Oncogenic BRAF regulates melanoma proliferation through the lineage specific factor MITF. PLoS One 2008;3(7):e2734.

[180] Garraway LA, Widlund HR, Rubin MA, Getz G, Berger AJ, Ramaswamy S, et al. Integrative genomic analyses identify MITF as a lineage survival oncogene amplified in malignant melanoma. Nature 2005;436(7047):117—22.

[181] Gurney JG, Ross JA, Wall DA, Bleyer WA, Severson RK, Robison LL. Infant cancer in the U.S.: histology-specific incidence and trends, 1973 to 1992. J Pediatr Hematol Oncol 1997;19(5):428—32.

[182] Brodeur GM. Neuroblastoma: biological insights into a clinical enigma. Nat Rev Cancer 2003;3(3):203—16.

[183] Grimmer MR, Weiss WA. Childhood tumors of the nervous system as disorders of normal development. Curr Opin Pediatr 2006;18(6):634—8.

[184] Wakamatsu Y, Watanabe Y, Nakamura H, Kondoh H. Regulation of the neural crest cell fate by N-myc: promotion of ventral migration and neuronal differentiation. Development 1997;124(10):1953—62.

[185] Nelms BL, Labosky PA. Transcriptional control of neural crest development. Morgan an Claypool LIfe Sciences. San Rafael (CA); 2010.

[186] Weiss WA, Aldape K, Mohapatra G, Feuerstein BG, Bishop JM. Targeted expression of MYCN causes neuroblastoma in transgenic mice. EMBO J 1997;16 (11):2985—95.

[187] Lasorella A, Noseda M, Beyna M, Iavarone A. Id2 is a retinoblastoma protein target and mediates signalling by Myc oncoproteins. Nature 2000;407(6804):592—8.

[188] Slack A, Chen Z, Tonelli R, Pule M, Hunt L, Pession A, et al. The p53 regulatory gene MDM2 is a direct transcriptional target of MYCN in neuroblastoma. Proc Natl Acad Sci USA 2005;102(3):731—6.

[189] Chen L, Iraci N, Gherardi S, Gamble LD, Wood KM, Perini G, et al. p53 is a direct transcriptional target of MYCN in neuroblastoma. Cancer Res 2010;70 (4):1377—88.

[190] Jerome LA, Papaioannou VE. DiGeorge syndrome phenotype in mice mutant for the T-box gene, Tbx1. Nat Genet 2001;27(3):286—91.

[191] Calmont A, Ivins S, Van Bueren KL, Papangeli I, Kyriakopoulou V, Andrews WD, et al. Tbx1 controls cardiac neural crest cell migration during arch artery development by regulating Gbx2 expression in the pharyngeal ectoderm. Development 2009;136 (18):3173−83.

[192] Vitelli F, Morishima M, Taddei I, Lindsay EA, Baldini A. Tbx1 mutation causes multiple cardiovascular defects and disrupts neural crest and cranial nerve migratory pathways. Hum Mol Genet 2002;11 (8):915−22.

[193] Xu H, Morishima M, Wylie JN, Schwartz RJ, Bruneau BG, Lindsay EA, et al. Tbx1 has a dual role in the morphogenesis of the cardiac outflow tract. Development 2004;131(13):3217−27.

[194] Garg V, Yamagishi C, Hu T, Kathiriya IS, Yamagishi H, Srivastava D. Tbx1, a DiGeorge syndrome candidate gene, is regulated by sonic hedgehog during pharyngeal arch development. Dev Biol 2001;235 (1):62−73.

[195] Kirby ML, Waldo KL. Neural crest and cardiovascular patterning. Circ Res 1995;77(2):211−5.

[196] Schneider C, Wicht H, Enderich J, Wegner M, Rohrer H. Bone morphogenetic proteins are required in vivo for the generation of sympathetic neurons. Neuron 1999;24:861−70.

[197] Shah NM, Groves AK, Anderson DJ. Alternative neural crest cell fates are instructively promoted by TGFbeta superfamily members. Cell 1996;85 (3):331−43.

[198] Guillemot F, Lo LC, Johnson JE, Auerbach A, Anderson DJ, Joyner AL. Mammalian achaete-scute homolog 1 is required for the early development of olfactory and autonomic neurons. Cell 1993;75: 463−76.

[199] Stanke M, Junghans D, Geissen M, Goridis C, Ernsberger U, Rohrer H. The Phox2 homeodomain proteins are sufficient to promote the development of sympathetic neurons. Development 1999;126 (18):4087−94.

[200] Pattyn A, Morin X, Cremer H, Goridis C, Brunet JF. The homeobox gene Phox2b is essential for the development of autonomic neural crest derivatives. Nature 1999;399(6734):366−70.

[201] Amiel J, Laudier B, Attie-Bitach T, Trang H, de Pontual L, Gener B, et al. Polyalanine expansion and frameshift mutations of the paired-like homeobox gene PHOX2B in congenital central hypoventilation syndrome. Nat Genet 2003;33(4):459−61.

[202] Matera I, Bachetti T, Puppo F, Di Duca M, Morandi F, Casiraghi GM, et al. PHOX2B mutations and polyalanine expansions correlate with the severity of the respiratory phenotype and associated symptoms in both congenital and late onset central hypoventilation syndrome. J Med Genet 2004;41(5):373−80.

[203] Dubreuil V, Ramanantsoa N, Trochet D, Vaubourg V, Amiel J, Gallego J, et al. A human mutation in Phox2b causes lack of CO2 chemosensitivity, fatal central apnea, and specific loss of parafacial neurons. Proc Natl Acad Sci USA 2008;105(3):1067−72.

[204] Bachetti T, Matera I, Borghini S, Duca MD, Ravazzolo R, Ceccherini I. Distinct pathogenetic mechanisms for PHOX2B associated polyalanine expansions and frameshift mutations in congenital central hypoventilation syndrome. Hum Mol Genet 2005;14(13):1815−24.

[205] Trochet D, Hong SJ, Lim JK, Brunet J-F, Munnich A, Kim K-S, et al. Molecular consequences of PHOX2B missense, frameshift and alanine expansion mutations leading to autonomic dysfunction. Hum Mol Genet 2005;14(23):3697−708.

[206] Pattyn A, Morin X, Cremer H, Goridis C, Brunet J-F. The homeobox gene Phox2b is essential for the development of autonomic neural crest derivatives. Nature 1999;399(6734):366−70.

[207] Young HM, Ciampoli D, Hsuan J, Canty AJ. Expression of Ret-, p75NTR-, Phox2a-, Phox2b-, and tyrosine hydroxylase-immunoreactivity by undifferentiated neural crest-derived cells and different classes of enteric neurons in the embryonic mouse gut. Dev Dyn 1999;216(2):137−52.

[208] Pattyn A, Goridis C, Brunet J-F. Specification of the central noradrenergic phenotype by the homeobox gene Phox2b. Mol Cell Neurosci 2000;15(3):235−43.

[209] Weese-Mayer D, Marazita M, Berry-Kravis E, Patwari P. Congenital central hypoventilation syndrome. In: Pagon RA, Adam MP, Bird TD, et al., editors. GeneReviews™ [Internet] Seattle (WA). Seattle: University of Washington; 2004. p. 1993−2013 [Updated 2011 Nov 10].

[210] Gronskov K, Ek J, Brondum-Nielsen K. Oculocutaneous albinism. Orphanet J Rare Dis 2007;2(1):43.

[211] Newton JM, Cohen-Barak O, Hagiwara N, Gardner JM, Davisson MT, King RA, et al. Mutations in the human orthologue of the mouse underwhite gene (uw) underlie a new form of oculocutaneous albinism, OCA4. Am J Hum Genet 2001;69(5):981−8.

[212] Bi W, Deng JM, Zhang Z, Behringer RR, de Crombrugghe B. Sox9 is required for cartilage formation. Nat Genet 1999;22(1):85−9.

[213] Lefebvre V. The SoxD transcription factors—Sox5, Sox6, and Sox13—are key cell fate modulators. Int J Biochem Cell Biol 2010;42(3):429−32.

[214] Mori-Akiyama Y, Akiyama H, Rowitch DH, de Crombrugghe B. Sox9 is required for determination

of the chondrogenic cell lineage in the cranial neural crest. Proc Natl Acad Sci USA 2003;100(16):9360–5.

[215] Day TF, Guo X, Garrett-Beal L, Yang Y. Wnt/beta-catenin signaling in mesenchymal progenitors controls osteoblast and chondrocyte differentiation during vertebrate skeletogenesis. Dev Cell 2005;8(5):739–50.

[216] Akiyama H, Lyons JP, Mori-Akiyama Y, Yang X, Zhang R, Zhang Z, et al. Interactions between Sox9 and beta-catenin control chondrocyte differentiation. Gen Dev 2004;18(9):1072–87.

[217] Komori T, Yagi H, Nomura S, Yamaguchi A, Sasaki K, Deguchi K, et al. Targeted disruption of Cbfa1 results in a complete lack of bone formation owing to maturational arrest of osteoblasts. Cell 1997;89(5):755–64.

[218] Jiang X, Iseki S, Maxson RE, Sucov HM, Morriss-Kay GM. Tissue origins and interactions in the mammalian skull vault. Dev Biol 2002;241(1):106–16.

[219] Yoshida T, Vivatbutsiri P, Morriss-Kay G, Saga Y, Iseki S. Cell lineage in mammalian craniofacial mesenchyme. Mech Dev 2008;125(9–10):797–808.

[220] Gagan JR, Tholpady SS, Ogle RC. Cellular dynamics and tissue interactions of the dura mater during head development. Birth Defects Res C Embryo Today Rev 2007;81(4):297–304.

[221] Melville H, Wang Y, Taub PJ, Jabs EW. Genetic basis of potential therapeutic strategies for craniosynostosis. Am J Med Genet A 2010;152A(12):3007–15.

[222] Iseki S, Wilkie AO, Morriss-Kay GM. Fgfr1 and Fgfr2 have distinct differentiation- and proliferation-related roles in the developing mouse skull vault. Development 1999;126(24):5611–20.

[223] Johnson D, Iseki S, Wilkie AOM, Morriss-Kay GM. Expression patterns of Twist and Fgfr1, -2 and -3 in the developing mouse coronal suture suggest a key role for Twist in suture initiation and biogenesis. Mech Dev 2000;91(1–2):341–5.

[224] Orr-Urtreger A, Bedford MT, Burakova T, Arman E, Zimmer Y, Yayon A, et al. Developmental localization of the splicing alternatives of fibroblast growth factor receptor-2 (FGFR2). Dev Biol 1993;158(2):475–86.

[225] Yu K, Xu J, Liu Z, Sosic D, Shao J, Olson EN, et al. Conditional inactivation of FGF receptor 2 reveals an essential role for FGF signaling in the regulation of osteoblast function and bone growth. Development 2003;130(13):3063–74.

[226] Holmes G, Basilico C. Mesodermal expression of Fgfr2S252W is necessary and sufficient to induce craniosynostosis in a mouse model of Apert syndrome. Dev Biol 2012;368(2):283–93.

[227] Komatsu Y, Yu PB, Kamiya N, Pan H, Fukuda T, Scott GJ, et al. Augmentation of smad-dependent BMP signaling in neural crest cells causes craniosynostosis in mice. J Bone Mineral Res 2013;28(6):1422–33.

[228] Chen ZF, Behringer RR. twist is required in head mesenchyme for cranial neural tube morphogenesis. Gen Dev 1995;9(6):686–99.

[229] Connerney J, Andreeva V, Leshem Y, Mercado MA, Dowell K, Yang X, et al. Twist1 homodimers enhance FGF responsiveness of the cranial sutures and promote suture closure. Dev Biol 2008;318(2):323–34.

[230] Ishii M, Merrill AE, Chan Y-S, Gitelman I, Rice DPC, Sucov HM, et al. Msx2 and Twist cooperatively control the development of the neural crest-derived skeletogenic mesenchyme of the murine skull vault. Development 2003;130(24):6131–42.

[231] Merrill AE, Bochukova EG, Brugger SM, Ishii M, Pilz DT, Wall SA, et al. Cell mixing at a neural crest-mesoderm boundary and deficient ephrin-Eph signaling in the pathogenesis of craniosynostosis. Hum Mol Genet 2006;15(8):1319–28.

[232] Santiago A, Erickson CA. Ephrin-B ligands play a dual role in the control of neural crest cell migration. Development 2002;129(15):3621–32.

[233] Ting M-C, Wu NL, Roybal PG, Sun J, Liu L, Yen Y, et al. EphA4 as an effector of Twist1 in the guidance of osteogenic precursor cells during calvarial bone growth and in craniosynostosis. Development 2009;136(5):855–64.

[234] Twigg SRF, Kan R, Babbs C, Bochukova EG, Robertson SP, Wall SA, et al. Mutations of ephrin-B1 (EFNB1), a marker of tissue boundary formation, cause craniofrontonasal syndrome. Proc Natl Acad Sci USA 2004;101(23):8652–7.

[235] Wieland I, Jakubiczka S, Muschke P, Cohen M, Thiele H, Gerlach KL, et al. Mutations of the Ephrin-B1 gene cause craniofrontonasal syndrome. Am J Hum Genet 2004;74(6):1209–15.

[236] Marra KG, Rubin JP. The potential of adipose-derived stem cells in craniofacial repair and regeneration. Birth Defects Res C Embryo Today Rev 2012;96(1):95–7.

[237] Magalhaes MHCG, Barbosa da Silveira C, Moreira CR, Cavalcanti MGP. Clinical and imaging correlations of Treacher Collins syndrome: Report of two cases. Oral Surgery, Oral Medicine, Oral Pathology and Endodontology 2007;103:836–42.

[238] Raveh E, Papsin BC, Forte V. Branchio-oculo-facial-syndrome. International Journal of Pediatric Otorhinolaryngology 2000;53:149–56.

[239] Bassett AS, et al. Practical Guidelines for Managing Patients with 22q11.2 Deletion Syndrome. The Journal of Pediatrics 2011;159:332–39e1.

[240] Nagaraja S, Anslow P, Winter B. Craniosynostosis. Clinical Radiology 2013;68:284–92.

Human Neural Crest Cells and Stem Cell-Based Models

Erin Betters*, Barbara Murdoch*, Alan W. Leung*
and Martín I. García-Castro

OUTLINE

INTRODUCTION

Wilhelm His first introduced neural crest cells (NCC) to the scientific world in a

* Authors of equal contribution.

landmark study of chick development [1]. Following his work in aves, His invested considerable effort in investigating the embryonic development of other vertebrates, including humans (reviewed by [2–4]) (Figure 18.1). In the century and a half since His's pioneering

(A) (B)

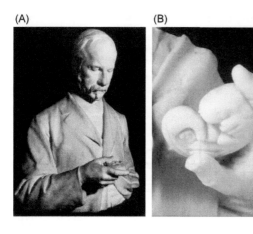

FIGURE 18.1 **Wilhelm His described the neural crest for the first time in 1868 using chick embryos and championed embryological studies and modeling in other organisms, including humans.** Photographs of a marble sculpture made in 1900 by Carl Sefner showing Wilhelm His contemplating a model of a human embryo (A), and a close up from above of the embryo held in his hand, which he classified as "embryo," seen from above, c. 1992, from Anatomisches Museum Basel. *Reprinted from [4], Copyright 2013, with permission from Elsevier.*

work, embryologists have continued to study NCC formation in a variety of model organisms, including the hagfish, lamprey, salamander, frog, fish, chick, and mouse (for review, see Chapters 2−4). Collectively, these studies in model systems have provided insight into the evolution, embryonic origin, and differentiation potential of NCC. Despite all we know about NCC development in lower vertebrates and amniotes, the early formation and migration patterns of the human neural crest (NC) remain elusive.

NCC are thought to be vertebrate-specific (although tunicates possess cells with some similarities) [5,6], arise early during embryogenesis at the edge of the neural plate, and become incorporated into the dorsal portion of the neural tube, from where they emigrate after undergoing an epithelial-to-mesenchymal transition. Upon their exit from the neural tube, NCC embark on extensive and stereotypic migration patterns, eventually stopping and differentiating into a bewildering variety of derivatives at their final destinations. Remarkably, derivatives of the NC include classic ectodermal cell types such as melanocytes, neurons (sensory, sympathetic, and parasympathetic), and glia (Schwann cells and olfactory ensheathing cells), in addition to a wide range of mesenchymal derivatives (e.g.,

head cartilages and bones including both endochondral and intramembranous, smooth muscles in the cardiac outflow tracts, and other connective tissues). Although NCC are often referred to as stem cells due to their proliferative and differentiative capacities, these traits are not homogenously distributed among all NCC. Instead, a great deal of heterogeneity has been identified within this cell population, and some NCC contribute to multiple and varied derivatives, while others are less potent or restricted to only single derivatives [7−11]. Furthermore, mixed populations of NCC with varied differentiation competence seem to coexist in both premigratory and differentiation stages. In this regard, an additional level of complexity has emerged, as mechanisms for reversible conversions from restricted to more multipotent precursors as has been suggested from culture experiments of NC derivatives [12,13]. The precise root of this variation among NCC, however, remains unknown, but unlocking its molecular nature is of great interest for both basic and clinical science.

Research in model organisms has established ample conservation in NC development and has yielded a list of common signaling molecules and transcription factors involved in NCC formation. These molecules have been arranged in "cascades" or "networks" of gene regulation responsible for NC development [14–18]. Although such regulatory networks typically synthesize information on NC formation across an array of model organisms, it is important to recognize that specific differences exist for each system (i.e., frog, mouse, and chick; reviewed by [19]). For example, although transcription factors of the Snail family play a major role in NC development in the frog and chick [20,102], accumulating evidence in the mouse challenges the function of Snail proteins in mammalian NC formation [21,22]. Furthermore, a recent study characterized an ectodermal role for FGF signaling in avian NC development, differing from the proposed mesodermal role of FGF in frog NC development [23]. Interesting differences between model organisms and humans have been recently noted in other early developmental events. Soon after proliferation, early mammalian embryonic cells differentiate into in or out groups of cells, with the outer cells, or trophectoderm (TE), contributing to extraembryonic components of the placenta, and the inner cells, or inner cell mass (ICM), generating the embryo proper. In mouse, separation between TE and ICM depends heavily on two key transcription factors Cdx2 and Oct4 (respectively); however, outstanding differences in their regulation and mechanism of action in rabbits, cows, and humans strongly suggest that the mouse might not be very representative for these early stages of mammalian development [24]. Additional evidence comes from studies analyzing the next differentiation event for ICM into either hypoblast or epiblast cells, which in the mouse rely on FGF/MAPK signaling, but instead are sensitive to MAPK but not to FGF in bovine embryos, and more importantly, MAPK had no effect on this human lineage decision ([25,103]). Another report looking further down the differentiation events analyzed neuroectoderm specification, and suggests that the transcription factor Pax6 is necessary and sufficient for neuroectoderm

specification in human embryos and embryonic stem cells, a role not shared with rodents [26]. These studies provide strong evidence that different wiring of signaling and transcription factors are employed in early human development. Whether differences in the contribution of signaling pathways and transcription factors exist in NC formation between humans and other model species remains unknown. The fact that developmental differences exist between different organisms is remarkably important and must be considered when evaluating human embryonic development. In fact, such differences lay behind the tragic outcome of thalidomide use to treat morning sickness during human pregnancies. Children born to women using thalidomide during early pregnancy often exhibited severe limb abnormalities; however, thalidomide exposure in rodents had no effect on embryonic development [27,28]. These studies show how imperative it is that we not only continue to expand our research on embryonic development in model organisms but also invest heavily in the study of human embryogenesis and NC biology.

Over the years, morphological studies of human NC development have been based largely on the histological analysis of embryonic sections. More recent molecular analyses of human NC development have been performed, but these lag behind the progress made in model organisms. Additional insights into human NC biology have come from surrogate models based on human embryonic stem cells (hESCs) and induced pluripotent stem cells (iPSCs). These approaches have enabled the generation of NC precursors, stem cells, or progenitors (hereafter referred to collectively as neural crest precursors or NCPs) from hESCs and iPSCs. Such stem cell work carries great diagnostic and therapeutic promise and may provide insight into human NC development. In the absence of specific methodology, it remains difficult to establish the extent to which these models replicate the precise developmental processes occurring *in vivo* in human embryos. However, these cell culture methods offer genuine and vital experimental tools that undoubtedly advance our understanding of processes in human biology. Of interest, variations in the usages of transcription factors and signaling pathways have been identified in the maintenance and differentiation of hESC cultures compared to ESC cultures in other species—for instance, the mouse. These variations may be underpinned by species-specific differences and could potentially help establish precise hypotheses to test with the limited number of precious human embryo specimens. With both *in vitro* and *in vivo* approaches, we can certainly gain a better understanding of how human NCC form and differentiate and improve therapeutic strategies for NC-associated diseases.

Here, we describe what is known about human NC formation, focusing upon marker expression analyses performed in human embryos or NC-associated human cell lines. We additionally review how NCPs are generated from hESCs or iPSCs and comment on how such *in vitro* culture methods may benefit from NC marker expression analyses performed in human sections.

Embryonic Development of the Human Neural Crest

Historic collections of human embryos, such as the Carnegie Collection at the National Museum of Health and Medicine (Silver Spring, Maryland) or the more recent Kyoto Collection from Kyoto University (Kyoto, Japan) allow scientists to evaluate human development (e.g., see [29–32]). In fact, these specimens have made a possible comprehensive staging system for human embryos during the embryonic period (up to 8 weeks of *in utero* development), one that relies on a combination of morphological characteristics such as somite

number, neuropore closure and limb and organ formation [29]. Although the embryos contained in the Carnegie Collection provide a remarkable resource for histological and structural studies, research carried out in these specimens is constrained by the processing and age of the tissue samples. Many of these embryos were collected in the early twentieth century (see [29]) and have not been subjected to contemporary molecular analysis. Despite these caveats, embryo sections within the Carnegie Collection (and others) have allowed for a thorough histological description of human NC formation [32]. In a review of nearly 200 historic specimens, human NCC were identified in embryo sections by unique morphological characteristics such as cell shape. Migratory cranial NCC were seen emerging from the level of the midbrain as early as Carnegie Stage (CS) 9, and trunk NCC began their migration slightly later at CS10 (both stages occurring between 3–3.5 weeks, from 25 to 28 days of in utero development, and characterized by the appearance of 1–3 somite pairs in CS9 and 4–8 somite pairs in CS10). In chick embryos, cranial NC emigration has been reported on or after HH9 or the seventh somite stage, but not in earlier embryos. In the mouse, emigration can be seen at Theiler stages 12–13 or 8–8.5d, which display from 1 to 7 pairs of somites). Although most cranial NCC were noted to end their migration by CS13 (4 weeks), cessation of trunk human NCC migration was not identified. Remarkably, migratory pathways followed by human NCC closely mirror those documented in model organisms (for NC migration patterns in the chick and mouse, see [33–37]).

Neural Crest Marker Expression in Human Embryos

Despite the thorough morphological description of early human NC migration, we know very little about what NC markers are expressed in human embryos, how specifically such markers label the human NC, and at what developmental time points these genes are active. In fact, gene expression patterns in human embryos have only recently begun to be explored. Given the caveats with working with historical specimens, marker expression analyses have relied on recent donations of fresh embryonic tissues. The inherent ethical concerns of working with human embryos, paired with the technical difficulties of acquiring such samples, have effectively created a research bottleneck. As a result, few groups have chosen to undertake human NC gene expression studies. Furthermore, the limited specimens available for research typically correspond to 4 weeks of gestation or later, as pregnancies are rarely detected before this point. These specimens are more advanced than required for the optimum study of early stages of NC induction, specification, and cranial migration. In fact, it is likely that in many of these embryos anterior NCC migration has already ceased. Therefore, most NC marker expression studies carried out in human embryos have characterized NC derivatives (i.e., the dorsal root ganglia, DRG), but not premigratory or migratory NCC.

Neural Crest Markers in Human Cell Lines and Embryonic Fragments

Despite the difficulties of investigating the early events of human NC development, a recent study reported a RT-PCR-based expression profile of NC markers contained within cranial segments of CS12–14 (approximately 3.5–5 weeks old) human embryos [38]. This work demonstrated the expression of SOX9, PAX3, MSX1, FOXD3, and SNAI1 in these NC-populated regions. A separate study generated NCC lines from CS10–14 (3–5 weeks old) neural tubes and performed an RNA expression analysis through SAGE ([39]; see [29] for

approximate ages). This work confirmed the presence of SOX9, PAX3, MSX1, FOXD3, and SNAI1 in human NCC, and additionally noted the expression of SNAI2, SOX10, and p75 (NTR) in these cells. This research has given us a glimpse into human NC formation; however, both studies are constrained by certain caveats. For example, evidence from the mouse suggests that non-NCC also emigrate from neural tube explants [40]. In addition, these studies cannot specifically determine the exact locations where NC markers are expressed in the human embryo (i.e., in what cells or structures, in what specific territory, or in relation to other markers).

Other work has sought to elucidate the expression of NC-associated markers directly in human embryo sections. Importantly, such research provides a context of expression in relation to other embryonic structures and enables the direct identification of NCC, their precursors, and their derivatives. Yet, as above, this research carries with it certain limitations. Much of this work has evaluated the expression of single markers in human embryos; however, no individual marker uniquely identifies NCC in any system (human or model organism). As NC markers are typically associated with other developmental processes, studies that evaluate the expression of NC markers in human embryos do not always decisively focus on NC development. Those studies that do explicitly describe NC marker expression in human NCC or NC-derivatives typically rely on older embryos (8–9 weeks) at the cusp of the embryonic/fetal divide, due to their prevalence compared to younger CS9–CS13 embryos (3–4 weeks). Here, we provide a brief overview of the expression patterns observed for well-established NC markers—SOX9, SOX10, PAX3, PAX7, p75(NTR), and AP-2α—during the human embryonic period. Although the studies described below use a mix of CS or "week-old" embryonic staging, for consistency we include approximate CSs (as above; using the criteria set forth in [29]) for embryos identified in studies as being between 4 and 8−9 weeks of age.

NC-marker expression has been best characterized in the DRGs of human embryos. Between 8 and 9 weeks of development (CS22−23, or the early fetal period)—presumably after NC migration has come to an end—SOX9, SOX10, PAX3, p75(NTR), and AP-2α signals have all been demonstrated within the DRGs or their associated nerves [41−44]. At slightly younger stages of 5−6 weeks (CS15−17), both HNK-1 and SOX10 expression have been described in the DRGs, with SOX10 signals also appearing in the sympathetic and cranial ganglia, in addition to presumptive enteric NCC populating the gut [45,46]. In embryonic sections from younger human embryos (4 weeks, or approximately CS11−13), SOX10 + cells have also been observed flanking the neural tube, an expression pattern suggestive of migratory NCC [46]. At similar early stages (4 weeks, or CS11−13), additional SOX10 expression—along with p75(NTR)—has been reported in presumptive migratory NCC surrounding the dorsal aorta and populating portions of the gut [47]. Outside of components of the peripheral nervous system or potential migratory NCC, SOX9 + (CS13, CS15, and CS18), SOX10 + (6 weeks; CS17), or PAX3 + (8−9 weeks; CS23 or early fetal period) cells additionally occur in the NC-populated craniofacial mesenchyme [41,44,46]. PAX3 signal is specifically observed in the regions that will go on to form the connective tissue of the tongue and jaws, and SOX9 expression has been noted in the future palate region [41,44].

Until recently, the above marker expression studies, paired with the histological descriptions of O'Rahilly and Müller [32], comprised the majority of our knowledge of *in vivo* human NC development. This work revealed only minimal information about NC marker expression in early migratory or premigratory

NCC. However, a recent study assessed the expression of multiple NC markers in four human embryos between CS12 and CS18 (4−7 weeks) [48]. Although rostral human embryo segments or human NCC lines were previously shown to express a number of NC-related markers in their RNA profiles [38,39], the Betters et al. [48] study was the first to morphologically demonstrate p75, AP-2α, PAX7, and SOX9 in migrating—and SOX10, SOX9, PAX7, and PAX3 in premigratory—NCC human embryos from CS12 to CS14 (4 weeks; see Figure 18.2). In addition, this work described the novel expression of NC markers in several NC derivatives or territories populated by these cells, such as PAX7 signal in the craniofacial mesenchyme flanking the olfactory placode or mandibular process at CS13 or CS18, respectively. Collectively, this *in vivo* expression analysis, which was performed in few precious specimens assumed to represent "normal" embryonic development, has revealed critical aspects of early human NC formation. Importantly, this research provides expression profiles for (1) both the premigratory and migratory NC at the posterior end of the embryo, (2) migrating and/or differentiating NCC (and NC-associated derivatives) in the trunk, and (3) cranial NCC at various stages of differentiation.

Data related to neural crest development from different model organisms has been incorporated into an NC-gene regulatory network (NC-GRN) model. This integrative model suggests a progressive acquisition of markers, with Wnt, FGF, Notch, and BMP signaling molecules starting a cascade of gene expression [19]. The first tier of transcription factors referred to as neural plate border specifiers includes AP-2α, Zic1, Pax3/7, Msx1, and Dlx5 genes. The second tier includes Snail1/2, FoxD3, Sox9, and Sox10. This NC-GRN has been heavily derived from the early induction of cranial NC, and at present the shape of its architecture at different antero-posterior

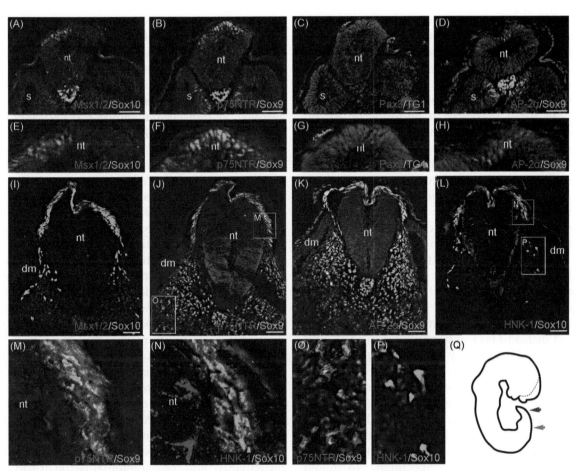

FIGURE 18.2 NCC markers are expressed in prospective premigratory and migratory NCC at CS12. Sox10 (A), Sox9 (B, D), and Pax3 (C) signal occurs in the dorsal neural tube of caudal trunk sections, prior to the appearance of the hind limb. These cells likely represent premigratory NCC. Images in (E–H) are magnifications of the dorsal neural tubes shown in (A–D). We do not observe definitive Msx1/2, p75NTR, or AP-2! signal in premigratory NCC (A, B, D). The monoclonal antibody TG1 has previously been used to recognize other migratory cells and was thus here tried, but no signal can be associated with premigratory NCC (C, G). In slightly rostral sections, Sox10 (I, L), Sox9 (J–K), p75NTR (J), and AP-2! (K) are expressed in migratory NCC situated dorsal, or dorso-lateral to, the neural tube. The majority of AP-2!+ migratory NCC appear to co-express Sox9 (K). Panels (M–P) are magnifications of boxed regions shown in (J, L). Although robust p75NTR signal is co-expressed with Sox9 in migratory NCC adjacent to the neural tube (M), or subadjacent to the dermo-myotome (O), this cell population (when identified via Sox10 expression) does not express HNK-1 in similar sections (N, P); migratory NCC generally appear HNK-1 negative. Image in (Q) is a tracing of the CS12 whole mount, prior to sectioning; the dotted line marks the contra-lateral side of the unfused neural tube. Color-coded arrows represent the axial levels of depicted sections as follows: purple (A–H), blue (I–P). Scale bars in all panels represent 100 μm. Note: Levels in (A–D) are modified separately from those in (I–L) to optimize signals in different embryonic regions. dm, dermomyotome; nt, neural tube; s, somite.

regions remains unknown. In the expression study of human NC markers by Betters et al., Pax3, (and perhaps Pax7), along with Sox9 and Sox10, label premigratory NC in caudal regions (hind limb) of the human embryo (CS12−14). A few sections more anteriorly, these same markers, in addition to AP-2α, accompany early migratory NC. Msx1 was only found in migratory NC of older embryos (CS15). The patterns for AP-2α, Pax3/7, and Sox9/10 are broadly consistent with those reported in caudal regions of model organisms. However, at least three specific differences at equivalent levels are noted: (1) the exclusion of AP-2α from premigratory NC resembles chick, but differs from murine NC development, (2) Msx1 is thought to label premigratory and migratory NC both in chick and mice, yet no signal was found in premigratory and early migratory NC on CS12−14, and instead Msx1 only appeared in CS15, and (3) Pax7 is necessary for avian NC development and is robustly expressed in both premigratory and migratory NCC in the chick, inclusive in caudal territories [49−51]. In contrast, Pax7 expression in the caudal regions of mouse embryos is limited, similar to the expression observed in migratory trunk NCC within a human CS13 specimen. No model for NC-GRN corresponding to this region has been generated, and therefore we do not know if the sequential progression proposed to operate in early/cranial NC development also operates in late/caudal NC development. Further comparative analyses of the NC-GRN between human and model organisms awaits a more thorough and comprehensive characterization of profiles of expression in equivalent regions/stages across different organisms. In this context, the human expression profiles will require additional information from models of hNC development (in vitro) to test functional interactions between these molecules.

Perhaps the most important aspects of in vivo human NC marker expression—and the ones that have the greatest relevance to NC-associated hESC or iPSC therapies (see below)—are the expression patterns of HNK-1 and p75(NTR) described in human embryonic sections. Within human embryos at CS12−13, most migratory NCC fail to express HNK-1, although in very few instances HNK-1 was co-expressed with either SOX10 or PAX7 [48]. This is in sharp contrast to the robust HNK-1 signal commonly seen in avian NCC [34]; however, NC-related expression of HNK − 1 in mouse and rat embryos remains controversial [35,45,52,53]. Compared to HNK-1, p75 (NTR) was more broadly expressed in human NCC; however, p75(NTR) identifies only a subset of migratory human NCC when compared to the expression of SOX9, AP-2α, or SOX10 in this cell population in similar sections [48]. Further complications emerge from the fact that both HNK-1 and p75—like other NC markers—are robustly expressed in neural and mesodermal tissues. In similar sections, both HNK-1 and p75 signal occur in the lateral and ventral regions of the human neural tube at CS12−18, and p75 + cells are further observed in components of the potential musculature and/or sclerotomal tissue at similar stages [48].

Given the observed expression patterns of HNK-1, p75, Pax7, and other NC markers in human embryos, it seems that even combinations of markers thought to be "bona fide" reporters of NC status may not exclusively reflect—or identify—human NCC. Further ESC studies offer a great opportunity to identify more selective markers, which then should be validated in human embryonic sections. Additional expression analyses at CS12−18, as well as at yet-uncharacterized embryonic stages (e.g., CS9−CS11 or CS14), will help to improve our understanding of basic human NC biology.

Neural Crest Marker Expression: Insights into Neural Crest-Associated Human Diseases

By understanding where and when NC markers appear in human embryos, researchers can speculate how the misregulation of specific genes can result in NC-associated diseases (for review of these diseases, see [54]). For example, cleft lip and palate are two of the most recognizable pathologies resulting from defects in NC development. These disorders typically result from the failure of bilateral NC-populated facial processes (e.g., maxillary, palatal, and frontonasal/nasal) to properly fuse during embryonic development. In chick, researchers can experimentally recapitulate cleft lip and cleft palate by ablating portions of the cranial NC [55]. Interestingly, unpublished work from our laboratory has noted the expression of MSX1/2 in the facial mesenchyme of a CS15 (5-week) human embryo, and the additional expression of SOX9, SOX10, AP-2α, PAX3, and PAX7 have all been reported at some stage within the craniofacial mesenchyme of human embryos [41,44,46,48]. Given these expression patterns, it is very likely that the above NC-associated markers play a role in the formation of the human palate, face, and jaws. This is corroborated by much evidence, including (1) recent work demonstrating that some human cases of orofacial clefting are caused by Sox9 misregulation [44,56], (2) the observation that specific maternally inherited PAX3 and PAX7 SNPs [57]—and PAX7 SNPs identified in sibling-pair or case-parent trios— appear to be associated with human orofacial clefts [58–60], and (3) the fact that AP-2, PAX3, or MSX1 mutations/allelic markers have been identified in individuals with cleft lip or palate ([61–64]; for further discussion, see [48]). As we gain a better understanding of what NC markers are expressed in the craniofacial mesenchyme and other territories during human embryogenesis, we may identify new markers potentially involved in the manifestation of cleft lip, cleft palate, and other NC-associated diseases.

SURROGATE MODELS OF HUMAN NEURAL CREST DEVELOPMENT

The tissues needed for the study of the earliest events in human NC development are remarkably difficult to find. Thus, we remain largely ignorant of the early events governing human NC development. However, an excellent alternative to study human NC development has emerged in the form of multipotent stem cells, either as hESCs or iPSCs. As discussed below, both hESCs and iPSCs can be directed to form cells resembling NCC (referred to as NCPs), which are identified on the grounds of gene expression and differentiation potential.

Generating Neural Crest Precursors from Human Embryonic Stem Cells

First derived by Thomson et al. [65], hESCs are isolated from the ICM of blastocyst-staged embryos. When cultured under the appropriate conditions, hESCs can proliferate indefinitely and maintain their potential to form any cell type in the body, and hence are said to be "pluripotent." As proof of this, after injection into immune-deficient mice, hESCs produce benign tumors termed teratomas that contain cells derived from each of the three germ layers (ectoderm, mesoderm, and endoderm) [65]. Importantly, NCPs obtained from hESCs can generate NC terminal derivatives, strongly supporting their NC character. These derivatives include neurons and glia of the peripheral nervous system [66–74], melanocytes [70,75,76], and mesenchymal cells such as osteoblasts and chondrocytes [67,70,74,77]. Due to their vast proliferative and differentiative capacity, hESCs hold great promise for cell therapies, drug discovery, and disease modeling—especially when it comes to the human NC and NC-associated pathologies [78].

Three general strategies for the production of NCPs from hESCs have evolved. The first reports used either stromal cell co-cultures [66,68,69] or relied on the production of intermediate structures such as embryoid bodies, neural precursor cells, or neural rosettes [38,67–70,73,75,77,79,80]. The unknown signaling contributions made by feeder cells, embryoid bodies, or neural rosettes however have hindered these approaches to efficiently dissect the mechanisms involved in human NCP formation. Additionally, NCPs are often found in cultures as a subpopulation among several central nervous system derivatives. However, in some cases, the formation of NCPs from neural rosettes has been related to "normal" developmental processes, including dorsal neural tube patterning and emigration [70,80].

The third strategy uses adherent cultures that provide more defined and in some cases more homogeneous induction of NCPs [72,76,81–84]. These approaches are particularly suited toward evaluating the molecular mechanisms that drive early human NC induction and differentiation from hESC derivatives. A recent report enabled the differentiation of hES in low-density cultures without exogenous cytokines, apparently letting the hES drift into minimal mesoderm and instead mainly producing ectodermal derivatives including non-neural, placodal, crest and neural markers [84]. Paired with this feeder-free approach, the inhibition of both Activin/Nodal and BMP signaling in hESC cultures can yield colonies containing 15–40% p75(NTR)+ presumptive NCC after approximately 12 days in culture [81,82]. Interestingly, a more recent study produced NC-like stem cells from hESCs in a

single step, without the need for cell depletion or enrichment via FACS [83]. Menendez et al. [83] protocol requires the simultaneous activation of the canonical Wnt pathway paired with the blockade of SMAD signaling, achieved through BIO (a small mimetic molecule) and SB431542 treatment, respectively. SB431542 is a small molecule used in previous studies to stimulate NCPs from hESCs [81,82]. Treatment of hESCs with BIO and SB431542 resulted in a highly enriched population of cells with NC characteristics. After 12 days, 88–96% of the cells expressed p75(NTR) + /HNK-1 +, and these co-positive cells displayed high levels of AP-2 protein and several additional NC-associated transcripts (PAX3, SOX9, SOX10, BRN3A, ZIC1) [83]. The latest modification of this strategy incorporated both dual smad inhibition and Wnt signaling activation. Here, BMP and TGFß inhibition were provided for 3 and 4 days, respectively, while Wnt induction (via GSK3ß inhibition) was provided from day 2 onward. After 11 days of differentiation 50–60% of these cells express SOX10 and display many other neural crest markers [76]. In contrast to Mernendez et al. [83], continuous activation of Wnt from day 0 onward blocked SOX10 induction in differentiating hESC. The different results may be due to the choice of markers (p75/HNK1 versus SOX10) and/or small molecule Wnt activators (BIO versus CHIR99021).

hESC Derived Neural Crest Precursors as a Model for Development

It is hopped that the *in vitro* differentiation of hESCs could replicate the endogenous intermediate steps that occur during *in vivo* development. Identifying these early NCC precursors is important to understand the normal progression of NC-development as well as for possible clinical applications. Characterizing the specific signaling pathways and transcription factors involved and their mechanisms of action will improve our understanding of human NC biology. These studies could also promote the yield and increase the speed of NC formation in cultures and may identify novel diagnostic and therapeutic targets.

In addition to the signaling pathways and transcription factors, key regulatory elements responsible for the activation or repression of transcription play a fundamental role in the orchestration of development, ensuring a spatiotemporal specificity. Specifically for NC, these enhancer or repressor elements display temporal regulations and are active in distinct intermediate developmental stages during the course of NC development. Recent examples have identified regulatory elements of the early neural plate border specifier Pax7, as well as for the later neural crest specifier Sox10 in avian NC development [18,85]. These elements are likely predictive of changes of cellular competence to signaling cues. Identifying and characterizing these enhancer elements has already led to a better understanding of the temporal controls and the sequence specific binding factors that cooperate to form the regulatory networks leading to NC formation. In this context, over 3000 distinct putative enhancer elements have been recently identified in hESC-derived NCPs that are active only in NCPs but not in hESC [86]. These putative enhancers are genomic regions 1−200 kb distal to the nearest transcriptional start sites that are bound by the histone acetyltransferase p300 and enriched with histone 3 lysine 4 monomethylation (H3K4me1) and histone 3 lysine 27 acetylation (H3K27ac) chromatin marks. Functional annotations of the closest transcripts associated with these newly identified elements suggest that they may represent elements that are active in the cranial NC-derived mesenchyme. Supporting this notion, the hESC-NCPs derived from their protocol are *HOX*-negative and *DLX1/2*-positive, and have

upregulated cartilage (*COL2A1*, *BGN*) and bone (*COL1A1*, *COL1A2*) transcripts [79,86]. However, we still do not know the specific elements associated with distinct stages, especially those for the naïve premigratory or multipotent early migratory populations, of human NCC. Other more recently identified chromatin marks may be incorporated into the existing schematics to promote the predictive power for the temporal activities of NC developmental enhancers.

Neural Crest Derivatives from Neural Crest Precursors

A variety of methods have been used to characterize derivatives arising from NCPs, including cell morphology, gene expression, and histological and immunological assays. Sympathetic and sensory neurons have been detected via the co-expression of peripherin and tyrosine hydroxylase, or peripherin and Brn3a, respectively [66–70,72]. In addition, peripheral glial cells expressing a subset of individual markers (S100, S100β, and GFAP, and in some cases markers indicative of myelin) have been identified, but only after extensive culture initiated with either embryoid bodies or neurosphere intermediates [67,70,71,73]. Melanocytes are typically identified by their production of pigments or the expression of MelanA, dopachrome tautomerase (DCT) or microphthalmia-associated transcription factor (MITF) [70,75]. Finally, mesenchymal derivatives are identified using a variety of methods, each specific to the cell type being assayed. These methods include oil red O staining for adipocytes, collagen II and alcian blue staining for chondrogenic cells, and Alizarin red or alkaline phosphatase staining for osteogenic cells [67,70,73,77,83].

Interestingly, NCC are composed of a heterogeneous population of cells, each with a unique differentiation potential. A wealth of evidence has identified pluripotent NCPs both *in vivo* and *in vitro*, either through the clonal analysis of cultured avian, amphibian, or murine cells, or through live-labeling approaches within whole mount embryos [7–11]. Most mammalian (including human) stem cell studies evaluating NC potency have embraced the original test devised by Anderson and colleagues to identify rat neural crest stem cells [87]. In this study, prospectively isolated p75+ cells were plated at clonal density, or the density where one colony arises from the division of a single cell. Individual colonies are tested via immunocytochemistry for their ability to produce three different cell types—neurons, glia, and myofibroblasts. A colony producing all three of these cell types is considered multipotent, whereas more restricted colonies only exhibit one or two of these different cell types. Using this assay, the potential of NCPs derived from hESCs has revealed clonal progeny of varying combinations of neurons, glia, and myofibroblasts, underscoring the heterogeneity of the (p75+) precursors arising from these protocols [67,69,70].

Transplantation of hESC Derivatives of the Neural Crest: Advances and Limitations

Despite both multipotent NCC and a wide range of differentiated cell types being shown to arise from hESCs *in vitro*, only limited cell types are found in NC-associated territories after hESCs or NCPs are transplanted into chick or mouse embryos [67,69,70,83]. Explanations for this lack of robust engraftment include that these human cells do not favorably survive, migrate, and/or differentiate in response to the endogenous chick or mouse embryonic environments. The paucity of *in vivo* human studies that address the expression of signaling receptors, ligands, and

extracellular matrix molecules further complicates the successful and robust engraftment of hESCs or NCPs into model organism embryos. Additionally, studies using hESC derivatives for transplantation generally fail to characterize NCPs prior to transplant. Further analysis at the cellular level for the expression of cadherins, integrins, and growth factor receptors may help to improve chimeric assays, as well as hESC-based *in vitro* assays.

To solve the problem of poor NCP contribution to *in vivo* embryonic structures following transplantation, human cells may have to be primed to activate specific molecules like GTPases of the Rho family, which are required for cell motility [88,89]. Alternatively, hESC derivatives could be co-transplanted with supporting cells to enhance the overall engraftment of human cells into mouse or chick chimeras. Finally, transplantation assays may need to assess the processing of cells (i.e., single, dissociated, or cell clumps) prior to transplantation, the number of cells injected, and the method and site of injection. Furthermore, the dependence upon migration could be avoided by the transplantation of cells directly into specific sites for differentiation, a technique that has greatly improved the repopulating function of hESC blood derivatives [90].

iPSC Contributions to Our Understanding of Human NC Development

iPSCs are somatic cells that have been reverted to an embryonic stem cell-like state after reprogramming with the expression of specific transcription factors (OCT3/4, SOX2, MYC, and KLF4; or OCT3/4, SOX2, NANOG, and LIN28) [91–94]. Interestingly, several studies suggest that hESC and iPSCs operate under similar signaling conditions to generate NCPs [72,81–83,95]. Like hESCs, iPSCs can be differentiated to form NC derivatives, and

NCPs can also be isolated from iPSC cultures through p75 and/or HNK-1 expression [73,81,82,95,96]. In addition, iPSC technology is extremely exciting because (1) iPSCs, like ESCs, can be differentiated to form any cell type, (2) unlike ESCs, the generation of iPSCs does not require human embryos, thus avoiding many ethical concerns (reviewed in [97]), (3) iPSCs can be derived from patients with specific diseases, allowing for the screening of potential therapeutic interventions at the cellular level, (4) patient-derived iPSCs can provide an unlimited number of cells for tissue repair and regeneration, and (5) patient-derived iPSCs can bypass concerns regarding immunological rejection [73,81,82,95,96,98]. Models of developmental abnormalities, or the significance of genetic background on the development of cancer, can also be addressed with iPSCs. Human iPSCs and disease-specific iPSCs will be covered in detail later in this book (Chapter 21); therefore, we will only briefly comment on them here and readers are referred to the corresponding chapter for further information.

Using Stem Cells to Model and Treat Neural Crest-Associated Diseases

Of the few studies that have used iPSCs to model NC development and disease, most have focused on the production of cells from the peripheral nervous system. Much is known regarding the differentiation of peripheral neurons and glia from iPSCs and hESCs, a prerequisite to studying disease phenotypes that are usually only manifested in specific cell types [99]. For example, using a pool of small molecules to manipulate different signaling pathways, TUJ1+ neurons can be rapidly produced from hESC precursors within 10 days in culture [72]. By 15 days, peripheral neurons involved with pain reception (nociceptors) arise via a Sox10+ precursor and are detected much faster than estimates of

in vivo development. Additionally, peripheral glia (Schwann cells) derived from iPSCs and hESCs have been shown to myelinate rodent neurons in vitro [73] and to promote neuron regeneration in vivo [96]. Methods for the scalable production of NC progenitors and differentiated Schwann cells allows them to be frozen, thawed, and expanded [73], thus producing unlimited cell numbers and banks of tested stocks prior to cell therapy and drug discovery [99].

Perhaps the most exciting disease modeling studies have utilized disease-specific iPSCs. A few NC-associated diseases, including familial dysautonomia and pigmentation disorders, have been recently modeled using iPSCs. Familial dysautonomia is a rare but fatal peripheral neuropathy that causes a loss of autonomic and sensory neurons. iPSC-derived NCPs from familial dysautonomia patients mimicked disease phenotypes of reduced tissue-specific levels of IKBKAP splice variants, as well as defects in neurogenic differentiation and migration [100]. The iPSC-based model of familial dysautonomia has also been used to screen for compounds that can rescue IKBKAP expression [95,100], demonstrating the promise of this in vitro approach and its potential to accelerate drug discovery. Patients with Chediak–Higashi and Hermansky–Pudlak syndromes carry characteristic mutations in selected genes within the biogenesis pathway of lysosome-related organelles, including melanomes in the melanocytes. These patients either display abnormal retention (Chediak–Higashi, LYST gene mutations) or deficiency in the formation (Hermansky–Pudlak, AP3B1 and HPS1 gene mutations, among others) of melanosome vesicles. To model these disorders, a defined protocol was developed to generate relatively pure populations of melanocytes from hESC-derived NCPs [76]. Notably, melanocytes derived from patient-specific iPSC displayed defects in the extent of melanocyte pigmentation, and the number, the size, and the morphology of melanosomes that nicely correlated with the corresponding disease phenotypes in Chediak–Higashi and Hermansky–Pudlak patients [76] again demonstrating the feasibility of the iPSC approach for NC-associated disease modeling.

A recent study has reported the establishment of a mutant hESC line that carries a mutation in TWIST1 (A129P), which is responsible for a NC-related disorder Saethre–Chotzen syndrome [101] characterized by the premature fusion of certain skull bones. While iPSC bypass the embryo bottleneck, they are restricted to somewhat more benign phenotypes, and instead hESC derived from excess pre-implantation embryos that have been genetically tested may enable the generation of lines carrying genetic changes that may lead to more severe (including lethal) phenotypes.

Although great progress has been achieved using hESCs or iPSCs, whether these cells will perform the same functions after transplantation into a human—or whether they will alleviate degenerative disease phenotypes—remains to be seen. A better understanding of the *in vivo* human microenvironment, including its region-specific and developmental changes, will help to improve future research.

The Relevance of Embryonic Marker Expression Studies to hESC/iPSC-Based Research

As a cautionary note, even though p75 and HNK-1 are not exclusive to NCPs, most studies have relied on their expression; p75 and HNK1 are convenient cell surface markers and are often used for cell sorting and enrichment. Importantly, a recent study of migratory NCC in the caudal trunk of human embryos suggests that p75 labels only a subset of NCC, as well as several cells outside of the NC (see above; [48]). Compared to p75, HNK-1 labels an even more restricted number of migratory

NCC in human embryos and is also co-expressed with p75 in ventral neural tube cells [48]. Hence, *in vivo* p75+ and HNK1+ cells within human embryos appear more indicative of a small subset of migratory NCC and also identify cells of the ventral neural tube. While the axial restriction of sections assessed in this study prevents definitive conclusions, it is remarkable that, *in vitro*, both p75+ and p75− cells can express equivalent levels of Sox10 and differentiate into presumptive NC-derivatives with similar efficiency ([70]; Murdoch and Garcia-Castro, unpublished observations). A more detailed analysis recently showed that P75+/HNK1+ sorted cells derived from hES low-density cultures promoting ectodermal derivatives displayed enriched expression by PCR of multiple NC markers; however, alternative populations including P75-hi HNK− and P75-low HNK1+ provide very similar profiles [84].

These studies suggest the need to consider alternative or additional markers for the identification of specific subsets of NCPs from hESC or iPSC cultures, as such markers may better define clonal progenitors with more refined cell fates. It would be interesting to examine hESC or iPSC derivatives *in vitro* for proteins at the cellular level that have been identified *in vivo* as being indicative of earlier human premigratory NCC, such as Sox9, Sox10, and Pax3 [48].

In the human, the temporal emergence of different NC progenitors may lead to the discovery of phenotypically more restricted cell types. Where p75 and/or HNK-1 are used as markers, future studies should incorporate more detailed analyses of hESC- or iPSC-derived precursors. These analyses may include testing (1) single p75+ cells (as opposed to clonal assays) for their ability to produce specific differentiated NCC types, to examine the heterogeneity of the population and how this changes over time in culture (note that technical limitations currently prevent this analysis), (2) varying the number of p75+ cells initially placed in differentiation assays, thereby better determining the role of cell density during differentiation, and (3) the production of differentiated cell types from outside of the NC (e.g., central nervous system neurons and glia) to better characterize the potential of p75+ progenitors.

CONCLUSIONS AND PERSPECTIVES

In vivo and *in vitro* studies have provided us with a glimpse of early human NC development, and more recent gene expression analyses have yielded clues as to what roles certain markers play in the development of NC-associated diseases. However, studies in human embryos are limited not only by the rarity of tissue but also the fact that most specimens correspond to 4 weeks of *in utero* development or older (CS12 and later); these embryos can only provide limited information about early NC events, including premigratory and early migratory stages. Complementing this *in vivo* research are hESC and iPSC systems, which can serve as *in vitro* models for the earliest inductive and specification events of human NC development. Human embryonic stem cells and iPSCs also carry with them great therapeutic potential and may one day be successfully used to treat NC-associated pathologies such as familial dysautonomia.

Despite this great clinical promise, iPSCs and hESC models have not yet been perfected, and transplantations of hESC-derivatives into avian or murine embryos often demonstrate poor engraftment. As we continue to learn more about the *in vivo* development of human NCC—and the markers expressed in these cells and their surrounding environments—iPSC and hESC systems and therapies will improve. For example, although p75 and HNK-1 are routinely used to enrich for NCPs from hESC or

iPSC preparations, *in vivo* data suggests the need to better characterize the differentiation potential of *in vitro*-derived presumptive NCC expressing these markers. Through the combined effort of *in vivo* research in human embryos and *in vitro* studies using hESCs or iPSCs, we can gain a clearer picture of all stages of human NCC development, and ultimately provide better therapies for NC-associated diseases.

References

[1] His W. Untersuchungen über die erste Anlage des Wirbeltierleibes. Die erste Entwicklung des Hühnchens im Ei. Leipzig: F. C. W. Vogel; 1868.

[2] Hopwood N. Producing development: the anatomy of human embryos and the norms of Wilhelm His. Bull Hist Med 2000;74:29—79.

[3] Hopwood N. A history of normal plates, tables and stages in vertebrate embryology. Int J Dev Biol 2007;51:1—26.

[4] Hopwood N. Anatomist and embryo: a portrait sculpture. Lancet 2013;381:286—7.

[5] Jeffery WR. Ascidian neural crest-like cells: phylogenetic distribution, relationship to larval complexity, and pigment cell fate. J Exp Zool B Mol Dev Evol 2006;306:470—80.

[6] Jeffery WR, Chiba T, Krajka FR, Deyts C, Satoh N, Joly JS. Trunk lateral cells are neural crest-like cells in the ascidian *Ciona intestinalis*: insights into the ancestry and evolution of the neural crest. Dev Biol 2008;324:152—60.

[7] Le Douarin NM, et al. The stem cells of the neural crest. Cell Cycle 2008;7:1013—9.

[8] Garcia-Castro M. Early origin and differentiation capacity of the neural crest. In: Chimal-Monroy J, editor. Topics in animal and plant development: from cell differentiation to morphogenesis. Kerala, India: Transworld Research Network; 2011. p. 55—74.

[9] Achilleos A, Trainor PA. Neural crest stem cells: discovery, properties and potential for therapy. Cell Res 2012;22:288—304.

[10] Dupin E, Sommer L. Neural crest progenitors and stem cells: from early development to adulthood. Dev Biol 2012;366:83—95.

[11] Kaltschmidt B, et al. Adult craniofacial stem cells: sources and relation to the neural crest. Stem Cell Rev 2012;8:658—71.

[12] Prince S, Wiggins T, Hulley PA, Kidson SH. Stimulation of melanogenesis by

tetradecanoylphorbol 13-acetate (TPA) in mouse melanocytes and neural crest cells. Pigment Cell Res 2003;16:26—34.

[13] Kormos B, Belso N, Bebes A, Szabad G, Bacsa S, Szell M, et al. *In vitro* dedifferentiation of melanocytes from adult epidermis characterisation of the fibroblast growth factor dependent transcriptome in early development 2011;6:e17197.

[14] Mayor R, et al. Development of neural crest in *Xenopus*. Curr Top Dev Biol 1999;43:85—113.

[15] Aybar MJ, Mayor R. Early induction of neural crest cells: lessons learned from frog, fish and chick. Curr Opin Genet Dev 2002;12:452—8.

[16] Sauka-Spengler T, Bronner-Fraser M. A gene regulatory network orchestrates neural crest formation. Nat Rev Mol Cell Biol 2008;9:557—68.

[17] Meulemans D, Bronner-Fraser M. Gene-regulatory interactions in neural crest evolution and development. Dev Cell 2004;7:291—9.

[18] Betancur P, et al. Assembling neural crest regulatory circuits into a gene regulatory network. Annu Rev Cell Dev Biol 2010;26:581—603.

[19] Stuhlmiller TJ, Garcia-Castro MI. Current perspectives of the signaling pathways directing neural crest induction. Cell Mol Life Sci 2012;69:3715—37.

[20] Nieto MA, et al. Control of cell behavior during vertebrate development by Slug, a zinc finger gene. Science 1994;264:835—9.

[21] Jiang R, et al. The Slug gene is not essential for mesoderm or neural crest development in mice. Dev Biol 1998;198:277—85.

[22] Murray SA, Gridley T. Snail family genes are required for left-right asymmetry determination, but not neural crest formation, in mice. Proc Natl Acad Sci USA 2006;103:10300—4.

[23] Stuhlmiller TJ, Garcia-Castro MI. FGF/MAPK signaling is required in the gastrula epiblast for avian neural crest induction. Development 2012;139:289—300.

[24] Berg DK, Smith CS, Pearton DJ, Wells DN, Broadhurst R, Donnison M, et al. Trophectoderm lineage determination in cattle. Dev Cell 2011;20:244—55.

[25] Kuijk EW, van Tol LT, Van de Velde H, Wubbolts R, Welling M, Geijsen N, et al. The roles of FGF and MAP kinase signaling in the segregation of the epiblast and hypoblast cell lineages in bovine and human embryos. Development 2012;139:871—82.

[26] Zhang X, et al. Pax6 Is a Human Neuroectoderm Cell Fate Determinant. Cell Stem Cell 2010;7:90.

[27] Fratta ID, et al. Teratogenic effects of thalidomide in rabbits, rats, hamsters, and mice. Toxicol Appl Pharmacol 1965;7:268—86.

[28] Chung F, et al. Thalidomide pharmacokinetics and metabolite formation in mice, rabbits, and

multiple myeloma patients. Clin Cancer Res 2004;10: 5949—56.

[29] O'Rahilly R, Müller F, Streeter GL. Developmental stages in human embryos : including a revision of Streeter's "Horizons" and a survey of the Carnegie collection. Washington, D.C.: Carnegie Institution of Washington; 1987306 p., 1 leaf of plates pp.

[30] Saitsu H, et al. Development of the posterior neural tube in human embryos. Anat Embryol (Berl) 2004;209:107—17.

[31] Yamada S, et al. Phenotypic variability in human embryonic holoprosencephaly in the Kyoto collection. Birth Defects Res A Clin Mol Teratol 2004;70:495—508.

[32] O'Rahilly R, Müller F. The development of the neural crest in the human. J Anat 2007;211:335—51.

[33] Johnston MC. A radioautographic study of the migration and fate of cranial neural crest cells in the chick embryo. Anat Rec 1966;156:143—55.

[34] Bronner-Fraser M. Analysis of the early stages of trunk neural crest migration in avian embryos using monoclonal antibody HNK-1. Dev Biol 1986;115: 44—55.

[35] Serbedzija GN, et al. Pathways of trunk neural crest cell migration in the mouse embryo as revealed by vital dye labelling. Development 1990;108:605—12.

[36] Lumsden A, et al. Segmental origin and migration of neural crest cells in the hindbrain region of the chick embryo. Development 1991;113:1281—91.

[37] Serbedzija GN, et al. Vital dye analysis of cranial neural crest cell migration in the mouse embryo. Development 1992;116:297—307.

[38] Pomp O, et al. PA6-induced human embryonic stem cell-derived neurospheres: a new source of human peripheral sensory neurons and neural crest cells. Brain Res 2008;1230:50—60.

[39] Thomas S, et al. Human neural crest cells display molecular and phenotypic hallmarks of stem cells. Hum Mol Genet 2008;17:3411—25.

[40] Zhao H, et al. An *in vitro* model for characterizing the post-migratory cranial neural crest cells of the first branchial arch. Dev Dyn 2006;235:1433—40.

[41] Terzic J, Saraga-Babic M. Expression pattern of PAX3 and PAX6 genes during human embryogenesis. Int J Dev Biol 1999;43:501—8.

[42] Josephson A, et al. GDNF and NGF family members and receptors in human fetal and adult spinal cord and dorsal root ganglia. J Comp Neurol 2001;440:204—17.

[43] Gershon TR, et al. Temporally regulated neural crest transcription factors distinguish neuroectodermal tumors of varying malignancy and differentiation. Neoplasia 2005;7:575—84.

[44] Benko S, et al. Highly conserved non-coding elements on either side of SOX9 associated with Pierre Robin sequence. Nat Genet 2009;41:359–64.

[45] Tucker GC, et al. Expression of the HNK-1/NC-1 epitope in early vertebrate neurogenesis. Cell Tissue Res 1988;251:457–65.

[46] Bondurand N, et al. Expression of the SOX10 gene during human development. FEBS Lett 1998;432:168–72.

[47] Fu M, et al. HOXB5 expression is spatially and temporarily regulated in human embryonic gut during neural crest cell colonization and differentiation of enteric neuroblasts. Dev Dyn 2003;228:1–10.

[48] Betters E, et al. Analysis of early human neural crest development. Dev Biol 2010;344:578–92.

[49] Lacosta AM, et al. Novel expression patterns of Pax3/Pax7 in early trunk neural crest and its melanocyte and non-melanocyte lineages in amniote embryos. Pigment Cell Res 2005;18:243–51.

[50] Basch ML, et al. Specification of the neural crest occurs during gastrulation and requires Pax7. Nature 2006;441:218–22.

[51] Otto A, et al. Pax3 and Pax7 expression and regulation in the avian embryo. Anat Embryol (Berl) 2006;211:293–310.

[52] Nagase T, et al. Roles of HNK-1 carbohydrate epitope and its synthetic glucuronyltransferase genes on migration of rat neural crest cells. J Anat 2003;203:77–88.

[53] Kizuka Y, Oka S. Regulated expression and neural functions of human natural killer-1 (HNK-1) carbohydrate. Cell Mol Life Sci 2012;69:4135–47.

[54] Etchevers HC, et al. Molecular bases of human neurocristopathies. Adv Exp Med Biol 2006;589:213–34.

[55] van Limborgh J, et al. Cleft lip and palate due to deficiency of mesencephalic neural crest cells. Cleft Palate J 1983;20:251–9.

[56] Jakobsen LP, et al. Pierre Robin sequence may be caused by dysregulation of SOX9 and KCNJ2. J Med Genet 2007;44:381–6.

[57] Sull JW, et al. Maternal transmission effects of the PAX genes among cleft case—parent trios from four populations. Eur J Hum Genet 2009;17:831–9.

[58] Prescott NJ, et al. Identification of susceptibility loci for nonsyndromic cleft lip with or without cleft palate in a two stage genome scan of affected sib-pairs. Hum Genet 2000;106:345–50.

[59] Murray JC. Gene/environment causes of cleft lip and/or palate. Clin Genet 2002;61:248–56.

[60] Beaty TH, et al. A genome-wide association study of cleft lip with and without cleft palate identifies risk variants near MAFB and ABCA4. Nat Genet 2010;42:525–9.

[61] Pasteris NG, et al. Discordant phenotype of two overlapping deletions involving the PAX3 gene in chromosome 2q35. Hum Mol Genet 1993;2:953–9.

[62] van den Boogaard MJ, et al. *MSX1* mutation is associated with orofacial clefting and tooth agenesis in humans. Nat Genet 2000;24:342–3.

[63] Vieira AR, et al. MSX1 and TGFB3 contribute to clefting in South America. J Dent Res 2003;82:289–92.

[64] Milunsky JM, et al. TFAP2A mutations result in branchio-oculo-facial syndrome. Am J Hum Genet 2008;82:1171–7.

[65] Thomson JA, et al. Embryonic stem cell lines derived from human blastocysts. Science 1998;282:1145–7.

[66] Pomp O, et al. Generation of peripheral sensory and sympathetic neurons and neural crest cells from human embryonic stem cells. Stem Cells 2005;23:923–30.

[67] Lee G, et al. Isolation and directed differentiation of neural crest stem cells derived from human embryonic stem cells. Nat Biotechnol 2007;25:1468–75.

[68] Brokhman I, et al. Peripheral sensory neurons differentiate from neural precursors derived from human embryonic stem cells. Differentiation 2008;76:145–55.

[69] Jiang X, et al. Isolation and characterization of neural crest stem cells derived from in vitro-differentiated human embryonic stem cells. Stem Cells Dev 2009;18:1059–70.

[70] Curchoe CL, et al. Early acquisition of neural crest competence during hESCs neuralization. PLoS One 2010;5:e13890.

[71] Ziegler L, et al. Efficient generation of Schwann cells from human embryonic stem cell-derived neurospheres. Stem Cell Rev 2011;7:394–403.

[72] Chambers SM, et al. Combined small-molecule inhibition accelerates developmental timing and converts human pluripotent stem cells into nociceptors. Nat Biotechnol 2012;30:715–20.

[73] Liu Q, et al. Human neural crest stem cells derived from human ESCs and induced pluripotent stem cells: induction, maintenance, and differentiation into functional Schwann cells. Stem Cells Transl Med 2012;1:266–78.

[74] Menendez L, Yatskievych TA, Antin PB, Dalton S. Wnt signaling and a Smad pathway blockade direct the differentiation of human pluripotent stem cells to multipotent neural crest cells. PNAS 2011;108:19240–5.

[75] Fang D, et al. Defining the conditions for the generation of melanocytes from human embryonic stem cells. Stem Cells 2006;24:1668–77.

[76] Mica Y, Lee G, Chambers SM, Tomishima MJ, Studer L. Modeling neural crest induction, melanocyte specification, and disease-related pigmentation defects in hESCs and patient-specific iPSCs. Cell Rep 2013;3:1140–52.

[77] Zhou Y, Snead ML. Derivation of cranial neural crest-like cells from human embryonic stem cells. Biochem Biophys Res Commun 2008;376:542–7.

[78] Yu J, Thomson JA. Embryonic stem cells: regenerative medicine. National Institutes of Health, Bethesda: Department of Health and Human Services; 2006. pp. 1–12.

[79] Bajpai R, et al. CHD7 cooperates with PBAF to control multipotent neural crest formation. Nature 2010;463:958–62.

[80] Cimadamore F, et al. Human ESC-derived neural crest model reveals a key role for SOX2 in sensory neurogenesis. Cell Stem Cell 2011;8:538–51.

[81] Chambers SM, et al. Highly efficient neural conversion of human ES and iPS cells by dual inhibition of SMAD signaling. Nat Biotechnol 2009;27:275–80.

[82] Lee G, et al. Derivation of neural crest cells from human pluripotent stem cells. Nat Protoc 2010;5:688–701.

[83] Menendez L, et al. Wnt signaling and a Smad pathway blockade direct the differentiation of human pluripotent stem cells to multipotent neural crest cells. Proc Natl Acad Sci USA 2011;108:19240–5.

[84] Mengarelli I, Barberi T. Derivation of multiple cranial tissues and isolation of lens epithelium-like cells from human embryonic stem cells. Stem Cells Transl Med 2013;2:94–106.

[85] Vadasz S, Marquez J, Tulloch M, Shylo NA, Garcia-Castro MI. Pax7 is regulated by cMyb during early neural crest development through a novel enhancer. Development 2013;140:3691–702.

[86] Rada-Iglesias A, Bajpai R, Prescott S, Brugmann SA, Swigut T, Wysocka J. Epigenomic annotation of enhancers predicts transcriptional regulators of human neural crest. Cell Stem Cell 2012;11:633–48.

[87] Morrison SJ, et al. Prospective identification, isolation by flow cytometry, and in vivo self-renewal of multipotent mammalian neural crest stem cells. Cell 1999;96:737–49.

[88] Perris R, Perissinotto D. Role of the extracellular matrix during neural crest cell migration. Mech Dev 2000;95:3–21.

[89] Liu JP, Jessell TM. A role for rhoB in the delamination of neural crest cells from the dorsal neural tube. Development 1998;125:5055–67.

[90] Wang L, et al. Generation of hematopoietic repopulating cells from human embryonic stem cells independent of ectopic HOXB4 expression. J Exp Med 2005;201:1603–14.

[91] Takahashi K, Yamanaka S. Induction of pluripotent stem cells from mouse embryonic and adult fibroblast cultures by defined factors. Cell 2006;126: 663–76.

[92] Takahashi K, et al. Induction of pluripotent stem cells from adult human fibroblasts by defined factors. Cell 2007;131:861–72.

[93] Yu J, et al. Induced pluripotent stem cell lines derived from human somatic cells. Science 2007;318:1917–20.

[94] Park IH, et al. Generation of human-induced pluripotent stem cells. Nat Protoc 2008;3:1180–6.

[95] Lee G, et al. Large-scale screening using familial dysautonomia induced pluripotent stem cells identifies compounds that rescue IKBKAP expression. Nat Biotechnol 2012;30:1244–8.

[96] Wang A, et al. Induced pluripotent stem cells for neural tissue engineering. Biomaterials 2011;32: 5023–32.

[97] Nishikawa S, et al. The promise of human induced pluripotent stem cells for research and therapy. Nat Rev Mol Cell Biol 2008;9:725–9.

[98] Yamanaka S. Strategies and new developments in the generation of patient-specific pluripotent stem cells. Cell Stem Cell 2007;1:39–49.

[99] Saha K, Jaenisch R. Technical challenges in using human induced pluripotent stem cells to model disease. Cell Stem Cell 2009;5:584–95.

[100] Lee G, et al. Modelling pathogenesis and treatment of familial dysautonomia using patient-specific iPSCs. Nature 2009;461:402–6.

[101] Frumkin T, Malcov M, Telias M, Gold V, Schwartz T, Azem F, et al. Human embryonic stem cells carrying mutations for severe genetic disorders. In vitro cellular and developmental biology. Animal 2010;46:327–36.

[102] Aybar MJ, et al. Snail precedes slug in the genetic cascade required for the specification and migration of the Xenopus neural crest. Development 2003;130:483–94.

[103] Roode M, Blair K, Snell P, Elder K, Marchant S, et al. Human hypoblast formation is not dependent on FGF signalling. Dev Biol 2012;361:358–63.

Neural Crest Stem Cell: Tissue Regeneration and Repair

Pedro A. Sanchez-Lara[a,b] and Hu Zhao[b]

[a]Children's Hospital Los Angeles, Department of Pediatrics & Pathology and Laboratory Medicine, Keck School of Medicine, University of Southern California, Los Angeles, CA, USA, [b]Center for Craniofacial Molecular Biology, Ostrow School of Dentistry, University of Southern California, Los Angeles, CA, USA

INTRODUCTION

As conventional surgical treatments for congenital and acquired craniofacial problems continue to make progress, the final functional and cosmetic outcomes can be varied, unpredictable, and sometimes unsatisfactory, mostly because of complications, infections and scar tissue. The promise of regenerative medicine brings new energy and hope for improved outcomes by replacing damaged or absent tissues with healthy regenerated tissue.

The concept of craniofacial neural crest stem cell covers two different aspects. The first is the neural crest stem cell during embryonic development. It is well known that migrating and post-migrated cranial neural crest cells (CNCC) are multipotential stem cells that are capable of forming various types of tissue. They give rise to craniofacial structures of

413

various tissue types, including bone, cartilage, neuron, glial cells, melanocyte, adipocyte, adrenal medulla, and smooth muscle [1]. The second aspect of the CNCC stem cells concept are stem cells residing in adult tissue derived from the CNCC. They are tissue-specific stem cells being responsible for supporting adult tissue turnover and injury repair. Although the adult CNCC stem cells are definitely derived from the prenatal CNCC stem cells, it remains unknown whether they are the continuation of those prenatal CNC stem cells or they are specified populations formed through lineage restriction during postnatal development. In this chapter, we will mainly focus on these adult tissue-specific stem cells.

The most important value of the stem cells in the adult tissue is their potential clinical applications on repairing or regenerating tissue and organs. An open question for many stem cell studies is this: "What is the advantage of using tissue-specific stem cells over using non-tissue-specific multipotential stem cells for tissue engineering?" It is well known that same type of tissue around the body may have different origins. For example, craniofacial bones are mainly derived from CNCC, whereas bones of other regions are derived from mesoderm. Future investigators will further define the stem cell properties from various developmental origins and the unique impact they have on the tissues in which they are residing. Do we have to use CNCC-derived stem cells to repair or regenerate craniofacial tissue defects if non-CNCC-derived stem cells can do the same work better? This is a difficult question to answer before we have more knowledge on the CNCC-derived stem cells. However, we can classify the CNCC-derived stem cells into two categories based on their translational advantages over other stem cells. The first category is CNC unique stem cells. They possess unique properties that no other stem cells have, such as dental mesenchymal stem cells (MSCs) and periodontal

ligament (PDL) MSCs. Obviously, these stem cells are irreplaceable in the tissue engineering applications. The second category includes CNCC stem cells that are not unique to CNCC. Some of them have advantages over other stem cells to tissue engineering applications, whereas others may not. Most of CNCC-derived stem cells fall into this category. For example, MSCs taken from the mandible bone or gingival have advantages over MSCs taken from other tissue in repairing craniofacial defects because of their easy access and strong proliferation ability.

GINGIVAL MSC

The human oral mucosa is highly active in terms of cell turnover and regeneration, which suggests the existence of one or multiple types of stem cell populations. Recently, a stem cell population was identified from the lamina propria of adult human oral mucosa. This population was identified by positive expression of embryonic stem cell markers Oct4, Sox2, Nanog and p75. These cells were localized *in vivo* to some cord-like structures. They are highly proliferative *in vitro* and are able to differentiate into tissue of mesodermal (osteoblast, chondrocyte, and adipocyte), endodermal (endothelium), and ectodermal lineages (neuronal cells). Surprisingly, when transplanted into nude mice and treated with dexamethasone, these cells were able to form tumors containing mixed types of tissue. This study indicates caution needs to be taken when applying stem cells for tissue engineering [2]. In addition to this, in 2003, Tran et al. [3] reported the transdifferentiation of bone marrow MSCs into buccal epithelial cells in human patients. By tracing the Y-chromosome of the bone marrow MSC male donor in the female recipient patients, they were able to localize the distribution of donor bone marrow MSC cells on the buccal epithelial cells of the recipients. 1.8% of the recipients' cheek epithelial cells were from the donor MSC origin, and this was detectable at 56–1964 days after the procedure [3].

Regenerative therapy is always aiming at reducing wound-healing time and reducing scar formation. Wound healing of the skin is comprised of three phases: the coagulation/early inflammation phase, late inflammation phase, and proliferative phase [4]. Although the oral mucosa healing goes through the same three phases, it always heals with an accelerated rate and reduced scar formation [5]. Fibroproliferative scars such as keloid and hypertrophic scars are rarely seen in the oral cavity [6]. The only exception is the hard palate of the mouse, which heals at a much slower rate than any other area of the oral mucosa [7]. This unique property of the oral mucosa is critical for any tissue engineering study. The first reason for the difference between oral mucosa healing and skin healing processes is the distinctive inflammatory response to the injury. The ratio of TGF-beta1 to TGF-beta3 is much lower in the oral mucosa than in the skin [8]. In addition, fewer inflammatory cells infiltrate the mucosa wound at the initial stage, and fewer inflammatory cytokines and chemokines are activated in the wound. Also, angiogenesis is less active in the oral mucosa wound than the skin, so oral wound healing is quite similar to fetal skin wound healing [9].

To date, no satisfactory FDA-approved therapy is available for the treatment of scarring. Some reagents have been shown to possess anti-scarring effects. Topical hyaluronic acid and saponin may reduce the scar formation by stimulating hyaluronic acid production [10]. Some TGF-beta3 formulation and neutralizing antibody to TGF-beta1, 2 have been shown to be effective on reducing scar formation [11]. Decorin can limit the duration of TGF-beta effects on inflammation and fibrosis [12]. Other factors, including TNF-alpha, Platelet-Derived Growth Factor (PDGF), Fibroblast Growth Factor (FGF), Vascular Endothelial Growth Factor (VEGF), Insulin-Like Growth Factor (ILGF), and Epidermal Growth Factor (EGF), have also been shown to have various effects on preventing scar formation [13].

DENTAL MSC

Human teeth consist of enamel, dentin, tooth pulp, and cementum covering the root surface. Surrounding the tooth is the PDL, which supports the tooth. Unlike the bone, hard tissue in tooth does not undergo renewal after its formation, except the dentin, which can regenerate itself internally upon injury, suggesting the existence of stem cell populations within the tooth pulp. One of the first dental-related stem cells identified was named DPSCs (dental pulp stem cells) [14]. DPSCs are capable of differentiating into multiple types of tissue, including odontoblast, bone, adipocyte, and neurons [14–16]. In addition, stem cells from human exfoliated deciduous teeth pulp (SHED) were also identified as possessing multipotential differentiation ability [15]. Both SHED and DPSC were shown to be able to generate tissue resembling human tooth pulp under appropriate conditions [17–20]. Efforts have been made to rebuild the teeth *in vitro* by combining tooth pulp-derived stem cells with proper scaffold materials [17,21,22]. In the Ohazama study, different types of non-dental derived mesenchymal cells, including ES cells, neural stem cells, and adult bone marrow MSCs, were mixed with embryonic oral epithelium cells. The mesenchyme-epithelium cell mixtures were then delivered into the kidney capsule of adult mice in an effort to recapitulate the classical dental epithelium-mesenchyme interactions that initiate and direct tooth development. All the mixtures resulted in the development of tooth-like structures and surrounding bone. This experiment indicated that it is possible to regenerate a tooth by mimicking the natural developmental process [22]. Other studies have used DPSCs or SHED for treatment of disease of non-dental tissue such as muscle dystrophies, critical size bony defect, spinal cord damage, corneal injury, and even systemic lupus erythematosus [23–28].

PDL MSC

Periodontal diseases affect at least 15% of the human adult population, with the symptoms and signs of periodontal soft tissue loss and subsequent supporting bone resorption that leads to loss of teeth [29]. Current therapeutic approaches include the use of guided tissue regeneration, bioactive grafting materials, and application of bioactive molecules to induce regeneration, but the overall effects of these approaches are relatively modest and limited in practical applications. Regenerating the periodontium has always been a challenge in the treatment of periodontal diseases due to the complex structures of the periodontium, consisting of cementum, PDL, gingival, and supporting bone. Its regeneration, therefore, will require either multiple cell populations or a multipotential stem cell population.

PDL is unique among all the ligament and tendon tissues of the body because it is the only soft tissue connecting two distinct hard tissues [30]. The PDL suspends the tooth like a cushion to transduce the mechanical load from the teeth evenly onto the supporting bone. Early observations conducted on different animal models demonstrated that the periodontal tissues possess some regeneration activity, suggesting the existence of stem cell population within the periodontium [31–35]. After depletion of various periodontal tissues, not only the PDLs but also cementum and alveolar bone can be regenerated, suggesting the presence of multipotential stem cell

populations [33–35]. Studies conducted on the human periodontum indicate the presence of a putative PDL stem cell population (PDLSCs) [36]. They are positive for MSC markers, including STRO-1 and CD146, and are able to differentiate into ostelbast, adipocyte, and cementoblast. The *in vitro* expanded human PDLSCs can contribute to the periodontal tissue regeneration when transplanted into the immunocompromised mice [36]. PDLSC were shown to be multipotential. *In vitro*, they can differentiate into various tissues of mesodermal (adipocyte, osteoblast, and chondrocyte), ectodermal (neuron) and endodermal lineages (hepatocyte) [37].

Cells of non-dental origins have also been tested for periodontal tissue regeneration. In 2004, Kawaguchi et al. transplanted *ex vivo* expanded bone marrow MSC into recipient dogs with periodontal defects. After a month, the transplanted cells were found to be able to repair the defective periodontal tissue, including cementoblast, PDL, and bone. This study suggested that bone marrow MSC can be used as a source for periodontal tissue regeneration [38]. Their following study indicated that this regeneration could be enhanced by brain-derived neurotrophic factor (BDNF). Presence of BDNF increased expression level of multiple bone and periodontal tissue related markers, including OPN, BMP2, collagen I, ALPase, and VEGF [39].

Cell sheet engineering has emerged as a novel alternative approach for periodontal tissue engineering without the disruption of both critical cell surface proteins such as ion channels, growth factor receptors, and cell-to-cell junction proteins. PDL cells can be isolated from an extracted tooth and can be cultured on temperature-responsive culture dishes at 37°C. Transplantable cell sheets can be harvested by reducing the temperature to 20°C, to be transplanted into the bony defect [40]. This was found to produce an obvious cementum layer and Sharpey's fibers [41]. Cell sheet engineering therefore allows for tissue regeneration by either direct transplantation of cell sheets to host tissues or the creation of three-dimensional structures via the layering of individual cell sheets. By avoiding the use of any additional materials such as carrier substrates or scaffolds, the complications associated with traditional tissue engineering approaches, such as host inflammatory responses to implanted polymer materials, can be avoided. Cell sheet engineering thus presents several significant advantages and can overcome many of the problems that have previously restricted tissue engineering with biodegradable scaffolds [42].

TMJ STEM CELLS

The temporomandibular joint (TMJ) is comprised of both osseous and cartilaginous structures. It can deteriorate due to injuries, osteoarthritis, or rheumatoid arthritis. The cartilage tissue has a limited capacity of intrinsic repair, so even minor lesions of injury may lead to progressive damage. Severe TMJ lesions need surgical replacement of the mandibular condyle [43]. Currently, only a few studies on TMJ tissue engineering have been conducted in animal models. In one study, bone marrow MSCs were isolated from the long bone marrow and expanded *in vitro* under either osteogenic or chondrogenic culture conditions. The expanded osteogenic and chondrogenic cells were mixed with PEGDA hydrogel and seeded onto an adult human cadaver mandible condyle in two stratified yet integrated layers. These bi-layer constructs were then placed under nude mice skin for culture. After 4 weeks of implantation, de novo formation of human condyle-like structures were observed replicating the relevant shape and dimensions. Chondrocytes and osteocytes of donor origin were identified in separated layers, and the two cell types infiltrate into each other's territory resembling the native condition. However, both chondrogenic

and osteogenic layers showed suboptimal maturation, possibly due to an insufficient amount of cells [44,45]. The same group also constructed a mandibular condyle scaffold by using CAD/CAM techniques and combined it with autologous bone marrow MSC cells. The construct was then transplanted into minipig TMJs for function evaluation. Evaluation and analysis after 1 and 3 months indicated bone regeneration of condyle shape and normally administrated masticatory function [46].

BONE MARROW MESENCHYMAL STEM CELLS (BMMSCs)

Current clinical approaches for reconstructing craniofacial defects include autologous bone grafts, allogeneic bone grafts, and prosthetic grafts such as titanium frameworks [47−50]. Stem cell-based strategies are currently a hot topic in craniofacial bone tissue engineering. Various cell sources have been used for repairing craniofacial bony defects. Bone marrow MSC taken from the iliac crest is the most commonly used cell source. Bone marrow MSCs have the ability of multipotent differentiation into bone, cartilage, tendon, muscle, adipose tissue, and neuronal tissue [51−53]. BMMSCs collected from the illiac crest have been reported to reconstruct jaw defect in various clinical studies [54,55].

Alveolar bone also contains the BMMSCs. They are essentially different from iliac crest BMMSCs in many aspects. Alveolar bone marrow MSCs are originated from CNCC, whereas iliac crest BMMSCs are from the mesoderm [56,57]. Some congenital diseases, such as Cherubism [58], Treacher Collins syndrome [59], craniofacial fibrous dysplasia [60], and hyperparathyroid jaw tumor syndrome [61], affect only craniofacial bones but not long bone, indicating that craniofacial bone development differs from that of mesoderm-derived bone formation. In addition to that, CNCC-derived

BMMSCs also possess different differentiation ability as the mesoderm-derived BMMSCs.

In a study by Chung et al. [57], alveolar bone MSCs were compared with long bone MSCs. Both populations can undergo multipotential differentiation into bone, cartilage, and adipose tissue. Alveolar bone MSCs are more responsive to osteogenic inductive signals like BMP and TGF-beta. In addition, following transplantation into nude mice, alveolar bone MSCs formed bone with densely packed lamella structures that were separated by abundant connective tissues. Interestingly, unlike bone originating from BMMSCs, bone derived from alveolar bone MSCs did not contain prominent hematopoietic components. These features may be a reflection of the intrinsic characteristic of craniofacial bone from intramembranous ossifications. In another study [62], alveolar bone MSCs were compared with long bone MSCs on their immunomodulation properties. Alveolar bone MSCs obtained from mice showed a stronger suppressive effect on the proliferation of anti-CD3 antibody-activated T-cells, along with high levels of NO production, when stimulated with IFNγ. Also, alveolar bone MSCs were capable of maintaining naïve splenocyte (including T-cell) survival more effectively than long bone MSCs.

Similar comparison studies were also performed on human subjects. Akintoye et al. [63] investigated skeletal site-specific phenotypic and functional differences between craniofacial bone-derived and long bone-derived human MSCs. Compared with long bone MSCs cells, craniofacial MSCs proliferated faster with less senescence, expressed higher levels of alkaline phosphatase, and demonstrated more bone formation ability *in vitro*. Both populations of MSCs showed similar affinity to titanium surface. But titanium-attached craniofacial MSCs were more osteogenically responsive than long bone MSCs [64]. In their follow-up study, human craniofacial MSCs showed more susceptible osteogenic response to BMP2 or bisphosphonates stimulation than long bone MSCs [65,66]. Craniofacial bone MSC survived higher radiation doses and recovered quicker than long bone MSC [67]. Alveolar BMSCs obtained from individuals of different ages indicated that the donor age has little effect on their gene expression pattern [68].

ADIPOSE MSC

Adipose-derived mesenchymal stem cells (AMCs) have also been used for repairing the craniofacial bone defects. AMCs are readily obtained via lipo-aspirate and grow up easily *in vitro*. They are multipotential and capable of forming different types of tissue including muscle, bone, neural and chondrocyte tissues [69–71]. AMCs taken from human sources were shown to be able to form bone when seeded in HA-TCP scaffolds and transplanted into immunocompromised mice [71]. In a clinical experiment conducted by Cowan et al., AMCs were expanded *in vitro* and seeded in apatite-coated PLGA scaffolds. The construct was then transplanted into a human patient to repair a critical size calvaria bone defect.

CNCC also give rise to the adipose tissue in the craniofacial region [72]. However, no study has been conducted to compare the CNCC-derived adipose MSC with other AMCs.

CONCLUSIONS AND PERSPECTIVES

The impact of tissue engineering and potential applications of stem cells to reconstruct different dental, oral, and craniofacial tissues and structures extends well beyond craniofacial and dental practices. It is to be hoped that future stem cell-based therapeutics will replace allograft and autologous tissue grafts, while improving long-term function and eliminating donor site morbidity.

References

[1] Achilleos A, Trainor PA. Neural crest stem cells: discovery, properties and potential for therapy. Cell Res 2012;22(2):288–304.

[2] Marynka-Kalmani K, et al. The lamina propria of adult human oral mucosa harbors a novel stem cell population. Stem Cells 2010;28(5):984–95.

[3] Tran SD, et al. Differentiation of human bone marrow-derived cells into buccal epithelial cells in vivo: a molecular analytical study. Lancet 2003;361 (9363):1084–8.

[4] Nauta A, Gurtner GC, Longaker MT. Wound healing and regenerative strategies. Oral Dis 2011;17(6):541–9.

[5] Whitby DJ, Ferguson MW. The extracellular matrix of lip wounds in fetal, neonatal and adult mice. Development 1991;112(2):651–68.

[6] Wong JW, et al. Wound healing in oral mucosa results in reduced scar formation as compared with skin: evidence from the red Duroc pig model and humans. Wound Repair Regen 2009;17(5):717–29.

[7] Graves DT, et al. IL-1 plays a critical role in oral, but not dermal, wound healing. J Immunol 2001;167 (9):5316–20.

[8] Schrementi ME, et al. Site-specific production of TGF-beta in oral mucosal and cutaneous wounds. Wound Repair Regen 2008;16(1):80–6.

[9] Mak K, et al. Scarless healing of oral mucosa is characterized by faster resolution of inflammation and control of myofibroblast action compared to skin wounds in the red Duroc pig model. J Dermatol Sci 2009;56(3):168–80.

[10] Mast BA, et al. Hyaluronic acid is a major component of the matrix of fetal rabbit skin and wounds: implications for healing by regeneration. Matrix 1991;11(1):63–8.

[11] Rhett JM, et al. Novel therapies for scar reduction and regenerative healing of skin wounds. Trends Biotechnol 2008;26(4):173–80.

[12] Jarvelainen H, et al. A role for decorin in cutaneous wound healing and angiogenesis. Wound Repair Regen 2006;14(4):443–52.

[13] Lawrence WT. Physiology of the acute wound. Clin Plast Surg 1998;25(3):321–40.

[14] Gronthos S, et al. Postnatal human dental pulp stem cells (DPSCs) in vitro and in vivo. Proc Natl Acad Sci USA 2000;97(25):13625–30.

[15] Miura M, et al. SHED: stem cells from human exfoliated deciduous teeth. Proc Natl Acad Sci USA 2003;100(10):5807–12.

[16] Huang GT, Gronthos S, Shi S. Mesenchymal stem cells derived from dental tissues vs. those from other sources: their biology and role in regenerative medicine. J Dent Res 2009;88(9):792–806.

[17] Cordeiro MM, et al. Dental pulp tissue engineering with stem cells from exfoliated deciduous teeth. J Endod 2008;34(8):962–9.

[18] Sakai VT, et al. SHED differentiate into functional odontoblasts and endothelium. J Dent Res 2010;89 (8):791–6.

[19] Demarco FF, et al. Effects of morphogen and scaffold porogen on the differentiation of dental pulp stem cells. J Endod 2010;36(11):1805–11.

[20] Casagrande L, et al. Dentin-derived BMP-2 and odontoblast differentiation. J Dent Res 2010;89(6): 603–8.

[21] Young CS, et al. Tissue engineering of complex tooth structures on biodegradable polymer scaffolds. J Dent Res 2002;81(10):695–700.

[22] Ohazama A, et al. Stem-cell-based tissue engineering of murine teeth. J Dent Res 2004;83(7):518–22.

[23] Kerkis I, et al. Early transplantation of human immature dental pulp stem cells from baby teeth to golden retriever muscular dystrophy (GRMD) dogs: local or systemic? J Transl Med 2008;6:35.

[24] Seo BM, et al. SHED repair critical-size calvarial defects in mice. Oral Dis 2008;14(5):428–34.

[25] Monteiro BG, et al. Human immature dental pulp stem cells share key characteristic features with limbal stem cells. Cell Prolif 2009;42(5):587–94.

[26] Ishkitiev N, et al. Deciduous and permanent dental pulp mesenchymal cells acquire hepatic morphologic and functional features in vitro. J Endod 2010;36 (3):469–74.

[27] Nosrat IV, et al. Dental pulp cells produce neurotrophic factors, interact with trigeminal neurons in vitro, and rescue motoneurons after spinal cord injury. Dev Biol 2001;238(1):120–32.

[28] Yamaza T, et al. Immunomodulatory properties of stem cells from human exfoliated deciduous teeth. Stem Cell Res Ther 2010;1(1):5.

[29] Mase J, et al. Cryopreservation of cultured periosteum: effect of different cryoprotectants and preincubation protocols on cell viability and osteogenic potential. Cryobiology 2006;52(2):182–92.

[30] McCulloch CA, Lekic P, McKee MD. Role of physical forces in regulating the form and function of the periodontal ligament. Periodontol 2000;24:56–72.

[31] Karring T, Nyman S, Lindhe J. Healing following implantation of periodontitis affected roots into bone tissue. J Clin Periodontol 1980;7(2):96–105.

[32] Nyman S, et al. Healing following implantation of periodontitis-affected roots into gingival connective tissue. J Clin Periodontol 1980;7(5):394–401.

[33] Nyman S, et al. The regenerative potential of the periodontal ligament. An experimental study in the monkey. J Clin Periodontol 1982;9(3):257–65.

[34] Nielsen IM, Ellegaard B, Karring T. Kielbone in healing interradicular lesions in monkeys. J Periodontal Res 1980;15(3):328–37.

[35] Parlar A, et al. New formation of periodontal tissues around titanium implants in a novel dentin chamber model. Clin Oral Implants Res 2005;16(3):259–67.

[36] Seo BM, et al. Investigation of multipotent postnatal stem cells from human periodontal ligament. Lancet 2004;364(9429):149–55.

[37] Seo BM, et al. Recovery of stem cells from cryopreserved periodontal ligament. J Dent Res 2005;84 (10):907–12.

[38] Kawaguchi H, et al. Enhancement of periodontal tissue regeneration by transplantation of bone marrow mesenchymal stem cells. J Periodontol 2004;75(9):1281–7.

[39] Takeda K, et al. Brain-derived neurotrophic factor enhances periodontal tissue regeneration. Tissue Eng 2005;11(9–10):1618–29.

[40] Huang S-Y, Zhang D-S. Periodontal Ligament Cell Sheet Engineering: A New Possible Strategy to Promote Periodontal Regeneration of Dental Implants. Rhode Island: Dental Hypotheses Westerly; 2011.

[41] Flores MG, et al. Periodontal ligament cell sheet promotes periodontal regeneration in athymic rats. J Clin Periodontol 2008;35(12):1066–72.

[42] Yang J, et al. Cell sheet engineering: recreating tissues without biodegradable scaffolds. Biomaterials 2005;26 (33):6415–22.

[43] Sarnat BG, Laskin DM. The temporomandibular joint: a biological basis for clinical practice. 4th ed. Philadelphia, PA: Saunders; 1992 xix, p. 505.

[44] Alhadlaq A, Mao JJ. Tissue-engineered neogenesis of human-shaped mandibular condyle from rat mesenchymal stem cells. J Dent Res 2003;82(12):951–6.

[45] Alhadlaq A, Mao JJ. Mesenchymal stem cells: isolation and therapeutics. Stem Cells Dev 2004;13 (4):436–48.

[46] Mao JJ, et al. Craniofacial tissue engineering by stem cells. J Dent Res 2006;85(11):966–79.

[47] Marchac D. Split-rib grafts in craniofacial surgery. Plast Reconstructive Surg 1982;69(3):566–7.

[48] Shenaq SM. Reconstruction of complex cranial and craniofacial defects utilizing iliac crest-internal oblique microsurgical free flap. Microsurgery 1988;9 (2):154–8.

[49] Goodrich JT, Argamaso R, Hall CD. Split-thickness bone grafts in complex craniofacial reconstructions. Pediatr Neurosurg 1992;18(4):195–201.

[50] Cowan CM, et al. Adipose-derived adult stromal cells heal critical-size mouse calvarial defects. Nat Biotechnol 2004;22(5):560–7.

[51] Friedenstein AJ, et al. Heterotopic of bone marrow. Analysis of precursor cells for osteogenic and hematopoietic tissues. Transplantation 1968;6(2):230–47.

[52] Deans RJ, Moseley AB. Mesenchymal stem cells: biology and potential clinical uses. Exp Hematol 2000;28 (8):875–84.

[53] Caplan AI, Bruder SP. Mesenchymal stem cells: building blocks for molecular medicine in the 21st century. Trends Mol Med 2001;7(6):259–64.

[54] Yamada Y, et al. Autogenous injectable bone for regeneration with mesenchymal stem cells and platelet-rich plasma: tissue-engineered bone regeneration. Tissue Eng 2004;10(5–6):955–64.

[55] Meijer GJ, et al. Cell based bone tissue engineering in jaw defects. Biomaterials 2008;29(21):3053–61.

[56] Chai Y, et al. Fate of the mammalian cranial neural crest during tooth and mandibular morphogenesis. Development 2000;127(8):1671–9.

[57] Chung IH, et al. Stem cell property of postmigratory cranial neural crest cells and their utility in alveolar bone regeneration and tooth development. Stem Cells 2009;27(4):866–77.

[58] Ueki Y, et al. Mutations in the gene encoding c-Abl-binding protein SH3BP2 cause cherubism. Nat Genet 2001;28(2):125–6.

[59] Dixon J, Trainor P, Dixon MJ. Treacher Collins syndrome. Orthod Craniofac Res 2007;10(2):88–95.

[60] Riminucci M, et al. The histopathology of fibrous dysplasia of bone in patients with activating mutations of the Gs alpha gene: site-specific patterns and recurrent histological hallmarks. J Pathol 1999;187(2):249–58.

[61] Simonds WF, et al. Familial isolated hyperparathyroidism: clinical and genetic characteristics of 36 kindreds. Medicine (Baltimore) 2002;81(1):1–26.

[62] Yamaza T, et al. Mouse mandible contains distinctive mesenchymal stem cells. J Dent Res 2011;90(3):317–24.

[63] Akintoye SO, et al. Skeletal site-specific characterization of orofacial and iliac crest human bone marrow stromal cells in same individuals. Bone 2006;38(6):758–68.

[64] Akintoye SO, et al. Comparative osteogenesis of maxilla and iliac crest human bone marrow stromal cells attached to oxidized titanium: a pilot study. Clin Oral Implants Res 2008;19(11):1197–201.

[65] Osyczka AM, et al. Age and skeletal sites affect BMP-2 responsiveness of human bone marrow stromal cells. Connect Tissue Res 2009;50(4):270–7.

[66] Stefanik D, et al. Disparate osteogenic response of mandible and iliac crest bone marrow stromal cells to pamidronate. Oral Dis 2008;14(5):465–71.

[67] Damek-Poprawa M, et al. Human bone marrow stromal cells display variable anatomic site-dependent

response and recovery from irradiation. Arch Oral Biol 2010;55(5):358–64.

[68] Han J, et al. Collection and culture of alveolar bone marrow multipotent mesenchymal stromal cells from older individuals. J Cell Biochem 2009;107(6):1198–204.

[69] Zuk PA, et al. Multilineage cells from human adipose tissue: implications for cell-based therapies. Tissue Eng 2001;7(2):211–28.

[70] Gimble J, Guilak F. Adipose-derived adult stem cells: isolation, characterization, and differentiation potential. Cytotherapy 2003;5(5):362–9.

[71] Hicok KC, et al. Human adipose-derived adult stem cells produce osteoid in vivo. Tissue Eng 2004;10 (3–4):371–80.

[72] Billon N, et al. The generation of adipocytes by the neural crest. Development 2007;134(12):2283–92.

Functional Significance of Cranial Neural Crest Cells During Tooth Development and Regeneration

Carolina Parada[a], Yang Chai[a] and Paul Sharpe[b]

[a]Center for Craniofacial Molecular Biology, Herman Ostrow School of Dentistry, University of Southern California, Los Angeles, CA 90033, USA

[b]Department of Craniofacial Development and Stem Cell Biology, Dental Institute, Kings College London, Guy's Hospital, London Bridge, London, SE1 9RT, UK

INTRODUCTION

One of the key features of craniofacial development is the formation of cranial neural crest cells (CNCC). The specification, delamination, and migration, proliferation, survival, and ultimate fate determination of the CNCC play important roles in regulating craniofacial development. Unlike cells from the trunk neural crest, CNCC give rise to a wide array of cell types during embryonic development. For example, CNCC form most of the hard tissues of the head, such as bone, cartilage, and teeth. During craniofacial development, CNCC migrate ventrolaterally and establish contact with the pharyngeal ectoderm, mesoderm, and endoderm. Subsequently, the CNCC-derived mesenchyme proliferates and contributes to formation of a series of discrete swellings known as branchial arches (BAs) [1,2]. Once in the BAs, a Hox-negative distal-less homeobox (Dlx) code provides CNCC with patterning information and intra-arch polarity along the dorsoventral (DV)/proximodistal axis. This polarity is essential for the establishment of the maxillary and mandibular prominences from the first BA and is achieved through the differential expression of Dlx genes. The subdivision of the first BA is primarily achieved with two Dlx combinations: *Dlx1/2* for the maxillary and *Dlx1/2/5/6* for the mandibular prominence [3–5]. The maxillary and mandibular prominences together with the frontonasal prominence constitute the primitive face and surround the stomodeum, or primordial mouth. The frontonasal prominence contributes significantly to the formation of the nose and the upper lip. The maxillary process gives rise to part of the upper lip, the maxillary bone, and the secondary palate, whereas the mandibular prominence forms the mandible and part of the tongue [6]. Maxillary and mandibular CNCC contribute significantly to the development of upper and lower teeth, respectively. Here, we review current concepts regarding the contribution of CNCC to tooth development.

TOOTH DEVELOPMENT

Origin of Tooth Cells

Tooth development is a complex process that involves a series of reciprocal interactions between the epithelium and the mesenchyme, similar to the development of other ectodermal organs. These interactions are mediated by major signaling pathways, members of which act reiteratively at diverse stages during the development of these organs. Once the first BA is formed, the oral epithelium thickens to become the dental lamina, the first morphological evidence of tooth development. At this stage, the dental epithelium possesses the competence to initiate and instruct tooth development. The epithelium then invaginates into the underlying CNCC-derived mesenchyme to form the tooth bud. As the mesenchyme condenses around the invaginating epithelium, it acquires the potential to promote odontogenesis. Next, the epithelium grows and folds to form a cap and then a bell-like structure. An epithelial organizing center that develops at the tip of the late bud, known as the enamel knot, regulates these events. At the late-bell stage, the condensing mesenchyme entirely surrounds the invaginating epithelium. Cell differentiation occurs during the late-bell stage. The CNCC-derived dental mesenchymal cells differentiate into dentin-secreting odontoblasts, and the neighboring epithelial cells located in close contact with the dental mesenchyme differentiate into enamel-producing ameloblasts [7]. Root formation occurs after crown development is completed. While development of the tooth crown is relatively well understood, much less is known regarding the molecular mechanisms directing root development. Dentin comprises most of the root, which normally contains root canals filled with a highly vascularized and loose dental pulp.

Cell mapping using the two-component genetic system *Wnt1-Cre;R26R* for indelibly marking the progeny of the CNC shows that CNCC contribute significantly to the formation of condensed dental mesenchyme, dental papilla, odontoblasts, dentine matrix, pulp, cementum, periodontal ligaments, and alveolar bone [8]. Regarding the origin of the epithelium, although in the mouse oral cavity it appears to be an ectodermal derivative, it has dual molecular features. Recently, Ohazama and colleagues [9] demonstrated that molar teeth, but not incisors, derive from epithelium that shares molecular features with the pharyngeal endoderm (Figure 20.1A). This finding is consistent with the phenotype of Ikka-null mice, which have abnormal incisors and other ectodermal organs but normal molars [10]. A recent study performed in transgenic axolotl showed that oral teeth are derived from both ectoderm and endoderm and, moreover, demonstrated that individual teeth may have a mixed ecto/endodermal origin, at least in this species. Despite the enamel epithelia having different embryonic sources, oral teeth in the axolotl exhibit remarkable developmental uniformity in morphology [11]. This suggests a dominant role for the neural crest mesenchyme over the epithelia in tooth initiation and, from an evolutionary point of view, that an essential factor in tooth evolution was the odontogenic capacity of NCC. Because the teeth of axolotl larvae are all monocuspid, the influence of the

endoderm in defining tooth identity cannot be studied in this model. Another illustration of the dominant neural crest role comes from transplants of mouse neural tube into chick embryos. The mouse CNCC respond to signals from the chick oral ectoderm and form tooth-like structures [12]. In mice, the extension of genes expressed in the developing pharyngeal endoderm into the proximal oral epithelium might reflect a requirement for endodermal signals for hard tissue patterning in the developing face, specifically for molar identity [9].

Patterning the Dentition and Instructive Signals for Patterning

During tooth bud formation, epithelial thickenings invaginate into the underlying CNCC-derived mesenchyme, which condenses around these epithelial buds. At this stage, the odontogenic potential switches from the epithelium to the mesenchyme [13], which correlates with

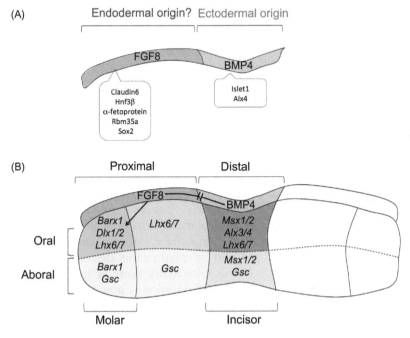

FIGURE 20.1 **Early patterning of the mouse dentition.** (A) Differential gene expression in the dental epithelium along the proximal–distal axis, suggesting a double origin from endoderm and ectoderm. (B) Diagram showing an isolated mandibular arch with distinct gene expression domains in the CNCC-derived mesenchyme along the proximal–distal and oral–aboral axes.

changes in the expression pattern of Bmp4 [14]. Recent results demonstrate that the mesenchymal condensation that subsequently drives tooth formation is induced by epithelial morphogens in an unexpected way. FGF8 and SEMA3F, which are produced at early stages by the dental epithelium, attract and repulse mesenchymal cells, causing them to pack tightly together. These mechanical compaction-induced changes in cell shape induce the odontogenic transcription factors *Pax9* and *Msx1* as well as *Bmp4*. Mechanical compression of the mesenchyme is also sufficient to induce tooth-specific cell fate switching, which underscores the role of mechano-transduction during developmental processes [15].

Bmp signaling is pivotal for morphogenesis of the dental organ. Deficiency of *BmprIa* in CNCC leads to an arrest of tooth development at the bud/early cap stages. Defective tooth development is accompanied by the downregulation of BMP-responsive genes and decreased cell proliferation levels in the dental mesenchyme [16]. Bmp signaling regulates the activation of expression of *Msx1* and *Pax9* in the developing tooth mesenchyme [17]. In turn, multiple lines of evidence suggest that these two transcription factors play an important role in the maintenance of mesenchymal *Bmp4* expression, which ultimately drives progression of tooth development from the bud to the cap stage [18,19]. For instance, deficiency of *Msx1* or *Pax9* in mice causes arrested tooth development at the bud stage, which is associated with downregulation of *Bmp4* expression in the mesenchyme (Figure 20.2) [20−22]. In $Msx1^{-/-}$ mice, the addition of recombinant Bmp4 protein rescues $Msx1^{-/-}$ mutant mandibular first molar tooth germs to the late-bell stage in explant cultures [23,24]. Bmp4-releasing beads placed in contact with isolated dental epithelium induce localized expression of primary enamel knot marker p21 [25]. Transgenic *Bmp4* expression driven by an *Msx1* gene promoter also partially rescues $Msx1^{-/-}$ mutant first molar tooth germs to the cap stage with the formation of a primary enamel knot [26].

Although it has long been believed that BMP4 plays an essential role in the bud to cap stage transition, this was based on the indirect evidence from Msx1 mutant studies mentioned above. Recently, genetic deletion of BMP4 using $Wnt1-cre:Bmp4^{fl/fl}$ mice showed, somewhat surprisingly, that although BMP signaling is essential for the bud to cap transition in mandibular molars, it is not required for maxillary molar or incisor development [27]. A critical factor in explaining these results is the transcription factor Osr2, whose expression is upregulated in the molar tooth mesenchyme when BMP signaling is lost (Figure 20.2). Osr2 acts to antagonize Msx1-mediated activation of mesenchymal signaling pathways required for the bud to cap transition. The reduction in Osr2 might lead to sufficient activation of as-yet-unidentified Msx1-induced signals to drive the bud to cap transition in BMP-less incisors and maxillary molars as observed in $Osr2^{+/-}$; $Wnt1-Cre;Bmp4^{fl/fl}$ mice. Regardless of the mechanism, these results highlight the differing molecular control of incisor versus molar and maxilla versus mandibular morphogenetic pathways, consistent with earlier discoveries demonstrating the roles of Dlx proteins and activin βA in regulating tooth morphogenesis (see below) [28,29].

Type of Teeth

Mammalian dentition contains teeth with different shapes, situated in different regions of the jaw. Mice have two groups of teeth, incisors and molars, whereas humans possess incisors, canines, premolars, and molars. It is clear that individual morphology and position within the arch are two essential factors defining types of teeth, but despite many years of investigation, the fundamental question of how these different tooth shapes are specified remains unanswered. In part, this reflects the fact that at the onset of

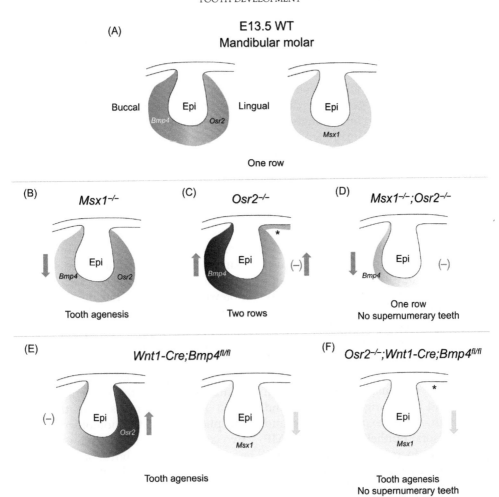

FIGURE 20.2 Bmp4-Msx1 and Osr2 in tooth morphogenesis and sequential tooth formation. (A) In E13.5 wild-type (WT) mandibular molars, *Bmp4* and *Osr2* are expressed in complementary gradients, with *Bmp4* mRNA being more concentrated on the buccal side (blue) and *Osr2* on the lingual side (red). *Msx1* is expressed homogenously along this axis (green). (B) In *Msx1*$^{-/-}$ mice, *Bmp4* expression is downregulated whereas *Osr2* expression is comparable to that of wild-type embryos. *Msx1* deficiency results in absence of the first molar. (C) *Osr2*-deficient mice display supernumerary teeth lingual to their molars in a "second row" (asterisk). This phenotype is due to an expansion of *Msx1* expression and its feedback activator, *Bmp4*, to the lingual side. (D) In *Msx1*$^{-/-}$;*Osr2*$^{-/-}$ mice, *Bmp4* expression is downregulated but sufficient to drive morphogenesis of the first molars, and no supernumerary teeth are initiated. (E) *Wnt1-Cre;Bmp4*$^{fl/fl}$ mice exhibit developmental arrest of the first molar at the bud stage, which is associated with upregulation of *Osr2* and downregulation of *Msx1* expression. (F) In *Osr2*$^{-/-}$;*Wnt1-Cre;Bmp4*$^{fl/fl}$ mice, molar development progresses only to the cap stage. Although supernumerary tooth development is initiated lingual to the first molar (asterisk) in these mutants, it does not progress to form teeth. Epi, dental epithelium.

morphogenesis few genes are expressed in a molar- or incisor-specific pattern. The obvious exception to this is the molar-specific transcription factor Barx1 [17]. It is still not understood how the intermediate morphologies such as canines and premolars are generated during development and evolution. Interestingly, the position of different types of teeth within the jaws is invariant regardless of species. Therefore, the formation of different tooth types may be closely related to the patterning of the jaws. In general terms, tooth and jaw patterning is controlled by the specific temporal and spatial expression of homeobox-containing genes in the CNCC-derived mesenchyme. The dental CNCC-derived mesenchyme is tightly patterned not only along the distal—proximal axis but also along the oral—aboral (rostral—caudal), DV (maxillary-mandibular arch) and lingual—buccal axes in a complex fashion that includes the action of the dental epithelium [7,30].

Many studies have shown that before tooth initiation, the mandibular epithelium is divided into the proximal domain, which expresses *Fgf8* and is the origin of molars, and the distal domain, which expresses *Bmp4* and gives rise to incisors. Epithelial FGF and BMP act antagonistically to restrict *Barx1* and *Dlx2* expression to the proximal domain of the first arch mesenchyme and *Msx1*, *Msx2*, and *Alx4* expression to the distal domain (Figure 20.1B) [7,30]. The biological significance of such a regional molecular specification has been demonstrated through inhibition of BMP signaling. This results in ectopic *Barx1* expression in the distal mesenchyme, producing a morphological change of tooth shape from incisor to molar configuration, indicating a change in tooth identity [17].

Another important factor determining the position of teeth in the arches is the establishment of the oral—aboral (rostral—caudal) axis, in which FGF and BMP are also involved. FGF signals influence the establishment of the oral—aboral axis through the regulation of *Lhx*

and *Gsc* expression. Postmigratory CNCC located in the oral (rostral) region of the mandibular arch express *Lhx6* and *Lhx7* [31]. In the aboral (caudal) region, *Gsc* is expressed in an *Lhx6/7*-negative area (Figure 20.1B) [32]. This subdivision along the rostral—caudal axis is established by FGF8 from the ectoderm and is achieved by the differential response of CNCC to the FGF8 signal according to their proximity to it [32]. Surprisingly, double homozygous $Lhx6^{-/-};Lhx7^{-/-}$ mice lack molars. The absence of molars in $Lhx6^{-/-};Lhx7^{-/-}$ mice is due to a failure of specification of only the molar mesenchyme. Despite molar agenesis, *Lhx6/7*-deficient animals have normal incisors that are flanked in the maxilla by a supernumerary pair of incisor-like teeth [33]. This finding is noteworthy because the dentition is also patterned along the DV axis, based on differential gene expression between the maxillary and mandibular mesenchyme. There is also evidence that the nature of the molecular signaling in the upper and lower jaws varies. The dental formula is the same in both arches in mice and in humans, but the shapes and morphologies of the homologous teeth in the two jaws are distinctive from each other.

As mentioned above, a Dlx code defines maxillary and mandibular structures, including teeth. Interestingly, maxillary molars in *Dlx1/2* double mutant mice fail to develop, whereas all the other teeth appear normal [28,34]. Loss of function of *Dlx1/2* affects only the CNCC-derived mesenchyme, which develops into cartilage instead of teeth. *Dlx1* or *Dlx2* may have redundant functions with *Dlx5/6* in the mandible because lower molars are not affected in $Dlx1^{-/-}; Dlx2^{-/-}$ mice [28]. Similarly, deficiency of *activin-A* does not affect the development of all teeth equivalently. Maxillary molars are unaffected in *activin-A* mutant mice that lack all other teeth [29]. The lack of a maxillary molar defect in *activin-A* mice suggests that the molecular regulation of maxillary molar development is different to that of other teeth. *Dlx1/2* may be a crucial part of the alternative genetic pathway used to establish a maxillary molar field within the CNCC-derived mesenchyme. The early genetic studies that revealed different molecular pathways controlling tooth formation in the maxilla and mandible, together with the recent tantalizing results indicating different qualitative signaling requirements, clearly highlight fundamental differences in the way the tooth develops in the maxilla versus the mandible. How, and more intriguingly why, this is so will provide major new insights into the positional control of tooth morphogenesis.

Number of Teeth

Another important aspect of the dental formula is the size of the dental fields. Mice have one incisor and three molars per quadrant. However, numerous mouse models exhibiting supernumerary teeth have been reported and are useful for understanding how tooth number is controlled molecularly. Supernumerary teeth can be roughly classified into two groups: (1) *de novo* supernumerary teeth that develop directly from the primary tooth germs or dental lamina, and (2) supernumerary teeth that form from vestigial tooth rudiments. Examples of the first group are Apc loss-of-function, β-catenin gain-of-function [35], and Sp6 (Epiprofin) loss-of-function mice [36,37]. The formation of supernumerary teeth via Apc loss-of-function is non-cell-autonomous. A small number of Apc-deficient cells is sufficient to induce contiguous wild-type epithelial and mesenchymal cells to contribute to the formation of new teeth. Interestingly, Msx1, which is necessary for endogenous tooth formation, is not required for supernumerary tooth formation in these APC mutants [35]. Moreover, the extra teeth that formed did so in the context of a mass of tissue that resembled an odontoma, which is not surprising given the role of canonical Wnt signaling in tumor formation. Furthermore, it has been

demonstrated that Fgf8 is a direct target of Wnt/beta-catenin signaling in this process. Sp6-deficient mice develop multiple supernumerary teeth in both the incisor and molar regions. This phenotype is due to the late appearance of multiple non-proliferating enamel knot-like structures formed ectopically in the enamel organ of primary tooth germs, which then grow and branch into the jaw mesenchyme [38].

Most reported mouse supernumerary teeth do not arise *de novo* but rather belong to the second group. Large vestigial tooth buds have been described at the mesial limit of the developing molars in both jaws; these buds either disappear or, in the case of the mandible, fuse with the mesial aspect of the first molar [39–41]. In addition, numerous small and transient primordia are found within the upper diastema and between the developing maxillary and mandibular incisors. These are eliminated by programmed cell death, which is an important early determinant of tooth number [42]. These vestigial tooth rudiments have the potential to develop into supernumerary diastema teeth [43]. Mice with mutations in a number of major signaling pathways exhibit supernumerary diastema teeth from vestigial buds. These mutant models include the following:

1. EDA: mice with overexpression of Eda or Edar [44–46], and Tabby mice with mutation of Eda [47,48]. The development of supernumerary molars in K14-Eda mice is marked by ectopic *Shh*-expressing placodes, located where transient upregulation of Shh is sometimes observed in wild-type mice. This ectopic Shh signaling may promote the development of the rudimentary dental placode into a fully erupted tooth [44,49]. Supernumerary teeth in homo/hemizygous Tabby mice may arise due to the segregation of the distal premolar vestige from the molar dentition

and thus represent an evolutionary throwback (atavism) [48].

2. FGF: *Sprouty2*- and *Sprouty4*-null mutant mice. The diastema epithelial buds in *Spry2*-null mice contain a functional enamel knot and are hypersensitive to FGF signaling from the mesenchyme, thereby maintaining *Shh* expression, which presumably enables them to develop into teeth [50]. On the other hand, SPRY4 functions to repress diastema bud development by preventing FGF signaling from maintaining *Fgf3* expression in the dental mesenchyme [50]. In $Spry2^{+/-};Spry4^{-/-}$ mice, supernumerary teeth are the result of the secondary splitting of the incisor primordium and *Shh* expression domain after normal development during the initial stages [51].

3. SHH: hypomorphic Polaris and *Wnt1-Cre; Polaris* conditional mutant mice and *Gas1*-null mutant mice [52]. The transient *Shh* expression in these mutant mice is augmented in vestigial tooth germs distal to the first molar, which progress to form supernumerary teeth. In wild-type mice, only transient upregulation and weak expression are found in this region and tooth development fails to progress. Thus, only those vestigial tooth germs with higher levels of Shh transcriptional activity proceed to form tooth germs [44]. SHH appears to be crucial for cell survival of the vestigial teeth.

4. WNT: *Sostdc1*-null and *Lrp4* hypomorphic mice. These mice display supernumerary teeth in both incisor and molar regions [53–55]. Inactivation of Sostdc1, an inhibitor of both BMP and WNT, leads to elevated Wnt signaling, increased proliferation, and continuous development of vestigial tooth buds to form supernumerary teeth [53]. *Lrp4* and *Sostdc1* mutant mice exhibit identical tooth phenotypes including not only supernumerary incisors and molars but also fused molars. This is consistent with the function of Lrp4 to inhibit canonical Wnt signaling by binding to Sostdc1 [55].

These studies suggest that the function of the epithelium in controlling the number of teeth in mice is more prominent than that of the CNCC-derived mesenchyme, although they act together to promote supernumerary tooth formation.

Successional Teeth

Most mammals, including humans, have two sets of teeth during their lifetimes and replace their deciduous teeth with a permanent dentition. The permanent teeth result from the extension of the dental lamina on the lingual side of the primary teeth or from the posterior growth of the molar dental lamina [56]. The dental lamina gradually degenerates when the crown of the permanent tooth has formed. Interestingly, in patients with cleidocranial dysplasia syndrome (CCD), multiple supernumerary teeth form, derived from the permanent teeth and representing successional teeth. Some researchers have proposed that this pool of supernumerary teeth represents a third dentition [57]. CCD is an autosomal-dominant disorder produced by the loss of function of one allele of the *RUNX2* gene [58]. The human phenotype is consistent with the existence of lingual buds in front of the upper molars in *Runx2* knockouts and *Runx2/Runx3* double knockouts, as well as in *Runx2* heterozygotes [59]. Runx2 encodes a runt-containing transcription factor that is widely expressed in the mesenchymal compartment of the tooth germ. Wang and colleagues speculate that Runx2 inhibits secondary tooth formation in wild-type mice and, thus, that Runx2 is an inhibitor of tooth cycling [59]. Most likely, this action occurs through the inhibition of Shh signaling, which can prevent the budding of successional teeth. Although tooth replacement does not occur in mice, formation of the second and third molars has some similarities with successional tooth replacement. Specifically, the second molar forms directly from the posterior region of the first molar. The first and second molar then separate after formation of the cervical loops. The junction where this separation occurs is labeled by the presence of an undifferentiated epithelium, characterized by the expression of SHH and absence of BMP and WNT signaling. Downregulation of Shh or upregulation of BMP and WNT activities in this region lead to altered differentiation of the inner enamel epithelium and fusion of the tooth germs [55].

Because mice have only one dentition, they may not represent the best model for studying tooth replacement. Other models have been used to evaluate successional tooth formation, including fish and reptiles such as lizards and snakes, which replace their teeth continuously throughout life, as well as some mammals with secondary teeth, such as ferrets [60–62]. Without a doubt, however, the best model for studying tooth replacement is the human embryo. With the availability of human embryos for research purposes, combined with the huge advances in human genome and epigenome analysis, we are poised to advance our understanding of tooth replacement in the future. A key question regarding the mechanism of tooth replacement is how the outer epithelium of the primary tooth germ becomes regionalized on its lingual aspect to produce a cervical extension of the dental lamina. In the ferret, Sostdc1 is expressed in the elongating successional dental lamina at the interface between the lamina and the deciduous tooth, as well as the buccal side of the dental lamina, suggesting that Sostdc1 may play a role in defining the identity of the dental lamina [62]. Similarly, successional tooth formation in snakes and lizards is dependent on the activity of the WNT signaling pathway and the regulation of epithelial stem cell fate [60]. Reactivation of a competent dental lamina

appears to be crucial for both replacement teeth and supernumerary formation [43,62]. These studies also demonstrate that the epithelium possesses important information regarding the number of teeth to be formed within the same dentition and in different dentitions.

Number of Rows

Mammals have only one row of teeth in each jaw, whereas certain non-mammalian species, including many fish and snakes, have multirowed dentitions. The number of rows is dependent on the establishment of the lingual–buccal (medial–lateral) axis of the maxillary and mandibular arches during embryonic development. The BMP4-MSX1 pathway from the neural crest-derived component of the tooth germ is involved in the regulation of this process. In Osr2-null mutant mouse embryos, supernumerary tooth germs develop directly from the oral epithelium lingual to their molar tooth germs [63]. This is similar to the formation of a second tooth row in multirowed fish and differs from the successional or supernumerary tooth formation described in diverse vertebrate species (see above) [63,64]. Expression of several odontogenic factors, including Bmp4, Pitx2, Shh, Msx1, and Lef1, is upregulated or expanded in $Osr2^{-/-}$ embryos, consistent with the formation of additional teeth. Interestingly, deletion of Msx1 in the Osr2-mutant background prevents the development of a new row of teeth, suggesting that Msx1 is required for expansion of the odontogenic field in $Osr2^{-/-}$ mice and that the antagonistic actions of Msx1/Bmp4 and Osr2 pattern mammalian teeth into a single row. In fact, during odontogenesis, Osr2 demonstrates a gradient of expression within the odontogenic molar mesenchyme that predominates in the lingual region, complementary to the expression pattern of Bmp4 (Figure 20.2) [63]. Interestingly, in $Osr2^{-/-}$;Wnt1-Cre;Bmp4$^{fl/fl}$ mice, supernumerary tooth development is initiated lingual to the first molar but does not progress to form teeth. Moreover, the *Msx1* expression domain is not expanded, unlike in $Osr2^{-/-}$ embryos (Figure 20.2). The lack of expansion of *Msx1* to the mesenchyme underlying the supernumerary tooth placode explains the subsequent failure of supernumerary tooth morphogenesis. These findings suggest that mesenchymal Bmp4 signaling is required to propagate *Msx1* expression during sequential tooth formation [27]. The exact molecular mechanism involved in the crosstalk between Osr2 and the Bmp4-Msx1 pathway is currently unknown. Mikkola proposed that Osr2 may repress the expression of *Bmp4* directly, or indirectly through an intermediate molecule like *Msx2*, or it may regulate Msx1 activity at the post-transcriptional level [63]. Recent results have shown that Osr2 acts downstream of Pax9 and interacts with both Msx1 and Pax9 to pattern the tooth developmental field [27,65].

CNCC AND DENTAL STEM CELLS

After tooth morphogenesis is complete, CNCC in the teeth differentiate and their cell fate is controlled by the context-dependent integration of extrinsic and intrinsic signals. CNCC give rise to odontoblasts, dentine matrix, pulp tissue, cementum, and periodontal ligaments of the teeth in the adult dentition [8]. During dentinogenesis, CNCC-derived odontoblasts secrete predentin and dentin following terminal differentiation [8,66]. Odontoblast terminal differentiation is controlled in part by the inner enamel epithelium and is also supported by matrix-mediated interactions [66,67]. Several studies have shown that the TGFβ superfamily and WNT contribute to odontoblast terminal differentiation [68]. Exogenous TGFβ1, BMP2, BMP4, and BMP7 can induce odontoblast differentiation and dentin formation in dental papilla cells *in vitro* [69]. Furthermore, loss of TGFβ signaling in *Wnt1-Cre;Tgfbr2^{fl/fl}* mice results in abnormal

dentin formation [70]. Accordingly, ablation of *Smad4* in the dental mesenchyme leads to a defect in odontoblast differentiation. Instead of forming dentin, ectopic bonelike structures develop in *Osr2-Cre;Smad4^{fl/fl}* mice [69], similar to those observed in transgenic mice in which *Runx2* is overexpressed in odontoblasts [71,72]. In *Osr2-Cre;Smad4^{fl/fl}* mice, the canonical WNT signaling pathway is upregulated. Thus, proper CNCC fate choice during odontogenesis appears to be dependent on the interplay between TGFβ/BMP and other signaling pathways, including WNT [69]. Consistent with the studies described above, postmigratory CNCC maintain mesenchymal stem cell (MSC) characteristics and are able to differentiate into specific cell types according to environmental conditions. Recently, an *in vitro* culture system for CNCC was developed to investigate the proliferation and differentiation potential of postmigratory CNCC within the first BA through the analysis of *Wnt1-Cre;R26R* mouse embryos. Specifically, cells from E9.5 and E10.5 were studied because they provide more undifferentiated CNCC than at later stages [73]. After CNCC were purified, they demonstrated a robust proliferative capability and maintained their undifferentiated state. Specific culture conditions led to their differentiation into particular cell types, including neurons, Schwann cells, myofibroblasts, and osteoblasts, faithfully mimicking the differentiation process of postmigratory CNCC *in vivo* [73].

Strong evidence supports the existence of stem cells in adult teeth. Mouse incisors grow continuously throughout life, supported by the division of dental epithelial stem cells that reside in the cervical loop region. TGFβ signaling-mediated mesenchymal–epithelial interactions control these dental epithelial stem cells [74]. Deficiency of TGFβ type 1 receptor in the CNCC-derived dental mesenchyme affects the proliferation of transit-amplifying cells and the maintenance of dental epithelial stem cells. Incisors of *Wnt1-Cre;Alk5^{fl/fl}* mice

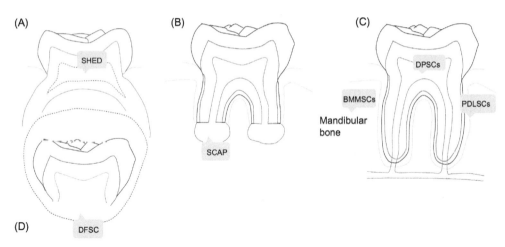

FIGURE 20.3 **Stem cells in adult teeth.** (A) Primary tooth, (B) Partially erupted permanent tooth, (C) erupted permanent tooth, and (D) Non-erupted permanent tooth. SHED: stem cells from human exfoliated deciduous teeth; DFSCs: dental follicle stem cells; SCAP: stem cells from the apical papilla; BMSCs: bone marrow-derived MSCs from orofacial bone; DPSCs: dental pulp stem cells; PDLSCs: periodontal ligament stem cells. *Source: Modified from Ref. [75].*

lose their ability to continue growing *in vitro*, which is due to downregulation of FGF signals in the dental mesenchyme [74]. Similarly, a discrete niche of CNCC-derived MSCs that is present at the cervical end of the mouse incisor provides a constant supply of progenitor cells during growth. During tooth development, CNCC-derived mesenchymal cells are stimulated to differentiate into various cell populations that form the developing tooth. Clearly, a small group of mesenchymal cells must evade such signals and remain multipotent cells. The mechanisms that regulate this are currently unknown. In the adult dental pulp of both incisors and molars, dental pulp stem cells (DPSCs) reside in a perivascular niche (Figure 20.3) [76,77]. Some pericytes differentiate into odontoblasts during tooth growth and also in response to damage *in vivo* [77]. In addition to the DPSCs [78], other MSC populations have been isolated from human CNCC-derived dental tissues such as the apical papilla (stem cells from the apical papilla, SCAP), the periodontal ligament (periodontal ligament stem cells, PDLSCs) [77], and the dental follicle (dental follicle precursor cells,

DFPCs) (Figure 20.3) [79]. Stem cells from human exfoliated deciduous teeth (SHED) have also been identified. Besides their dentinogenic potential, subpopulations of DPSCs, SCAP, DFPCs, and SHED also possess adipogenic and neurogenic differentiation capacities based on their adipocyte- and neuron-like cell morphologies and expression of relevant gene markers. They also exhibit osteogenic, chondrogenic and myogenic differentiation *in vitro* [80]. PDLSCs also exhibit osteogenic, adopogenic, and chondrogenic characteristics under defined culture conditions [81]. Differentiation of PDLSCs into neurons has not been reported so far. Also, a population of immature DPSCs of deciduous teeth (IDPSC) has been reported. IDPSC express embryonic stem cell (ESC) markers such as Oct-4, Nanog, SSEA-3, SSEA4, TRA-1-60, and TRA-1-81, as well as several other MSC markers and have the potential to differentiate into smooth and skeletal muscles, neurons, cartilage, and bone under chemically defined culture conditions [82]. The presence of MSCs with a CNCC origin is promising in the field of regenerative medicine [83], not only for repairing missing or defective dental tissues but also for use in the treatment of diverse life-threatening diseases.

CNCC-Derived Stem Cells in Tooth and Alveolar Bone Regeneration

Among the many tissues that originate from CNCC, tooth and alveolar bone have an interdependent relationship. Alveolar bone provides the support for a functional dentition, and it can be reshaped based on the precise needs of the tooth. Development of the alveolar bone involves CNCC-derived mesenchymal cell condensation to form the dental sac, osteoid deposition, and mineralization. A well-defined intramembranous bony socket, consisting of differentiating osteoblasts that are enriched in bone matrix, is built around the developing tooth. Clinically, the loss of a tooth leads to the absorption of alveolar bone, suggesting that there is a close relationship between alveolar bone and tooth development [84]. Therefore, to design a biological solution for tooth regeneration, it is critical to regenerate not only the tooth itself but also alveolar bone with which it is fully integrated to provide support for the newly formed tooth [85]. To do so, it is necessary to understand the molecular mechanisms that regulate CNCC differentiation into osteoblasts and the formation of alveolar bone.

Postmigratory CNCC have been observed to maintain stem cell characteristics even into adulthood [85]. These cells express MSC markers such as CD90.2 and SSEA4 that are also expressed by ESC lines (The International Stem Cell Initiative). CNCC share more characteristics with ESC than with bone marrow mesenchymal stem cells (BMMSCs). Interestingly, the expression pattern of some MSC markers is modified during development. In MSCs from the mandible of adult mice, the expression level of CD90.2 and SSEA4 is decreased and that of CD29, CD44, Sca-1, and CD49e is increased compared with MSCs from embryonic stages. Therefore, properties of CNCC-originated stem cells change during their contribution to the development of craniofacial structures [85]. Postmigratory CNCC are more responsive to *in vitro* osteogenic induction than BMMSCs, consistent with previous findings showing that neural crest-derived progenitor cells possess increased osteogenic capacity and enhanced osteogenesis compared with mesoderm-derived progenitor cells [86]. In addition, following transplantation into hosts, CNCC are able to form bone with densely packed lamella structures that are separated by abundant connective tissues. Unlike bone originating from BMMSCs, CNCC-derived bone does not contain prominent hematopoietic components [85], which is characteristic of craniofacial bone derived from intramembranous ossification. The differences in the histological appearance of bone formed by postmigratory CNCC and BMMSCs might be the result of intrinsic differences in embryological origin and functional demands at each skeletal site [87]. The CNCC that form craniofacial hard tissue do not express Hox genes, which may account for the difference in cellular differentiation potential in comparison to other neural crest cells [87]. Moreover, in repair responses, MSCs differentiate into hard tissue-forming cells and a major source of MSCs are perivascular cells (pericytes) that are neural crest-derived in the craniofacial region [88].

Long bones are physiologically adapted to support body weight, contain more bone marrow, and contribute more to hematopoiesis. In contrast, the CNCC-originated mandible and maxilla are part of the craniofacial complex and contain less bone marrow but offer protection to vital structures including the brain, dentition, and neurovascular bundles [89]. Long bones contain more abundant bone marrow because skeletal bone formation results from endochondral ossification, in which hypertrophic chondrocytes mineralize their surrounding matrix and attract blood vessels. The proper formation of bone marrow thus requires the normal development of skeletal bones. Several studies of mutant mice in which hematopoiesis is defective as a consequence of primary defects in bone development have implicated osteoblasts in the formation and function of the bone marrow hematopoietic stem cell environment [90]. In addition, cells involved in bone formation play a role supporting hematopoiesis, and specialized osteoblasts lining the bone marrow function to maintain and regulate hematopoietic stem cells [90]. During endochondral ossification, hypertrophic chondrocytes express BMP2 and BMP6. Deficiency of these two genes in mice results in reduction of trabecular bone volume and suppressed bone formation [90,91]. Accordingly, exogenous BMP2 induces increased bone marrow along with increased bone formation in transplants of CNCC and hydroxyapatite/tricalcium phosphate. Thus, the concerted action of BMP2 and CNCC results in increased formation of hematopoietic components in the bone matrix, consistent with previous studies showing that bone marrow development in the rat mandible occurs in a BMP2 dose-dependent fashion [92]. These findings support the hypothesis that the microenvironment influences how postmigratory CNCC differentiate and regenerate tissues.

Postmigratory CNCC appear to be crucial for tooth germ survival. Well-formed supporting structures, especially the alveolar bone, are also required for the proper development of teeth. Subcutaneous transplantation of tooth germs and postmigratory CNCC results in apparently normal tooth development [85]. CNCC contribute to the dental pulp and the bone adjacent to the tooth. In contrast, transplants of BMMSCs and tooth germs fail to form a normal tooth and the supporting structures are also altered. Thus, it appears that CNCC-derived bone, with the unique features of craniofacial bone, is required for the

survival of tooth germs. Differences in supporting structures based on differences in embryonic origin between CNCC and BMMSCs might explain the abnormal tooth development in BMMSC-tooth transplants [85]. As discussed above, various transcription factors and signaling molecules, including BMP, FGF, Activin, Hedgehog, and Wnt family members, participate in tooth development [7,17,93]. Among these, BMPs are key signals [13,94]. During jaw bone development, BMP activity plays a vital role in the formation of alveolar bone via a Bmp/Msx signaling cascade [84,95]. In loss-of-function studies that inhibit BMP activity using Noggin, kidney capsule transplantation of Noggin-treated tooth germs gave rise to keratinized cysts [84]. Similarly, inactivation of Smad4 in CNCC results in abnormal tooth development in tooth germ transplants. However, administration of BMP4 into transplants of BMMSCs and tooth germs failed to promote normal tooth development [85], suggesting that BMP signaling in CNCC is necessary, but not sufficient, to support tooth formation. Remarkably, there is an intrinsic difference between CNCC and BMMSCs in supporting tooth development, and postmigratory CNCC have unique properties essential for tooth development. Postmigratory CNCC create a link between alveolar bone and tooth development as a functional unit. These cells have the ability to support an organ survival environment, also known as an organ niche, which can provide the proper conditions for tooth germ survival.

CONCLUSIONS AND PERSPECTIVES

The presence of neural crest during development is pivotal for vertebrate specific features, such as the sensory ganglia, craniofacial skeleton, and teeth. The appearance of these features was fundamental in vertebrate evolution because it facilitated a predatory lifestyle. Accordingly, the diversity in craniofacial and dental morphologies in vertebrates has been shown to be highly dependent on the patterning and development of the postmigratory neural crest populating the first BA. The contribution of NCC to early tooth development has been extensively demonstrated during the last three decades. A complex and tightly regulated molecular network including transcription factors and signaling molecules that reiteratively appear during odontogenesis provides the neural crest-derived mesenchyme with the potential to drive tooth morphogenesis and differentiation. This is evident from several disease models in which tooth formation is compromised due to defective migration, proliferation, or differentiation of the NCC. Despite all the knowledge of early odontogenesis, the function of NCC in the regulation of late events in tooth development is yet to be completely understood, particularly in dental root formation. In addition to the developmental features of the NCC, it is now obvious that some of their derivatives in the adult tooth possess stem cell properties. This discovery is of great importance for regenerative therapies not only for oral diseases but also for life-threatening disorders.

Acknowledgments

We are grateful to Drs. Julie Mayo and Bridget Samuels for critical reading of this manuscript. Studies in Paul Sharpe's laboratory are supported by the MRC. Studies in Yang Chai's laboratory are supported by the National Institute of Dental and Craniofacial Research, NIH (DE012711, DE014078, DE020065, and DE022503).

References

[1] Bronner-Fraser M. Neural crest cell migration in the developing embryo. Trends Cell Biol 1993;3:392—7.

[2] Selleck MA, Scherson TY, Bronner-Fraser M. Origins of neural crest cell diversity. Dev Biol 1993;159:1—11.

[3] Depew MJ, Simpson CA, Morasso M, Rubenstein JLR. Reassessing the Dlx code: the genetic regulation of branchial arch skeletal pattern and development. J Anat 2005;207:501—61.

[4] Jeong J, Li X, McEvilly RJ, Rosenfeld MG, Lufkin T, Rubenstein JLR. Dlx genes pattern mammalian jaw primordium by regulating both lower jaw-specific and upper jaw-specific genetic programs. Development 2008;135:2905—16.

[5] Minoux M, Rijli FM. Molecular mechanisms of cranial neural crest cell migration and patterning in craniofacial development. Development 2010;137: 2605—21.

[6] Moore KL, Persaud TVN. The developing human. Philadelphia, PA: Saunders; 2008.

[7] Tucker A, Sharpe P. The cutting-edge of mammalian development; how the embryo makes teeth. Nat Rev Genet 2004;5:499—508.

[8] Chai Y, Jiang X, Ito Y, Bringas P, Han J, Rowitch DH, et al. Fate of the mammalian cranial neural crest during tooth and mandibular morphogenesis. Development 2000;127:1671—9.

[9] Ohazama A, Haworth KE, Ota MS, Khonsari RH, Sharpe PT. Ectoderm, endoderm, and the evolution of heterodont dentitions. Genesis 2010;48:382—9.

[10] Hu Y, Baud V, Delhase M, Zhang P, Deerinck T, Ellisman M, et al. Abnormal morphogenesis but intact IKK activation in mice lacking the IKKalpha subunit of IkappaB kinase. Science 1999;284:316—20.

[11] Soukup V, Epperlein H-H, Horacek I, Cerny R. Dual epithelial origin of vertebrate oral teeth. Nature 2008;455:795—8.

[12] Mitsiadis T, Chéraud Y, Sharpe PT, Fontaine-Pèrus J. Development of teeth in chick embryos following mouse neural crest transplantations. Proc Natl Acad Sci USA 2003;100:6541—5.

[13] Kollar EJ, Baird GR. The influence of the dental papilla on the development of tooth shape in embryonic mouse tooth germs. J Embryol Exp Morphol 1969;21:131—48.

[14] Vainio S, Karavanova I, Jowett A, Thesleff I. Identification of BMP-4 as a signal mediating secondary induction between epithelial and mesenchymal tissues during early tooth development. Cell 1993;75: 45—58.

[15] Mammoto T, Mammoto A, Torisawa Y-S, Tat T, Gibbs A, Derda R, et al. Mechanochemical control of mesenchymal condensation and embryonic tooth organ formation. Dev Cell 2011;21:758—69.

[16] Li L, Lin M, Wang Y, Cserjesi P, Chen Z, Chen Y. Bmpr1a is required in mesenchymal tissue and has limited redundant function with Bmpr1b in tooth and palate development. Dev Biol 2011;349:451—61.

[17] Tucker AS, Matthews KL, Sharpe PT. Transformation of tooth type induced by inhibition of BMP signaling. Science 1998;282:1136—8.

[18] Peters H, Balling R. Teeth: where and how to make them. Trends Genet 1999;15:59—65.

[19] Ogawa T, Kapadia H, Feng JQ, Raghow R, Peters H, D'Souza RN. Functional consequences of interactions between Pax9 and Msx1 genes in normal and abnormal tooth development. J Biol Chem 2006;281:18363—9.

[20] Satokata I, Maas RL. Msx1 deficient mice exhibit cleft palate and abnormalities of craniofacial and tooth development. Nat Genet 1994;6:348—56.

[21] Peters H, Neubeuser A, Kratochwil K, Balling R. Pax9-deficient mice lack pharyngeal pouch derivatives and teeth and exhibit craniofacial and limb abnormalities. Genes Dev 1998;12:2735—47.

[22] Zhang Z, Song Y, Zhao X, Zhang X, Fermin C, Chen Y. Rescue of cleft palate in Msx1-deficient mice by transgenic Bmp4 reveals a network of BMP and Shh signaling in the regulation of mammalian palatogenesis. Development 2002;129:4135—46.

[23] Bei M, Kratochwil K, Maas RL. BMP4 rescues a non-cell-autonomous function of Msx1 in tooth development. Development 2000;127:4711—8.

[24] Chen Y, Bei M, Woo I, Satokata I, Maas R. Msx1 controls inductive signaling in mammalian tooth morphogenesis. Development 1996;122:3035—44.

[25] Jernvall J, Aberg T, Kettunen P, Keränen S, Thesleff I. The life history of an embryonic signaling center: BMP-4 induces p21 and is associated with apoptosis in the mouse tooth enamel knot. Development 1998;125: 161—9.

[26] Zhao X, Zhang Z, Song Y, Zhang X, Zhang Y, Hu Y, et al. Transgenically ectopic expression of Bmp4 to the Msx1 mutant dental mesenchyme restores downstream gene expression but represses Shh and Bmp2 in the enamel knot of wild type tooth germ. Mech Dev 2000;99:29—38.

[27] Jia S, Zhou J, Gao Y, Baek JA, Martin JF, Lan Y, et al. Roles of Bmp4 during tooth morphogenesis and sequential tooth formation. Development 2013;140: 423—32.

[28] Thomas BL, Tucker AS, Qui M, Ferguson CA, Hardcastle Z, Rubenstein JL, et al. Role of Dlx-1 and Dlx-2 genes in patterning of the murine dentition. Development 1997;124:4811—8.

[29] Ferguson CA, Tucker AS, Christensen L, Lau AL, Matzuk MM, Sharpe PT. Activin is an essential early mesenchymal signal in tooth development that is required for patterning of the murine dentition. Genes Dev 1998;12:2636—49.

[30] Chai Y, Maxson RE. Recent advances in craniofacial morphogenesis. Dev Dyn 2006;235:2353—75.

[31] Grigoriou M, Tucker AS, Sharpe PT, Pachnis V. Expression and regulation of Lhx6 and Lhx7, a novel subfamily of LIM homeodomain encoding genes, suggests a role in mammalian head development. Development 1998;125:2063—74.

[32] Tucker AS, Yamada G, Grigoriou M, Pachnis V, Sharpe PT. Fgf-8 determines rostral-caudal polarity in the first branchial arch. Development 1999;126: 51—61.

[33] Denaxa M, Sharpe PT, Pachnis V. The LIM homeodomain transcription factors Lhx6 and Lhx7 are key regulators of mammalian dentition. Dev Biol 2009;333: 324—36.

[34] Qiu M, Bulfone A, Ghattas I, Meneses JJ, Christensen L, Sharpe PT, et al. Role of the Dlx homeobox genes in proximodistal patterning of the branchial arches: mutations of Dlx-1, Dlx-2, and Dlx-1 and −2 alter morphogenesis of proximal skeletal and soft tissue structures derived from the first and second arches. Dev Biol 1997;185:165—84.

[35] Wang XP, O'Connell DJ, Lund JJ, Saadi I, Kuraguchi M, Turbe-Doan A, et al. Apc inhibition of Wnt signaling regulates supernumerary tooth formation during embryogenesis and throughout adulthood. Development 2009;136:1939—49.

[36] Nakamura T, Unda F, de-Vega S, Vilaxa A, Fukumoto S, Yamada KM, et al. The Krüppel-like factor epiprofin is expressed by epithelium of developing teeth, hair follicles, and limb buds and promotes cell proliferation. J Biol Chem 2004;279:626—34.

[37] Talamillo A, Delgado I, Nakamura T, de-Vega S, Yoshitomi Y, Unda F, et al. Role of Epiprofin, a zinc-finger transcription factor, in limb development. Dev Biol 2010;337:363—74.

[38] Jimenez-Rojo L, Ibarretxe G, Aurrekoetxea M, de Vega S, Nakamura T, Yamada Y, et al. Epiprofin/Sp6: a new player in the regulation of tooth development. Histol Histopathol 2010;25:1621—30.

[39] Peterkova R, Churava S, Lesot H, Rothova M, Prochazka J, Peterka M, et al. Revitalization of a diastemal tooth primordium in Spry2 null mice results from increased proliferation and decreased apoptosis. J Exp Zool B Mol Dev Evol 2009;312B: #292—308.

[40] Prochazka J, Pantalacci S, Churava S, Rothova M, Lambert A, Lesot H, et al. Patterning by heritage in mouse molar row development. PNAS 2010;107: 15497—502.

[41] Witter K, Lesot H, Peterka M, Vonesch JL, Mísek I, Peterková R. Origin and developmental fate of vestigial tooth primordia in the upper diastema of the field vole (Microtus agrestis, Rodentia). Arch Oral Biol 2005;50:401—9.

[42] Peterkova R, Lesot H, Peterka M. Phylogenetic memory of developing mammalian dentition. J Exp Zool B Mol Dev Evol 2006;306:234—50.

[43] Tummers M, Thesleff I. The importance of signal pathway modulation in all aspects of tooth development. J Exp Zool B Mol Dev Evol 2009;312B:309—19.

[44] Kangas AT, Evans AR, Thesleff I, Jernvall J. Nonindependence of mammalian dental characters. Nature 2004;432:211—4.

[45] Pispa J, Mustonen T, Mikkola ML, Kangas AT, Koppinen P, Lukinmaa PL, et al. Tooth patterning and enamel formation can be manipulated by misexpression of TNF receptor Edar. Dev Dyn 2004;231: 432—40.

[46] Tucker AS, Headon D, Courtney J-M, Overbeek P, Sharpe PT. The activation level of the TNF-family receptor, Edar, determines cusp number and tooth number during tooth development. Dev Biol 2004;268: 185—94.

[47] Charles C, Pantalacci S, Tafforeau P, Headon D, Laudet V, Viriot L. Distinct impacts of Eda and Edar loss of function on the mouse dentition. PLoS One 2009;4:e4985.

[48] Peterková R, Lesot H, Viriot L, Peterka M. The supernumerary cheek tooth in tabby/EDA mice—a reminiscence of the premolar in mouse ancestors. Arch Oral Biol 2005;50:219—25.

[49] Pummila M, Fliniaux I, Jaatinen R, James MJ, Laurikkala J, Schneider P, et al. Ectodysplasin has a dual role in ectodermal organogenesis: inhibition of Bmp activity and induction of Shh expression. Development 2007;134:117—25.

[50] Klein OD, Minowada G, Peterkova R, Kangas A, Yu BD, Lesot H, et al. Sprouty genes control diastema tooth development via bidirectional antagonism of epithelial—mesenchymal FGF signaling. Dev Cell 2006;11:181—90.

[51] Charles C, Hovorakova M, Ahn Y, Lyons DB, Marangoni P, Churava S, et al. Regulation of tooth number by fine-tuning levels of receptor-tyrosine kinase signaling. Development 2011;138:4063—73.

[52] Ohazama A, Haycraft CJ, Seppala M, Blackburn J, Ghafoor S, Cobourne M, et al. Primary cilia regulate Shh activity in the control of molar tooth number. Development 2009;136:897—903.

[53] Ahn Y, Sanderson BW, Klein OD, Krumlauf R. Inhibition of Wnt signaling by Wise (Sostdc1) and negative feedback from Shh controls tooth number and patterning. Development 2010;137:3221—31.

[54] Murashima-Suginami A, Takahashi K, Kawabata T, Sakata T, Tsukamoto H, Sugai M, et al. Rudiment incisors survive and erupt as supernumerary teeth as a result of USAG-1 abrogation. Biochem Biophys Res Commun 2007;359:549—55.

[55] Ohazama A, Johnson EB, Ota MS, Choi HY, Porntaveetus T, Oommen S, et al. Lrp4 modulates extracellular integration of cell signaling pathways in development. PLoS One 2008;3:e4092.

[56] Cobourne MT, Sharpe PT. Making up the numbers: the molecular control of mammalian dental formula. Semin Cell Dev Biol 2010;21:314—24.

[57] Jensen BL, Kreiborg S. Development of the dentition in cleidocranial dysplasia. J Oral Pathol Med 1990;19: 89—93.

[58] Mundlos S, Otto F, Mundlos C, Mulliken JB, Aylsworth AS, Albright S, et al. Mutations involving the transcription factor CBFA1 cause cleidocranial dysplasia. Cell 1997;89:773—9.

[59] Wang XP, Aberg T, James MJ, Levanon D, Groner Y, Thesleff I. Runx2 (Cbfa1) inhibits Shh signaling in the lower but not upper molars of mouse embryos and prevents the budding of putative successional teeth. J Dent Res 2005;84:138—43.

[60] Handrigan GR, Leung KJ, Richman JM. Identification of putative dental epithelial stem cells in a lizard with life-long tooth replacement. Development 2010;137: 3545—9.

[61] Handrigan GR, Richman JM. A network of Wnt, hedgehog and BMP signaling pathways regulates tooth replacement in snakes. Dev Biol 2010;348: 130—41.

[62] Järvinen E, Tummers M, Thesleff I. The role of the dental lamina in mammalian tooth replacement. J Exp Zool B Mol Dev Evol 2009;312B:281—91.

[63] Zhang Z, Lan Y, Chai Y, Jiang R. Antagonistic actions of Msx1 and Osr2 pattern Mammalian teeth into a single row. Science 2009;323:1232—4.

[64] Mikkola ML. Controlling the number of tooth rows. Sci Signal 2009;2:e53.

[65] Zhou J, Gao Y, Zhang Z, Zhang Y, Maltby KM, Liu Z, et al. Osr2 acts downstream of Pax9 and interacts with both Msx1 and Pax9 to pattern the tooth developmental field. Dev Biol 2011;353:344—53.

[66] Ruch J, Lesot H, Bègue-Kirn C. Odontoblast differentiation. Int J Dev Biol 1995;39:51—68.

[67] Thesleff I, Keranen S, Jernvall J. Enamel knots as signaling centers linking tooth morphogenesis and odontoblast differentiation. Adv Dent Res 2001;15: 14—8.

[68] Lohi M, Tucker AS, Sharpe PT. Expression of Axin2 indicates a role for canonical Wnt signaling in development of the crown and root during pre- and postnatal tooth development. Dev Dyn 2010;239:160—7.

[69] Li J, Huang X, Xu X, Mayo J, Bringas P, Jiang R, et al. SMAD4-mediated WNT signaling controls the fate of cranial neural crest cells during tooth morphogenesis. Development 2011;138:1977—89.

[70] Oka S, Oka K, Xu X, Sasaki T, Bringas Jr P, Chai Y. Cell autonomous requirement for TGF-beta signaling

during odontoblast differentiation and dentin matrix formation. Mech Dev 2007;124:409—15.

[71] Li S, Kong H, Yao N, Yu Q, Wang P, Lin Y, et al. The role of runt-related transcription factor 2 (Runx2) in the late stage of odontoblast differentiation and dentin formation. Biochem Biophys Res Commun 2011;410: 698—704.

[72] Miyazaki T, Kanatani N, Rokutanda S, Yoshida C, Toyosawa S, Nakamura R, et al. Inhibition of the terminal differentiation of odontoblasts and their transdifferentiation into osteoblasts in Runx2 transgenic mice. Arch Histol Cytol 2008;71:131—46.

[73] Zhao H, Bringas P, Chai Y. An in vitro model for characterizing the post-migratory cranial neural crest cells of the first branchial arch. Dev Dyn 2006;235: 1433—40.

[74] Zhao H, Li S, Han D, Kaartinen V, Chai Y. Alk5-mediated transforming growth factor beta signaling acts upstream of fibroblast growth factor 10 to regulate the proliferation and maintenance of dental epithelial stem cells. Mol Cell Biol 2011;31:2079—89.

[75] Egusa H, Sonoyama W, Nishimura M, Atsuta I, Akiyama K. Stem cells in dentistry—Part I: stem cell sources. J Prosthodont Res 2012;56:151—65.

[76] Shi S, Gronthos S. Perivascular niche of postnatal mesenchymal stem cells in human bone marrow and dental pulp. J Bone Miner Res 2003;18:696—704.

[77] Feng J, Mantesso A, De Bari C, Nishiyama A, Sharpe PT. Dual origin of mesenchymal stem cells contributing to organ growth and repair. PNAS 2011;108: 6503—8.

[78] Miura M, Gronthos S, Zhao M, Lu B, Fisher LW, Robey PG, et al. SHED: stem cells from human exfoliated deciduous teeth. PNAS 2003;100:5807—12.

[79] Morsczeck C, Goetz W, Schierholz J, Zeilhofer F, Kuehn U, Moehl C, et al. Isolation of precursor cells (PCs) from human dental follicle of wisdom teeth. Matrix Biol 2005;24:155—65.

[80] Huang GT, Gronthos S, Shi S. Mesenchymal stem cells derived from dental tissues vs. those from other sources: their biology and role in regenerative medicine. J Dent Res 2009;88:792—806.

[81] Yu S, Long J, Yu J, Du J, Ma P, Ma Y, et al. Analysis of differentiation potentials and gene expression profiles of mesenchymal stem cells derived from periodontal ligament and Wharton's jelly of the umbilical cord. Cells Tissues Org 2012;Dec(14).

[82] Kerkis I, Kerkis A, Dozortsev D, Stukart-Parsons GC, Gomes Massironi SM, Pereira LV, et al. Isolation and characterization of a population of immature dental pulp stem cells expressing OCT-4 and other embryonic stem cell markers. Cells Tissues Org 2006;184: 105—16.

[83] Achilleos A, Trainor PA. Neural crest stem cells: discovery, properties and potential for therapy. Cell Res 2012;22:288—304.

[84] Zhang Z, Song Y, Zhang X, Tang J, Chen J, Chen Y. Msx1/Bmp4 genetic pathway regulates mammalian alveolar bone formation via induction of Dlx5 and Cbfa1. Mech Dev 2003;120:1469—79.

[85] Chung I-H, Yamaza T, Zhao H, Choung P-H, Shi S, Chai Y. Stem cell property of postmigratory cranial neural crest cells and their utility in alveolar bone regeneration and tooth development. Stem Cell 2009; 27:866—77.

[86] Leucht P, Kim JB, Amasha R, James AW, Girod S, Helms JA. Embryonic origin and Hox status determine progenitor cell fate during adult bone regeneration. Development 2008;135:2845—54.

[87] Gronthos S, Akintoye SO, Wang CY, Shi S. Bone marrow stromal stem cells for tissue engineering. Periodontol 2006;41:188—95.

[88] Etchevers HC, Vincent C, Le Douarin NM, Couly GF. The cephalic neural crest provides pericytes and smooth muscle cells to all blood vessels of the face and forebrain. Development 2001;128:1059—68.

[89] Charbord P, Tavian M, Humeau L, Péault B. Early ontogeny of the human marrow from long bones: an immunohistochemical study of hematopoiesis and its microenvironment. Blood 1996;87:4109—19.

[90] Wilson A, Trumpp A. Bone-marrow haematopoietic-stem-cell niches. Nat Rev Immunol 2006;6:93—106.

[91] Kugimiya F, Kawaguchi H, Kamekura S, Chikuda H, Ohba S, Yano F, et al. Involvement of endogenous bone morphogenetic protein (BMP) 2 and BMP6 in bone formation. J Biol Chem 2005;280:35704—12.

[92] Arosarena O, Collins W. Comparison of BMP-2 and -4 for rat mandibular bone regeneration at various doses. Orthod Craniofac Res 2005;8:267—76.

[93] Thesleff I, Sharpe P. Signalling networks regulating dental development. Mech Dev 1997;67:111—23.

[94] Plikus MV, Zeichner-David M, Mayer JA, Reyna J, Bringas P, Thewissen JG, et al. Morphoregulation of teeth: modulating the number, size, shape and differentiation by tuning Bmp activity. Evol Dev 2005;7: 440—57.

[95] Nie X, Luukko K, Kettunen P. Bmp signalling in craniofacial development. Int J Dev Biol 2006;50:511—21.

Using Induced Pluripotent Stem Cells as a Tool to Understand Neurocristopathies

John Avery[a], Laura Menendez[a], Michael L. Cunningham[b,c], Harold N. Lovvorn III[d] and Stephen Dalton[a]

[a]Department of Biochemistry and Molecular Biology, University of Georgia, [b]Department of Pediatrics, University of Georgia- Athens, GA 30602, Vanderbilt University- Nashville, TN 37232, University of Washington, Seattle 98195, USA, [c]Seattle Children's Research Institute, Seattle 98101, USA, [d]Department of Pediatric Surgery, Vanderbilt University Medical Center

INTRODUCTION

Technologies that allow induced pluripotent stem cells (iPSCs) to be generated through cell reprogramming [1,2] have led to the development of powerful tools that can be used to model human disease and as platforms for drug discovery [3,4]. Disease modeling using iPSCs offers advantages over more traditional animal-based models and has real potential to advance molecular understanding of diseases that are poorly understood. Most notably, patient-derived iPSC models are likely to more completely and faithfully recapitulate the human

Neural Crest Cells.
DOI: http://dx.doi.org/10.1016/B978-0-12-401730-6.00022-3

441

disease state in cases where the disease is cell autonomous. This is particularly relevant where disease pathogenesis is multifactorial and dependent on the patient's genetic background. Access to patient-derived cells also allows for potential differences between individuals to be assessed. iPSCs are also well suited to modeling disease-related events that occur during development because iPSCs can be differentiated to a wide range of cell types by tightly controlled protocols. If disease progression is associated with a specific developmental defect, differentiation models should capture this and facilitate a more detailed molecular understanding that may then lead to new pathways of therapeutic intervention. As knowledge of progenitor differentiation pathways is gained from this technology, the potential for tissue engineering to repopulate defective cell types becomes more promising.

NEURAL CREST CELLS AND NEUROCRISTOPATHIES

Neural crest cells (NCC) arise from the neural plate border and migrate to diverse targets throughout the embryo where they differentiate and incorporate into functional tissue [5–7]. Defects associated with the emergence of NCC from the neural plate border, their migration throughout the embryo, or their differentiation can contribute to a broad spectrum of defined diseases or syndromes, collectively known as neurocristopathies. This chapter will focus on how iPSC-based technologies can be used to gain a better understanding of neurocristopathies.

NCC originate from four discrete segments (cranial, cardiac, vagal, and trunk) along the neural tube's rostro–caudal axis. Figure 21.1 illustrates the corporal locations that NCC are

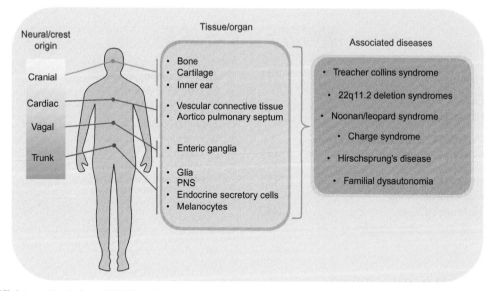

FIGURE 21.1 **Depiction of NCC origin along the rostro–caudal axis of the neural tube with corresponding tissues and associated diseases in humans.** In vertebrates, the neural tube gives rise to neural crest cells, which then migrate from distinct regions along the neuraxis to specific areas of the organism where they differentiate into an extremely diverse array of tissues. As a consequence of NCC multipotency, perturbations in NCC differentiation, migration, proliferation, and/or survival may lead to a wide variety of associated diseases, termed neurocristopathies, which may primarily be limited to one type of tissue, such as the ENS observed in Hirschsprung's disease, or affect many tissue types as observed in LEOPARD syndrome. PNS, peripheral nervous system.

targeted as they migrate away from different regions of the neural tube. Several prevalent neurocristopathies are listed that have specific relationships with the different populations of NCC. Most of these arise from improper specification of neural crest cells, defective migration, compromised proliferation, and/or decreased NCC survival. Most of these aspects of neurocristopathies are likely to be cell autonomous and can therefore be approached using iPSCs as a tool.

Cranial NCC give rise to the bulk of the bone and cartilage that form scaffolding of the head and face and contribute to ganglia, smooth muscle, connective tissue, and even pigment cells [7]. NCC emanating from the cardiac segment contribute to heart development by forming connective tissue associated with the great vessels, the aorticopulmonary septum, and smooth muscle cells of the great arteries, as well as celiac, superior, mesenteric, and aortic ganglia [8]. Vagally derived NCC establish the ganglia of the enteric nervous system (ENS) and contribute to neurons of the parasympathetic nervous system [7,8]. Trunk NCC establish the dorsal root ganglia, contribute to the lower sympathetic nervous system and the peripheral nervous system, and give rise to secretory cells of the endocrine system and to pigment cells of the skin [7,8]. Because of the multipotent nature of NCC, neurocristopathies may take many and varied forms. These include craniofacial defects (Treacher Collins syndrome), cardiac defects (CHARGE syndrome, Noonan/Leopard syndrome, 22q.11.2 deletion syndromes), ENS defects (Hirschsprung's disease), peripheral nervous system defects (familial dysautonomia), and melanocyte defects (Waardenburg syndrome, melanoma, piebaldism) (Figure 21.1). This chapter examines several conditions of NCC origin in which iPSCs have the potential to serve as tools to understand detailed aspects of disease progression. There are only a few reports describing the use of iPSCs for modeling neurocristopathies, but as will be seen, these have provided encouraging outcomes that support their use as a tool in this important area.

The strategy for generating iPSCs from patients with neurocristopathies and their subsequent characterization *in vitro* is similar to that of "disease in a dish" modeling for other clinical conditions. Our laboratory uses Sendai virus particles to deliver the four reprogramming factors Oct4, Sox2, Klf4, and c-Myc to recipient donor cells that over 3 weeks are reprogrammed to a pluripotent state. For details about reprogramming procedures, see Menendez et al. [9]. Once iPSC colonies are obtained, they are genotyped by fluorescence *in situ* hybridization (FISH) and subjected to G-band analysis to confirm that genetic aberrations have not been introduced during the reprogramming procedure. iPSC colonies are then transitioned to culture in chemically defined media that is compatible with differentiation into lineages such as neural crest. Cell stocks are frozen at different stages to provide a tiered stock of low passage cells [9].

METHODS FOR NCC DIFFERENTIATION FROM PLURIPOTENT STEM CELLS

Differentiation of human embryonic stem cell (hESCs) and iPSCs toward a neural crest cell fate offers great opportunities to study important aspects of early embryonic development and also to elucidate molecular mechanisms related to human neurocristopathies. A general scheme for generation of patient-derived iPSCs is shown in Figure 21.2. The first reports of human cells differentiating to NCC involved co-culture on PA6 stromal cells, but this was relatively inefficient [10], and obtaining highly enriched populations required fluorescence-activated cell sorting (FACS) [11]. $Sox9^+$ $Ap2^+$ cells in these cultures, however, could be further differentiated to peripheral

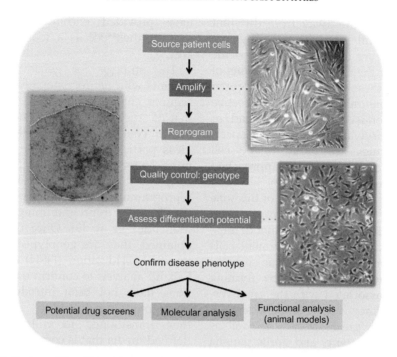

FIGURE 21.2 **Patient-specific cells, such as dermal fibroblasts, are collected via biopsy or operative incision.** The isolated fibroblasts are amplified in culture prior to reprogramming (red box). Various transduction methods have been developed to introduce reprogramming factors; iPSC induction via Sendai virus delivery allows for transient expression of *OCT4*, *SOX2*, *KLF4*, and *c-MYC* that yields colonies of reprogrammed pluripotent cells (green box) within 3 weeks when kept on a feeder layer of irradiated mouse embryonic fibroblasts. Reprogrammed cells are adapted to feeder-free culture and are genotyped for quality control to ensure no genetic aberrations are introduced during reprogramming. The patient-specific iPSCs are then transitioned to defined media to assess their capacity to differentiate into cells that are representative of the endoderm, mesoderm, and ectoderm, such as NCC (blue box). Genetic and functional assays may demonstrate that these cells are disease-specific cells, which can authentically replicate the "disease in a dish" and are powerful tools for disease modeling and high-throughput drug screening.

neurons through a neurosphere intermediate stage, demonstrating that they had developmental potential comparable to their embryonic counterparts [12]. A second method used co-culture of hESCs and the mouse stromal line MS5 to generate neural rosettes [13]. This approach represents the adaptation of existing methods where contaminating NCC were isolated from neural rosette cultures. A shortfall of this method is that it is inefficient and requires a labor-intensive FACS step.

More recently, methods have been developed that eliminate the need for stromal co-cultures, instead using chemically defined media supplemented with small molecules and/or growth factors [14–16]. Most of these methods, however, are inefficient and often still require a FACS step. To this point, no method had been developed for efficient, lineage-specific differentiation of human pluripotent cells to a NCC fate. This is obviously important if the development of neurocristopathies is to be understood at the molecular level. To solve this problem, we developed a protocol using chemically defined medium and two small molecule inhibitors, SB431542 (Lefty/Activin/TGFB inhibitor) and BIO (GSK3 Inhibitor/ Wnt activator) [9,17]. This method generates NCC cultures without other contaminating cell types under feeder-free conditions, within a relatively short time frame (10–15 days). NCC produced by this method have similar developmental potential to that described by other methods but do not require FACS isolation or co-culture steps. It is unclear if any of these methods generate NCC that are patterned as is seen along the rostro–caudal axis during embryogenesis or if they have broad NCC developmental potential. The paucity of molecular markers that discriminate between cranial, cardiac, vagal, and trunk identities have made this question difficult to address. NCC derived from human pluripotent cells can, however, be further differentiated into peripheral neurons, adipocytes, bone, cartilage, and smooth muscle, indicating that they possess cranial, cardiac, and perhaps vagal NCC differentiation potential.

Both hESCs and iPSCs may be utilized to balance the *in vivo* and *in vitro* information garnered from neural crest development in model organisms. While many molecular markers and processes of neural crest development are shared between human and model organisms such as *Xenopus*, chick, and mouse, there are vast differences that require examination of human model systems [18]. In recent years, significant progress has been made in complementing our knowledge of human neural crest development and its distinctive variances. These advancements were achieved through exhaustive morphological analysis of early human neural crest development [19], molecular profiling of human neural tube explants [20], and marker analyses defining cranial and trunk regions from intact early human embryos [21]. Despite these developments, stem cell technologies are recognized as much-needed tools that have the capacity to address limitations of studying the human model and further our understanding of molecular mechanisms that drive normal and aberrant neural crest development [18].

NEUROCRISTOPATHIES

Familial Dysautonomia

Familial dysautonomia (FD), also known as Riley–Day disease, is a rare disease in the general population but one that affects 1:3700 births from Jewish individuals of eastern European ancestry [22,23]. Most notably, it impacts the autonomic and sensory nervous systems, resulting in an age-increasing reduction of dorsal root ganglion and superior cervical sympathetic ganglion neurons. FD symptoms include the inability to produce tears, hypotonia, and speech and growth developmental delays, and only 50% of patients reach age 30 [23]. FD patients also suffer from cardiac problems due to postural hypotension without compensatory tachycardia and episodic hypertension, as well as chronic kidney disease caused by the inability to regulate renal hemodynamics [23,24]. FD is an autosomal recessive disease where approximately 99% of affected patients have point mutations in the IκB kinase complex-associated protein (IκBKAP) gene, resulting in splicing defects and reduced IκBKAP expression [25].

In 2009, the Studer laboratory described the first use of iPSCs to model a neurocristopathy following the characterization of FD patient-derived cells [26]. FD patient-derived iPSCs (FD-iPSCs) differentiate to NCC but show migratory defects and tissue-specific differences between mutant and wild type IkBKAP transcripts, similar to that seen in FD patient-derived tissue. Even though FD-NCC differentiate to peripheral neurons, they do so at a reduced efficiency. Microarray analysis comparing wild type and FD-NCC identified 89 differentially expressed genes. Some of these are known to be involved in peripheral neurogenesis (SLC17A6, MAP4, INA, STMN2, and ASCL1) and were downregulated in FD-NCC relative to controls. Pharmacological evaluation of FD-NCC was also used to evaluate the effects of potential therapeutics such as the plant hormone kinetin, which elevates the expression of wild type IκBKAP. Under sustained exposure to kinetin, FD-NCC showed elevated expression of wild type IκBKAP and increased differentiation to peripheral neurons. These observations illustrate how iPSC models can be used broadly for drug screening applications and therapeutic development. More generally, the study also shows the power of iPSCs as a tool to recapitulate and understand neurocristopathies.

Hirschsprung's Disease

Hirschsprung's disease (HSCR) is a developmental disorder arising from malformation of the ENS within the hindgut [27,28]. At the tissue level, HSCR is characterized by aganglionosis in the lower intestine resulting in loss of peristaltic activity due to the absence of peripheral neuron innervation (Figure 21.3). Proper peristaltic activity requires the coordination of gut smooth muscle action via enteric nervous signaling. This signaling is propagated by plexuses of ganglia beneath the longitudinal muscle and within the intestinal submucosa. Normal proximal ganglionic colon is shown in Figure 21.3A (uppermost portion), contrasted against the contracted portion of aganglionic distal colon, which is unable to relax in the absence of neural crest-derived ganglia (lower portion of Figure 21.3A). Histologically, differences between ganglionic

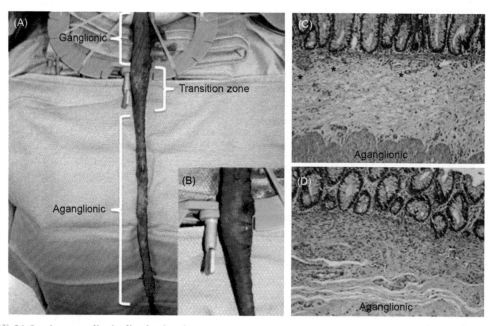

FIGURE 21.3   **An aganglionic distal colon from a Hirschsprung's disease patient.** (A) Operative photograph depicting long-segment aganglionic colon, which is contracted relative to the more proximal normal caliber colon. (B) Enlargement of operative photograph of the transition zone between normal and aganglionic colon. The normal ganglionated colon is at the top of the photo; the transition zone is the middle tapered segment of colon and on microscopy has rare ganglia; and the contracted aganglionic colon appears at the bottom of the image. (C) High power (400×) photomicrograph of normal ganglion cells populating the proximal resected colon (ganglia are denoted by blue stars). The layer of ganglia appears between the outer and inner muscle layers of the colon. (D) Low power (200×) photomicrograph of aganglionic segment. Submucosa in this region contains numerous hypertrophic nerve twigs.

and aganglionic colon are apparent: ganglia (blue stars) are visible within the ganglionic portion of the proximal colon between the longitudinal and circular muscle layers (Figure 21.3C), but are absent from within the submucosal layer and exterior to the circular muscle layer in the aganglionic portion (Figure 21.3D). Diagnosis of HSCR is typically made postnatally due to intestinal obstruction associated with the failure to pass meconium within the first 48 hours of life. Patients also display abdominal distension, vomiting, and neonatal enterocolitis [27]. Depending on the degree of severity for this disease, other individuals may be diagnosed later in infancy or even into adulthood with severe constipation, chronic abdominal distension, vomiting, and failure to thrive [27,29].

Hirschsprung-associated aganglionosis presumably results from the migration failure of vagally and sacrally derived NCC to reach their target in the developing hindgut between gestational weeks 5 and 12 [27,30]. The incidence of HSCR is approximately 1:5000 live births, although the degree of aganglionosis and corresponding disease severity is variable and only presents as an isolated trait (aganglionosis not associated with other underlying causes) in around 70% of cases, whereas the remaining cases are observed as associated malformations of complex syndromes [27,31]. HSCR displays strong sexual dimorphism, with males being affected up to two to four times more often/severe than females [27]. There may also be ethnic disparities in terms of HSCR predisposition [32].

In addition to Hirschsprung's syndromic or isolated nature, the disease may arise through familial or sporadic modes and may be further categorized into two subtypes depending on the degree of aganglionosis. Short-segment HSCR (S-HSCR) is the most common and affects the rectum and a short portion of the colon. Long-segment HSCR (L-HSCR) affects longer tracts of the colon and in severe cases,

presents as total colonic (approximately 10% of cases) or total intestinal aganglionosis [33,34]. Although the vast majority of patients with HSCR are sporadic, it can also be inherited. Isolated HSCR transmission is non-Mendelian, implying that it is a multifactorial disease. The dissimilar characteristics of syndromic versus isolated HSCR can be striking: cases of isolated HSCR display lower penetrance in that genotypic abnormality leading to aganglionosis in one individual that may not cause the same diseased phenotype in another individual, isolated cases also present with the greatest variability in the extent of aganglionosis, and isolated cases display the highest degree of male gender bias [27,33].

Genetic heterogeneity in HSCR has been demonstrated with at least 10 specific genes suspected of contributing to pathogenesis [27]. The two most commonly implicated genes in isolated HSCR are *RET* (rearranged during transfection) and *EDNRB* (endothelin receptor type B), with *RET* suffering coding sequence mutations in around 50% and 15–20% of familial versus sporadic HSCR cases, respectively [27]. In addition to *RET* and *EDNRB*, mutations have been found within *GDNF*, *EDN3*, *ECE1*, *SOX10*, *ZFHX1B*, and *PHOX2B*. In the context of ENS development, these genes encode proteins that form a network of interrelated signaling pathways that are responsible for the development of enteric ganglia from NCC; collectively and in coordinated fashion, these proteins regulate the survival, proliferation, migration, and differentiation of NCC in the developing ENS. Furthermore, HSCR has been associated with at least 32 syndromes and shows a strong association with Goldberg–Shprintzen and BRESHEK—brain abnormalities, retardation, ectodermal dysplasia, skeletal malformation, Hirschsprung's disease, ear/eye anomalies, and kidney dysplasia [27]. Chromosomal abnormalities in HSCR cases are less common, although up to 10% of all HSCR cases are associated with trisomy 21 [27].

Although several mouse models have identified genetic factors (such as *RET*) associated with HSCR and can recapitulate many features of human HSCR, the full genetic pathogenesis of this disease remains incomplete [35]. HSCR provides an excellent model for multigenic disease study and, like many congenital disorders, lends itself extremely well to the use of iPSC technology. Although HSCR is likely to be multifactorial in nature, elucidating its pathogenesis can be approached with iPSC technology and various *in vitro* approaches, including differentiation and migration assays. Hirschsprung's disease-specific iPSCs have already been generated and differentiated to neural crest in support of this general premise (Figure 21.4). Amenability to manipulation, scalability of disease-specific cell types, the capacity to focus on single or multiple targets simultaneously, and the rapidity of phenotype assessment make iPSCs a potent tool for studying Hirschsprung's disease.

While mutations in *RET* coding sequences account for 50% of familial and 15−20% of sporadic HSCR cases, most familial HSCR cases show association with the *RET* locus by linkage analysis. Investigation into non-coding mutations has identified a single nucleotide polymorphism (SNP) within intron 1 that is far more frequently associated with HSCR than observed coding sequence mutations [36,37]. The detailed understanding of *RET* and its role in Hirschsprung's disease allows for a targeted approach to be used in iPSC-based models of the disease that are likely to generate important information. Additional evidence demonstrates that *RET* plays a role in HSCR symptoms associated with other congenital syndromes such as congenital central hypoventilation syndrome (CCHS). Alterations in RET expression or stability therefore appear to lie at the nexus for complex modes of HSCR pathogenesis. As previously mentioned, patients with PHOX2B mutations display the HSCR phenotype. The vast majority of CCHS

patients also carry PHOX2B mutations [36]. Importantly, transmission of a *RET* hypomorphic allele in CCHS patients increases the risk for HSCR, implying an epistatic interaction between PHOX2B and RET in HSCR [36]. Data generated in mice show that PHOX2B is required for RET expression in enteric NCC and that mutant forms of PHOX2B are linked to compromised RET expression and, thus, aganglionosis beyond the foregut [36,38]. Furthermore, Down syndrome patients that concomitantly display HSCR also carry hypomorphic *RET* alleles [27,36]. In mouse models of HSCR, reduced RET expression in *Ret*−/− mice (mice with a knocked-in RET [9] isoform that has reduced expression) recapitulates many of the most common features of HSCR, including distal colon aganglionosis and incomplete penetrance, and confers a similar sex bias to that observed in humans [36,39]. While mutations in other proteins such as EDNRB display colonic aganglosis, especially in closed populations such as Mennonite communities, the importance of RET mutations cannot be understated. Fully 5% of all HSCR cases involve only mutations in EDNRB and its ligand EDN3, which typically present with S-HSCR; however, in mouse studies concomitant mutations in both *Ret* and *Ednrb* lead to almost complete aganglionosis [34,40]. In fact, mutation of the *RET* gene is now thought to be requisite in the majority of HSCR cases [36].

Together, these factors suggest that RET occupies a critical intersection between the monogenic and complex forms of HSCR pathogenesis. However, a complete molecular description of intestinal aganglionosis remains to be determined. Patient-specific iPSCs (see Figure 21.4) should prove an invaluable tool to determine these associations. As proof of concept, reprogramming of Hirschsprung's patient fibroblasts to iPSCs and their efficient differentiation to NCC are shown in Figure 21.4. Fibroblasts from an HSCR affected adult male with two affected children were collected

Hirschsprung's hiPSCs

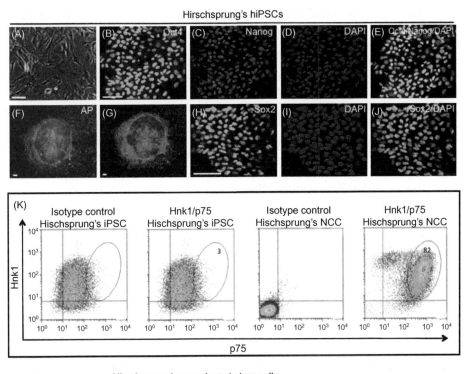

Hirschsprung's neural crest stem cells

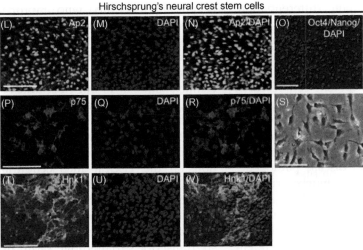

FIGURE 21.4 **Hirschsprung's disease (HSCR) iPSCs and neural crest cells (NCC).** (A) Bright field image of HS fibroblasts before reprogramming. Immunocytochemistry analysis of HSCR iPSCs shows they are positive for pluripotent markers such as OCT4, NANOG (B−E), and SOX2 (H−J). (F) Alkaline phosphatase (AP) staining of a reprogrammed HSCR iPSC colony. (G) Bright field image of a reprogrammed HSCR colony. (K) Flow cytometry analysis of HSCR iPSCs and HSCR NCC at day 20 showing an increase of p75 + /HNK1 + population in NCC. Double positive cells are shown in the circled area. HSCR NCC are positive for AP2 (L−O), p75 (P−R), and HNK1 (T−V) but negative for OCT4 and NANOG (O). (S) Bright field image of HSCR NCC at day 20. Scale bars, 100 μM.

(Figure 21.4A) and reprogrammed according to Menendez et al. 2013 [9]. Pluripotency was demonstrated following establishment of stem cell colonies by live alkaline phosphatase (AP) staining and immunofluorescence staining of three pluripotent markers—OCT4, NANOG, and SOX2 (Figure 21.4B–J). These iPSCs were then differentiated toward a NCC fate according to the method developed by Menendez et al. [17]. Differentiation from iPSCs to NCC was confirmed by flow cytometry (Figure 21.4K) using antibodies against the low-affinity nerve growth factor receptor (p75) and HNK1, a glucoronyl transferase involved in neuronal cell surface protein glycosylation and neuronal cell adhesion. NCC differentiation was further confirmed by immunofluorescence using antibodies against AP-2B (activating enhancer binding protein 2 beta), p75 and HNK1 (Figure 21.4L–N, P–R, T–V). Increased staining for these markers correlated with loss of OCT4 and NANOG expression (Figure 21.4O). Figure 21.4S demonstrates typical NCC morphology upon differentiation. These cells are well suited to serve as an experimental platform to evaluate pathogenic mechanisms of HSCR.

Beyond replicating a "disease in a dish," a major application of pluripotent cell research is to cure human disease through transplantation of functional cells derived from embryonic or reprogrammed cells. As a route to achieving this goal, it is also critical to understand the underlying mechanisms of disease pathogenesis. Often it is necessary for human pluripotent cells to be xenografted into model organisms to understand their behavior in an *in vivo* context, or as in the case of NCC, their proper proliferation, migration, and subsequent colonization of target tissues. However, reintroduction of pluripotent cells into model systems, such as Ret $-/-$ mice, may encounter hurdles that are a consequence of non-cell autonomous effects. The obstacle of non-cell autonomous effects with regards to NCC

migration and ENS development was highlighted by Bogni et al. [41], who demonstrated that murine intestinal ENS progenitors could be transplanted into Ret-deficient hosts; however, while transplanted cells were able to fully colonize the entire length of the gastrointestinal tract in wild type mice, transplanted cell migration was halted at the proximal foregut in Ret $-/-$ mice [41]. Thus, it must be appreciated that surrogate disease systems, such as in RET $-/-$ mice for HSCR, for example, may display complications arising from factors extrinsic to deficiencies in the cell type under study and co-dependent on proper microenvironments and signaling milieus of target tissues.

Treacher Collins Syndrome

Treacher Collins syndrome (TCS) is a rare condition affecting 1:50,000 live births, although some perinatal deaths have been reported, and is typically associated with improper craniofacial architecture arising from insufficient generation of cranially fated NCC (Figure 21.5) [42]. TCS was initially described by Treacher Collins in 1900 and clinically presents with a range of characteristics, including hypoplasia of the facial bones, cleft palate, external and middle ear anomalies, and defects in brain development [42–48]. Multiple studies indicate that the majority of TCS cases can be attributed to mutations in the *TCOF1* gene on chromosome 5 [49–53]. *TCOF1* encodes the serine/alanine-rich nucleolar protein, treacle, [49,54,55].

The *TCOF1* gene comprises 26 exons and can be differentially spliced to generate up to six transcript variants [56]. Over 200 different mutations have been identified in *TCOF1*, and many of these have implicated roles in TCS pathology, including deletions, insertions, missense, nonsense, and splicing mutations [42,57]. Deletions of 1–41 nucleotides in length are the most common class, and only exon 24

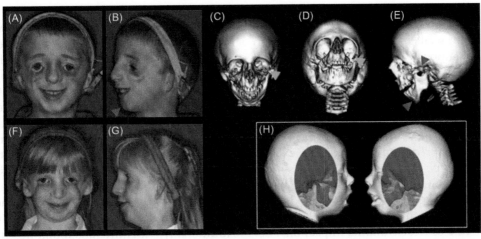

FIGURE 21.5 **Examples of hereditary and sporadic conditions affecting craniofacial structures derived from cranial neural crest.** (A–G) Clinical and 3D computerized tomography (CT) scan images of Treacher Collins syndrome patients demonstrating lateral oblique orbital clefts (green arrowheads), micrognathia (red arrowheads), zygomatic arch clefts (blue arrowhead), facial soft tissue deficiency, microtia, and auditory canal atresia (A–E). Images (F) and (G) demonstrate the phenotypic variability in TCS in the sister of (A). Note the less severely affected facial soft tissue and ear involvement. Middle ear ossicular chain malformations necessitate the use of bone conduction hearing aids (A, F). Craniofacial microsomia (H) is a sporadic condition affecting the same neural crest derivatives as in TCS but usually presenting with asymmetric involvement. Windows in 3D reconstructions of soft and skeletal tissues demonstrates the asymmetric mandibular dysplasia and zygomatic arch clefting.

has been reported as a mutational hot spot, with a common 5-bp deletion occurring with a frequency of 17% of TCS cases [57]. Half of all TCS cases have no previous family history and arise from *de novo* mutation [42,58]. Diagnosing TCS may often be confounded by the fact that a high degree of both inter- and intrafamilial phenotypic variations are commonly observed [59,60]. In many cases, TCS phenotypes are mild and go undiagnosed until an offspring or sibling displaying a more severe phenotype is diagnosed [42].

Mouse models of TCS have demonstrated that normal *Tcof1* expression is critical to the formation, proliferation, and survival of NCC [61,62]. These studies, in conjunction with Gonzales and colleagues [63], established that Tcof1 participates in ribosome biogenesis and maturation, and that defects in Tcof1 result in a diminished pool of mature ribosomes and compromised protein synthesis. The current view of TCS is that compromised ribosome biogenesis contributes to reduced numbers of emerging NCC at the time of neural tube closure due to decreased proliferation and increased apoptosis [42,62,63]. Jones and colleagues [64] demonstrated that *Tcof1* haploinsufficiency in mice provokes p53-dependent apoptosis within the neuroepithelium of the neural crest, thus diminishing the NCC progenitor pool. As a consequence, fewer proliferating NCC emanate from the branchial arches to form the proper craniofacial architecture [64].

Collectively, data suggest that mutations within *Tcof1* result in aberrant NCC formation and survival. Subsequently, an insufficient number of cranially fated NCC are available to create the scaffold on which the head and face are constructed. This is a simplified scenario, as the high degree of variability in disease severity suggests that a myriad of factors contribute to TCS. Likewise, the complete picture of the molecular function of the treacle protein is not well characterized. In addition to the aforementioned association in ribosome biogenesis, treacle may play other roles depending on its subcellular localization. Recently, Barlow and colleagues demonstrated that *Tcof1* may act as a modifier of *Pax3* during ENS development, wherein compound haplo-insufficient *Pax3:Tcof1* mice display an exacerbated phenotype of colonic agangliosis—this implies that TCOF1 may also play a role in HSCR pathogenesis [35]. Animal models of TCS alone may be inadequate to fully characterize the role of treacle and its impact on TCS. A primary reason for this is that treacle conservation among vertebrates is low. For example, amino acid sequence identity between mice and human is only 62% and between *Xenopus* and human only 19% [54,63]. Mechanistic studies on human embryos are obviously not an option, making patient-specific iPSC technology an attractive tool for understanding the molecular pathogenesis of TCS. Furthermore, parallel studies between parent-, offspring-, and sibling-derived iPSCs and derivative NCC are likely to yield a critical understanding relating to phenotype variability. Because TCS is thought to be a monogenic disease, mechanistic studies utilizing patient-specific iPSCs should be straightforward.

Currently, the Dalton laboratory has successfully generated patient-specific iPSCs from three different patients, two of whom are siblings (Figure 21.5). Both the clinical and 3D computerized tomography (CT) scan images demonstrate several characteristics typical of TCS patients, including the variability of phenotype among family members: both exhibit the undersized jaw (micrognathia; Figure 21.5A, B, E–G and red arrowheads), clefting of the inferolateral orbit (Figure 21.5A–D, F, G and green arrowheads), as well as the zygomatic arch (Figure 21.5E, blue arrowheads). The male child exhibits more severe soft tissue

involvement as noted by the increased downward slope of the eyes, eyelid coloboma, and microtia (small ear) (Figure 21.5A, B). The external and middle ear malformations common to TCS necessitate the use of bone conducting hearing aids (Figure 21.5A, B, F, G).

Fibroblasts from these patients have been reprogrammed to iPSCs and differentiated to NCC via the method presented in Menendez et al. [9]. Studies to determine the nature of NCC defects are ongoing. Specifically, flow cytometric analyses of apoptosis are being employed to determine if similar phenomena observed in mice may be recapitulated *in vitro*. Additionally, identifying apoptotic maxima during differentiation to NCC will guide future experiments to determine the role of treacle specific to its subcellular localization and involvement in NCC development.

CHARGE Syndrome

CHARGE syndrome was defined in 1981 as an acronym for a collection of symptoms: coloboma of the eye, heart defects, atresia of the nasal choanae, retardation of growth and/or development, genital and/or urinary abnormalities, and ear abnormalities and deafness [65]. CHARGE is an autosomal dominant disease affecting 1:10,000 births. Around 65% of these cases are a result of mutations in *CDH7*, an ATP-dependent chromatin remodeler gene [66]. Due to the nature of CHARGE-associated symptoms, it was originally assumed to be a neurocristopathy, but it was not until 2010 that direct evidence for this was obtained, using hESCs and their differentiation to NCC [67].

CHARGE syndrome is an excellent example of how pluripotent stem cells have been used to gain a deeper insight into the molecular basis of a poorly understood condition. Using differentiation models, *CDH7* was identified as a key regulator of NCC formation and function [67]. Genes impacted by *CHD7* include *TWIST*,

SLUG, and *SOX9*, all of which are critical for NCC migration. CHD7 functions by binding BRG1, a member of the PBAF chromatin-remodeling complex. Once in a complex, CDH7 and PBAF regulate expression of SOX9 and TWIST by binding H3K4me1-marked enhancer regions that control chromosome architecture and transcriptional activity [67]. The use of patient-derived iPSCs (CHARGE-iPSCs) will be critical to unravel the molecular complexities of CHARGE syndrome. For example, CHARGE-iPSCs from patients harboring CDH7 mutations can be differentiated to NCC to test if migration and/or differentiation is affected. Overexpression of *CDH7* should rescue the expression of genes like *SOX9* and *TWIST*, as well as the migration problems. Jongmans and colleagues [66] reported 67 different *CDH7* mutations in 69 CHARGE patients with no corresponding genotype/phenotype relationship. The use of iPSCs might help find such correlation, since they might result in different levels of *CHD7* expression during NCC differentiation, which could correlate with different NCC defects.

Noonan and LEOPARD Syndromes

LEOPARD syndrome is an acronym of its most common symptoms: lentigines, ECG conduction abnormalities, ocular hypertelorism, pulmonary stenosis, abnormal genitalia, retardations of growth, and deafness [68]. Noonan syndrome symptoms include dwarfism, hypertrophic cardiomyopathy, and craniofacial malformations and affects 1:1000–1:2500 births [69,70]. Around 50% of cases for Noonan syndrome and 80–90% of LEOPARD syndrome are caused by an allelic dominant mutation in *PTPN11*, a gene encoding the tyrosine phosphatase SHP2 [71,72]. In Noonan patients, most *PTPN11* mutations are gain of function, promoting protein stabilization at the structural level and elevated mitogen-activated

protein (MAP) kinases ERK1/2 activity [71]. In mice, this leads to reduced osteoblasts differentiation from NCC and, consequently, craniofacial defects. Treatment of mice with the ERK1/2 inhibitor U0126 restored normal craniofacial development [73]. In contrast to Noonan syndrome, LEOPARD syndrome mutations in *PTPN11* are frequently associated with loss of function. Mice with reduced SHP2 expression have a corresponding decreased ERK1/2 activity resulting in loss of NCC migration to the heart and cranial areas [74]. Complete loss of SHP2 in NCC results in lower numbers of cells reaching the outflow tract cushions and resulting in septation defects. Cranial NCC migrated properly but failed to differentiate to osteoblasts, without any increase of apoptosis.

Induced pluripotent stem cells from two LEOPARD syndrome patients (LEOPARD iPSCs) have been generated and used to study differentiation to cardiomyocytes [75]. Comparison of LEOPARD iPSCs to those from normal patients showed significant differences in the phosphoproteosome as well as decreased ability to respond to FGF activation of ERK. As described above, proper regulation of ERK is very important for NCC differentiation since either too much or too little ERK signaling leads to craniofacial and cardiac problems. LEOPARD and Noonan iPSCs will be very useful to study NCC differentiation and to verify if the defects seen in mouse are comparable to humans.

22q11.2 Deletion Syndromes

22q11.2 deletion syndromes, which include DiGeorge syndrome, velocardiofacial syndrome, and conotruncal anomaly face syndrome, affect 1:4000 to 1:1600 births. Fully 95% of cases are caused by the *de novo* deletion of a 3 Mb region in chromosome 22 at band q11.2 [76,77]. There are more than 180 different clinical features associated to this spectrum of diseases, some of the most common being heart defects, palatal anomalies, vascular anomalies, psychiatric disorders, learning difficulties, and immune disorders. The q11.2-deleted region contains 30–40 genes of developmental importance, including *COMT, UFD1L, TBX1,* and *DGRC8.* Cardiac defects for this syndrome have been associated with migration and survival defects in cardiac NCC (cNCC).

TBX1 encodes the transcription factor T-box 1 that, although not expressed in NCC, is necessary to regulate signaling molecules from the pharyngeal ectoderm and endoderm such as *GBX2, SLIT,* and *FGF8.* Without these signals, mouse cNCC migration to the fourth pharyngeal arch fails, resulting in pharyngeal abnormalities that resemble those seen in 22q11.1 syndrome [78]. Another important gene for NCC in the 22q11.2 region is *DGCR8.* This gene encodes a protein critical for microRNA (miRNA) processing, and its conditional ablation in mouse NCC results in severe cardiovascular malformations [79]. Contrary to *TBX1* deficiency, these defects are due to an increase of apoptosis of cNCC just before they reach the outflow tract [79]. Patient-specific iPSCs could be differentiated to neural crest and then implanted in mice to study migration to the branchial arches. Similar phenotypes to that seen in mouse models may be seen such as migration defects (*TBX1* mutants) or increased apoptosis (*DGCR8* mutations). It would be interesting to see if NCC from these patients show an increase of apoptosis *in vitro,* and if so, if it can be rescued by overexpressing *DGCR8.*

Waardenburg Syndrome

Waardenburg syndrome (WS) is an autosomal dominant disease affecting 1:42,000 births. WS symptoms include deafness, constipation, partial albinism, dystopia canthorum, and, in some cases, atrial septal defects and spina bifida [80]. Patients are classified as type I, II, III,

or IV depending on the combination of symptoms they present. Up to six genes have been linked to the disease: *PAX3* in 90% of type I and III, *MITF* in type II (15%), *EDN3* and *EDNRB* in type II (<5%) and type IV (20–30%), *SOX10* in type II (15%) and type IV (50%), and *SNAI2* in type II (<5%) [81]. Approximately 70% of patients with type II and 15–35% of type IV patients do not exhibit mutations in these six genes. Most cases of type I WS are due to loss of function mutations in *PAX3,* an important regulator of early neural crest specification. There are more than 70 reported point mutations of *PAX3* in type I WS, with no correlation between them and the severity of the syndrome [81]. Most mutations affect the *PAX3* DNA binding domain or its homeodomain. In mouse models, *PAX3* mutations lead to increased apoptosis that could be rescued by p53 mutations [82]. Since *PAX3* is a gene expressed very early during neural crest specification, differentiation of WS-iPSCs (iPSCs derived from WS patients) to NCC might be the only way to study the effects these multiple mutations have in early human NCC differentiation. For example, iPSCs from patients with different mutations could show different degrees of apoptosis. Experiments with mouse fibroblast show that mutations in PAX3 not only affect the DNA binding ability but its nuclear localization [83]; therefore, iPSCs could show a similar defect in WS patients. Type IV WS, also referred to as Waardenburg-Shah syndrome, presents an association of hypopigmentation and Hirschsprung's disease. Around 50% of those patients carry mutations in SOX10, and mouse models have shown it is very important for NCC differentiation to glial cells and melanocytes [81,84,85]. Both *PAX3* and *SOX10* control *MITF* expression, a gene required for melanocyte differentiation. WS-iPSCs could be differentiated to NCC and then to melanocytes. Since both *PAX3* and *SOX10* regulate early melanocyte differentiation, iPSCs offer the opportunity to study early steps, as opposed to the study of mature melanocytes. Furthermore, in those WS-iPSCs that present defects in melanocyte differentiation and *MITF* expression, microarray analysis could point to some unknown *MITF* regulator, which could be mutated in those patients and be the cause of the phenotypic variability or explain WS cases with unknown mutation. It is important to point out that the methods utilized to differentiate NCC into melanocytes are currently inefficient—an important technical barrier that must be overcome in order to study pathologies such as WS [86].

Piebaldism

Piebaldism is characterized by the absence of melanocytes in patches of skin and hair and by the presence of a white forelock in around 90% of patients. Piebaldism is a rare autosomal dominant disorder in which approximately 75% of cases are due to mutations in the *KIT* gene. The KIT protein is a receptor tyrosine kinase, and it activates multiple signaling cascades such as phosphatidylinositol 3′-kinase (PI3′-kinase), Src Family Kinase (SFK), MAP kinase, and phospholipase C and D pathways [87]. KIT is a receptor for the stem cell growth factor expressed in NCC-derived melanoblasts and it is required for their migration [88]. Melanoblasts also show increased apoptosis under conditions where KIT signaling is compromised [89].

Although piebaldism is mostly considered a cosmetic defect, the study of iPSCs from these patients could help understand how deregulation of *KIT* affects melanocyte differentiation. For example, they could help understand if, in humans, *KIT* mutations result in an increase of apoptosis during melanoblast differentiation, a migration defect of the NCC, or a combination of both, and microarray analysis could point to new *KIT* targets in human melanocytes.

Piebaldism iPSCs can also be used similarly to FD-iPSCs for drug screens. Piebaldism patient-derived melanocytes could be screened for activation of KIT downstream pathways, and those drugs could be potentially used in topical ointments to treat piebaldism.

CONCLUSIONS AND PERSPECTIVES

Development of technology for the derivation of patient-specific pluripotent cells represents a major leap forward for modeling and potentially treating human disease [85]. These tools will be a valuable part of deciphering disease pathogenesis and are likely to uncover aspects of human disease that would be unanticipated from other approaches. Several reports have already described the use of iPSC technology to model neurocristopathies. One potential outcome of this is the development of new therapeutics from iPSC-based drug screens. Although animal models have been critical to reach our current level of understanding, patient-derived iPSC models are likely to take this to the next level particularly in the area of personalized medicine.

References

[1] Takahashi K, Tanabe K, Ohnuki M, Narita M, Ichisaka T, Tomoda K, et al. Induction of pluripotent stem cells from adult human fibroblasts by defined factors. Cell 2007; 131(5):861−72.
[2] Yu J, Vodyanik MA, Smuga-Otto K, Antosiewicz-Bourget J, Frane JL, Tian S, et al. Induced pluripotent stem cell lines derived from human somatic cells. Science 2007;318(5858):1917−20.
[3] Grskovic M, Javaherian A, Strulovici B, Daley GQ. Induced pluripotent stem cells–opportunities for disease modelling and drug discovery. Nat Rev Drug Discov 2011;10(12):915−29.
[4] Trounson A, Shepard KA, DeWitt ND. Human disease modeling with induced pluripotent stem cells. Current Opinion in Genet & Development 2012;22(5):509−16.

[5] LaBonne C, Bronner-Fraser M. Induction and patterning of the neural crest, a stem cell-like precursor population. J Neurobiol 1998;36(2):175–89.

[6] Le Douarin NM, Calloni GW, Dupin E. The stem cells of the neural crest. Cell Cycle 2008;7(8):1013–9.

[7] Achilleos A, Trainor PA. Neural crest stem cells: discovery, properties and potential for therapy. Cell Res 2012;22(2):288–304.

[8] Hall BK. The neural crest and neural crest cells: discovery and significance for theories of embryonic organization. J Biosci 2008;33(5):781–93.

[9] Menendez L, Kulik MJ, Page AT, Park SS, Lauderdale JD, Cunningham ML, et al. Directed differentiation of human pluripotent cells to neural crest stem cells. Nat Protoc 2013;8(1):203–12.

[10] Pomp O, Brokhman I, Ben-Dor I, Reubinoff B, Goldstein RS. Generation of peripheral sensory and sympathetic neurons and neural crest cells from human embryonic stem cells. Stem Cells 2005;23(7):923–30.

[11] Jiang X, Gwye Y, McKeown SJ, Bronner-Fraser M, Lutzko C, Lawlor ER. Isolation and characterization of neural crest stem cells derived from in vitro-differentiated human embryonic stem cells. Stem Cells Dev 2009;18(7):1059–70.

[12] Pomp O, Brokhman I, Ziegler L, Almog M, Korngreen A, Tavian M, et al. PA6-induced human embryonic stem cell-derived neurospheres: a new source of human peripheral sensory neurons and neural crest cells. Brain Res 2008;1230:50–60.

[13] Lee G, Kim H, Elkabetz Y, Al Shamy G, Panagiotakos G, Barberi T, et al. Isolation and directed differentiation of neural crest stem cells derived from human embryonic stem cells. Nat Biotechnol 2007;25(12):1468–75.

[14] Chambers SM, Fasano CA, Papapetrou EP, Tomishima M, Sadelain M, Studer L. Highly efficient neural conversion of human ES and iPS cells by dual inhibition of SMAD signaling. Nat Biotechnol 2009;27(3):275–80.

[15] Lee G, Chambers SM, Tomishima MJ, Studer L. Derivation of neural crest cells from human pluripotent stem cells. Nat Protoc 2010;5(4):688–701.

[16] Curchoe CL, Maurer J, McKeown SJ, Cattarossi G, Cimadamore F, Nilbratt M, et al. Early acquisition of neural crest competence during hESCs neuralization. PloS one 2010;5(11):e13890.

[17] Menendez L, Yatskievych TA, Antin PB, Dalton S. Wnt signaling and a Smad pathway blockade direct the differentiation of human pluripotent stem cells to multipotent neural crest cells. Proc National Acad. Sci U S A 2011;108(48):19240–5.

[18] Stuhlmiller TJ, Garcia-Castro MI. Current perspectives of the signaling pathways directing neural crest induction. Cellular Mol Life Sci: CMLS 2012;69(22):3715–37.

[19] O'Rahilly R, Muller F. The development of the neural crest in the human. J Anatomy 2007;211(3):335–51.

[20] Thomas S, Thomas M, Wincker P, Babarit C, Xu P, Speer MC, et al. Human neural crest cells display molecular and phenotypic hallmarks of stem cells. Human Mol Genet 2008;17(21):3411–25.

[21] Betters E, Liu Y, Kjaeldgaard A, Sundstrom E, Garcia-Castro MI. Analysis of early human neural crest development. Dev biology 2010;344(2):578–92.

[22] Axelrod FB. Familial dysautonomia: a review of the current pharmacological treatments. Expert Opin pharmacother 2005;6(4):561–7.

[23] Slaugenhaupt SA, Blumenfeld A, Gill SP, Leyne M, Mull J, Cuajungco MP, et al. Tissue-specific expression of a splicing mutation in the IKBKAP gene causes familial dysautonomia. American J Human Genet 2001;68(3):598–605.

[24] Rekhtman Y, Bomback AS, Nash MA, Cohen SD, Matalon A, Jan DM, et al. Renal transplantation in familial dysautonomia: report of two cases and review of the literature. Clinical J Am Soc Nephrology: CJASN 2010;5(9):1676–80.

[25] Lee G, Studer L. Modelling familial dysautonomia in human induced pluripotent stem cells. Philos Transactions R Soc London Ser B, Biol Sci 2011;366(1575):2286–96.

[26] Lee G, Papapetrou EP, Kim H, Chambers SM, Tomishima MJ, Fasano CA, et al. Modelling pathogenesis and treatment of familial dysautonomia using patient-specific iPSCs. Nature 2009;461(7262):402–6.

[27] Amiel J, Sproat-Emison E, Garcia-Barcelo M, Lantieri F, Burzynski G, Borrego S, et al. Hirschsprung disease, associated syndromes and genetics: a review. J Med Genet 2008;45(1):1–14.

[28] Whitehouse FR, Kernohan JW. Myenteric plexus in congenital megacolon; study of 11 cases. Arch Intern Med (Chic) 1948;82(1):75–111.

[29] Parc R, Berrod JL, Tussiot J, Loygue J. [Megacolon in adults. Apropos of 76 cases]. Annales de gastroenterologie et d'hepatologie 1984;20(3):133–41.

[30] Kenny SE, Tam PK, Garcia-Barcelo M. Hirschsprung's disease. Semin Pediatric Surgery 2010;19(3):194–200.

[31] Mundt E, Bates MD. Genetics of Hirschsprung disease and anorectal malformations. Semin Pediatric Surgery 2010;19(2):107–17.

[32] Torfs C. An epidemiological study of Hirschsprung disease in a multiracial California population. The Third International Meeting: Hirschsprung Disease and Related Neurocristopathies. Evian, France; 1998.

[33] Brooks AS, Oostra BA, Hofstra RM. Studying the genetics of Hirschsprung's disease: unraveling an oligogenic disorder. Clin Genet 2005;67(1):6—14.

[34] Heanue TA, Pachnis V. Enteric nervous system development and Hirschsprung's disease: advances in genetic and stem cell studies. Nat Rev Neurosci 2007;8(6):466—79.

[35] Barlow AJ, Dixon J, Dixon M, Trainor PA. Tcof1 acts as a modifier of Pax3 during enteric nervous system development and in the pathogenesis of colonic aganglionosis. Human Mol Genet 2013.

[36] Burzynski G, Shepherd IT, Enomoto H. Genetic model system studies of the development of the enteric nervous system, gut motility and Hirschsprung's disease. Neurogastroenterology Motil : Official J Eur Gastrointestinal Motil Soc 2009;21(2):113—27.

[37] Emison ES, McCallion AS, Kashuk CS, Bush RT, Grice E, Lin S, et al. A common sex-dependent mutation in a RET enhancer underlies Hirschsprung disease risk. Nature 2005;434(7035):857—63.

[38] Pattyn A, Morin X, Cremer H, Goridis C, Brunet JF. The homeobox gene Phox2b is essential for the development of autonomic neural crest derivatives. Nature 1999;399(6734):366—70.

[39] Uesaka T, Nagashimada M, Yonemura S, Enomoto H. Diminished Ret expression compromises neuronal survival in the colon and causes intestinal aganglionosis in mice. J Clin Invest 2008;118(5):1890—8.

[40] Barlow A. de Graaff E, Pachnis V. Enteric nervous system progenitors are coordinately controlled by the G protein-coupled receptor EDNRB and the receptor tyrosine kinase RET. Neuron 2003;40(5):905—16.

[41] Bogni S, Trainor P, Natarajan D, Krumlauf R, Pachnis V. Non-cell-autonomous effects of Ret deletion in early enteric neurogenesis. Development 2008;135 (18):3007—11.

[42] Trainor PA. Craniofacial birth defects: The role of neural crest cells in the etiology and pathogenesis of Treacher Collins syndrome and the potential for prevention. Am J Med Genet Part A 2010;152A (12):2984—94.

[43] Milligan DA, Harlass FE, Duff P, Kopelman JN. Recurrence of Treacher Collins' syndrome with sonographic findings. Military Med 1994;159 (3):250—2.

[44] Poswillo D. The pathogenesis of the Treacher Collins syndrome (mandibulofacial dysostosis). British J Oral Surg 1975;13(1):1—26.

[45] Cohen J, Ghezzi F, Goncalves L, Fuentes JD, Paulyson KJ, Sherer DM. Prenatal sonographic diagnosis of Treacher Collins syndrome: a case and review of the literature. Am J Perinatology 1995; 12(6):416—9.

[46] Teber OA, Gillessen-Kaesbach G, Fischer S, Bohringer S, Albrecht B, Albert A, et al. Genotyping in 46 patients with tentative diagnosis of Treacher Collins syndrome revealed unexpected phenotypic variation. Eur J Human Genet: EJHG 2004;12(11):879—90.

[47] Stovin JJ, Lyon Jr. JA, Clemmens RL. Mandibulofacial dysostosis. Radiology 1960;74:225—31.

[48] Treacher Collins E. Case with symmetrical congenital notches in the outer part of each lower lid and defective development of the malar bones. Trans Ophthalmological Soc U K 1900;20:90.

[49] Wise CA, Chiang LC, Paznekas WA, Sharma M, Musy MM, Ashley JA, et al. TCOF1 gene encodes a putative nucleolar phosphoprotein that exhibits mutations in Treacher Collins Syndrome throughout its coding region. Proc National Acad Sci U S A 1997;94 (7):3110—5.

[50] Sakai D, Trainor PA. Treacher Collins syndrome: unmasking the role of Tcof1/treacle. International J Biochem & Cell Biol 2009;41(6):1229—32.

[51] Trainor PA, Dixon J, Dixon MJ. Treacher Collins syndrome: etiology, pathogenesis and prevention. Eur J Human Genet: EJHG 2009;17(3):275—83.

[52] Jabs EW, Li X, Coss CA, Taylor EW, Meyers DA, Weber JL. Mapping the Treacher Collins syndrome locus to 5q31.3——q33.3. Genomics 1991;11(1):193—8.

[53] Jabs EW, Li X, Lovett M, Yamaoka LH, Taylor E, Speer MC, et al. Genetic and physical mapping of the Treacher Collins syndrome locus with respect to loci in the chromosome 5q3 region. Genomics 1993;18(1):7—13.

[54] Dixon J, Hovanes K, Shiang R, Dixon MJ. Sequence analysis, identification of evolutionary conserved motifs and expression analysis of murine tcof1 provide further evidence for a potential function for the gene and its human homologue, TCOF1. Human Mol Genet 1997;6(5):727—37.

[55] Isaac C, Marsh KL, Paznekas WA, Dixon J, Dixon MJ, Jabs EW, et al. Characterization of the nucleolar gene product, treacle, in Treacher Collins syndrome. Mol Biol Cell 2000;11(9):3061—71.

[56] Bowman M, Oldridge M, Archer C, O'Rourke A, McParland J, Brekelmans R, et al. Gross deletions in TCOF1 are a cause of Treacher-Collins-Franceschetti syndrome. European J Human Genet: EJHG 2012;20 (7):769—77.

[57] Splendore A, Fanganiello RD, Masotti C, Morganti LS, Passos-Bueno MR. TCOF1 mutation database: novel mutation in the alternatively spliced exon 6A and update in mutation nomenclature. Human Mutation 2005;25(5):429—34.

[58] Jones KL, Smith DW, Harvey MA, Hall BD, Quan L. Older paternal age and fresh gene mutation: data on additional disorders. J Pediatrics 1975;86(1):84—8.

[59] Dixon MJ, Marres HA, Edwards SJ, Dixon J, Cremers CW. Treacher Collins syndrome: correlation between clinical and genetic linkage studies. Clin Dysmorphology 1994;3(2):96–103.

[60] Marres HA, Cremers CW, Dixon MJ, Huygen PL, Joosten FB. The Treacher Collins syndrome. A clinical, radiological, and genetic linkage study on two pedigrees. Arch Otolaryngology–Head & Neck Surgery 1995;121(5):509–14.

[61] Dixon J, Brakebusch C, Fassler R, Dixon MJ. Increased levels of apoptosis in the prefusion neural folds underlie the craniofacial disorder, Treacher Collins syndrome. Human Mol Genet 2000;9(10):1473–80.

[62] Dixon J, Jones NC, Sandell LL, Jayasinghe SM, Crane J, Rey JP, et al. Tcof1/Treacle is required for neural crest cell formation and proliferation deficiencies that cause craniofacial abnormalities. Proc National Acad Sci U S A 2006;103(36):13403–8.

[63] Gonzales B, Henning D, So RB, Dixon J, Dixon MJ, Valdez BC. The Treacher Collins syndrome (TCOF1) gene product is involved in pre-rRNA methylation. Human Mol Genet 2005;14(14):2035–43.

[64] Jones NC, Lynn ML, Gaudenz K, Sakai D, Aoto K, Rey JP, et al. Prevention of the neurocristopathy Treacher Collins syndrome through inhibition of p53 function. Nature Med 2008;14(2):125–33.

[65] Pagon RA, Graham Jr. JM, Zonana J, Coloboma Yong SL. congenital heart disease, and choanal atresia with multiple anomalies: CHARGE association. J Pediatrics 1981;99(2):223–7.

[66] Jongmans MC, Admiraal RJ, van der Donk KP, Vissers LE, Baas AF, Kapusta L, et al. CHARGE syndrome: the phenotypic spectrum of mutations in the CHD7 gene. J Med Genet 2006;43(4):306–14.

[67] Bajpai R, Chen DA, Rada-Iglesias A, Zhang J, Xiong Y, Helms J, et al. CHD7 cooperates with PBAF to control multipotent neural crest formation. Nature 2010;463(7283):958–62.

[68] Keyte A, Hutson MR. The neural crest in cardiac congenital anomalies. Differentiation; Res Biol Diversity 2012;84(1):25–40.

[69] Tullu MS, Muranjan MN, Kantharia VC, Parmar RC, Sahu DR, Bavdekar SB, et al. Neurofibromatosis-Noonan syndrome or LEOPARD Syndrome? A clinical dilemma. J Postgraduate Med 2000;46(2):98–100.

[70] Romano AA, Allanson JE, Dahlgren J, Gelb BD, Hall B, Pierpont ME, et al. Noonan syndrome: clinical features, diagnosis, and management guidelines. Pediatrics 2010;126(4):746–59.

[71] Tartaglia M, Mehler EL, Goldberg R, Zampino G, Brunner HG, Kremer H, et al. Mutations in PTPN11, encoding the protein tyrosine phosphatase SHP-2, cause Noonan syndrome. Nature Genet 2001;29(4):465–8.

[72] Digilio MC, Sarkozy A, Pacileo G, Limongelli G, Marino B, Dallapiccola B. PTPN11 gene mutations: linking the Gln510Glu mutation to the "LEOPARD syndrome phenotype". Eur J Pediatrics 2006;165(11):803–5.

[73] Nakamura T, Gulick J, Pratt R, Robbins J. Noonan syndrome is associated with enhanced pERK activity, the repression of which can prevent craniofacial malformations. Proc National Acad Sci U S A 2009;106(36):15436–41.

[74] Nakamura T, Gulick J, Colbert MC, Robbins J. Protein tyrosine phosphatase activity in the neural crest is essential for normal heart and skull development. Proc National Acad Sci U S A 2009;106(27):11270–5.

[75] Carvajal-Vergara X, Sevilla A, D'Souza SL, Ang YS, Schaniel C, Lee DF, et al. Patient-specific induced pluripotent stem-cell-derived models of LEOPARD syndrome. Nature 2010;465(7299):808–12.

[76] Oskarsdottir S, Vujic M, Fasth A. Incidence and prevalence of the 22q11 deletion syndrome: a population-based study in Western Sweden. Archives of disease in childhood 2004;89(2):148–51.

[77] Shprintzen RJ. Velo-cardio-facial syndrome: 30 Years of study. Dev Disabilities Res Rev 2008;14(1):3–10.

[78] Calmont A, Ivins S, Van Bueren KL, Papangeli I, Kyriakopoulou V, Andrews WD, et al. Tbx1 controls cardiac neural crest cell migration during arch artery development by regulating Gbx2 expression in the pharyngeal ectoderm. Development 2009;136(18):3173–83.

[79] Chapnik E, Sasson V, Blelloch R, Hornstein E. Dgcr8 controls neural crest cells survival in cardiovascular development. Developmental biology 2012;362(1):50–6.

[80] Banerjee AK. Waardenburg's syndrome associated with ostium secundum atrial septal defect. J R Soc Med 1986;79(11):677–8.

[81] Pingault V, Ente D, Dastot-Le Moal F, Goossens M, Marlin S, Bondurand N. Review and update of mutations causing Waardenburg syndrome. Human Mutation 2010;31(4):391–406.

[82] Wang XD, Morgan SC, Loeken MR. Pax3 stimulates p53 ubiquitination and degradation independent of transcription. PloS one 2011;6(12):e29379.

[83] Corry GN, Raghuram N, Missiaen KK, Hu N, Hendzel MJ, Underhill DA. The PAX3 paired domain and homeodomain function as a single binding module in vivo to regulate subnuclear localization and mobility by a mechanism that requires base-specific recognition. J Mol Biol 2010;402(1):178–93.

[84] Mollaaghababa R, Pavan WJ. The importance of having your SOX on: role of SOX10 in the development of neural crest-derived melanocytes and glia. Oncogene 2003;22(20):3024–34.

[85] Southard-Smith EM, Angrist M, Ellison JS, Agarwala R, Baxevanis AD, Chakravarti A, et al. The Sox10 (Dom) mouse: modeling the genetic variation of Waardenburg-Shah (WS4) syndrome. Genome Res 1999;9(3):215–25.

[86] Clewes O, Narytnyk A, Gillinder KR, Loughney AD, Murdoch AP, Sieber-Blum M. Human epidermal neural crest stem cells (hEPI-NCSC)–characterization and directed differentiation into osteocytes and melanocytes. Stem Cell Rev 2011;7(4):799–814.

[87] Lennartsson J, Ronnstrand L. Stem cell factor receptor/c-Kit: from basic science to clinical implications. Physiol Rev 2012;92(4):1619–49.

[88] Ezoe K, Holmes SA, Ho L, Bennett CP, Bolognia JL, Brueton L, et al. Novel mutations and deletions of the KIT (steel factor receptor) gene in human piebaldism. A J Human Genet 1995;56(1):58–66.

[89] Ito M, Kawa Y, Ono H, Okura M, Baba T, Kubota Y, et al. Removal of stem cell factor or addition of monoclonal anti-c-KIT antibody induces apoptosis in murine melanocyte precursors. J Invest Dermatology 1999;112(5):796–801.

Index

Printed and bound by CPI Group (UK) Ltd, Croydon, CR0 4YY

08/05/2025

01864982-0002